TANDEM ORGANIC REACTIONS

TANDEM ORGANIC REACTIONS

TSE-LOK HO
Department of Chemistry
National Taiwan University

A Wiley–Interscience Publication
JOHN WILEY & SONS, INC.
New York • Chichester • Brisbane • Toronto • Singapore

Dedicated to
Professor George A. Olah
with Admiration and Friendship

Copyright © 1992 by John Wiley & Sons, Inc.

All rights reserved. Published simultaneously in Canada.

Reproduction or translation of any part of this work
beyond that permitted by Section 107 or 108 of the
1976 United States Copyright Act without the permission
of the copyright owner is unlawful. Requests for
permission or further information should be addressed to
the Permssions Department, John Wiley & Sons, Inc.

Library of Congress Cataloging in Publication Data:

Ho, Tse-Lok.
    Tandem organic reactions / Tse-Lok Ho.
       p.   cm.
    "A Wiley–interscience publication."
    Includes index.
    1. Organic compounds—Synthesis. 2. Chemical reactions.
  I. Title.
QD262.H62 1992
547.2—dc20   92-10637
   ISBN 0-471-57022-2   CIP

Printed and bound in the United States of America
10 9 8 7 6 5 4 3 2 1

CONTENTS

PREFACE

A long-range mission of chemical synthesis is to provide materials on demand. In view of dwindling resources in the world the economic factor looms ever larger in the future. Consequently, the highest possible efficiency of synthetic operations must be attained, and in that regard the number of steps must be kept to a minimum. However, it is also critical that the yield of each step should approach 100%.

To meet this enormous challenge the chemist must devise the shortest routes which incorporate high-yielding reactions. He/she must recognize the possibility of combining two or more reactions in any unit operation. To be able to achieve the latter possibility, without incurring substantial sacrifice in yields, is always desirable because it is dispensed with the isolation of the intermediate.

In synthesis design by retrosynthetic analysis a useful procedure is the tactical combinations of transforms. In it, reactions that can be used together in a specific sequence are grouped into packages and each of these multistep packages is considered loosely as one transform. The most powerful tactical cominations are those in which one transform acts on a synthetic target to produce a full retron for the second, and so on. In the extreme case the synthetic reactions are tandem reactions.

As one of the basic tools for synthesis planning is a library of known reactions, the chemist is served better if the library is well equipped. To start a library one may assemble individual reactions from textbooks and various compendia. However, these sources do not usually supply adequate information on tandem reactions. I hope the present volume will provide at least a framework to build on.

Tandem reactions are rarely indexed as such in the chemical literature, therefore gleaning them practically requires examination of individual publications. As this task is far beyond my capacity to undertake, my knowledge and recollection become the arbitrator for the contents of this book. On the other hand, I have restricted examples of certain pericyclic reactions and rearrangement of electron-deficient species because these have been reviewed adequately. Reactions in the area of heterocylic chemistry are also limited; furthermore, I have been quite selective in dealing with serial publications of the same themes, only representatives of them are included.

The importance of tandem reactions is evident. In addition to their pragmatic value, they have an aesthetic appeal. It is my opinion that a survey of these reactions, not necessarily exhaustive, would serve to rivet the attention of many chemists to their merits and utility. This attention should also stimulate design and discovery of new sets of tandem reactions. It is even more urgent that young students of organic chemistry should be told of this prospect at an early stage.

To complete this monograph I owe the diligence of Miss Jia-Ling Chu who transformed my rough sketches into neat formulas.* The quiet forbearance of my family during a most trying period of our lives is deeply appreciated.

TSE-LOK HO

* Reaction scheme on p. 362 is reproduced from *J. Am. Chem. Soc.* **95**: 5769 (1973) with permission of the American Chemical Society.

1

INTRODUCTION

Tandem reactions are those which occur in succession. In the chemical literature there have been explicit mention of certain combinations of reactions as tandem reactions even though these reactions were effected separately. On the other hand, many authentic tandem reactions are not considered as such because of their familiarity. In view of this confusion we wish to define tandem reactions as combinations of two or more reactions whose occurrence is in a specific order, and if they involve sequential addition of reagents the secondary reagents must be integrated into the products. Thus the aldol condensation–dehydration sequence catalyzed by a base is considered tandem; however, if the dehydration step requires the addition of another reagent, then the two reactions are individual ones. On the other hand, the vicinal functionalization of a conjugated double bond by the Michael addition and alkylation of the enolate is classified as a tandem process despite the stepwise manner of reagent introduction. Both reagents contribute parts of the product.

We do make exception to include certain nontandem reaction sequences because there is possibility to modify them into tandem processes.

Numerous organic transformations are the result of tandem reactions. But in view of their commonness many of them will not be included in this book. Thus, coverage is extended to only the truly tandem processes which lead to skeletal changes rather than merely functional group transformations. In this context the discussion of the formation of 1,2-diol monoformate from an olefin by the action of performic acid is outside the scope of this book, even if its tandem nature, that is epoxidation followed by formolysis, is apparent.

Tandem reactions are composed of ordinary reactions, therefore there is no need for new or special knowledge to understand them. However, the "secondary" reaction, for which the structural prerequisite is absent in the initial

1

substrate, must be triggered by the first reaction. In a sense the prior reaction may be equated to a deblocking step for revealing a functional group, yet it achieves far more toward the synthetic goal.

Reactions that require in situ generation of reactive species are tandem reactions. From this perspective whole classes or subclasses of ordinary reactions belong to them, for example Mannich reactions, Diels–Alder reactions of benzynes, cycloadditions of ketenes, carbene/nitrene insertions.

The organization of tandem organic reactions is not easy, due to the fact that some of them involve more than two consecutive transformations. It seems to this author most logical they be classified according to the last reaction. Such organization would be also of assistance to readers who wish to add the relevant transforms to his/her reference library reserved for retrosynthetic analysis.

2

ALDOL CONDENSATION

One of the most important CC bond forming reactions is the aldol condensation of two carbonyl compounds [Nielsen, 1968]. Related process are the Knoevenagel, Doebner, and Perkin reactions.

The condensation of two aldehydes is very facile. Although all steps are reversible, equilibrium favors the aldol. Sometimes dehydration occurs after the condensation, especially at higher temperatures, to give an unsaturated aldehyde directly. Mixed aldol reactions used to cause problems, but they can now be made chemoselective by techniques other than base catalysis [Mukaiyama, 1982]. Tremendous progress on stereocontrol of the reaction [Heathcock, 1981; Braun, 1987] has also been made.

Some other significant results relevant to the aldol chemistry include the generation of enolates under neutral conditions, for example by decarballyl-oxylation of the corresponding β-keto esters [Nokami, 1989a].

erythro: threo = 1:5

There are several tandem reactions which are terminated by an aldol reaction. For example, multiple aldol condensations may occur when a molecule contains several properly spaced carbonyl groups [Snider, 1983b].

3

2.1. MICHAEL–ALDOL TANDEMS

This combination has a long history in the famous Robinson annulation [Gawley, 1976] which consists of a base-catalyzed Michael reaction between a ketone enolate and an enone, an intramolecular aldol reaction, and subsequent dehydration. The enone may be generated in situ [Cornforth, 1949].

Despite certain problems of this reaction the Robinson annulation has served organic synthesis well over the years. It remains to be a method for consideration in the construction of a six-membered ring compounds, especially cyclohexenones. Because of the importance of this reaction many of its associated problems have been scrutinized to provide solutions by ingenious modification of substrates and reaction conditions. Thus, polymerization of vinyl ketones may be suppressed by introducing a readily removable silyl substituent to the α-position, regiochemistry is controlled by specific blocking or activation of the Michael donor. It has also been demonstrated that the Robinson annulation may be conducted under acidic conditions. In certain cases reaction at high pressure [Dauben, 1983] may prove advantageous.

The annulation of mesityl oxide with methyl vinyl ketone constitutes an attractive route to piperitenone [Beereboom, 1966]. The self-condensation of mesityl oxide to form isoxylitone may be restricted considerably by conducting the reaction under heterogeneous conditions (KOH in THF).

Innumerable syntheses are hinged on the Robinson annulation [Ho, 1988]. Noteworthy are those employing the Robinson annulation to construct an internal ring. The presence of a functionalized chain at the α′-carbon of the vinyl ketone generally simplifies a synthetic route. In the case of an approach to cortisone the A-ring assembly presented difficulties pertaining to stereoselectivity. By reversing the alkylation steps, that is incorporating the A-ring component in the enone for the annulation of a CD-ring synthon and effecting angular methylation after an intermediate containing the BCD-ring system was secured, the desired product became predominant [Stork, 1956; Barkley, 1956].

Another solution to this problem is a double annulation of a 1,3-cyclopentanedione which bears a vinyl ketone sidechain [Danishefsky, 1971].

An excellent synthetic intermediate is the Wieland–Miescher ketone which is formed by reaction of 2-methyl-1,3-cyclohexanedione with 4-diethylamino-2-butanone in the presence of pyridine. The activated 1,3-diketone undergoes Robinson annulation with ease. Interestingly, a more complex reaction path was observed with 1,3-cyclohexanedione itself in the presence of two equivalents of the Michael acceptor [Balasubramanian, 1974]. The formation of a hexalone was due to retro-Claisen cleavage of the six-membered ring of the 1:2 adduct which was followed by two aldol reactions.

2,5-Cyclohexadienones have been obtained from reaction with methyl ethynyl ketone [Woodward, 1950]. The yields are low, owing to the preferential formation of the (E)-enone intermediates which cannot cyclize.

Another aspect of the Robinson annulation which should be mentioned is that conjugated dienones and trienones may be used as the Michael acceptors, although the product yields tend to decrease progressively. Thus, a 19-norsteroid synthesis via C-ring formation [Eschenmoser, 1953] has been accomplished.

The intramolecular Michael–aldol combination for the formation of bridged ring system, as imposed by relative location of the two ketone groups, provided a precursor of the tricyclic ketone for the synthesis of zizaene [Alexakis, 1978].

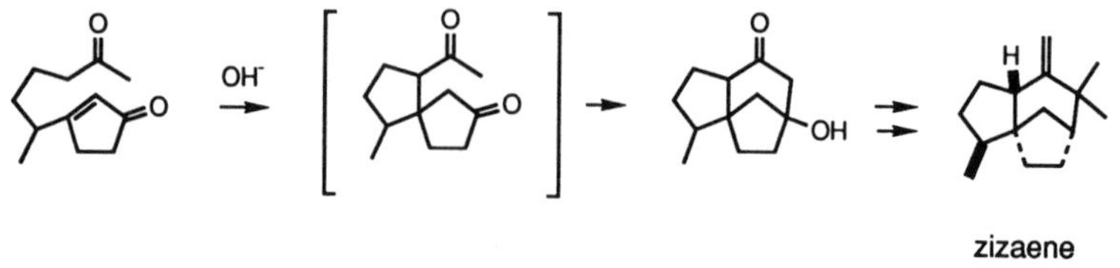

zizaene

A method for annulation of carbonyl compounds which also incorporates a butenolide moiety in the same operation is particularly useful for the synthesis of pertinent natural products. For example, it served to construct the cyclic skeleton of paniculide-A [Kido, 1982].

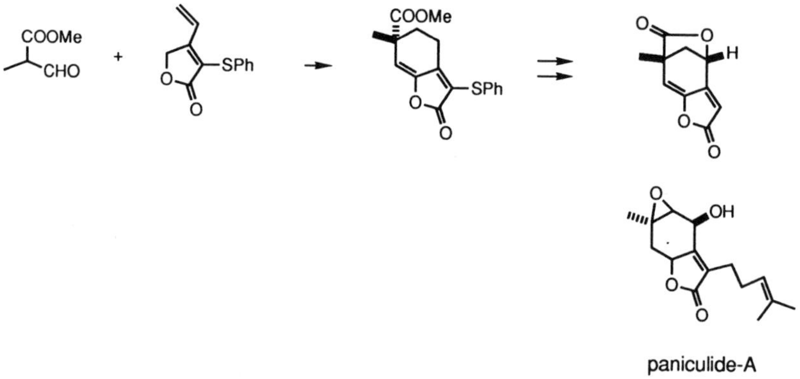

paniculide-A

An alternative ring construction by a Michael–aldol reaction tandem follows the 3C + 3C mode instead of the 2C + 4C combination. This alternative method most frequently requires α-methylenecycloalkanones as the acceptor species. A B-ring annulation for the establishment of a tricyclic steroid synthon [Hajos, 1973] is representative.

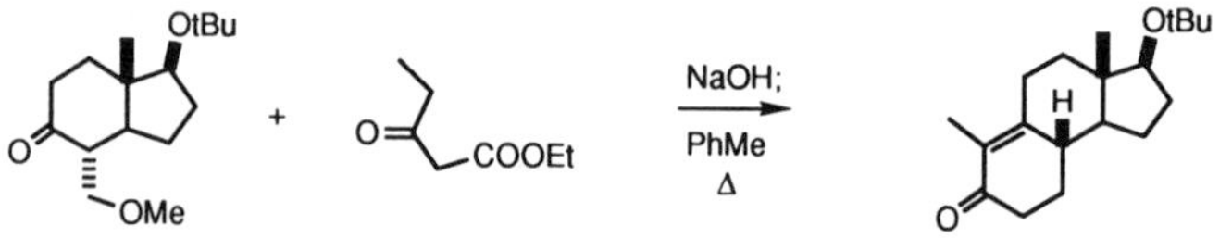

It should be noted that the Stork annulation of enamines with α,β-enones also has a 3C + 3C version [Meyer, 1986].

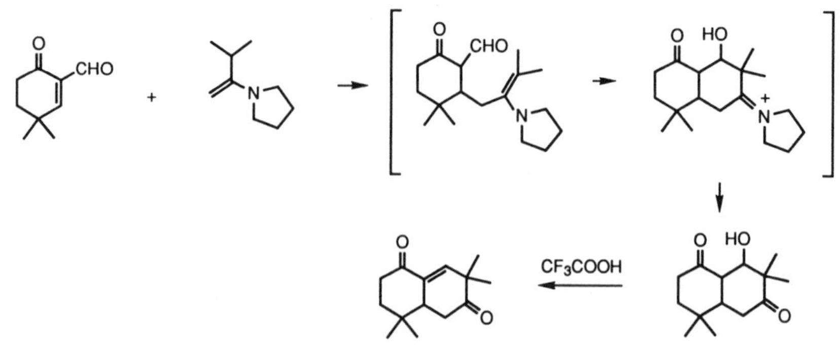

The *trans*-fused, angularly methylated hydrindanone system of steroids has been assembled by a novel approach based on intramolecular Michael–aldol reactions [Stork, 1982b,c]. Actually, adrenosterone was synthesized starting from this mode of assembly. The crucial stereochemistry at the ring juncture was dictated by the preferred conformation with chelation of oxygen atoms of both the donor and the acceptor moieties to the metal ion.

R = H, X = H₂
R = OMe, X = (OCH₂)₂

An intramolecular Michael–aldol tandem was the penultimate step of a shinjulactone-C synthesis [Collins, 1990]. The transannular bond formation process was initiated by deprotonation at C-5; the attack on C-13 brought C-1 and C-12 within bonding distance. This latter aldolization is a favorable reaction since two α-diketone systems, and hence dipole–dipole interactions, are removed.

shinjulactone-C

A synthon for forskolin [Koft, 1987] has been obtained from an acetylacetoxy β-ionone by way of a one-step Michael–aldol reactions and elimination of acetic acid. A very important aspect of this approach is that the resulting ketone possesses the relative configurations of C-1 and C-10 corresponding to the natural product.

forskolin

Two rings were created from a naphthalene derivative with appended sidechains which contain a Michael donor and a Michael acceptor, respectively, thus effectively providing a tetracyclic intermediate for anthracycline synthesis [Uno, 1984].

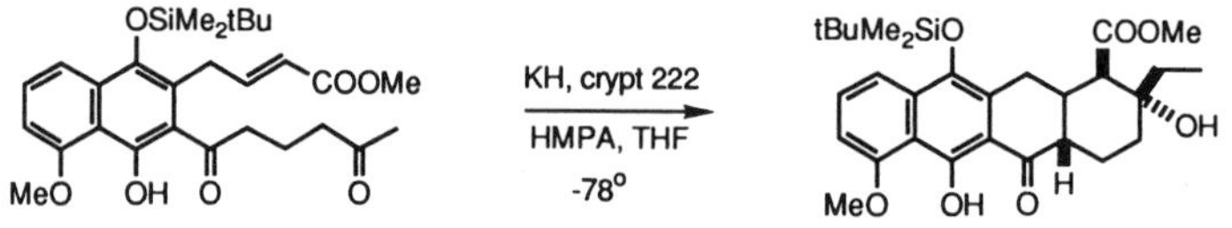

The Michael–aldol tandem also constitutes an excellent method for the construction of bridged ring systems [Danishefsky, 1969]. This theme reaction formed the basis for syntheses of patchouli alcohol, seychellene [K. Yamada, 1979], and picrotoxinin [Niwa, 1984].

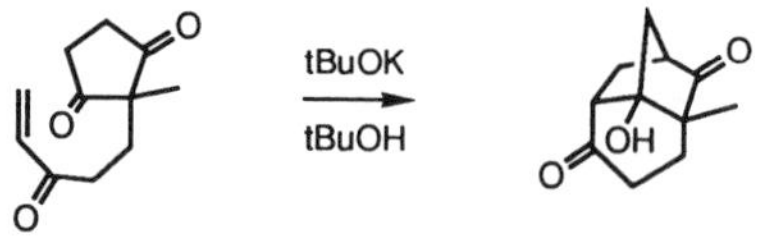

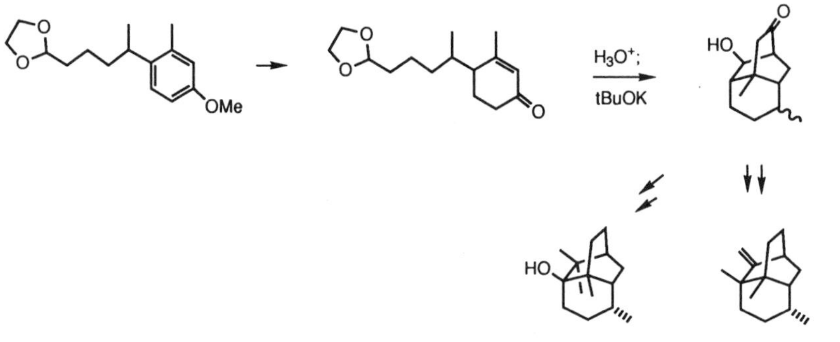

patchouli alcohol   seychellene

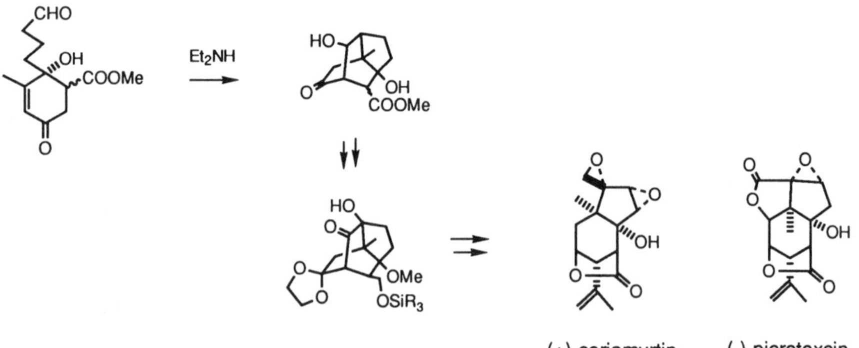

(+)-coriamyrtin   (-)-picrotoxcin

A 2C + 3C annulation via the Michael–aldol tandem results [Rajagopalan, 1989] when 2-arylsuccinimides are treated with base and enones.

Titanium(IV)-catalyzed condensation of the bis(trimethylsilyl) ether of methyl acetoacetate with β-alkoxy enones constitutes a convenient method for synthesis of substituted methyl salicylates [Chan, 1980]. Generally, the Michael reaction takes precedence over the aldol-type reaction and it is of interest to be able to control the regiochemistry by varying the electrophilic component. Thus the monoacetal of a β-dicarbonyl substance reacts initially at the C=O group.

The process embodying in situ generation of donor species by Michael addition of organometals to α,β-unsaturated carbonyl compounds and their intramolecular interception by properly placed acceptors has become a very effective ring formation method. It constitutes a key step in an elaboration of coriamyrtin [K. Tanaka, 1982] whereby the functional groups attaching to the six-membered ring were completely established in the two hydrindane products.

Although in one of the isomers the two new stereocenters are opposite to those present in the target molecule, these centers are epimerizable after slight modification. In other words, the other isomer is usable.

coriamyrtin

2-Aryl-1,3-dithian-2-yllithiums are sufficiently soft to behave as Michael donors toward various α,β-unsaturated carbonyl substances. This property has been exploited in the Michael–aldol tandem approach to certain lignans such as podorhizol [Ziegler, 1975] and galactin [Mpango, 1980].

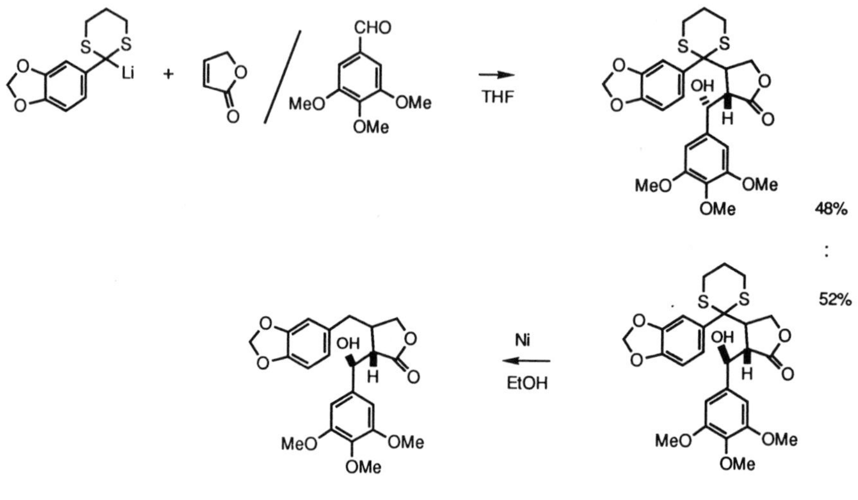

48%

:

52%

podorhizol

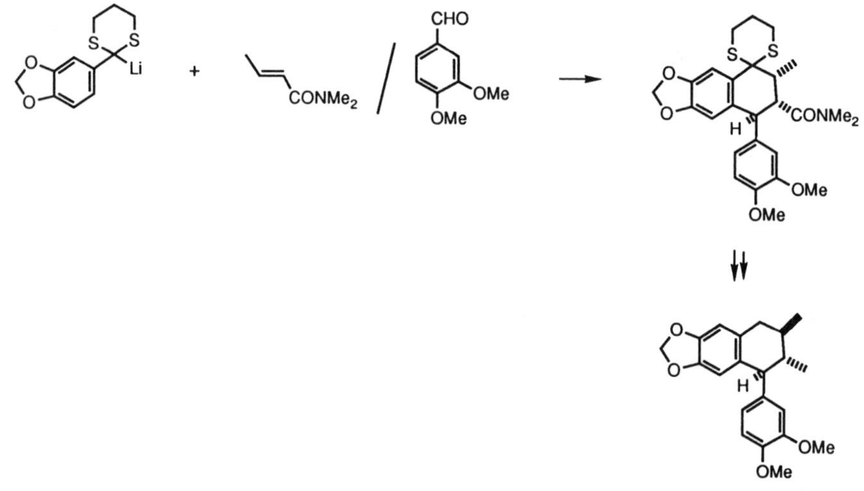

galactin

It is also possible to incorporate an organometallic donor with attenuated reactivity in the same molecule containing the acceptor carbonyl for the aldol condensation [Gommans, 1987]. The organometallic species must be a reactive Michael donor while being mild enough not to destroy the carbonyl.

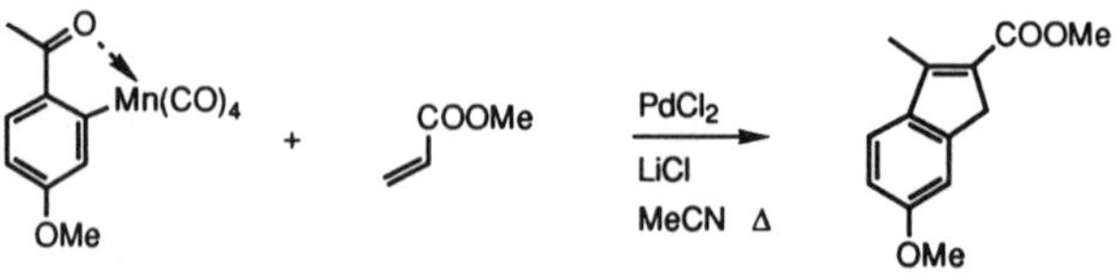

Organocuprate reagents attack CO ligands of complexed metal carbonyls. The ensuing acylmetals may act as Michael donors. A Michael–aldol route to a bicyclic ketone via acetyldemetalation and subsequent trapping has been witnessed [Rosenblum, 1990].

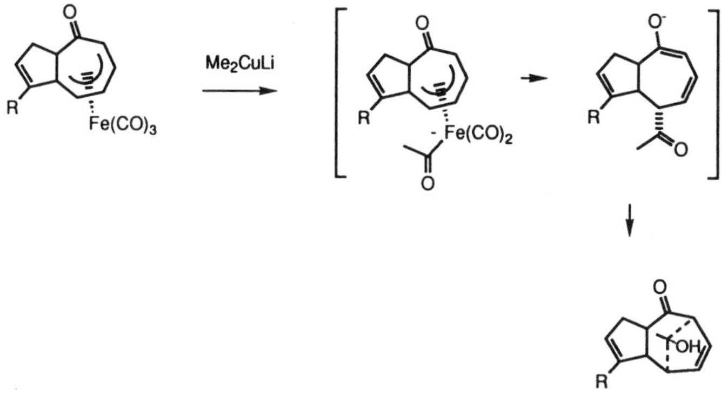

Besides construction of condensed ring systems, spirocyclic skeletons are also accessible by such a technique [Näf, 1975]. Also related to this Michael–aldol approach is the generation of cyclopentadienide species (donors) via a Michael reaction, its application having been illustrated in the syntheses of β-vetivone [Büchi, 1976] and $\Delta^{9(12)}$-capnellene [Y. Wang, 1990].

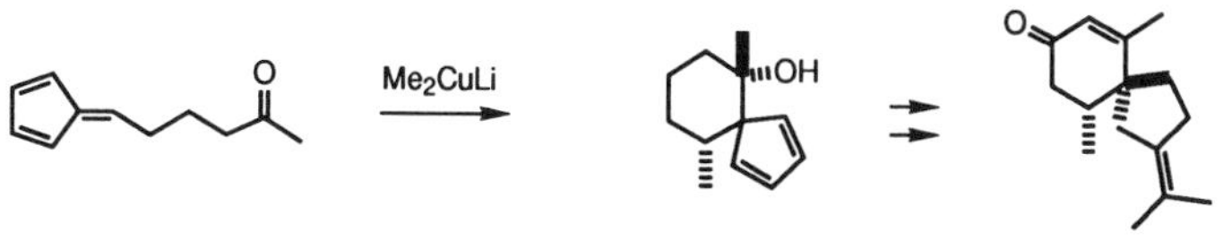

β-vetivone

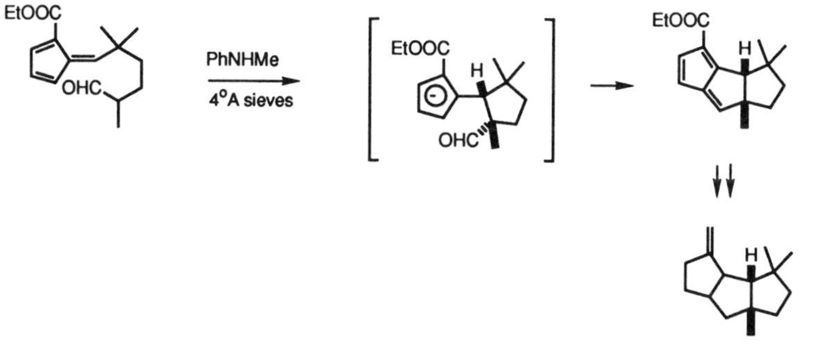

$\Delta^{9(12)}$-capnellene

The formal intramolecular Michael reaction in which the donor is generated by samarium(II) reduction of an alkyl iodide and trapping of the enolate adduct with an aldehyde has some potential in the area of diquinane synthesis [Curran, 1991b]. The Michael addition requires two electrons to complete, apparently single electron transfer occurs in rapid succession in such a process.

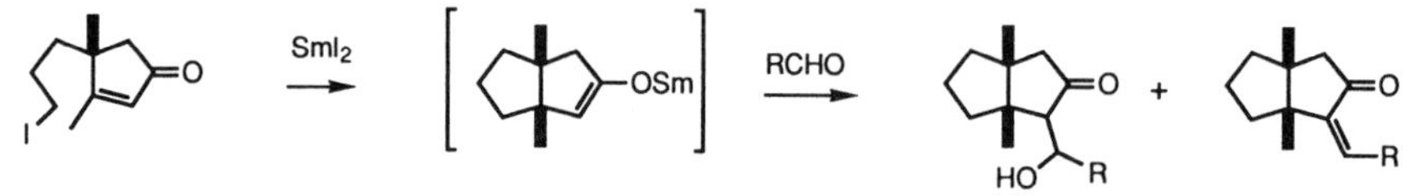

It has been observed that in some cases the Tishchenko reaction of the aldehyde was faster than the aldol condensation.

Ring formation by aldol condensation with two preceding Michael reactions offers efficiency and novelty. For example, in a model study for the synthesis of aflavinine [Danishefsky, 1985a] two rings were created in a one-pot process.

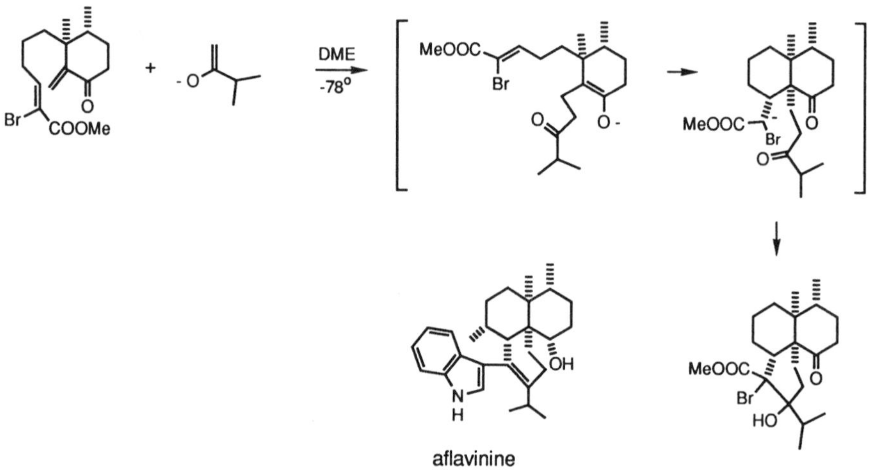

aflavinine

The termolecular version of the Michael–Michael–aldol annulation provides isophthalic esters [Boekelheide, 1969].

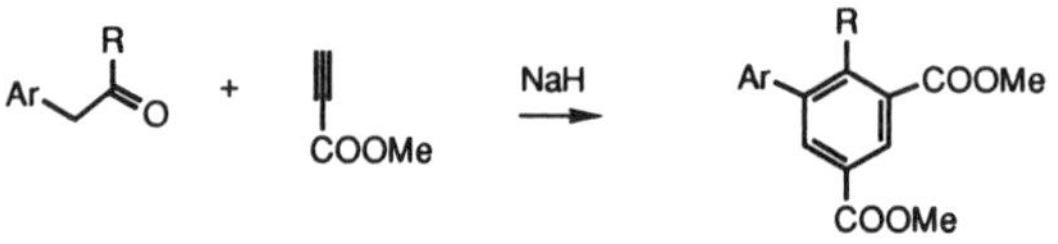

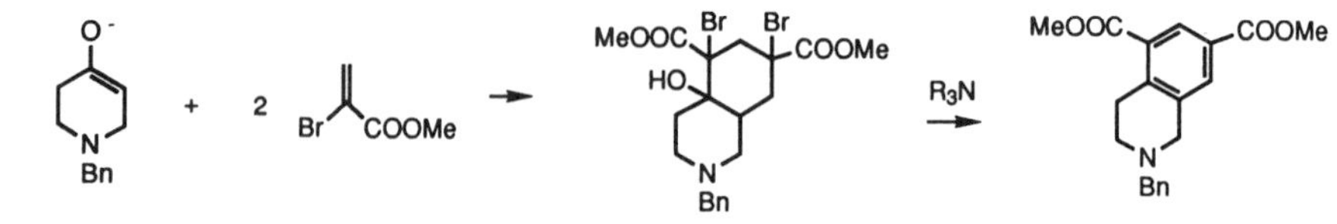

When methyl vinyl ketone or a simple acrylic ester is employed as the Michael acceptor the annulation gives rise to cyclohexane derivatives [Wakselman, 1973; Posner, 1986b]. Remarkably, in the former case the Cu(I) species present suppressed the Robinson annulation while promoting another Michael reaction.

The synthetic value of multicomponent annulation depends on the controllability of reactants to enter certain phases of the reaction sequence. It has been possible to accomplish the following transformation by operating at low temperatures [Posner, 1986a].

An unexpected double Michael reaction and aldolization tandem involving the reaction of a kinetic enolate and two molecules of methyl acrylate is depicted in the following equation.

By generating the Michael donor by means other than enolization, additional synthetic possibilities may be developed. Thus, an access to mesocyclic substances [Posner, 1988] may exploit the conjugative addition of a triorganotin anion to a cyclic α,β-unsaturated carbonyl compound as a prelude to the annulation, since the γ-hydroxystannane product is liable to oxidative fragmentation.

Cyclization of 2-ene-1,4-diones has been effected by the addition of methoxide ion [Piancatelli, 1980]. The reaction giving rise to 4-methoxy-2-cyclopentenones lacks regiocontrol except for symmetrical substrates and those in which one of the ketone groups is nonenolizable.

One of the dimers formed by treatment of prostaglandin-B with alkali has a pentacyclic structure [Toda, 1982]. Apparently, a reflexive Michael reaction between two molecules, initiated by attack of the hydroxyl group at C-15, forms a tetrahydrofuran derivative which is followed by a Michael–aldol reaction tandem.

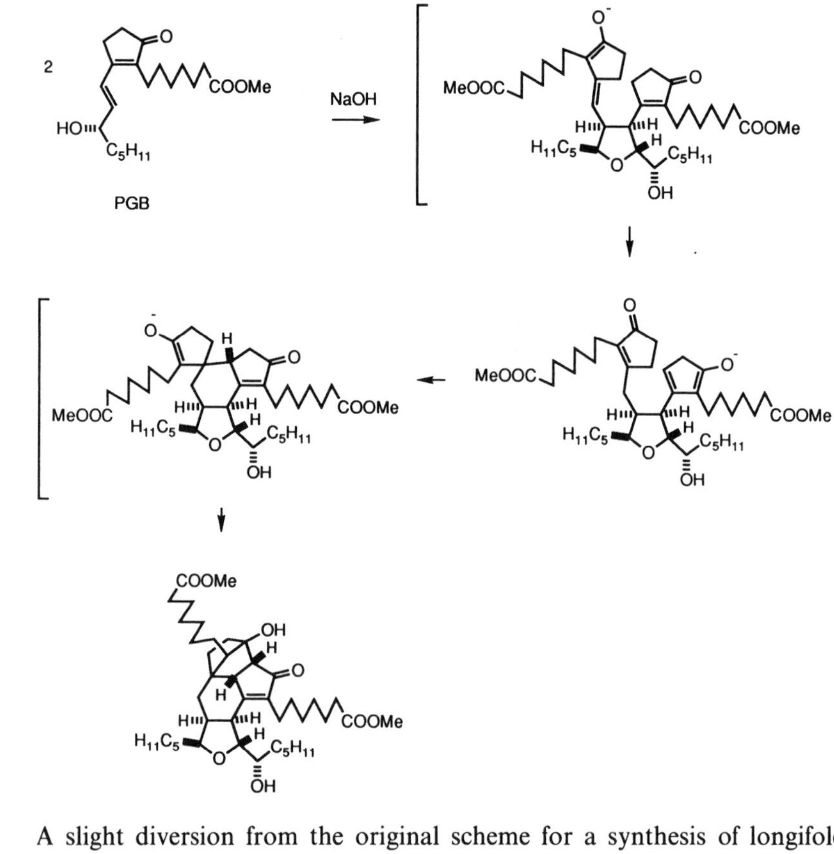

A slight diversion from the original scheme for a synthesis of longifolene [McMurry, 1972] was due to an intramolecular aldol condensation upon establishment of the gem-dimethyl subunit by conjugate addition to an enone system. Fortunately, the carbocycle formed by this aldol step could be reopened by a Grob fragmentation at a subsequent stage.

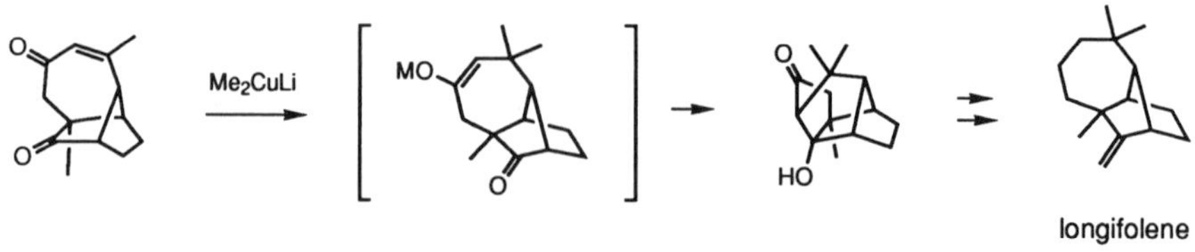

Certain bridged α-sulfonyl γ-lactones undergo Grob fragmentation and aldolization which is initiated by conjugate addition of methoxide ion to the highly strained (E)-cycloalkene ester intermediate [Kende, 1989].

The reaction of β-heterosubstituted carbonyl compounds with Michael acceptors can lead to six-membered unsaturated heterocycles. The Michael donor may be present in a latent form if it can be generated in situ [McIntosh, 1977].

A neat approach to emetine [D.E. Clark, 1962] exploited the hidden symmetry of the molecule. Disconnection of the C-ring of emetine together with functional group addition operation and retro-Michael transform reveals a 1,3-bis(tetrahydroisoquinolin-1-yl)acetone. The Michael reaction of this ketone with methyl vinyl ketone is expected to form a 1:2 adduct, but since there is only one acceptor group for the subsequent aldol condensation the other nitrogen atom remains in a protected form and the secondary amine will be uncovered later by removal of the methyl vinyl ketone unit.

Tertiary amines may be used as Michael donors, as demonstrated in a general and convenient synthesis of 2-(hydroxylalkyl)-2-propenoic esters [H.M.R. Hoffmann, 1985]. This DABCO-catalyzed coupling of aldehydes with methyl acrylate apparently involves the formation of a zwitterionic adduct of DABCO and the acrylic ester and subsequent reaction with the aldehyde. The termolecular adducts decompose with regeneration of the catalyst. The overall reaction is a Michael–aldol–elimination tandem.

A convergent synthesis of resistomycin [T.R. Kelly, 1985a] started from Friedel–Crafts acylation of a 9,10-dihydroanthracene derived from emodin. The Friedel–Crafts reaction set up the acid-catalyzed Michael- and Claisen-type condensations.

resistomycin

It is interesting to note that the acidity of the reaction medium for the first stage of the annulation sequence is very crucial to the regiochemistry. Omitting dichloroethane shifted the acylation site to the other terminal ring due to protection of the more nucleophilic nucleus by protonation.

δ-Hydroxy-α,β-enones also fragment (vinylogous retro-aldol fission) on base treatment. A complex transformation may ensue when additional functionalities are unveiled from the retroaldol fission. For example, the generation of a new enone system from the deannulation of a hydrindenone permitted an intramolecular. Michael addition and vinylogous aldol condensation to follow [Dauben, 1977a].

The Darzens reaction combines an intramolecular O-alkylation with an aldol (Knoevenagel) condensation. The Darzens reaction can be initiated by a Michael reaction [D.R. White, 1975] to allow formation of cycloalkene oxides in one operation. A potential A-ring synthon for aklavinone has been obtained by this method [Bauman, 1984].

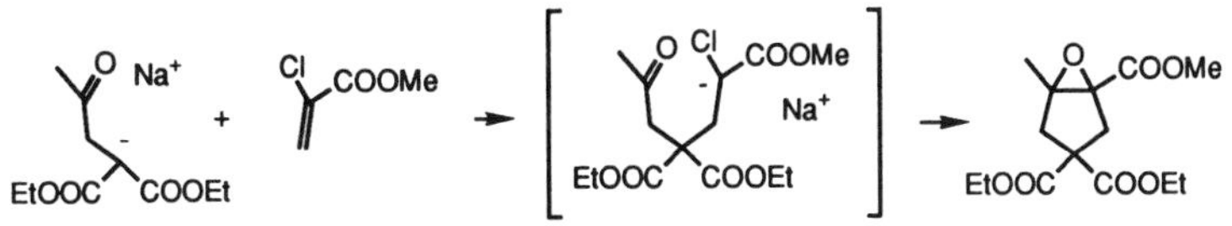

Conjugated nitroalkenes are excellent Michael acceptors, therefore the Michael–Henry reaction tandem is analogous to the Michael–aldol process.

The condensation of β-nitrostyrenes with β-dicarbonyl compounds furnishes directly 3-aryl-5-hydroxy-4-nitrocyclohexanones [Ehrig, 1975] which are obvious synthetic intermediates of certain Amaryllidaceae alkaloids [Weller, 1982].

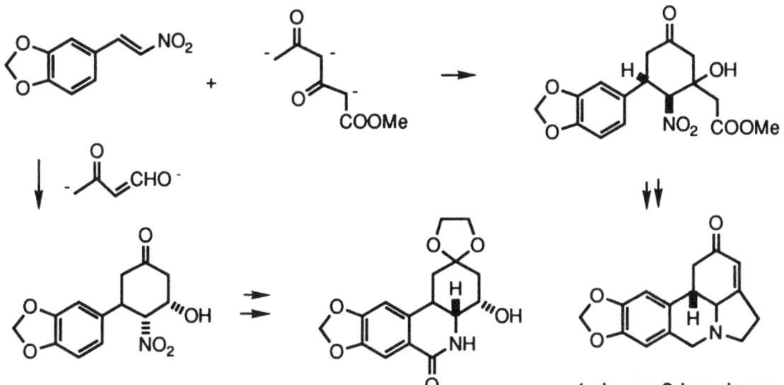

1-deoxy-2-lycorinone

With a monoenamine of an α-diketone as the Michael donor and Henry acceptor the reaction with a nitroalkene gives a cyclopentanone derivative [Felluga, 1991].

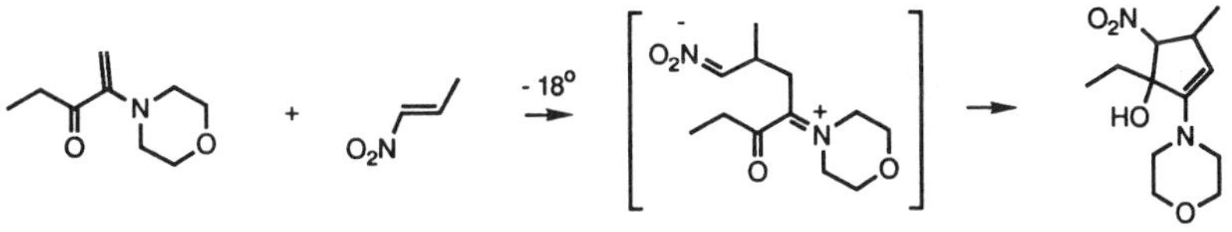

Electroreduction of a hexalindione effected a Michael reaction. Further reduction led to the bisenolate which is sterically disposed to form an aldol [Margaretha, 1982].

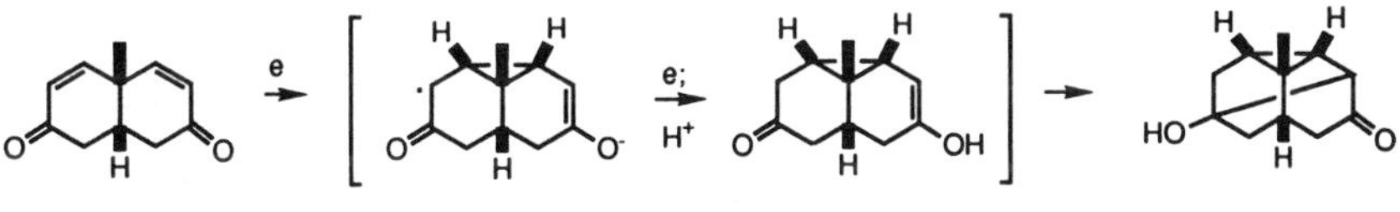

A rearrangement which may be regarded as tandem retro-Homo Michael–aldol reactions is also mentioned here. It is essentially the latter part of a cyclopentannulation protocol featuring an initial cyclopropane formation. A synthetic approach to specionin [Hudlicky, 1990] embodying this rearrangement gives a clear illustration of the process and an indication of its synthetic potential.

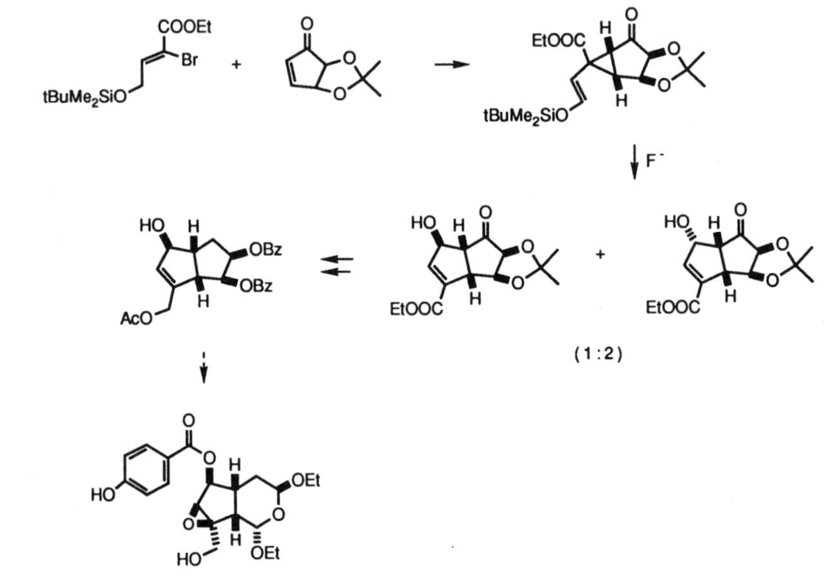

specionin

2.2. CLAISEN–ALDOL AND MULTIPLE ALDOL CONDENSATIONS

A rather general method for the synthesis of phenolic compounds related to polyketide natural products involves Claisen and aldol condensations The simplest examples are the formation of methyl orsellinate from methyl acetoacetate and diketene [T. Kato, 1972] and that of *tert*-butyl orsellinate 4-methyl ether [Stockinger, 1976].

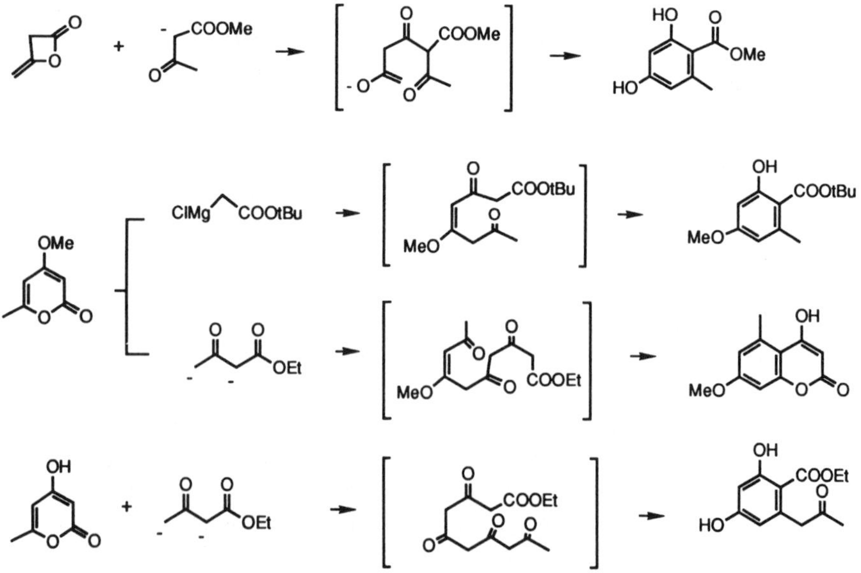

The reaction of 4-methoxy-6-methyl-α-pyrone with the dianion of acetoacetate furnishes a coumarin. Interestingly, the unprotected triacetic lactone reacts in a different manner, giving a resorcylate ester [Stockinger, 1976]. This Claisen–aldol approach can be extended to the synthesis of various homologs.

Anthracene derivatives can be rapidly assembled via condensation of a 2,4-pentanedione dianion with a dialkyl 3-oxoglutarate in which the ketone group is protected or in a reduced form [Harris, 1976].

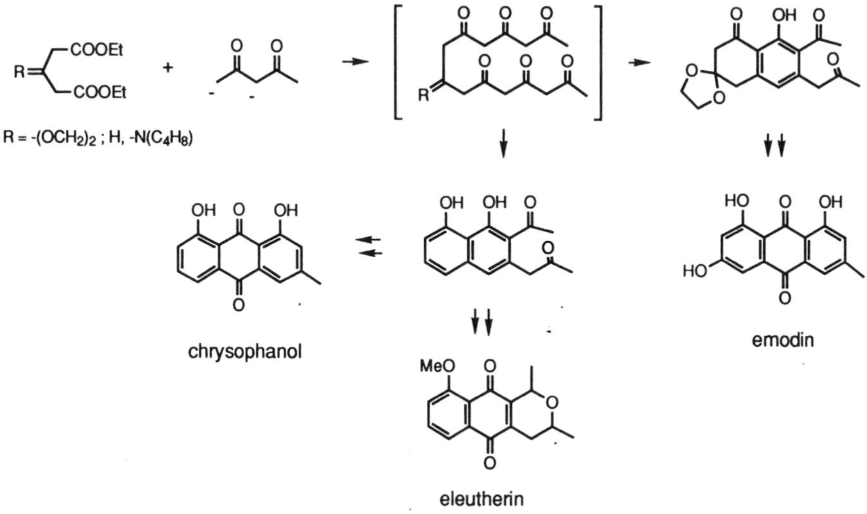

chrysophanol

emodin

eleutherin

Another strategy of polyaromatic synthesis is based on a Claisen–aldol tandem with homophthalate esters [M. Yamaguchi, 1990]. Formation of angular isomers (e.g. phenanthrenes vs anthracenes) is disfavored due to peri-interactions.

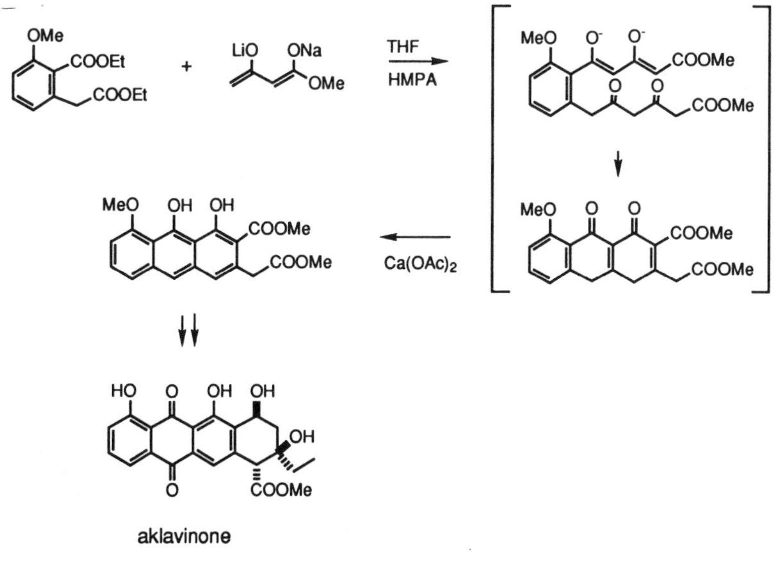

aklavinone

The groundwork of poly-β-carbonyl compounds was laid by the study of 2,4,6-heptanetrione [Collie, 1922] and this and related studies led to the formulation of the polyketide theory for the biogenesis of many aromatic substances.

2,4,6-Heptanetrione cyclizes in acid or under strongly alkaline media to give orsinol; however, on treatment with barium hydroxide dimerization (aldol tandem) occurs to yield a naphthalenediol.

The bimolecular tandem aldol reactions has also been applied to a synthesis of emodin [Mühlemann, 1951].

emodin

Long-chain poly-β-carbonyl compounds tend to undergo cyclization via aldol tandems. For example, a naphthalene derivative has been obtained from the self-condensation of a hexaone [Wittek, 1973].

The preparation of 6-methylsalicyclic acid by the cyclization protocol necessarily involves an intermediate in which the oxygen function at C-4 is deleted. An approach is by a sequence of inter- and intramolecular aldol condensations [Harris, 1974].

Perhaps the most complex molecules prepared thus far by the Claisen–aldol approach is pretetramide and 6-methylpretetramide. Because of the great number of β-dicarbonyl subunits, many strategies may be considered.

The $[3 + (2 \times 2) + 1 + 2]$ approach [Gilbreath, 1988] starts from a bifunctional chain of three C_2 units. Extension at each terminus a C_4 (i.e. $2 \times C_2$) chain is attended by cyclization. After proper modification of the naphthalene derivative thus produced, another C_2 fragment, and finally a C_4 group, are added in succession. Since a primary amide is needed in pretetramide, an isoxazole derivative is the most useful C_4 group to use in the final condensation.

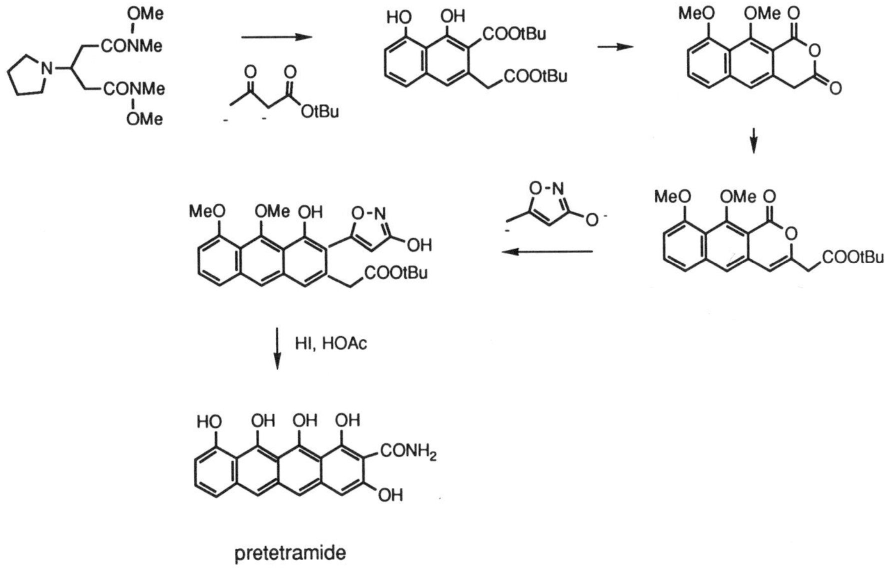

pretetramide

Based on a condensation of homophthalate esters with a dianion of methyl acetoacetate, a $[5 + (2 \times 2) + 1]$ route and a $[5 + 2 + 3]$ route to pretetramide have been developed [Haris, 1988b]. The former method can be modified to gain access to 6-methylpretetramide [Harris, 1988a].

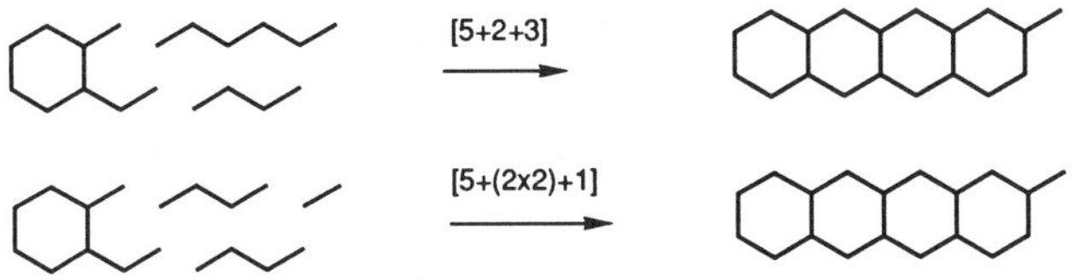

The annulation pathway of an α-tetralone leading to a B-aromatic tricyclic synthon of D-homosteroids [M.J.T. Robinson, 1957] is not clear. However, a Claisen–aldol route is a distinct possibility in view of the hindered nature of the α-tetralone carbonyl. In other words, formation of a β-keto ester preceded the aldol condensation.

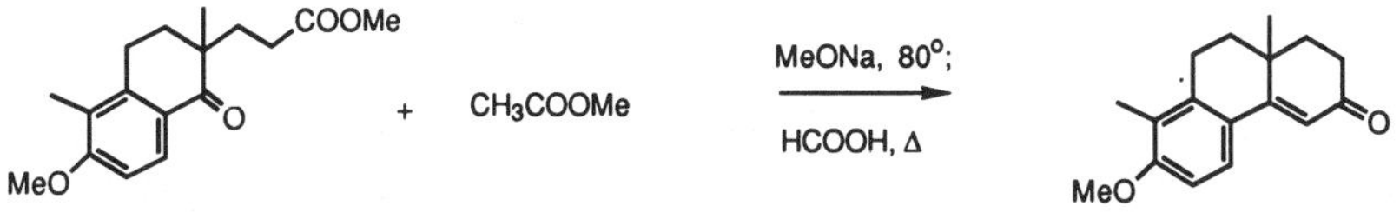

Tandem aldol condensations have a long history in the service of synthesis. Both self-condensation of dicarbonyl substances (e.g. formation of 2,5-dimethyl-1,4-benzoquinone from biacetyl [Pechmann, 1888]) and mixed aldolizations are well-known. Only a few examples from the latter category need mention [Dilthey, 1930; Prelog, 1947; Woodward, 1950b; Ried, 1953].

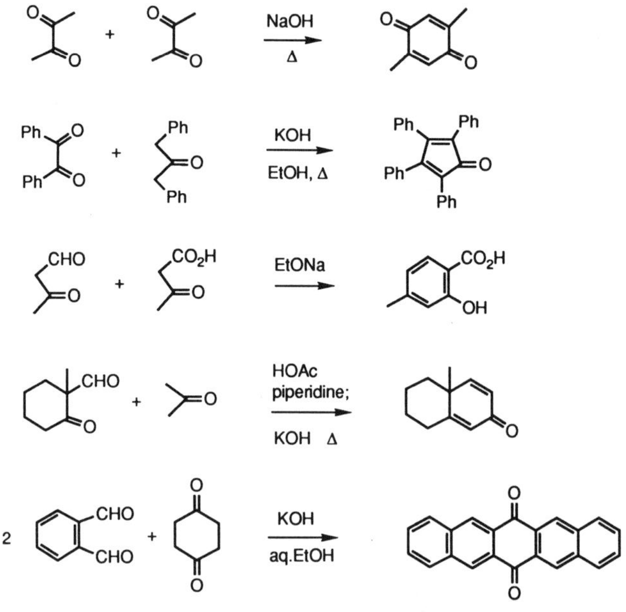

6,13-pentacenequinone

The synthesis of 8-oxabicyclo[3.2.1]octane-3-ones from 1,4-dicarbonyl compounds is a formal double aldol reaction [Molander, 1989, 1991].

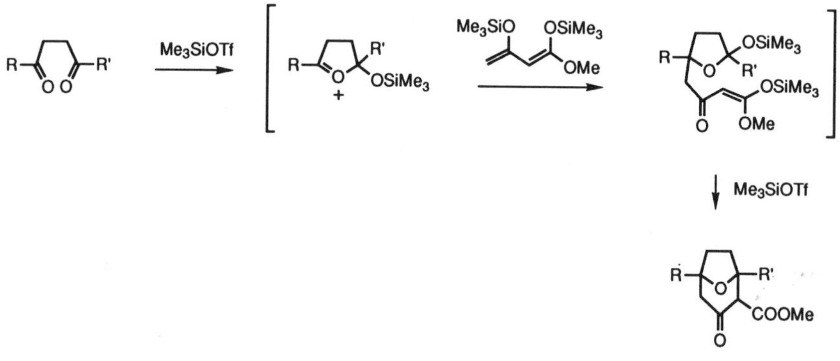

2.3. RETRO-ALDOL–ALDOL TANDEMS

Since the aldol condensation is reversible the retroaldol fission is a common occurrence under the same conditions. The combination of retro-aldol fission and aldol condensation in a different fashion is of some synthetic significance.

For example, the kinetic formation of bridged ring systems in the Robinson or other annulation methods may be corrected by such a reaction tandem. The conversion of the bridged ketol from a Grignard reaction on an enol lactone to a precursor of estrone [Bucourt, 1967] is a case in point.

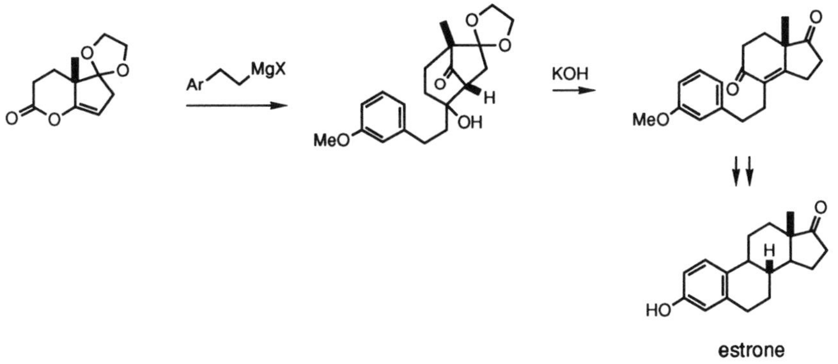

estrone

Triticone-A and triticone-B are interconvertible and racemized in solution, undoubtedly via a retro-aldol–aldol reaction sequence [Sugawara, 1988].

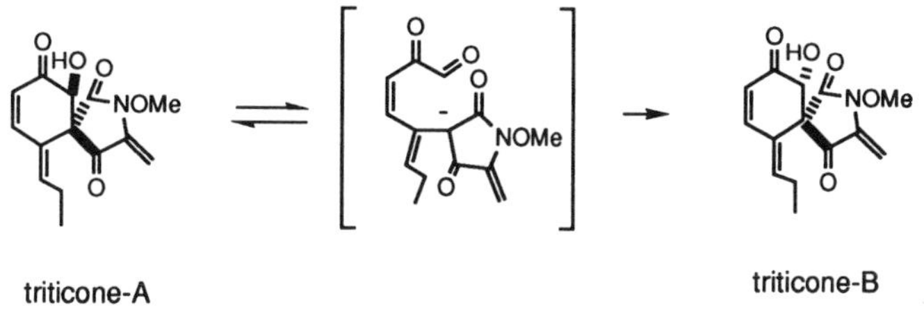

triticone-A   triticone-B

There are many intriguing transformations of natural products and their derivatives which are relevant to the topic under discussion. A representative of these is the generation of an aromatic compound by base treatment of the compound obtained from saponification and periodate cleavage reaction of oxonitine [Wiesner, 1963].

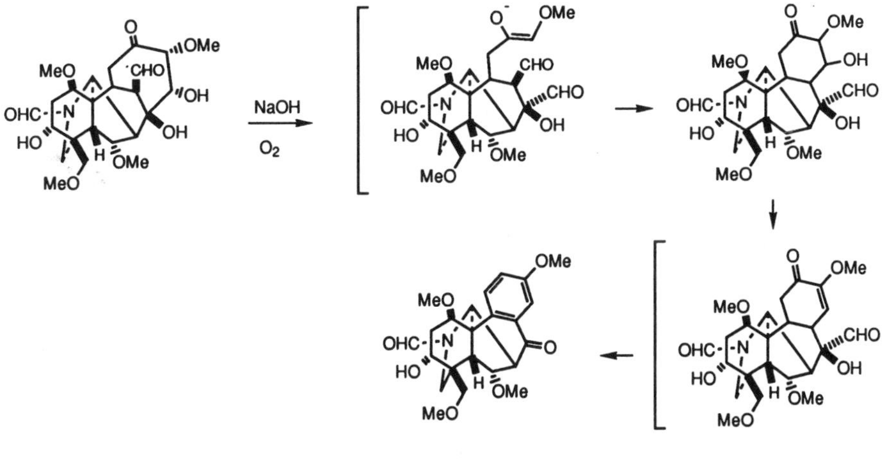

The rearrangement of tenulin into desacetylneotenulin [Herz, 1962] has been clarified. The formation of the latter compound is readily described by a mechanism consisting of retro-aldol fission, double bond migration, and vinylogous aldol condensation at the donor site which is the β-carbon of the original cyclopentenone system.

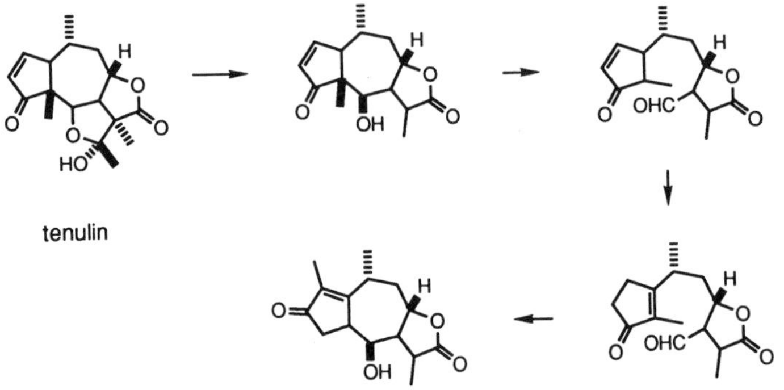

tenulin

The vigorous saponification of forskolin yielded a cycloheptanone [Saksena, 1985]. The inversion of the C-1 configuration indicates a hidden retro-aldol–aldol reaction sequence.

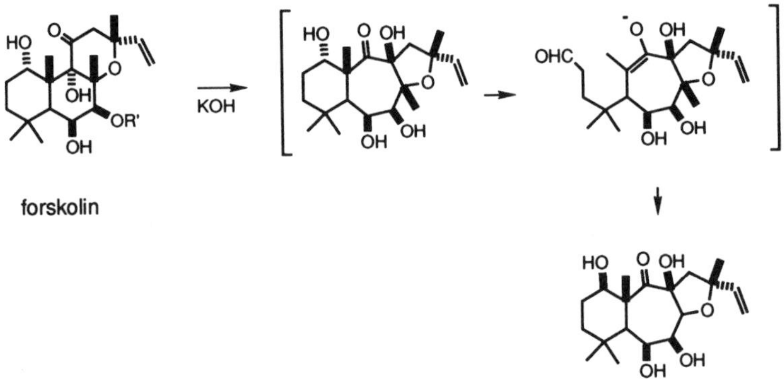

forskolin

The epimerization of the A-ring hydroxyl group of gibberellins, for example GA_1 methyl ester, by treatment with dilute alkali involves seco intermediates [Cross, 1961; MacMillan, 1967].

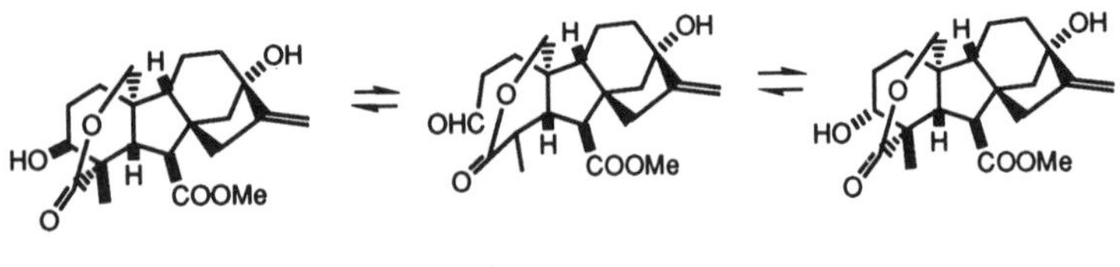

16-Acetoxy-17α-hydroxy-20-keto steroids undergo D-homoannulation in the presence of base [Wendler, 1960]. This transformation involves cleavage and reclosure of the D-ring.

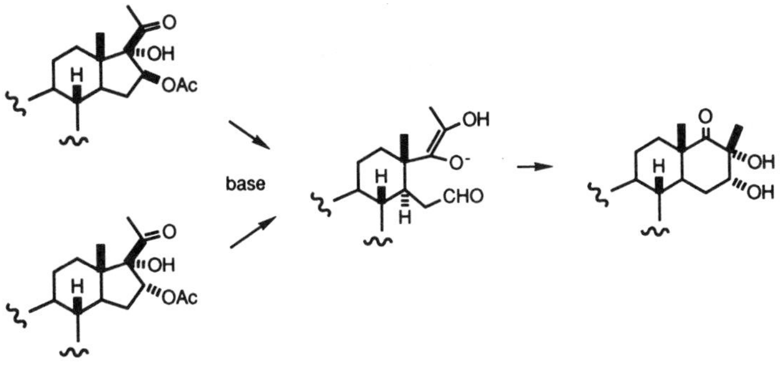

By virtue of a thermal equilibration via a retro-aldol–aldol reaction tandem the major but otherwise useless product for a synthesis of (-)-phytuberin [Findlay, 1980], obtained from kinetic hydroxymethylation of dihydrocarvone, can be converted into a mixture enriched in the desired epimer.

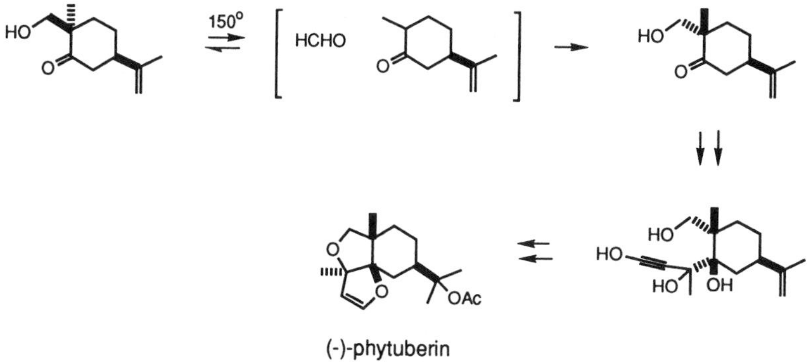

(-)-phytuberin

It is interesting to note that O-methylation of both epimeric 4-hydroxy-2-adamantanones resulted in the same product [Duddeck, 1979]. Apparently the equatorial alcohol underwent methylation more rapidly, whereas the axial isomer equilibrated with the former compound via the retro-aldol–aldol reaction circuit.

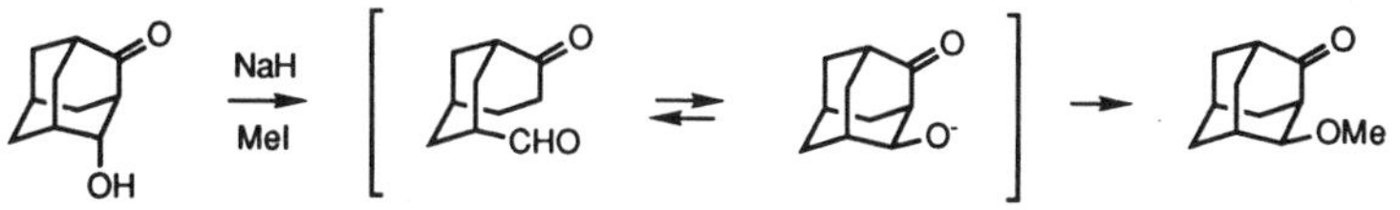

β-Ketol systems in which the hydroxyl group is attached to a small ring are very susceptible to retro-aldol fragmentation. Under basic conditions the cleavage products most frequently would recyclize [Trost, 1972].

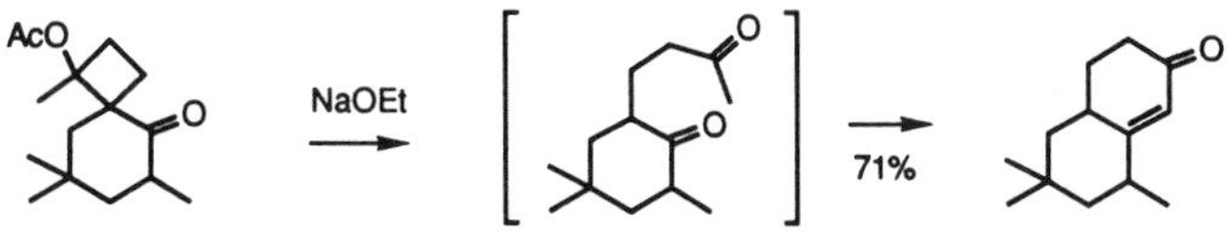

A spirocyclic β-lactone in which the β-carbon is also β- to a cyclohexanone carbonyl was converted into a bridged ring system on treatment with aniline [T. Kato, 1978], as a result of transacylation, retro-aldol cleavage, and an intramolecular aldolization. The final enamination of the β-ketoamide is a trivial reaction.

On treatment with mercury(II) perchlorate in an aqueous medium, the photocycloadducts of certain cyclohexenones and allene undergo mercurio-hydroxylation, retro-aldol, demercuration, and alfol condensation in sequence [Duc, 1981].

During a synthesis of longifolene [Oppolzer, 1978b] which was based on an intramolecular de Mayo reaction the unraveling of the bridged seven-membered ring by alkali treatment of the tertiary acetate also led to aldolization after the retro-aldol fission. In this case, steric proximity of the two ketone groups favors the observed reaction. This undesirable transformation was avoided by changing the de Mayo reaction substrate to an enol carbonate so that the ring cleavage step proceeded under nonbasic conditions.

On the other hand, the retro-aldol–aldol reaction tandem which occurred en route to daucene [Seto, 1985] served the purpose of selective protection of one of the two carbonyl functions.

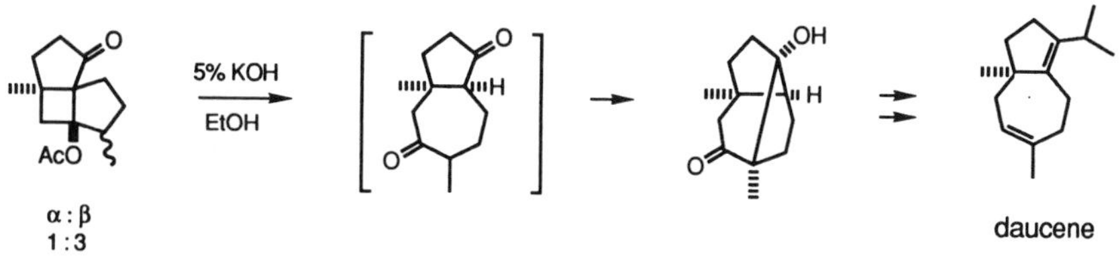

α : β
1 : 3

daucene

A short synthesis of hirsutene [Disanayaka, 1985] based on the intermolecular de Mayo reaction to create a 5:8-fused diketone intermediate directly was unfortunately marred by the predominant formation of the undesirable regioisomer.

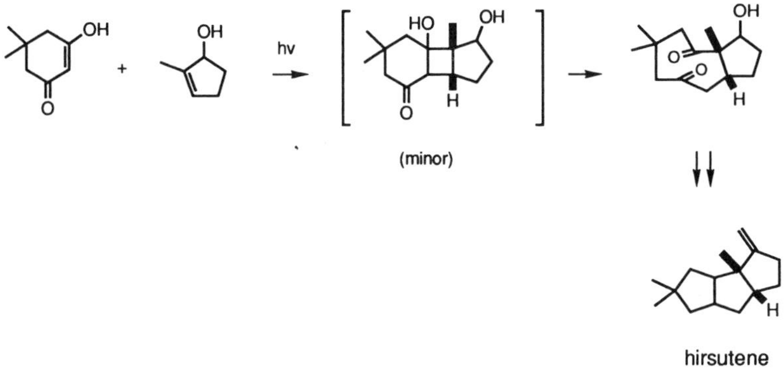

(minor)

hirsutene

The same reaction tandem is also crucial to the elaboration of atisine [Guthrie, 1966] and ishwarane [R.B. Kelly, 1972], among many other natural products. The precursors are available from photocycloaddition and proper modification of the adducts.

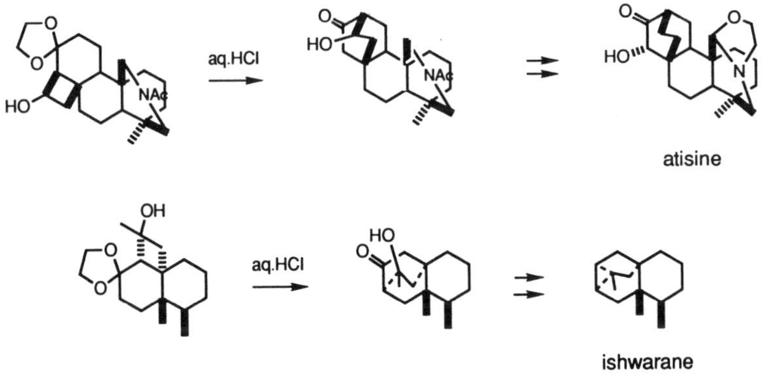

atisine

ishwarane

2-Acetoxy- and 2-alkoxycyclopropyl ketones are even more readily converted into cyclopentenones via the retro-aldol–aldol reaction sequence. Many natural products containing a five-membered ring have been synthesized by this method: β-vetivone and cuparene [Wenkert, 1978], and jasmonoids [Wenkert, 1970; McMurry, 1971].

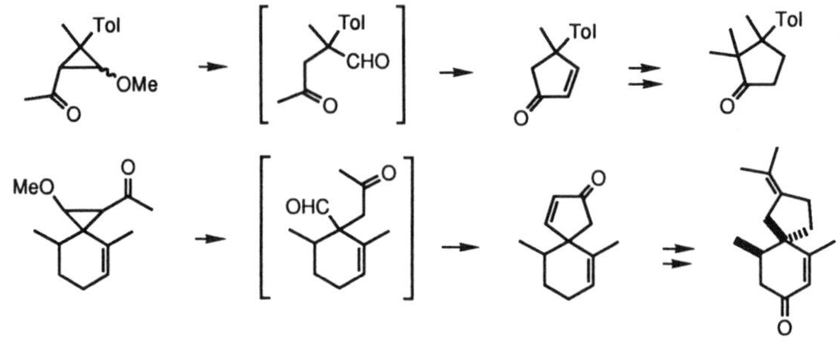

β-vetivone

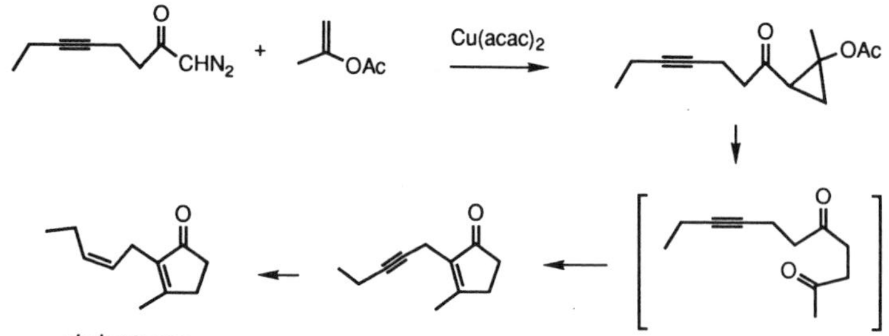

cis-jasmone

The cyclopropyl ketones may be obtained conveniently by carbenoid addition to enol ethers or enol esters.

Cyclopropylcarboxylic esters bearing an enol ether substituent at C-2 are susceptible to ring opening on exposure to Lewis acids. Spontaneous cylization of the resulting switterions is also expected [Harvey, 1991a].

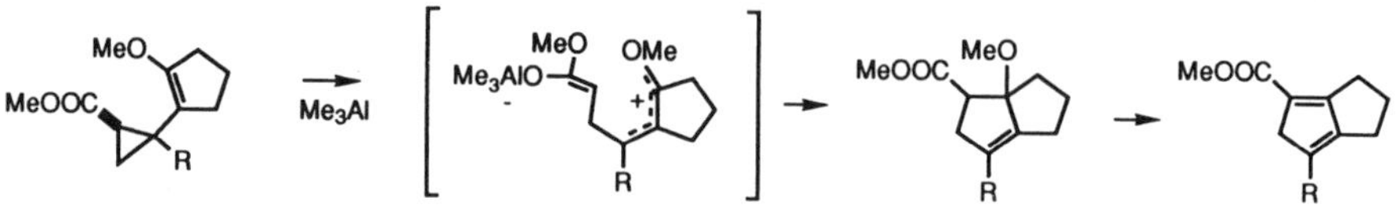

Acetolysis of the *endo*-tosylate of 1,5-dimethyl-9-oxobicyclo[3.3.1]nonan-2-ol [J. Martin, 1972] afforded two products with rearranged skeletons, besides others. A condensed bicyclic enone can be viewed as arising from the 1,2-acyl migration, hydration, retro-aldol and aldol reaction tandem.

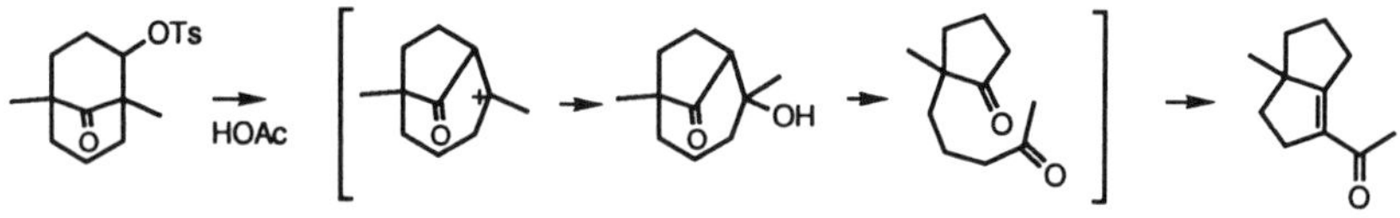

2.4. MANNICH–ALDOL TANDEMS

Several reports on annulation consist of tandem Mannich–aldol reactions. Thus, the liberation of di(3-formylpropyl)amine from the bis(dimethylacetal) in the presence of a β-keto ester led to an indolizidine. Elaokanine A can be acquired by proper treatment of the condensation product [Gribble, 1988].

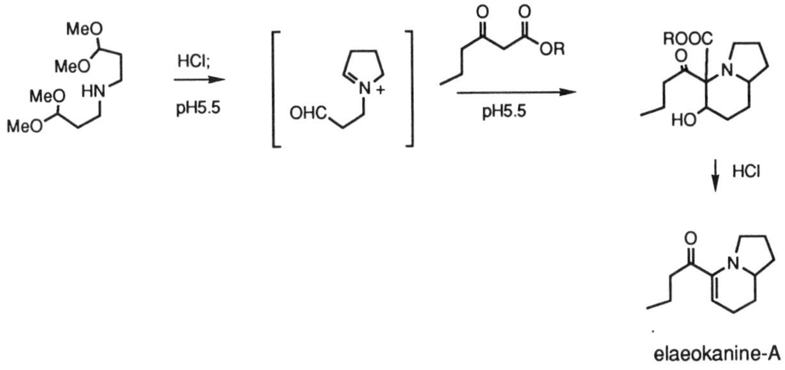

elaeokanine-A

Elaboration of the tetracyclic skeleton of the olivacine-type alkaloids via closure of the internal benzene ring is a very intriguing process, in view of the complete loss of a tryptamine sidechain [Takano, 1979b]. The first CC bond formation is a Mannich reaction, and the cyclization step may be regarded as an enamine-mediated intramolecular aldolization. Aromatization of the newly formed ring provides the driving force for the sidechain elimination.

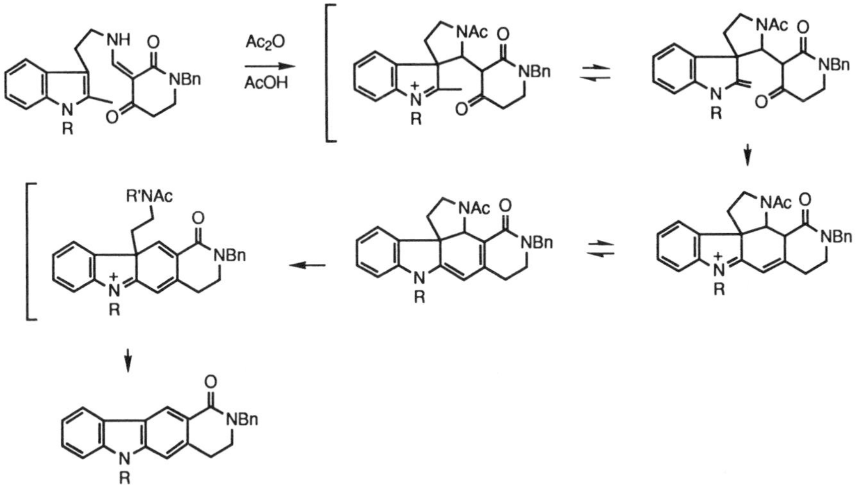

2.5. REARRANGEMENT–ALDOL TANDEMS

As oxy-Cope rearrangement produces enol intermediates, aldol condensation may be induced if another carbonyl group is present or co-produced in this prior step. The direct isolation of two tricycline enones from pyrolysis of 2-vinyl-3-propenyl-2,3-bornanediol [Leriverend, 1970] attests to the facility of the intramolecular aldolization of the nascent mesocyclic 1,6-diones. The minor product has been converted into β-patchoulene.

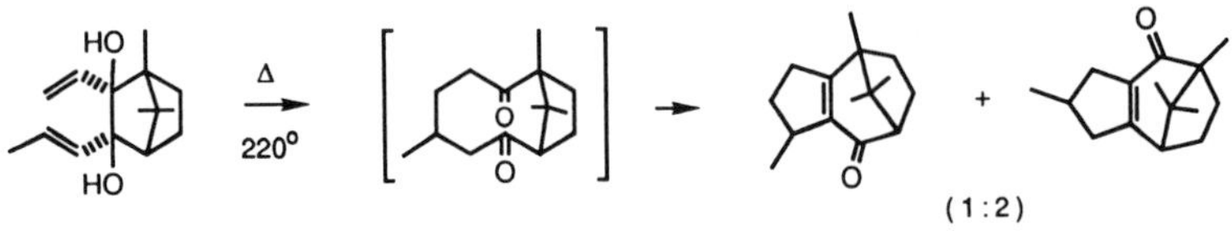

(1:2)

Another useful annulation method consists of vinylogous aldolization following an anionic oxy-Cope rearrangement [Raju, 1980; Sathyamoorthi, 1989, 1990].

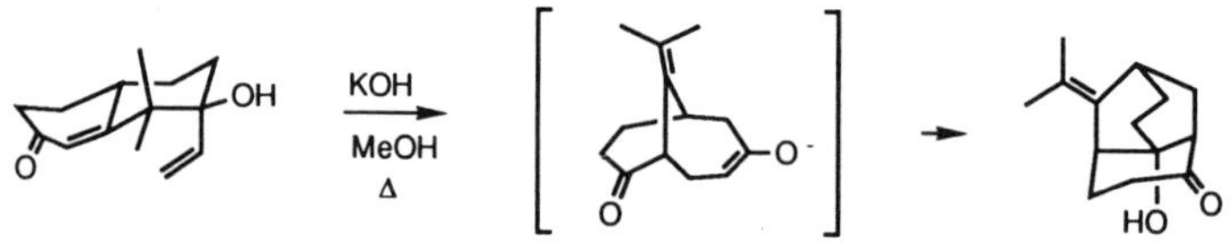

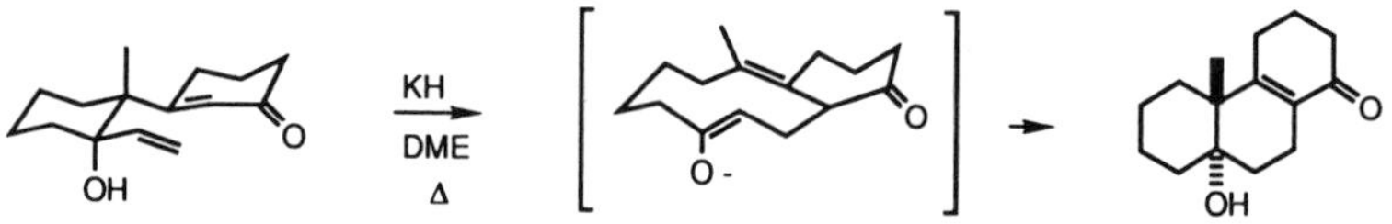

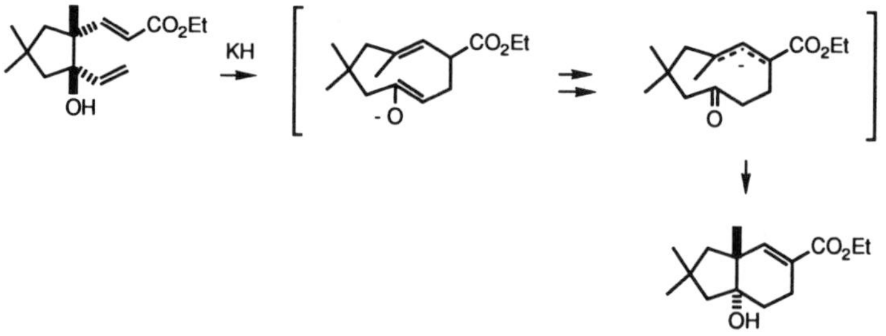

A further variation involves an aldol-type condensation which is promoted by hydroxylation of an anionic oxy-Cope rearrangement product [Paquette, 1989a]. Transannular addition of the conjugate base of the α-hydroxy ketone generates the carbanionic species.

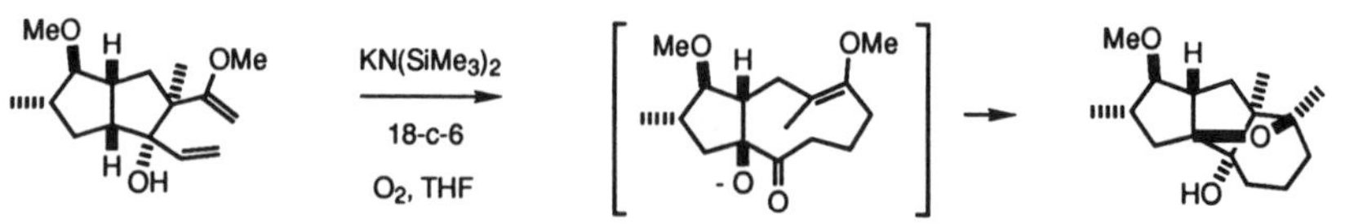

A convenient method for the synthesis of bicyclo[3.3.0]octa-2,6-diene derivative [Mandai, 1991a] is by a tandem Claisen rearrangement–aldol condensation.

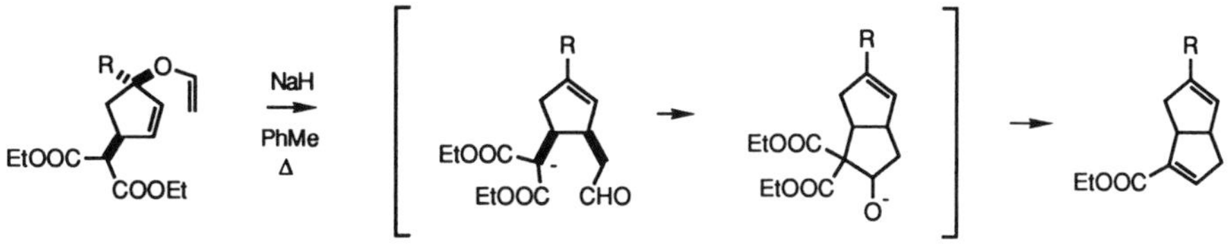

2.6. ALDOL-TYPE REACTIONS IN TANDEM WITH OTHER PROCESSES

During a synthesis of coriolin [Koreeda, 1983] the construction of a diquinane intermediate was accomplished by a one-pot reaction featuring an alkylation and an intramolecular aldol reaction.

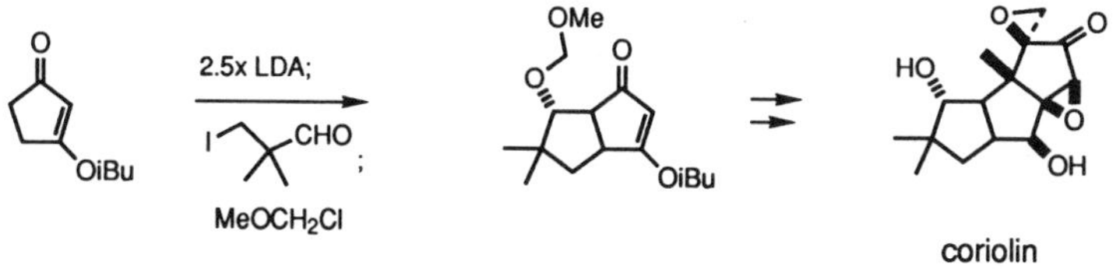

coriolin

A different combination of alkylation aldol condensation was employed in the formation of the pseudoguaianolide skeleton from a cyclodecatrione [Kretchmer, 1976]. It is impossible to determine which step (aldol or alkylation) took place first, although the intramolecular aldolization should be very favorable. It should be noted that the acetic ester sidechain is *cis* to the angular methyl group. This relative configuration corresponds to that required for the elaboration of damsin.

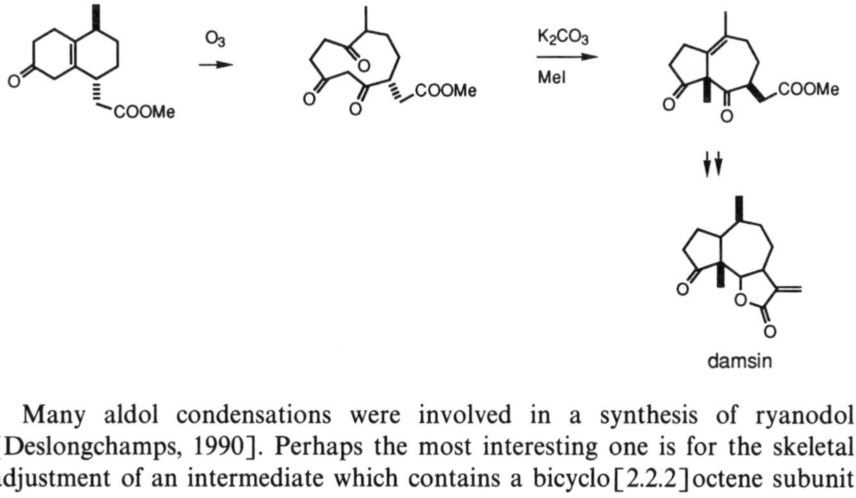

damsin

Many aldol condensations were involved in a synthesis of ryanodol [Deslongchamps, 1990]. Perhaps the most interesting one is for the skeletal adjustment of an intermediate which contains a bicyclo[2.2.2]octene subunit to one possessing the hydrindanone framework. The transformation occurred on ozonolysis (and in situ aldolization).

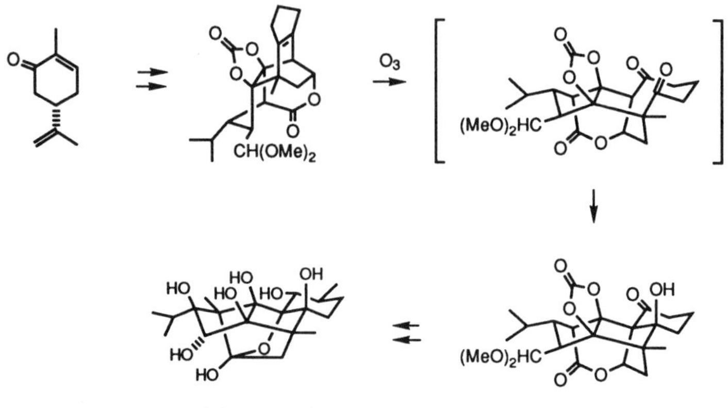

(+)-ryanodol

An aldol condensation in combination with Friedel–Crafts alkylation constitutes an excellent strategy for the convergent and flexible synthesis of A-aromatic steroids [Mander, 1981]. After linkage of preexisting A- and D-ring components with a C_6-chain, cyclization of the above-mentioned tandem reactions completes the assembly of the molecular framework.

The equilenin analog has been synthesized by cyclization of a slightly different substrate [Daniewski, 1985]. In this version the opening of the cyclopropane ring by hydrogen bromide presumably initiated the aldol condensation.

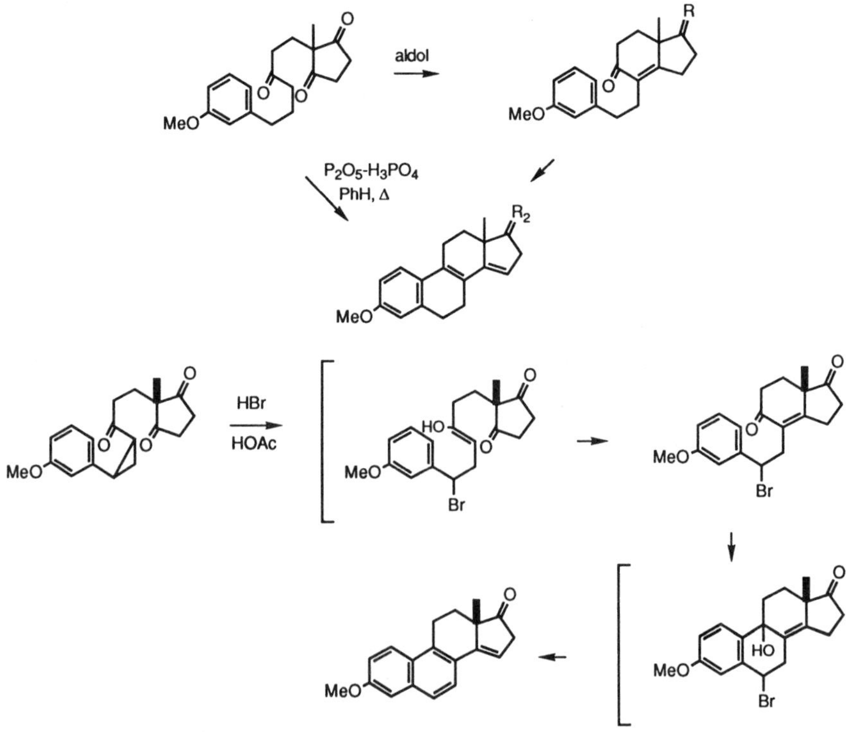

3

MICHAEL REACTIONS

The conjugate addition of a compound possessing an active hydrogen atom (proton donor) to an activated unsaturated double or triple bond is known as the Michael reaction [Bergmann, 1959a]. Besides alcohols, amines, thiols, etc., substances with an active C–H component behave as Michael donors. These include carbonyl compounds, alkyl sulfones, organonitriles, and nitroalkanes, and the Michael reactions involving them are very important synthetic procedures because they achieve CC bond formation and increase functionality. In other words, more complex carbon frameworks are thereby constructed.

Generally, a Michael reaction comprises deprotonation, conjugate addition, and protonation of the conjugate base of the adduct. As all steps are reversible, it is imperative to choose reaction conditions and the addends judiciously to drive the equilibria toward formation of the adduct.

Base-catalyzed Michael reactions can be carried out catalytically or after complete deprotonation of the Michael donors. The two processes are under thermodynamic control and kinetic control, respectively. Variants of the Michael reaction that do not involve deprotonation complement the conventional methods. In most cases these involve breakdown of enol derivatives such as enol silyl ethers, or replacement of the enolate anions with enamines. Besides their application to base sensitive systems there are certain other advantages such as that of employing 2-siloxyfurans as synthetic equivalents for the γ-anions of butenolides, illustrated in a synthetic route to mitomycin-A [Fukuyama, 1987].

Organocopper reagents are suitable Michael donors for enones, and reactions involving these components have gained enormous popularity in synthesis.

Given the preeminence of the Michael reaction one can imagine its increased value when coupled with another reaction in tandem.

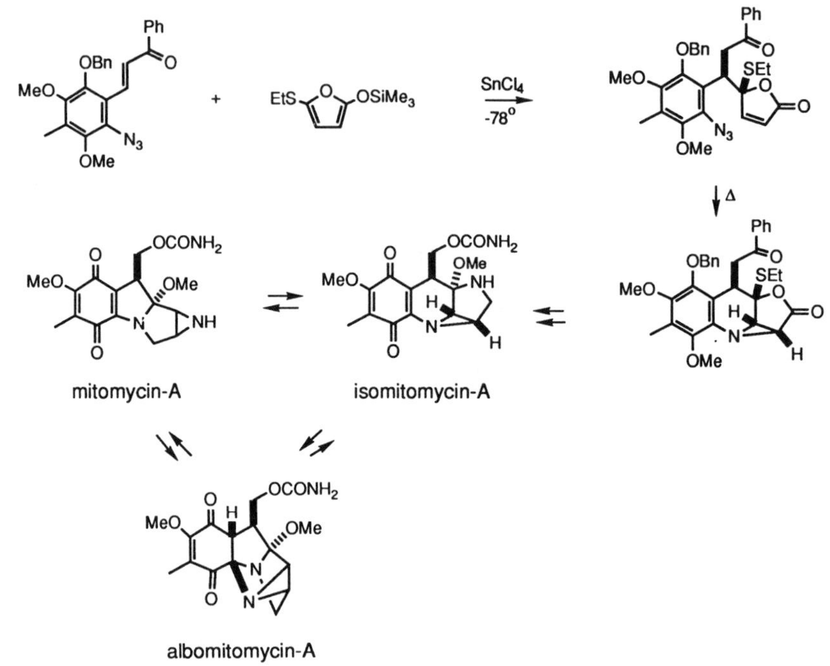

mitomycin-A   isomitomycin-A

albomitomycin-A

3.1. MICHAEL REACTIONS INITIATED BY OTHER REACTIONS

Cyclopentene formation by intramolecular opening of (E)-2-alkenylcyclopropane-1,1-dicarbo-xylic esters is sterically impossible. However, an alternative reaction pathway available to these compounds is that of a reversible Michael reaction which permits the change of (E) to (Z) configuration [Danishefsky, 1975]. It should be noted that the initial Michael donor actually is a homo-Michael aceptor, and the overall reaction is formally a vinylcyclopropane to cyclopentene rearrangement.

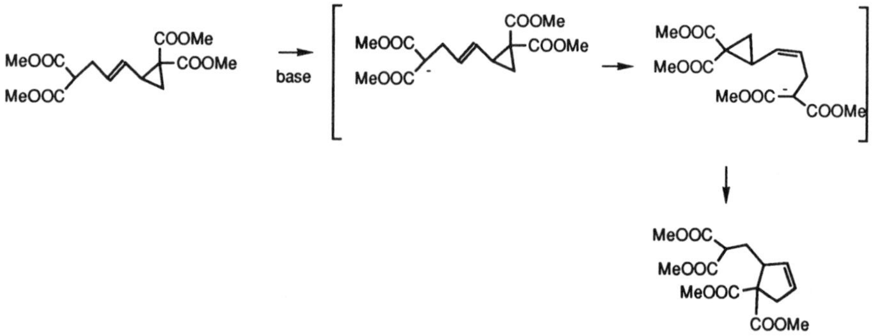

Another intramolecular homo-Michael cyclization [Deb, 1990] is also structurally dependent. It requires an aryl or some other carbenium ion stabilizing group at the 2-position of the cyclopropyl ketone.

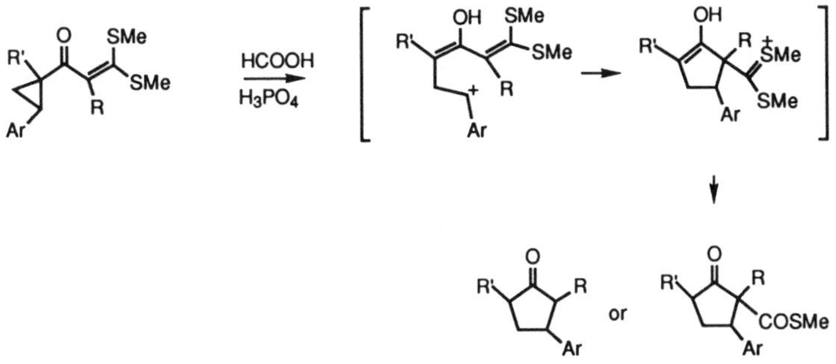

Of some biosynthetic relevance is the reaction course (but not the reagent used) for the transformation of a neolignan skeleton of the epiburchellin type into the epiguainin analog [Angle, 1990] on exposure to fluoride ion. Desilylation caused generation of a quinonemethide species which regained stability by an intramolecular Michael addition. It is also significant that in the presence of a Lewis acid, formation of a substance with the futoenone skeleton is favored.

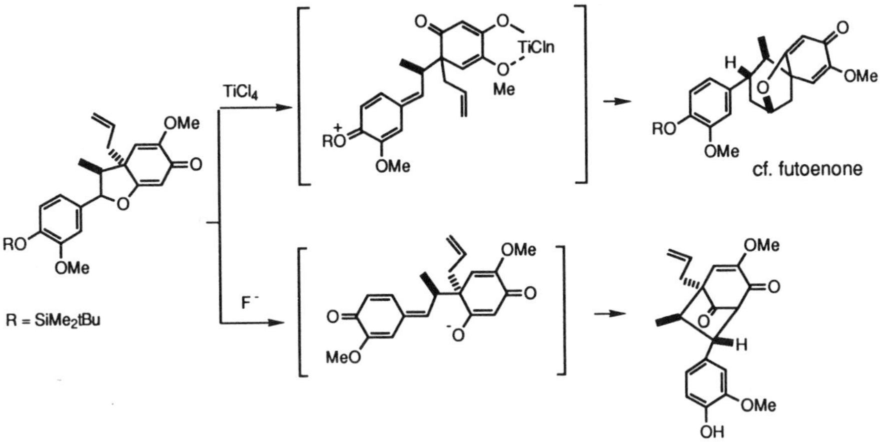

Both condensed and spirocyclic systems have been acquired by intramolecular Michael reactions of ketone enolate species which were generated by Pd-catalyzed decarboallyloxylation of the appropriate β-keto esters [Nokami, 1989b]. Nonbasic conditions are maintained in these reactions.

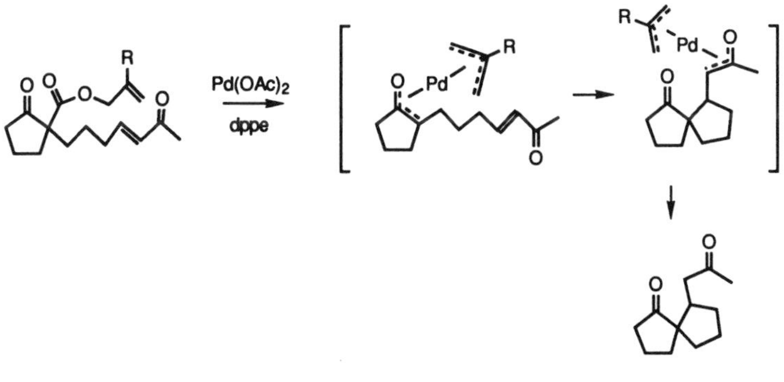

The reduction of anemonin with sodium amalgam yields isotetrahydroanemonin. The structure of the latter compound has been clarified [Stamos, 1977]. It is evident that the transient γ-anion of the butenolide half from cleavage of the cyclobutane ring added to the other butenolide in a Michael fashion.

Skeletal rearrangement of a cage ketol on treatment with base was observed [Mehta, 1981]. Retro-aldol fragmentation also caused the establishment of a conjugated ketone substructure to allow an intramolecular Michael reaction to take place.

Exposure of the Diels–Alder adduct of 1-methoxy-4-methyl-1,3-cyclo-hexadiene and methyl vinyl ketone to acid afforded 10-methyl-3,7-decalindione and two bridged ketols [Birch, 1966]. The intermediate for all these compounds is a cyclohexenone generated by retro-aldol fission. Two divergent pathways, that is intramolecular Michael reaction and aldolization, then led to the observed products.

Desilylation of methyl 2-trimethylsiloxy-2-vinylcyclopropanecarboxylate by fluoride ion in the presence of a nitroalkane led to the 6-nitro-4-oxoalkanoic ester [Zschiesche, 1988]. Since fluoride ion is capable of deprotonating nitroalkanes, it served a double role in the present transformation.

Generation of an ester enolate by the retro-homoaldol fission has also been coupled to an intramolecular Michael reaction [Marino, 1988] in which the acceptor is an unsaturated sulfone. The stereochemical outcome of this tandem process is such that the product represents a potential intermediate for synthesis of compactin.

An allylic alcohol derived from the saturated ketone unit of the Wieland–Miescher ketone is prone to rearrange on treatment with base. Two reaction pathways have been identified [Uma, 1986] and both lead to the same product. In aprotic solvents an anionic oxy-Cope rearrangement is preferred; however, in methanol the molecule would fragment to afford a dienolate and a vinyl ketone. Intramolecular Michael reaction then brings about the closure of an eight-membered ring.

Analogous reactions have been observed with homologous ring systems (6:5- and 5:5-fused systems) and an propargyl alcohol may replace the allyl alcohol subunit.

The peristylane skeleton can be readily formed from a C_{14}-pentaquinanedione on Claisen–Schmidt condensation [Eaton, 1977]. Because of the spatial proximity of the functionalities the enone intermediate cannot avoid an intramolecular Michael reaction.

Spiroannulation has been achieved by reaction of a bicyclic enol lactone with an ynamine [Ficini, 1981]. Acylation of the ynamine leads to a charged β-imino enone which is an excellent Michael acceptor (also a Mannich acceptor).

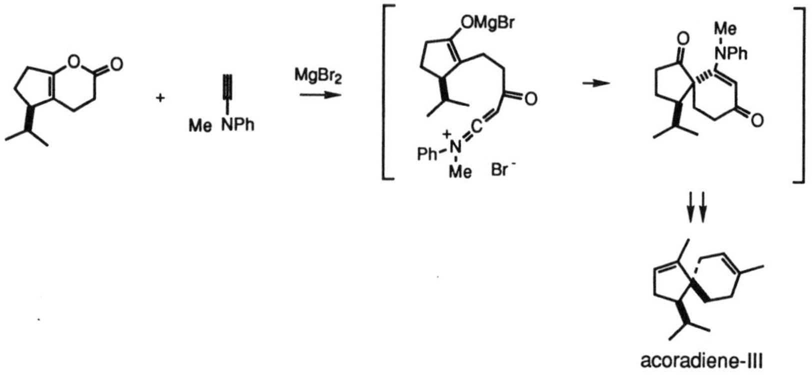

acoradiene-III

Alkaloids such as ochotensimine may originate from the berberine skeleton by rearrangement. It is of interest that the spirocyclic system can be created by treatment of a dibenzoquinolizidine containing a catechol subunit [Shamma, 1969]. Opening of the C-ring results in a quinonemethide species which is capable of undergoing intramolecular Michael reaction, by virtue of the presence of the phenolate ion in an ortho position of the carbonyl group.

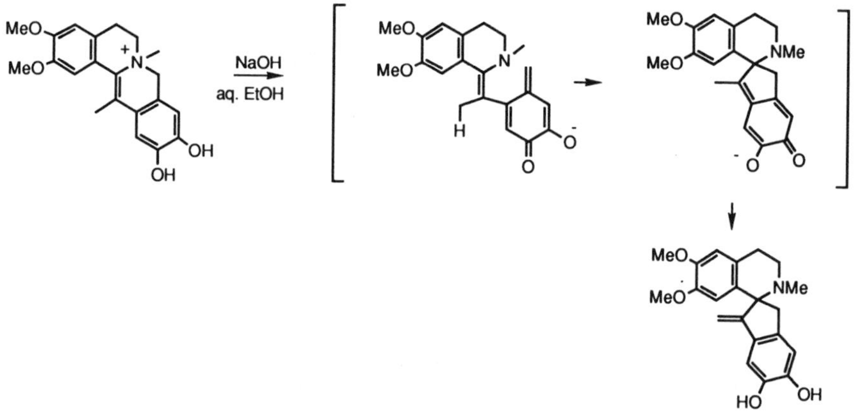

Benzoquinone generated by oxidation may be trapped by Michael donor present in the same molecule [Ulrich, 1977]. However, the cyclization is very sensitive to the acidity (reactivity) of the Michael donor.

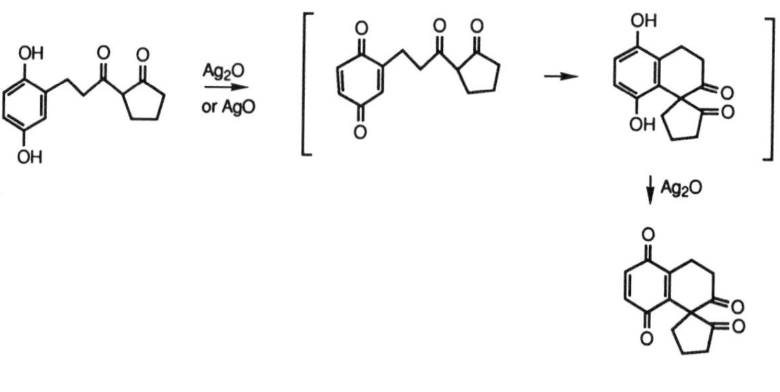

Acylmetallic species are naked acyl anion equivalents that are able to undergo conjugate addition. Fortunately, these transient acylmetals are quite easy to prepare in situ by rearrangement of σ-bonded metal carbonyl compounds, which in turn are obtainable by displacement reactions. A ring-forming reaction by this method is illustrated in a synthesis of (2-oxocyclopentane)acetonitrile and ester [Cooke, 1977].

R = CN, COOEt

Interestingly, a skeletal rearrangement of an alkyltetracarbonylferrate intermediate resulting from the reaction of an aziridinium ion with disodium tetracarbonylferrate has been observed [Overman, 1992]. Due to the presence of an oxindole moiety which is capable of inducing fragmentation, it becomes possible to establish a polycyclic system containing a conjoint dioxo functionality by an intramolecular Michael reaction.

Free radicals may act as Michael donors. For example, CC bond formation via in situ generation of free radicals from organomercurials by reduction with sodium borohydride and trapping with Michael acceptors is an established method. A tandem reaction involving oxidative cleavage of a cyclic ketone to give a free radical with an ester terminus, and subsequent Michael reactions [Karim, 1988] is useful in view of the extensive modification of the carbon framework of the addends.

An unexpected product from *p*-nitration of an anisole arose from a tandem Michael reaction [Beeley, 1983].

A tandem reaction involving reduction and intramolecular Michael reaction was crucial to a synthesis of ibogamine [Büchi, 1966]. When the synthesis was planned on the basis of a Diels–Alder reaction to create the azabicyclo[2.2.2]octane skeleton a formal dyotropic rearrangement at the stage of C-ring formation that united the indole nucleus and the azabicycle was perhaps not anticipated. Fortunately, an acetyl chain which originated from methyl vinyl ketone and destined to become the ethyl group helped rectify the skeletal impropriety.

ibogamine

Imines react with 1-methoxy-3-trimethylsiloxy-1,3-butadiene in the presence of zinc chloride. Direct hydrolysis of the resulting β-amino ketones with aqueous acid brings about ring closure which involves an intramolecular Michael reaction [Kunz, 1989].

The bridged system derived from an anionic oxy-Cope rearrangement of a spirocyclic hydroxy dienone led to a perhydrophenalenedione [Rao, 1982], as a result of equilibration of the enone and the enolate components to give a species disposed to undergo an intramolecular Michael addition.

As indicated before, the reversibility of the Michael reaction can be either an advantage or disadvantage. In the following, a few cases in which the retro-Michael–Michael reaction sequence intervenes are described. Thus, in a synthesis of (−)-norsecurinine [Jacobi, 1989] the epimeric furano ketone could be made useful via equilibration before the establishment of the bridged ring system.

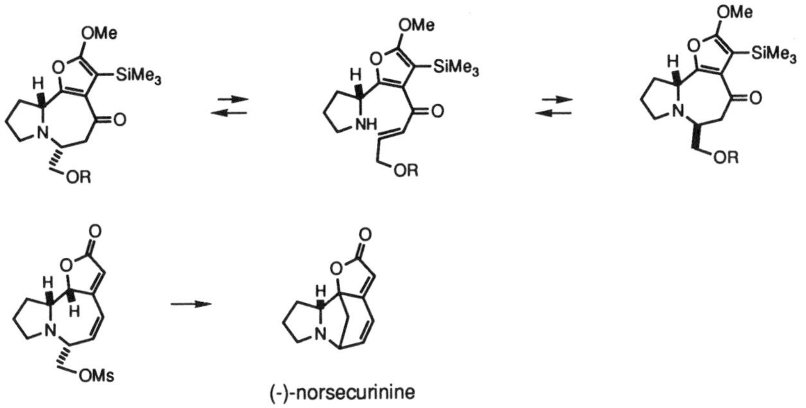

(-)-norsecurinine

Similarly, the hetero-Diels–Alder cycloaddition approach to cephalotaxine [Burkholder, 1988] via an *anti-cis* and an *anti-trans* 1,2-oxazine was benefitted by the stereochemical rectification of the minor *anti-trans* isomer at the α-diketone stage by means of a retro-Michael–Michael reaction tandem.

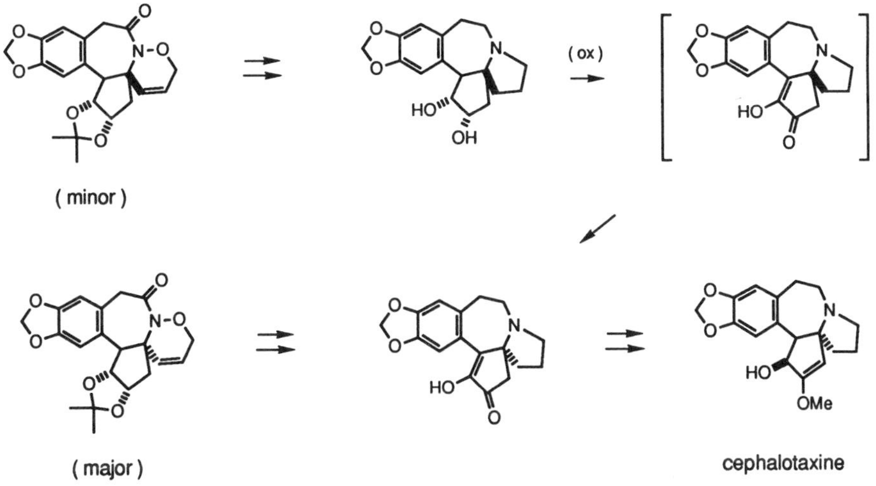

(minor)

(major)   cephalotaxine

On the other hand, an attempt at synthesizing cephalotaxine [Dolby, 1972] based on another method was thwarted by the intervention of the same reaction sequence that transformed the desired skeleton into one with a quinolizidine nucleus.

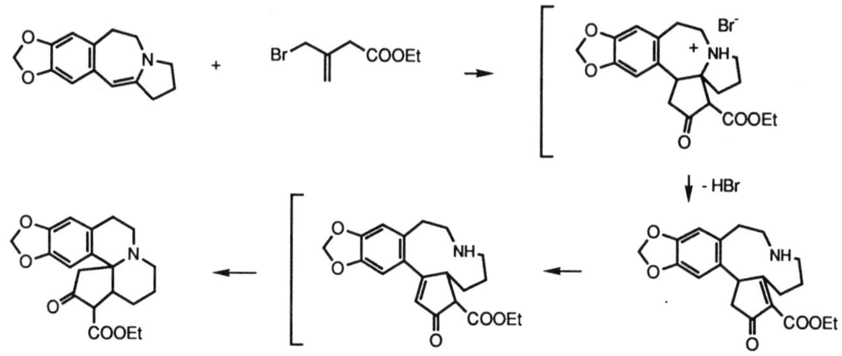

The biomimetic synthesis of erysodienone [Mondon, 1966; Gervay, 1966] via intramolecular phenolic coupling has been accomplished. In fact, the pathway involves formation of a dibenzoazonine from which erysodienone is obtained upon further oxidation and intramolecular Michael reaction.

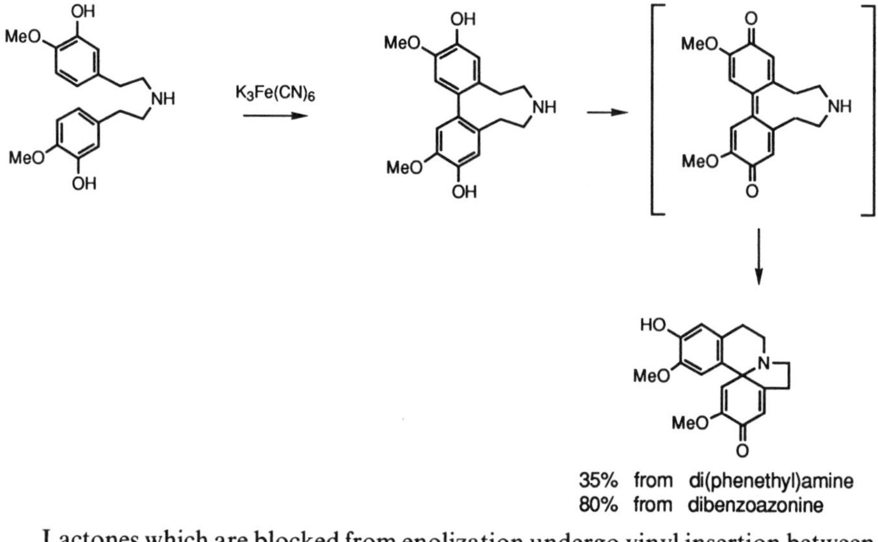

35% from di(phenethyl)amine
80% from dibenzoazonine

Lactones which are blocked from enolization undergo vinyl insertion between the O-C(acyl) bond when reacted with alkynyllithiums in the presence of hexamethylphosphoric triamide [S.L. Schreiber, 1984]. Apparently such conditions favor ring opening of the adducts as well as intramolecular Michael addition.

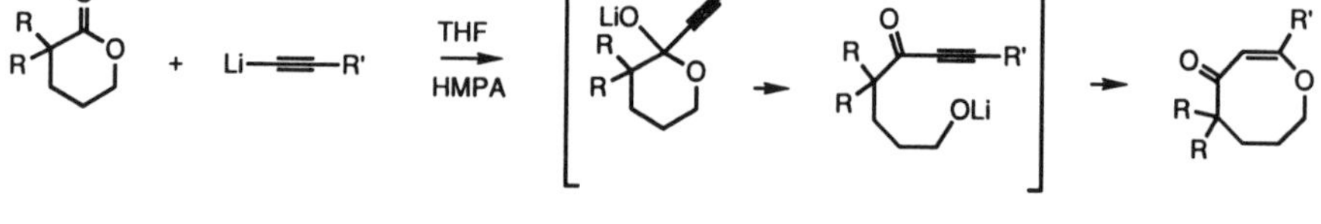

Treatment of *N*-(4-nitrobenzyl)- and *N*-phenacyltetrahydroisoquinolinium salts with potassium *t*-butoxide led to benzazepines [S. Smith, 1984]. This transformation may be considered as proceeding via generation of the quinonemethide intermediates and reclosure of the heterocycle by an intramolecular Michael reaction.

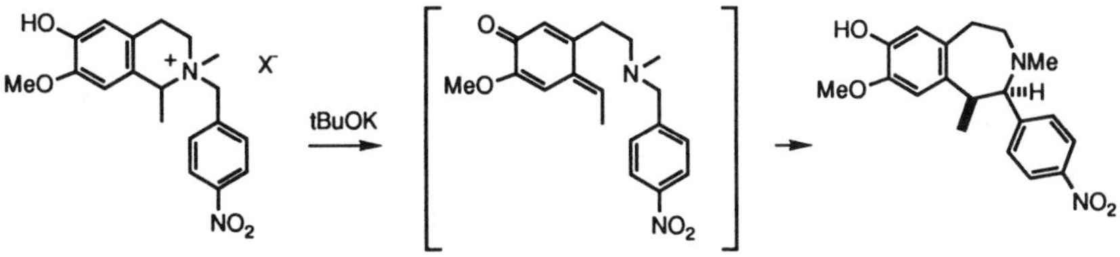

It is of interest to mention a few retro-Michael reactions here. First, a synthetically significant regiocontrolled spirannulation is via intramolecular carbenoid cycloaddition to cyclic β,γ-unsaturated ketals. The following acid treatment whereby the cyclopropane ring suffers cleavage is a formal retro-Michael fragmentation. Applications of such a process can be found in the syntheses of agarospirol [LaFontaine, 1980] and α-chamigrene [J.D. White, 1981].

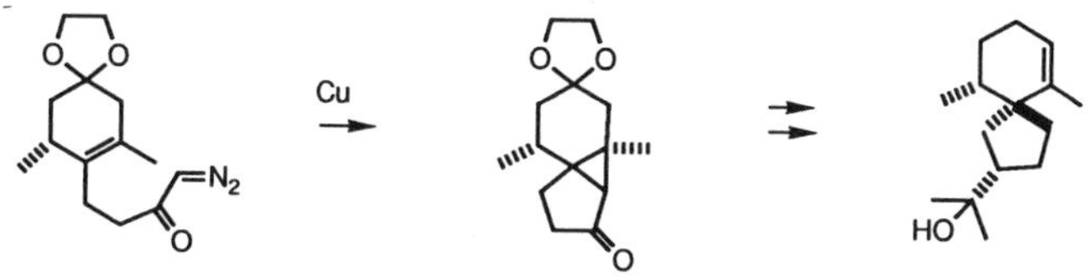

agarospirol

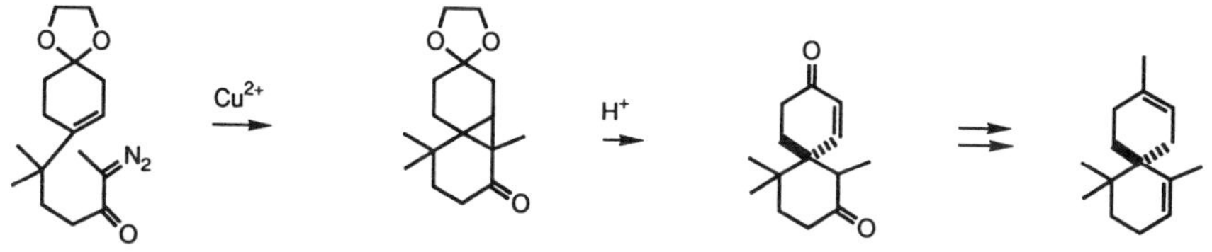

α-chamigrene

A reaction sequence involving alternate Michael addition and its retrograde reaction may be postulated to rationalize the ring expansion/rearrangement [Fayos, 1977] which is shown here.

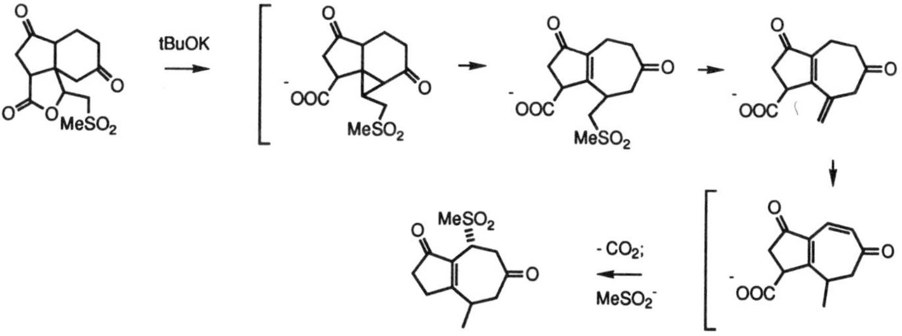

A facile retro-Michael elimination of malonic ester from *o*-bromomethyl-benzylidenemalonates on reaction with hydrazines has been observed [Sha, 1991]. This elimination is preceded by the formation of tetrahydrophthalazines.

3.2. DOUBLE MICHAEL REACTIONS

Two Michael reactions in tandem form two bonds in one step. An α-functionalization of β-unsubstituted acrylic esters which cannot undergo enolization makes use of a double Michael reaction. Thus, certain tertiary amines and phosphines catalyze the dimerization of acrylic esters by forming zwitterions which then add to a second molecule of the acrylic ester. Proton transfer, either intramolecular or intermolecular, triggers an elimination and regeneration of the catalyst. Trisdimethylaminophosphine appears to be an excellent catalyst for the preparation of 2-methyleneglutaric esters [Amri, 1989].

The double Michael reaction is a very useful process if the reactions are controlled to afford the desired products. Thus, cross-conjugated dienones can react with Michael donors containing an active XH_2 fragment to give directly six-membered ring compounds [Marvel, 1949]. It is interesting that if one of the activating groups is also a good leaving group, 1,3-elimination of the cyclic ketones may be induced [Britten-Kelly, 1981].

R = H, Me

The instability of cyclopentadienone makes it very difficult to use it as a discrete double Michael acceptor. However, certain 4-substituted 2-cyclo-pentenones may be employed instead, provided that the initial Michael adducts would undergo elimination under the reaction conditions. 4-Acetoxy-2-cyclopentenone is such a cyclopentadienone surrogate [Osterhun, 1977].

The convenient synthesis of bicyclo[3.3.0]octane-3,7-diones by condensation of acetonedicarboxylic esters with α-dicarbonyl compounds is known as the Weiss reaction [A.K, Gupta, 1991]. It is reasonable that the mechanism consists of two aldolization reactions to give 4-hydroxy-2-cyclopentenone intermediates, and two Michael reactions intervenient by an in situ dehydration step.

E = COOMe

Formation of a cyclohexenone from an active methylene compound and a cross-conjugated enynone via double Michael reactions was the focal point of a concise synthesis of griseofulvin [Stork, 1964]. It should be noted that the *trans*-addition across a triple bond that results in a (Z) relationship between the donor atom and the ketone group is critical to the ring closure, assuming the first step involves such a reaction. The stereospecific formation of griseofulvin also indicates a kinetically controlled process, because epigriseofulvin is the more stable isomer.

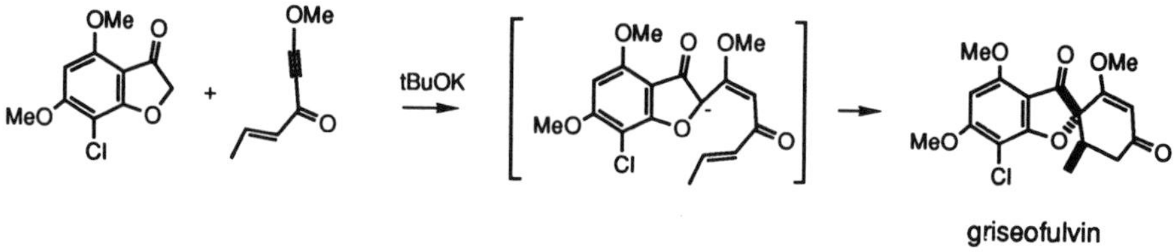

griseofulvin

As model for aklavinone synthesis [Cava, 1984] a 83% yield of a tetracyclic ketronitrile was obtained by a double Michael reaction of phenylthioacetonitrile with an enone derived from chrysazin.

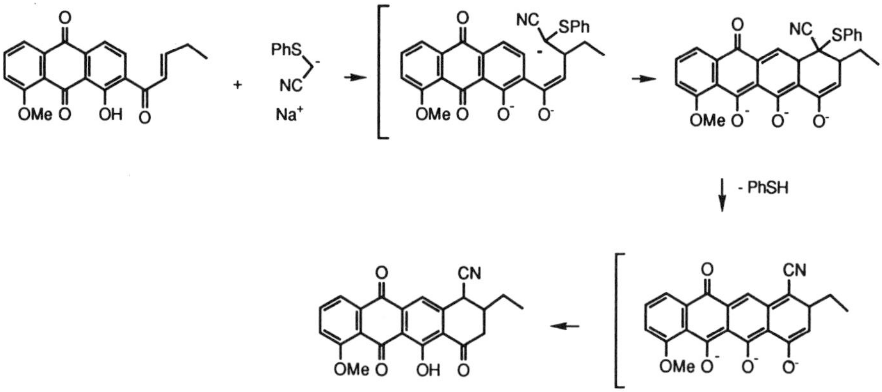

Nitroalkanes are very acidic and their capacity for acting as Michael donors are well documented. It is possible to effect double Michael reactions, as shown in the formation of 4-nitrocyclohexanones from their reaction with cross-conjugated dienones [Richter, 1987].

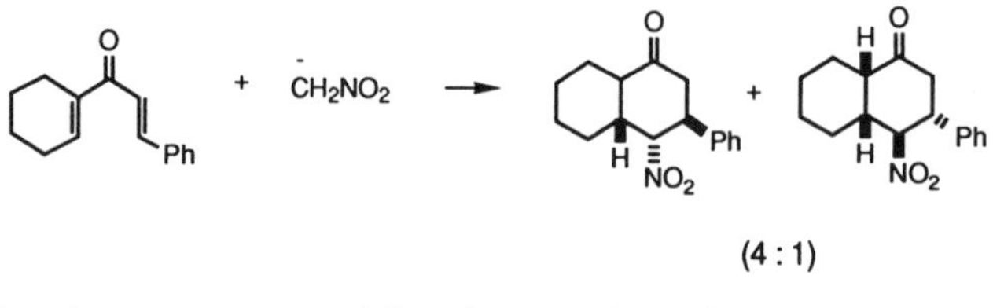

(4 : 1)

A β-nitro ketone generated by the reaction of a vinyl ketone with nitrite anion may add to a second molecule of the Michael acceptor. The reaction can be adjusted in such a way that elimination of nitrous acid occurs at this stage to furnish a Δ^2-ene-1,6-dione as the final product [Miyakoshi, 1982].

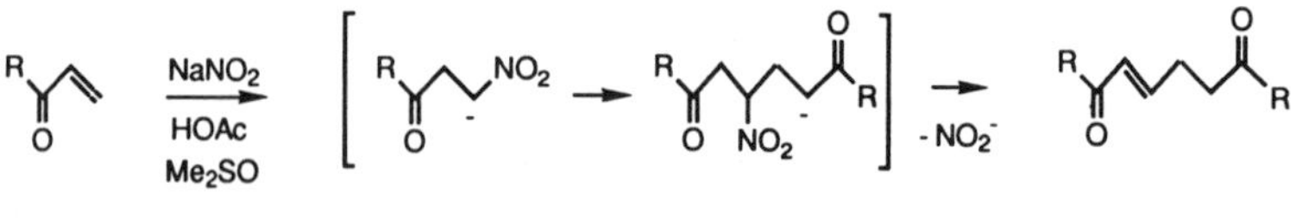

Polynitroaromatic compounds from Meisenheimer complexes with various nucleophiles. Michael reactions beyond this point can lead to bridged ring systems [Strauss, 1973, 1974].

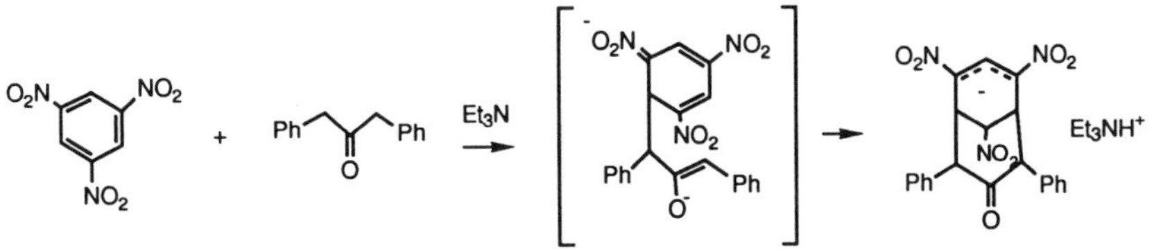

Nitroalkenes are typical Michael acceptors also. A tandem strategy [Barco, 1990] proves to be very efficient for the assembly of pyrrolidines and it has enabled the completion of an enantioselective synthesis of (−)-α-kainic acid [Barco, 1991].

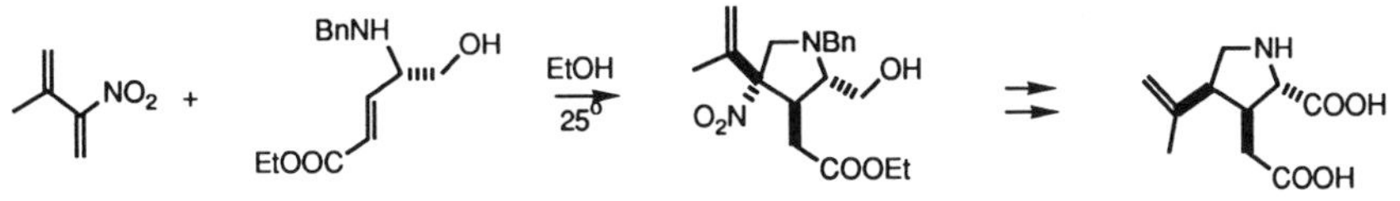

(-)-α-kainic acid

The double Michael addition approach to cyclic systems has many other ramifications. One such example is the elaboration of the tricyclic skeleton of protoemetine from a tetrahydroisoquinoline precursor [Hirai, 1986]. When the tetrahydroisoquinoline is substituted at C-1 by a γ-crotonic ester unit its adduct with methyl vinyl ketone is capable of undergoing an intramolecular Michael reaction. (Note the report described a two-step reaction with the second step catalyzed by pyrrolidine. It is conjectured that a tandem reaction is achievable by using a tertiary amine as catalyst.) An analogous cyclization provided useful precursors of heteroyohimbine alkaloids [Rosenmund, 1990].

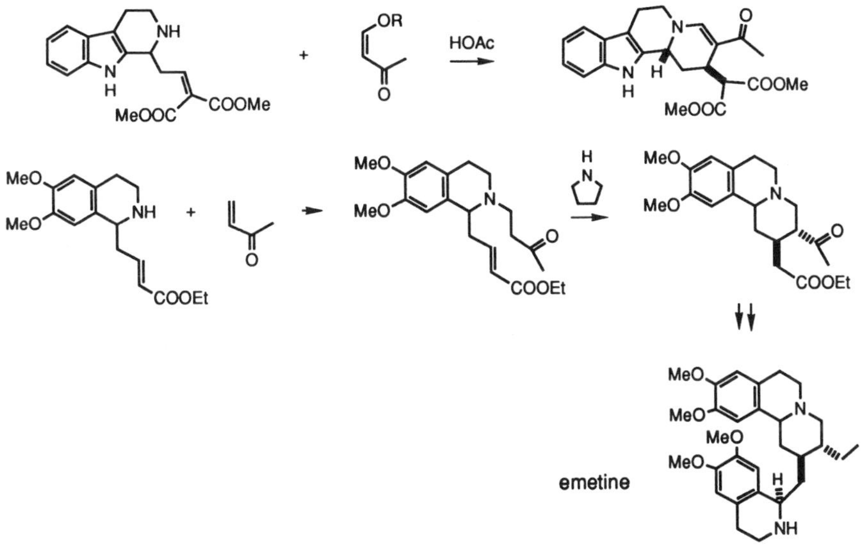

emetine

2,6-Octadienoic diesters have been caused to cyclize by the Kharasch reaction [Saito, 1989]. It is significant that substituents at C-4,5 have profound effects on the steric course of the initial attack by the nucleophile and that of the cyclization by an intramolecular Michael reaction.

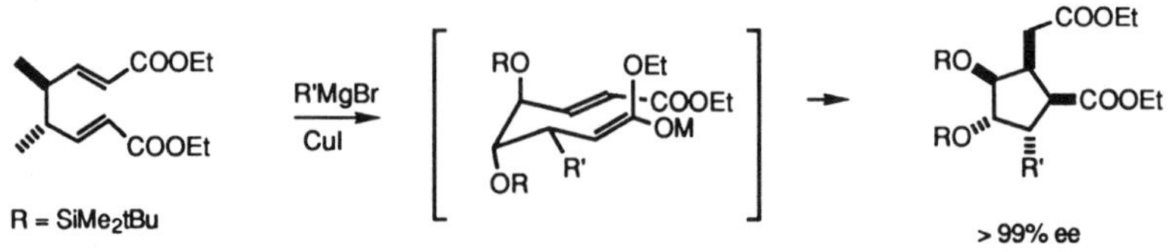

A second type of double Michael reactions occurs with β-substituted α,β-unsaturated compounds in which the β-substituent is liable to elimination after the first Michael addition. Particularly interesting examples are those involving spiroannulation [Canonne, 1988; Wender, 1988b].

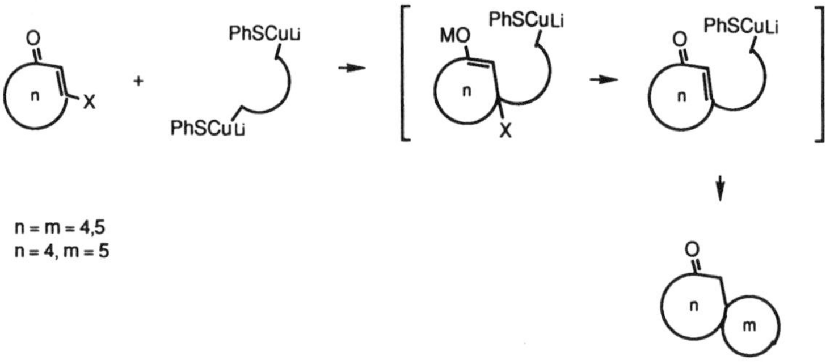

A variant of the above reaction is restricted to Michael acceptors in which the α-carbon is branched and the other β-carbon bears a leaving group. The first Michael addition is followed by elimination to generate a new acceptor system. Introduction of two identical or different groups, that is by using one or two Michael donors, is possible [Dunham, 1971; Mitra, 1979; T. Takahashi, 1981a].

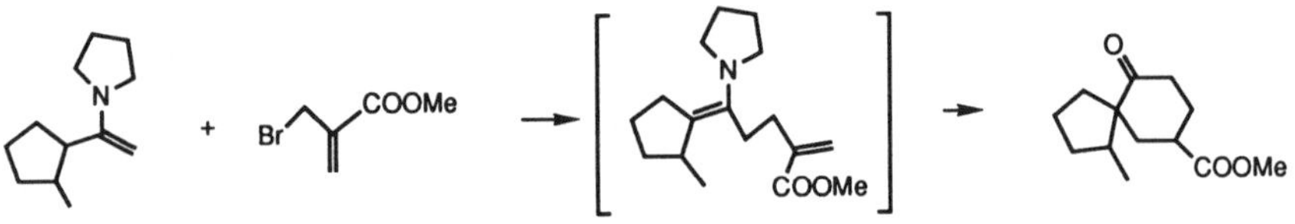

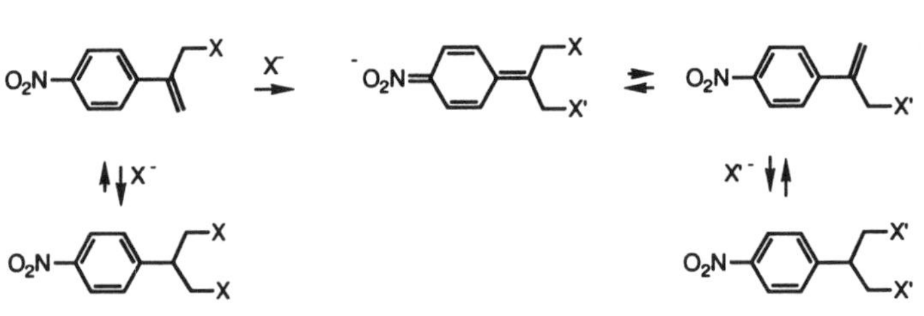

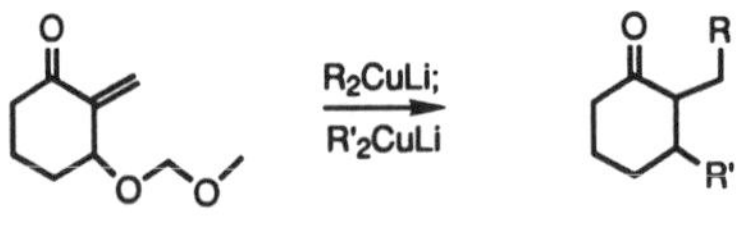

A third type of double Michael reactions involves dienones or more extended conjugate systems as Michael acceptors. The prerequisite for a dienone to undergo such reactions is that the β-carbon atom must be a poorer acceptor than the δ-carbon. Furthermore, the formation of a six-membered ring by reaction of a dienone with an acetonedicarboxylic ester has stereochemical implications when the dienone is imbedded in, for example, another ring system. Because the second Michael reaction is intramolecular the product will have a *cis* ring juncture.

An exploitation of these properties of the double Michael reactions succeeded in syntheses of occidentalol [Irie, 1978; Mizuno, 1980] and furanoeremophilane [Irie, 1978].

occidentalol

When a molecule contains two separate Michael acceptors, providing their separation is not too far as to preclude a second, intramolecular Michael reaction from an initial adduct with a divalent Michael donor, ring formation may result (!). 2,3-Bisbenzenesulfonyl-1,3-butadiene is such a double Michael acceptor that its reaction with primary amines leads to 3,4-bisbenzenesulfonyl-pyrrolidines [Padwa, 1990b]. The 5-endo-trig cyclization is somewhat surprising, but it is the only reaction pathway available to the adduct.

(major)

(Note that the major products from corresponding reactions of the dicarboalkoxy-1,3-butadienes are the unsaturated lactams, generated via a 5-exo-trig cyclization and double bond migration.)

3.3. REFLEXIVE MICHAEL REACTIONS

A special kind of double Michael reactions between a kinetic enolate of an enone (cross-conjugated dienolate anion) and a Michael acceptor is called reflexive Michael reactions. In reflexive Michael reactions each reactant serves as both acceptor and donor, the second stage of the process forms a six-membered ring, and from cyclic donors the overall result is the emergence of bridged ring systems [Lee, 1973; Kraus, 1977].

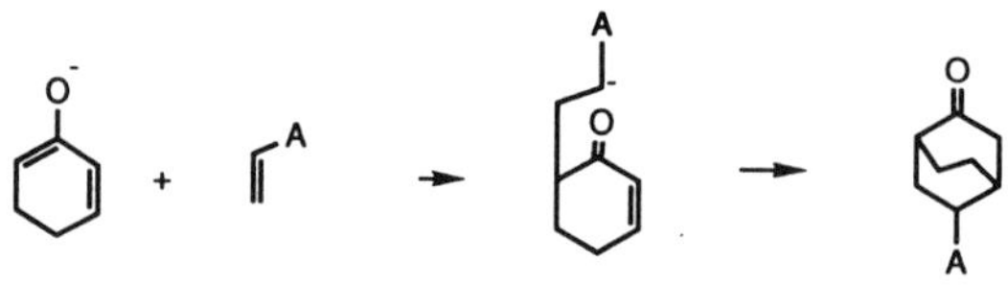

While the reflexive Michael reactions are complementary to the Diels–Alder reactions, they have certain advantages. For example, lower reaction temperatures are required, and generally there are much less side reactions due to diene polymerization and elimination. Thus, in the course of synthesizing khusitone [Hagiwara, 1988] and khusilal [Hagiwara, 1989] the superiority of the reflexive Michael reactions versus the Diels–Alder method was evident.

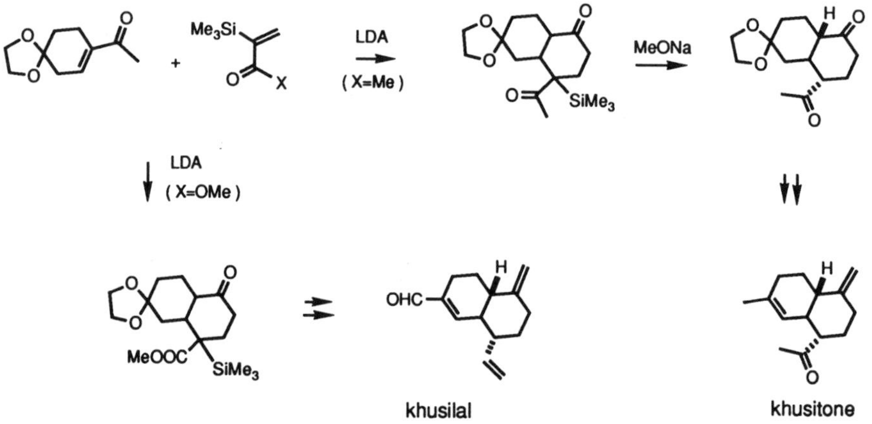

khusilal

khusitone

Subtle stereochemical effects are shown in the following reactions [Bateson, 1991], although regiochemistry is the same for both the reflexive Michael reactions and the Diels–Alder cycloaddition.

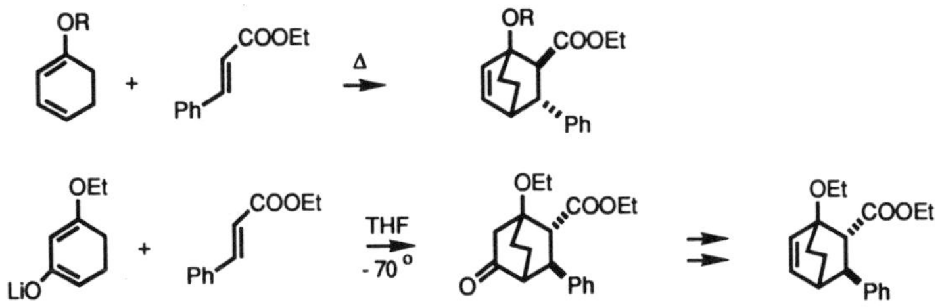

In certain cases the thermal cycloaddition fails to provide products, or it gives unacceptable mixture of isomers. Thus, two parallel experiments have shown the stereospecific and stereorandom annulations, respectively involving a kinetic enolate and the corresponding dienol ether [Spitzner, 1989].

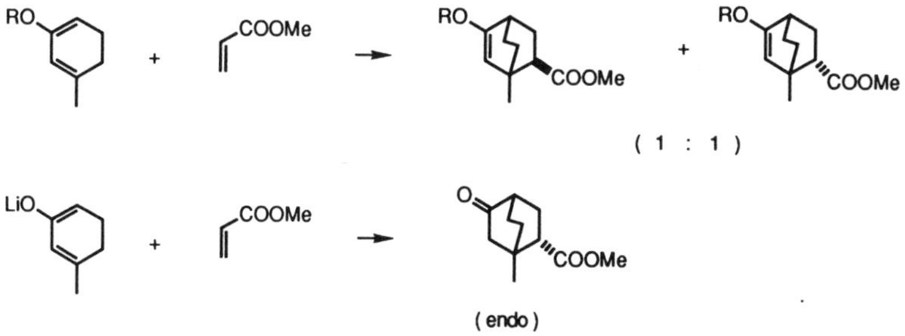

(1 : 1)

(endo)

Many elegant syntheses have been developed based on the reflexive Michael reactions. In a synthesis of pleuromutilin [Gibbons, 1982] such reactions served to provide a tricyclic scaffold from which the cyclooctane component was elaborated. Classical approaches for cyclization would be difficult on account of the dense array of asymmetric centers and the additional nonbonding interactions present in the transition states, due to the existing hydrindanone moiety.

The ideal steric disposition of a reactive funtionalities in the reflexive Michael reaction product permitted construction of a latent eight-membered ring in the form of a lateral cyclohexene subunit. Subsequent fragmentation removed the extra CC bond.

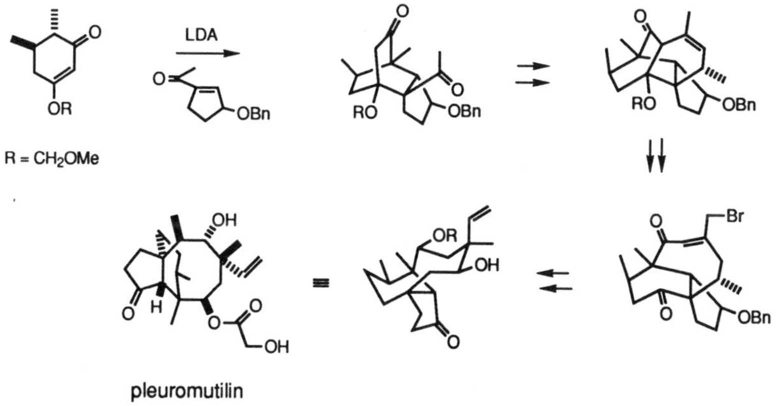

R = CH$_2$OMe

pleuromutilin

This very effective transitory annulation strategy [Ho, 1988] also formed the basis of a sanadaol synthesis [Nagaoka, 1987]. The extra ring was attached to a bicyclo[2.2.2]octanone, which was assembled by reflexive Michael reactions, and upon Grob fragmentation the original [2.2.2] framework was partially destroyed, revealing a bicyclo[4.3.1]decane skeleton.

sanadaol

It proved expedient to employ the reflexive Michael reactions to start a synthesis of eriolanin [Roberts, 1981] despite the requirement of a double bond instead of an ethanone bridge.

eriolanin

The building block for the DE-ring system suitable for coupling with 6-methoxytryptamine to give a viable precursor of reserpine is a cyclohexane-acetaldehyde. Since the aldehyde chain is *cis* to the oxygen function at C-18, the two functionalities could be derived from an ethanone bridge placed across the cyclohexane ring by Baeyer–Villiger reaction and semireduction of the lactone. Thus the subtarget for a reserpine synthesis [Stork, 1989] is a bicyclo[2.2.2]octanone which must be substituted with an ester and a functionalized one-carbon unit in the *syn* configuration and a potential methoxy group *anti* to the ketone bridge. Such substitution patterns of the bicyclic compound suggest a reflexive Michael reaction route.

According to this plan a β-substituted methyl (E)-acrylate in which the β-substituent is convertible to a methoxy group should be employed. A 2-furyldimethylsilyl group was found to be serviceable.

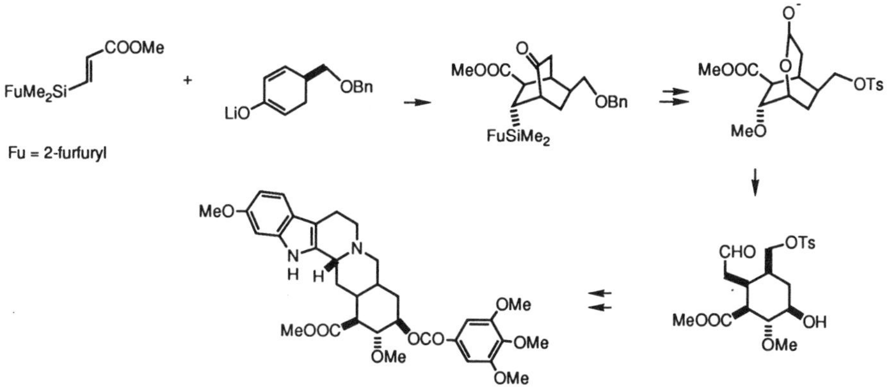

reserpine

Intramolecular reflexive Michael reactions to form the complete network of (+)-atisirene [Ihara, 1986] is a very satisfying achievement. The substrate, derived from an optically active Wieland–Miescher ketone, cyclized in 92% yield on treatment with lithium hexamethyldisilazide. A lithium ion chelation by both the dienolate and the ester carbonyl oxygen atoms favored the endo transition state.

The use of this procedure in a synthesis of atisine [Ihara, 1988b] is a logical extension.

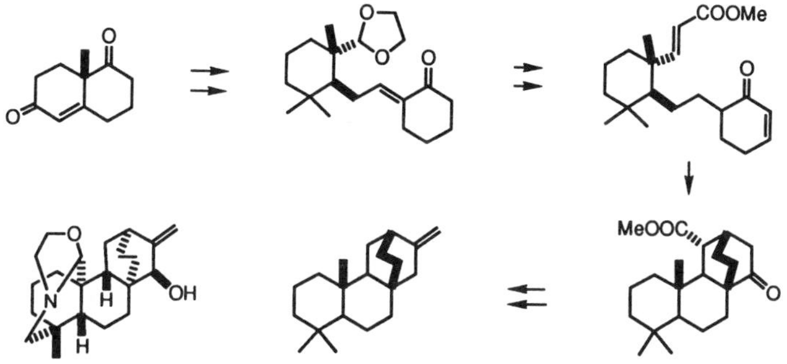

atisine   atisirene

Reflexive Michael reactions of vinylogous amides proceed very readily. A useful precursor of the kopsane alkaloids has been assembled [Schinzer, 1989].

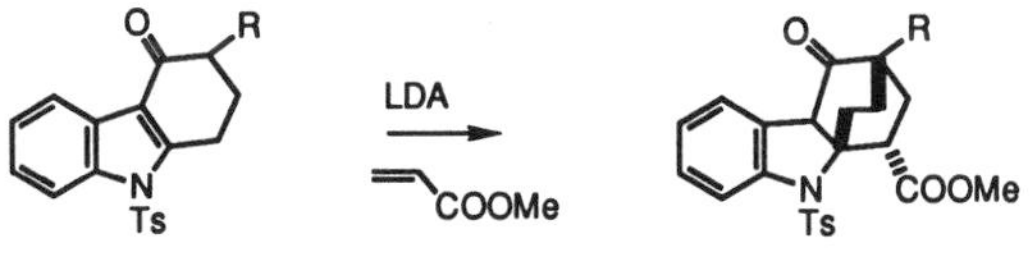

In other circumstances intramolecular reflexive Michael reactions fail under the enolization conditions. It may be possible to induce such reactions by heating the substrates with a mixture of chlorotrimethylsilane, zinc chloride and triethylamine. This protocol was crucial to the realization of a synthesis of pentalenene and pentalenic acid [Ihara, 1988a]. The fact that silyl enol ethers were not the primary products ruled out the Diels–Alder reaction pathway.

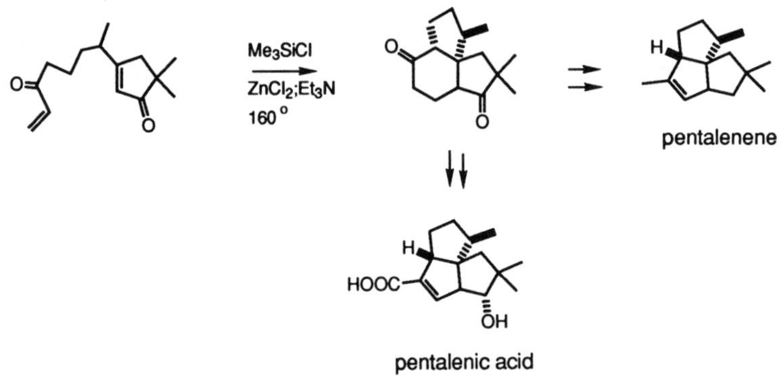

pentalenene

pentalenic acid

Three isomeric estranes were obtained by this method [Ihara, 1989]. Analogous reaction conditions may be used to construct indolizidines and quinolizidines, as demonstrated by a synthesis of epilupinine [Ihara, 1987]. It should be noted that an *N*-acryloylpiperidine was formed when the amide was treated with various bases. In other words, the simple Michael reaction aborted further involvement of the functionalities.

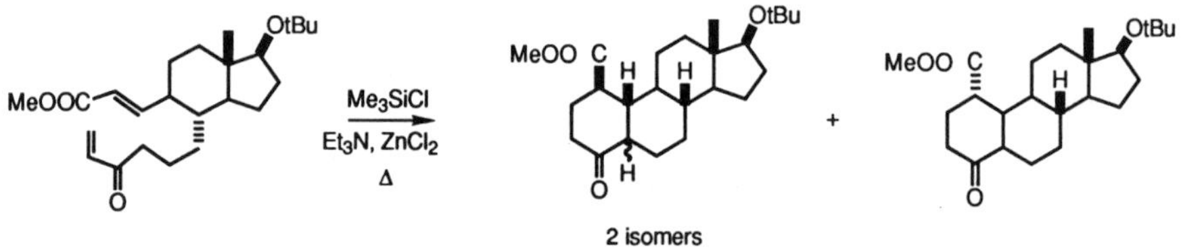

2 isomers

In a synthesis of (−)-tylophorine [Ihara, 1990] the reflexive Michael cyclization was performed under extremely mild conditions, that is by treatment with *t*-butyldimethylsilyl triflate and triethylamine at −78°C.

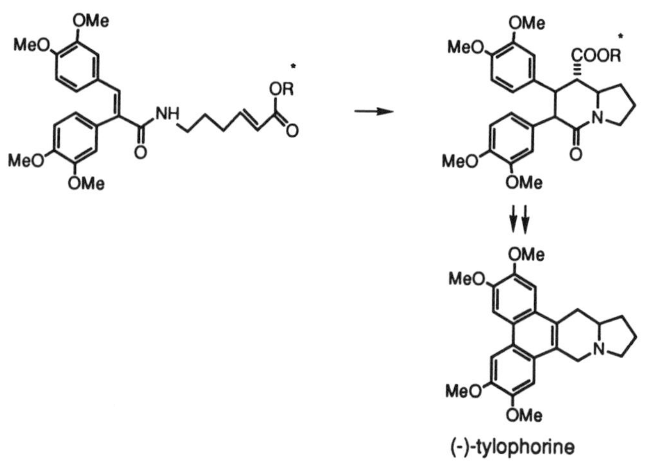

(-)-tylophorine

A synthesis of *trans*-hydrindanones by intramolecular reflexice Michael reactions [Ihara, 1991b] is stereoselective, presumably involving zwitterionic intermediates. The presence of a phenylthio group at the β-position of the enone moiety appears to be critical for the cyclization under the silylating conditions.

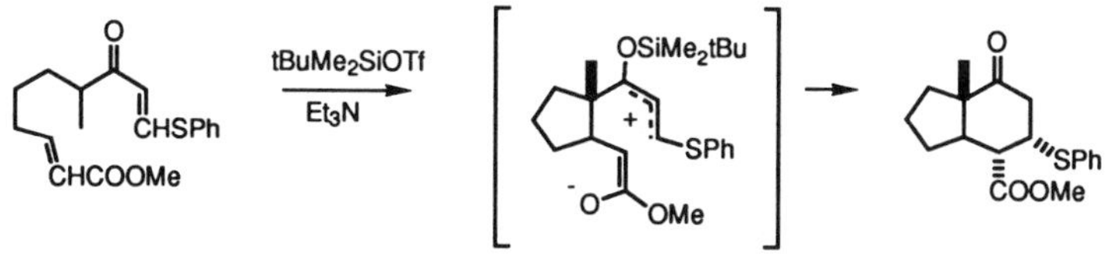

Five-membered rings can be assembled by reflexive Michael reactions of bifunctional donors. Of particular interest are those reactions involving donors in which an activating group is also a good leaving group [Beak, 1986].

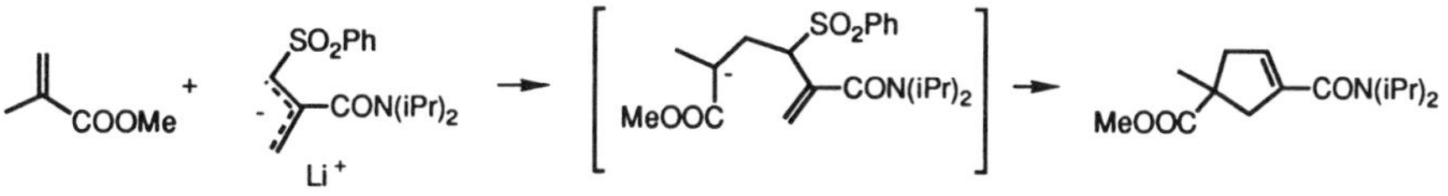

A three-component reflexive Michael reaction actually started with Michael addition of an ester enolate to a cross-conjugated dienone [Thanupran, 1986. Also the cyclotrimerization of protoanemonin [Bigorra, 1985] is a case of Michael addition-initiated triple Michael reactions.

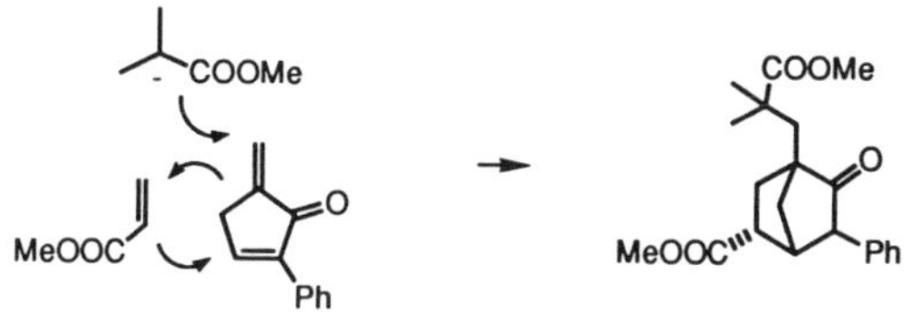

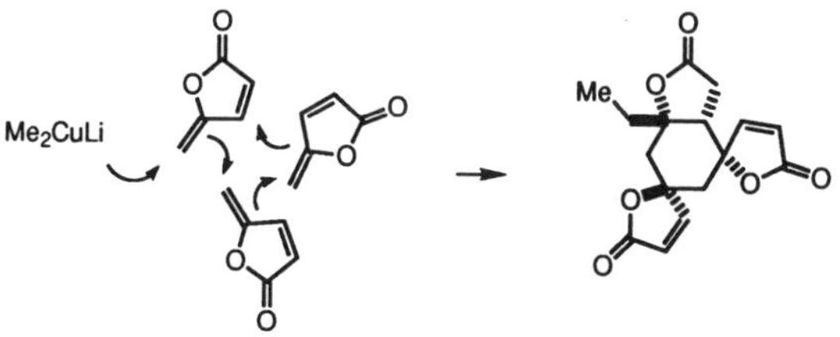

A most concise scheme for the construction of the homoisotwistane system is based on three Michael reactions, a reflexive Michael reaction pair, and a cyclizing addition [Hagiwara, 1985].

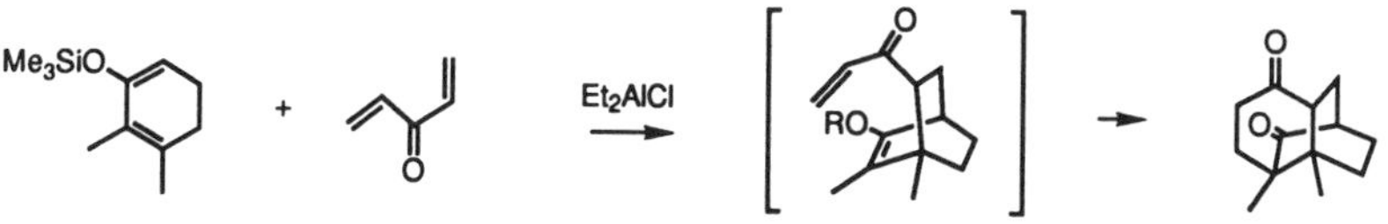

In water, sodium 3,5-hexadienoate reacts with 2,6-dimethyl-*p*-benzoquinone to afford a 1 : 2 cycloadduct which has a very complex structure [Grieco, 1983]. This product is formed by a triply reflex Michael reaction between the normal Diels–Alder adduct and another molecule of 2,6-dimethyl-*p*-benzoquinone.

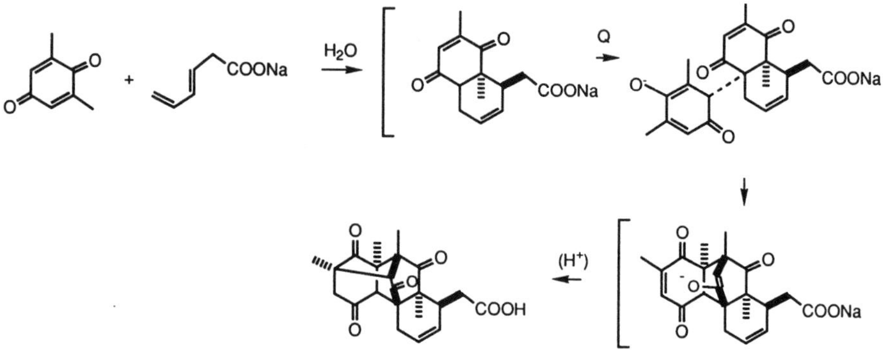

4

TANDEM VICINAL
DIFUNCTIONALIZATION OF
ALKENES AND ALKYNES

4.1. MICHAEL REACTION/ELECTROPHILATION

Many organic reactions belong the category of tandem vicinal difunctionalization; some of them are well known, for example epoxidation, Diels–Alder reaction, 1,3-dipolar cycloaddition, [2 + 2] photocycloaddition. In this chapter we shall discuss tandem vicinal additions of two carbon chains to a double bond.

An important development in recent years concerns the stereoselective introduction of two carbon chains to an α,β-unsaturated carbonyl compound. It is of course extremely difficult to achieve controlled reaction to an unactivated substrate in view of the high activation energy for the addition step and, even if possible, the adduct would be so reactive that polymerization would complicate the situation.

The most common process of the formal 1,2-addition involves Michael reaction under aprotic conditions and capture of the resulting enolate with an appropriate electrophile.* Sometimes a stepwise protocol is implemented, that is the enolate is converted into enol derivatives and resubmitted to α-functionalization. An extensive compilation of these reactions [Chapdelaine, 1990] has appeared recently, therefore only a few representative cases are given here.

The reactivity profile of organocopper reagents is compatible with enecarbonyl substances. These reagents behave as Michael donors and they do not generally attack isolated carbonyl groups. Furthermore, the enolate species generated from such a reaction is capable of reaction with various other

---

* It is proposed to use the term "electrophilation" for such reactions which encompass alkylation, acylation, silylation, etc.

electrophiles, enabling the tandem vicinal difunctionalization of a conjugated double bond.

Mechanistically the Michael addition step consists of oxidative *trans* addition of a d^{10} cuprate to the substrate to produce a transient copper(III) (d^8) intermediate, and a subsequent reductive *cis* elimination to generate the CC bond at the β-position of the enecarbonyl compound.

The second step pertaining to reaction of the enolate species with an electrophile to complete the tandem functionalization is similar to the familiar C-alkylation, notwithstanding the involvement of single electron transfer prior to bond formation with electrophiles of low reduction potentials (e.g. RI). The reactions with electrophiles of higher reduction potentials are S_N2-type.

The Michael addition process is of course subject to steric and stereoelectronic control inherent to the substrate structure. However, the final product is liable to some equilibration, depending on the nature of the enolate, its counterion, reaction conditions, etc., although the thermodynamically more stable (*trans*) products usually prevail.

Both stoichiometric and catalytic organocopper reagents have been successfully employed in the tandem *vic*-functionalization scheme. In a catalytic reaction the active species is generated in situ from a Grignard reagent and a small amount of a copper salt.

Organocopper reagents are activated on solubilization by the presence of organophosphorus or organosulfur ligands. Another advantage of this technique is that the reagent is derived from only one equivalent to the precursor. Further variations include the use of mixed homocuprate reagents (RR′CuLi) and heterocuprate reagents (RCuXLi) in the Michael step in place of homocuprate reagents (R$_2$CuLi).

A short, stereocontrolled route to 11-oxoequilenin methyl ether [Lentz, 1978] is representative of the synthetic value of such methods.

A regiocontrolled synthesis of substituted pyridines [T.R. Kelly, 1985b] has been developed on the basis of the tandem process. With organocuprate reagents derived from α-lithio-N,N-dimethylhydrazones and trapping of the enolates with acyl cyanides the condensation products readily cyclize to afford N-dimethylamino-1,4-dihydropyridines which aromatize on acid treatment.

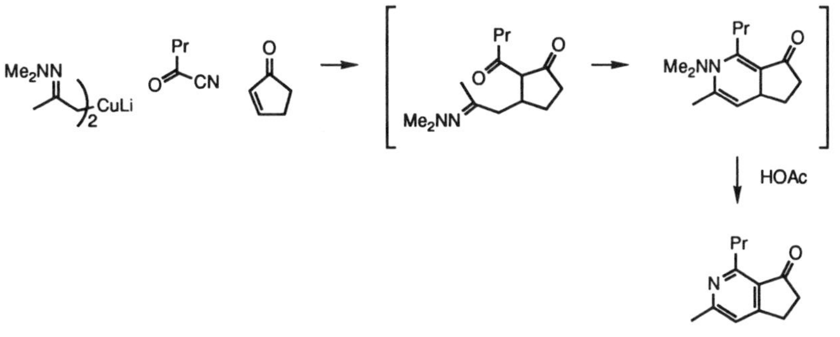

Although the softness of organocopper compounds makes them the primary Michael donors to initiate the *vic*-difunctionalization, occasionally other organometallic species may be employed. Further synthetic uses of the difunctionalized products depend on the structural features of the chains. Thus, a method for benzannulation of enones has been developed [Posner, 1990].

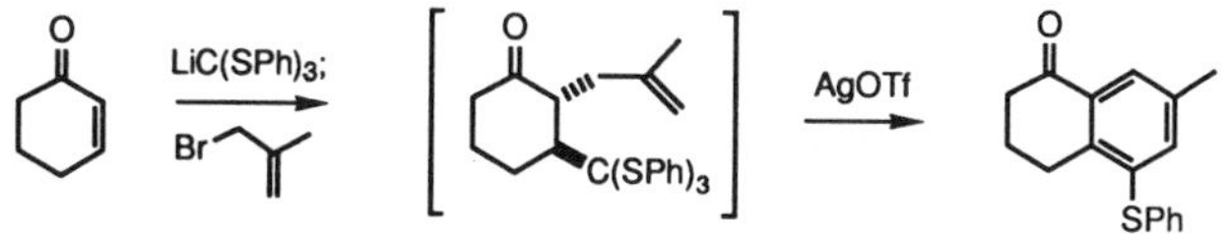

An oxazoline group at C-2 of a naphthalene nucleus activates C-1 toward attack by organometallic reagents in the Michael reaction sense. Quenching the lithiated adducts with an alkylating agent completes the stereoselective additive dialkylation of the naphthalene [Robichaud, 1991]. Analogously, the 1-oxazolinyl group in the naphthalene nucleus mediates similar reactions, including asymmetric annulation by intramolecular tandem alkylations [Meyers, 1989].

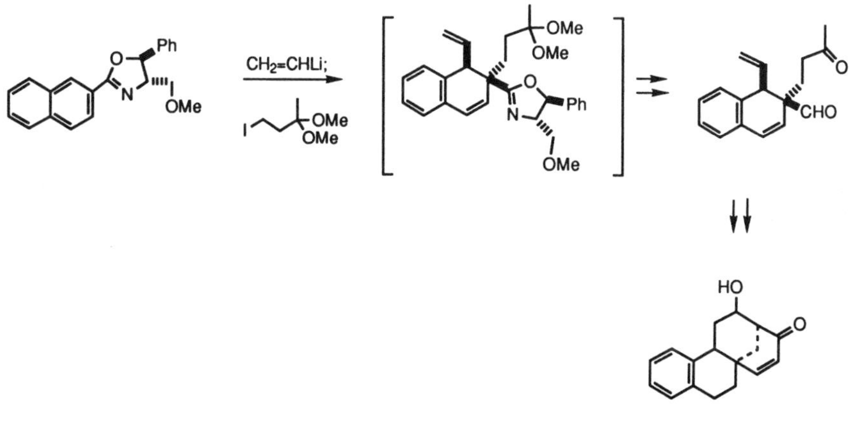

The coupling the tandem *vic*-difunctionalization with the Robinson annulation magnifies further the utility of the latter process [Boeckman, 1973]. Of course, the *vic*-difunctionalization is a double Michael reaction.

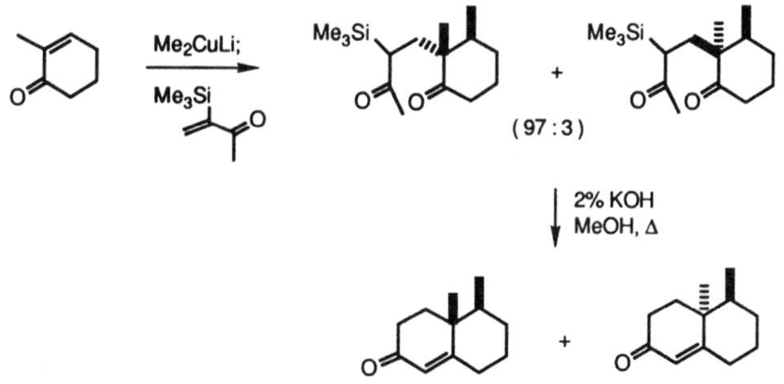

The employment of in situ generated acylnickel carbonyls to effect Michael addition [Corey, 1969b], with trapping of the resulting enolate ions by alkyl halides constitutes a novel pathway to 1,4-dioxygenated substances. This protocol is the basis for a concise approach to naphthoquinone antibiotics such as nanaomycin-A [Semmelhack, 1985b].

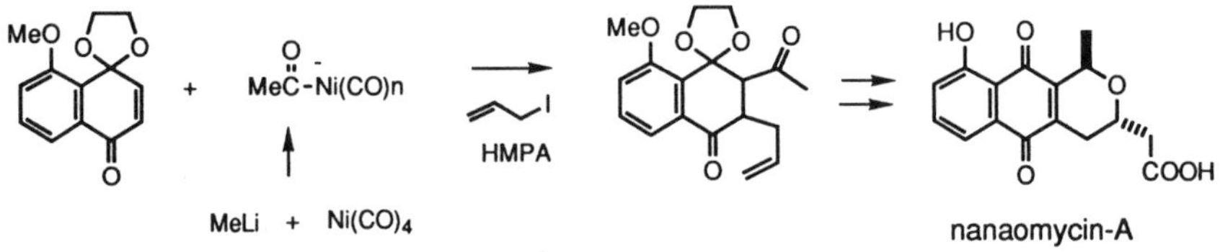

Complexation of α,β-unsaturated carbonyl compounds with an iron carbonyl moiety enhances its Michael acceptor characteristics. Furthermore, alkylation of the Michael adducts may lead to carbonyl insertion [B.W. Roberts, 1980].

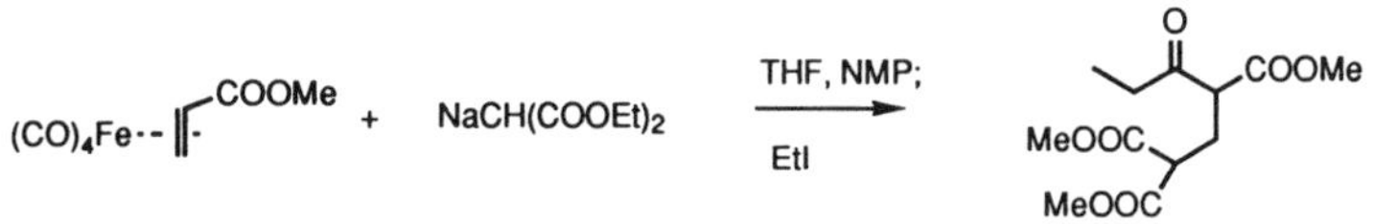

The presence of another electrophilic center in the Michael acceptor molecule may induce cyclization upon formation of the initial adduct [Cooke, 1979; Little, 1982]. Interestingly, a remarkable solvent effect has been observed [M. Yamaguchi, 1985] on the stereochemistry, and diastereomers which differ in one of the three newly created contiguous asymmetric centers may be obtained as the major products.

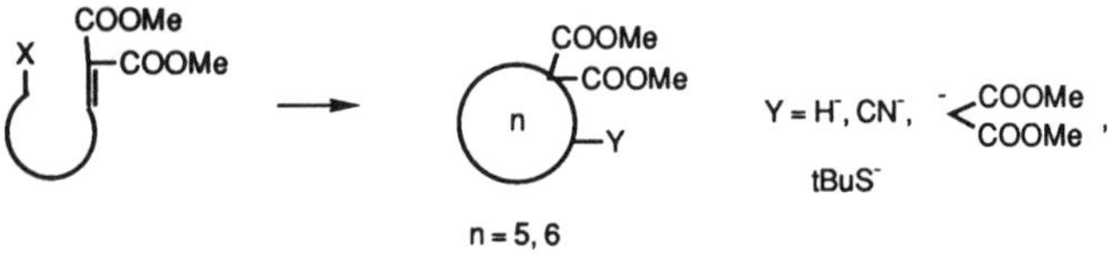

The protocol is effective for bisannulation [Lavallée, 1987] when the initial Michael donor, the acceptor, and the alkylating agent exist in the same molecule.

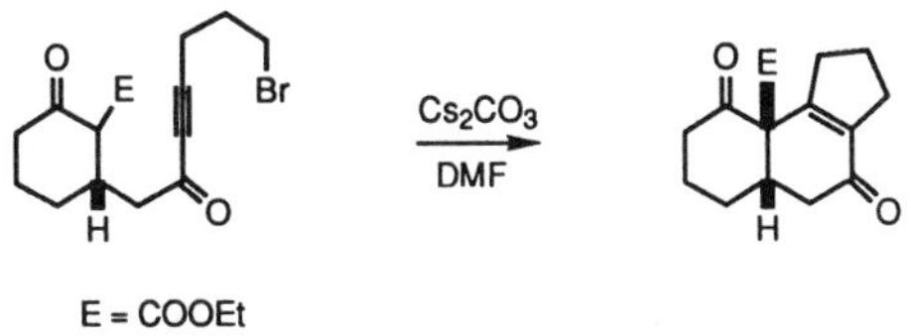

E = COOEt

On the other hand, compounds with both Michael donors and acceptors which are noninteracting offer special opportunity for ring formation. For example, five-membered ring compounds have been synthesized from dimethyl (E)-5-(methoxycarbonyl)-2-hexenedioate and enones [Bunce, 1987]. In this reaction the donor atom constituting part of a malonate appears to provide the best results, whereas the corresponding bis(sulfone) undergoes elimination instead of Michaeal reaction. Reflexive Michael reactions which belong to the same category have been discussed in Section 3.3.

It should be emphasized that the electrophile is not limited to an alkylating agent. Thus the trapping reaction may be a Claisen condensation [Tada, 1982].

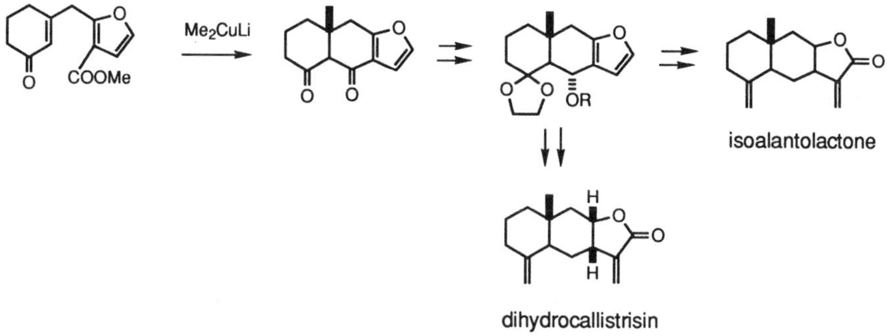

isoalantolactone

dihydrocallistrisin

It is perhaps surprising that strained ring systems have been prepared by the method. For example, a bridged lactone is obtained when γ-(2-tosyloxyethyl)butenolide is exposed to an alkyllithium in the presence of the copper(1) bromide-dimethyl sulfide complex [Hizuka, 1989].

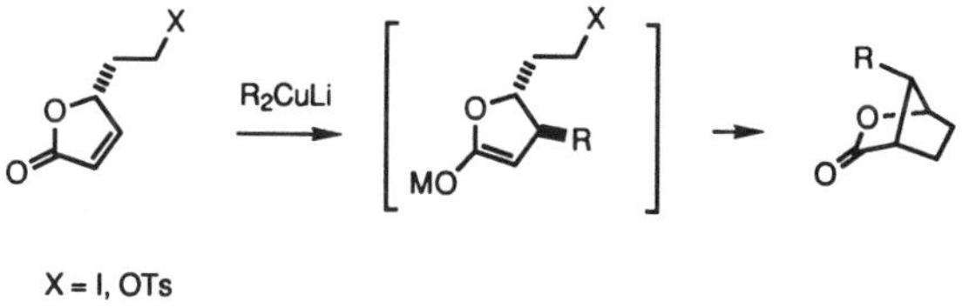

X = I, OTs

The three-membered ring is very readily formed by a similar reaction [Torii, 1977].

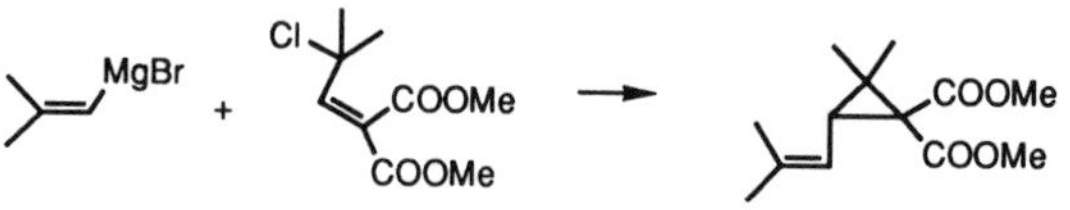

The reaction of 1,1-dipolar species with Michael acceptors can result in three-membered ring compounds which include epoxides. Cyclopropanation is now frequently performed with the aid of sulfonium [Corey, 1967] and phosphonium ylides [Grieco, 1972; Krief, 1988]. α-Lithiosulfones are also useful reagents [Krief, 1985].

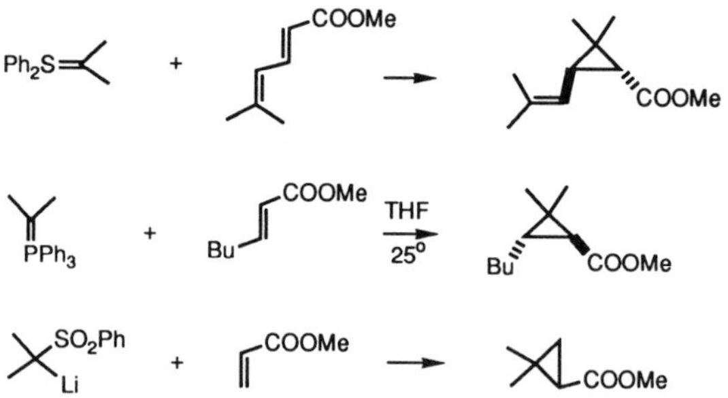

Conjugated nitroalkenes are latent 1,1-dipoles. Thus, a spirocyclic product was produced when 1-nitromethylcyclohexene oxide was treated with methyl cyanoacetate in the presence of a base [Yokomori, 1991]. The nitroalkene was generated in situ.

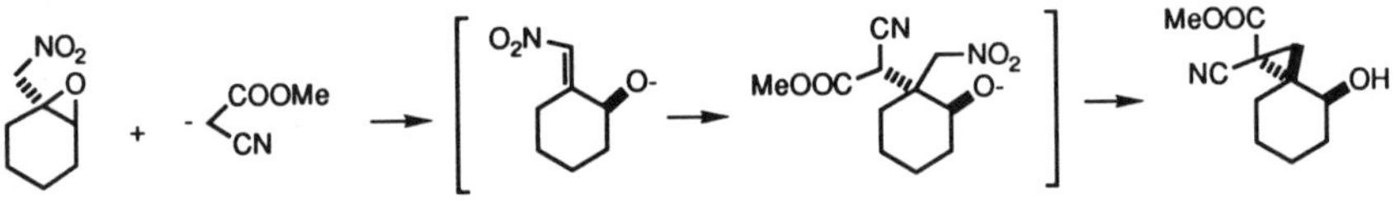

Cyclopropanation involving α,β-unsaturated 1,1-dipolar species has further implications in annulation of a cyclopentene moiety. As illustrated in a convergent approach to retigeranic acid [Hudlicky, 1988], deprotonated α-bromoacrylic esters behave excellently in that role.

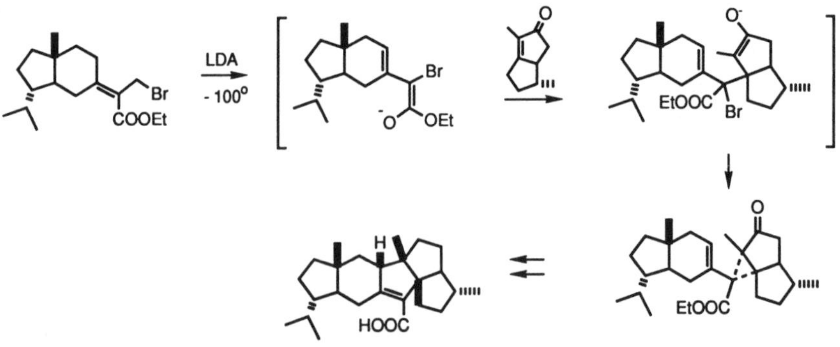

retigeranic acid-A

Dimerizative cyclopropanation of methyl α-bromoacrylate by the action of nucleophiles has been reported [Joucla, 1985].

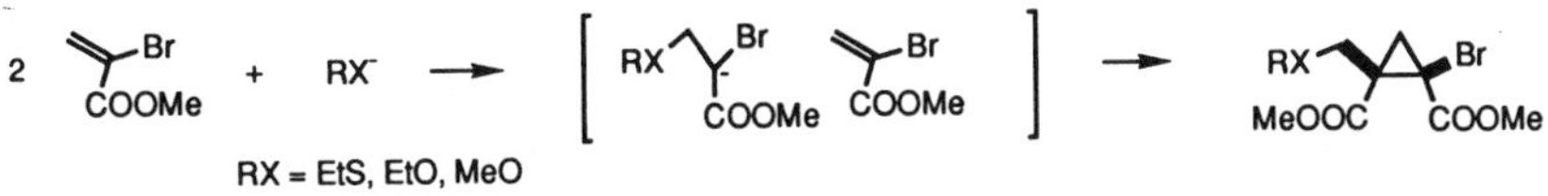

RX = EtS, EtO, MeO

The formation of three CC bonds in one step via reflexive Michael reaction and intramolecular alkylation is an extremely valuable method. Its many uses include the assembly of the bicyclo[3.1.0]hexanone skeleton [Hagiwara, 1991] and the key step in a synthesis of ishwarane [Hagiwara, 1980] (cf. [Spitzner, 1982]).

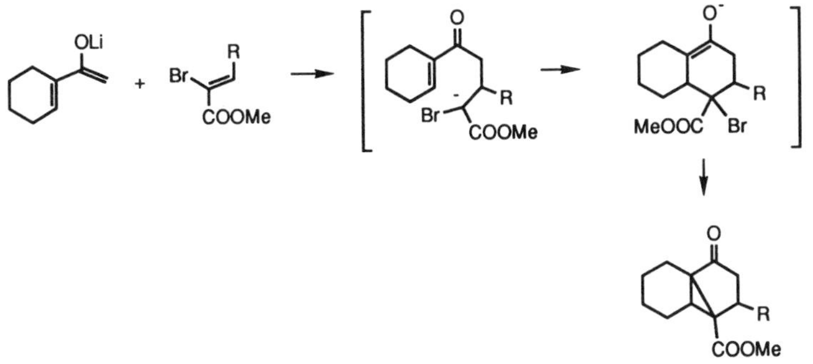

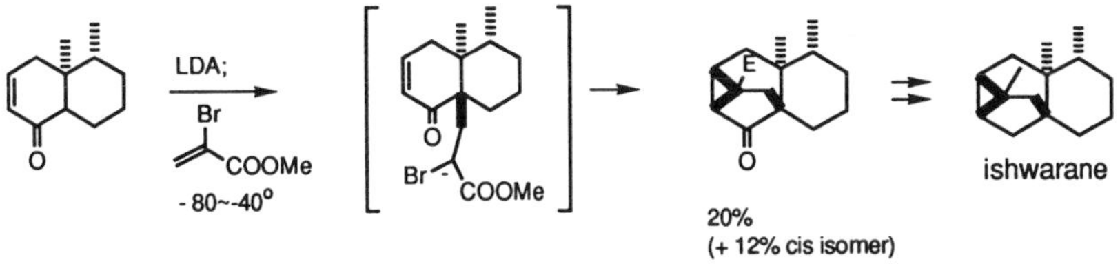

20%
(+ 12% cis isomer)

ishwarane

Related to this mode of bisannulation is a synthesis of the mitosane ring system [Cory, 1983] in which an imine was involved in the formation of the second and third bonds.

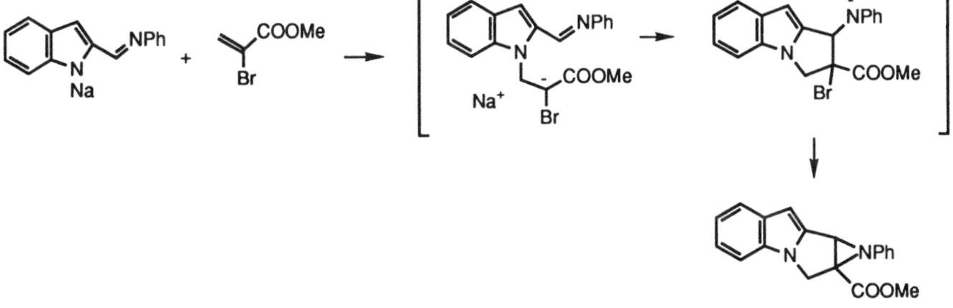

Complementary to an α-bromoacrylic ester a few other reagents have been employed in the reflexive Michael reaction–intramolecular alkylation combination. Among these reagents are alkenyltriphenylphosphonium bromides

[Cory, 1980], vinyl sulfones [Cory, 1984], and conjugated nitroalkenes [Cory, 1985]. The electron-withdrawing substituents play an activating role for the olefins to participate in the reflexive Michael reactions as well as that of the leaving group in the intramolecular alkylation step.

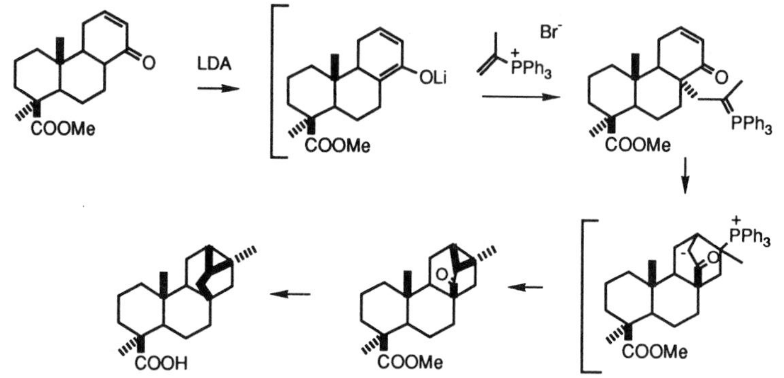

trachyloban-19-oic acid

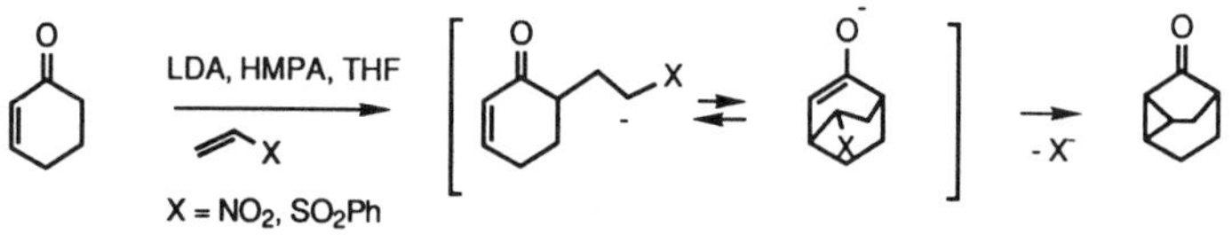

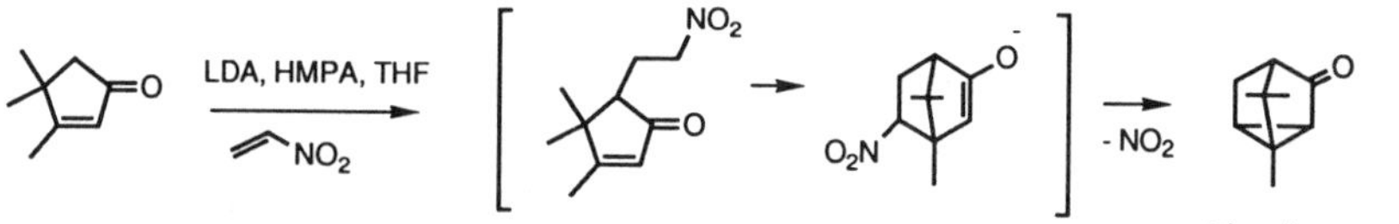

X = NO₂, SO₂Ph

tricyclanone

Metal carbenoids may also be used in cyclopropanation. Interestingly, the reaction of methyl acrylate with a Fischer carbene complex which contains an acetylenic linkage four bonds away has been shown to afford a cyclopentenyl-cyclopropylcarboxylic ester [Harvey, 1990]. However, serious limitations concerning the initial carbocycle formation have been noted, as an extremely poor yield of the homologous product was obtained.

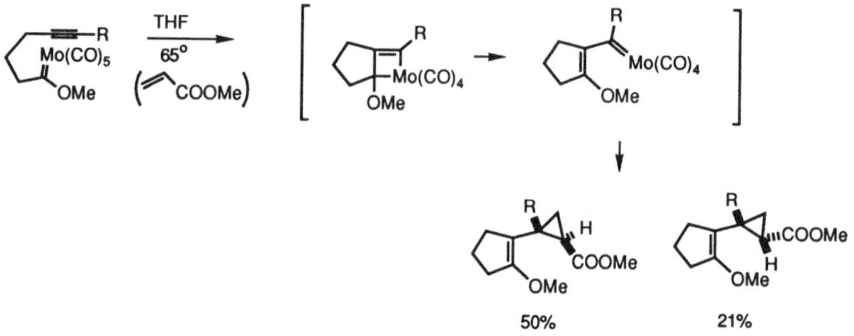

50%   21%

A more conventional route to such metal carbenoids is by decomposition of diazo compounds in the presence of a heavy metal compound. The relay bond formation which results in an intramolecular cyclopropanation [Hoye, 1990] is now shown.

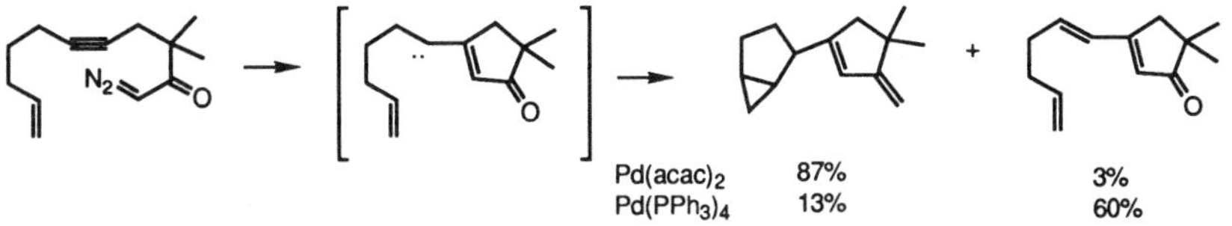

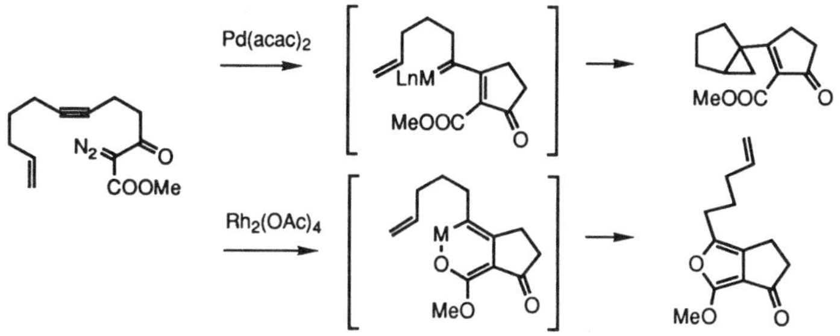

Pd(acac)$_2$	87%	3%
Pd(PPh$_3$)$_4$	13%	60%

It must be emphasized that cyclopropanation based on carbenoid addition has much less reactivity restriction on the part of the alkene, particularly for the intramolecular version. Generally the efficiency of such reactions is greatly enhanced by the use of Rh(II) catalysts.

It is of interest that in the next reaction shown the aromatic ring is more reactive than the vinyl group. The cycloheptatriene derivative is a potential intermediate for the synthesis of harringtonolide [Mander, 1991].

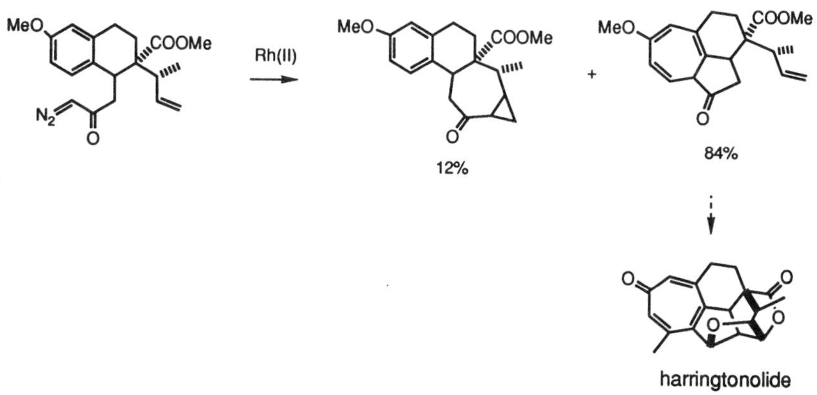

harringtonolide

The reactivity pattern is quite different for the metal-catalyzed cyclopropanation by consecutive displacement of a 2-ene-1,4-dicarbonate with malonate ester [Trost, 1987]. The chemoselectivity of the first step can be changed by variation of the metal template, for example Pd(0) to W(0).

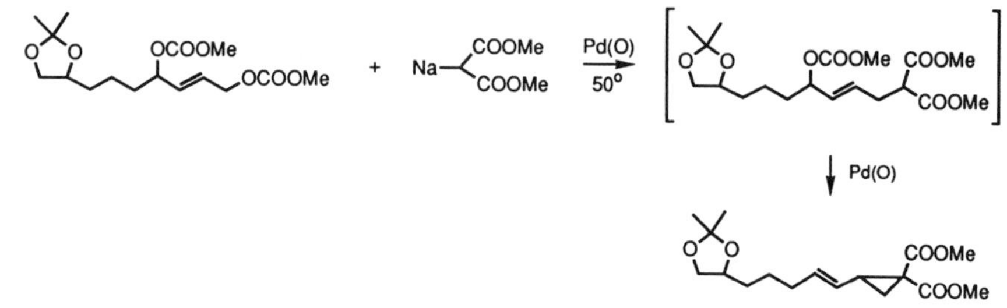

o-Benzologous 1,1-dipole equivalents have been in great demand for annulation owing to the discovery of many bioactive natural quinones and related substances. Prototypes of several such synthetic equivalents are shown below.

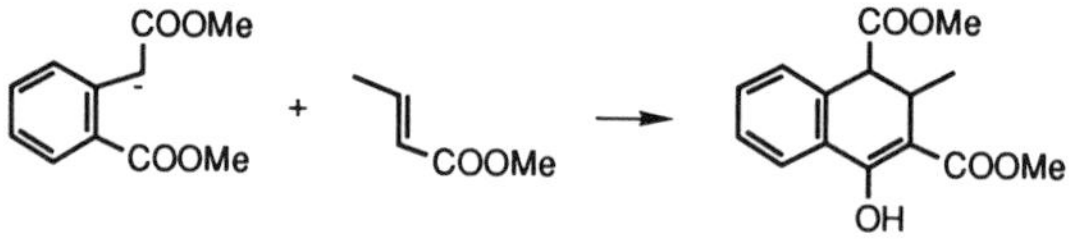

[Eisenhuth, 1965]

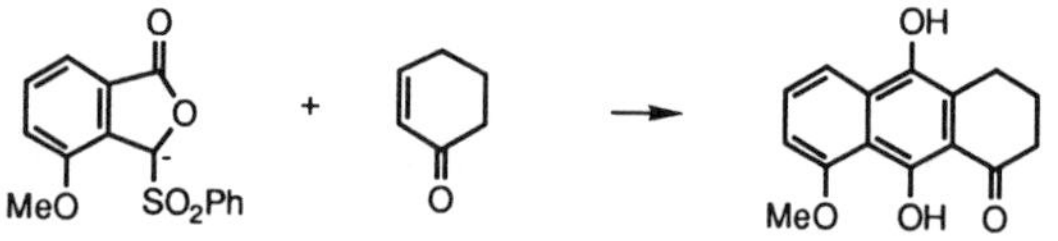

[Hauser, 1978]

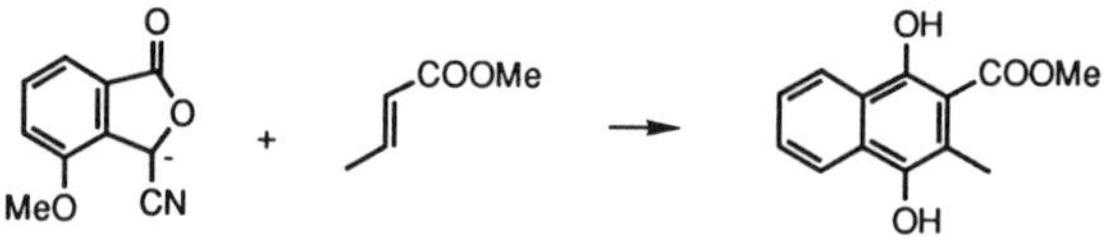

[Kraus, 1978]

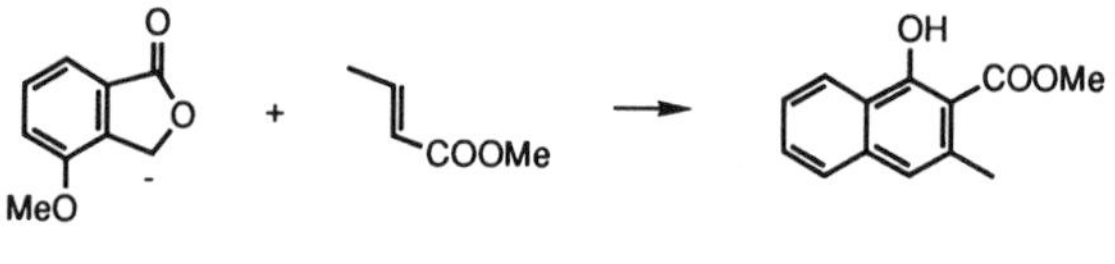

[Broom, 1978]

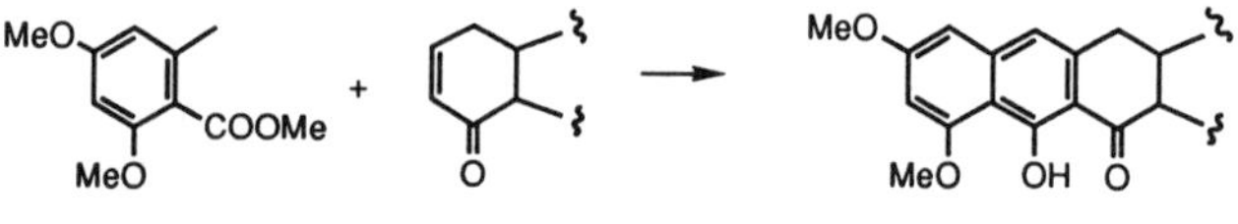

[Dodd, 1979]

While reactions of the sulfonyl and cyanophthalide systems lead directly to hydroquinones via the Michael reaction, Dieckmann condensation, and decomposition of the resulting cyanohydrin or α-hydroxysulfone anions, others require oxidation to reach the dioxygenated state.

Besides the common enecarbonyl compounds, benzynes have been employed in the annulation [Townsend, 1981].

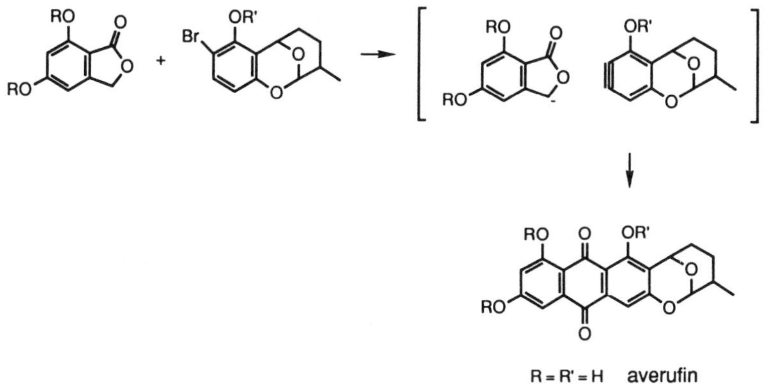

R = R' = H   averufin

Related to the cycloaddition the tandem electrophilic–nucleophilic additions to benzynes have been effected and the regiocontrol by an oxazolinyl group has also been delineated [Pansegrau, 1988].

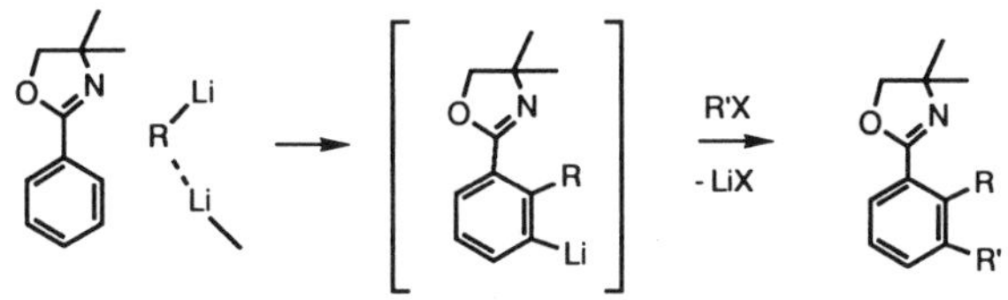

A variant of the Michael reaction–electrophilation is the use of conjugate sulfones to direct bond formation [Fuchs, 1986]. This acceptor group is complementary to the carbonyl and certain types of compounds may be more readily acquired. For example, sulfinic acid elimination from the products regenerates an alkene linkage which proved expedient in syntheses such as those of cephalotaxine [Burkholder, 1988] and morphine [Toth, 1988].

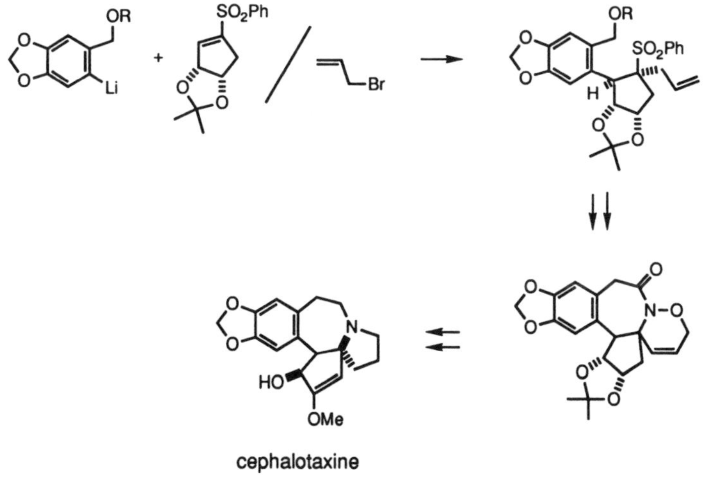

cephalotaxine

morphine

4.2. *vic*-DIFUNCTIONALIZATION OF TEMPORARILY ACTIVATED ALKENES

A Michael acceptor contains a CC unsaturation which is usually activated by an electron-withdrawing substituent. The synthetic value of alkene activation toward vicinal difunctionalization by attachment of a temporary auxiliary group is apparent because it is no longer necessary to deal with the substituent if its role is merely activating. In this respect the activation of double bonds using temporary devices is a welcome development.

The depletion of the π-electron cloud of an aromatic ring by complexation with a carbonylmetal species is now widely recognized. The sidechain of styrene becomes highly electrophilic when the aromatic ring is ligated to a tricarbonylchromium unit [Semmelhack, 1980], therefore *vic*-difunctionalization is possible.

Even simple alkenes such as propene have been activated (e.g. by Pd(II)) in situ to react with nucleophilic reagents. Subsequent carbonylation would lead to carboxylic acid derivatives [Hegedus, 1980].

A cycloacylation of alkenyl tosylates is illustrated in its application to the formation of the bridged ring system of aphidicolin [McMurry, 1979]. The tosylate underwent substitution with disodium tetracarbonylferrate, and after carbonyl insertion the acyliron species then reacted with the double bond. Upon protonation on workup the ketone precursor of aphidicolin was obtained.

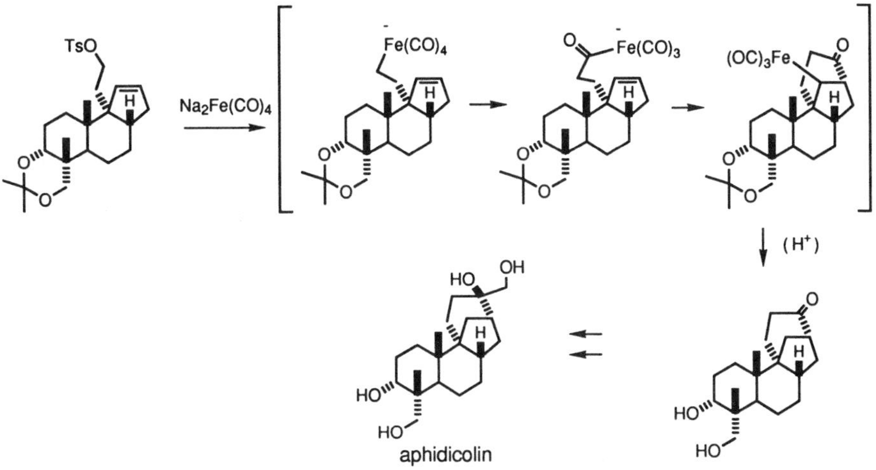

aphidicolin

vic-Difunctionalization of a diene may be achieved via metal–diene complexes. The tandem acylation–alkylation of 1,3-cyclohexadiene [Semmelhack, 1983] is an example.

Perhaps the stereoselective 1,2-addition to benzene is more remarkable [Kündig, 1983], in view of its destruction of the aromaticity.

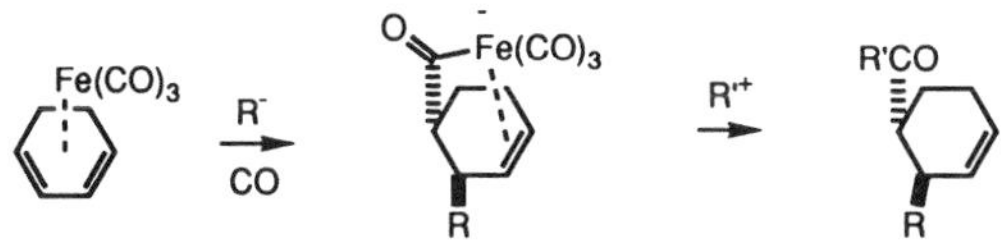

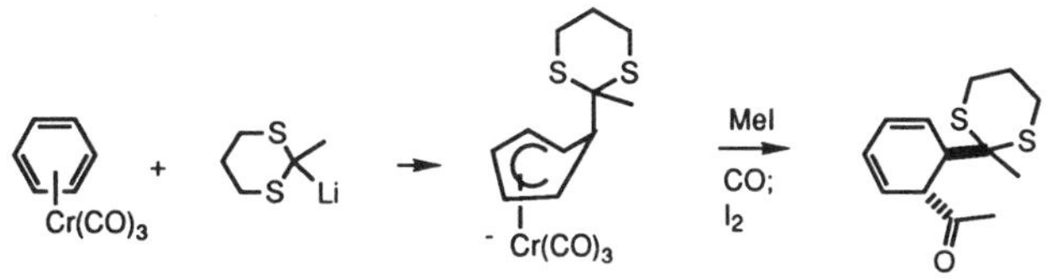

Sterically favored intramolecular cyclization of alkenyllithiums leads to new structures. Polycyclic products may be obtained [W.F. Bailey, 1989]. It is significant that the more strained *trans*-bicyclo[3.3.0]octane derivatives are formed in a highly stereoselective manner [W.F. Bailey, 1990].

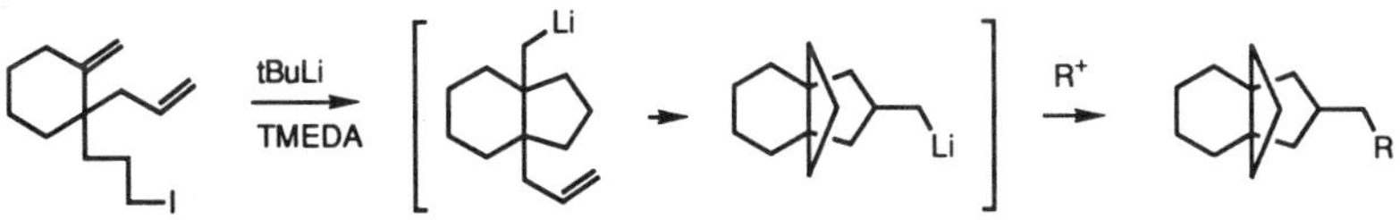

Substituted 1,3-dienes form magnesiacyclopentenes which react with α,ω-dihalides and bromo nitriles to afford cyclic products [Xiong, 1991.

Intramolecular alkylation with double bond transposition and in tandem with substitutive carbonylation constitute a useful method for simultaneous carbocyclization and lactonization [Semmelhack, 1981]. Thus, in one step a common structural unit of many biologically active compounds can be assembled.

Perhaps it should be mentioned that cationic cobalt complexes of conjugated dienes undergo twofold alkylation without activation of the system after the first group is attached [Barinelli, 1986]. However, the alkene products derive from a formal 1,4-addition.

4.3. PAUSON–KHAND CYCLIZATION AND DÖTZ REACTION

The cyclopentenone synthesis from an alkyne, an alkene in the presence of dicobalt octacarbonyl, is formally a *vic*-cycloacylation of the triple bond. This very useful reaction is known as the Pauson–Khand cyclization [Pauson, 1985; Schore, 1988a].

While many alkynes may be used, the scope of the Pauson–Khand cyclization is somewhat limited by the alkene. It seems that strained cycloalkenes are the best substrates. However, ethylene has been successfully incorporated under more vigorous conditions [Newton, 1983].

The intramolecular version of the Pauson–Khand cyclization is a valuable tool for the preparation of bicyclic structures and its application to the elaboration of polyquinane systems such as the perhydrotriquinacene system [Carceller, 1985] and several sesquiterpenes (e.g. coriolin [Exon, 1983]) has been extensively exploited.

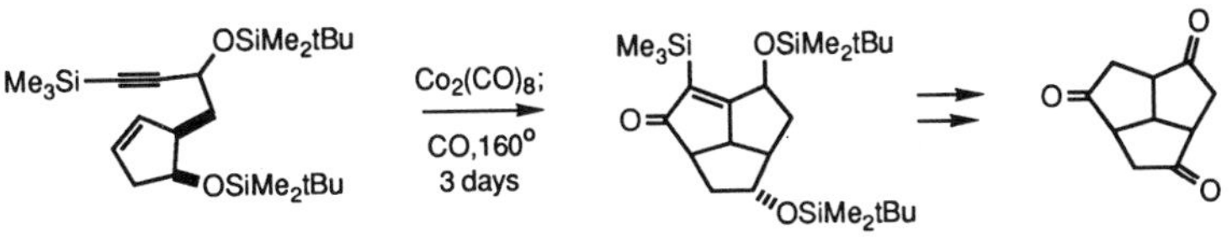

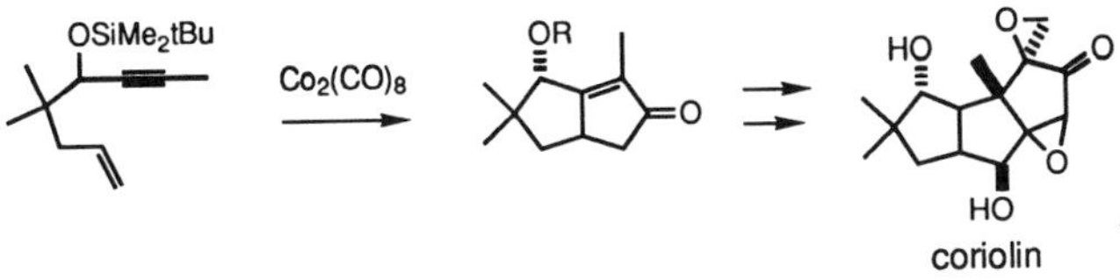

coriolin

Asymmetric bicyclization using enol ethers containing an alkyne terminus has been achieved [Castro, 1990]. The alkyl moieties of the enol ethers are chiral and diastereofacially biased, so that CC bond formation can only occur from one side of the enyne.

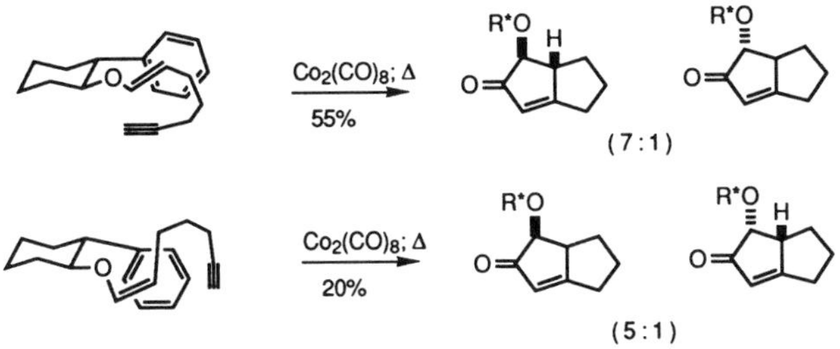

(7:1)

(5:1)

A formal synthesis of (+)-hirsutene attests to the effectiveness of the method.

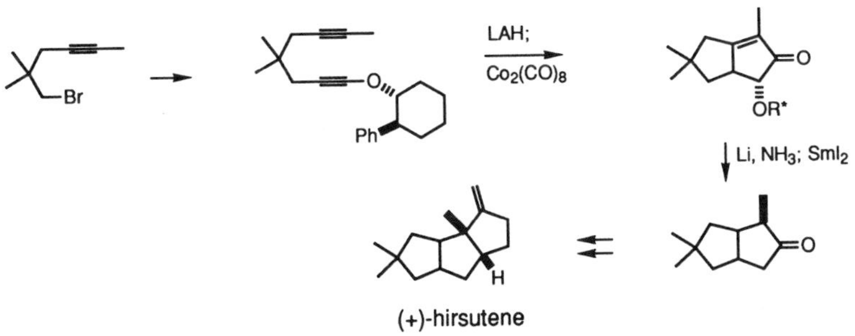

(+)-hirsutene

Both regiochemistry and stereochemistry are sterically controlled. For an unsymmetrical alkyne, olefin insertion into its derived dicobalt complex proceeds at the original alkyne carbon bearing a smaller group. The CC bond formation is directed toward the alkene carbon nearest the larger allylic substituent in order to avoid steric repulsion by the large $Co(CO)_3$ residue. It is remarkable that the difference in bulk between a methyl group and a methoxy substituent in the two allylic positions is responsible for the high regioselectivity of the annulation of a fused cyclobutene [Sampath, 1987].

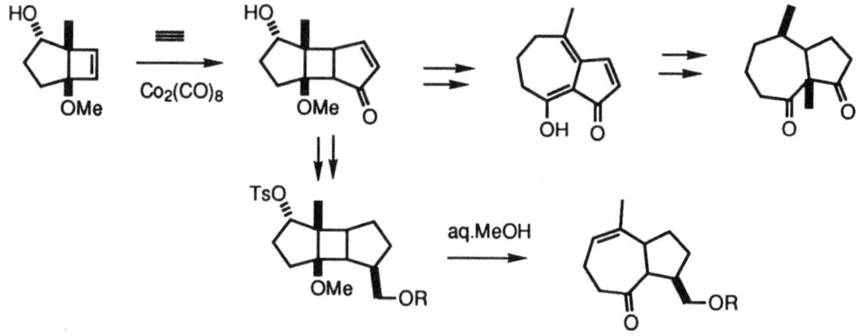

In the intramolecular Pauson–Khand cyclization the 1,2- and 1,3-stereoselectivity as influenced by the propargylic and allylic substituents, respectively, have been determined [Magnus, 1985].

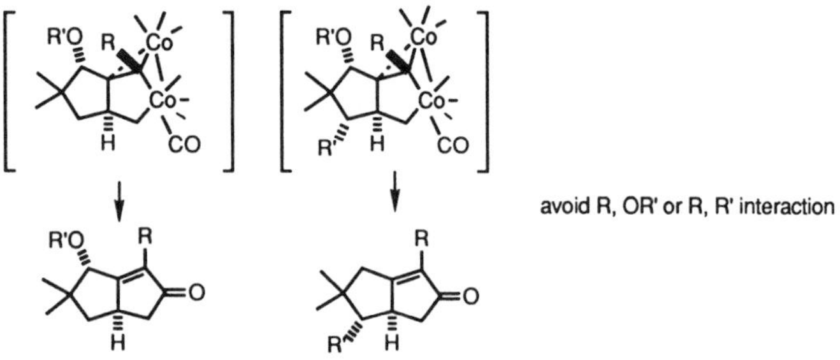

avoid R, OR' or R, R' interaction

Thus, in the synthesis of quadrone [Magnus, 1987] it is crucial to produce a diquinane with a *cis*-relationship between the angular methine hydrogen and the C_2-sidechain on the adjacent carbon atom, established upon alkene insertion of the C–Co bond. Blocking the endo side by the terminal siloxymethyl group was the deciding factor in favor of the desired product.

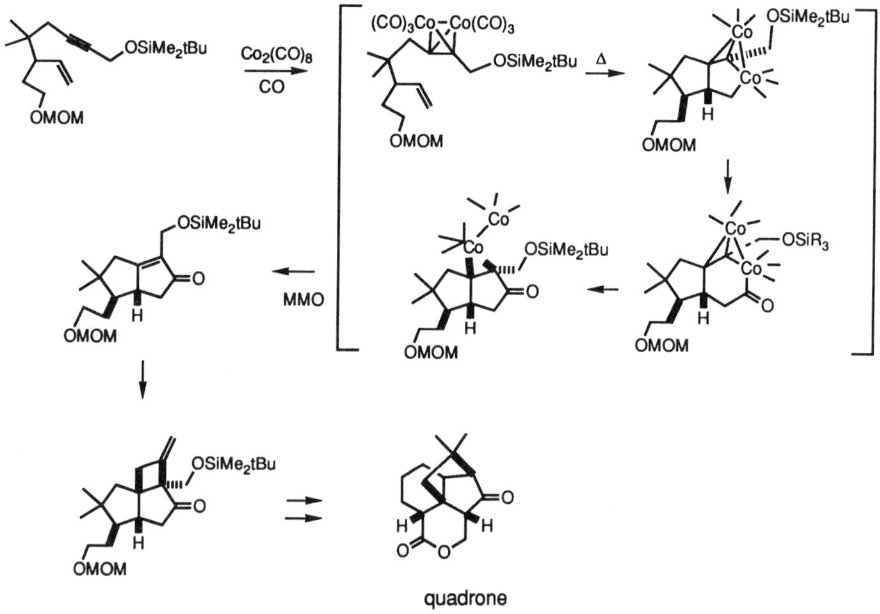

quadrone

The high stereoselectivity shown in the cyclization step of a pentalene synthesis [Schore, 1988b] has been ascribed to the adoption of a more product-like transition state for alkene insertion.

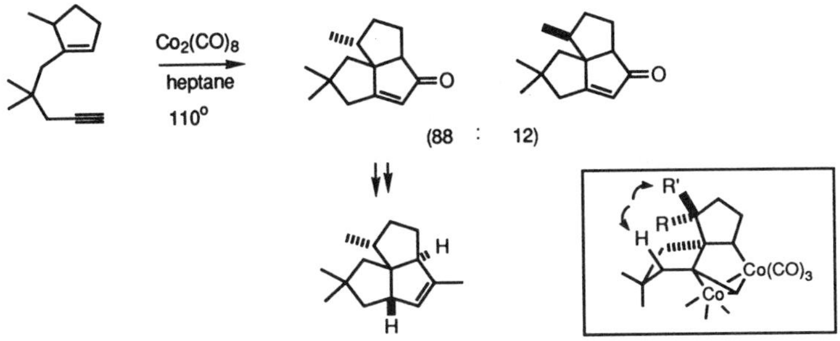

pentalenene

Reaction conditions for the Pauson–Khand cyclization have been modified in various manners. The most significant improvement is the promotion of an intramolecular reaction by treatment of the alkyne–cobalt complex with a tertiary amine N-oxide at room temperature [Shambayati, 1990].

3-Azabicyclo[3.3.0]9ct-5-en-7-ones are readily obtained by coupling the N-propargylation of allylamines with the Pauson–Khand cyclization [Jeong, 1991].

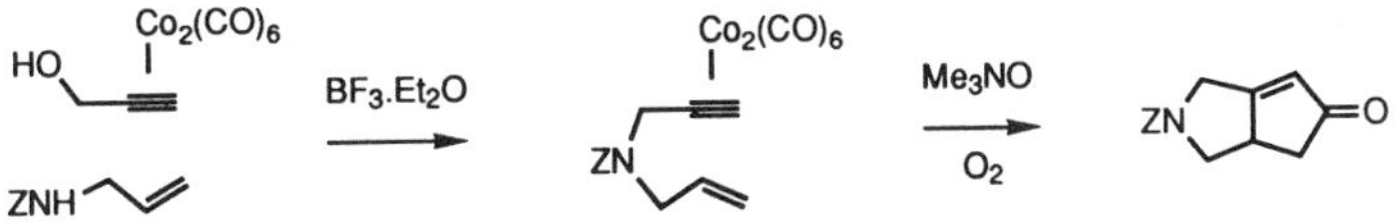

Alternative methods for the cyclopentenone formation include those mediated by zirconocene [Negishi, 1985] which provided particularly useful in the elaboration of the A-ring of tiglianes and daphnanes [Wender, 1990a].

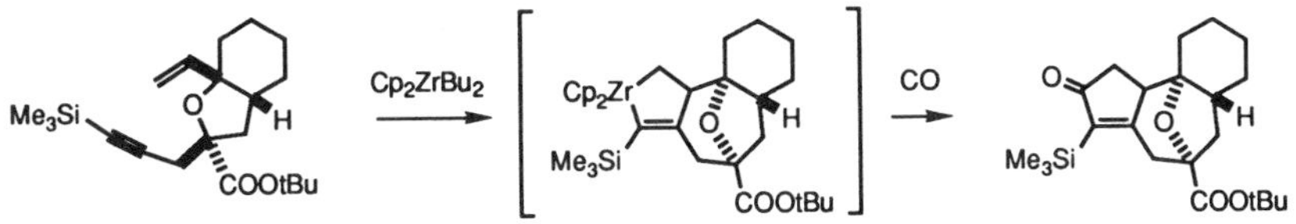

(Methylzircona)cycloalkenes also react with alkenes in a carbon monoxide atmosphere [Buchwald, 1989] to afford fused cyclopentenones.

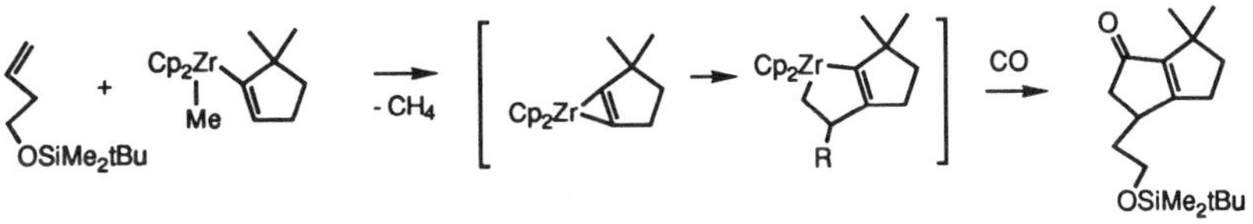

An isonitrile may be employed as the CO surrogate in a nickel(0)-promoted carbocyclization [Tamao, 1988].

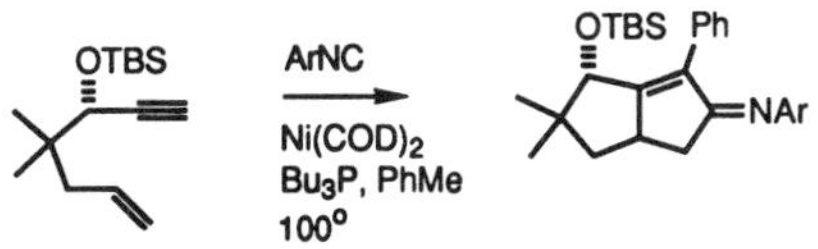

In the Dötz reaction an α,β-unsaturated alkoxycarbene pentacarbonyl-chromium complex undergoes annulation on treatment with an alkyne. The product is a π-complex of 1,4-hydroquinone monoether [Dötz, 1975, 1984].

Formation of three CC bonds occurs within the coordination sphere of the chromium atom. The initial cycloaddition generates a chromacyclobutene, from which a vinylketene complex is derived via ring opening and carbonylation. Subsequent electrocyclization involving the CC unsaturation originally present in the carbene complex leads to the cyclic structure.

In terms of synthetic value the regioselectivity shown by the reaction is a positive contribution. Generally the relative sizes of the alkyne substituents are important; the highest regioselectivity is observed in reactions with mono-substituted alkynes. The predominant isomer has the large substituent adjacent to the free hydroxyl group of the hydroquinone monoether.

Many elegant syntheses have been devised on the basis of this versatile reaction. For example, in the vitamin-K series [Dötz, 1983] the quinone system was created in only two steps from a simple carbene complex. By varying the enyne substrates, many congeners can be obtained. Vitamin-E has also been prepared.

Anthracyclinones can be approached via formation of the B- or C-ring. Thus, a highly efficient synthesis of a C-seco intermediate paved a smooth way to daunomycinone [Wulff, 1984].

The C-ring formation strategy is demonstrated by a synthesis of 11-deoxydaunomycinone [Wulff, 1988]. In both approaches the closure of the fourth ring did not cause any regiochemical problems, because only one site is open for the intramolecular acylation.

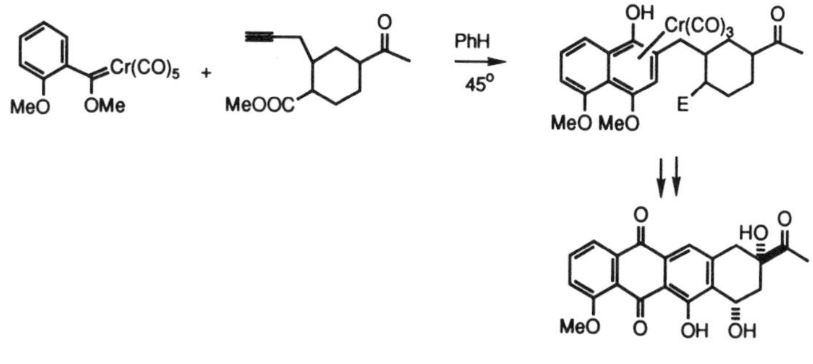

11-deoxydaunomycinone

Variants of these routes have also been developed [Dötz, 1988].

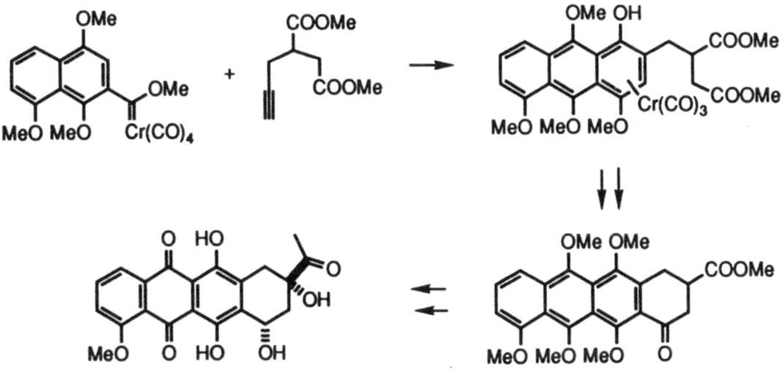

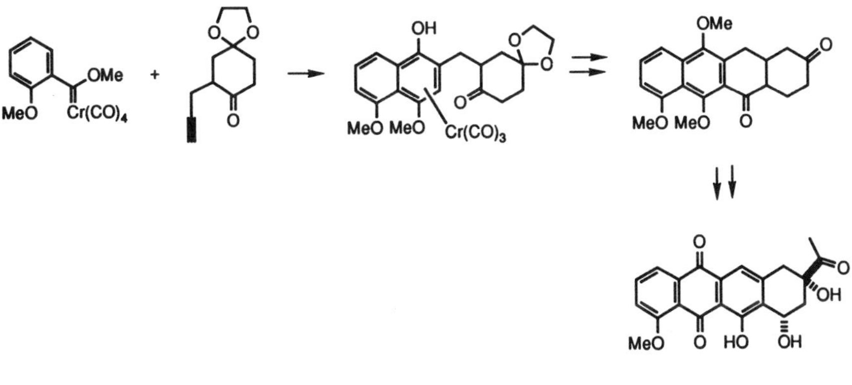

11-deoxydaunomycinone

At least in one instance the reaction of a chromium carbene complex with an enyne gives rise to a fused cyclobutanone [Wulff, 1985]. The synthetic potential of this process is evident.

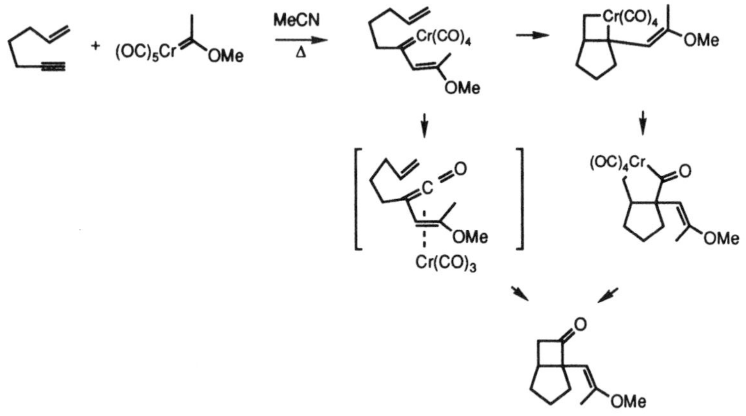

Another route to *p*-quinones based on vicinal diacylation of alkynes, via insertion to maleoyl- and phthaloylcobalt complexes and reductive elimination, is complementary to the Dötz reaction. The cobalt(III) complexes can be prepared from cyclobutenediones and benzocyclobutenediones by reaction with low-valent metal species. (Note that the CC bond to be cleaved must be a strained bond.)

Since alkyne coordination of the phosphine ligated cobalt center is energetically unfavorable, certain manipulations are required before the quinone formation can be effected, e.g., treatment of the chlorobisphosphinocobalt complexes with a silver(I) salt. The utility of this process is shown in a convergent synthesis of nanaomycin-A [South, 1984].

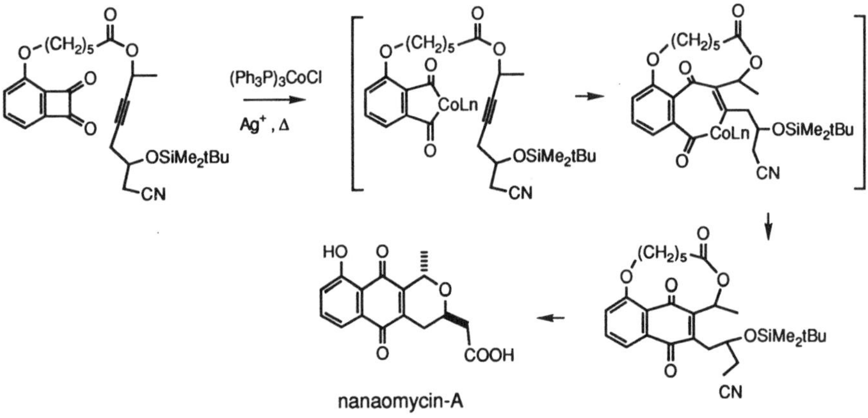

nanaomycin-A

4.4. INTRAMOLECULAR METALLO–ENE REACTIONS AND RELATED CC BOND FORMATION PROCESSES

The ene reaction is a thermal reaction of an olefin containing an allylic hydrogen (ene component) with an electron-deficient unsaturated compound (enophile) to form a 1:1 adduct via a cyclic six-electron transition state. Whether the reaction is intramolecular or intermolecular, it is a concerted suprafacial reaction.

On replacement of the allylic hydrogen of the substrate with a metal atom a metallo–ene reaction may result. However, it generally proceeds well only in the intramolecular version [Oppolzer, 1989]. In such cases the new organometallic species can then be functionalized, for example by carbonylation, alkylation, etc., to increase the complexity of the products. Significant syntheses have been devised on the basis of the metallo–ene processes.

For type-I reactions applications are found in approaches to protoilludene [Oppolzer, 1986c], sinularene [Oppolzer, 1982c], 12-acetoxysinularene [Oppolzer, 1984], and (+)-iridomyrmecine and (+)-δ-skytanthine [Oppolzer, 1986b].

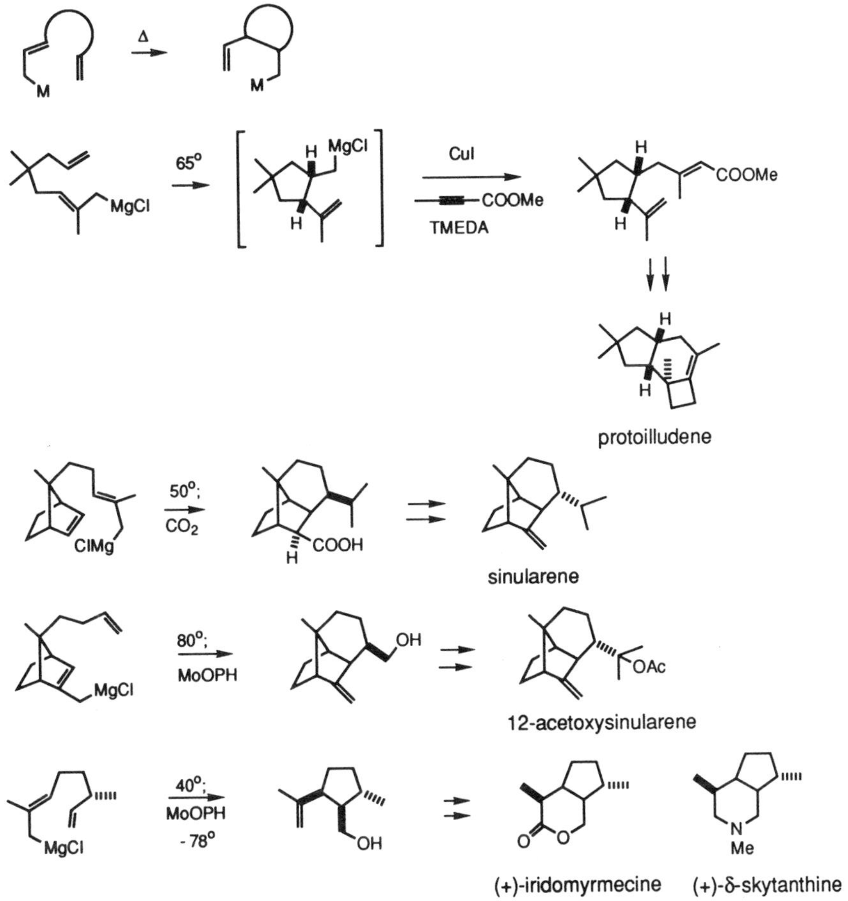

protoilludene

sinularene

12-acetoxysinularene

(+)-iridomyrmecine   (+)-δ-skytanthine

Two alternative methods for ring closure leading to the sinularene ring system are seen. The synthesis of (+)-iridomyrmecin and (+)-δ-skytanthine demonstrates stereodirecting effects of an existing stereocenter.

The possibility of iterative application of the method in the construction of polyquinane system is evident from the synthesis of $\Delta^{9(11)}$-capnellene [Oppolzer, 1982a]. Although a 3:2 mixture of stereoisomers was obtained at the closure of the central ring, the thermodynamically controlled intramolecular aldolization to be employed to complete the triquinane skeleton effectively obliviated the lack of diastereoselection in the metallo–ene reaction step.

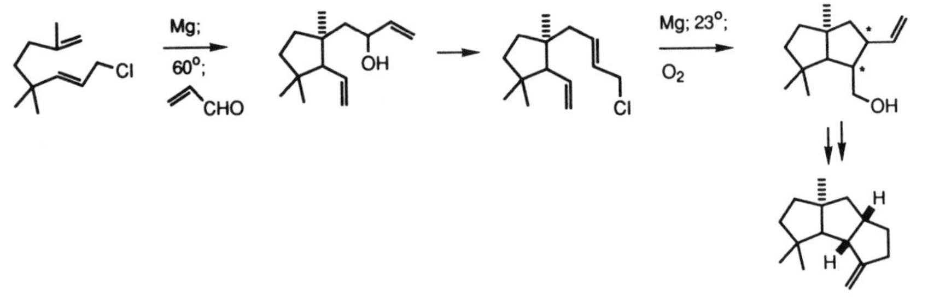

$\Delta^{9(12)}$ - capnellene

The type-II metallo–ene reaction is represented in the key steps for the syntheses of chokol-A [Oppolzer, 1986a] and (+)-khusimone [Oppolzer, 1982b].

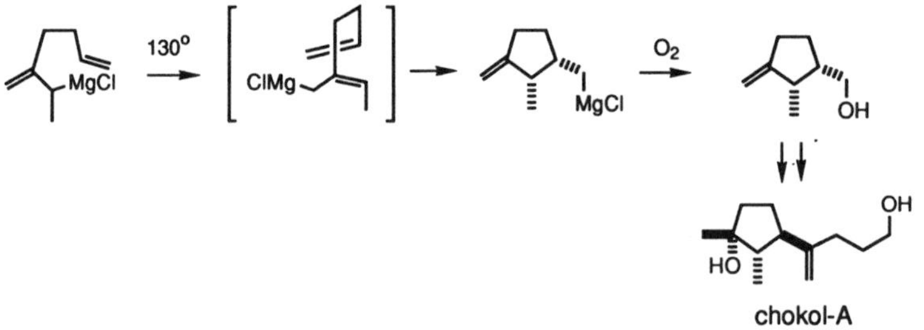

chokol-A

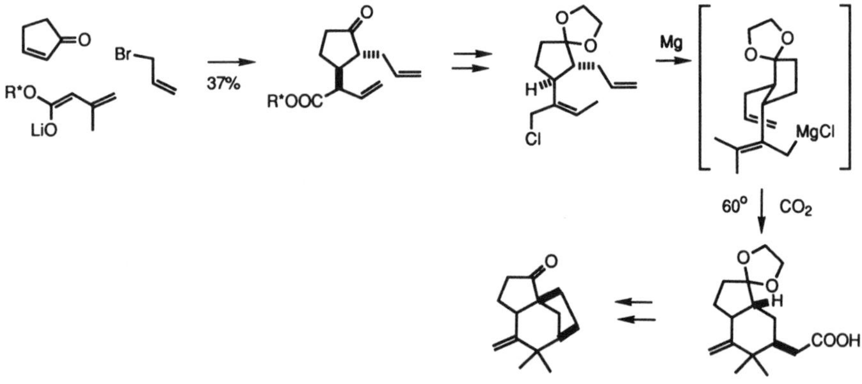

(+) - khusimone

The regio-, diastereo-, and enantioselective synthesis of (+)-khusimone is a best endorsement of the type-II magnesium–ene cyclization. It must be noted that substituents tend to influence the selection of transition states; the lack of diastereoselection during cyclization of a β-necrodol precursor [Oppolzer, 1986d] is fully understandable.

(1 : 1)

β-necrodol

Generally, the size of the developing ring dictates the ease of the cyclization:

Type-I metallo–ene reaction: 5 > 6 ≫ 7

Type-II metallo–ene reaction: 6 > 5 ∼ 7 ≫ 8

With respect to regioselectivity, the CC bond formation between the proximal sites of the ene and the enophile in the type-I process and that between the more highly substituted ene terminus and the proximal enophile site in the type-II situation is preferred.

Under kinetic control the type-I metallo–ene cyclizations usually afford five- and six-membered rings with *cis-vic* substituents from the reacting chains.

Catalytic intramolecular metallo–ene reactions are possible using Pd, Ni, and Pt complexes. The nickel reaction is particularly interesting in view of its *cis*-diastereoselectivity and the possibility of CO insertion.

Since Pd(0) species activate allyl esters toward alkylation, systems may be set up to realize an intramolecular metallo–ene reaction in tandem with the alkylation [Oppolzer, 1987].

A synthesis of (+)-3-isorauniticine [Oppolzer, 1991a] illustrates a further application of the tandem reaction in the area of natural products.

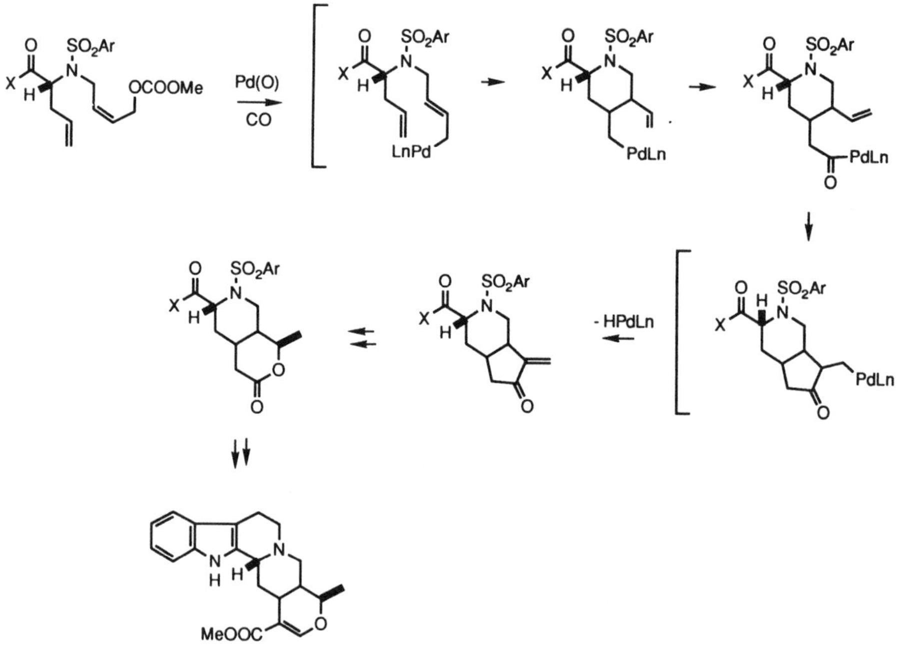

(+)-3-isorauniticine

Quite a few other metal atom transfer reactions accompanying CC bond formation have been discovered. For example, amidopalladation of N-acrylyl-o-allylanilides led to double cyclization [Danishefsky, 1983]. The palladium atom in the indoline intermediates migrated to the β-carbon atom of the acrylyl group upon formation of a new CC bond. A series of retro-hydrido and hydrido-palladations then followed.

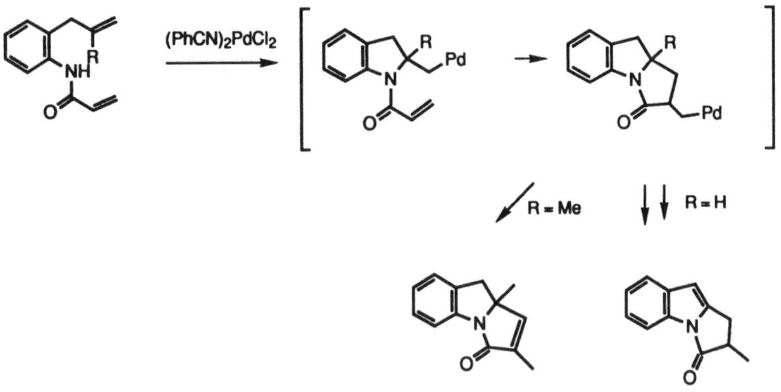

It is possible to form more than two rings in such Pd-catalyzed cascade carbometalation of polyunsaturated substrates [Abelman, 1988; Zhang, 1989,

1990; Carpenter, 1989; Trost, 1991a]. Juxtaposition of the unsaturation determines the ring system to be formed.

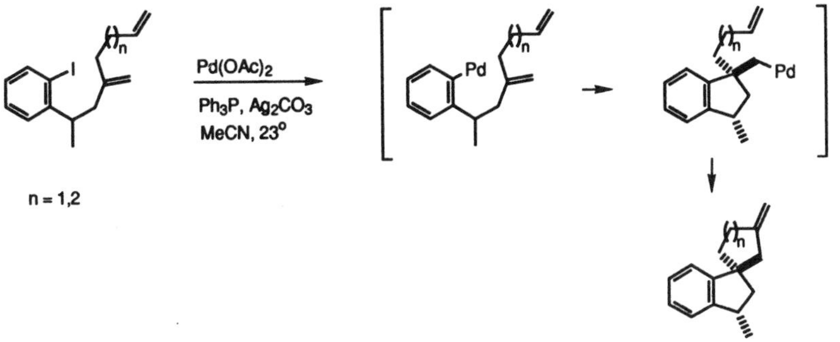

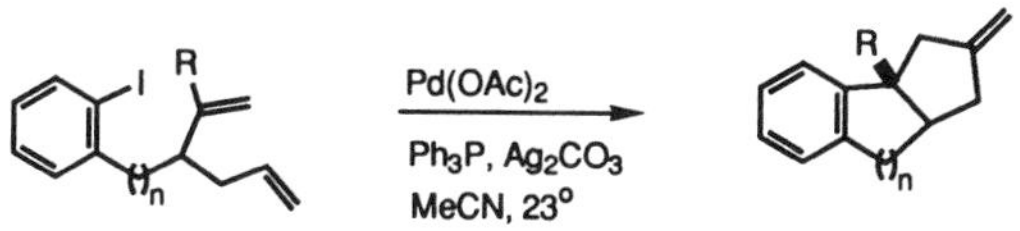

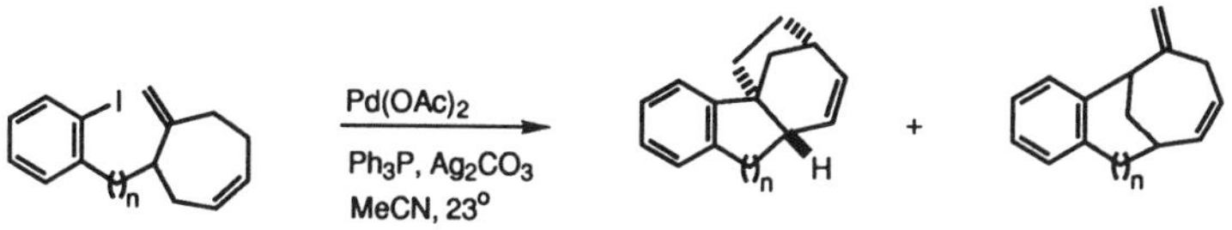

n = 1,2

n = 1,2   n = 1

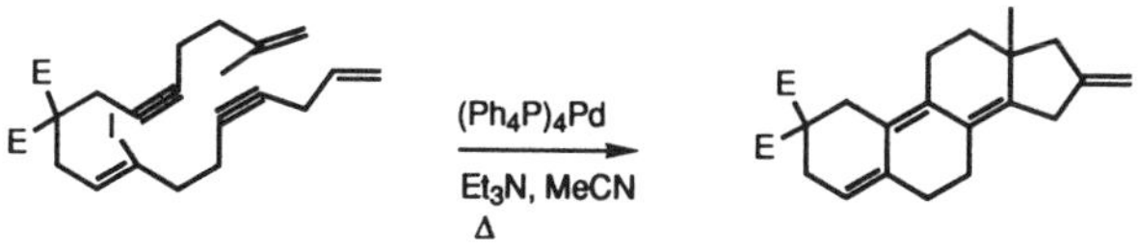

E = COOEt

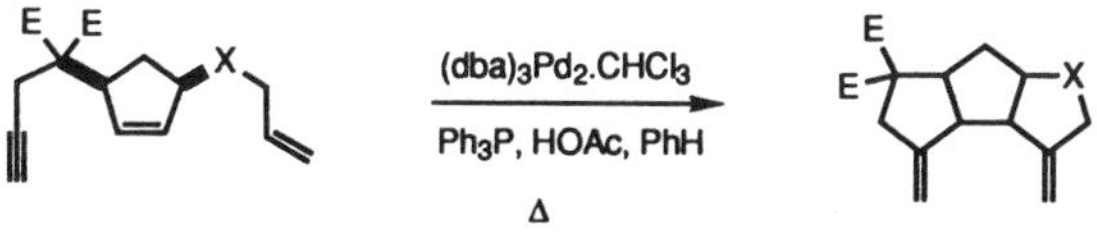

E = COOMe
X = CE$_2$, NTs

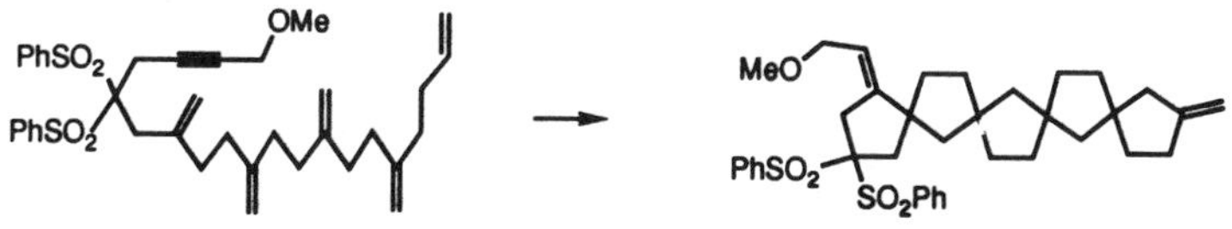

4.5. CATIONIC POLYENE CYCLIZATIONS

To the organic chemist the enzymatic cyclization of squalene is a most impressive and fascinating reaction. The Stork–Eschenmoser postulate of the stereochemical course stimulated an intense investigation into the stereoelectronic effects of the process and biomimetic synthesis of the tetracyclic skeleton to steroids. Taken apart, the cyclization involves electrophilic addition to one of the double bonds with the participation of the next double bond as the nucleophile, generating a new electrophile; the process continues until a terminator is encountered. In a stereoelectronically favorable parallel arrangement of the double bonds a polyene may undergo synchronous cyclization by which the all-(*E*) geometry of squalene is translated into a tetracyclic product with *trans-anti-trans-anti-trans* ring fusion.

In vitro polyene cyclizations catalyzed by protic acids were not promising due to indiscriminate protonation at various sites. Furthermore, the strongly acidic conditions are conducive to deprotonation of species in which only one, two, or three rings have been formed.

Allyl cations generated by formic acid catalyzed ionization of allyl alcohols are good initiators of polyene cyclizations [W.S. Johnson, 1976]. The formation of two new rings and three contiguous stereocenters was demonstrated and utilized earlier in a synthesis of fichtelite [W.S. Johnson, 1968a].

fichtelite

Three new rings including a seven-membered C-ring of a precursor of serratenediol were formed under very mild conditions [Prestwich, 1974]. The initiator being a symmetrical tertiary allylic cation eliminated any regiochemical problems.

serratenediol

The symmetrical 1,3-dimethylcyclopentenyl cation is valuable initiator for steroid synthesis, for example 16,17-dehydroprogesterone [W.S. Johnson, 1968b]. The impressive feat of creating five stereocenters in one step cannot be overemphasized.

As a terminator of the cationic polyene cyclization to construct directly the steroid skeleton an acetylene bond behaves impeccably. The vinyl cation may be trapped by weak nucleophiles such as acetonitrile. Nitroalkane and ethylene carbonate can also be used to capture the vinyl cation, as shown in a synthesis of testosterone benzoate [Morton, 1973a,b].

Allylsilane and propargylsilane moieties are efficient terminators of the polyene cyclization by virtue of the cation stabilizing properties of the silyl group, and the charge dissipation via desilylation. The propargylsilane function serves this purpose particularly well as it leaves an allenyl group in the steroid skeleton [R. Schmid, 1980] which is suitable for modification into a corticoid sidechain.

11α-Hydroxysteroids have also been obtained in an analogous fashion [W.S. Johnson, 1977]. The remarkable asymmetric induction is due to the unfavorable formation of the 11β-isomer via a sterically encumbered transition state. Enhancement of the rate of cyclization by a cation-stabilizing auxiliary at *pro*-C-9

[W.S. Johnson, 1987] is remarkable. This observation is important as the rate enhancement counterbalances the retardation by a hydroxyl group at the *pro*-C-11 position, which allows unwanted reactions to compete.

An intramolecular Friedel–Crafts alkylation terminated the cyclization leading to a pentacyclic precursor of alnusenone [Ireland, 1975].

alnusenone

Epoxides are excellent initiators of the cationic polyene cyclization (cf. squalene oxide). An approach to progesterone [van Tamelen, 1983] by a one-step formation of the BCD-ring subunit has been accomplished.

Many other functionalities have been tested for the role of an initiator. In a magnificent synthesis of cafestol [Corey, 1987c] the role was assigned to a cyclopropylcarbinyl cation conjugated to a double bond of the polyene system. Four stereocenters were established during the cyclization.

cafestol

Ionization at an alternative carbinyl position adjoining the cyclopropane ring led to a tetracarbocyclic intermediate of atractyligenin [Singh, 1987]. The tactical change was to facilitate implementation of the substitution pattern of ring-D.

The Prins reaction effects CC bond formation by electrophilic attack on an aldehyde by an alkene. Polyene aldehydes may undergo useful cyclizations (cf. steroid synthesis [Ziegler, 1981]).

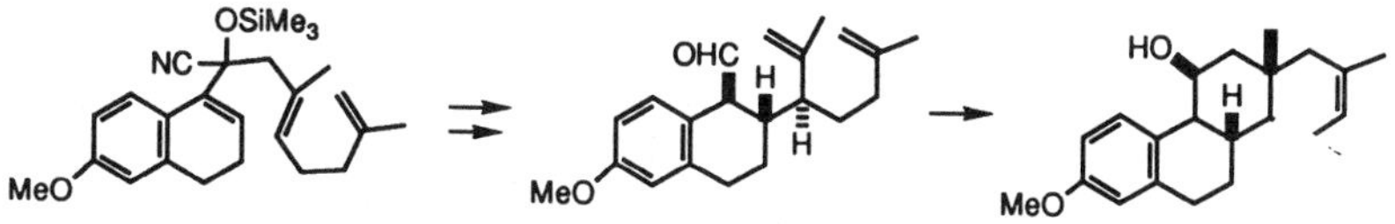

The suitability of an acetal function to induce cyclization via oxonium ion formation was affirmed some time ago. The coupling of this initiator with a vinylsilane terminator has the advantage of double activation of the polyene array. An illustration of this methodology is found in the construction of the nagilactone ring system [Burke, 1989].

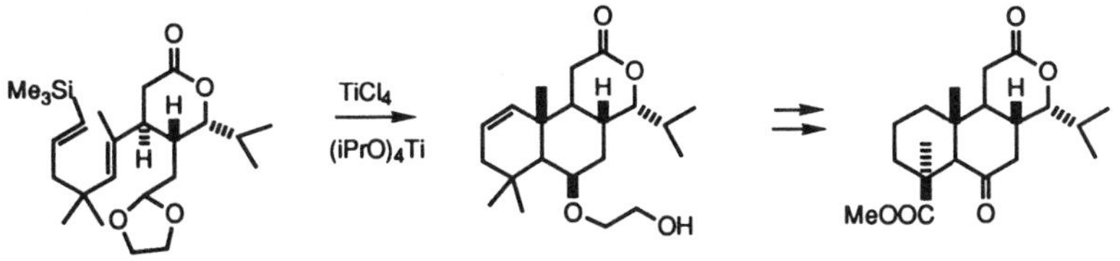

The acyliminum ion is another good electrophile. Formation of the hydroisoquinoline unit of O-methylpallidinine by the cyclization method simplified synthesis of the alkaloid [Kano, 1986].

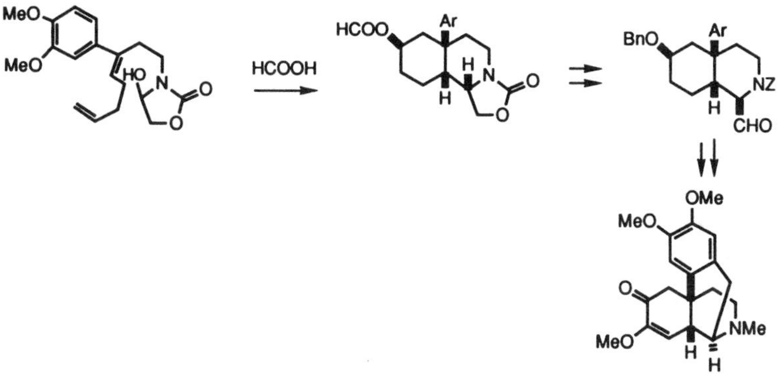

O-methylpallidinine

An elaboration of the octalin portion of dihydrocompactin [Burke, 1985] by intramolecular trapping of an acylium ion and with participation of a vinylsilane proved very effective. The double bond was created in the correct position.

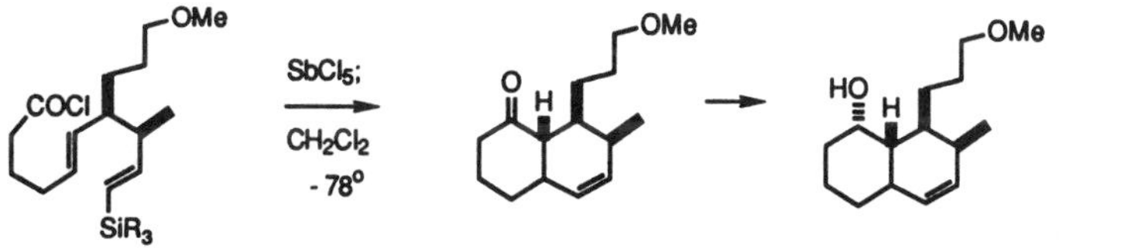

α-Diazo ketones under Lewis acid catalysis furnish formal α-acyl cations which may be intercepted intramolecularly by an alkene. Bicyclization is also possible [Smith, 1977].

Cyclization of those polyenes incorporating an iminium ion and an allylsilane end groups may provide a rapid entry into azapolycycles. A case in point is an application to a synthesis of yohimbone [Grieco, 1987].

yohimbone

A functional group may be introduced into the cyclic network via soft acid promoted cyclization of a polyene. Positive halogen reagents, mercury and selenium compounds are most frequently used for the purpose. For example, the reaction of a triene with mercury(II) and trifluoroacetate provided a *trans-β*-decalone from which aphidicolin was elaborated [Corey, 1980]. The 3α-hydroxyl group was derived from the mercurio pendant.

aphidicolin

4.6. MISCELLANEOUS *vic*-DIALKYLATIONS

Participation of an isolated double bond in double cyclization is represented by the formation of a tricyclic lactam from benzaldehyde and an unsaturated amide [Marson, 1991]. Another example is the trapping of an incipient α-sulfenylcarbenium ion with subsequent Friedel–Crafts alkylation. The ring formation is a crucial step in a concise synthesis of cuparene [Ishibashi, 1988]. Needless to say the wel-known preparation of β-tetralones from phenylacetyl chloride with alkenes under catalysis of aluminum chloride is an analogous process. For a recent example, see [Fleming, 1980].

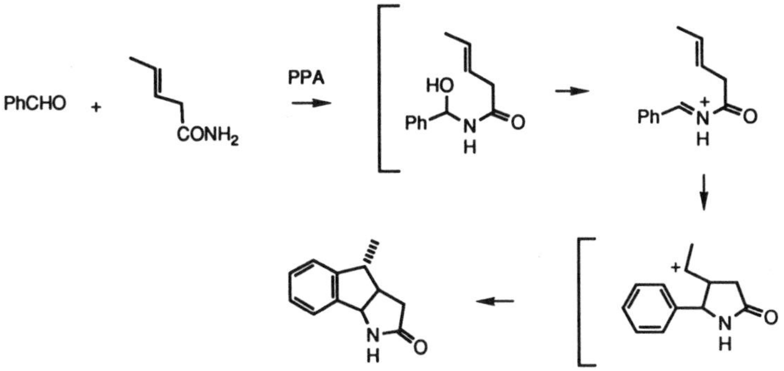

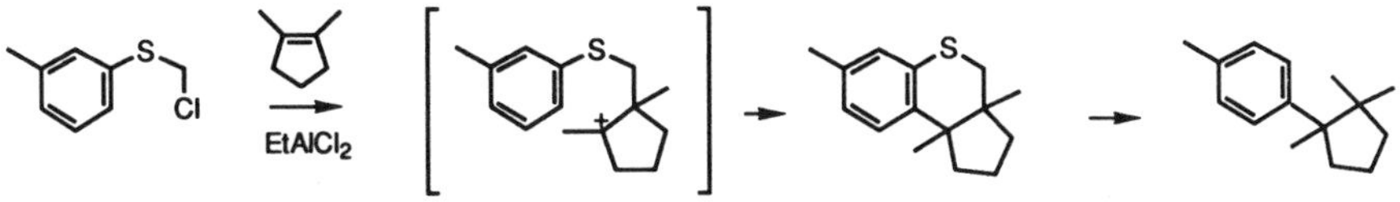

cuparene

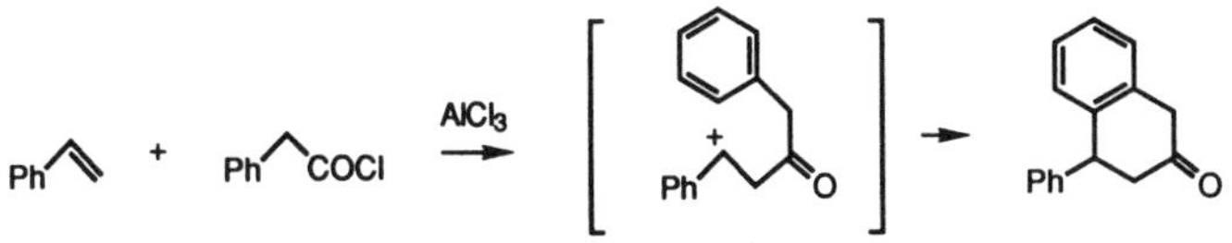

The transformation of kopsinine into kopsanone [Kuehne, 1985b] on heating in a methanol solution at 200°C is a very intriguing process. It has been proposed that the acylation involves a pentacyclic intermediate from which the intramolecular acylation is in tandem with the reestablishment of the C–N bond.

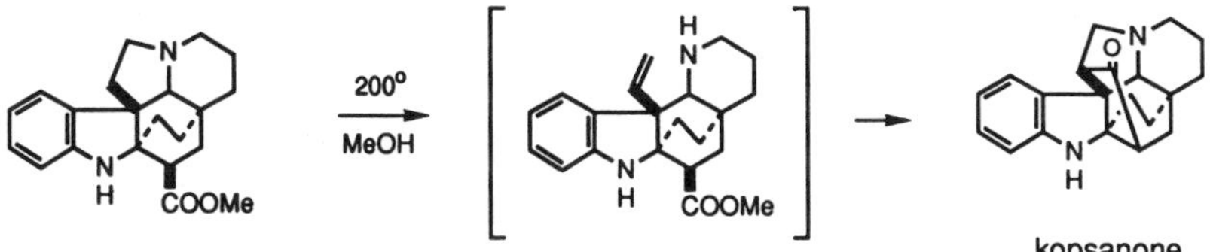

kopsinine

kopsanone

1,2-Dialkylation in the classical sense requires activation (i.e. deprotonation) at two adjoining carbon atoms. The simplest systems are succinic acid derivatives [Long, 1981].

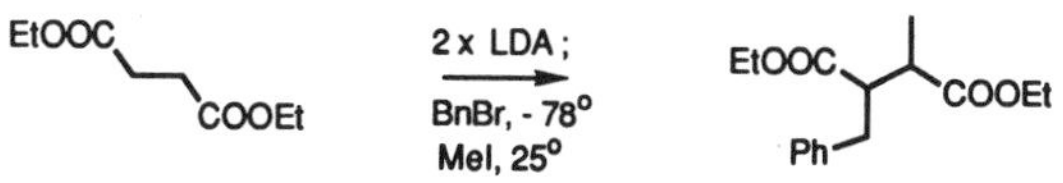

5

DIECKMANN AND CLAISEN CYCLIZATIONS

Bimolecular condensation of esters is known as the Claisen condensation. Dieckmann cyclization which is the intramolecular version involving diesters to form 2-oxocycloalkanecarboxylic esters [Schaefer, 1967] has not been rendered obsolete by new methodologies for the construction of 5-, 6-, and 7-membered rings. The creation of the all-important ketone functionality as part of the even more versatile β-keto ester is well appreciated by synthetic chemists. Its combination with another reaction to form a tandem usually facilitates many a synthetic endeavor.

The self-condensation of dialkyl succinates by exposure to an alkoxide base represents a Claisen–Dieckmann reaction tandem. The mixed condensation of certain diesters with an oxalic ester also belongs to the same category. In fact, the synthesis of camphoric acid [Komppa, 1903], and hence camphor itself, started from such a reaction.

Analogously, a keto ester may also react with an oxalic ester to form a cyclic triketo ester in which the newly established ring contains an α-diketone unit. The tandem reaction was one of the key steps in a celebrated synthesis of 6-demethyl-6-deoxytetracycline [Conover, 1962].

6-demethyl-6-deoxy-
tetracycline

5.1. MICHAEL–DIECKMANN AND MICHAEL–CLAISEN REACTION TANDEMS

Many cyclopentanone derivatives have been assembled by the very effective combination of Michael and Dieckmann reactions. Thus, for a synthesis of sarkomycin [Kodpinid, 1984], dimethyl itaconate with its double bond masked in a Diels–Alder adduct with anthracene underwent Michael reaction with methyl acrylate and the subsequent Dieckmann cyclization. The sensitive enone system was unveiled at the conclusion of all the necessary structural modifications on the spirocyclic intermediate.

α-Tetralones have been prepared from methyl *o*-toluate by the Michael–Dieckmann tandem [Tarnchompoo, 1987], by virtue of benzylic anion formation from such compounds.

In a synthesis of terramycin [Muxfeldt, 1979] the A-ring was assembled from an α-tetralone by a reaction sequence initiated by a Michael reaction. Formation of the B-ring involving a Claisen condensation may be accomplished without isolation of the tricyclic intermediate. This probably is a rare case that a β-keto amide serves as the donor in a Dieckmann cyclization.

The Michael–Dieckmann tandem was also the key step of an approach to 1-deoxylycorine [Muxfeldt, 1973]. The polyfunctionality of such a product indicates versatility of the methodology and its potential for further exploitation in solving other synthetic problems.

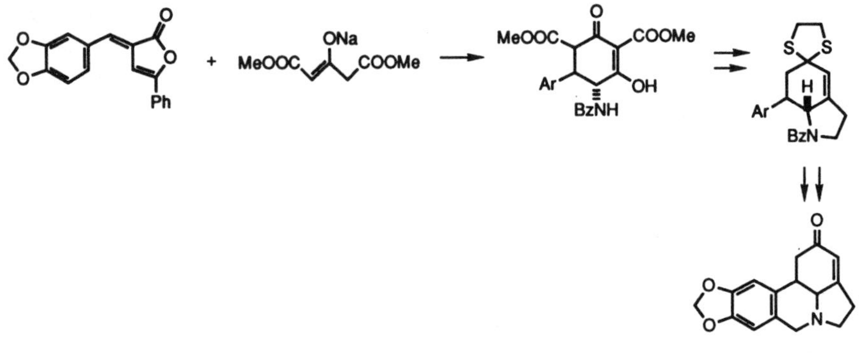

1-deoxylycorine

Dialkyl 2-hexenedioates are adequate Michael acceptors. Usually the Michael reaction would be followed by a Dieckmann cyclization. The synthesis of several naturally occurring cyclopentanoid compounds has exploited this reaction sequence, for example mitsugashiwalactone [Nugent, 1986] and (+)-dehydroiridodiol [M. Yamaguchi, 1986].

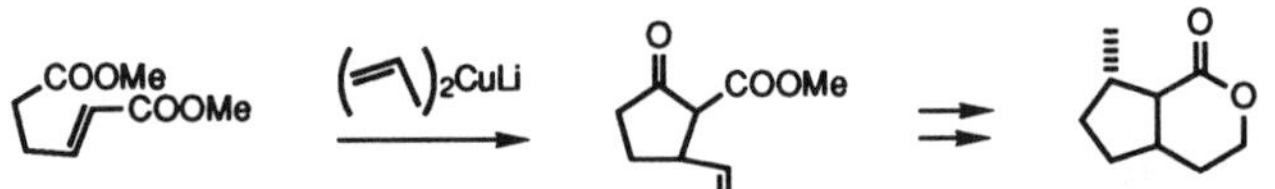

mitsugashiwa
lactone

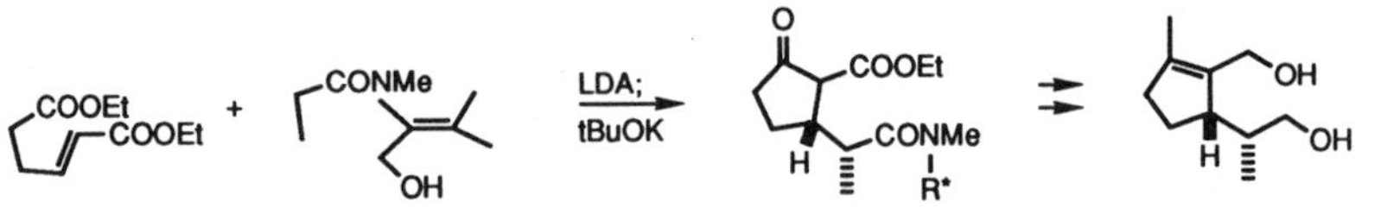

(+)dehydroiridodiol

A *formal* Michael–Dieckmann cyclization provided the key intermediate which contains all the carbon atoms of bilobalide [Corey, 1987a,b]. The synthesis of the complex terpenoid was thereby greatly facilitated. The critical requirement is the generation of a succinic ester α,α'-dianion.

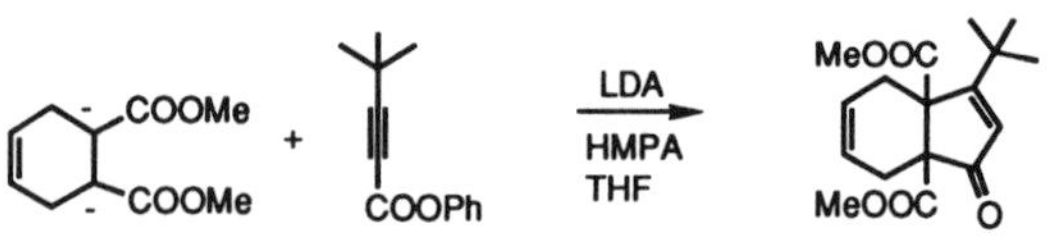

The tolerance of an ester group by organozinc reagents (cf. the well-known Reformatskii reagents) has made possible the use of such as ester β-carbanions in synthesis. A convenient method for the preparation of 2-cyclopentenone-2-carboxylic esters has been developed [Crimmins, 1990] on the basis of a Michael–Dieckmann tandem.

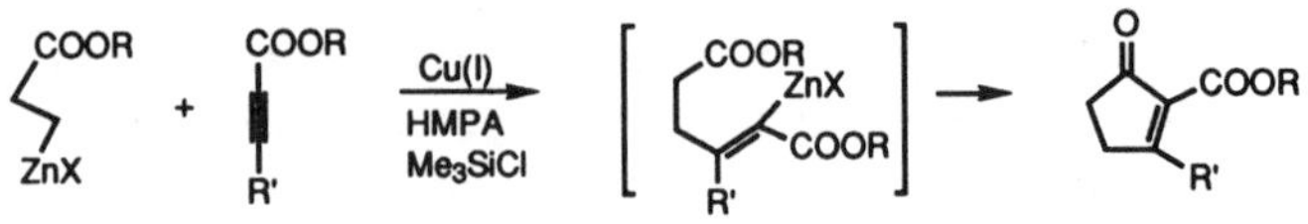

With heterosubstituted esters as Michael donors the variation of the theme makes available several heterocyclic systems. It has found use in the synthesis of alkaloids such as retronecine *Geissman*, 1962], norisotuboflavine [McEvoy, 1969], and camptothecin [Volkmann, 1971].

retronecine

norisotuboflavine

camptothecin

Equally accessible are the keto esters in which the heteroatom is either oxygen or sulfur. Interestingly, methyl thiolane-3-one-4-carboxylate [Woodward, 1946] has been identified as an α-acrylate anion equivalent [Baraldi, 1982] by virtue of its facile alkylation and retro-Dieckmann–Michael decomposition on treatment with dilute alkali.

Dieckmann cyclization succeeding two Michael reactions is also common. The reaction usually occurs when a Michael donor containing a highly activated methylene group is exposed to excess acrylic ester. The products are cyclohexanone derivatives (or 4-hetera analogs). A recent example [Ogura, 1986] is now depicted.

While this indicates double Michael reaction originating from the same donor atom, the serial version has also been developed [Murahashi, 1967; Olsson, 1977; Posner, 1989]. In order to conclude the annulation by a Dieckmann condensation the donor must be an ester.

The reaction of sodium cyanide with dimethyl maleate gives rise to a cyanocyclopentanonetricarboxylic ester [Michael, 1931] from which 4-oxocyclopentane-1,2-dicarboxylic acid is readily obtained. This acid and its diester are very important building blocks for various syntheses.

A bimolecular double Michael reaction coupled to Dieckmann condensation permitted formation of a perhydrindane system which is simultaneously bridged and spiroannulated to a cyclohexane [Danishefsky, 1973]. This carbon skeleton is the same as that of clovane except the stereochemistry at the ring juncture.

A unique Michael acceptor is an α-bromomethylacrylic ester which can be used twice. For example, its reaction with acetone α,α'-dianion would generate a cyclohexanonecarboxylic ester. In this manner adamantanedionetetracarboxylic esters are readily available from 4-oxocyclohexane-1,1,3,5-tetracarboxylic esters [Stetter, 1974]. By using the enamine donor an adamantanedionedicarboxylic ester has also been obtained [Stetter, 1968].

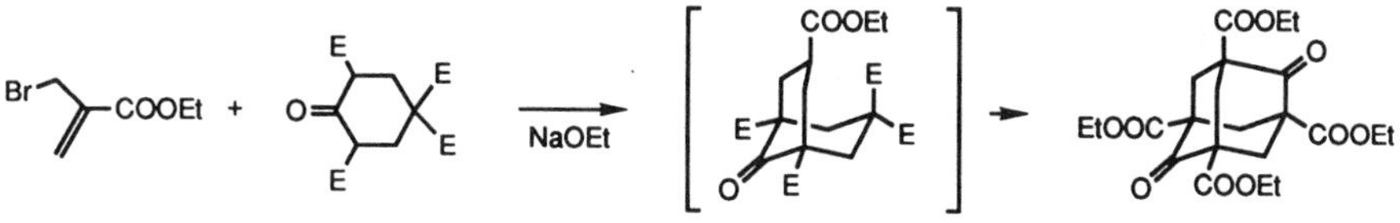

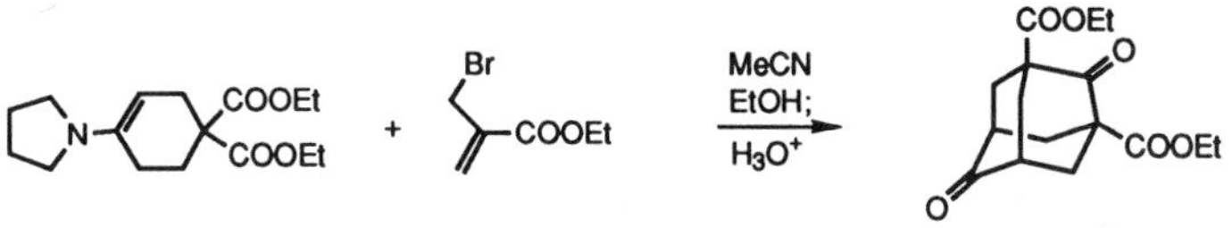

A formal Michael–Dieckmann cyclization tandem is that involving reaction of cyclopropenone ethyleneketal with α,β-enone α-carboxylic esters [Boger, 1984a] in which the ketal acts as a Michael donor and a Dieckmann acceptor.

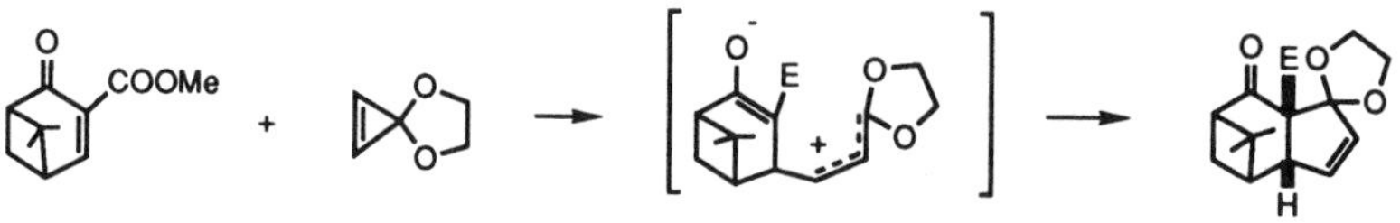

It is well known that cyclopropanes geminally substituted with acceptor groups are susceptible to attack by nucleophiles in the manner akin to a Michael reaction. The ring opening is an S_N2 reaction, causing inversion of configuration at the relevant carbon atom. When the donor is a malonic ester and the acceptor is a cyclopropane-1,1-dicarboxylic ester the product is a cyclopentanone-2,5-dicarboxylic ester, as expected from the tandem Dieckmann cyclization. The utility of this combination is evident from its incorporation into the elaboration of methyl jasmonate [Quinkert, 1982a], estrone [Quinkert, 1982b], and confertin [Quinkert, 1987].

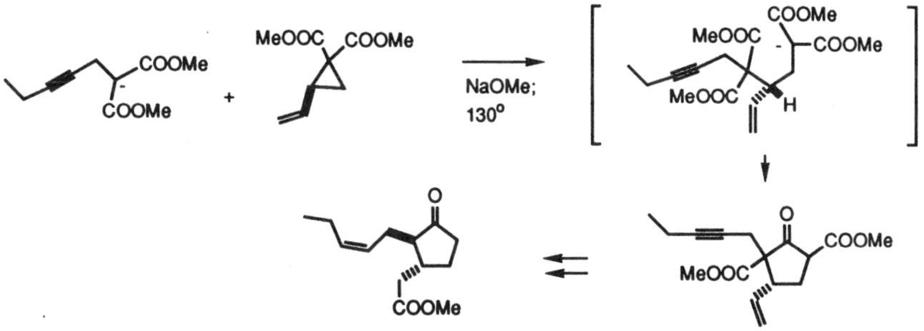

methyl jasmonate

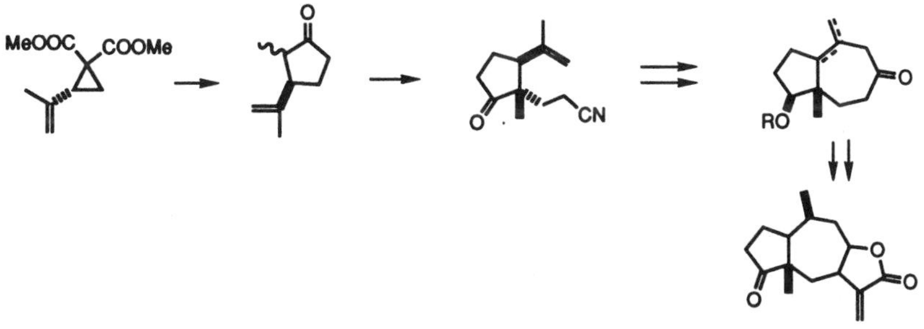

(+)-confertin

It is of particular interest to note the formation of hydropentalenone and hydrindanone derivatives via intramolecular ring opening and subsequent Dieckmann cyclization [Danishefsky, 1974].

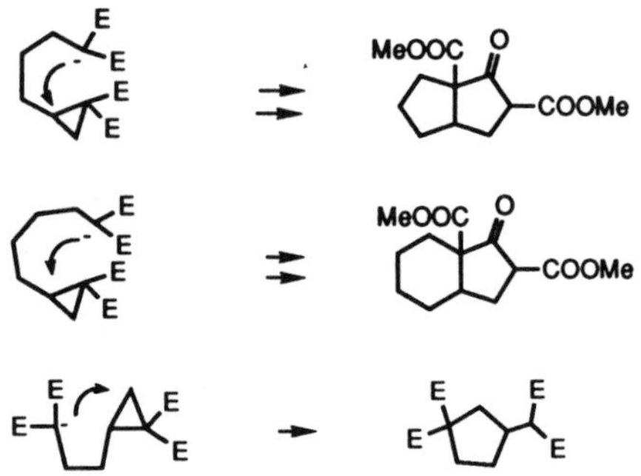

A reaction of long history is the formation of 5,5-dimethyl-1,3-cyclohexanedione from mesityl oxide and diethyl malonate [Vorländer, 1897]. The six-membered ring is formed by a Michael–Claisen reaction tandem; hydrolytic removal of an ester pendant results in the diketone.

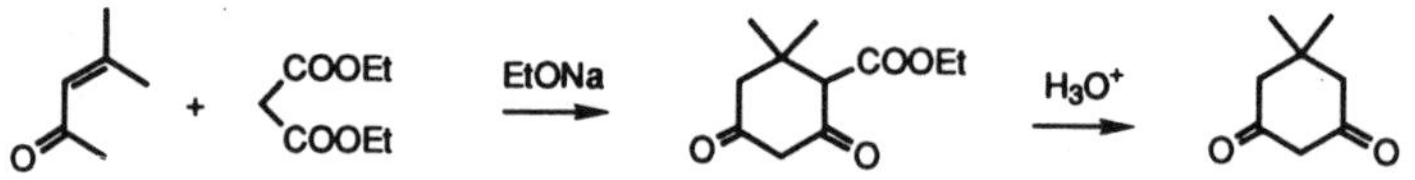

A slight structural variation of the product derived from methyl cinnamate and 2-acetyl-γ-butyrolactone is due to the presence of methoxide ion and the relative stability of the primary Michael–Claisen condensation product [Brimacombe, 1974].

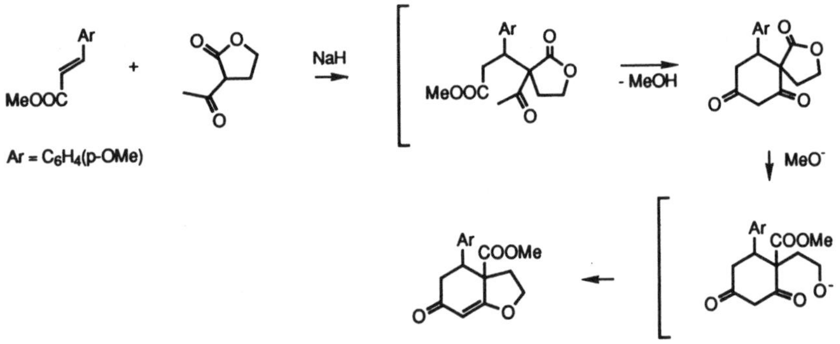

5.2. ALKYLATION–DIECKMANN AND RELATED TANDEM REACTIONS

Certain cycloalkanones have been acquired by reacting ester enolates with haloalkanoic or haloalkenoic esters, with the latter species acting as acceptor twice. In an approach to strigol [Heather, 1974] this methodology was specially rewarding because the resulting bicyclic β-keto ester developed a new donor site for the regioselective introduction of an acetic acid sidechain which eventually became part of the fused γ-lactone.

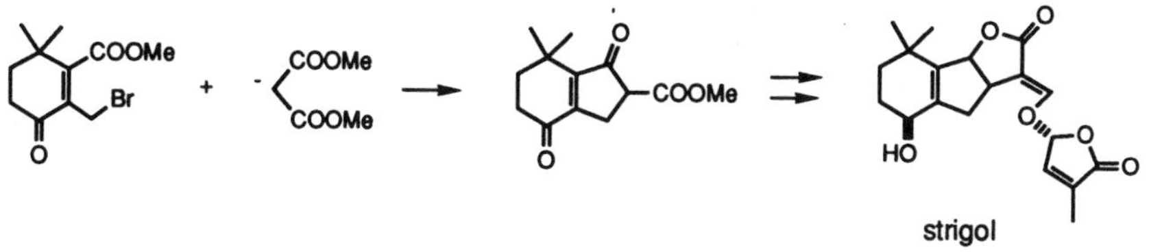

strigol

Two other versions of the alkylation–Dieckmann annulation [Funita, 1984; Misumi, 1984] are now presented.

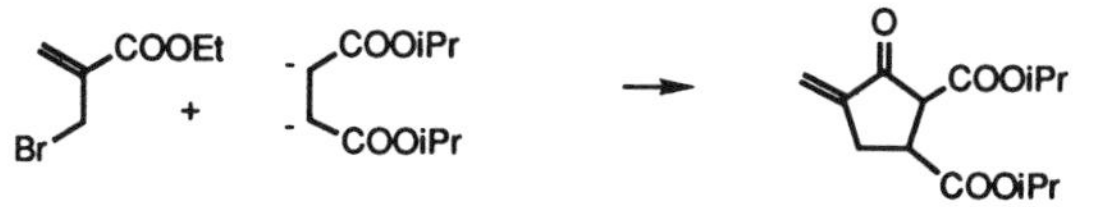

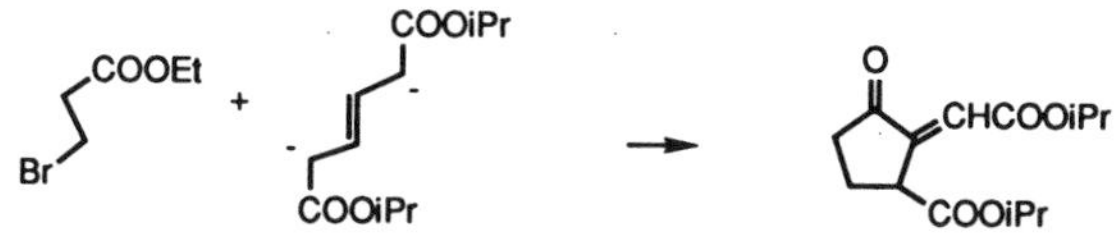

The succinic diester may be part of a ring system and naturally in such cases bi- or polycyclic keto esters are obtained [Noire, 1982; Garratt, 1986].

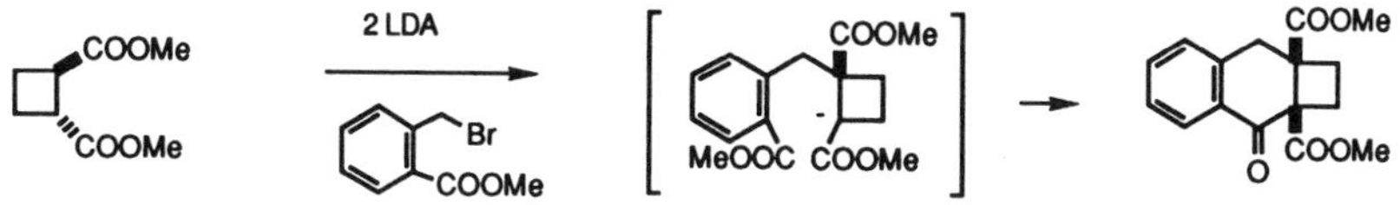

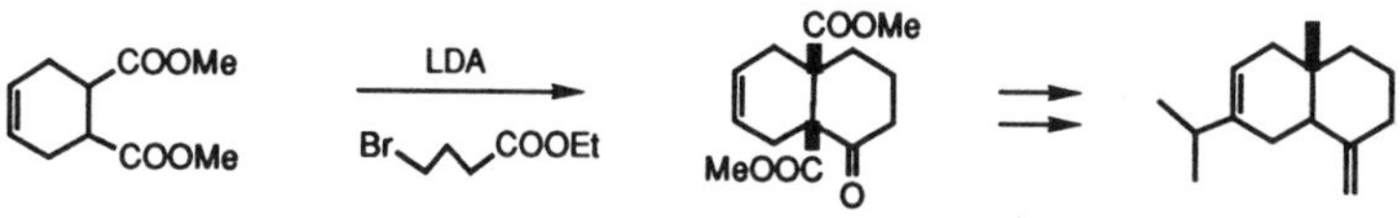

vetiselinene

N,N,N',N'-Bis(diethyl)succinamide can also be employed instead of the diester for the annulation [Mahalanabis, 1982].

A significant synthesis of a *trans*-hydrindanone bearing an angular methyl group which is characteristic of the CD-ring system of common steroids [Stork, 1960] involved a Dieckmann cyclization in tandem with a Favorskii rearrangement. It has been proposed that the cyclopropanone intermediate was the electrophile for the final step of the process.

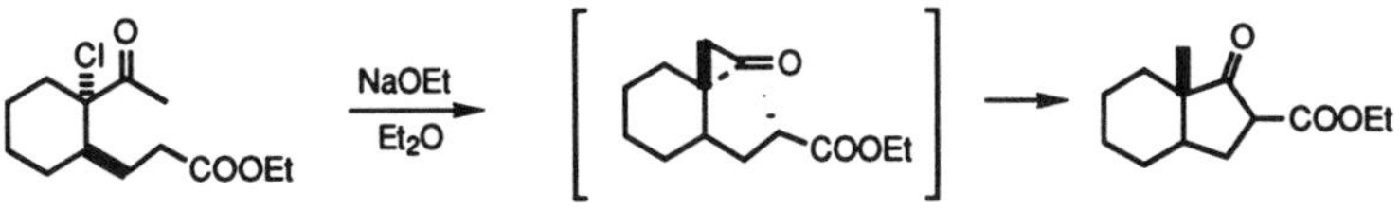

Also relevant to steroid synthesis is the formation of the ABC-ring skeleton by reaction of a keto diester with an alkoxide base [R. Robinson, 1937]. An aldol condensation set up the Dieckmann cyclization.

Reaction of ethyl 3-benzoylpropanoate with diethyl succinate in the presence of sodium hydride led to a cyclopentenone [Bergmann, 1959]. It is evident that the Stobbe condensation was followed by a Dieckmann cyclization.

In recent years a useful variant of the Dieckmann cyclization has been developed. This variation entails self-condensation of an ω-sulfonyl ester to produce an α-sulfonylcycloalkanone. Such a ketone presents a different set of reactions and therefore it may meet certain synthetic requirements better. An alkylation–Dieckmann tandem in the context of a maturone synthesis [Ghera, 1986] is now shown.

maturone

The acceptor ester group for this variant of Dieckmann cyclization may be juxtaposed by a preliminary alkylation step [Maurya, 1990].

The formation of 2-hydroxycarbazole on heating akuammicine in methanol [Edwards, 1961] was initiated by two retro-Mannich fissions. Intramolecular acylation of the pyridinium ylide then furnished the skeleton of the product.

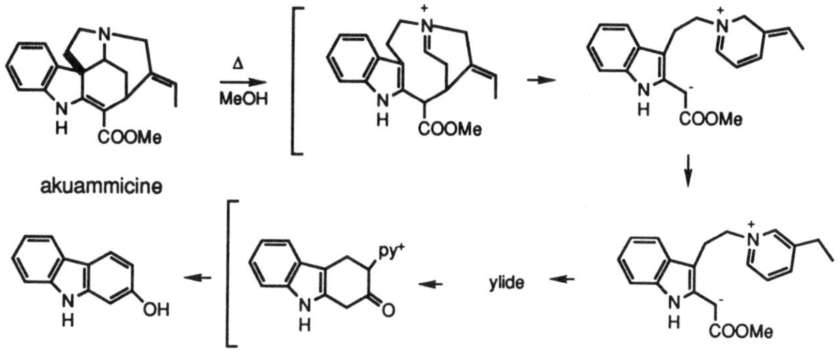

akuammicine

5.3. TANDEM REACTIONS TERMINATED BY CLAISEN CONDENSATION

We include in the Claisen condensation reactions which involve ketones and esters in the donor and acceptor roles, respectively. For reactivity reasons the usefulness of such reactions is generally limited to intramolecular condensation of keto esters. Naturally, double Claisen condensation of a ketone with an oxalic ester to form a 1,2,4-cyclopentanetrione was discovered very early. An application of the process to the preparation of a prostanoid synthon [Katsube, 1971; Sih, 1975] may be mentioned.

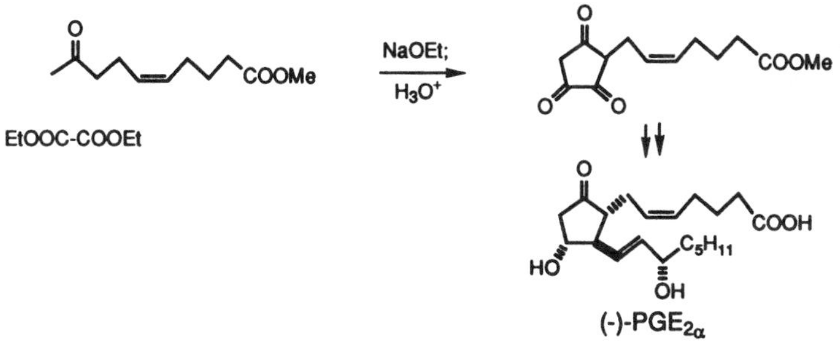

(-)-PGE$_{2\alpha}$

Under the O-methylation conditions (K$_2$CO$_3$, MeI, acetone) dimethyl homophthalate obtained from its monomethyl esters undergoes further Claisen condensations, enollactonization, aromatization, and phenol methylation [Murray, 1991].

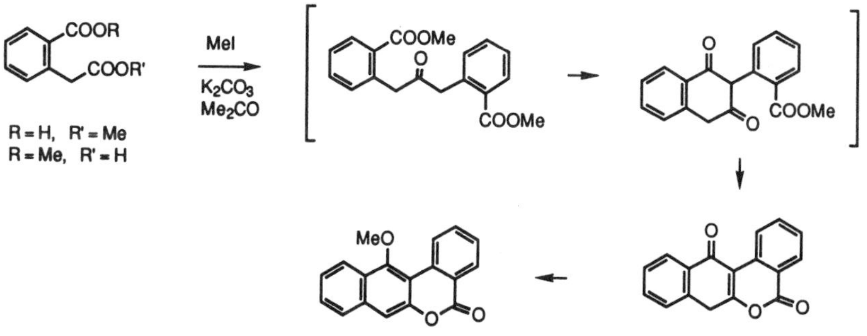

R = H, R' = Me
R = Me, R' = H

Keto esters may be acquired by Michael reaction, and if an ester group is properly appended cyclization of the adducts is likely to follow. To illustrate synthesis of β-diketones of both carbocyclic and heterocyclic skeletons [Schank, 1969; Winterfeldt, 1969] the following examples are presented.

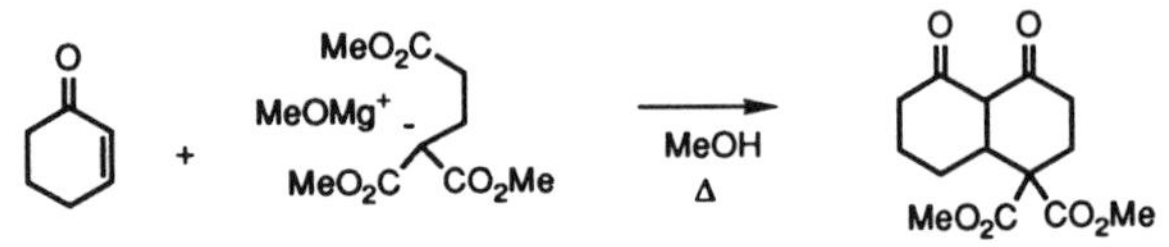

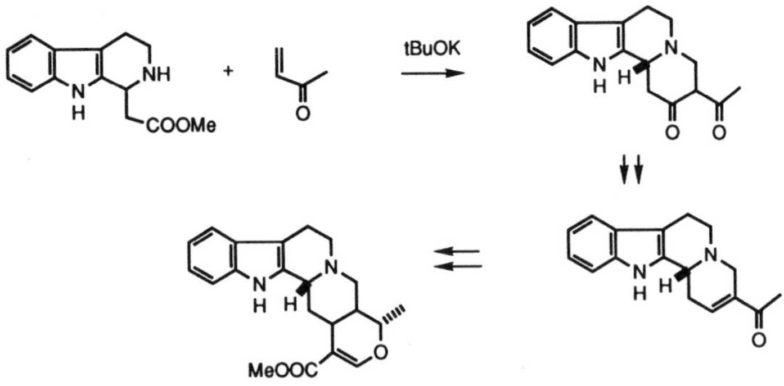

akuammigine

The key to a synthesis of dihydrocallistrisin [Tada, 1982] is the conjugate addition–Claisen condensation initiated by reaction of a 3-furfuryl-2-cyclohexenone with lithium dimethylcuprate.

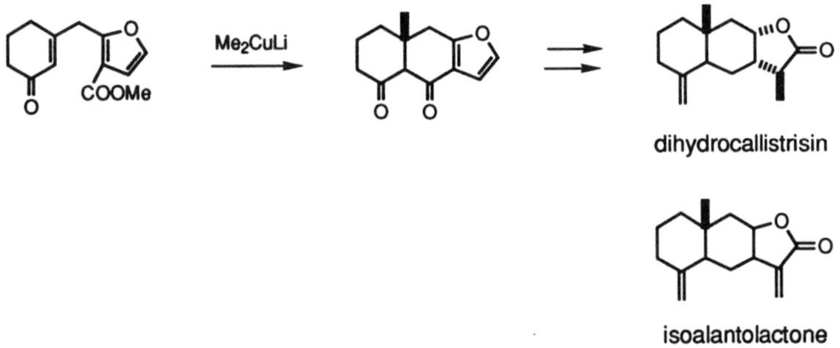

dihydrocallistrisin

isoalantolactone

Perhaps the most widely applied strategy for complex quinone synthesis is that which hinges on the Michael–Claisen tandem, especially with a phthalide 3-carbanion in which the C-3 also carries a leaving group. Thus, pachybasin has been obtained by the reaction of 3-cyanophthalide with 5-methyl-2-cyclohexenone, followed by oxidation [Kraus, 1983].

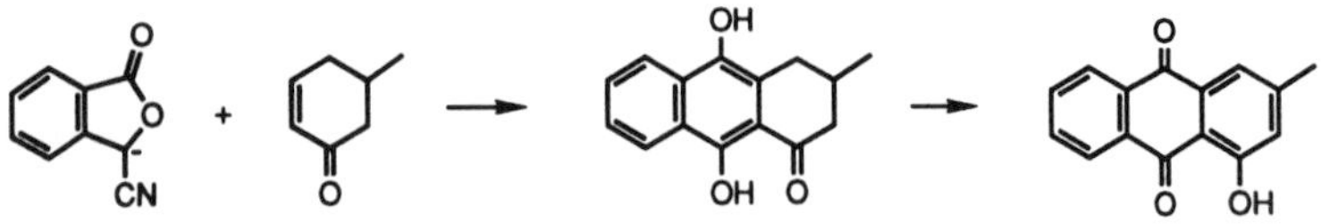

pachybasin

A more complex case pertains to the assembly of a precursor of daunomycinone [Dolson, 1981]. Certain homophthalic anhydrides may be employed as the Michael donor, as demonstrated by an approach to (−)-γ-rhodomycinone [Fujioka, 1989].

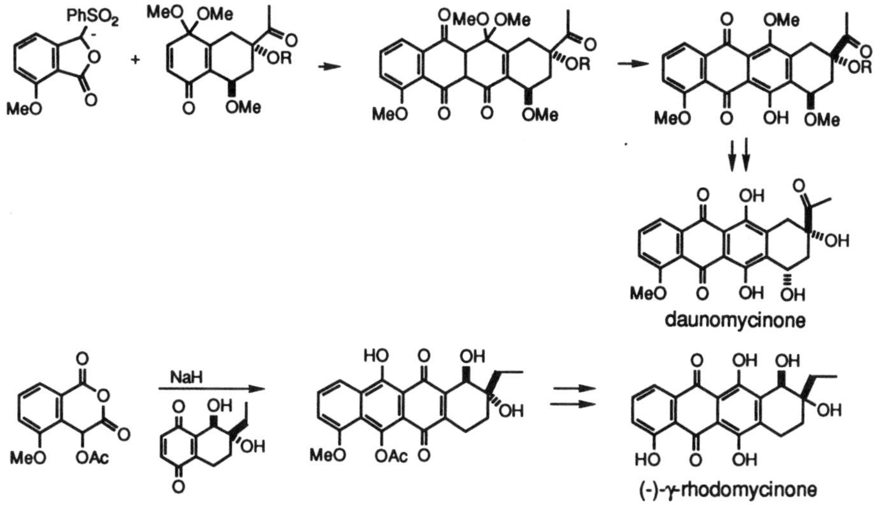

daunomycinone

(-)-γ-rhodomycinone

Quite different products are generated by the Michael–Claisen tandem involving homophthalic ester carbanions. For example, α-tetralones have been obtained from the condensation with acrylic esters [Eisenhuth, 1965].

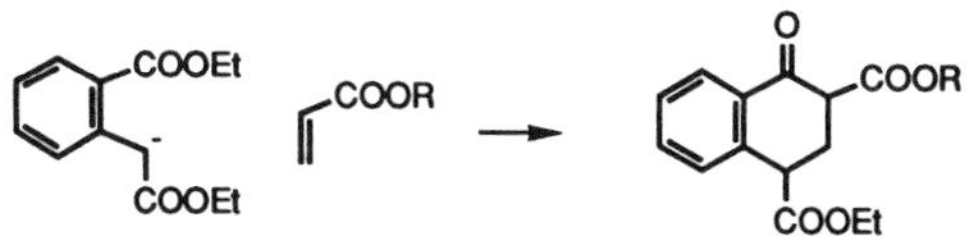

m-Dialkoxybenzannulation of p-quinones [Banville, 1974] by reaction with two molecules of 1,1-dialkoxyethylene has great significance in synthesis, in view of its simplicity and the substitution pattern corresponding to many naturally occurring aromatic compounds. The quinone poses as a Michael acceptor and its combination with the electron-rich keteneacetal precipitates the aldol-type condensation. However, the zwitterionic 1:2 adduct is sterically and entropically favorable for cyclization. The process is terminated by elimination of two molecules of the alcohol.

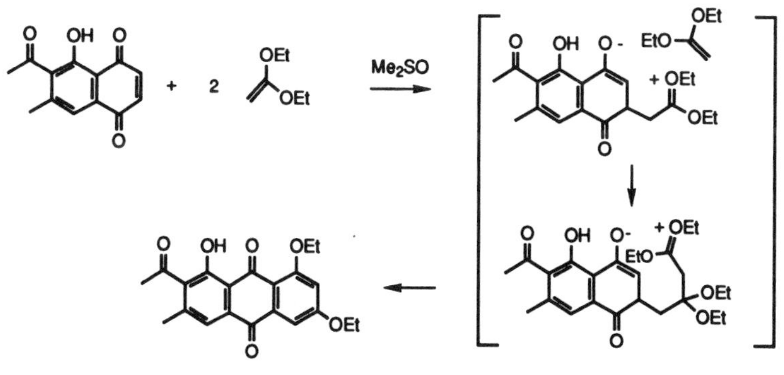

6

MANNICH REACTION

6.1. SCHIFF CONDENSATION–MANNICH CYCLIZATION

The Mannich reaction involves three components. An iminium species is formed from an amine (usually a secondary amine) and a carbonyl compound (Schiff condensation) and then it is intercepted by a donor species (enol). By definition the Mannich reaction is a bona fide tandem process. In view of its familiarity to organic chemists only certain significant aspects and applications are described in this book.

Azabicyclic structures are formed from self-condensation of aminodicarbonyl compounds. To be useful for synthesis, the substrates must be well-designed to assure the first step to be chemoselective, as shown in an efficient elaboration of lycopodine [Heathcock, 1982]. Here the imine formation is strongly biased toward formation of a six-membered ring (vs. 8-membered ring).

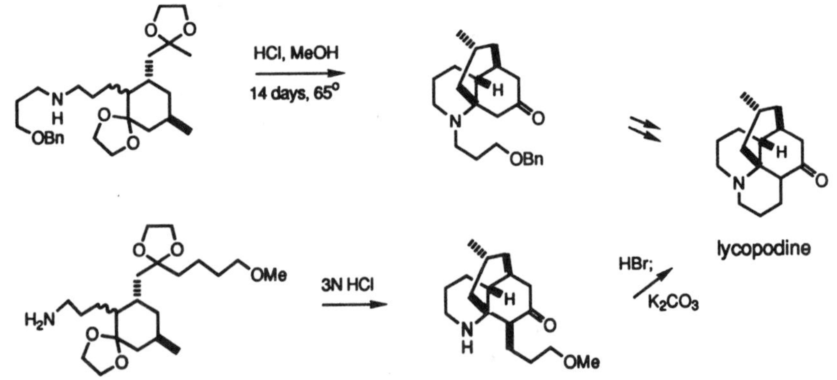

Another prominent aspect of this synthesis is the stereocontrol at the two new chirality centers created during the Mannich reaction. A cyclization pathway is available only to the iminium ion formed from the amino diketone with *cis–vic* sidechains. However, equilibration of the isomeric iminium species to the cyclizable entity would drive the reaction to completion.

The symmetrical nature of the molecule made the synthesis of trachelanthamidine [Takano, 1981) from an aminodiacetal very straightforward.

trachelanthamidine

In a quest for quinocarcinol [Danishefsky, 1985b] it was found expedient to construct the bridged ring system from a tetrahydroisoquinoline. The 1,2-relationship between the two nitrogen atoms of the molecule suggested a peptide intermediate in the synthetic pathway. The carbon chain to be introduced requires a carboxyl group which is translated into a γ-formylglutamic acid derivative as a building block. A Mannich reaction for the ring formation step from the glutamylisoquinoline was deemed applicable when necessary functionalities were properly modified.

quinocarcinol

(It is instructive to compare this approach to another synthesis of quinocarin [Fukuyama, 1988] based on the Mannich reaction.)

Retrosynthetic analysis of porantherine by a functional group interchange operation ($CH=CH$ to CH_2CO) reveals two Mannich retrons in sequence [Corey, 1974]. An azabicyclo[3.3.1]nonane containing two oxoalkyl sidechains is readily identified as a viable precursor for synthesis of the alkaloid. Accordingly, a synthesis was executed with guidance of these key reactions. (Another synthesis also relied on the Mannich cyclization [Ryckman, 1987].) The stepwise execution of the Mannich reactions steps was probably because the cyclization was disfavored in the presence of water. However, achieving the tandem reactions may be possible in view of the availability of acetal cleavage conditions that do not require water (e.g. using Me₃SiI).

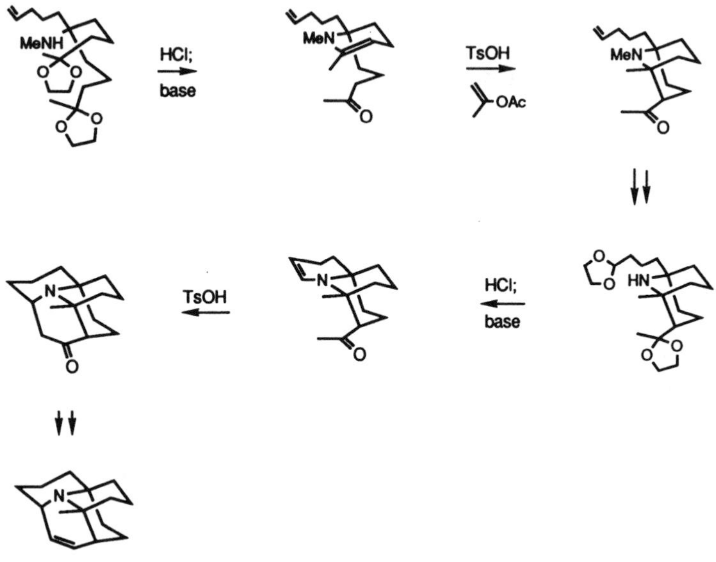

porantherine

Although amides rarely condense with carbonyl compounds except formaldehyde and a very few others, the very low equilibrium concentration of such species may still be trapped intramolecularly by a properly positioned donor. A tricyclic lactam thus formed was the template for eventual construction of (+)-methyl homodaphniphyllate [Heachcock, 1986].

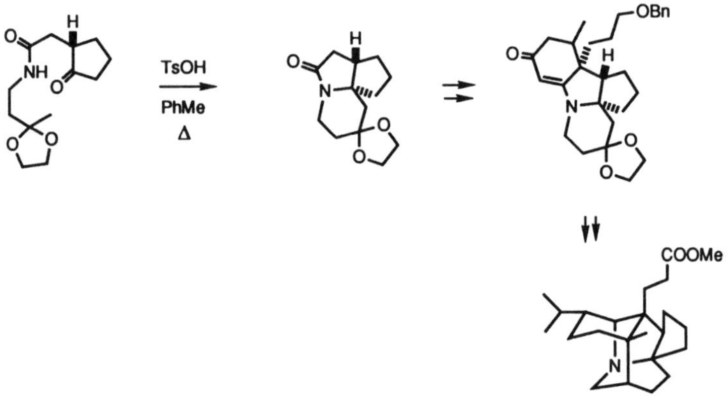

(+)-methyl homodaphniphyllate

Vinyl vicinal tricarbonyl compounds are versatile building blocks for synthesis because they present several electrophilic sites for bond formation.

Their reaction with amines give pyrrolidin-3-ones in one step via Michael addition and Schiff condensation. When the amine contains an enol ether group in a proper distance, Mannich cyclization may follow [Wasserman, 1990a].

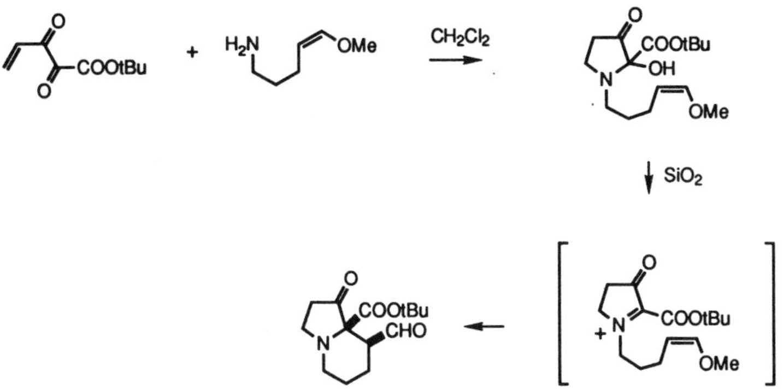

A well-positioned olefin linkage can act as a Mannich donor. To illustrate this possibility we can point out the spiroannulation step of a perhydro-histrionicotoxin synthesis [Evans, 1982].

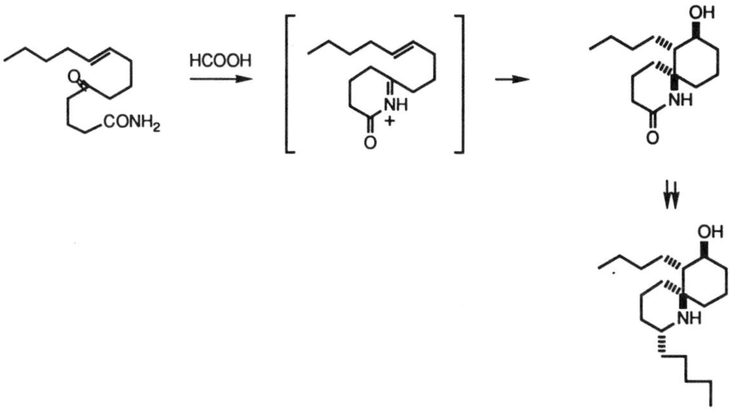

perhydrohistrionicotoxin

While on the subject of iminium ion capture with an olefin donor, it must be mentioned that a major development concerns the employment of silylalkenes. In addition to their higher reactivities the regioselectivity and stereocontrol of these processes afford much benefit to synthesis of compounds such as deplancheine [Overman, 1982].

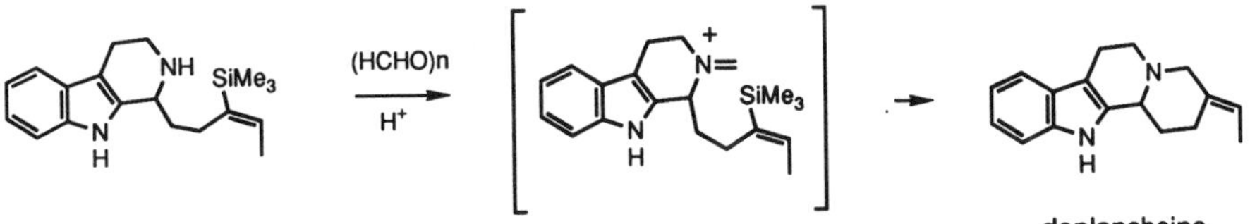

deplancheine

Exposure of 2-(2-[*1H*-indol-3-yl]ethylimino)cyclohexanone to hot concentrated sulfuric acid led to a pentacyclic product [Bobowski, 1981]. A Michael-type cyclization at the β-carbon atom of the indole nucleus is followed by a Mannich reaction.

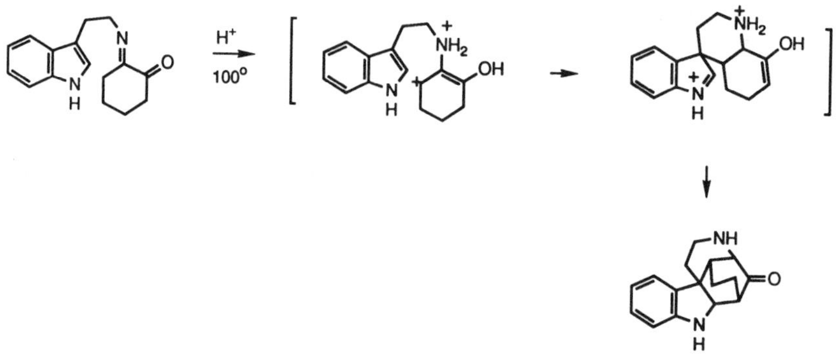

Mechanistically analogous is the reaction of *N*-(3,4-dimethoxyphenylacetyl)-tryptamine with phosphorus trichloride [Biswas, 1983].

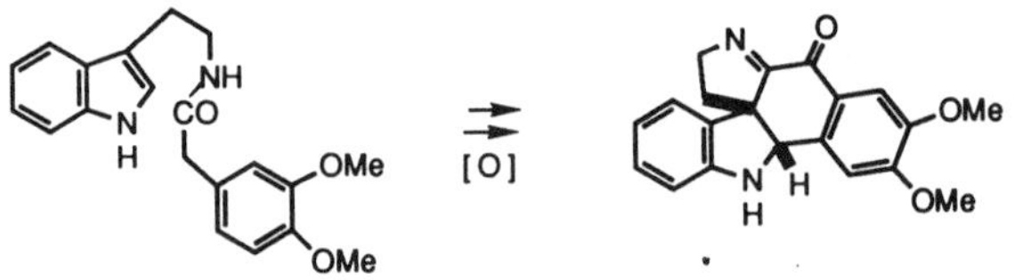

The Pictet–Spengler cyclization of *Ar*-oxygenated arylphenethylamines can be considered as a vinylogous version of Mannich reaction. By means of this process a concise assembly mode of the aromatic erythrina alkaloid skeleton has been developed [Haruna, 1976; Ito, 1978].

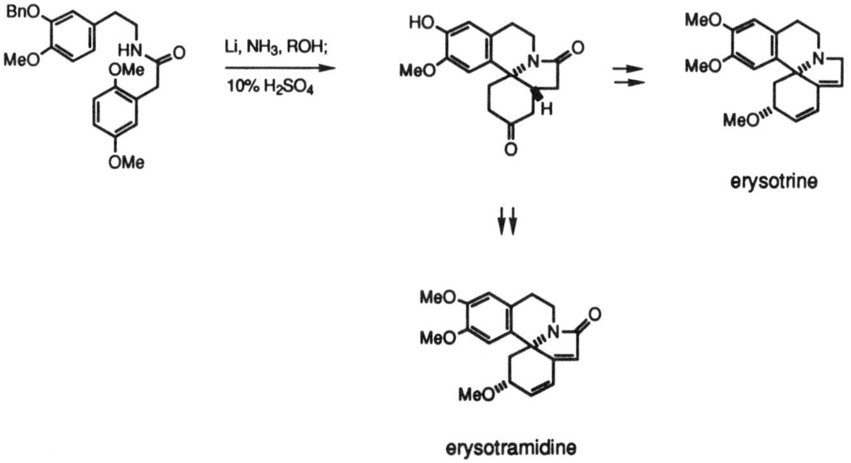

erysotrine

erysotramidine

Although not strictly within the realm of our discussion, the way by which two pyrrylmethane moieties were united during a synthesis of chlorophyll-*a* [Woodward, 1961; 1990] is still relevant to be considered. The likely generation of two different condensation products as the result of the unsymmetrical nature of the two reactants was a major obstacle. Other difficulties pertained to reactivities. Alkyldipyrrylmethanes are unstable and particularly sensitive to acids, whereas pyrrole-α-aldehydes are not very electrophilic, they being vinylogous amides. Thus, the condensation of two such dipyrrylmethanes has little chance to succeed.

Ingenuity showed in the solution of this problem. By identifying a β-aminoethyl chain as the latent vinyl group (cf. quinine synthesis [Woodward, 1945]) and using it to form a Schiff base with the aldehyde function of the other dipyrrylmethane, the condensation was rendered intramolecular. In practice the aldehyde was converted, via the simpler Schiff base with ethylamine, into a thioaldehyde. The advantage and necessity of this maneuver is that the thioaldehyde formed the desired Schiff base without the need for acid-catalysis.

phytylOOC   COOMe   chlorophyll-a

Despite radical differences from the Schiff condensation the generation of an active Mannich reaction substrate from a Claisen-type reaction of an oxindole is described here also. Thus after the Mannich cyclization of the β-amino ketone to afford a hexacycle using an ester chain as donor, a retro-Mannich fission involving the alternate bridge generated an intermediate which was capable of decarboxylation. The whole process was then terminated by yet another Mannich cyclization. The discovery of this process enabled the realization of a very short synthesis of aspidofractinine [Cartier, 1989].

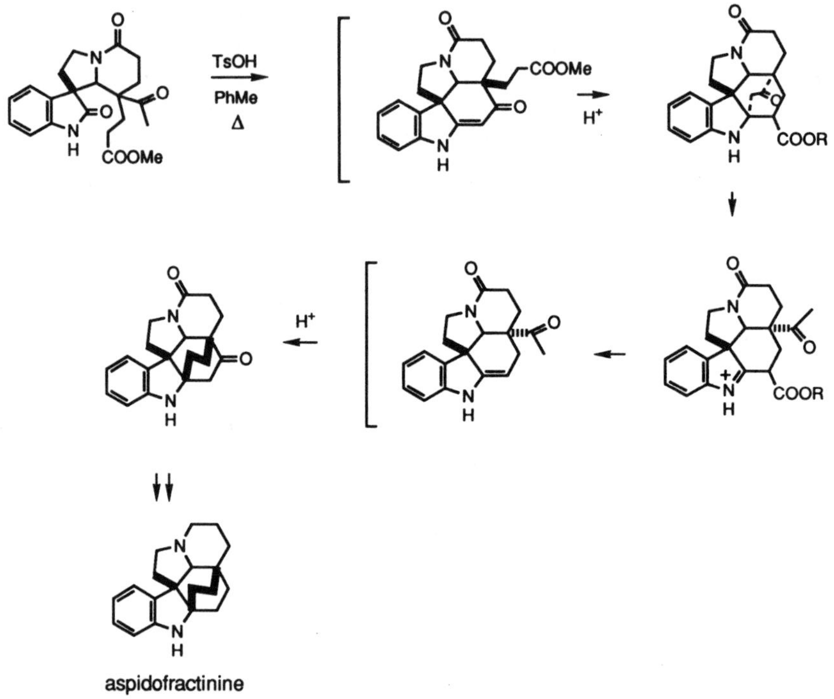

aspidofractinine

6.2. MANIFOLD MANNICH REACTIONS

The manifold Mannich reaction has played an important part in the development of organic synthesis. The Robinson tropinone synthesis [Robinson, 1917], accomplished by admixture of methylamine, succindialdehyde, and acetone, was a celebrated case that also provided an enormous stimulus and enchantment to the chemist in pursuit of efficient synthetic routes. The great improvement in yields of tropinone achieved subsequently [Schöpf, 1937] further attested to the usefulness of the Mannich reaction.

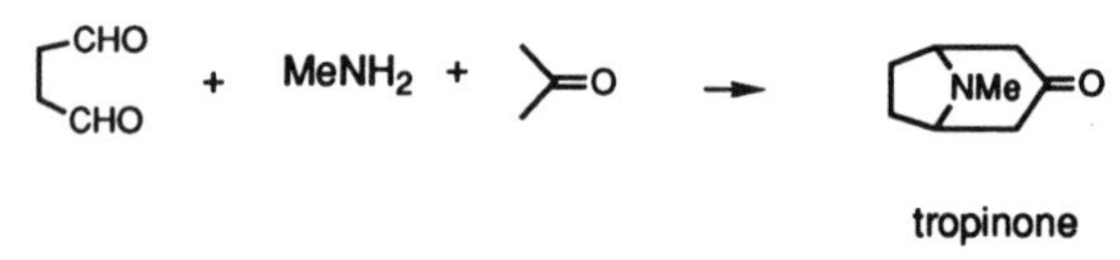

tropinone

The manifold Mannich reaction has attracted renewed interests in recent years. For example, a route to the ladybug defense substances [Stevens, 1979] that was based on the construction of the perhydro[9b]azaphenalene skeleton by the Mannich manifold is distinguished by its brevity. Two pairs of CC and CN single bonds were created in one operation. Furthermore, because of the reversibility of the reaction, control of the new stereocenters may be exercised.

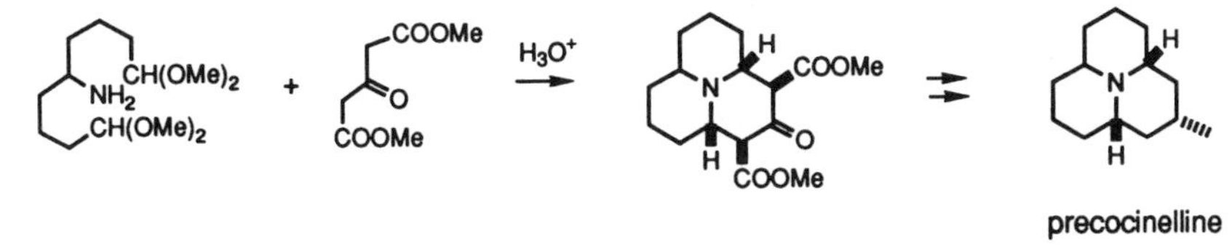

precocinelline

The fascinating stereoelectronic effects asserted the first CC bond formation to occur with a chair-like transition state in which the attacking donor would approach from the opposite face of the preexisting sidechain, while maintaining maximum orbital develop. The *syn* face approach to the alternative chair conformation of the protonated piperideine would have encountered a severe 1,3-diaxial interaction.

While the Robinson tropinone synthesis consists of consecutive Mannich reactions to form a bridged azabicyclic framework, a variation has enabled constructing three rings of lycopodine in one operation [Schumann, 1982], regardless the first stage of the condensation being more akin to a Michael addition than a Mannich reaction.

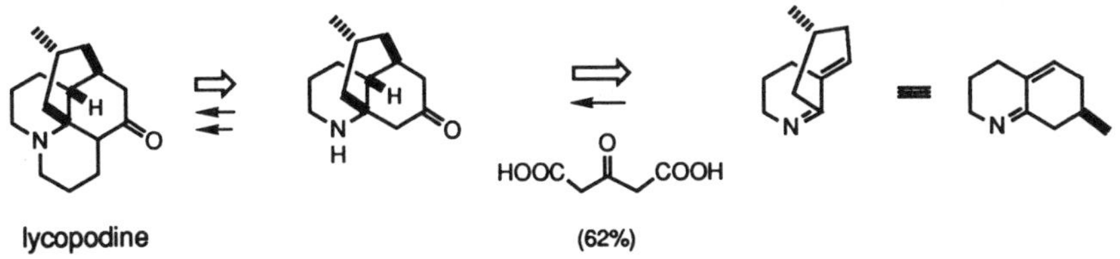

lycopodine   (62%)

A further variation in which 6-methylene-δ-valerolactam was employed as donor reagent simplified the assembly of certain diaza members of the *Lycopodium* alkaloids, including α-obscurine [Schumann, 1983].

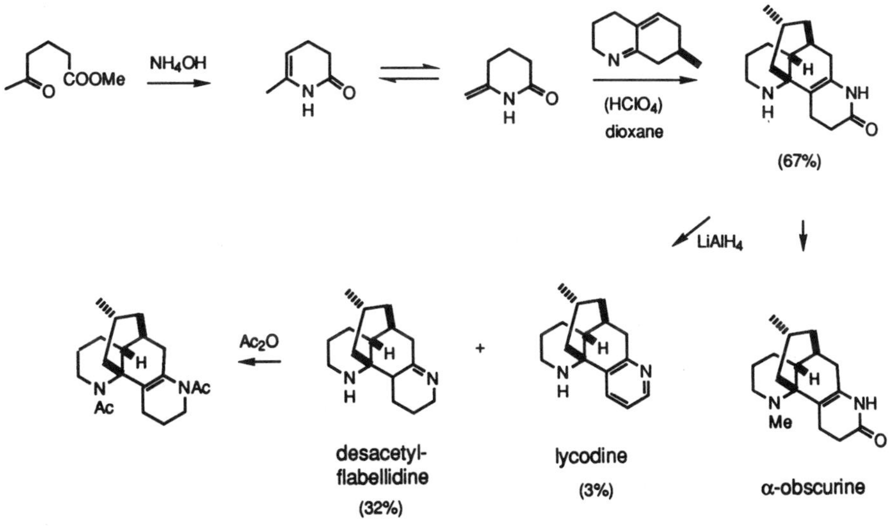

(67%)

LiAlH₄

desacetyl-
flabellidine
(32%)

lycodine
(3%)

α-obscurine

The autoxidation product of vincadifformine is a tetracyclic aldehydo oxindole. It is very unstable due to equilibration of various diastereomers by two different reversible retro-Mannich–Mannich sequences [Danieli, 1984].

vincadifformine

The formation of a 3-(aminoalkyl)indole from indole, an amine, and an aldehyde is generally categorized as a Mannich reaction. Such 3-(aminoalkyl)-indoles are excellent precursors of more complex 3-alkylindoles because the amino groups are readily displaced with a variety of nucleophiles. This propensity has been exploited in the formation of an aromatic ring by a double Mannich reaction which led to a synthetic intermediate of ellipticine [Besselievre, 1975].

ellipticine

Phloroglucinols can be capped very efficiently by refluxing with hexamethylene-tetramine in methanol [Risch, 1983]. The formation of 1-azaadamantanetriones is a threefold Mannich reaction.

1-Azaadamantan-4-one is equally easy to obtain [Black, 1981]. Thus, treatment of the ethyleneacetal of 4-oxocyclohexylmethylamine with formaldehyde in a 2% sulfuric acid solution afforded the compound in 53–56% yield.

Condensation of an N-methylisoquinolinium salt with nitromethane leads to a 2:1 adduct [Schleigh, 1969]. The structure suggests a multiple Mannich pathway.

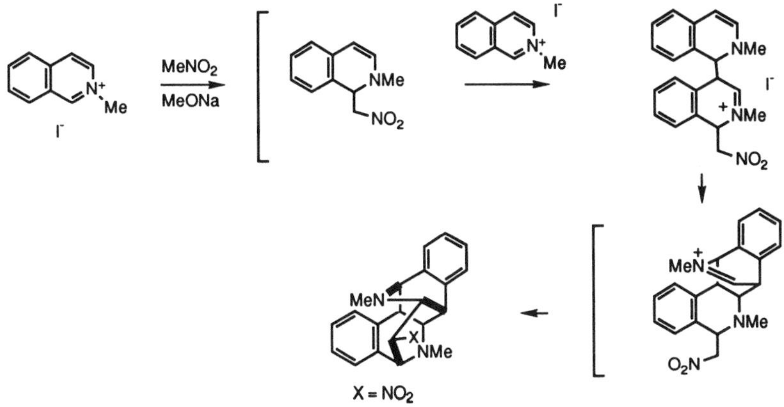

An efficient method for the assembly of the aspidosperma alkaloid skeleton is by treatment of an *N*-tryptophyl-3-(1,3-dioxoalkyl)piperidinone with trifluoroacetic anhydride [Jackson, 1987]. The bisannulation is a result of tandem Bischler–Napieralski cyclization/Mannich reaction.

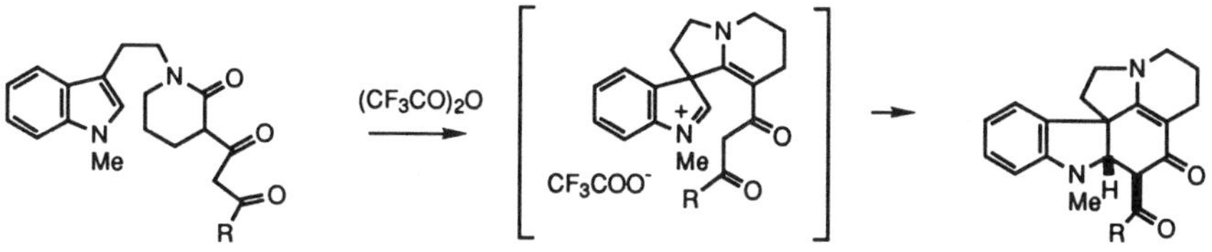

6.3. ALKYLATION OR ACYLATION–MANNICH CYCLIZATION

An excellent method for the synthesis of a levorotatory tricyclic ketone precursor of emetine [Openshaw, 1963] consists of a reaction between a dihydroisoquinoline and (3-oxobutyl)trimethylammonium iodide, and selective crystallization of the salt with (−)-10-camphorsulfonic acid. The annulation step involves a Michael addition and Mannich cyclization. (The Michael reaction is considered as an alkylation here.)

Because both of these reactions are reversible, the more soluble (+)-ketone can undergo racemization via a retro-Mannich step.

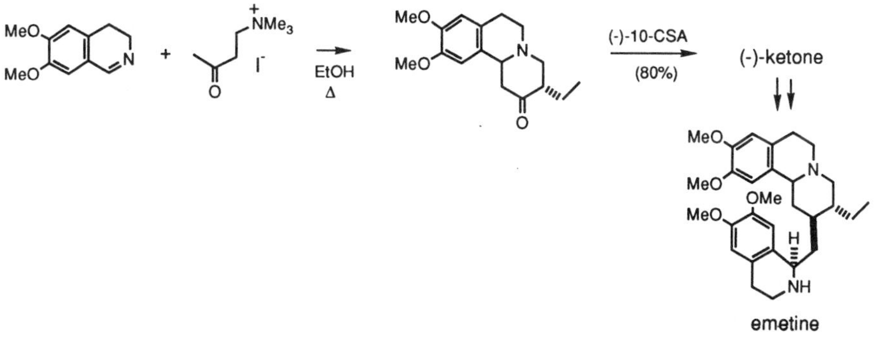

Further applications of the very efficient Michael–Mannich tandem reactions include a yohimbine synthesis [Szantay, 1965] and a dregamine synthesis [Kutney, 1978].

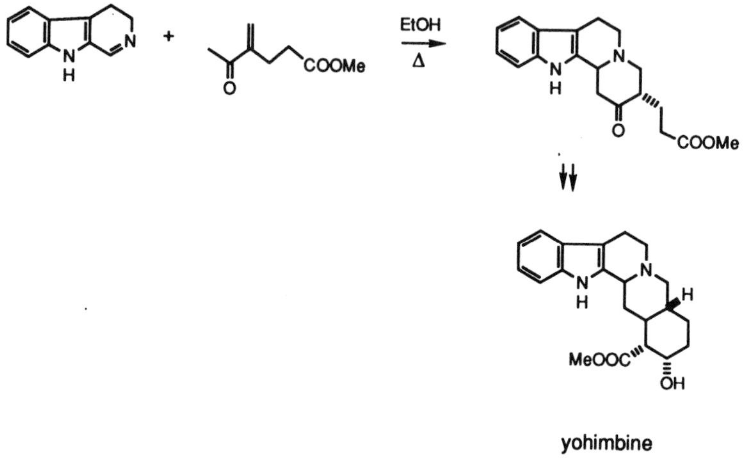

yohimbine

Secondary amides and lactams do not undergo Michael–aldol condensation. However, the desired transformation may be effected by converting a lactam into the thiolactim and reacting the latter compound with the proper enone in the presence of mercuric ion. In other words, replacing an aldol condensation by a Mannich reaction circumvents the lack of reactivity of the acceptor. Thiol elimination occurs after the Mannich cyclization [Takahata, 1986].

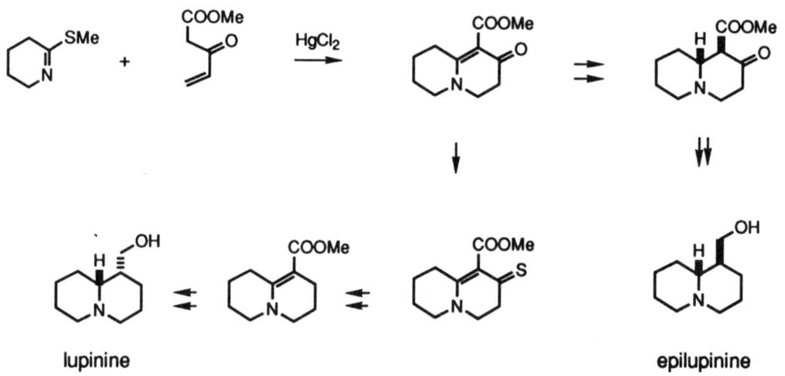

lupinine   epilupinine

The spontaneous resolution via salt formation was also a success in the acquisition of vincamine [Oppolzer, 1977]. The tetracyclic intermediate was obtained by an alkylation/Pictet–Spengler cyclization tandem. However, the resolution with (+)-malic acid proceeded by a retro-Mannich fragmentation–Mannich cyclization sequence. The Pictet–Spengler cyclization of indole derivatives may also be considered as a special version of the Mannich reaction.

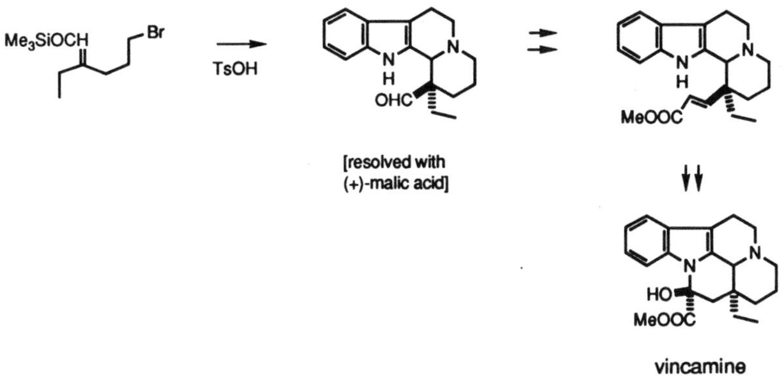

The formation of a bicyclo[2.2.1]hept-2-en-7-one by the condensation of an allylideneglycine ester with acrylonitrile [Bourhis, 1989] is an intriguing reaction. To this author a reasonable reaction course consists of Michael reaction and a tandem intramolecular acylation–Mannich cyclization. The same intermediate may be involved in the thermal isomerization of the bridged ring system to the condensed ring.

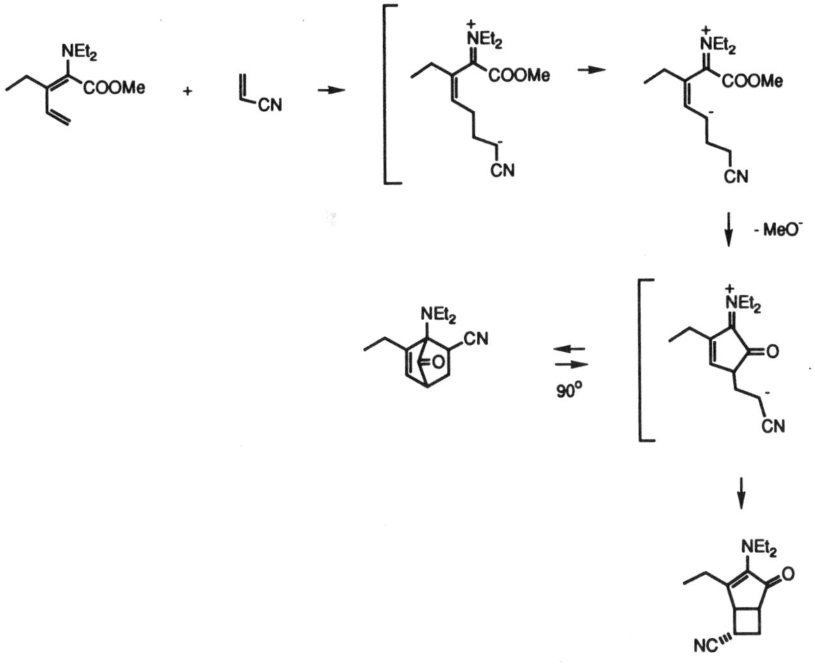

A Mannich-type reaction served to close the C-ring and complete the aspidosperma alkaloid framework in a double cyclization [Ogawa, 1987]. The product was an intermediate of kopsinine.

The reaction conditions are not similar to those normally employed in a Mannich reaction. However, the basic reagent was necessary for the alkylation step that generated the imine acceptor of the Mannich cyclization.

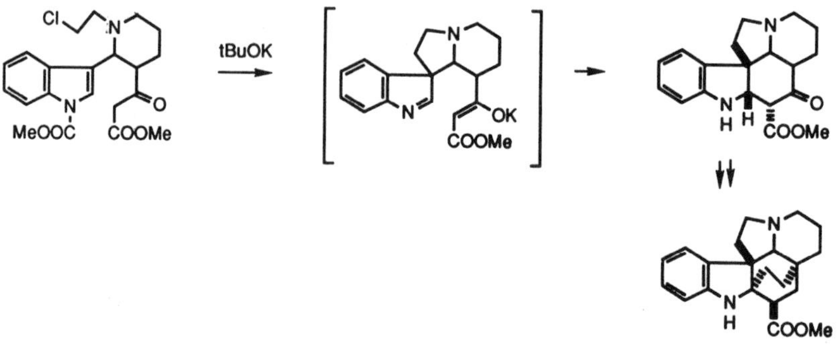

kopsinine

The Stork cyclohexenone synthesis from an enone and an enamine [Stork, 1963a] is a method which complements the Robinson annulation. Mechanistically. this annulation consists of a Michael–Mannich reaction tandem, followed by elimination of the secondary amine.

The utility of this method is pervasive [Gadamasetti, 1988] and only a very few syntheses that were based on it can be described here. Thus, the DE-component of yohimbine thereby assembled [Stork, 1972] was coupled with tryptophyl bromide to form the C-seco intermediate of the alkaloid.

A β-vetivone synthesis [Hutchins, 1984] demonstrates its applicability to the assembly of spirocyclic systems.

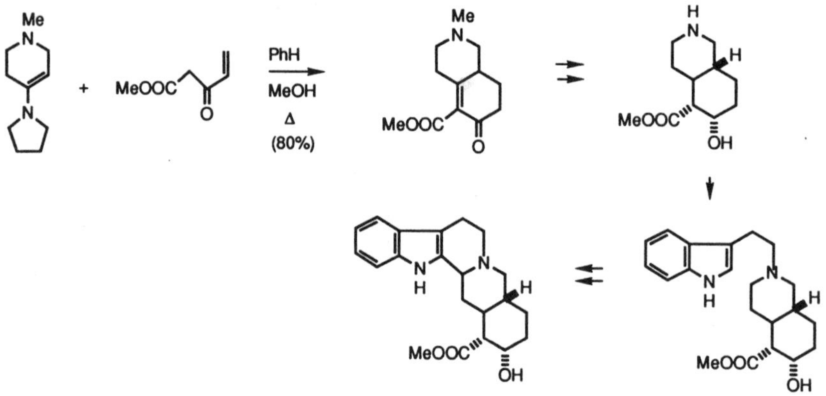

yohimbine

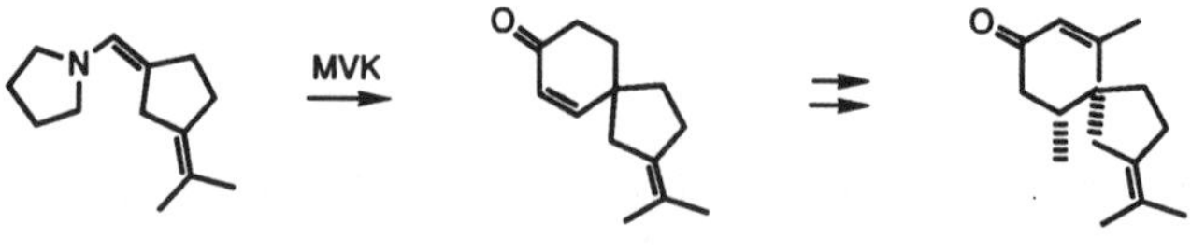

β-vetivone

A variation of Stork's yohimbine synthesis involved E-ring annulation from a zwitterion derived from a dihydropyridinium salt [Kuehne, 1991b]. The cyclization process took place at a vinylogous position to the iminium site.

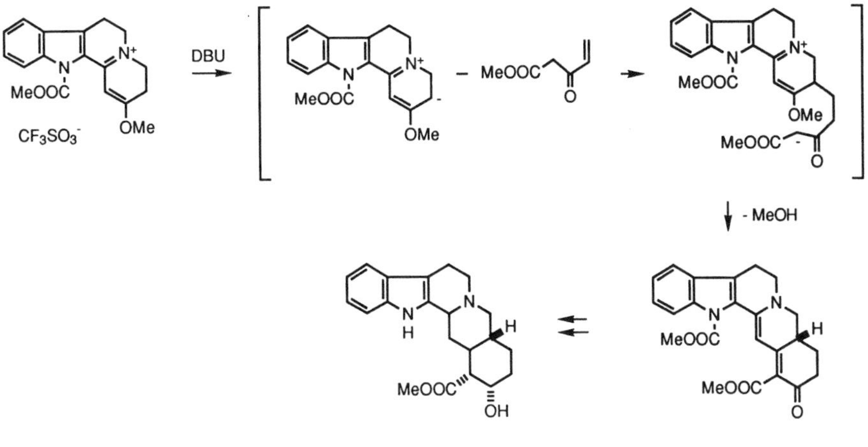

yohimbine

Extension of the Stork annulation using endocyclic enamines as donors has been rewarded handsomely. Specifically, many intricate azapolycyclic systems have been constructed in a most efficient fashion. These include mesembrine [Curphey, 1968; Stevens, 1968], precursors of cepharamine [Keely, 1970], elwesine [Stevens, 1972], and the Sceletium alkaloid-A$_4$ [Stevens, 1975b; Forbes, 1976]. The major problem pertaining to such syntheses seems to be that of the endocyclic enamine preparation. However, a highly reliable procedure that has evolved is the isomerization of cyclopropylimines.

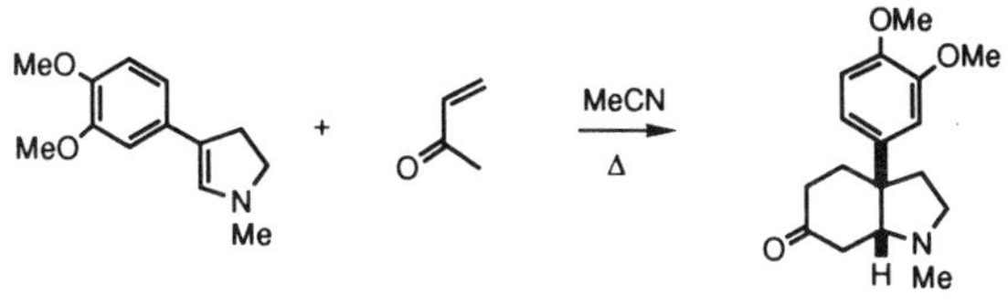

mesembrine

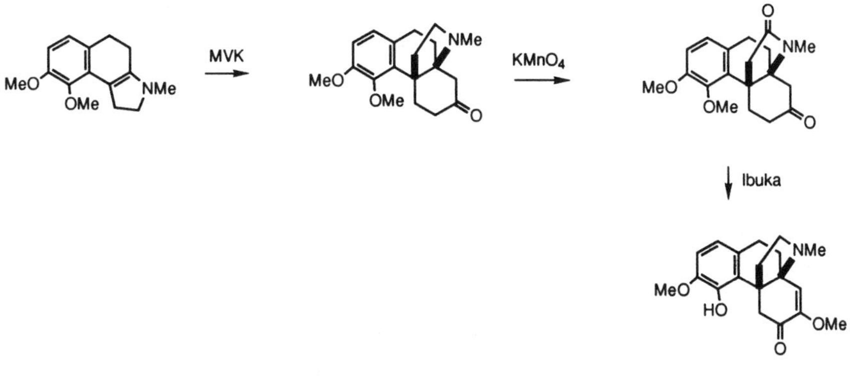

cepharamine

elwesine

Sceletium alkaloid -A$_4$

The last step of amine elimination in the Stork cyclohexenone synthesis can be suppressed in the cases involving endocyclic enamines. Such elimination is not as favorable entropically as that encountered in the more conventional reactions, and the intramolecular Michael addition can always intervene to restore the azacycle. (It is of interest to note that aspidospermine has been synthesized [Stork, 1963b] starting with a Stork annulation and via a lilolidine derivative. A perhydroquinoline intermediate was acquired by an intramolecular Michael addition.)

aspidospermine

In extending the synthetic use of vicinal tricarbonyl compounds formation of carbazoles has been demonstrated [Wasserman, 1990b]. The central ketone group of a 3-(3-indolyl)-2,3-dioxopropanoic ester is most reactive and its reaction with the enamine form of azomethanes generates intermediates which are susceptible to cyclization. Dehydration and enolization of the tricyclic products are favorable as the resulting carbazole derivatives are aromatic. The cyclization step is similar to the Pictet–Spengler reaction that is frequently employed in the synthesis of tetrahydro-β-carbolines, the difference is that in the present cases the nitrogen atom is placed in an exocyclic position.

One of the most delightful syntheses reported in recent years is the conversion of berberine into karachine [Stevens, 1983b] in one step. Three C bonds were formed in the sequence of Mannich, Michael, and Mannich reactions, the last two combined to form the cyclohexanone moiety.

karachine

Significantly, the final cyclization step required a boat transition state, which was sterically enforced in the present case.

Indoles are special but readily available endocyclic enamines. Their behavior as Mannich donors is well known. As an example demonstrating their suitability for the construction of 2-oxo-1,2,3,4-tetrahydrocarbazoles a synthesis of vindoline [Ando, 1975; cf. Pandit, 1984] may be cited. More recently, the reflexive Mannich reaction sequence was employed to create a pentacycle in one step in an elaboration of aspidospermidine [Wenkert, 1973; LeMenez, 1991].

vindoline

aspidospermidine

The dehydrative alkylation of an indoloquinolizidinone set up a sequence of retro-Mannich–Mannich reactions and resulted in the pentacyclic skeleton of the *Aspidosperma* alkaloids [Harley-Mason, 1967]. The product yielded aspidospermidine on reduction with lithium aluminum hydride.

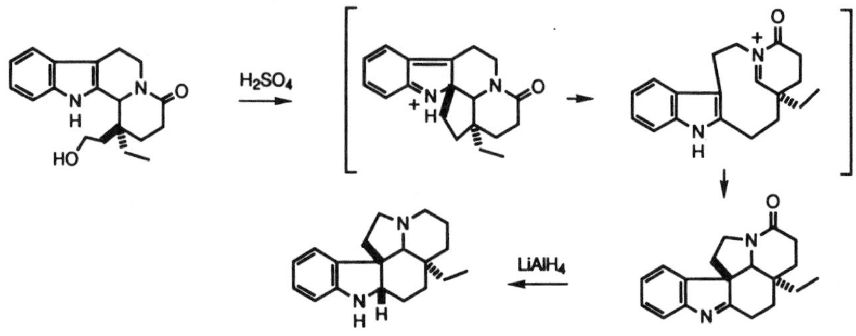

aspidospermidine

In the broadest sense, the attack of pyridinium ions (at C-4) with carbon nucleophiles may be considered as a Mannich reaction. A tandem acylation–intramolecular alkylation (Mannich reaction) occurred when harmalane was exposed to excess isonicotinoyl chloride [Naito, 1979].

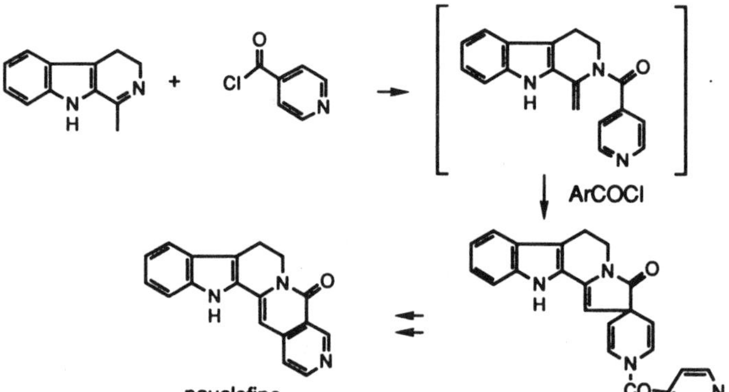

nauclefine

A convergent to guatambine and olivacine [Besselievre, 1976] is interesting as there was an exchange of a heterocycle for a carbocycle in the condensation step leading to a carbazole. Thus, alkylation of indole with the nascent 4-acetyl-1-methyl-1,2,5,6-tetrahydropyridine also caused double bond migration and an intramolecular Mannich reaction. However, fragmentation of the tetracyclic intermediate was a favorable reaction as it was accompanied by aromatization.

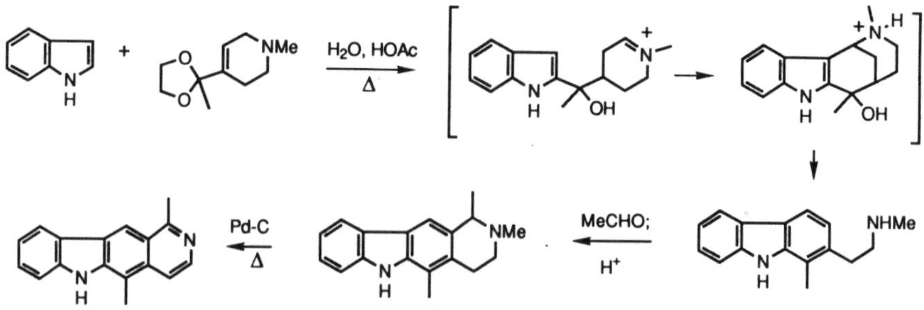

olivacine   guatambine

The erythrinane skeleton has been constructed from an α-methylsulfinyl-acetamide in one step [Tamura, 1982] via intramolecular trapping of the Pummerer rearrangement product and a Pictet–Spengler-type cyclization.

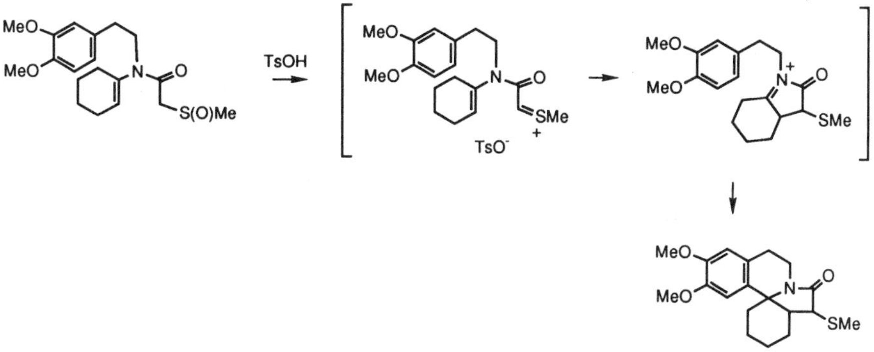

6.4. RETRO-MANNICH FRAGMENTATION–MANNICH CYCLIZATION

The reversibility of the Mannich reaction is often a crucial feature to consider during synthesis of nitrogenous substances, with respect to gaining access to one or more isomers. For example, the mold metabolites rugulovasine-A and -B may be interconverted by a fragmentation/recyclization pathway, and a synthesis [Rebek, 1980] that culminated in a rugulovasine-A also furnished rugulovasine-B upon equilibration.

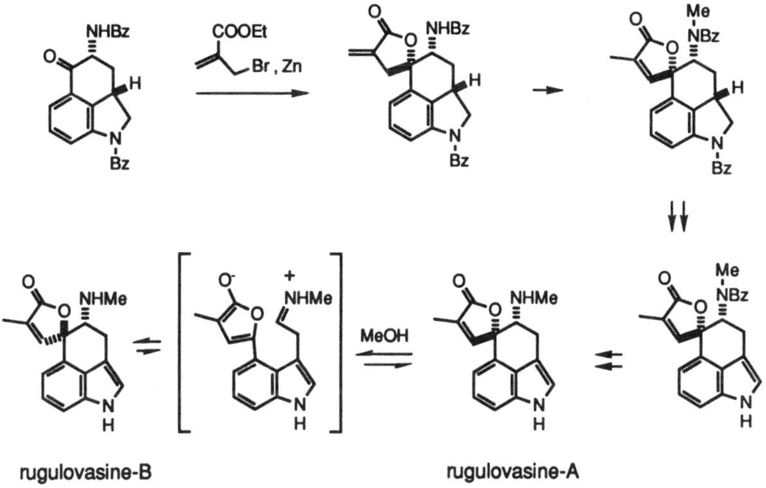

rugulovasine-B   rugulovasine-A

The thermal equilibration of indolyltetrahydropyridines [Natsume, 1984], which served as intermediates in a synthetic study o the aspidosperma alkaloids, proceeded by a retro-Mannich–Mannich reaction tandem. The nonequilibrat-ability of the corresponding N_b-hydroxyethyl derivatives is intriguing.

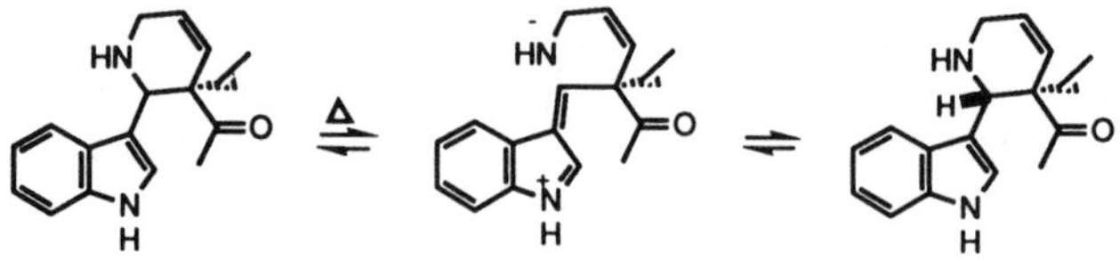

Similarly, the toxic fungal metabolites cyclopiamine-A and -B equilibrates via ring opening of the nitroindolizidine ring (retro-Mannich–Mannich tandem) [Bond, 1979].

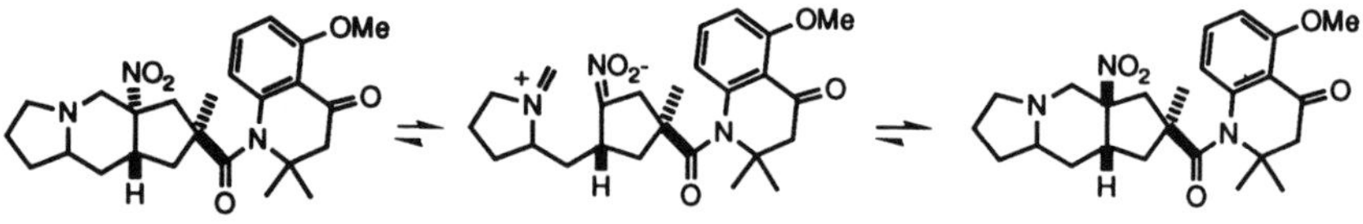

cyclopiamine-A   cyclopiamine-B

The acid-catalyzed isomerization of (−)-sophoramine into (+)-isosophoramine [Okuda, 1963] is represented by a transannular retro-Mannich fragmentation–Mannich cyclization tandem.

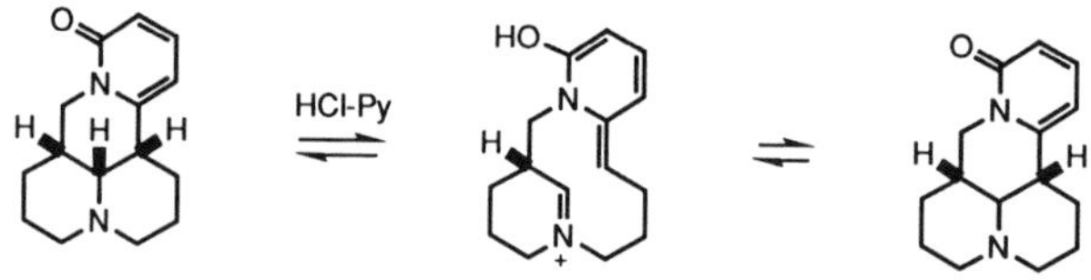

(-)-sophoramine   (+)-sophoramine

Epimerization of (19S)-vindolinine to the (19R)-isomer [Atta-ur-Rahman, 1986] has been rationalized by a mechanism as shown now.

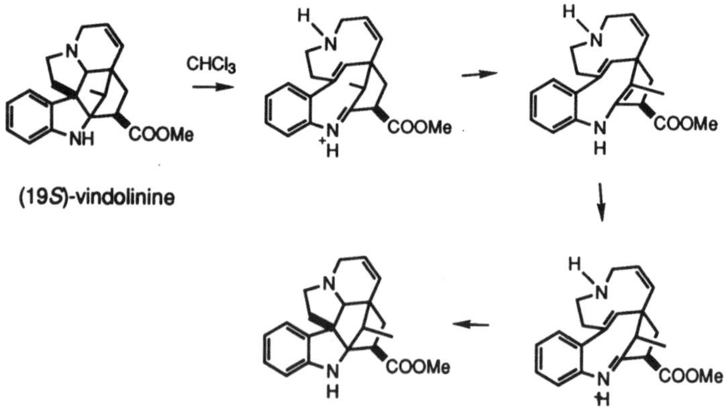

(19S)-vindolinine

(19R)-vindolinine

Condyfoline is partially transformed into tubifoline on heating [Schumann, 1963]. A mechanism consistent with the result suggests the occurrence of a tandem reaction series of retro-Mannich–Mannich reactions interposed by prototopic shift.

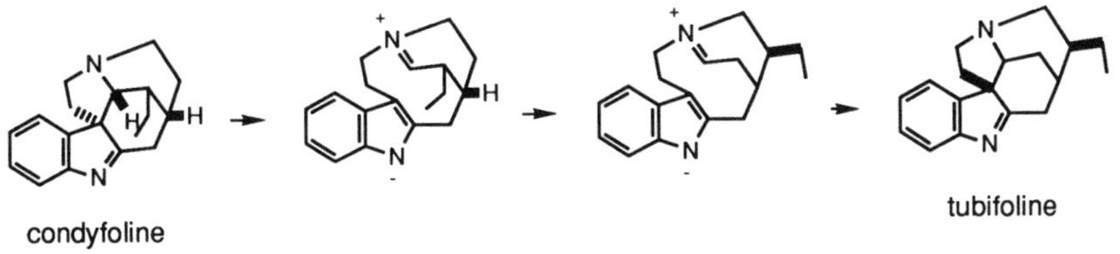

condyfoline   tubifoline

Aspidospermidine 16-oxime undergoes rearrangement to afford a condensed tetracycle [Hugel, 1974] under dehydrating conditions. Apparently a Beckmann fragmentation initiated the retro-Mannich deannulation–Mannich cyclization tandem.

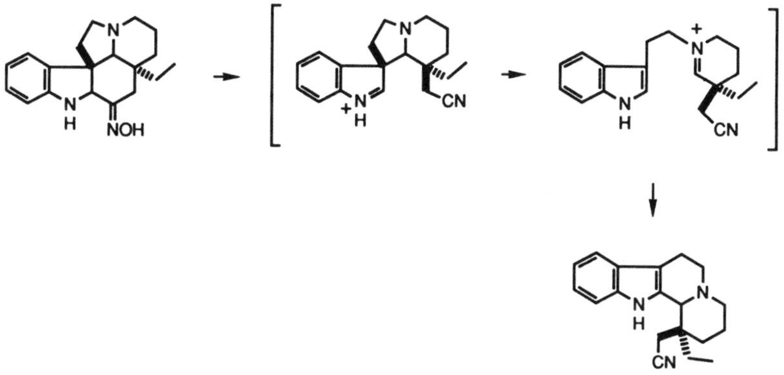

The akuammiline-to-akuammicine transformation [Olivier, 1965] on base treatment involved loss of an acetoxymethyl group, formation of a cyclopropane intermediate, retro-Mannich fragmentation, and Mannich cyclization.

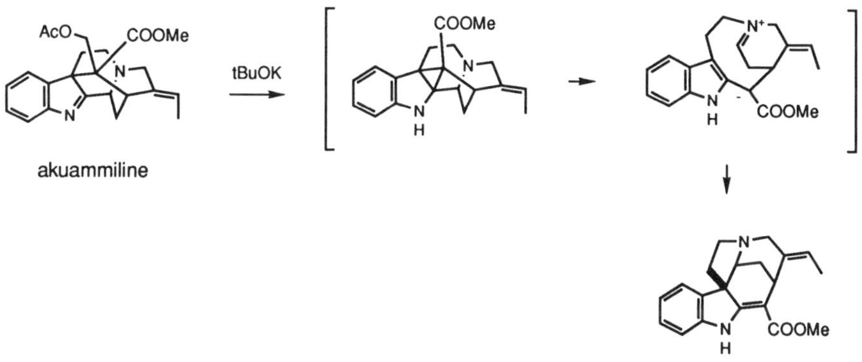

akuammiline

Reduction of the indolenine portion of aspidospermine-type compounds usually gives quebrachamine-type products. However, using Zn–Cu couple the reduction was found to proceed with a 1,2-alkyl shift. This observation is very useful in an approach to strempeliopine [Hajicek, 1981].

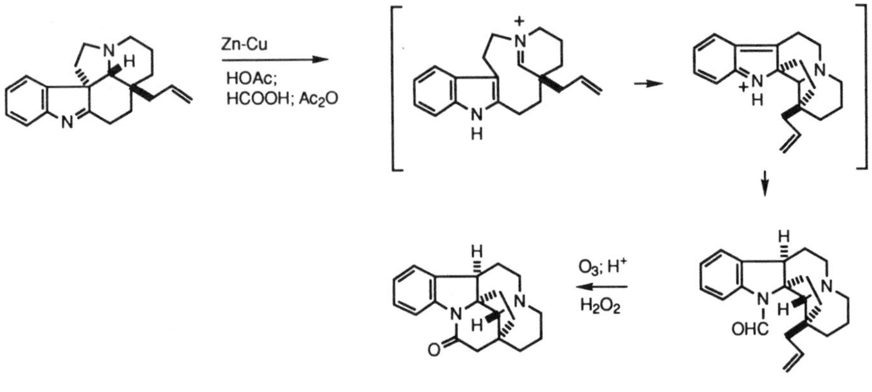

strempeliopine

The transformation apparently consisted of a retro-Mannich fragmentation followed by a Mannich cyclization of the chano intermediate in which CC bond formation involved the α-carbon atom of the indole nucleus. The new pentacyclic species underwent reduction.

The Fischer indolization of 1,3-disubstituted 4-piperidinone gives tetrahydropyrimidino[1,6-a]indole [Cattanach, 1968]. The indolenine intermediates cannot tautomerize but a retro-Mannich fission pathway is available to them. Consequently, fragmentation and reclosure of the heterocycle by the Mannich reaction take the course.

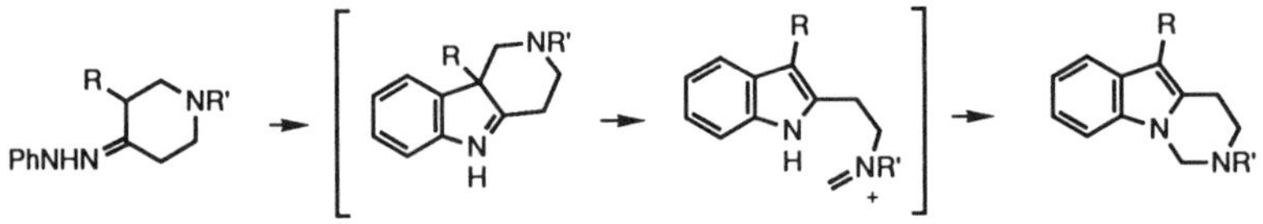

Heating (−)-tabersonine in acetic acid gave (+)-allocatharanthine as the major product [Muquet, 1972]. Retro-Mannich fragmentation, double bond migration, and carbocyclization account for the result.

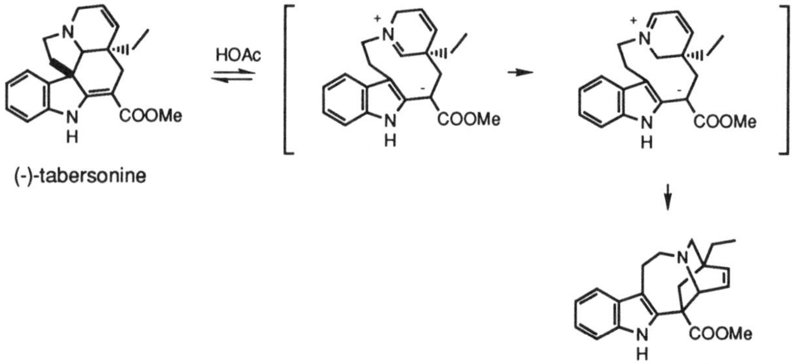

(-)-tabersonine

Rearrangement of a bicyclic amino ketol was observed during a synthesis of velbanamine [Büchi, 1970]. The occurrence of a retro-Mannich ring cleavage was favored by relief of ring strain. The product was formed via a vinylogous Mannich cyclization at an alternate site.

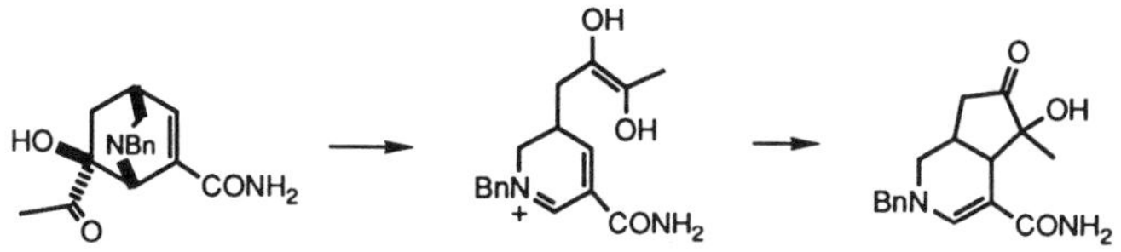

The synthesis of a Mannich base involving deliberate synthesis of a more highly strained isomer and equilibration of the latter compound is meritorious because the second step is thermodynamically driven. The retro-Mannich fragmentation relieves much strain of the molecule.

The intramolecular [2 + 2]photocycloaddition of an olefin and a β-amino enone is an excellent method for creating a fragmentation-prone Mannich base. The angular strain of the cyclobutane is dissipated by a ring cleavage action to produce a cyclic imine with a 3-oxoalkyl sidechain.

Although the Mannich cyclization of the fragmentation product generally requires a separate activation step of the nitrogen atom, thus making the retro-Mannich–Mannich reaction protocol inconsecutive, the strategy deserves mention here. And to illustrate the synthetic potential of this strategy routes to mesembrine [Winkler, 1988] and vindorosine [Winkler, 1990] are outlined now.

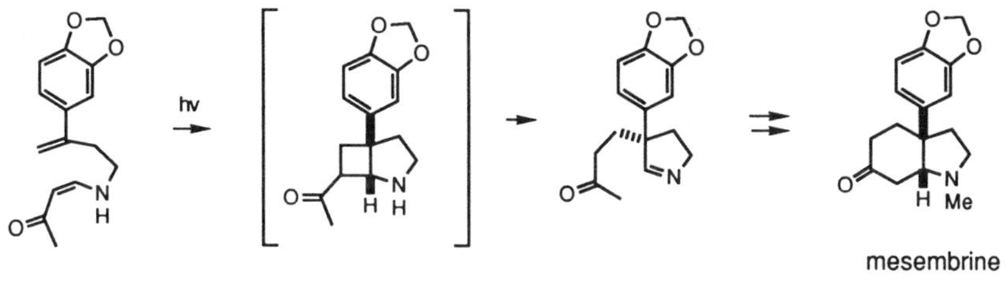

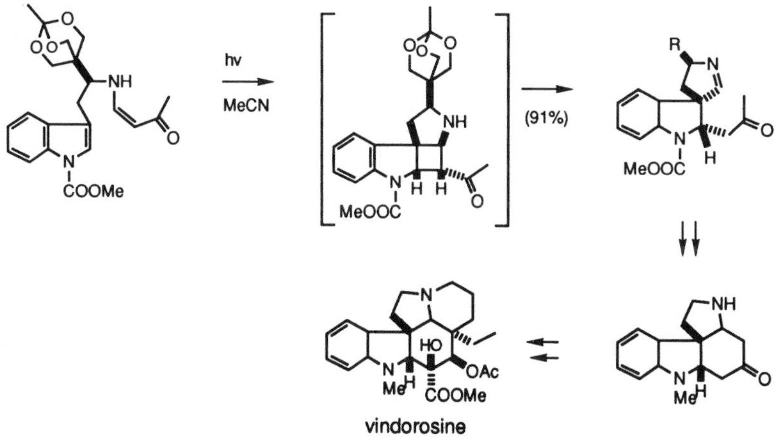

mesembrine

vindorosine

In a preparation of the indoloazepine intermediate which is of enormous synthetic utility for the aspidosperma alkaloids, the reaction of the 3-chloroindolenine isomer of a tetrahydro-β-carboline with thalliomalonate ester proved to be quite efficient [Kuehne, 1978]. The formal insertion actually proceeded by rearrangement, retro-Michael fragmentation, and a Mannich reaction.

vincadifformine

Exposure of certain 6-hydroxytetrahydroisoquinolines to lead(IV) acetate has resulted not only in the acetoxylation of the aromatic ring but also a rearrangement of the original oxygenation pattern [Hara, 1978]. The oxidation introduces an acetoxy group para to the hydroxyl function and the dienone is fragmentation-prone. Re-establishment to the tetrahydroisoquinoline nucleus from the iminium species is now energetically favorable.

6.5. MANNICH CYCLIZATION FOLLOWING OXIDATION OR REDUCTION

In addition to the classical two-staged condensation pathway to the Mannich reaction, generation of the ininium ion by other means may precede the CC bond formation. Examples of an oxidation-initiated Mannich cyclization are shown in the syntheses of tropinone [J. Hess, 1972] and (+)-anatoxin-*a* [Sardina, 1989] from proline derivatives.

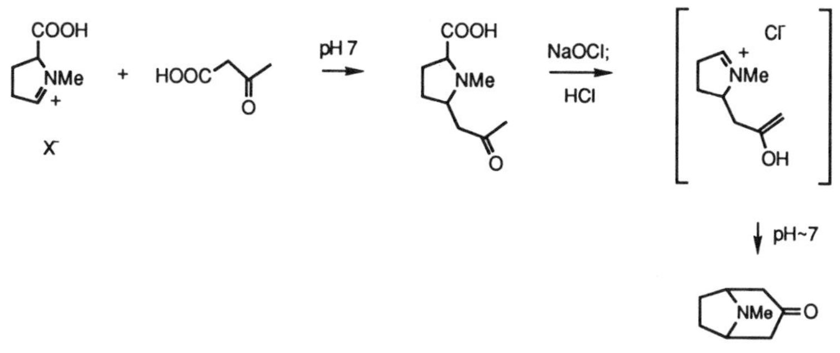

tropinone

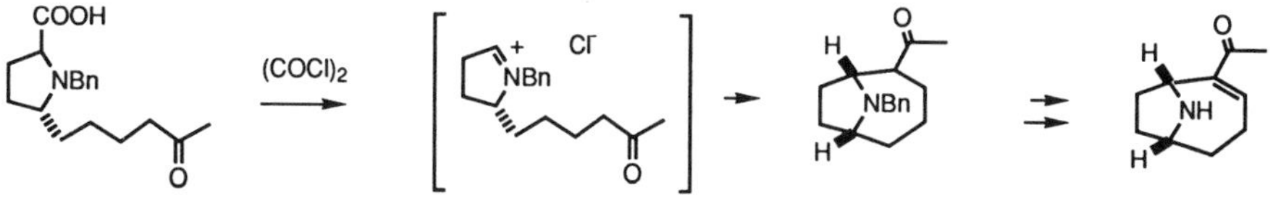

anatoxin-A

A biomimetic synthesis of ajmaline [van Tamelen, 1970] featured a tandem decarbonylative dehydration of a tryptophan moiety by reaction with dicyclohexylcarbodiimide and Mannich cyclization to form the quinuclidine ring system.

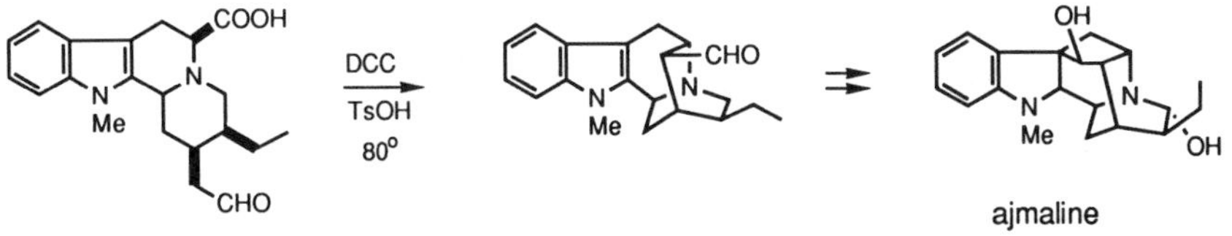

ajmaline

A biogenetically patterned approach to the sparteine bases [van Tamelen, 1960] involved two Mannich reaction steps, the second of which serving to establish the diazabicyclo[3.3.1]nonane nucleus by a two-staged intramolecular Mannich cyclization from intermediates created by oxidation of the tertiary amines.

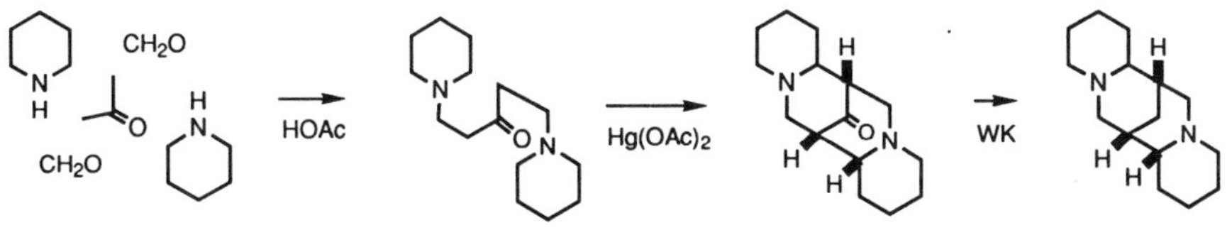

sparteine

A more obvious invocation of the Mannich reaction as a synthetic step toward coronaridine and dihydrocatharanthine is the presence of a β-amino ester subunit [Atta-ur-Rahman, 1980]. Employment of the transannular oxidative cyclization method was all the more favorable in view of the facility in acquiring the precursor.

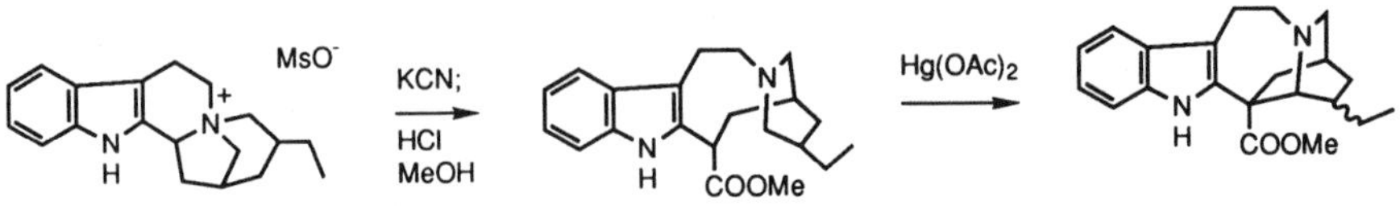

exo-Et   coronaridine
endo-Et   dihydrocatharanthine

On Swern oxidation a Mannich cyclization of 19-oxoaspidospermidine took place which resulted in 19-oxoaspidofractinine [Dufour, 1990].

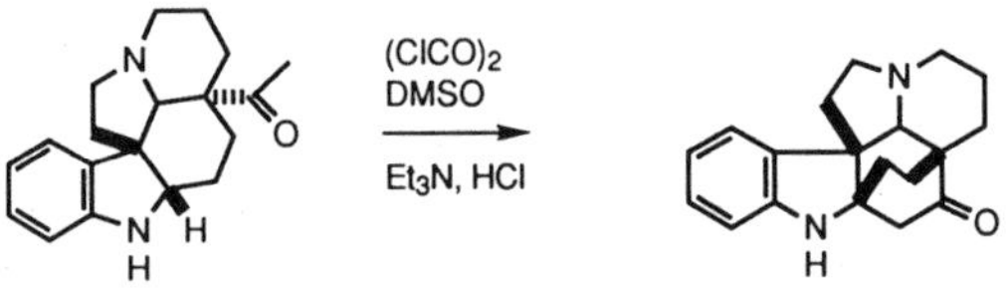

19-oxoaspidofractinine

Although the Mannich cyclization of acylinium ions that are generated by reduction of cyclic imides does not occur spontaneously under the reaction conditions, a casual mention of the reaction sequence is both worthwhile and relevant, in view of its facility and great utility for synthesis [Speckamp, 1985]. The following equation represents the key steps toward synthesis of vindorosine [Veenstra, 1981].

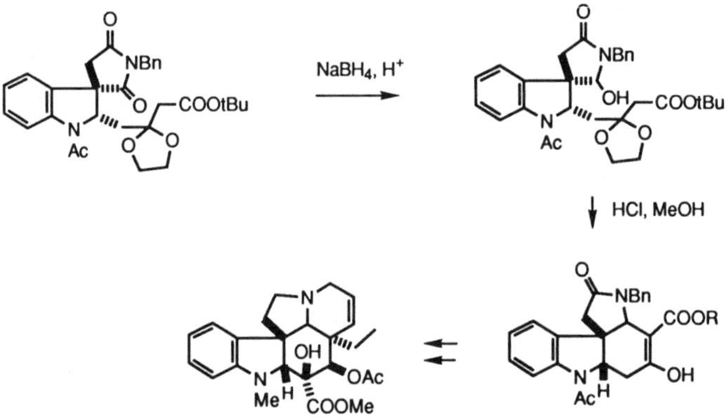

vindorosine

6.6. CATIONIC AZA-COPE REARRANGEMENT–MANNICH CYCLIZATION AND OTHERS

The aza-Cope rearrangement proceeds under very mild conditions, typically at or slightly above room temperature, with a high level of stereocontrol (which is characteristic of [3.3]-sigmatropic rearrangements). The many methods available for preparing iminium substrates also enhance the synthetic potentials of this rearrangement.

This rearrangement actually intervenes in the allylation of enamines, as indicated by the appearance of a transposed allyl group in the products [McCurry, 1974].

β-vetivone

Alkylation of an enamine with a propargyl bromide leads to an α-allenyl carbonyl compound, after mild hydrolysis [Corbier, 1970].

The ability to direct the equilibrium of the rearrangement is most valuable in the context of synthesis. Accordingly, an in situ capture of the rearranged iminium ion by a Mannich reaction in tandem with the cationic aza-Cope rearrangement would render the reaction irreversible. Various substituted 3-acylpyrrolidines have been synthesized by providing the necessary structural features [Overman, 1983b].

It has been found that the sigmatropic rearrangement occurs preferentially via a chair transition state (chair–boat preference >3 kcal/mol), in the same topographical sense as the normal Cope rearrangement [Doedens, 1988]. Generally, the rearrangement rates for substrates possessing an E-alkene moiety are much higher than those with a Z-alkene moiety. Such differences might reflect the influences of the chair–boat topographies.

An extreme case is an aza-Cope rearrangement of cinnamyl substrates where a complete bias towards adopting chair and boat transition states for the E- and Z-isomers, is indicated respectively [Steglich, 1984].

That a [3.3]-sigmatropic rearrangement intervenes in these reactions has received support from an examination of a chiral substrate. Generation of a racemic product rules out effectively the reaction pathway involving cyclization of the substrate to a 4-piperidinyl cation and pinacolic rearrangement of the latter species. The rearranged iminium intermediate undergoes rapid C–C single bond rotation before the intramolecular Mannich cyclization is also indicated, and the cyclization seems to occur preferentially via the E-iminium ion, with an approximately synclinal (or chairlike) geometry [Jacobsen, 1988].

It is significant that racemization of the sigmatropic rearrangement product is less likely in the ring-enlarging pyrrolidine annulation, due to conformational constraints of the intermediates with a medium-sized ring. Chiral polycyclic azacycles can then be synthesized by this method.

In terms of scope, hydrocycloalka[b]pyrrole systems in which the carbocycle containing 5, 6, and 7 atoms have been assembled [Overman, 1983a; 1985b].

After the establishment of its generality this tandem process has been applied in an ingenious manner to the synthesis of many complex alkaloids, such as perhydrogephyrotoxin [Overman, 1980a] and crinine [Overman, 1983c; 1985c].

perhydrogephyrotoxin

R = Bn, (R)-CHMePh

(±)- or (-)-crinine

In an approach to tazettine via pretazettine [Overman, 1989a] special attention must be devoted to the problem of oxygenation in the pyrrolidine ring. In the substrate the presence of an oxygen functionality in that subangular position is incompatible with the aza-Cope rearrangement–Mannich cyclization process, because it is prone to elimination. Consequently, in its place an acceptor substituent which is convertible to a hydroxyl group is required. A phenyldimethylsilyl group fulfills this need, as a configuration-conservative $SiMe_2Ph$ to OH conversion is known.

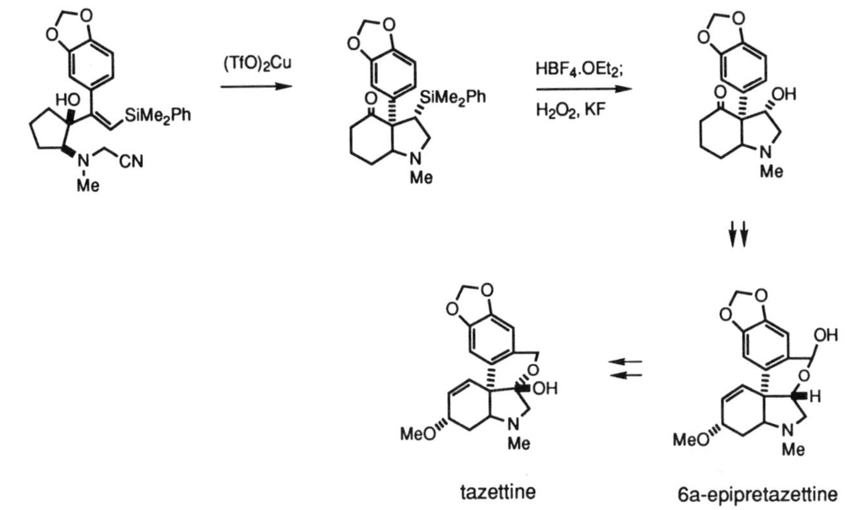

tazettine   6a-epipretazettine

An access of 16-methoxytabersonine [Overman, 1983d] affirms further the effectiveness of the aza-Cope rearrangement–Mannich cyclization route in the construction of very complex structures.

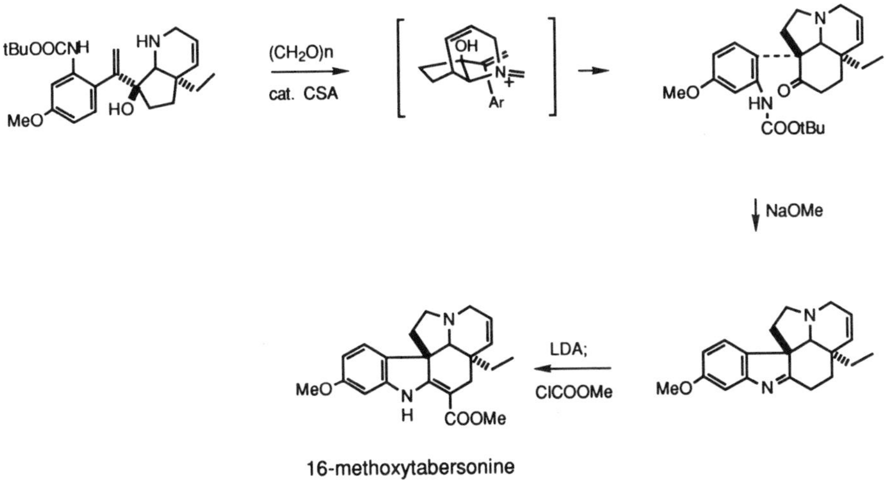

16-methoxytabersonine

An attempt at synthesizing strychnos alkaloids by the same strategy was thwarted, rather unfortunately, by the failure of E-ring formation [Overman, 1985a]. Apparently the prerequisite of a twist–boat conformation for the D-ring and an axial butenyl chain and the associated steric interactions is very difficult to meet. However, a modified scheme involving prior formation of the E-ring proved successful, as demonstrated by a synthesis of dehydrotubifoline [Fevig, 1991].

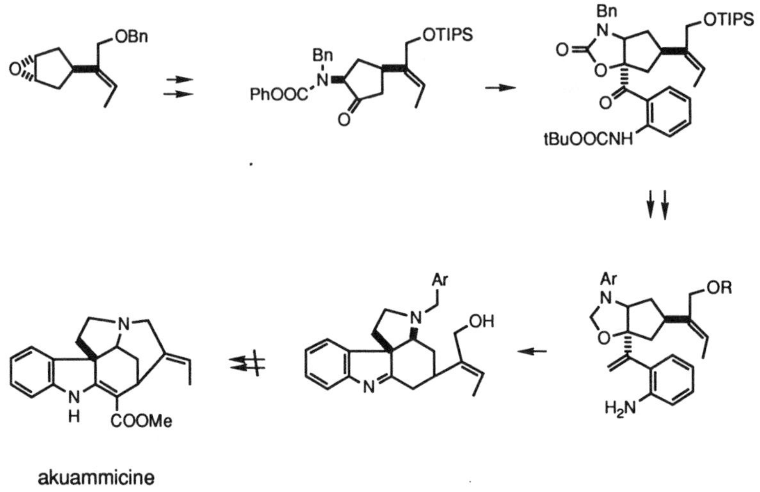

akuammicine

The *Melodinus* alkaloids are thought to arise in nature from aspidosperma bases via BC-ring disproportionation. The access of angularly arylated lilolidines by the aza-Cope rearrangement–Mannich cyclization method also opens a route to meloscine and epimeloscine [Overman, 1989b].

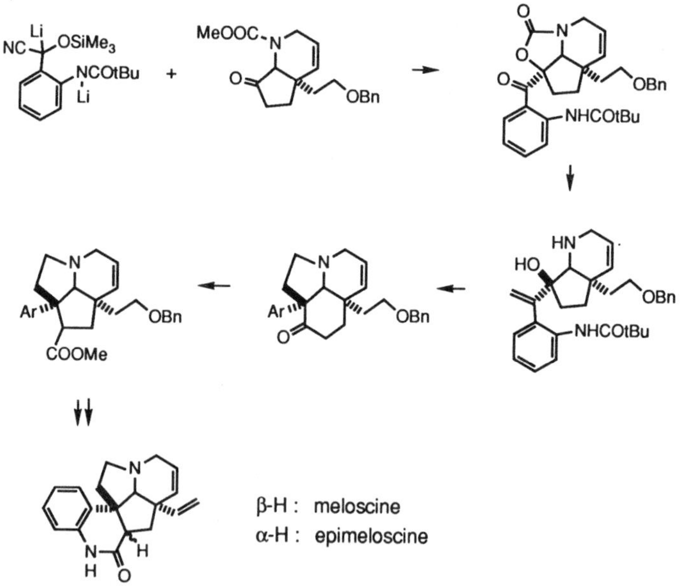

β-H : meloscine
α-H : epimeloscine

An approach to yohimbenone [W. Benson, 1979] via Birch reduction of 3-(*m*-methoxybenzyl)-1,2,3,4-tetrahydro-β-carboline and condensation with formaldehyde plausibly involved a cationic aza-Cope rearrangement. Closure of the D-ring can be regarded as a vinylogous Mannich cyclization.

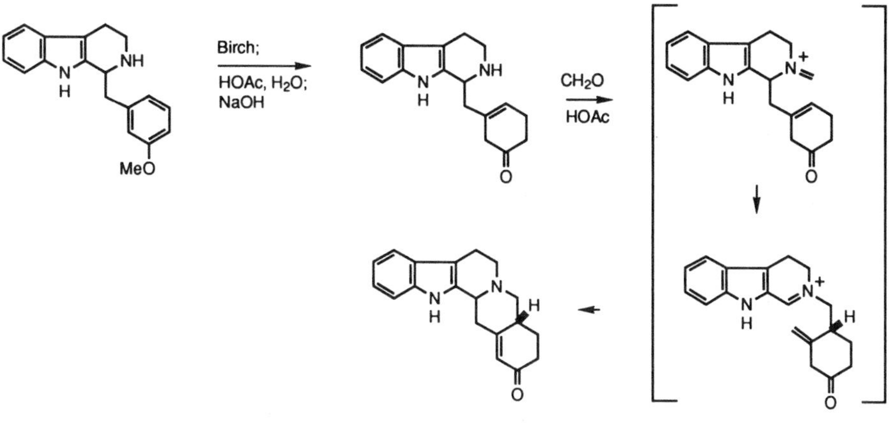

yohimbenone

Closely related to the cationic aza-Cope rearrangement–Mannich cyclization tandem is an assembly of the pyrrolizidine ring system [Hart, 1985]. The successful interception of the rearranged iminium ion by a trisubstituted double bond is the basis of a synthesis of (−)-hastanecine.

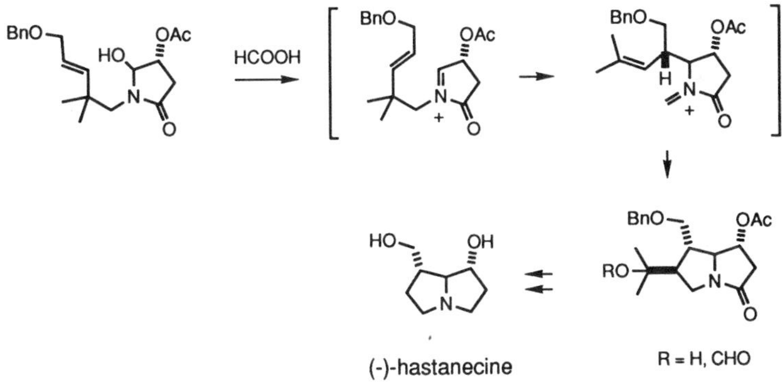

(-)-hastanecine   R = H, CHO

The intervention of an aza-Cope rearrangement is also evident in an intramolecular reaction of an acyliminium ion with an allene [Nossin, 1981].

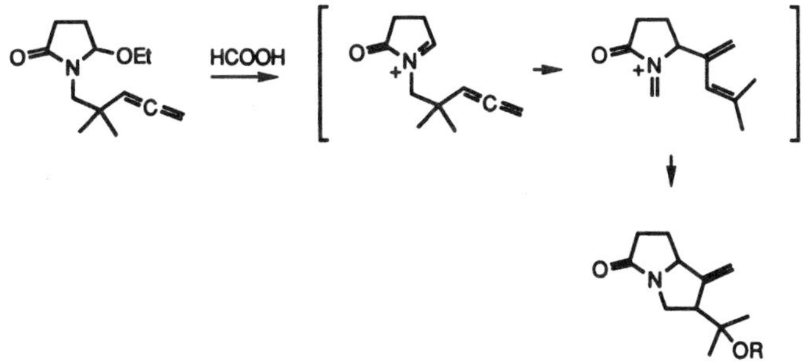

The photochemical rearrangement of oxotetrahydroberberinium salts into spiroisoquinoline derivatives [Nalliah, 1973] has been interpreted in terms of electrocyclic opening of the zwitterions and subsequent cyclization by a vinylogous Mannich reaction.

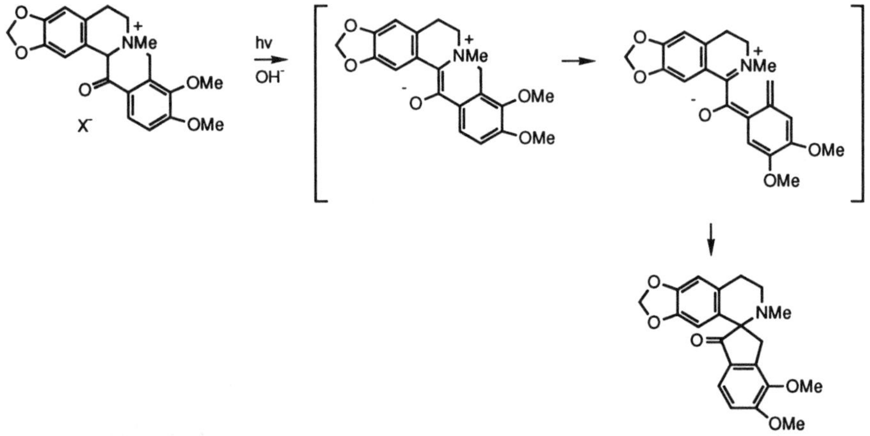

The Pictet–Spengler cyclization following a ring-enlarging rearrangement [Schell, 1983] is quite novel.

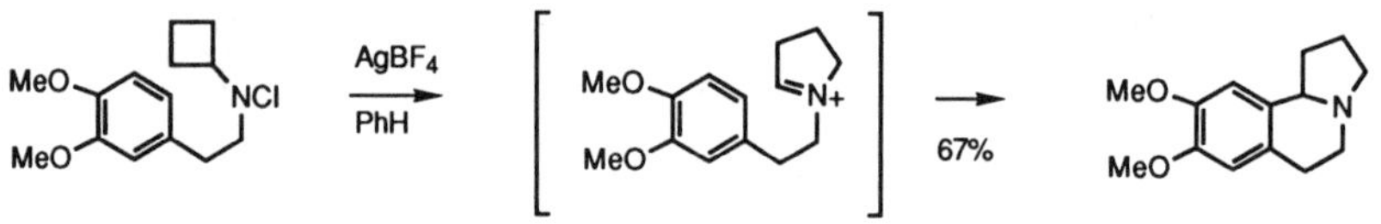

7

DIELS–ALDER REACTIONS

The Diels–Alder reaction [Carruthers, 1990] which features the formation of two new σ-bonds at the expense of two π-bonds to furnish an adduct with a six-membered hydroaromatic or heteroaromatic ring, is already one of the most efficient and hence popular reactions for synthesis. This versatility is further enhanced by its high regio- and stereo-selectivity, and the possibility of incorporating a great variety of functional groups in the addends.

With such attributes the Diels–Alder reaction in combination with another reaction in tandem would become an even more powerful tool for synthesis. To design tandem reactions of this kind either the diene or the dienophile component may be generated in situ and trapped by the reaction partner which is already present. Generally, the reaction tandem consists of an intramolecular Diels–Alder reaction and as such this step is favored by entropic factors. However, some Diels–Alder reactions are so facile (due to particularly high reactivity of the diene or dienophile) that an intermolecular reaction tandem may be effected. In these instances the reaction partners are introduced as trapping agents into the reaction media prior to the initial reactions. For example, benzyne generated in the presence of furan is intercepted readily to yield 1,4-dihydro-1,4-oxanaphthalene.

7.1. ELIMINATION/DIELS–ALDER REACTION TANDEMS

Some reactive dienes or dienophiles may be generated and captured in situ with appropriate reaction parameters to form Diels–Alder adducts. Formation of such intermediates by an elimination process thermally or under the influence of a catalyst has been well known, and such methods are particularly valuable in the acquisition of heterocycles.

A very interesting observation [Ohkita, 1991] is that a bicyclo[2.2.1]heptane-2-carbonitrile was produced by prolonged heating of acrylonitrile and 2-cyclopentenone ethyleneketal in acetonitrile. Most likely the minute amount of cyclopentadiene isomer of the spiroacetal in equilibrium was trapped by the dienophile. Naturally the more strained norbornene form of the Diels–Alder adduct would prefer the acetal structure.

Quaternary ammonium salts of Mannich bases undergo β-elimination readily. Thus, it is not surprising that a Diels–Alder adduct resulted when such an ammonium salt containing a conjugate diene subunit was boiled in water [Greengrass, 1985].

The Cope elimination is a common method for olefin synthesis. When the N-oxide is located in a sidechain of a cyclohexadienone an intramolecular Diels–Alder reaction could be effected in situ. Such was a crucial operation in an approach to seychellene [Fukamiya, 1973].

A synthesis of (−)-α-selinene [Caine, 1987] from (+)-carvone involved preparation of an acyclic intermediate whose intramolecular Diels–Alder reaction led to the bicyclic skeleton. Pyrolysis of the sulfoxide was succeeded by the cycloaddition.

α–selinene

Copyrolysis of 1-methyl-1-allyl-1-sila-2,4-cyclohexadiene and acetylene at 428°C led to 1-methyl-1-silabicyclo[2.2.2]octatriene [T.J. Barton, 1978]. *Si*-Methylsilabenzene was shown to be the intermediate.

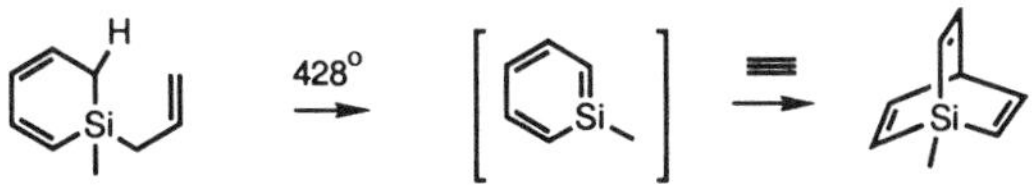

Synthetic schemes based on tandem dehydration and intramolecular Diels–Alder reaction have been explored. For example, a route to estrone [Jung, 1984] was hinged on C-ring formation by the cycloaddition. Tethering the dienophile to the latent diene by a three-carbon chain not only provides the elements of the D-ring, but also increases the reactivity and ensures regioselectivity.

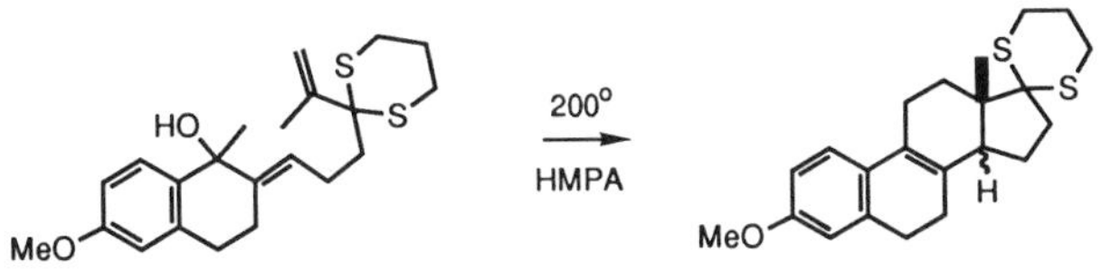

(*trans:cis* = 2.5 : 1)

An elegant route to 11-oxygenated steroids [Stork, 1981] was predicated on an intramolecular Diels–Alder reaction followed by cleavage of the cyclohexene and an aldol condensation. The Diels–Alder reaction was very facile as the conditions of dehydration of the allylic alcohol at −78°C to give the triene were sufficient to bring about it.

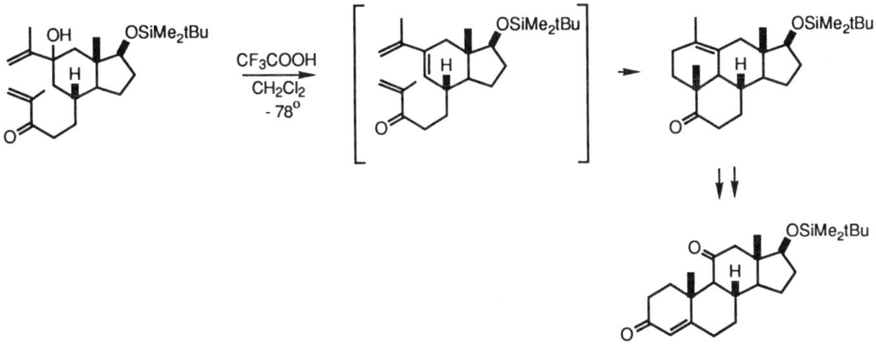

o-(1 – Hydroxyalkyl)phenols are susceptible to dehydration to form *o*-quinonemethides. As expected, intramolecular trapping of such species would lead to polycyclic chromanes [Talley, 1985].

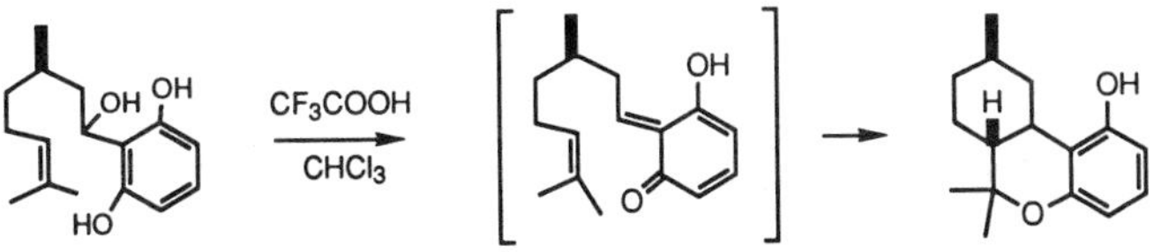

The sequence of intramolecular acetalization and 1,4-elimination of an alcohol to give isobenzofurans from *o*-(α-hydroxyalkyl)araldehyde acetals proceeds readily. In the presence of a dienophile the isobenzofuran may be transformed into benzo-7-oxabicyclo[2.2.2]heptanes. Synthetic utility of the combined reactions in lignan synthesis is quite evident [Rodrigo, 1980].

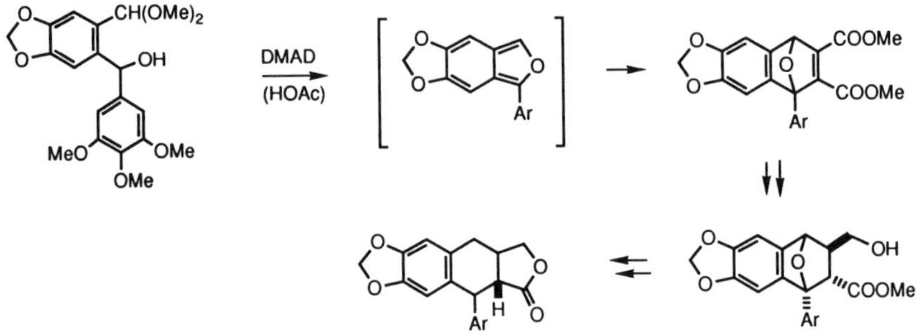

isodeoxypodophyllotoxin

An analogous intramolecular Diels–Alder reaction of an isobenzofuran was crucial to a synthesis of a resistomycin [Keay, 1982]. In this instance the precursor was an amino acetal. Remarkably the adduct was converted into the synthetic target by treatment with pyridinium hydrochloride. Four reactions (desilylation, aromatization, demethylation, and cyclization) occurred in this step.

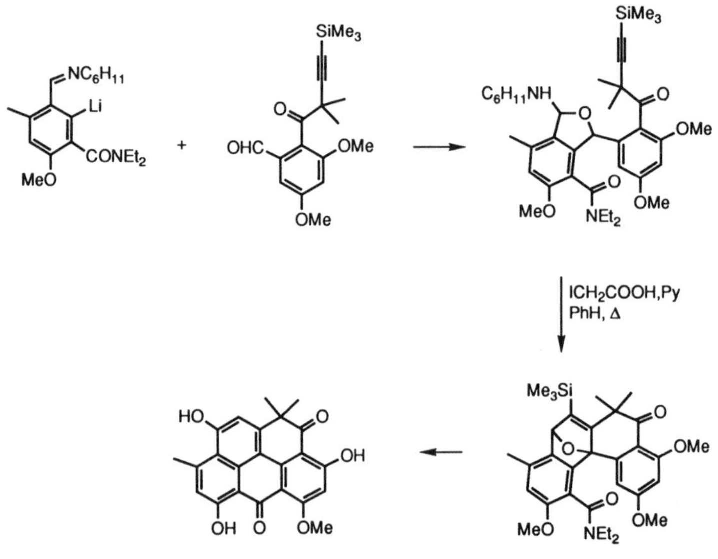

resistomycin

A rather unusual synthesis of ellipticine [Differding, 1985] involved amide dehydration to a ketenimine and subsequent cycloaddition to form the two internal rings.

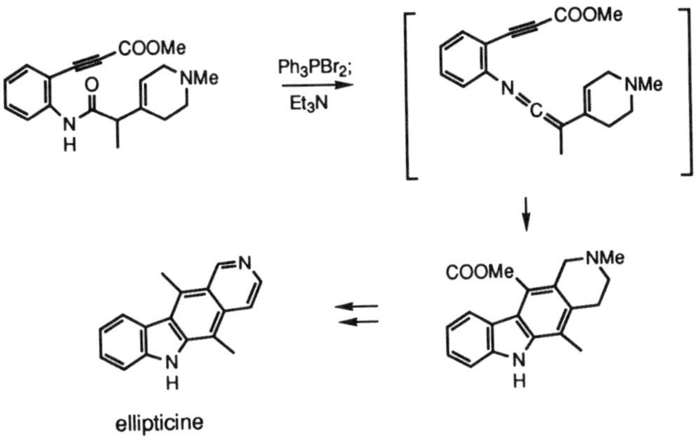

ellipticine

The Schiff condensation involves dehydration. The unsubstituted aldiminium ions thus prepared are capable of undergoing Diels–Alder reactions, an intramolecular version [Grieco, 1988] of which is shown now.

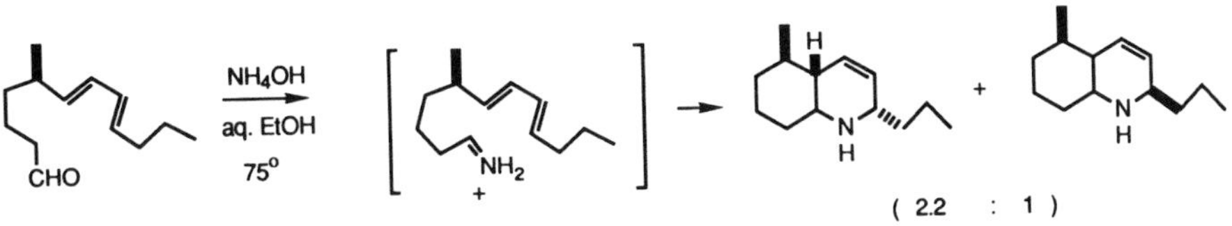

Unless structurally prohibited, enamines are preferably formed from aliphatic carbonyl compounds. In electronically and sterically favorable situations, enamine formation can cause reactions with functionalities present in the substrates. A felicitous result emerged from condensation of an indolazonine containing an acrylate moiety with butyraldehyde which engendered an intramolecular Diels–Alder reaction [Kuehne, 1991a]. An expeditious route to tubotaiwine was thereby revealed.

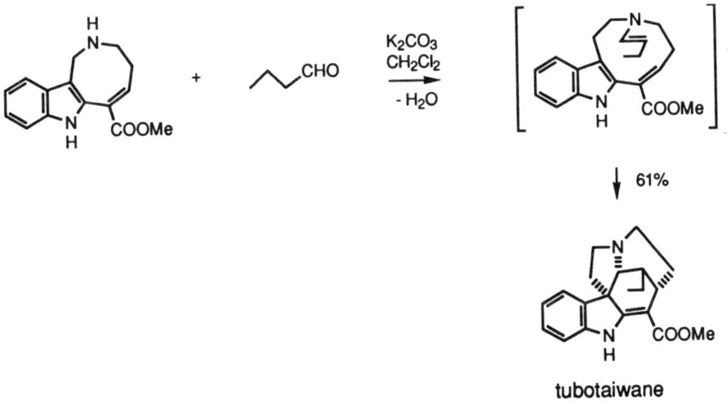

tubotaiwane

(The above process should be compared with a route to 20-epiibophyllidine [Jegham, 1989] in which an intramolecular Schiff condensation of an indoloazepine is followed by dequaternization by a retro-Michael reaction to furnish the precursor.)

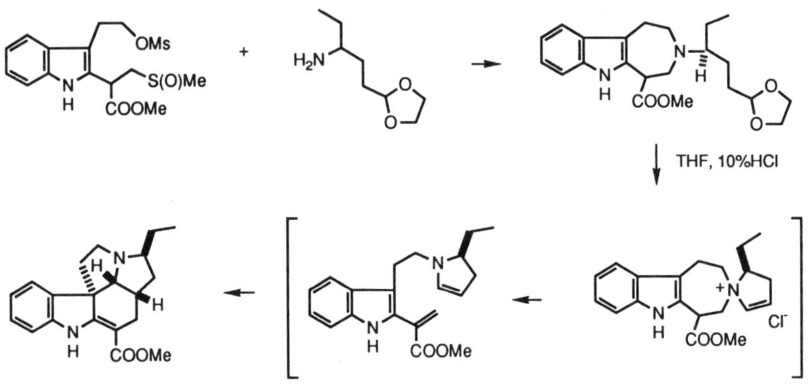

(+)-20-epiibophyllidine

Double condensation and dehydration take place when ammonia reacts with glutaraldehyde in a 1:2 ratio. A similar reaction is thought to be one of the key steps in an extremely efficient process that forms the pentacyclic skeleton of methyl homosecodaphniphyllate [Ruggeri, 1988]. It may also have some biosynthetic significance (cf. a biomimetic synthesis of protodaphniphylline [Piettre, 1990]).

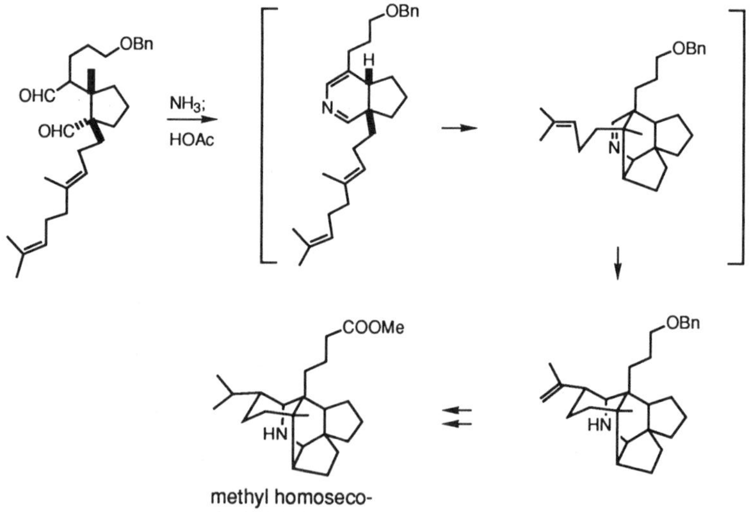

methyl homoseco-
daphniphyllate

In a formal synthesis of atisine [Shishido, 1989b] a hydrophenanthrene derivative bearing an angular cyano group was targeted as a key intermediate This cyano compound was prepared by an intramolecular Diels–Alder reaction, the substrate of which was in turn derived directly from a β-hydroxy silane.

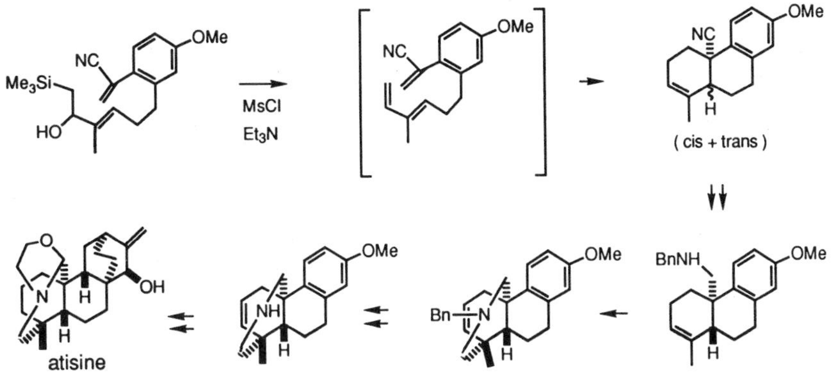

A hetero-Diels–Alder reaction was observed when an α-chloro ketoxime was released from its O-silyl derivative [Denmark, 1984]. A remote enol ether served as a trap for the unsaturated niroso compound.

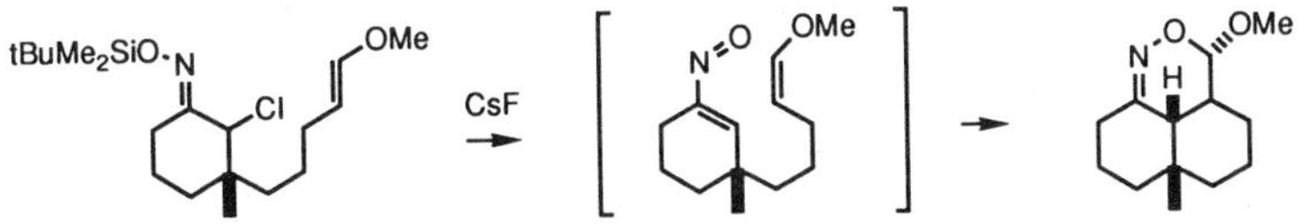

Dehydrochlorination of N′-4-pentenoyl-α-chlorohydrazones gives rise to 1,2-diazine derivatives [Gilchrist, 1983].

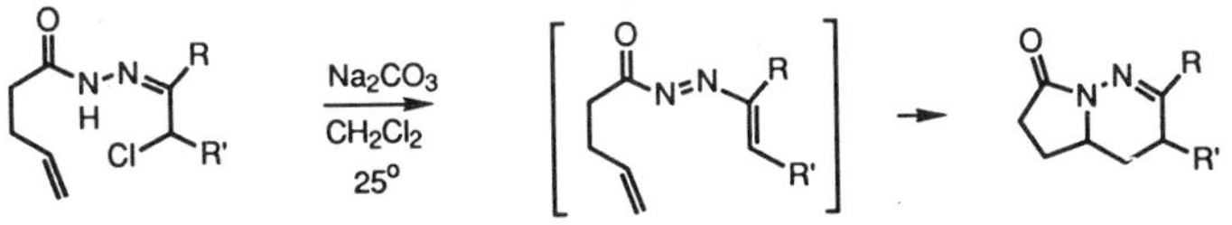

Benzynes are extremely reactive species. They are also known to participate in the Diels–Alder reaction, both inter- and intramolecularly. An intramolecular reaction of benzyne with an aromatic ring was observed [Houlihan, 1981] when an N-aryl-2,6-dichlorobenzamide was treated with butyllithium. The reaction is analogous to the formation of a symmetrical bridged benzobarrelene from 5-bromo[3.3]paracyclophane [Longone, 1976].

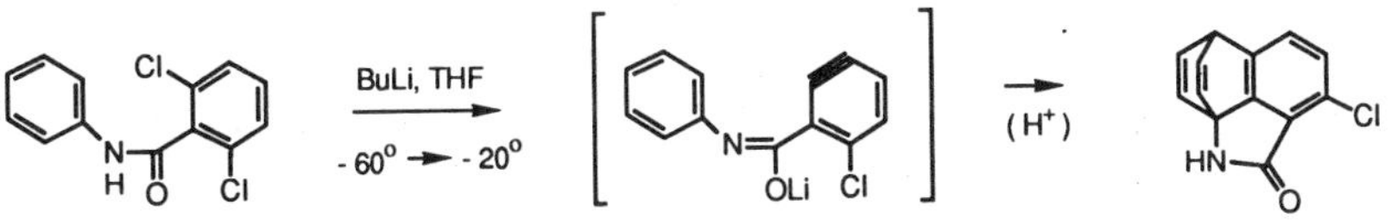

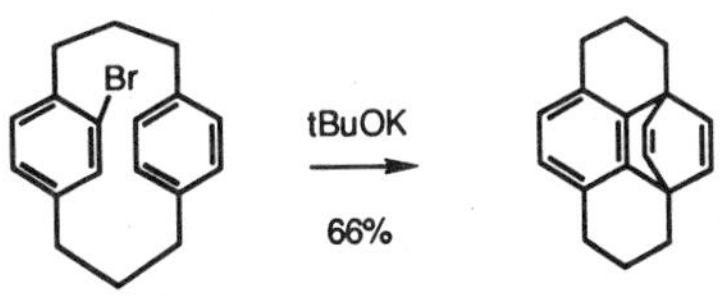

It is interesting to note that bridged benzobarrelenes are formed from diazonium salts of paracyclophanes [Mori, 1988]. Aryne intermediates are indicated.

Exploitation of the intramolecular Diels–Alder reaction of a nascent benzyne for synthesis of polycyclic aromatic compounds is illustrated by a synthesis of aristolactam [Estevez, 1989].

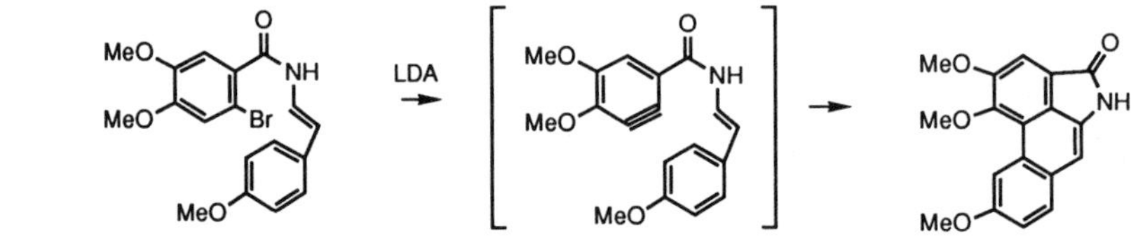

aristolactam

The very reactive 2,3-naphthoquinones have been generated and trapped by dienophiles [Jones, 1990].

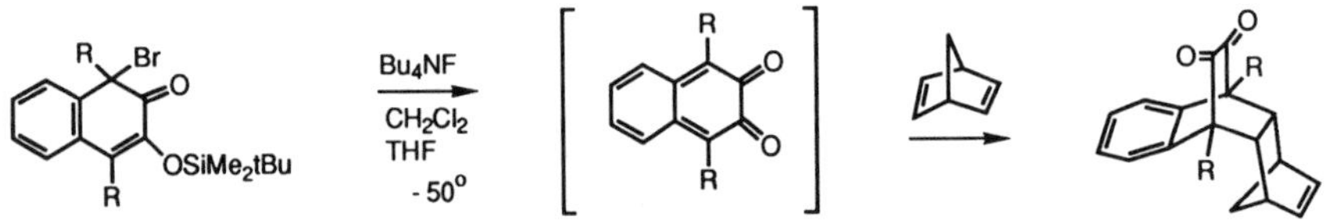

1,4-Elimination of benzylic substituents from *o*-xylyl derivatives furnishes the very reactive *o*-quinodimethanes. The latter species are readily trapped by dienophiles to form tetralins. The intramolecular version of the elimination–cycloaddition constitutes a highly efficient method for fabricating polycyclic structures.

Two slightly different schemes for the elaboration of estrone-type compounds have been developed [Djuric, 1980; Y. Ito, 1981]. The elimination step was initiated by the attack of a benzylic silicon group with fluoride ion.

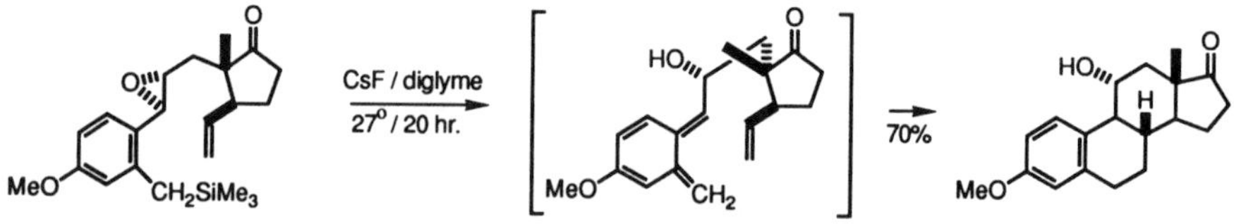

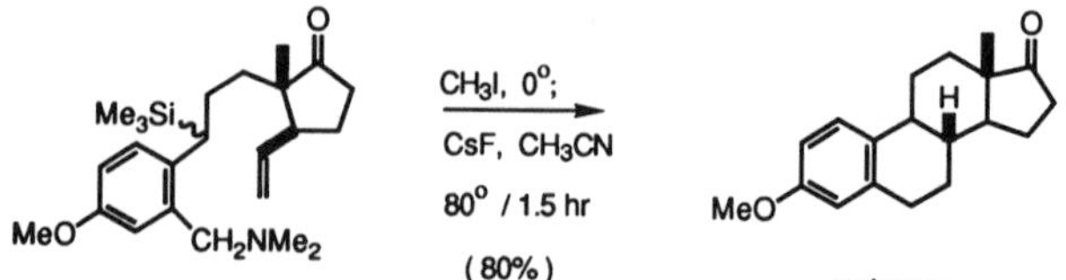

estrone
methyl ether

Further extension of this technique permitted the assembly of the tricyclic framework of gephyrotoxin [Y. Ito, 1983], demonstrating the accessibility of heterocycles by this method.

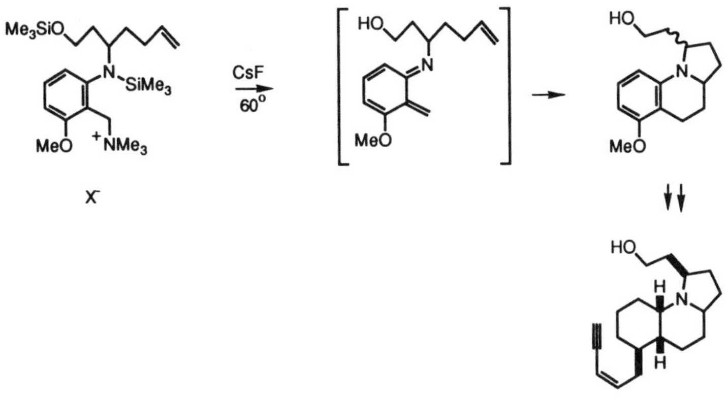

gephyrotoxin

The following equation shows adduct formation by an intramolecular cycloaddition of a benzyne as dienophile which was instrumental to a synthesis of mansonone-E [Best, 1981]. Note the method for the benzyne generation is more compatible to other functionalities present in the molecule.

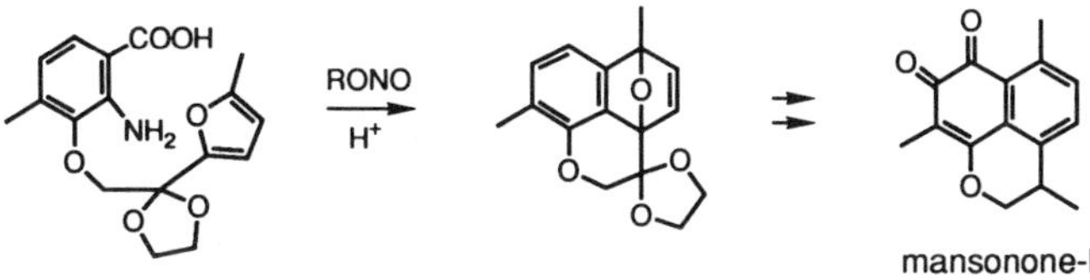

mansonone-E

Anatabine is a 6-substituted Δ^3-piperideine. By the presence of a double bond in a six-membered ring the structure beckons a Diels–Alder approach for its synthesis. Accordingly, the dienophile was formed by BF_3-catalyzed elimination of ethyl carbamate from the aminal derivative, and it was intercepted by butadiene [Quan, 1965].

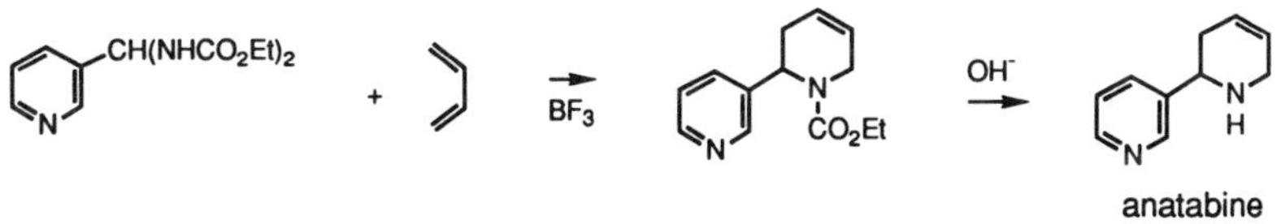

anatabine

In a synthesis of rutecarpine (and also evodiamine) [Kametani, 1976b] a mixed anhydride reacted with 3,4-dihydro-β-carboline at room temperature. The product is a formal Diels–Alder adduct of a ketene imine, although the intermediacy of the diene is suspect.

evodiamine   rutecarpine

Generally, intramolecular Diels–Alder reactions are more effective because the entropic factor can compensate to a large extent the electronic effects. Furthermore, regioselectivity can be predetermined for such reactions, as exemplified in a synthesis of anhydrocannabisativene [T.R. Bailey, 1984].

anhydrocannabisativene

The 1,3-relationship of the piperidine nitrogen atom and the sidechain oxygen suggests an anchoring technique to control regiochemistry. This approach was so successful that not only the required regiochemistry was developed, the quasi-boat transition state adopted by the cycloaddition (between the diene and the nascent imino ester) led to a *trans*-2,6-disubstituted Δ^3-tetrahydropyridine derivative corresponding to the target molecule.

The strategy as applied to δ-coniceine and tylophorine [Weinreb, 1979] is straightforward in stereochemical terms. On the other hand, two diastereoisomers were produced from a reaction that was designed to gain access to slaframine [Gobao, 1982]. Fortunately, the isomers differ only in the configuration of the secondary siloxy group, and both were useful synthetically.

δ-coniceine

tylophorine

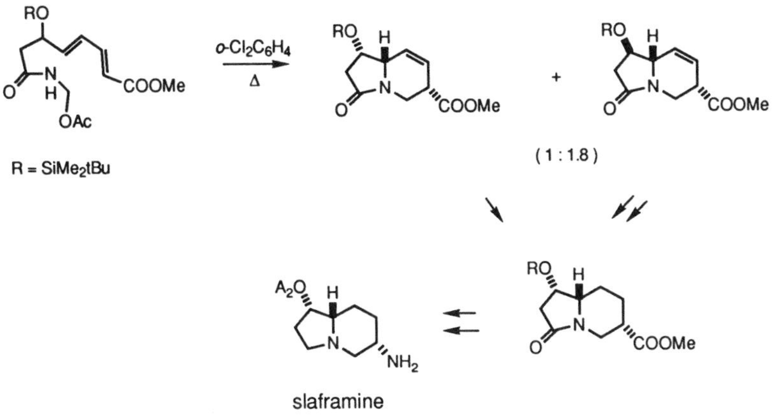

R = SiMe₂tBu

(1 : 1.8)

slaframine

A theoretically very significant molecule which contains both an anti-Bredt double bond and an anti-Bredt amide function has been prepared by the same technique [Shea, 1990]. The high reactivity of the double bond was demonstrated by its ready epoxidation on exposure to air.

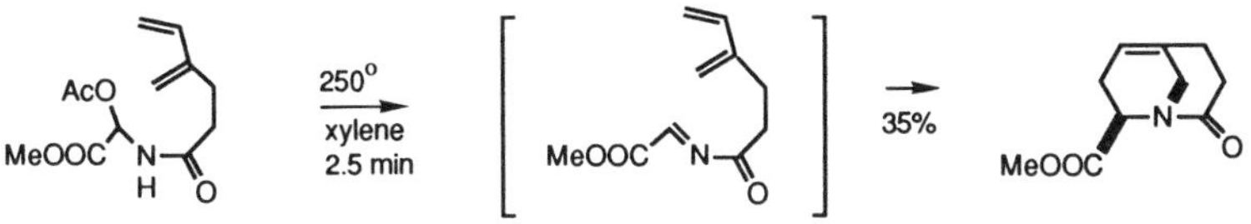

Another example concerns the assembly of the quinolizidine moiety of deoxynupharidine [Hwang, 1985]. Here the heteroatom is incorporated in the diene portion.

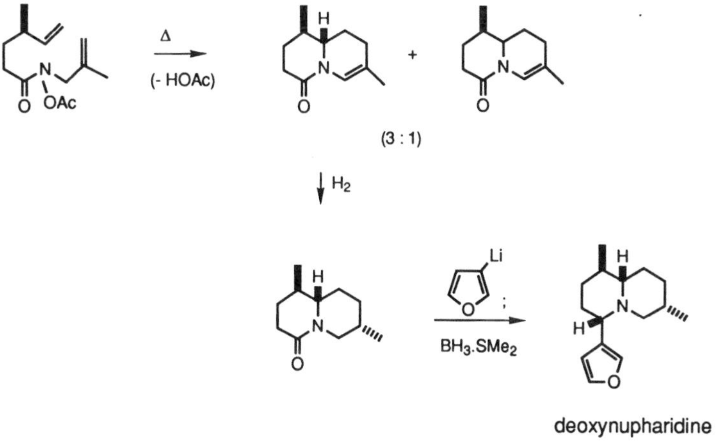

(3 : 1)

deoxynupharidine

Both the diene and dienophile components may be stored in latent forms that are sensitive to heat. Concurrent exposition of the components thermally would also induce the cycloaddition. Such is the case of an elaeokanine-A synthesis [Schmitthenner, 1980]. The elimination of sulfur dioxide from the sulfolene moiety is a cheletropic reaction.

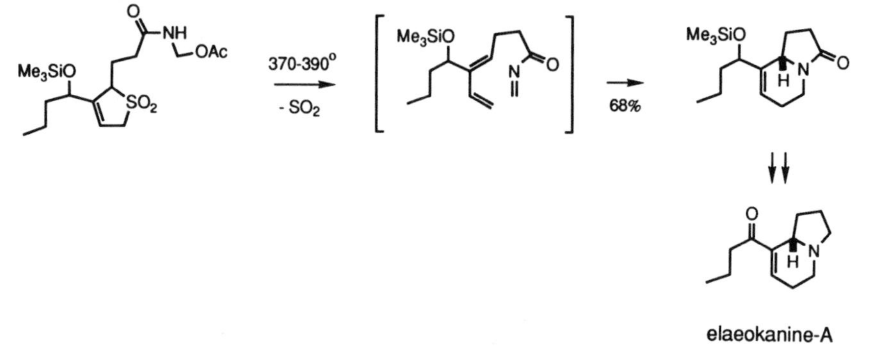

elaeokanine-A

Protection of 1,3-dienes as sulfolenes is a well-established method. The possibility of modifying the sulfolenes, for example by alkylation, has extended the utility of these substances. Synthetic plans for lupinine [Nomoto, 1985], aspidospermine [Martin, 1980], and lycorine [Martin, 1982] have been devised on the basis of such versatility.

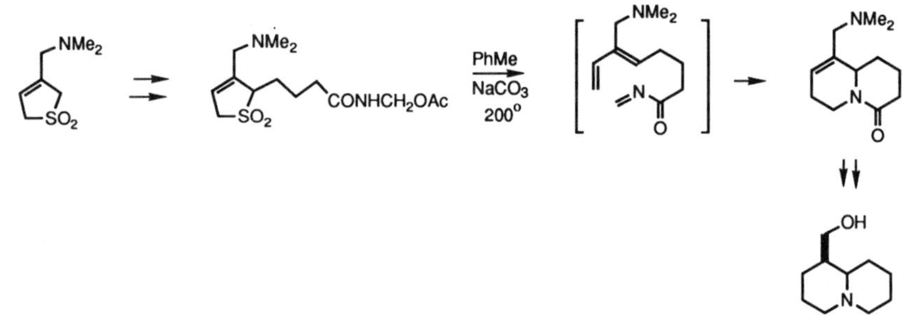

lupinine

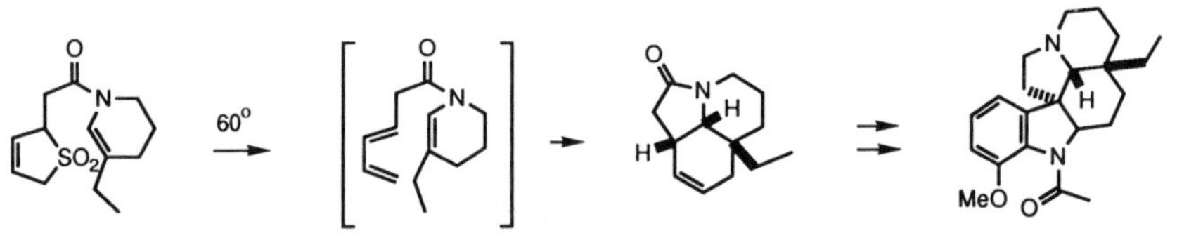

aspidospermine

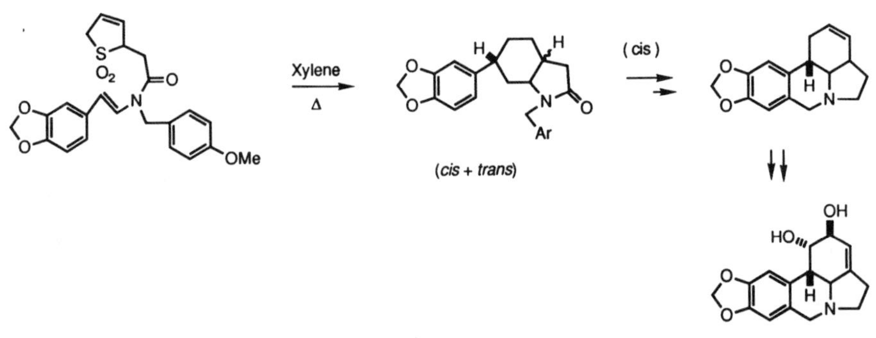

lycorine

Dihydroisobenzothiophen dioxides also undergo thermolysis. The products are *o*-quinodimethanes. Exploitation of this chemistry and the propensity of *o*-quinodimethanes to cycloadd to various dienophiles culminated in an elaboration of (+)-estradiol [Oppolzer, 1980; cf. Nicolaou, 1979], and also phyltetralin [Mann, 1982].

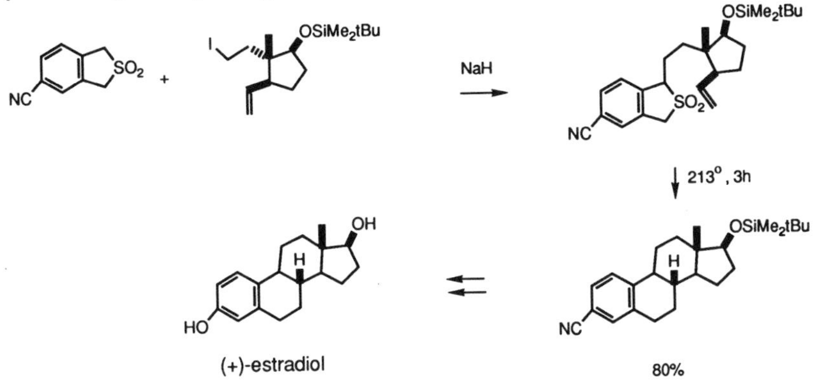

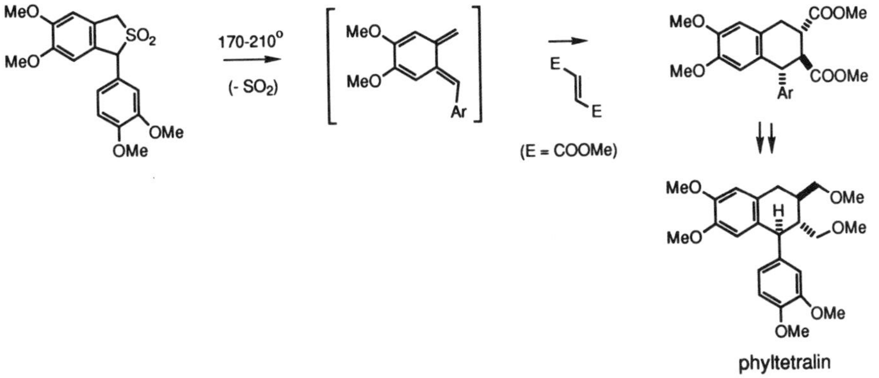

(+)-estradiol

80%

(E = COOMe)

phyltetralin

Three consecutive thermal reactions were the crux of a one-step preparation of a molecule possessing the iceane skeleton [Hamon, 1982]. These reactions are Diels–Alder reaction of a propellane with tetrachlorothiophen dioxide, cheletropic expulsion of sulfur dioxide, and intramolecular Diels–Alder reaction. The last cycloaddition only the *cis-syn-cis* isomer is able to partake.

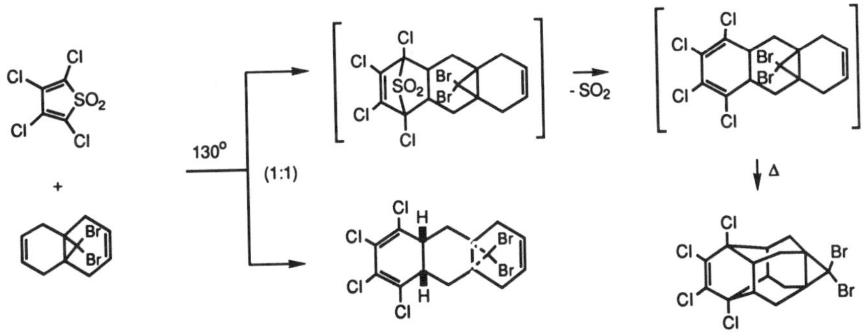

Thioaldehydes are unstable and an excellent method for their generation is by photochemical elimination of acetophenone from phenacyl sulfides. In the presence of a conjugate diene the nascent thioaldehyde may be trapped to afford dihydrothiopyran [Vedejs, 1988a].

cis : trans
X = H, H   2 : 1
X = O   3.5 : 1

lactone > 19 : 1

7.2. RETRO-DIELS–ALDER/DIELS–ALDER REACTIONS

The widely used cyclopentadiene is usually obtained from the dimer by thermolysis. Often a Diels–Alder reaction is carried out by generating the monomer in situ in the presence of dienophile. 1-Phenyl-2,3,4,5-tetramethyl-borole also exists in the dimeric form, it is very easy to effect its monomerization and subsequent Diels–Alder reaction due to the thermal lability of the dimer and the high reactivity of the borole toward simple dienophiles (e.g. ethylene, allene, butadiene, 2-butyne) [Fagan, 1988].

An in situ Diels–Alder reaction with anthracene was conducted to show the existence of benzocyclopropene quinone [Watabe, 1988] which was liberated from a tricyclic precursor.

Microwave irradiation induces a retro-Diels–Alder/Diels–Alder reaction sequence and hence racemization of vincadifformine [Takano, 1989].

(-)-vincadifformine

(+)-vincadifformine

In refluxing carbon tetrachloride, basketene formed an adduct with dimethyl azodicarboxylate [Allred, 1977]. Apparently the tricyclic isomer of basketene formed by a retro-Diels–Alder reaction was the intermediate.

25%

While accomplishing a Diels–Alder reaction consecutive to cheletropic extrusion of sulfur dioxide from a sulfolene is a very popular practice in a situation that warrants the use of a masked 1,3-diene, other versions of tandem reactions are also known.

The retro-Diels–Alder/Diels–Alder reaction sequence has many representatives. Thus, heating Δ^{18}-tabersonine in methanol at 145°C yielded andranginine and 21-epiandranginine [Andriamialisoa, 1975]. The formation of two epimers may indicate a nonconcerted nature of the cycloaddition.

andranginine

Thermolysis of an N-acyl-2-azabicyclo[2.2.1]hept-5-ene gives rise to cyclopentadiene and an N-acylimine. When a diene moiety is present in the acyl sidechain at a proper distance an intramolecular Diels–Alder could follow. δ-Coniceine has been obtained from such an adduct upon further reduction [Lasne, 1982].

δ-coniceine

In order to effect a synthesis of lysergic acid [Oppolzer, 1981a] by forming the CD-ring portion employing an intramolecular hetero-Diels–Alder reaction, the diene component which contains a carboxyl (better as an ester) group must be kept in a less reactive form during elaboration of the dienophile. An adequate device is a cyclopentadiene adduct which can be dissociated on pyrolysis.

In fact the masked diene was acquired by a Wittig reaction.

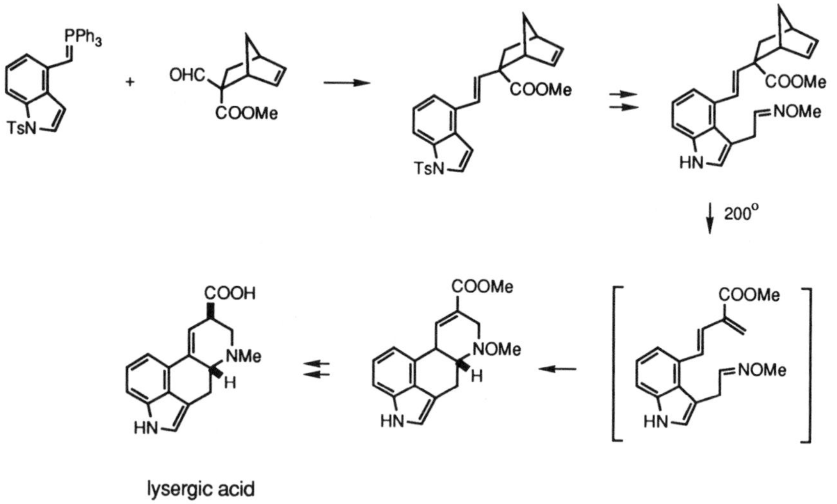

lysergic acid

α-Pyrones react with many dienophiles. The adducts usually undergo decarboxylation in situ to provide 1,3-cyclohexadienes. However, in the presence of excess dienophiles, bicyclo[2.2.2]-octenes are formed. This sequential Diels–Alder/retro-Diels–Alder/Diels–Alder process was the basis of a barrelene synthesis [Zimmerman, 1969a].

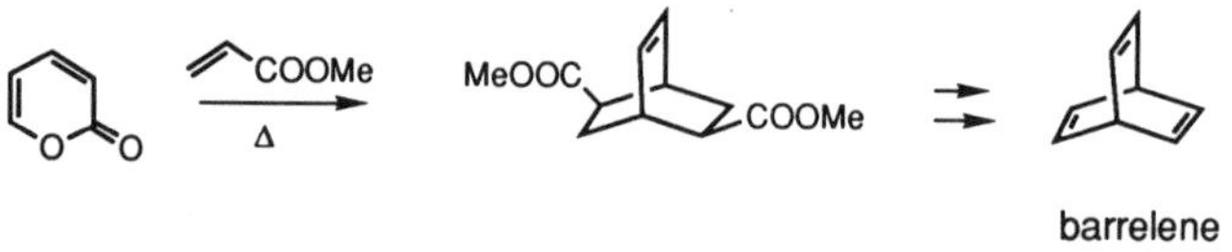

barrelene

When the dienophile contains two alkene linkages of vastly different dienophilic reactivities, its reaction with α-pyrone may lead to tricyclic product(s) in one step [Krantz, 1973; Swarbrick, 1991].

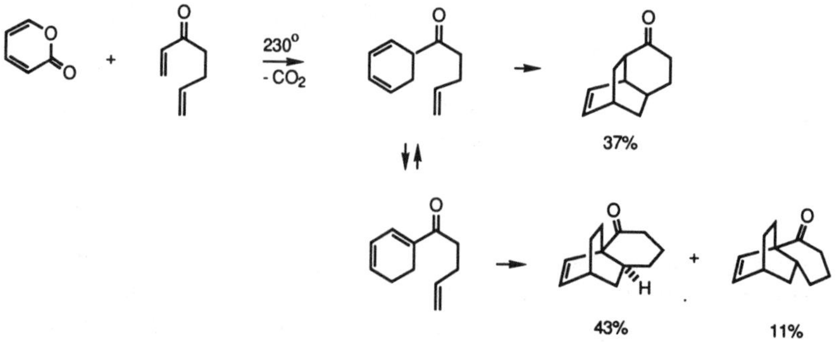

The generation of 4-methoxyisobenzofuran by thermal extrusion of carbon dioxide and benzene, and trapping the reactive intermediate with a bicyclic enedione cleared an entry to the anthracycline skeleton [Kende, 1977]. One of the products was converted into daunomycinone.

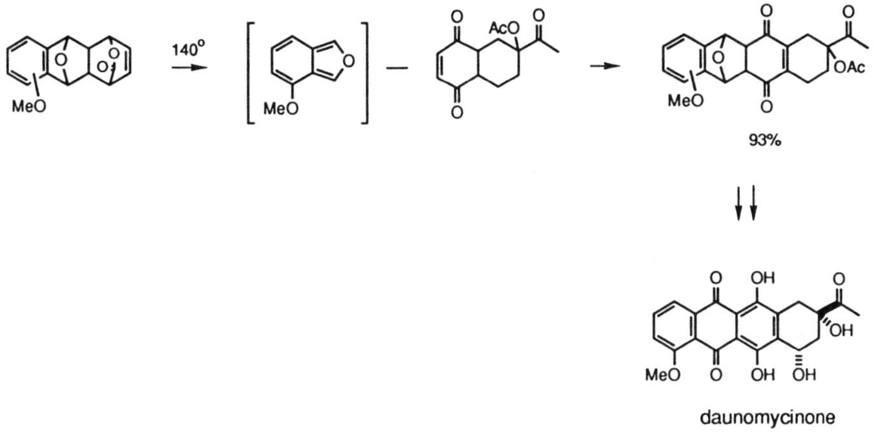

93%

daunomycinone

It should be noted that the decarboxylation of the α-pyrone adducts can be suppressed by conducting the Diels–Alder reactions at room temperature under high pressure [Gladysz, 1977].

3-Isochromanones undergo thermal decarboxylation to give o-quinodimethanes which isomerize to benzocyclobutenones [Spengler, 1977]. In the presence of dienophiles, the quinodimethanes are intercepted accordingly, and a route to 1-aryltetralin lignan lactones [Das, 1983] was based on such a reaction.

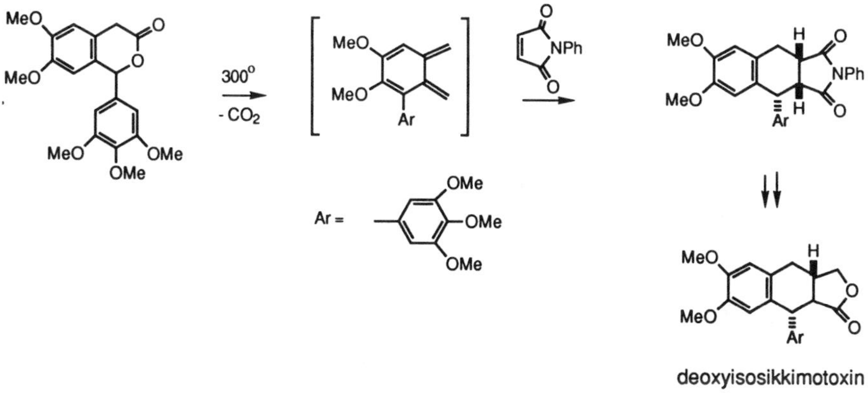

deoxyisosikkimotoxin

Rapid trapping of benzyne by an oxazole followed by extrusion of organonitrile and reaction of the resulting isobenzofuran with a second molecule of benzyne constitute a sequence of events occurring in refluxing dioxane when an oxazole is treated simultaneously with anthranilic acid and isoamyl nitrite [Reddy, 1980].

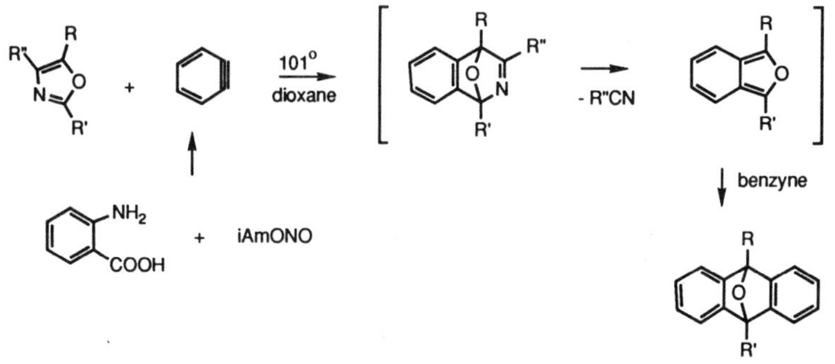

Coronafacic acid is amenable to assembly by an intramolecular Diels–Alder reaction, considering that the six-membered ring contains a double bond and the presence of a carbonyl group in a subangular position. In a route established on this premise [Ichihara, 1980] the diene component was unfolded from a cyclobutene moiety and the dienophile from a fulvene adduct. Both retro-Diels–Alder reaction and electrocycloreversion are orbital symmetry-allowed thermal reactions.

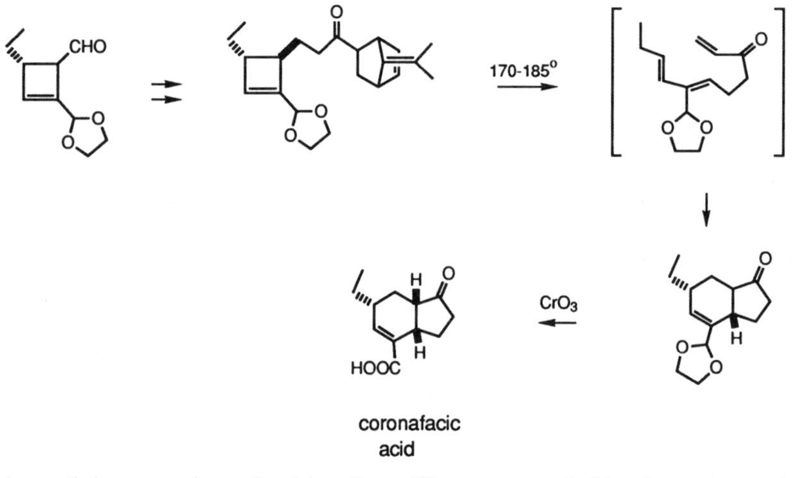

coronafacic
acid

A partial conversion of a bicyclic α-diketone to a bridged system perhaps proceeded via a retro-Diels/Diels–Alder reorganization [Y. Kitahara, 1976].

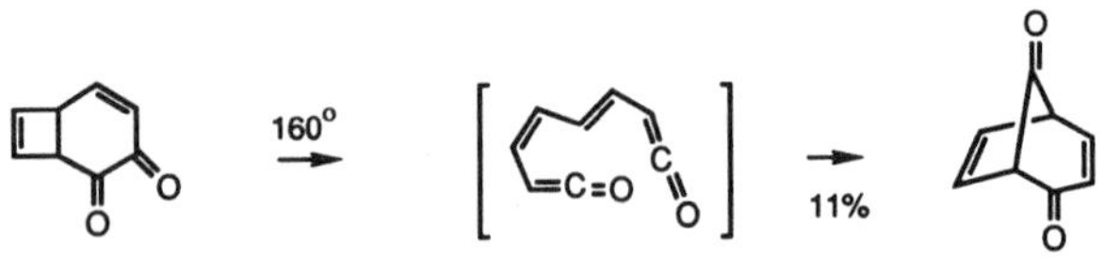

A synthetic route to heliotridine and retronecine [Keck, 1980] involving the retro-Diels–Alder and Diels–Alder reaction tandem is less obvious in the sense that the dienophile is composed of two heteroatoms. However, great advantages of this method include the preorganization of all the functionalities, the facile reductive N–O bond cleavage to give viable precursors of the necine bases.

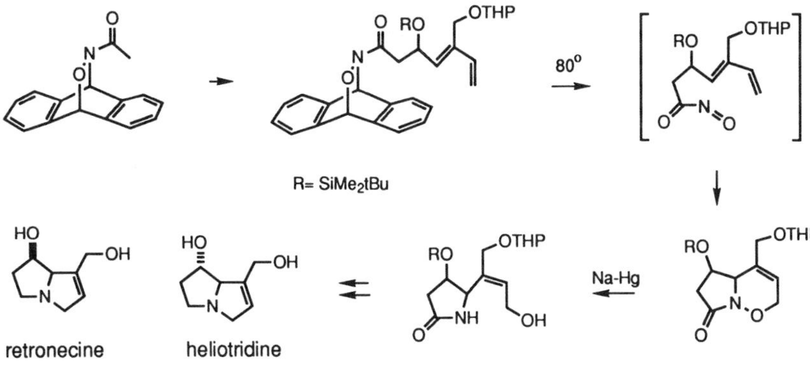

R= SiMe₂tBu

retronecine   heliotridine

Elimination of acetylene from a dioxa[2.2](1,4)-naphthalenophane precipitated an intramolecular Diels–Alder reaction [Halverson, 1982].

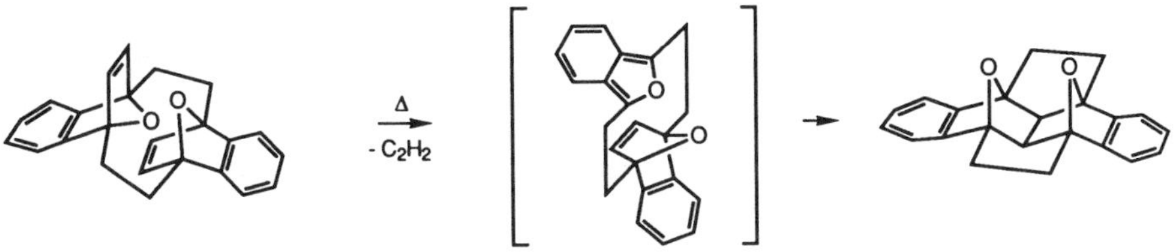

A retro-hetero-Diels–Alder fission was found to preempt a projected Claisen rearrangement [Burke, 1986]. The objective of this transformation was realized, however, as the fission was followed by an intramolecular Diels–Alder cycloaddition.

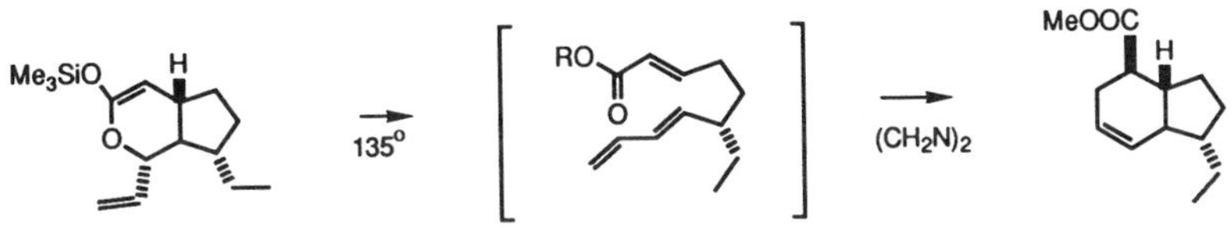

Cheletropic extrusion of carbon monoxide from adducts can also induce an intramolecular Diels–Alder reaction when such adducts carry an olefin linkage [Harano, 1988].

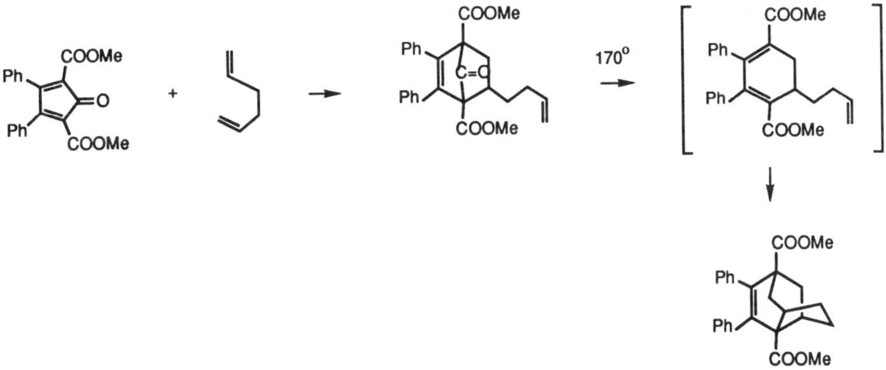

Many intermolecular versions of this reaction sequence are known. An application to the synthesis of the interesting hexaene [Lin, 1990] is indicated below.

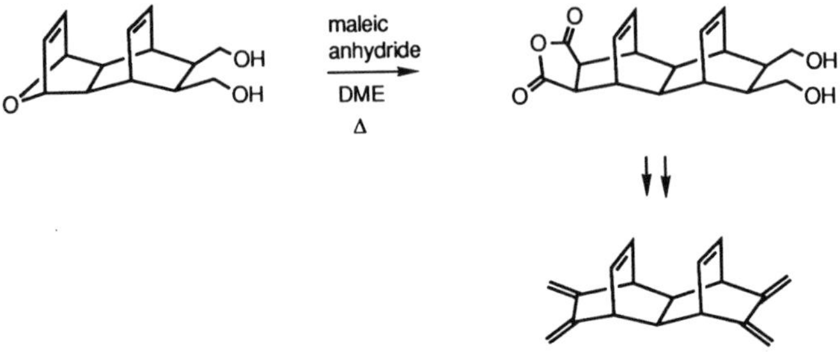

The Diels–Alder reaction of acyl ketenes with enol ethers affords 2-alkoxy-2,3-dihydro-*4H*-pyran-4-ones [Coleman, 1990]. A convenient method for generation of the ketenes is by pyrolytic elimination of acetone from certain dioxenones which may be considered as a retro-Diels–Alder reaction.

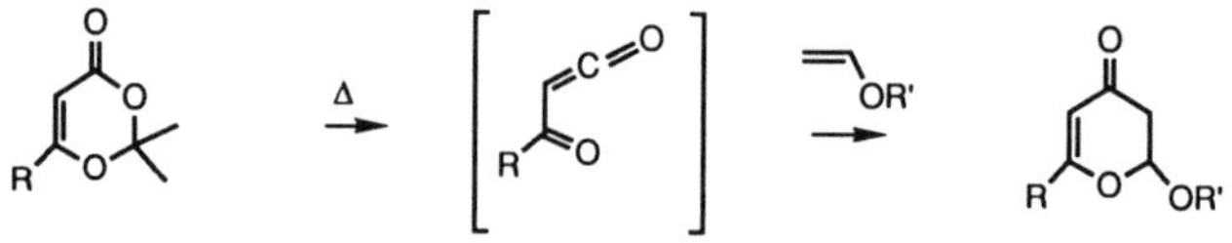

The net elimination of furan from an oxatetracyclic compound may have involved two retro-Diels–Alder reactions followed by recombination of two of the dissociated components [Cobb, 1977].

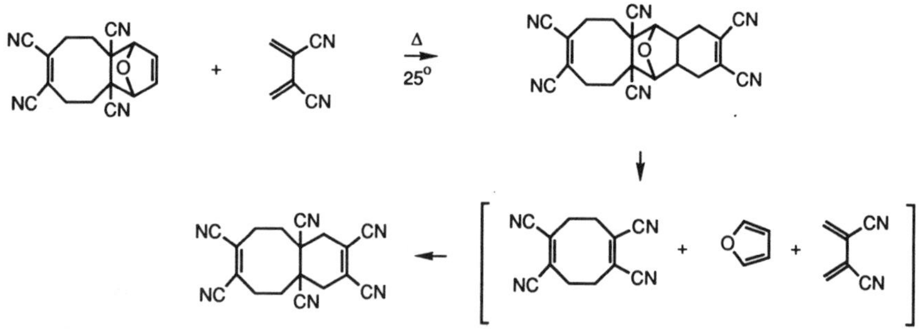

7.3. ELECTROCYCLIC/DIELS–ALDER REACTIONS

The major products of an aluminum chloride-catalyzed reaction of cyclododecene with methyl propynoate are two bicyclo[10.2.2]hexadecadienes [Snider, 1977]. They are the secondary Diels–Alder adducts from (*E,E*)-cyclotetradeca-1,3-diene which was formed by electrocyclic opening of the *trans*-fused [2 + 2]-cycloadduct.

13%
(ene adduct)

6%

28%

32%

A bridged ring system has been obtained by thermolysis of a linear tricycle [Oda, 1976]. The product may have been derived from electrocycloreversion, double bond isomerization, and intramolecular Diels–Alder reaction.

Electrocyclic opening of benzocyclobutenes leads to the very reactive *o*-quinodimethane derivatives. In the presence of a dienophile such an *o*-quinodimethane would be trapped and a Diels–Alder adduct is formed. This very powerful synthetic method has been investigated by many research groups and the results have been spectacular.

Inspired by the mass spectral fragmentation patterns of isoquinoline alkaloids which are dominated by the formal retro-Diels–Alder reaction pathway, synthetic plans based on the reverse, that is Diels–Alder cycloaddition, were formulated [Kametani, 1976a]. Necessarily surrogates for the *o*-quinodimethanes must be found.

Benzocyclobutenes fulfill this need because the angular strain of such compounds provides a strong driving force for ring opening. The conrotatory opening is allowed thermally. Indeed, heating a mixture of 6,7-dimethoxy-3,4-dihydroisoquinoline and 4,5-dimethoxybenzocyclobuten-1-ol gave rise to a tetracyclic product [Kametani, 1974b] which on sodium borohydride reduction resulted in xylopinine.

The 1-cyanobenzocyclobutene analog may also be used in the synthesis [Kametani, 1975b] but the intermediate bore an extra cyano group which was removed reductively.

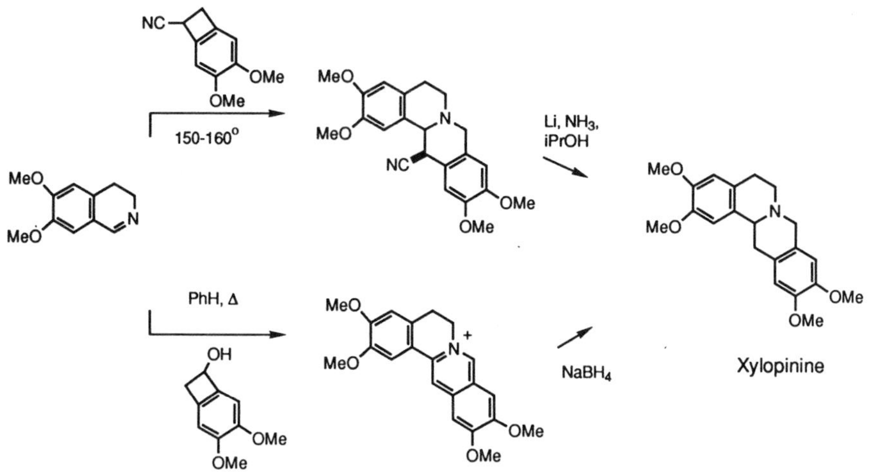

Thus, the *o*-quinodimethane intermediate may carry either an electron-withdrawing or electron-donating substituent.

By varying slightly the substitution pattern of the benzocyclobutene and the subsequent reduction conditions a route to emetine [Takano, 1978b] was developed.

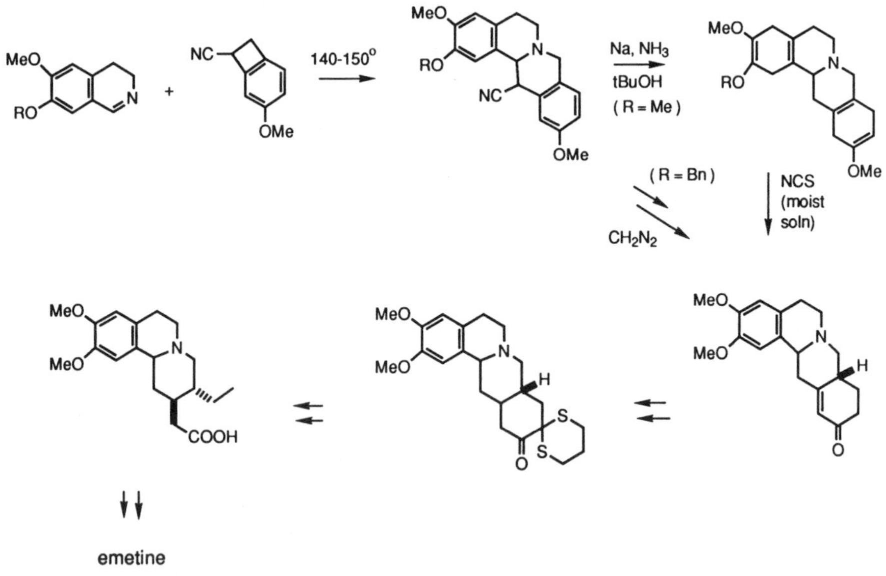

It should be noted that the dienophiles are not restricted to imines. A synthetic approach to cordiachrome-B [Watabe, 1987] relied on trapping of a quinodimethane with 3-methyl-2-cyclohexenone to create the tricyclic skeleton.

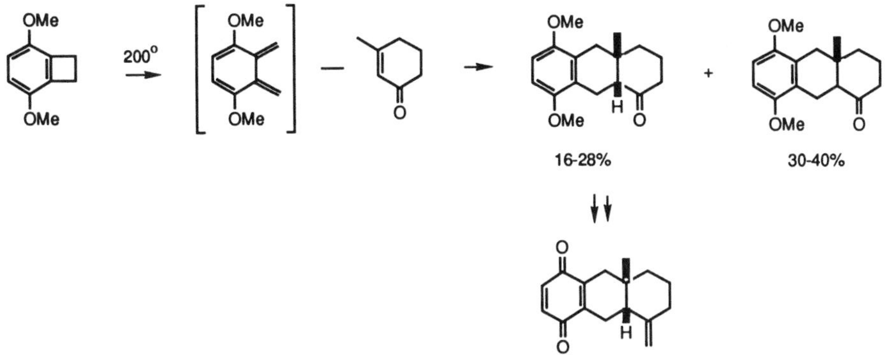

cordiachrome-B

This tandem process also constituted the key step for the construction of the carbon network of (+)-halenaquinone [Harada, 1988].

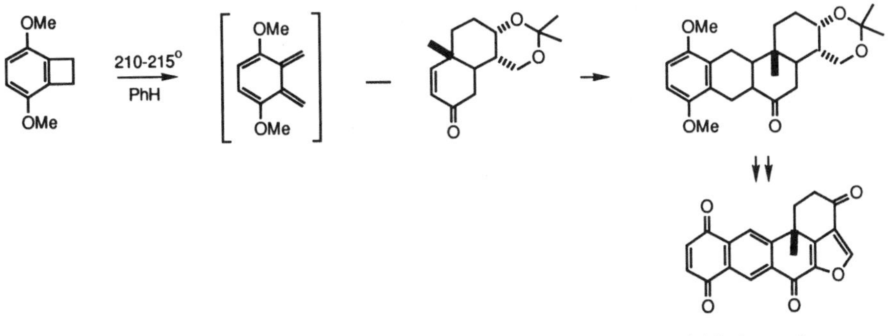

(+)-halenaquinone

The intramolecular interception of an o-quinodimethane has many intriguing ramifications. For example, this strategy allowed for the construction of a hydrophenanthrene intermediate of atisine [Kametani, 1976c] in a most concise manner. The cyano group in the four-membered ring was actually a required feature of the product. Its presence in the precursor also assisted attachment of the sidechain containing the dienophile.

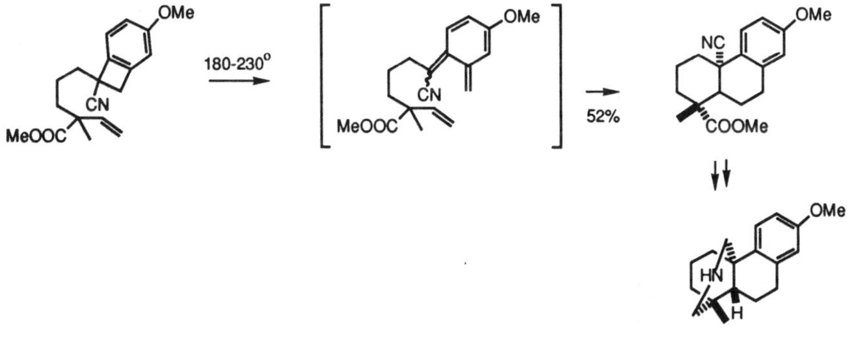

The angular cyano group of such adducts can serve different roles. For example, the corresponding primary alcohol of an analogous hydrophenanthrene underwent skeletal rearrangement upon tosylation to give an intermediate of pisiferin [Kametani, 1990].

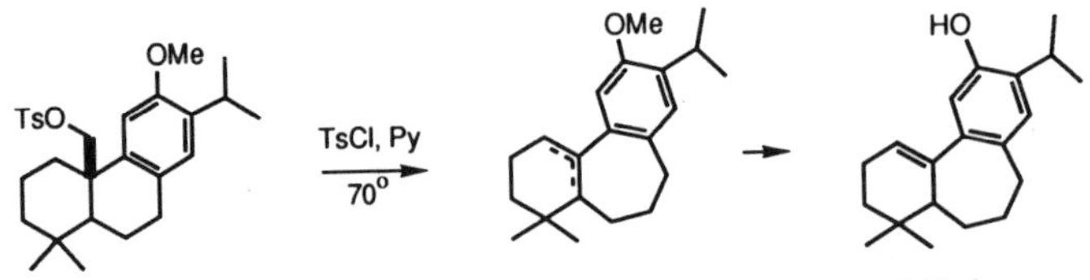

pisiferin

The effectiveness of a monosubstituted unactivated olefin as dienophile in these reactions has been thus demonstrated. With the exception of 1,1-dialkylbenzocyclobutenes which are not suitable substrates in view of the more favorable [1.5]-sigmatropic hydrogen migration process [Kametani, 1977b], numerous elegant synthesis have been executed on the basis of this strategy. In the steroid field earlier examples include approaches to estradiol [Kametani, 1978; Tsuji, 1981] and (+)-11-oxoestrone 3-methyl ether [Oppolzer, 1978].

It is remarkable that the (11-) oxime ether afforded exclusively a product with a *cis-anti-trans* ring skeleton [Oppolzer, 1977b].

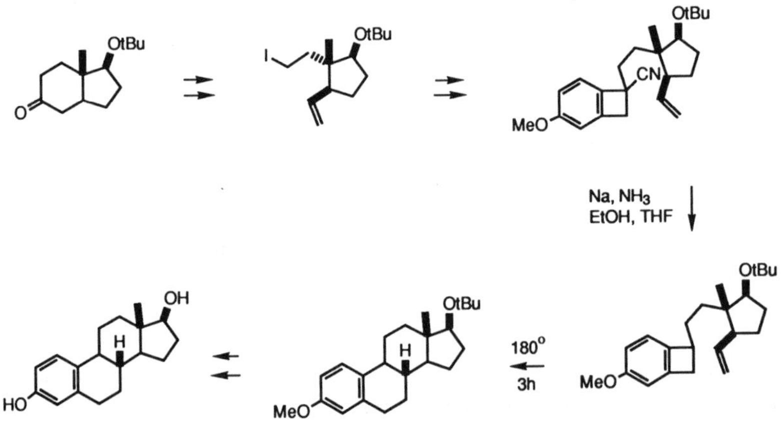

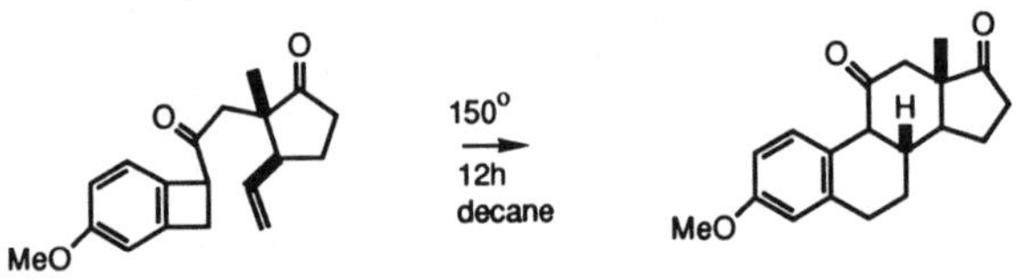

The bicyclo[2.2.1]heptenone building block which has been exploited extensively in the synthesis of many natural products has also found its way into estrone [Grieco, 1980].

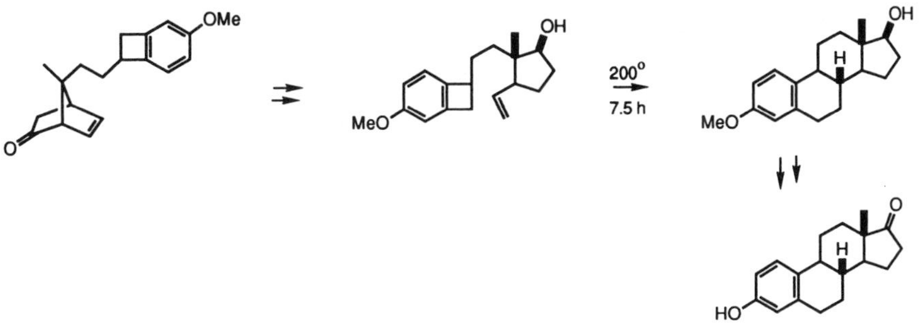

estrone

Perhaps the most impressive work in this area is that concerned with elaboration of the pentacyclic intermediates of alnusenone and friedelin [Kametani, 1977; 1978a].

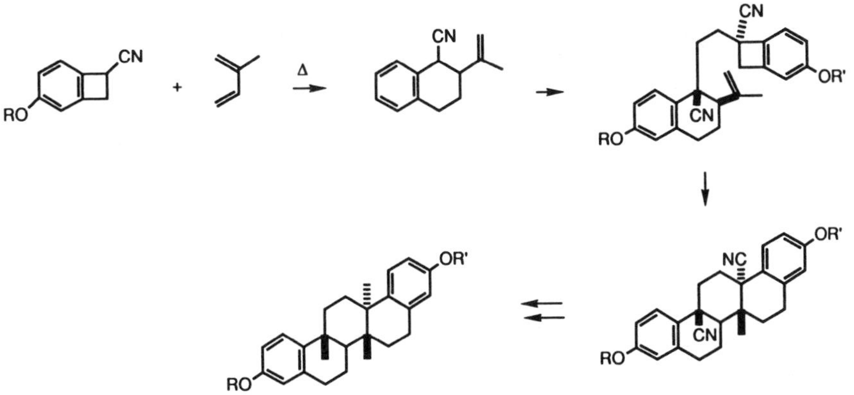

(+)-Chenodeoxycholic acid [Kametani, 1981b], hibaol [Kametani, 1979], a compound with the basic skeleton of the aphidicolanes [Kametani, 1980b], and a potential intermediate of klaineanone [Kametani, 1980a] are some of the other structures made available by following the general synthetic scheme.

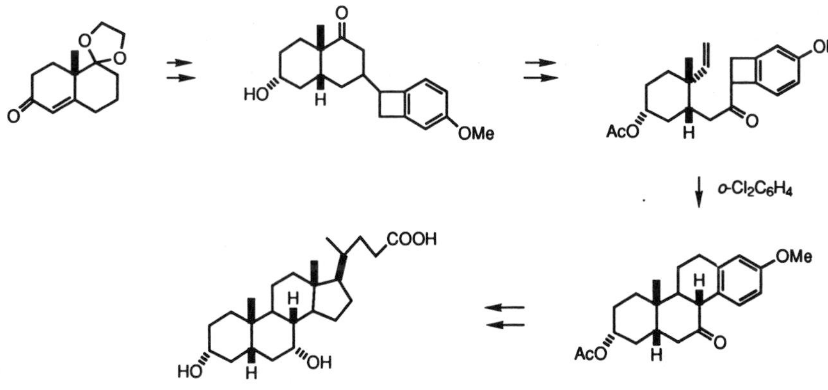

(+)-chenodeoxycholic acid

Amazingly, the whole or partial skeleton of steroids can be constructed in quite a variety of modes, as shown in an elaboration of intermediates containing the BCD-ring portion [Kametani, 1981a].

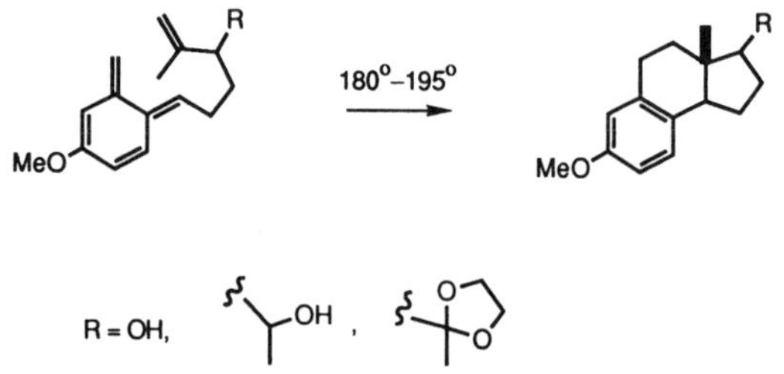

$$R = OH,$$

Needless to say, an alkyne is a suitable dienophile for intercepting *o*-quinodimethanes. Thus, in a synthesis of chelidonine [Oppolzer, 1971] a double bond was needed to introduce the secondary hydroxyl group, and it was created during the cycloaddition using an alkyne dienophile.

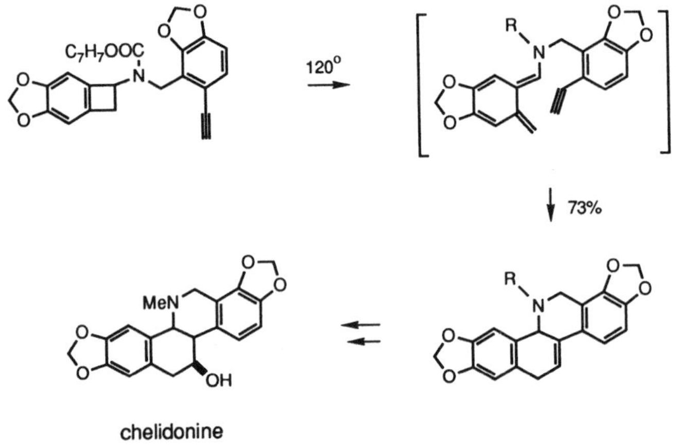

chelidonine

In connection with the chelidonine synthesis, the regiochemistry enforced by intramolecularity corresponds to an electronically disfavored transition state [Oppolzer, 1983].

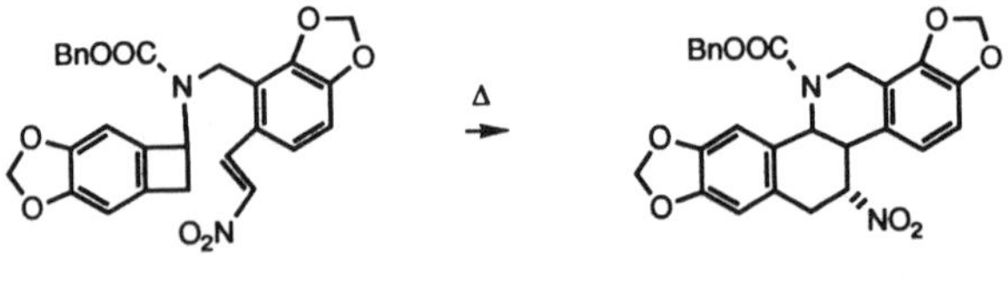

The conformational control of the transition state geometry in this type of intramolecular cycloaddition is illustrated by the following results [Oppolzer, 1974].

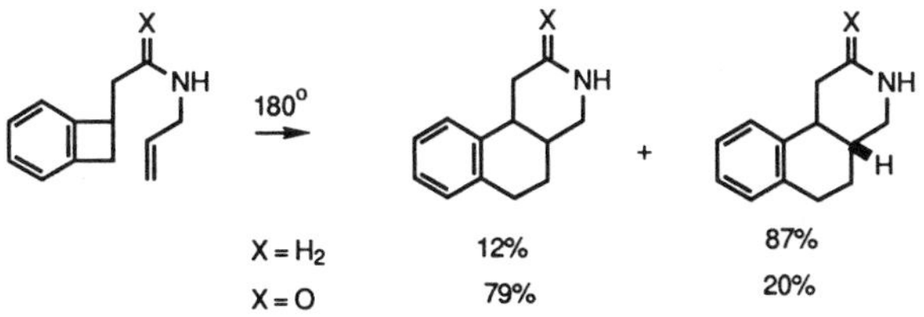

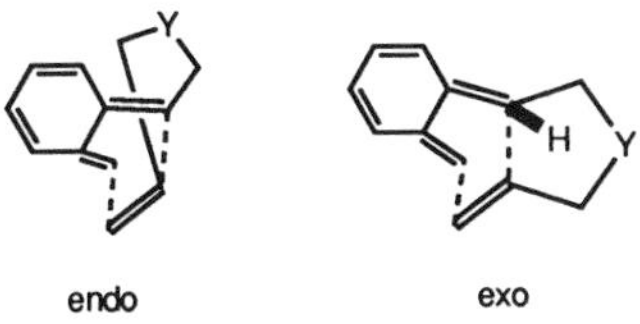

$X = H_2$ 12%   87%
$X = O$ 79%   20%

endo   exo

Podophyllotoxin is less stable than the epimer with a *cis*-fused lactone ring and this property has caused certain stereochemical problems in the synthesis. However, a neat solution is to enlist the electrocyclic opening/Diels–Alder cycloaddition tandem [Macdonald, 1986].

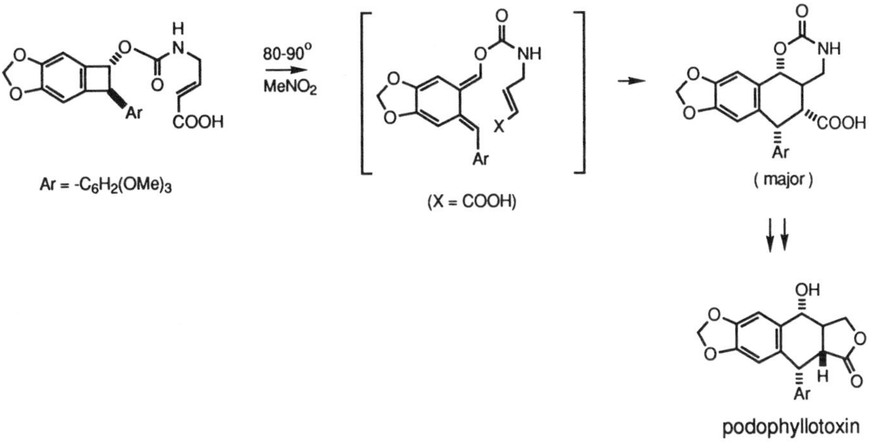

$Ar = -C_6H_2(OMe)_3$

$(X = COOH)$

(major)

podophyllotoxin

One of the most efficient pathways to multi-bridged cyclophanes involves pyrolysis of benzocyclobutenes [Boekhelheide, 1978]. The following equation shows an application of the method for the accomplishment of [2.2.2.2](1,2,4,5)-cyclophane.

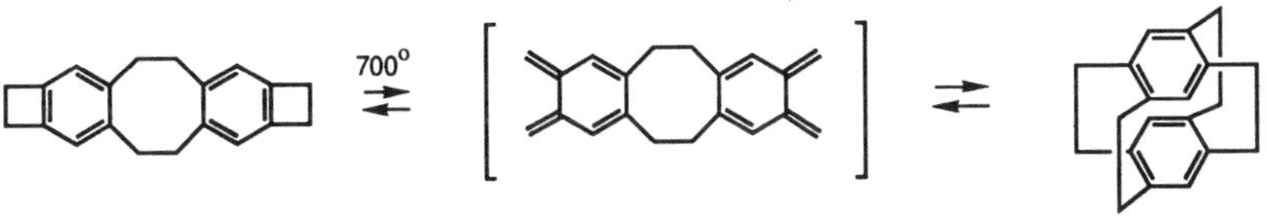

A novel procedure for the generation of *o*-quinodimethane is based on co-oligomerization of α,ω-dialkynes and monoalkynes in the presence of cyclopentadienylcobalt dicarbonyl, and it is readily adapted to the construction of A-aromatic steroids [Funk, 1977; 1979]. (For a route to B-aromatic steroids, see [Lecker, 1986].)

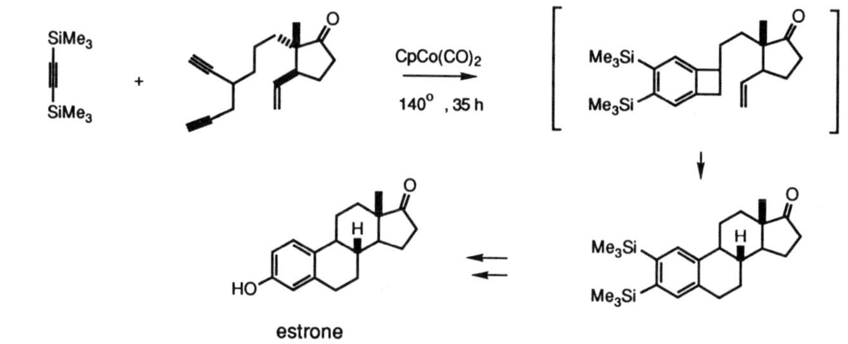

estrone

An anthracyclinone synthesis [Broadhurst, 1984] has enlisted *trans*-3,4-diacetoxybenzo-cyclobutene to provide the CD-ring. A 1-acetoxy-2-arylbenzo-cyclobutene has also been heated with a chiral butenolide to give an adduct which was eventually converted into epiisopodophyllotoxin [Choy, 1990].

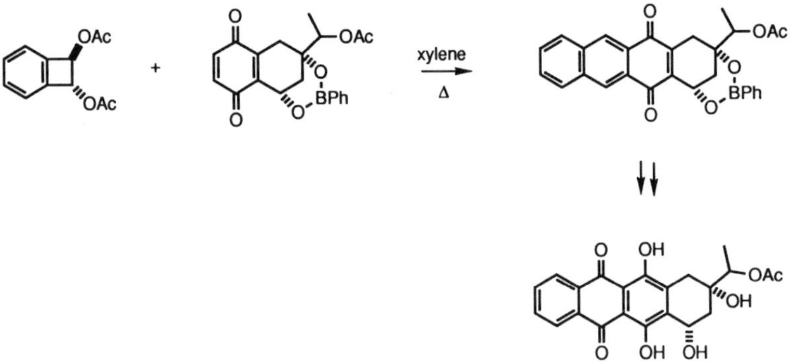

Benzocyclobutenediones are converted into bisketenes on uv irradiation. It has been shown that these ketenes react with benzoquinones, and on this basis a convenient route to islandicin and digitopurpone [Juung, 1977] has been developed. However, the yields of such reactions are very low.

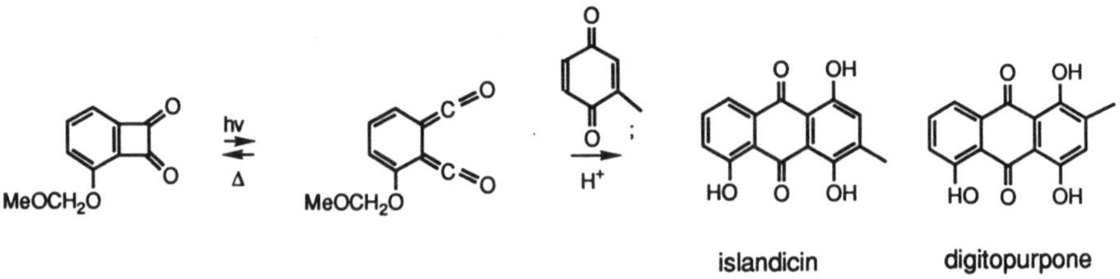

islandicin   digitopurpone

Cyclobutenes are synthetic equivalents of 1,3-butadienes. The limitation of their utility is due to mainly their accessibility. Interestingly, 2-trimethylsilyl- and 2-trimethylstannylmethylenecyclobutanes undergo themolysis to afford 2-trimethylsilylmethyl- and 2-trimethylstannylmethyl-1,3-butadienes, respectively [S. R. Wilson, 1979]. In the presence of dienophiles the latter compounds are captured.

The electrocyclic opening must be preceded by an allylic isomerization.

The equivalency of 1-diphenylphosphonylcyclobutene to 2-diphenyl-phosphonyl-1,3-butadiene in the Diels–Alder reaction has been demonstrated [Minami, 1986a].

A two-staged electrocyclic opening/Diels–Alder reaction was employed to build the bicyclic skeleton of coronafacic acid [Jung, 1981]. The ring cleavage occurred at ca. 100°C, but the intramolecular cycloaddition must be effected at a higher temperature. However, it is reasonable to assume that the two reactions can be conducted in tandem.

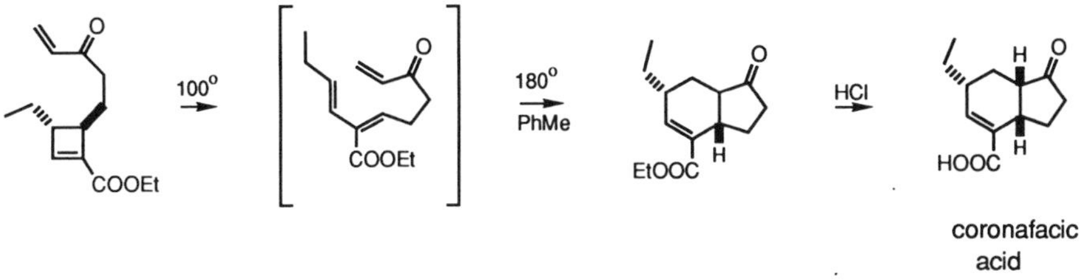

coronafacic
acid

Linear tetracyclic quinones have been acquired from a naphthoquinone and a cyclobutene which is fused to a six-membered ring [Boeckman, 1983]. Photocycloaddition was enlisted as method for the preparation of the ring fused cyclobutenes.

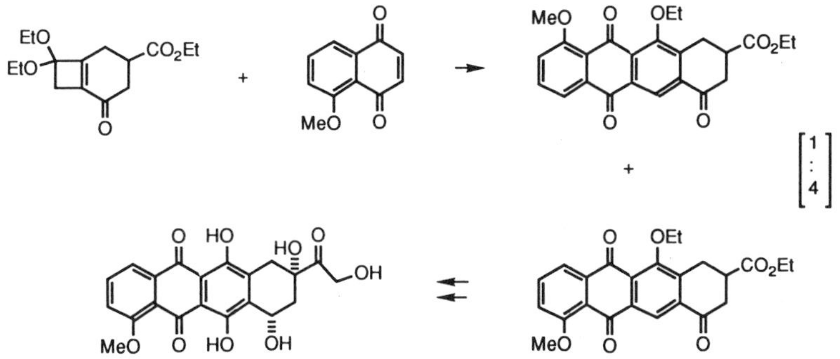

A dimeric hydrocarbon was produced on pyrolysis of the sodium salt of spiro[2.3]hexan-4-one tosylhydrazone [Wiberg, 1971]. The structure of this dimer clearly indicates the formation of bicyclo[2.2.0]hexene, its conversion into 1,2-dimethylenecyclobutane and a Diels–Alder reaction of the two species.

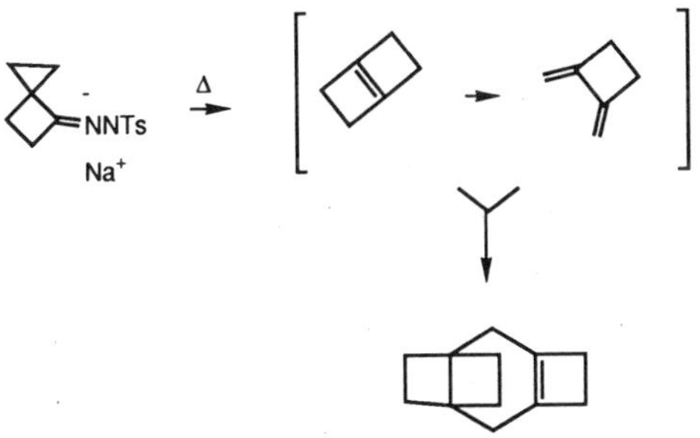

It should be emphasized that heteracyclobutenes are also subject to electrocycloreversion and intramolecular Diels–Alder reaction when opportunity presents itself, as illustrated in the pyrolytic formation of the indolizidinone system [Jung, 1991].

While many thermal reaction pathways are available to 2-allyloxy-2,4,6-cyclooctatrienone, it seems that the most favorable one is that involving a tandem electrocyclization–intramolecular Diels–Alder reaction [Y. Kitahara, 1976a].

Cyclooctatetraene is in a thermal equilibrium with the bicyclic triene (by a disrotatory ring closure). Although the concentration of the triene isomer is extremely small, it is in this form cyclooctatetraene undergoes Diels–Alder reaction with most of the commonly known dienophiles. The adduct with maleic anhydride is an excellent precursor of basketene [Masamune, 1966; Dauben, 1966].

basketene

In the same manner the tautomer of 1,6-methano[10]annulene may be trapped as the Diels–Alder adduct with dimethyl acetylenedicarboxylate [Vogel, 1965]. The latter compound gave benzocyclopropene on pyrolysis at 400°C.

Benzocyclopropene

The alkylation of cyclononatetraenide anion with tetrachlorocyclopropene was expected to afford a symmetrical molecule after an intramolecular

Diels–Alder reaction of the product. However, the Diels–Alder reaction occurred only after the rapid electrocyclization of a triene subunit [Mock, 1972].

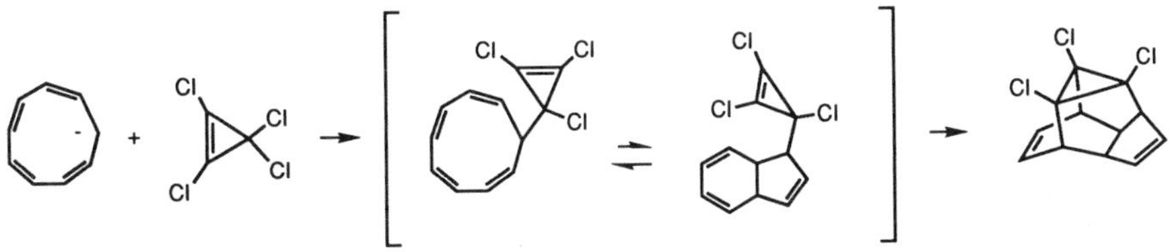

Opening of fused cyclobutenes may not be concerted due to the generation of (*E,Z*)-dienes. It is perhaps the case in which a capped [3]peristylane was formed by pyrolysis of two of the meta photocycloadducts of benzene and bicyclo[3.2.0]heptadiene [Srinivasan, 1972]. The conjugate diene intermediate participated in an intramolecular Diels–Alder reaction.

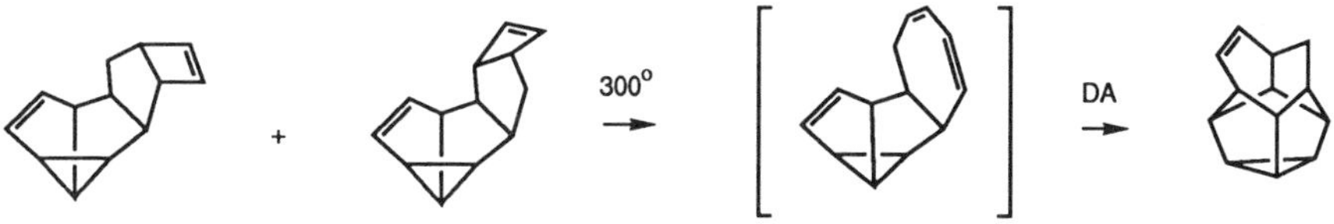

A unique way of generating the dienophilic benzynes via cyclization is by thermal decomposition of 3-azido-4-(trimethylsilyl)ethynyl-*o*-quinones [Chow, 1990]. Elimination of dinitrogen and carbon dioxide, and cyclization of the resulting enynyl ketenes can result in the phenol *O,C*-diyls only. However, the close proximity of the silyl group to the oxy radical enables the C to O transfer of the silyl residue.

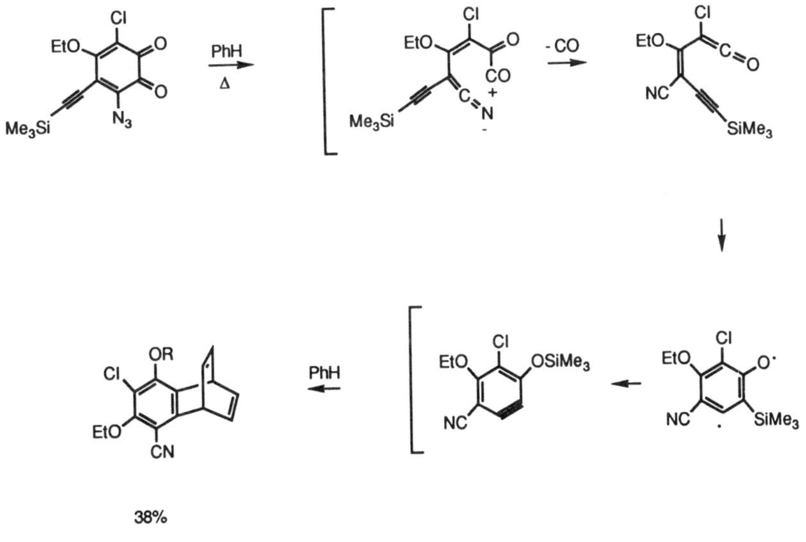

38%

Undoubtedly the most spectacular pericyclic reaction sequences employed by nature is that which generates the endiandric acids. Such reaction sequences have been duplicated in the laboratory. For example, Lindlar hydrogenation of a tetraenediyne followed by brief heating at 100° in toluene led to the isolation of endiandric acid-A methyl ester [Nicolaou, 1982a]. Four new rings and eight asymmetric centers evolved from an achiral open-chain precursor.

In the formation of endiandric acid-A derivative, two electrocyclizations were followed by an intramolecular Diels–Alder reaction in which the proximate olefin of the cyclohexadiene moiety combined with the sidechain diene.

endiandric acid-D   endiandric acid-E   endiandric acid-A

A pentaenediyne analog was similarly processed. The product mixture contained methyl esters of endiandric acid-B and a lesser amount of endiandric acid-C, in addition to the bicyclic precursors.

Endiandric acid-B is a vinylog of the acid-A, whereas endiandric acid-C has a different tetracyclic skeleton. The ultimate step of its formation is also an intramolecular Diels–Alder reaction, but the dienophile was the unsaturated ester.

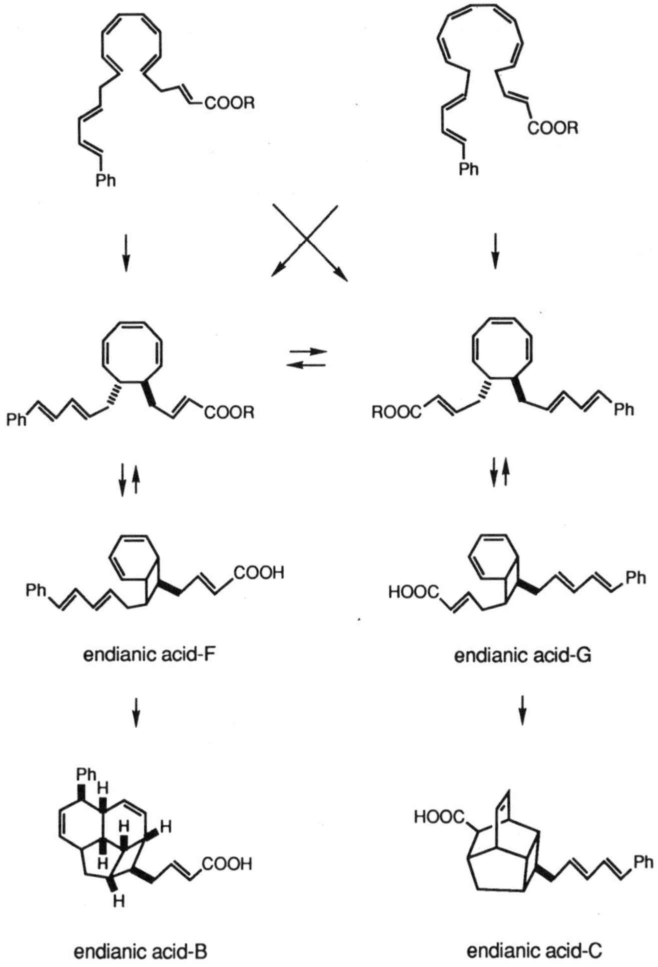

endianic acid-F

endianic acid-G

endianic acid-B

endianic acid-C

7.4. DIELS–ALDER REACTIONS PRECEDED BY OTHER PERICYCLIC REACTIONS

Two types of tandem Diels–Alder reactions may be conceived. In the one shown by the following equation which deals with a cross-conjugated enynone and a tetraene [Kraus, 1980] there is a spatial fixation in the first step.

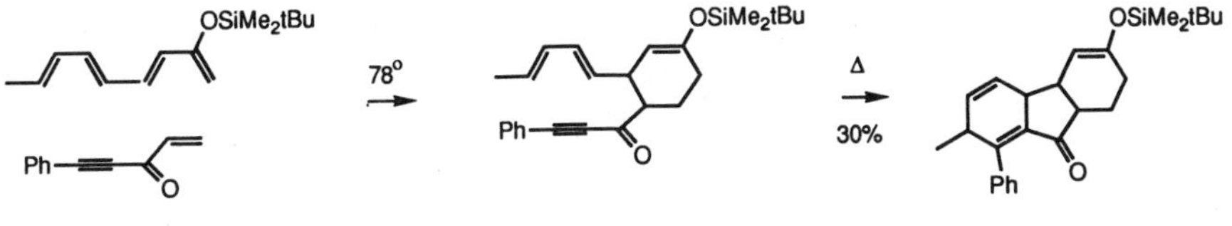

Actually cross-conjugated trienes have been shown to undergo the tandem reaction [Tsuge, 1983].

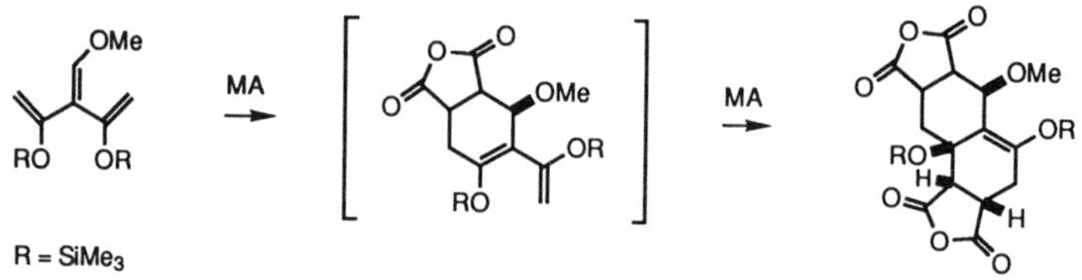

R = SiMe₃

On the other hand, domino Diels–Alder reactions involve the same dienophile [Paquette, 1978] or a nascent dienophile [Fessner, 1983]. It is amazing that the two domino Diels–Alder reactions shown afforded spectacular entries into dodecahedrane and pagodane, respectively.

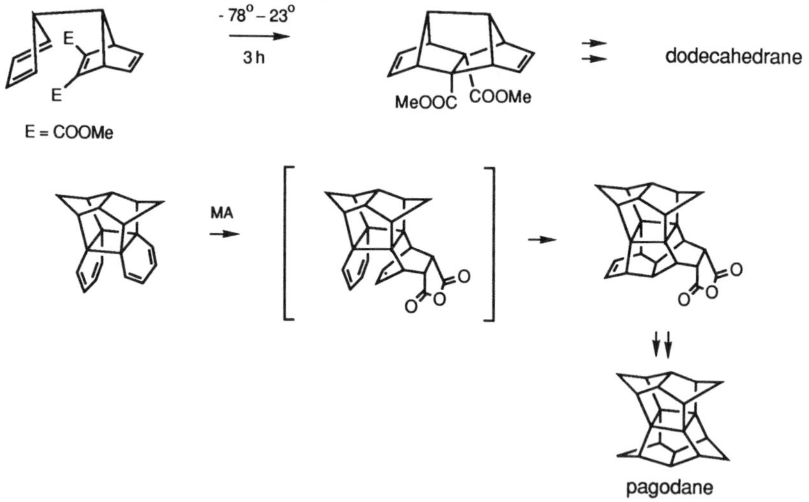

E = COOMe

pagodane

The domino Diels–Alder reaction of *N,N'*-dipyrrolylmethane with activated acetylenes [Visnick, 1985] clearly showed that the product can undergo rearrangement via the retro-Diels–Alder/Diels–Alder reaction sequence. Understandably, the more electron-deficient olefin of the bicycloheptadiene system in the initial adduct would participate in the formation of the kinetic product. Under equilibrating conditions, cycloreversion and the alternative intramolecular Diels–Alder reaction furnished the isomer in which the double bond in conjugation with the electron-deficient substituents was preserved.

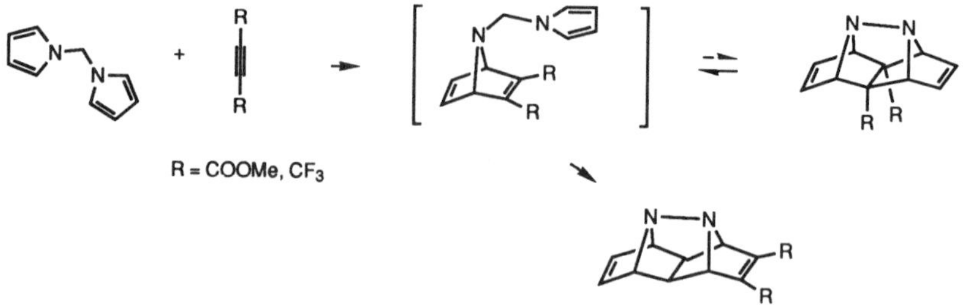

R = COOMe, CF₃

Two other intriguing domino Diels–Alder reactions are the following [Wasserman, 1966; Wynberg, 1971].

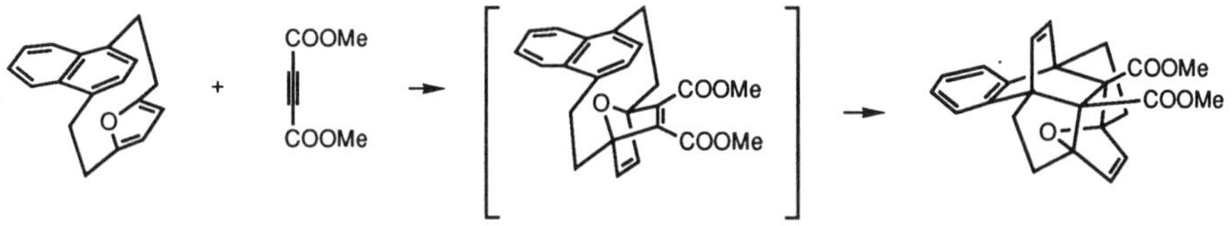

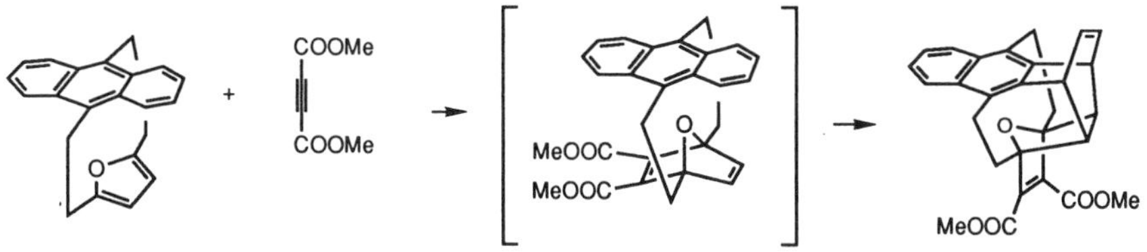

A most efficient synthesis of a molecular belt [Kohnke, 1989] consisted of subjecting a bisdiene and a bisdienophile to high pressure (10 kbars) in dichloromethane at 60°C. One cycloaddition process must have preceded the closure of the belt, considering that the establishment of such a molecule from two fragments is attended by a tremendous loss of entropy.

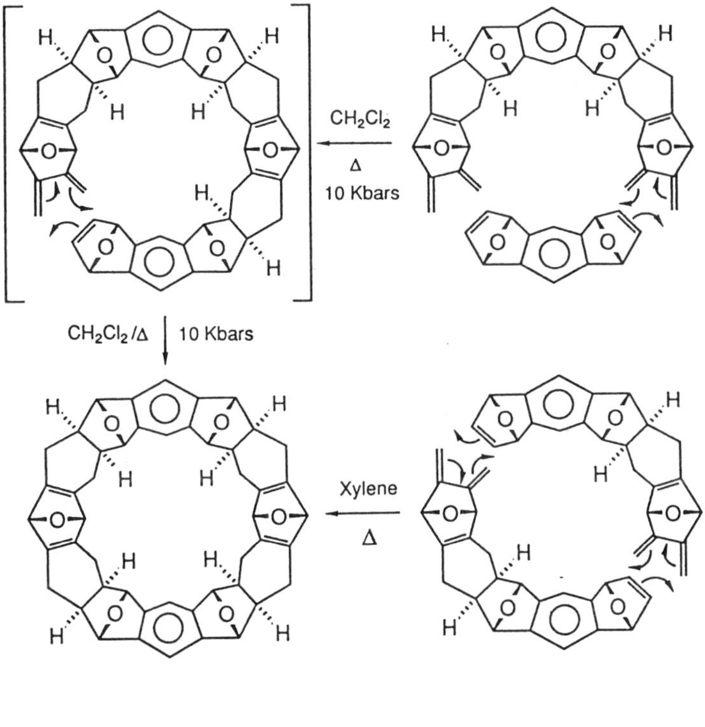

Dimerization of a propellatetraene imide [Sciacovelli, 1970] involved a Diels–Alder cycloaddition from the *syn* face of the imide molecule that acted as the diene, and the *anti* face with respect to the dienophilic molecule. This stereochemistry of the adduct is conducive to another, intramolecular, Diels–Alder reaction.

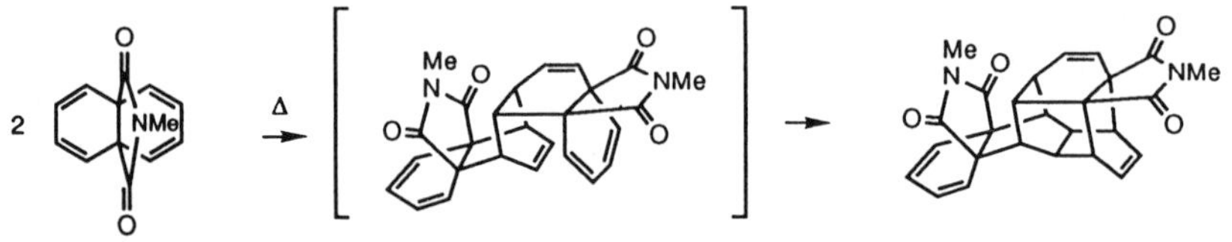

It is reasonable to assume that practically all the 2:1 adducts of 1,2,4-triazoline-3,5-diones and propellatetraenes containing two homoannular diene subunits resulted from tandem Diels–Alder reactions. Interestingly, through-space interactions between the nonreacting bridge of the propellane and the dienophile, when they exist, govern the stereochemical course of the cycloadditions [Ginsburg, 1974, 1983; Gleiter, 1979].

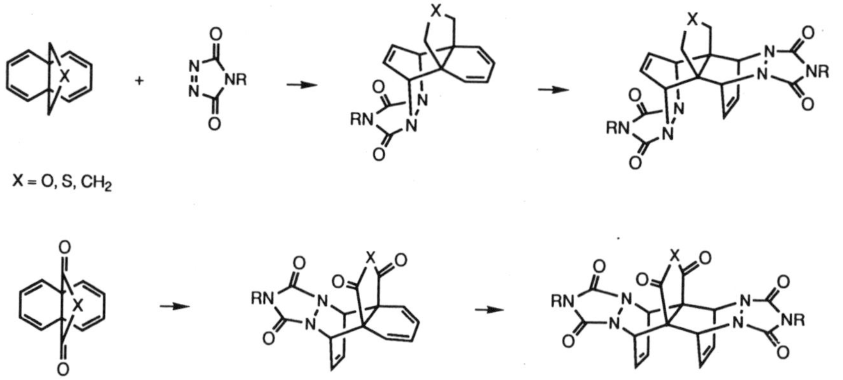

X = O, S, CH$_2$

The 1:2 adduct of benzene with maleic anhydride upon irradiation is now known to be derived from a [2 + 2]photocycloaddition followed by a thermally induced Diels–Alder reaction [Hartmann, 1974].

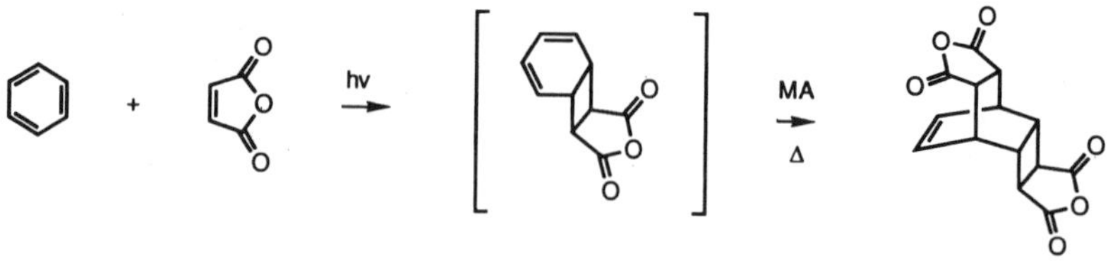

In sterically favorable situations a benzene ring can behave as dienophile. For example, the major 1:1 cycloadduct of 1,3-diphenylisobenzofuran and cyclooctatetraene is a cage compound, formed by a tandeom twofold Diels–Alder reaction [Saito, 1984].

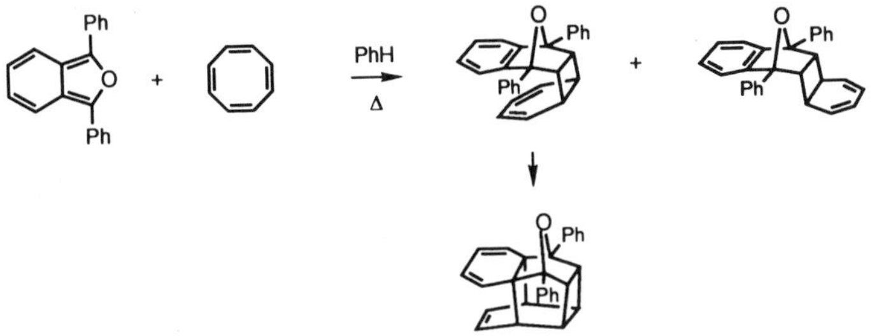

On heating at 270°C, o-(1E,5-hexadienyl)phenol was transformed into a chromane [Jug, 1972]. An o-quinonemethide which is connected to the phenol by [1.5]-sigmatropy is the reactive species which underwent the Diels–Alder reaction.

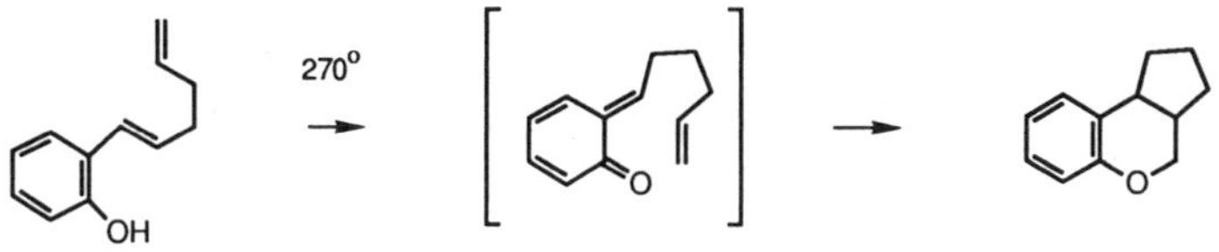

Photoenolization of o-alkylphenones is a sigmatropic rearrangement. A very interesting exploitation of this reaction is that which led to an estrone derivative [Quinkert, 1982b].

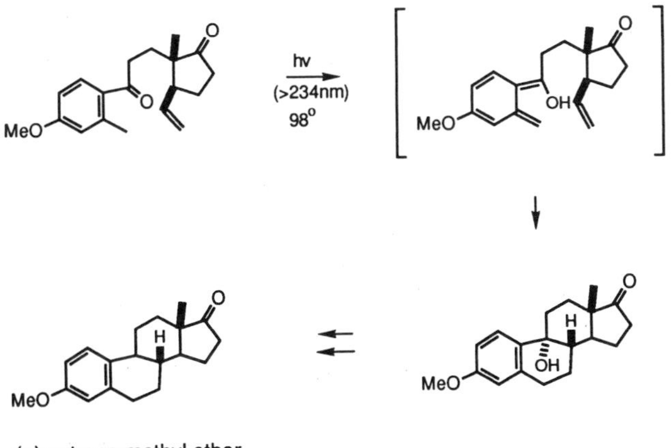

(+)-estrone methyl ether

Natural lignans such as dehydropodophyllotoxin, taiwanin-E, taiwanin-C, and justicidin-E have been obtained via Diels–Alder reaction of the proper photoenols [Arnold, 1973].

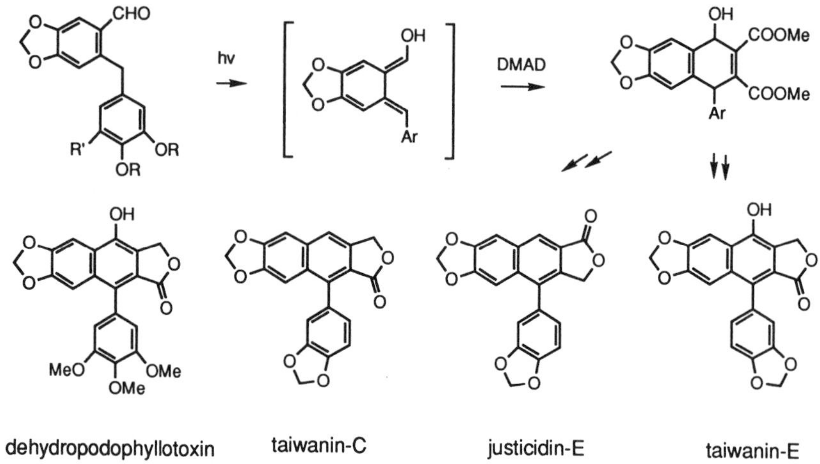

dehydropodophyllotoxin   taiwanin-C   justicidin-E   taiwanin-E

An interesting access to the rhoeadine-type alkaloids is by intramolecular trapping of the photoenols by an aldehyde [Prabhakar, 1977].

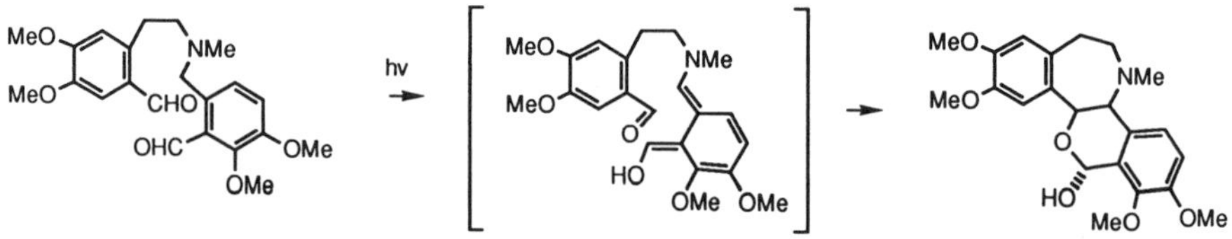

A less obvious application of the photoenol trapping is the elaboration of an acromelic acid analog [Hashimoto, 1990] from such a cycloadduct. The acetic acid sidechain derives from part of the diene and part of the dienophile.

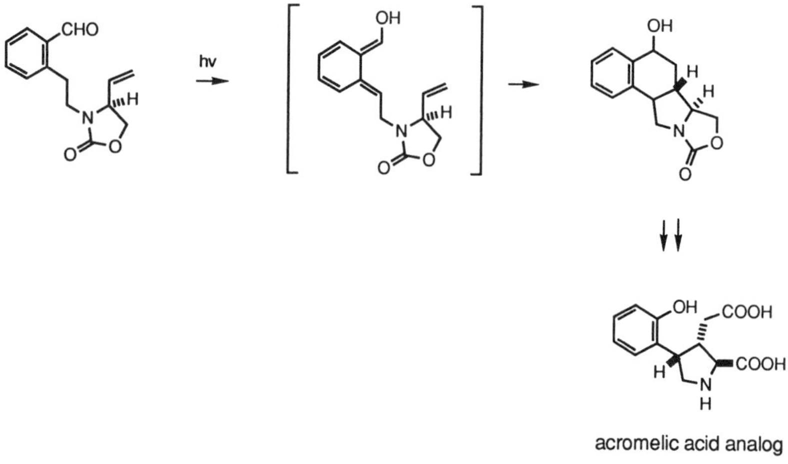

acromelic acid analog

Many dienophiles are capable of trapping the isoindene isomer to form benzobicyclo[2.2.1]heptenes. A significant use of this sigmatropic rearrangement/ Diels–Alder reaction is the formation of an adduct that was elaborated into the very complex diterpene alkaloid chasmanine [Tsai, 1977].

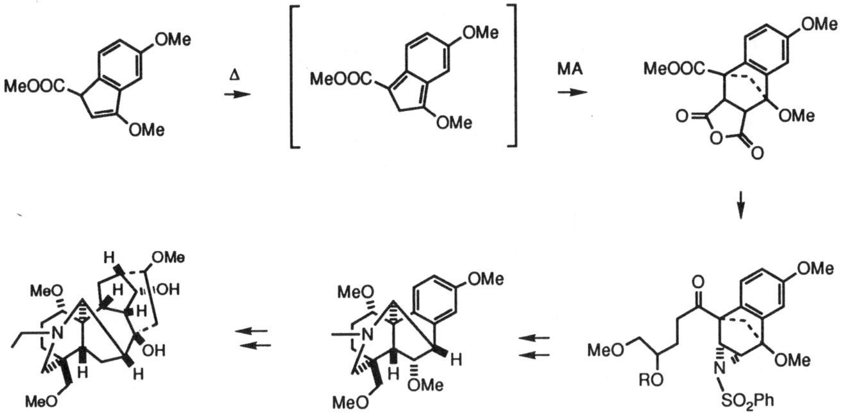

chasmanine

Isodicyclopentadiene may exist in two forms. The less stable isomer which is accessible in low concentration by [1.5]-hydrogen shift is more reactive and may be intercepted by less reactive dienophiles [Paquette, 1983b].

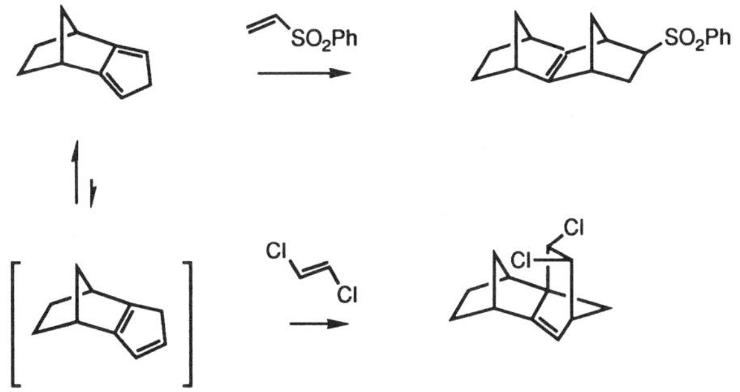

Spiroannulated isodicyclopentadienes are not very reactive toward tropone and their cycloadditions take place only after extensive carbon and hydrogen [1.5]-sigmatropy [Paquette, 1984].

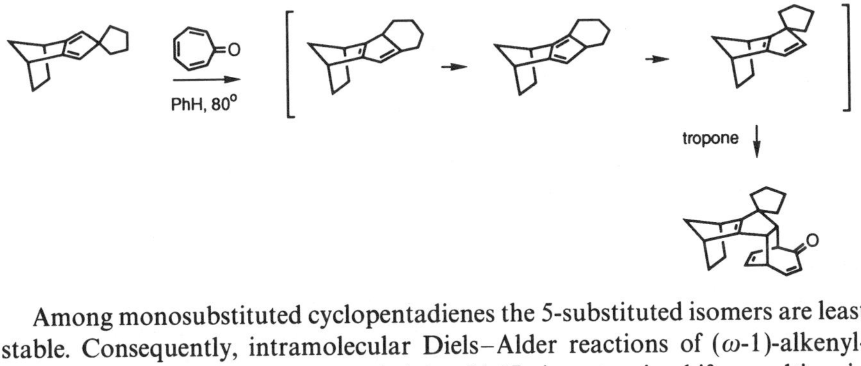

Among monosubstituted cyclopentadienes the 5-substituted isomers are least stable. Consequently, intramolecular Diels–Alder reactions of (ω-1)-alkenyl-cyclopentadienes often are preceded by [1.5]-sigmatropic shift, resulting in laterally fused norbornenes [Brieger, 1963].

Three thermally allowed reactions took place in tandem in the crucial process of skeletal construction of a precursor of 8,14-cedranediol [Landry, 1983]. These are retro-Diels–Alder decomposition, 1,5-sigmatropic shift (hydrogen), and intramolecular Diels–Alder reaction.

8,14-cedranediol

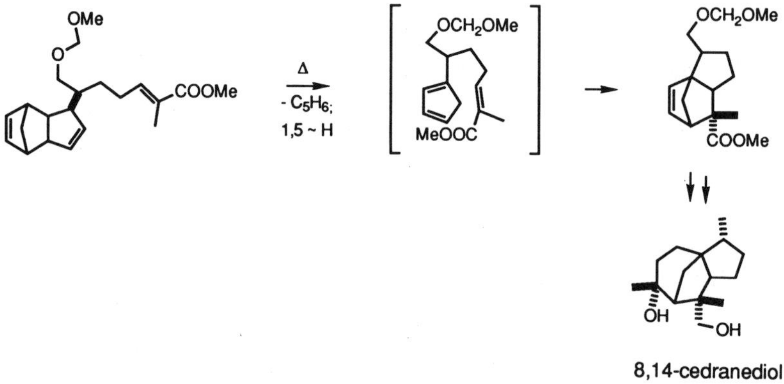

A furanoid diterpene derivative was found to undergo tandem [1.5]-hydrogen shift and intramolecular Diels–Alder reaction [Ghisalberti, 1974] in which one of the furan double bonds acted as the dienophile.

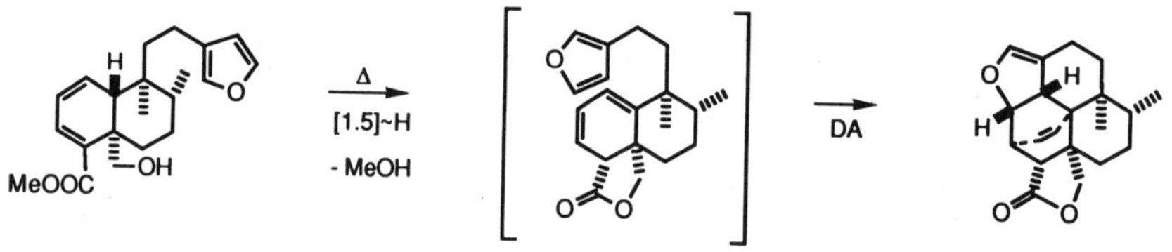

The formation of a hexacyclic hydrocarbon from the reaction of lithium cyclopentadienide with 1,8-dioodonaphthalene [Gronbeck, 1989] is due to cofacial alignment of the two cyclopentadiene rings in the primary product which is very favorable to mutual reaction. Thus, an intramolecular Diels–Alder condensation is expected after a [1.5]-hydrogen migration in one of the cyclopentadiene rings.

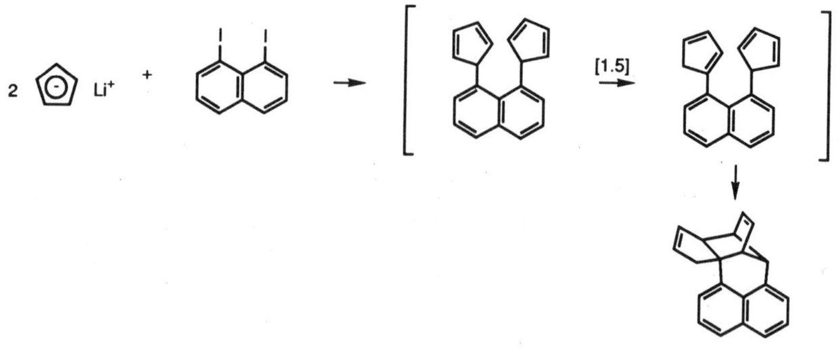

An example of intermolecular Diels–Alder reaction occurring immediately after [1.5]-sigmatropy is that between 4-hydroxy-3-vinylquinolone and 3-ethylidene-1,2,3,4-tetrahydroquinoline-2,4-dione [Kumaraswami, 1988]. This reaction appears to belong to the type of inverse electron demand.

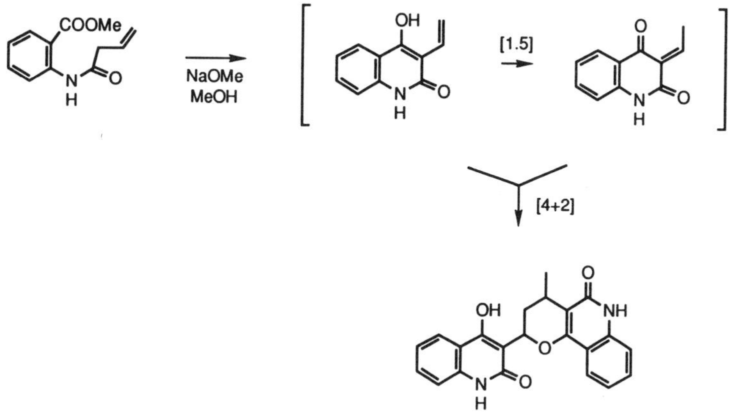

A [1.7]-hydrogen shift followed by a Cope rearrangement of 7-(3-butenyl)cycloheptatriene rendered the molecule reactive toward intramolecular Diels–Alder reactions. Two tricyclic dienes were formed by the process [Cupas, 1971].

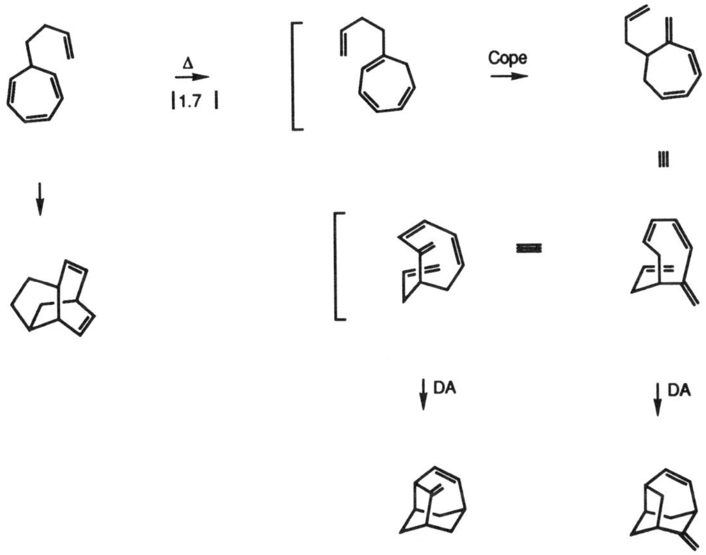

A 1:2 adduct of 2-isopropenyl-6-methoxybenzofuran and methyl propynoate was shown to be pentacyclic [De Capite, 1990]. Its formation via tandem Diels–Alder/ene/Diels–Alder reactions can be envisaged.

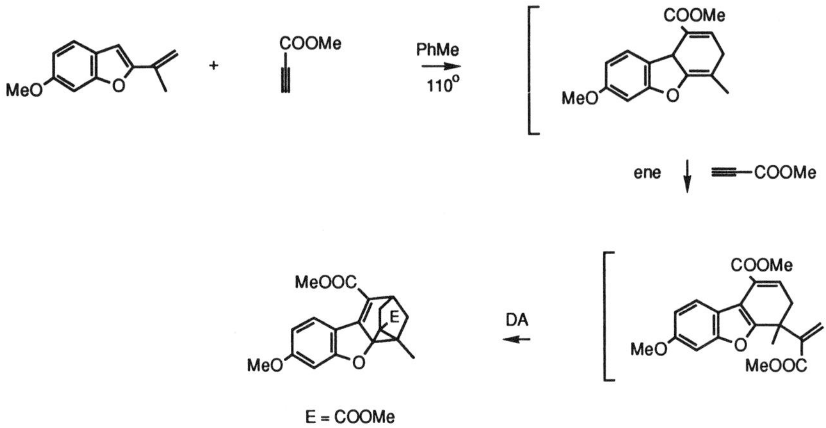

E = COOMe

A classical example of the two-step reaction is the formation of a 1:2 adduct of 1,4-cyclohexadiene and maleic anhydride [Alder, 1949].

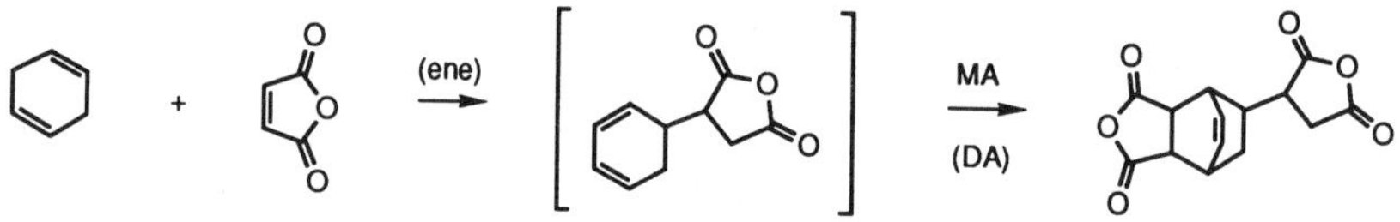

In recent times reactions by microwave irradiation have been scrutinized. Thus, the tandem ene/Diels–Alder reactions of 1,4-cyclohexadiene and methyl propynoate has been effected by such means [Giguere, 1987].

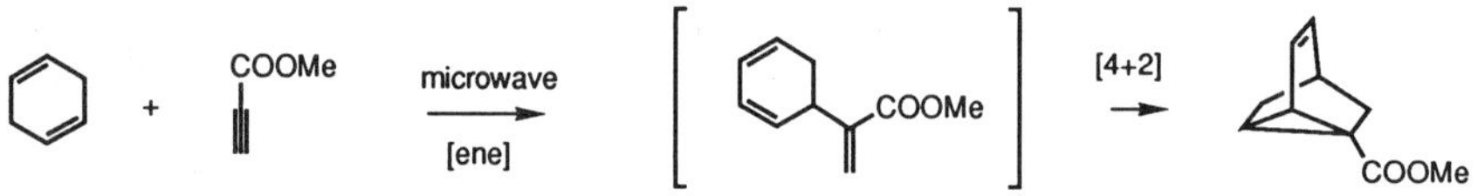

An interesting transformation resulted from pyrolysis of *N*-methyl-5-(1,3-butadien-2-yl)-pyrrolidone may be rationalized as involving a retro-ene reaction followed by an intramolecular Diels–Alder cycloaddition [Earl, 1982].

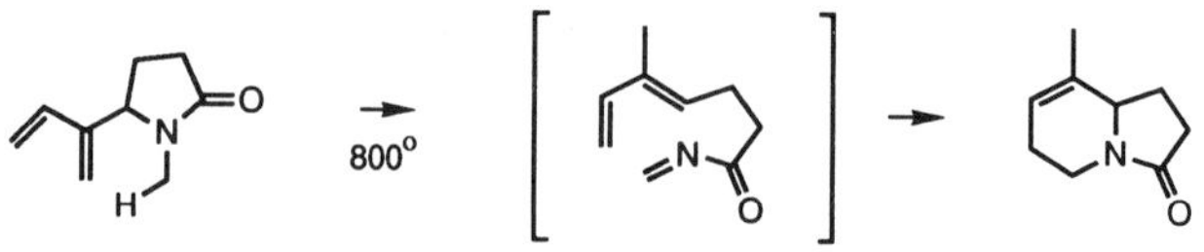

The formation of a *cis*-2,5-disubstituted Δ^3-dihydropyran from an ethylaluminum dichloride-catalyzed reaction of a 1,4-diene and formaldehyde is significant in that it provided a useful intermediate of pseudomonic acid [Snider, 1983c]. Apparently a Prins reaction took place, and the resulting alkoxyaluminum species delivered another molecule of formaldehyde to the conjugate diene in an activated form, through its complexation.

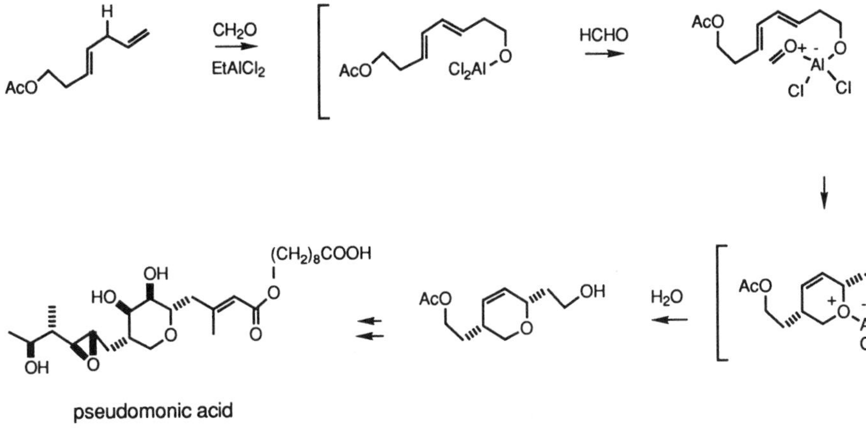

pseudomonic acid

On heating 9-(cyclopropen-3-yl)bicyclo[6.1.0]nona-2,4,6-triene furnished a hydrocarbon with a very different skeleton [Farnum, 1982]. The reaction course most likely involved a vinylcyclopropane rearrangement and a Diels–Alder reaction.

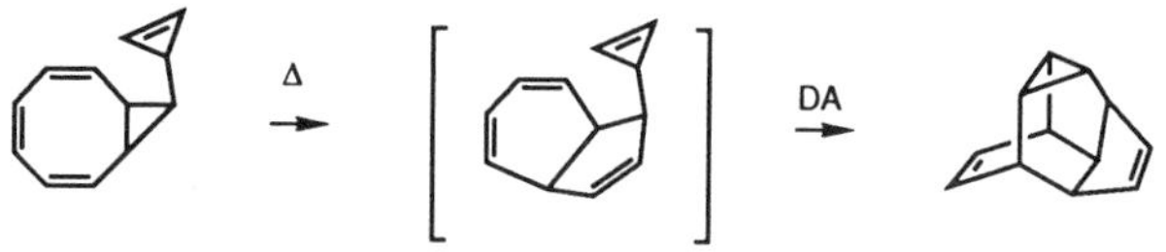

The cycloaddition of a spiro[2.4]hepta-4,6-diene with 1,3-diphenylisobenzo-furan was found to be preceded by a [1.5]-sigmatropic rearrangement of the spirocycle into a bicyclo[3.2.0]- heptadiene isomer [R.D. Miller, 1977].

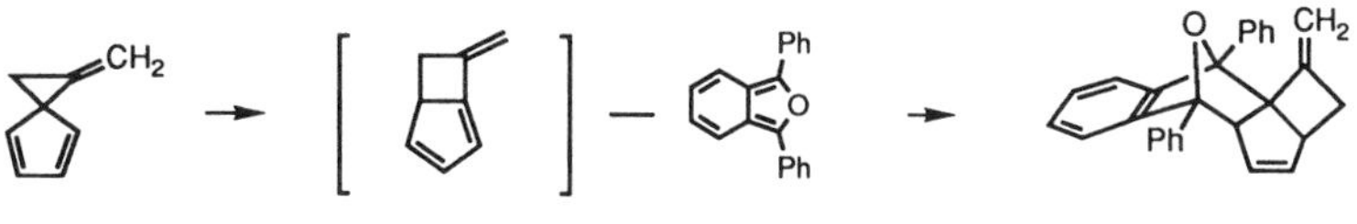

In the course of studying chorismate mutase inhibition a synthesis of transition state analogs was undertaken [Clarke, 1990]. Tricyclononenes have been secured through a tandem Cope rearrangement/Diels–Alder reaction.

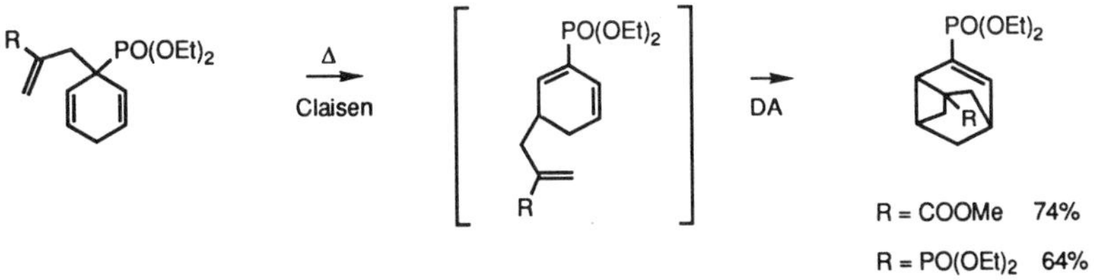

R = COOMe   74%

R = PO(OEt)₂   64%

Formation of molecules containing small rings is not unusual. For example, a Claisen rearrangement of an *O*-propargyltropolone was found to be followed by an intramolecular Diels–Alder reaction [Harrison, 1973].

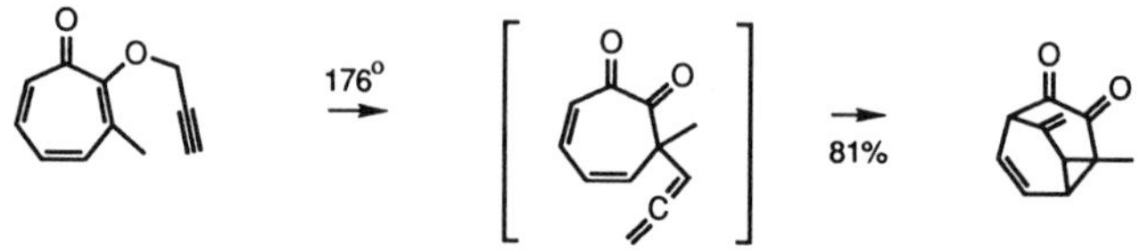

The intricate ring system of the morellins probably arise from a combination of aromatic Claisen rearrangement and intramolecular Diels–Alder reaction. This speculation has received strong support by an in vitro demonstration of the tandem reaction on a model xanthone [Quillinan, 1971].

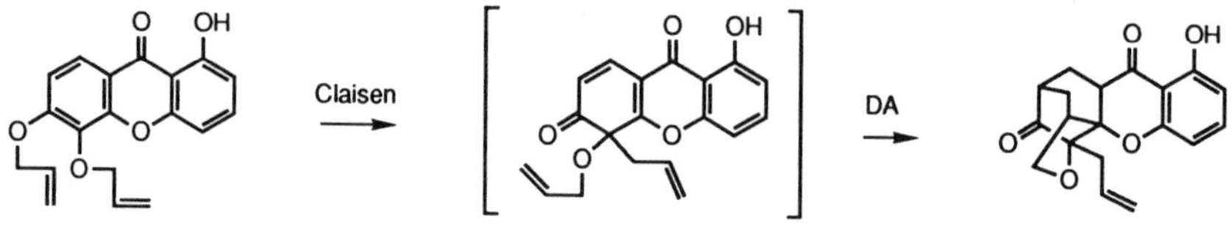

A linear version of the Claisen rearrangement-Diels–Alder reaction represents a valuable method for establishing the BCD-ring segment of the anthracyclines from a monocyclic phenolic derivative [Kraus, 1984; 1985].

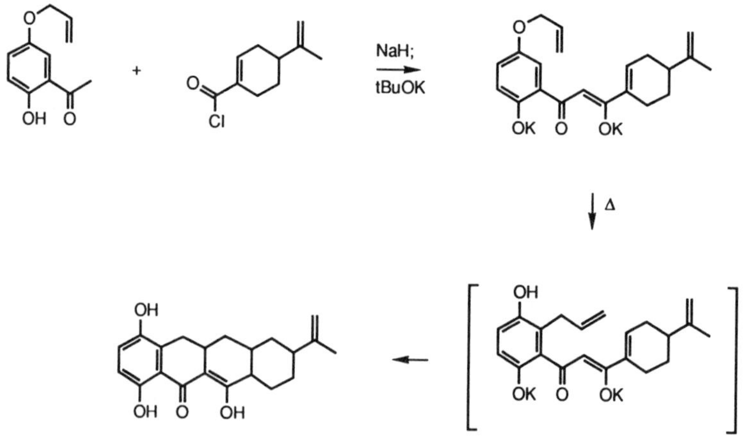

One of the products arising from thermolysis of a ketenimine proved to be a pentacyclic isoquinoline derivative [Molina, 1990]. The genesis of this compound is thought to proceed via electrocyclization, Claisen rearrangement, intramolecular Diels–Alder reaction, and aromatization of two rings via alkyl shift. The first two steps are independent of one another, but the Diels–Alder reaction must await the completion of both of them.

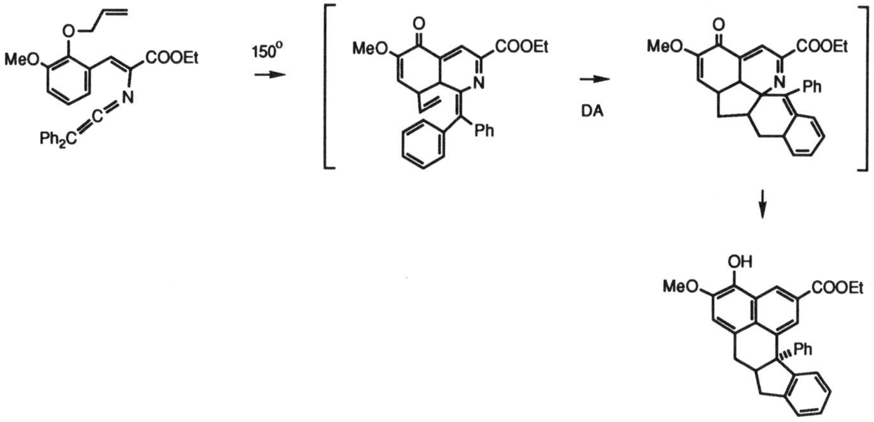

Each of the above processes depicts the generation of a dienophile by an aromatic Claisen rearrangement. It has also been demonstrated that the aliphatic Claisen rearrangement/intramolecular Diels–Alder reaction can be induced with the diene as the nascent moiety [Mulzer, 1991].

One of the two possibilities for the Claisen rearrangement, migration of the chain terminus is highly favored if the internal double bond is tri- or tetrasubstituted; otherwise there is nearly equal chance for both rearrangement pathways. Of course different produces arise from the overall transformation.

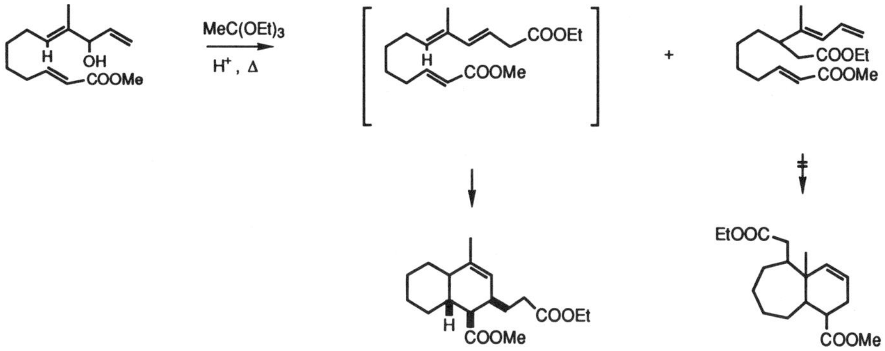

A tricyclic phosphine oxide was produced on prolonged heating of 1-allyloxy-1-methyl-2,4,6-triphenyl-λ^5-phosphorin [Schaffer, 1975]. It should be noted that the Claisen rearrangement-type intermediates did not arise from a concerted process (esr signals detected).

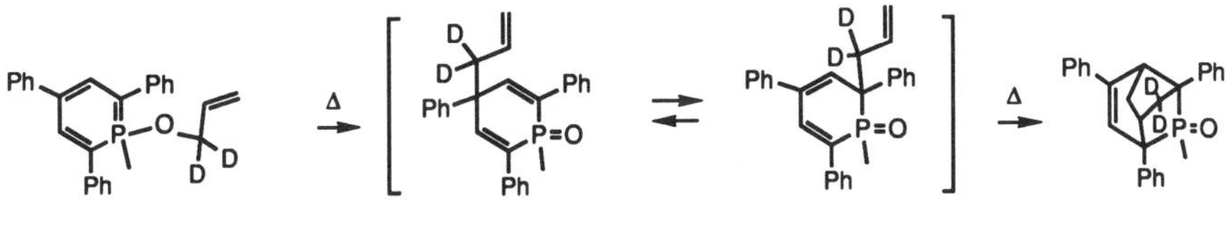

O-(2,4-alkadienyl)-*S*-(2-alkenyl) dithiocarbonates undergo facile transformation into tetrahydroisobenzothiophene derivatives [Harano, 1991] via two consecutive [3.3]-sigmatropic rearrangements and an intramolecular Diels–Alder reaction.

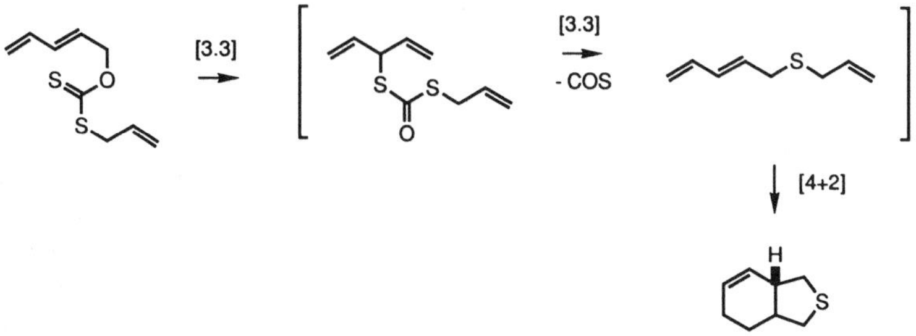

Finally, intramolecular Diels–Alder reactions by heating fulvene oxides have been observed [Näf, 1977]. The rearrangement of the fulvene oxides to 2,4-cyclohexadienones is somewhat unusual, and it may not be purely thermal reaction.

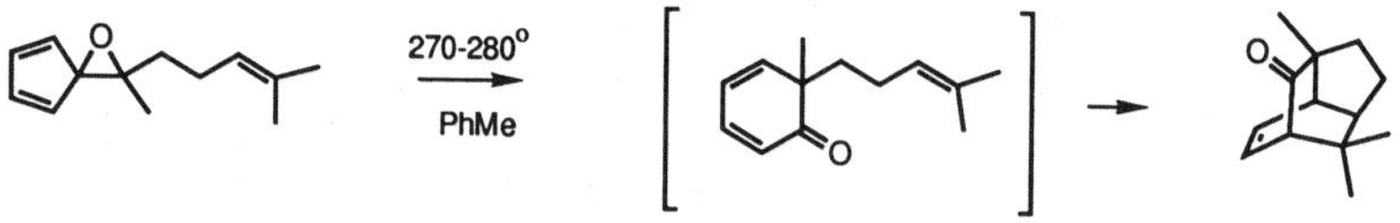

7.5. ACYLATION/DIELS–ALDER REACTION TANDEMS

Acylation of an imine shifts the double bond. In the case of a conjugated imine the reaction gives rise to a dienamide.

In the context of alkaloid synthesis the dienamination and subsequent trapping with a dienophile represents a powerful method of annulation. Novel routes to very complex indole bases have been developed, a prototype of which shown here is part of a synthetic scheme for aspidospermidine [Gallagher, 1982].

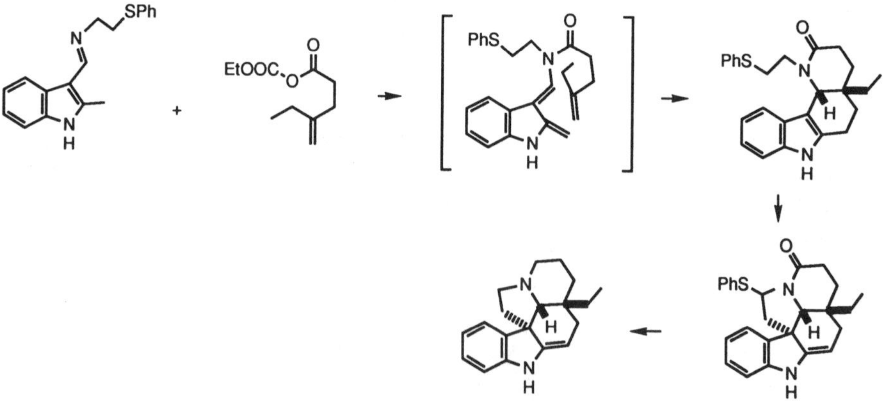

dehydroaspidospermidine

The mode of E-ring formation implies the primary pentacyclic intermediate contains a double bond at C-2. This feature is very beneficial to functionalization of the C-ring, and it permitted the use of simplest possible substrates for the acylation/Diels–Alder reaction steps. A synthesis of eburnamonine [Magnus, 1986] illustrates this point.

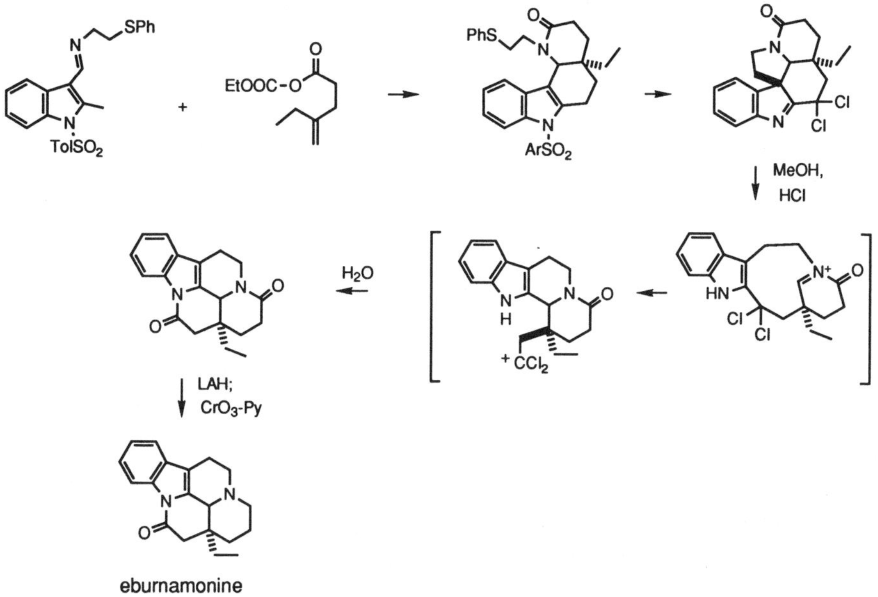

eburnamonine

A substrate with different distribution of functional groups has been used to successfully elaborate the core structure of staurosporine [Magnus, 1984].

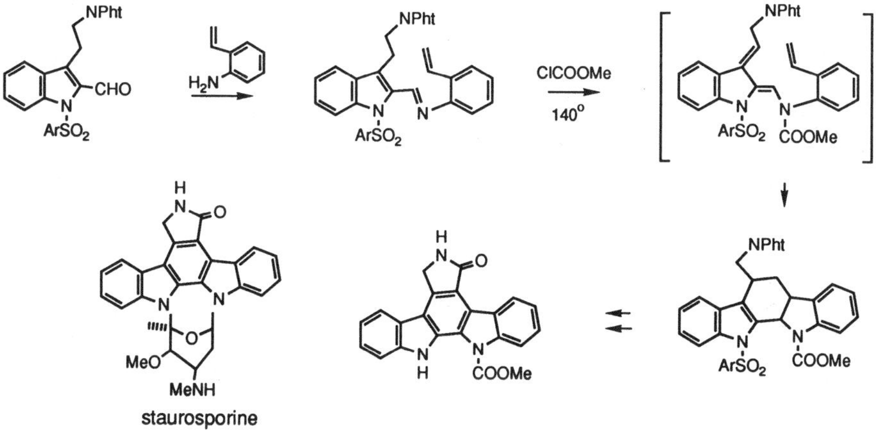

staurosporine

A stepwise N-acylation/Diels–Alder cycloaddition was employed in a formal synthesis of dendrobine [S.F. Martin, 1989].

Annulated N-acylpiperideines are readily prepared from N-trimethylsilyl-1-aza-1,3-dienes and acyl chlorides bearing an unsaturation [Uyehara, 1990]. The nitrogen atom is directly within the ring created by the cycloaddition instead of being in a lateral ring. The process has been applied to a synthesis of sedridine [Uyehara, 1991].

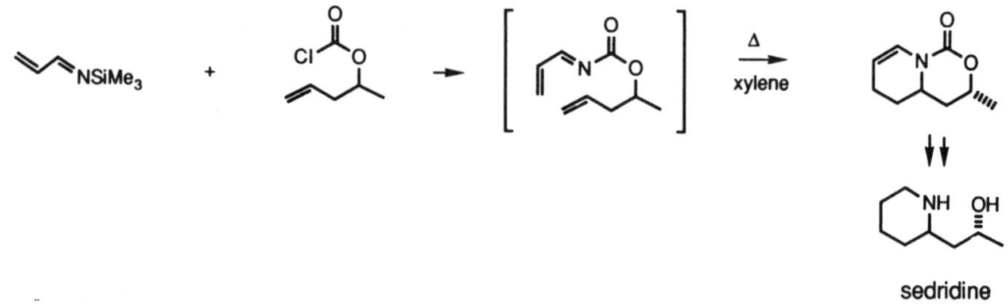

sedridine

7.6. OXIDATION OR REDUCTION INITIATED DIELS–ALDER REACTIONS

Acceptor substitution increases the reactivity of a dienophile in the normal Diels–Alder reaction. This change of reactivity pattern, which is also witnessed in the intramolecular Diels–Alder reactions, has been exploited many times in the synthesis of complex molecules, for example torreyol [Taber, 1979], sclerosporin [Kitahara, 1985], and biflora-4,10(19), 15-triene [Parker, 1986]. Thus, subjecting a precursor containing a conjugate diene and an allylic alcohol to an oxidation condition generally induces a spontaneous cycloaddition, if no excessive strain is introduced into the molecule.

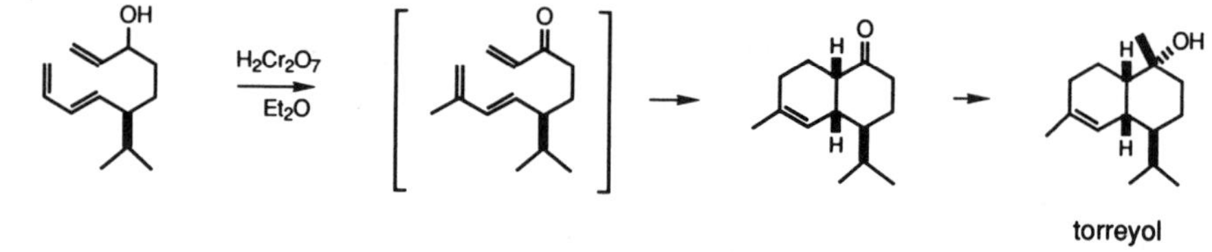

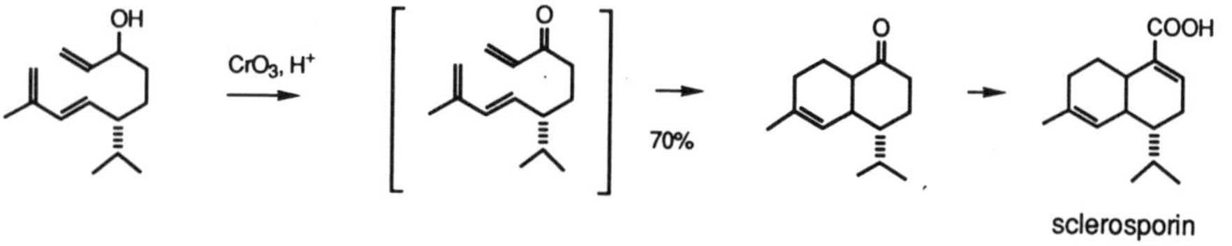

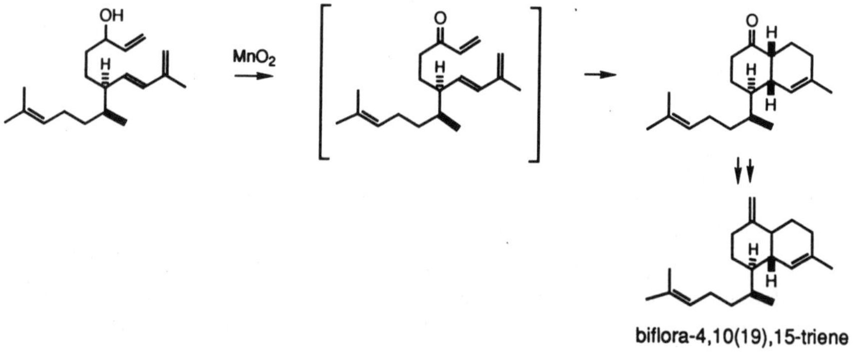

torreyol

sclerosporin

biflora-4,10(19),15-triene

Concerning the stereoselectivity of the cycloaddition, the existence of severe nonbonding interactions disfavored the chair transition states of the substrates, hence boat-like transition states were adopted. The predominant products were *cis*-octalones in which the subangular substituent is also *cis* to the angular hydrogen atoms.

An enedicarbonyl subunit generated from oxidation of a furan ring tethered to a cyclopentadiene reacts readily with the latter [H.-J. Wu, 1987]. The resulting annulated norbornene offers various options to structural modification. Oxidation of a furanophane caused the formation of a tetracyclic diketone [Wasserman, 1962].

A biomimetic synthesis of lachnanthocarpone [Bazan, 1978] consisted of oxidation of a catechol. The resulting *o*-quinone was intercepted intramolecularly by a diene moiety in the sidechain. Further oxidation of the adduct furnished the plant pigment.

lachnanthocarpone

Oxidative dimerization of a styrene also mimicked the biosynthesis of carpanone [Chapman, 1971], although the oxidant was definitely not the type used by nature. A hetero-Diels–Alder reaction of the dimer (in a higher oxidation state) ensued.

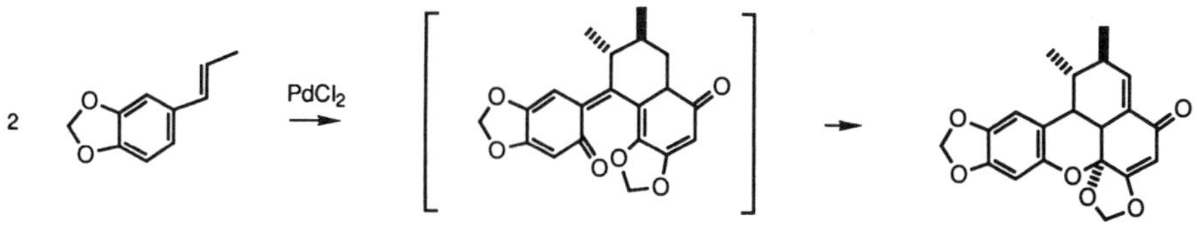

carpanone

Anodic oxidation of 4-hydroxycinnamic acids in methanol at low concentration leads to isoasatone-type products predominantly [Iguchi, 1979], apparently via Diels–Alder reaction of the 6,6-dimethoxy-2,4-cyclohexadienone intermediates and subsequent [2 + 2]-cycloaddition.

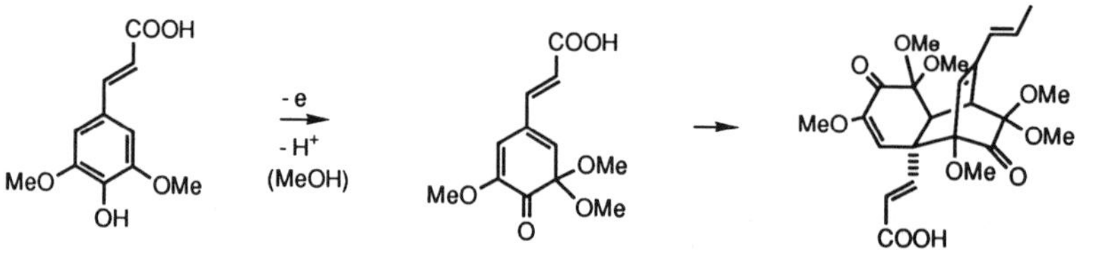

" asatone-type "

An analogous oxidation provides the skeleton of silydianin [Shizuri, 1986] when an appropriate dienophilic moiety is present in the substrate.

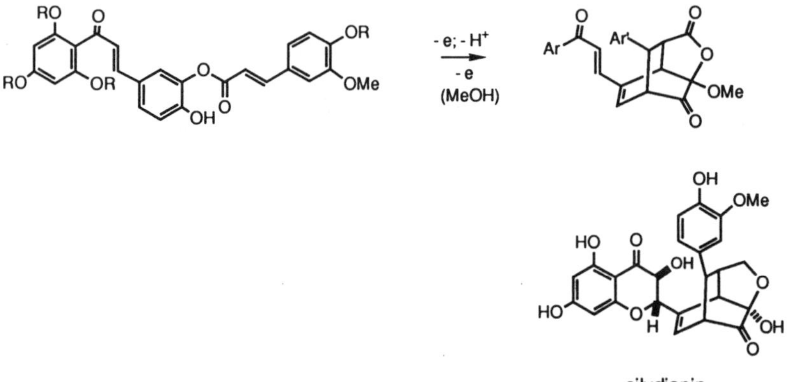

silydianin

Intramolecular interception of an *N*-acylnitroso group was a crucial step in the syntheses of gephyrotoxin-223AB [Iida, 1985] and (−)-nupharamine [Aoyagi, 1991]. For the former synthesis this cycloaddition established the piperidine ring and a potential leaving group four carbon atoms away from the nitrogen, facilitating the eventual cyclization of the pyrrolidine moiety. The *N*-acylnitroso functionality was generated in situ by oxidation of the hydroxamic acid.

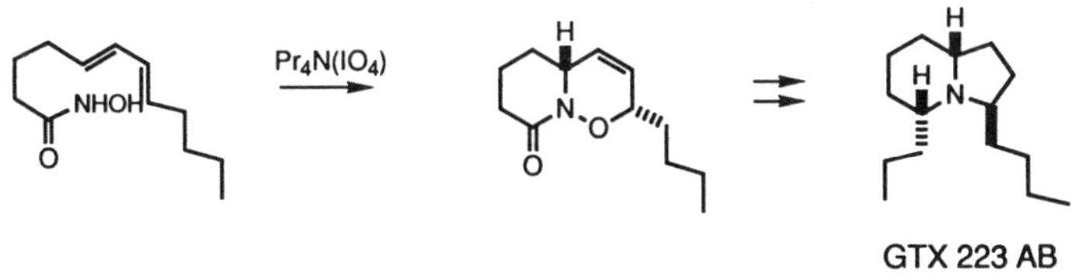

GTX 223 AB

Formation of the azepine ring of cephalotaxine [Burkholder, 1988] also employed the same technique.

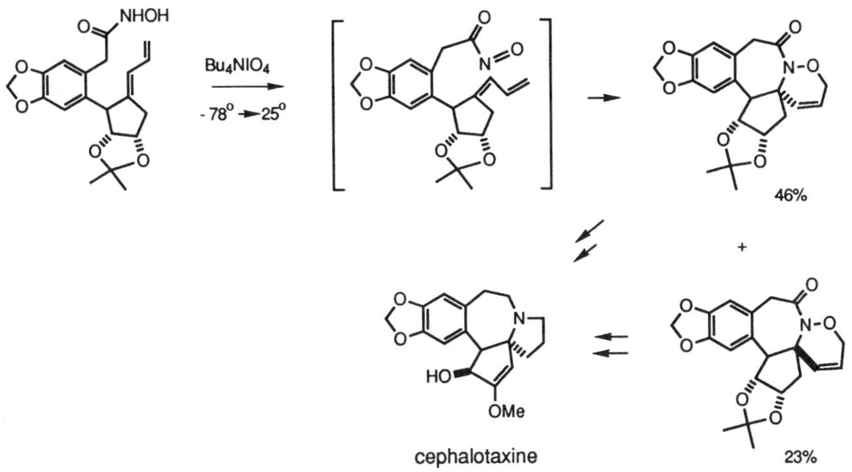

46%

+

cephalotaxine   23%

The internal photochemical redox reaction of an *o*-nitrobenzyl alcohol to give the corresponding nitroso phenone is the basis of a novel approach to mitomycin precursors [McClure, 1991], the nitroso function being trapped by a diene upon its generation.

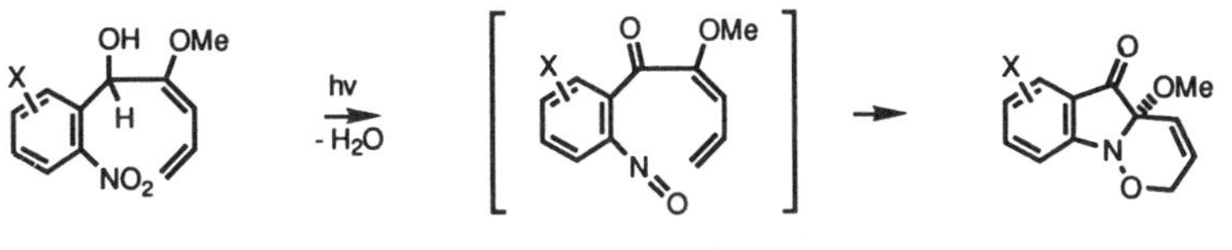

Oxidation of dihydropyrazolones containing a diene tether leads to cyclazine derivatives [Medina, 1987].

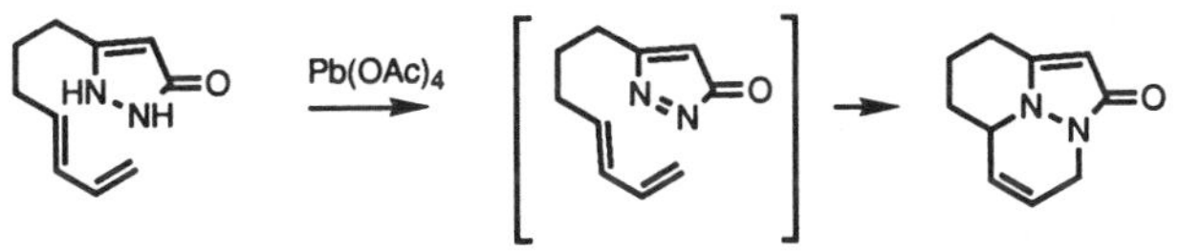

Peracid oxidation of 3,4-diaryl-6,7-dihydro-1-methyl-1*H*-1,2,5-triazepines leads to 2-hydroxy-1,2,5-triazabicyclo[3.2.1]oct-3-enes [Trepanier, 1973]. A mechanism involving nitrone ring cleavage with proton transfer and an intramolecular Diels–Alder reaction is consistent with the results.

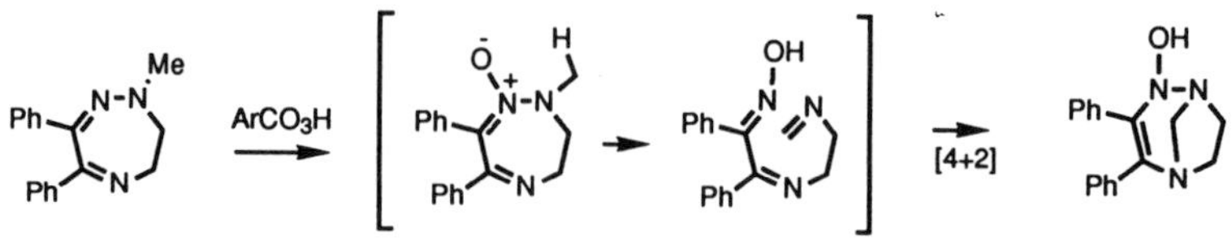

9,10-Antracenediol generated by reduction of anthraquinone under basic conditions reacts with active dienophiles [Koerner, 1991]. However, less reactive dienophiles undergo competitive reduction, resulting in lower yields of the adducts. The monoimine of anthraquinone behaves similarly with respective to reduction and trapping behavior.

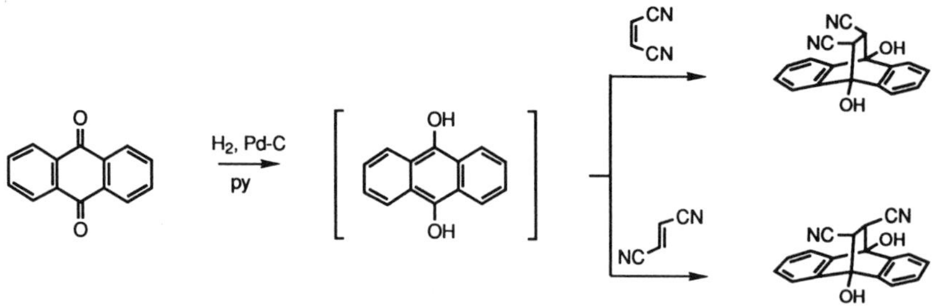

5-Acyloxy-8-hydroxy-1,4-naphthoquinones may undergo Diels–Alder reactions after internal transacylation (redox exchange of two rings). This behavior has important ramifications in the synthesis of anthracyclinone antibiotics [Kelly, 1977; 1980].

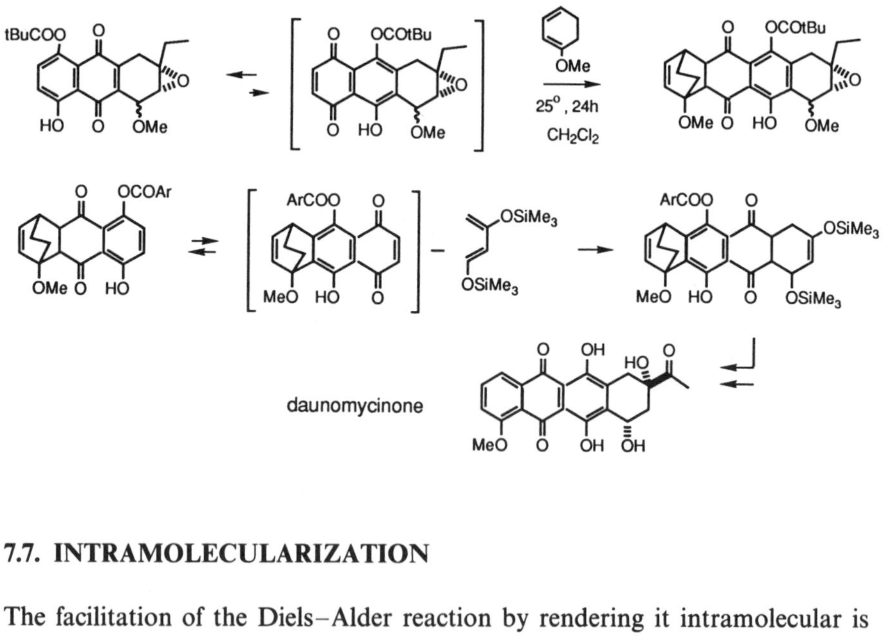

7.7. INTRAMOLECULARIZATION

The facilitation of the Diels–Alder reaction by rendering it intramolecular is due to pre-reduction of entropy. The dramatic effect of such a maneuver can be witnessed by the observed cycloaddition at room temperature [Greenlee,

1976] instead of at ca. 200°C [Boeckman, 1980], which was the key step in two synthetic pathways toward marasmic acid.

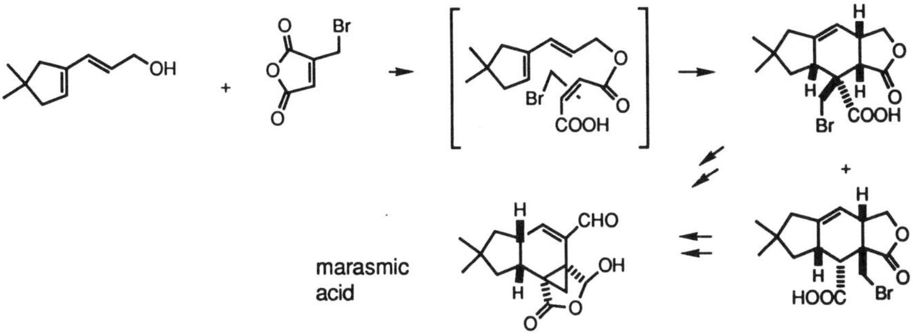

marasmic
acid

The same strategy has been applied to a synthesis of biflora-4,10(19), 15-triene [Grieco, 1986].

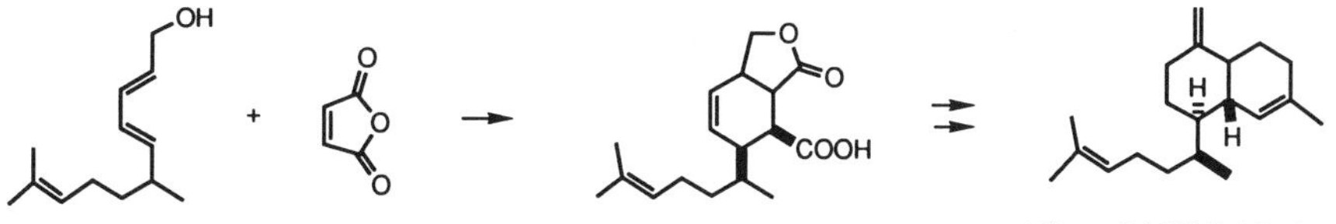

biflora-4,10(19),15-triene

Similarly, the formation of the hydroisoindolone core of (−)-aspochalasin-B [Trost, 1989] by heating a dimethyl alkylidenemalonate with 4-methyl-2,4-hexadienol must have been preceded by a transesterification.

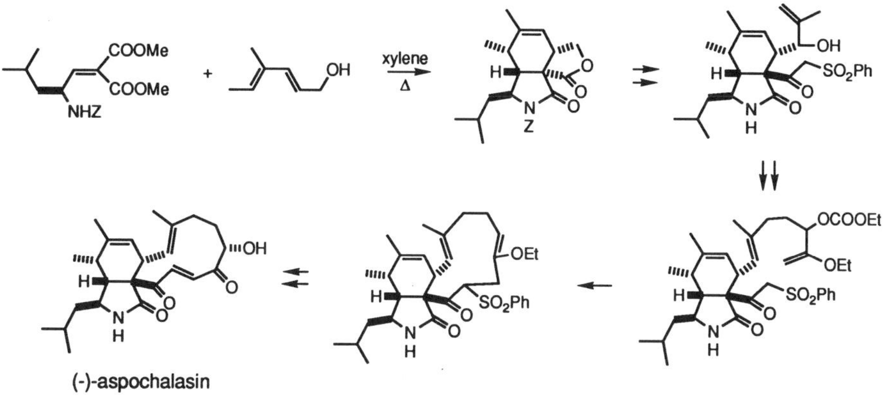

(-)-aspochalasin

Acylation of N-methyl-5-aryl-2,4-pentadienylamines with fumaryl ester chloride at 0°C led to the hydroisoindolones directly [Gschwend, 1972].

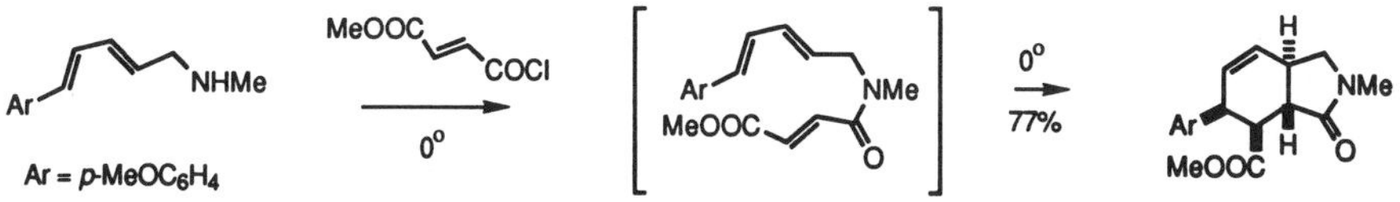

Ar = p-MeOC₆H₄

The replacement of an allylic chlorine atom of hexachlorocyclopentadiene with an allyloxy group caused a cycloaddition also [Jung, 1982].

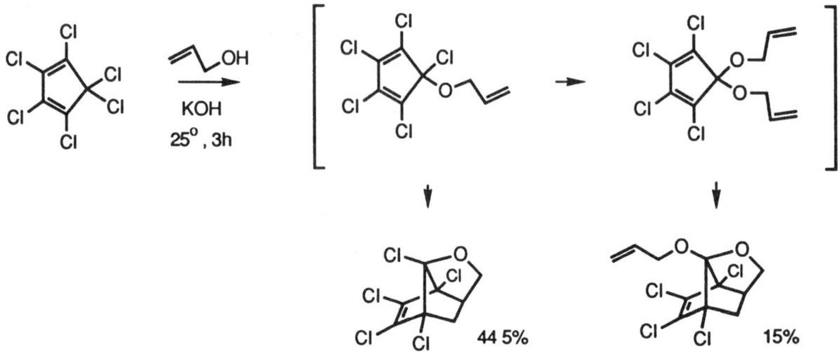

Nickelocene reacts with octachlorocycloheptatriene at room temperature to give a tricyclic compound [Moberg, 1974]. The unusual alkylation was followed by an intramolecular Diels–Alder reaction.

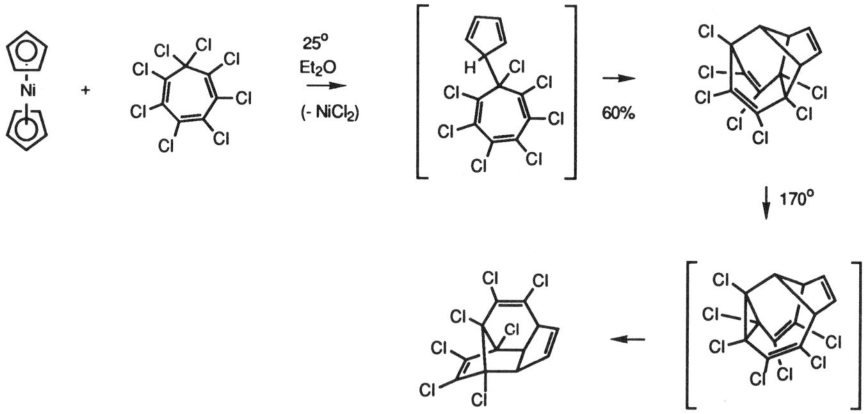

An intramolecular Diels–Alder reaction was found to occur in tandem with the alkylation of β-tropolone with diphenylcyclopropenium tetrafluoroborate [Takahashi, 1988].

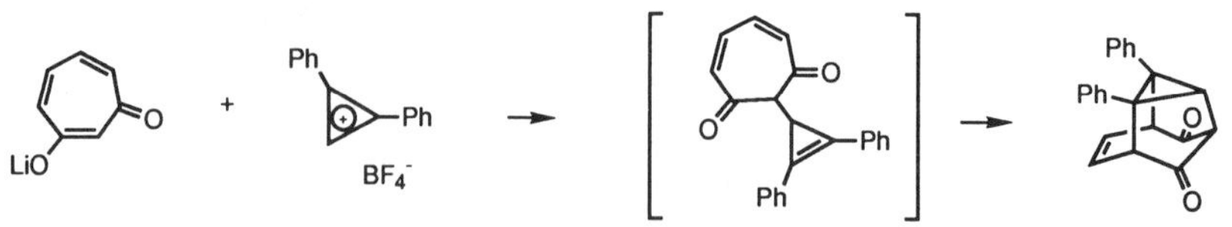

A general method for a one-step construction of the naphthalide lignan skeleton is by dimerization of a 3-arylpropynoic acid. Anhydride formation brings two molecules within distance for a facile cycloaddition. An example shown now pertains to a synthesis of helioxanthin [T.L. Holmes, 1970].

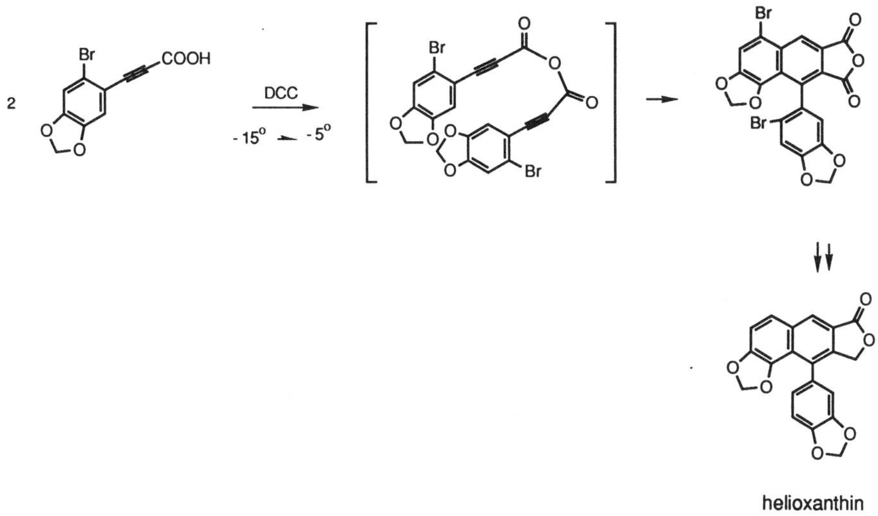

helioxanthin

(Note that much higher temperatures, e.g. $> 100°C$, seem to be required to effect an intramolecular Diels–Alder reaction of cinnamyl 3-arylpropynoates [Klemm, 1971].)

A novel approach to (+)-isovelleral [Bergman, 1990] called for an intramolecular Diels–Alder reaction to form the norcarene skeleton. The substrate was constructed from a diene amide by N → C transacylation with 2-methylcyclopropenyllithium.

The undesirable stereochemistry of the cycloaddition notwithstanding, the plan was successfully completed by virtue of a thermal equilibration via an intermediate with a seven-membered ring.

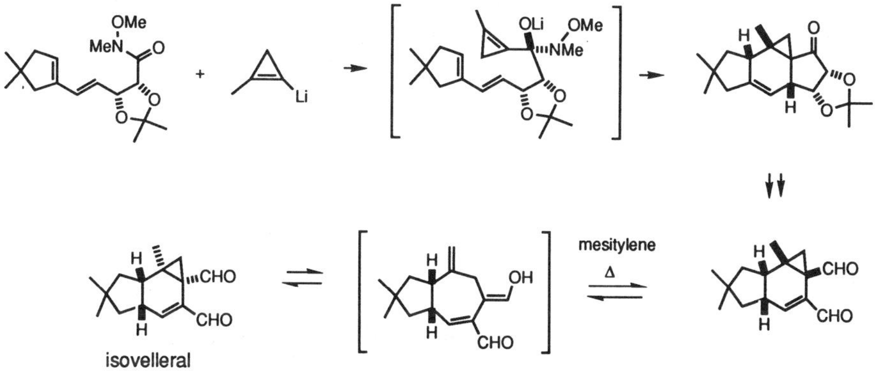

isovelleral

A model compound for dynemicin-A has been acquired by an intramolecular Diels–Alder reaction [Porco, 1990]. The substrate prepared by either macrolactonization or Pd(O)-catalyzed coupling, underwent the cycloaddition spontaneously at room temperature.

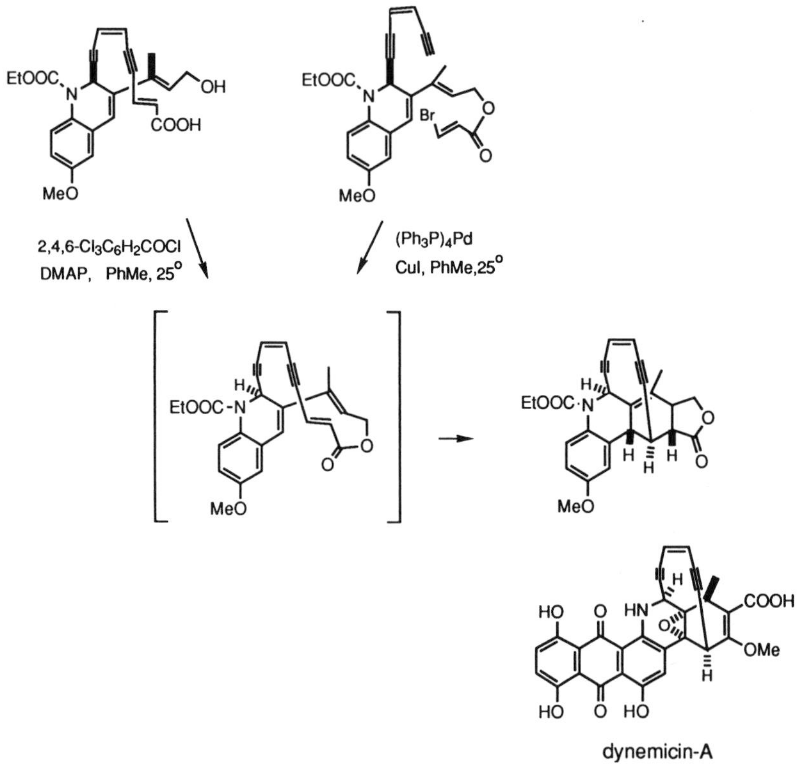

dynemicin-A

An intramolecular alkylation of a malonic ester did not permit the isolation of the 14-membered cyclic dienyne. Under the reaction conditions (85°C) a Diels–Alder reaction occurred [Y.-C. Xu, 1990].

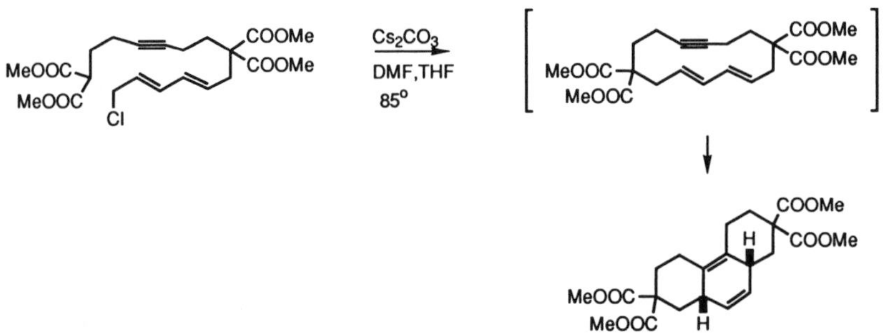

The aldol condensation of citronellal with 1,3-cyclohexanediones at moderate temperatures led directly to compounds related to hexahydrocannabinol [Tietze, 1982].

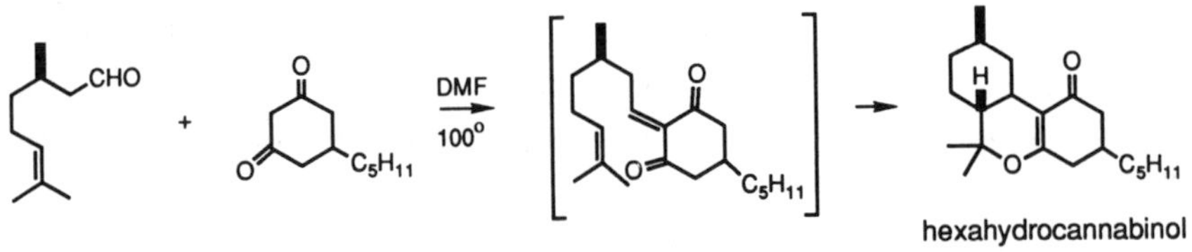

hexahydrocannabinol

In a (−)-lycoserone synthesis [Kuhnke, 1988] the iridoid skeleton was formed in the condensation step.

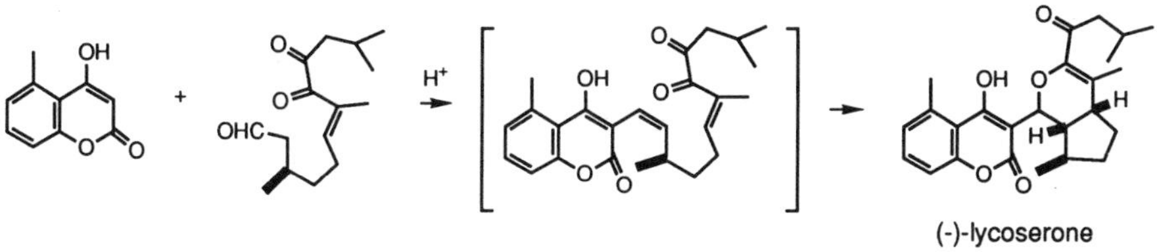

(-)-lycoserone

Reaction of phloroglucinol with citral led to a brichromane [Crombie, 1971]. An intramolecular hetero-Diels–Adler reaction was the bridging process to complete the tetracyclic skeleton.

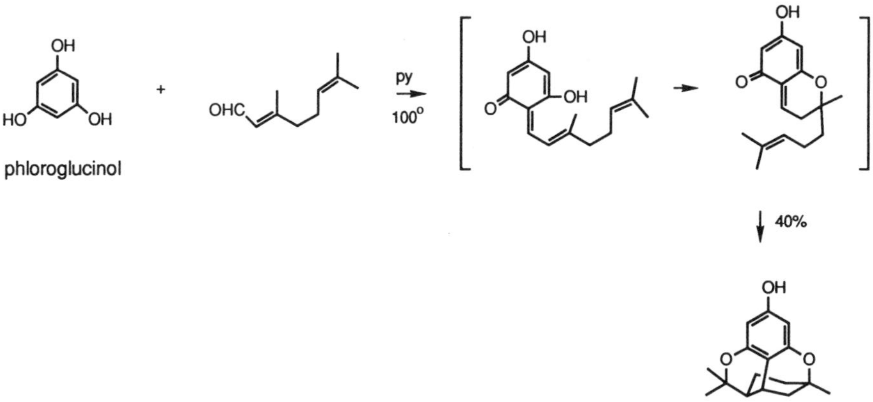

phloroglucinol

40%

The much more facile condensation/hetero-Diels–Alder reaction sequences involving Meldrum's acid as the aldolization donor has enabled the efficient access of (−)-kainic acid [Takano, 1988b] and furofuran lignans including (−)-sesamolin, (−)-sesamin, (−)acuminatolide [Takano, 1988a].

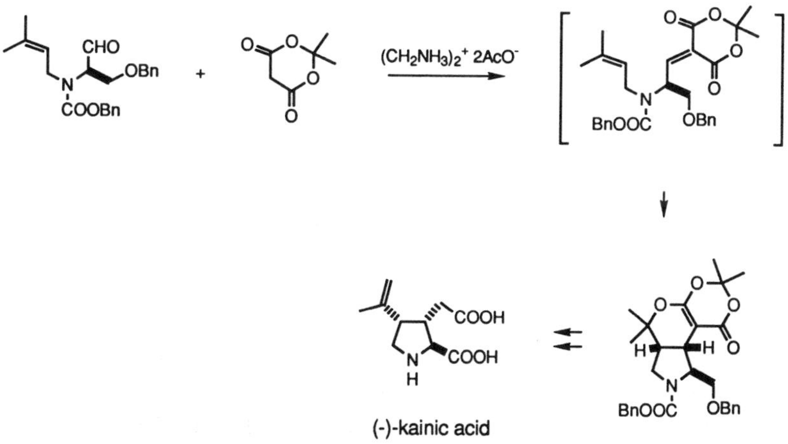

(-)-kainic acid

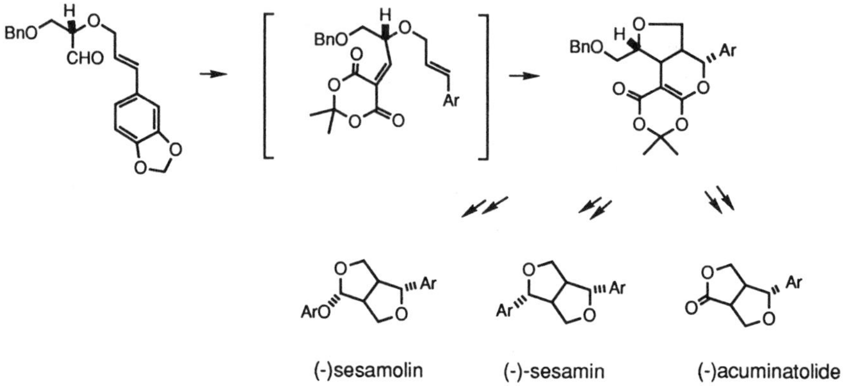

BnO
CHO
Ar
(-)sesamolin
(-)-sesamin
(-)acuminatolide

7.8. ASSORTED TRIGGERS FOR THE DIELS–ALDER REACTION

Photoenolization of a 2-acyl-1,4-benzoquinone provides a conjugate diene system. The monoenolization of a bis(2-acyl-1,4-benzoquinone) in which the two acyl groups are linked by a carbon chain can induce an intramolecular Diels–Alder reaction [Miyagi, 1984]. Necessarily the tether is of proper length so as to present a favorable steric situation.

hν
MeCN

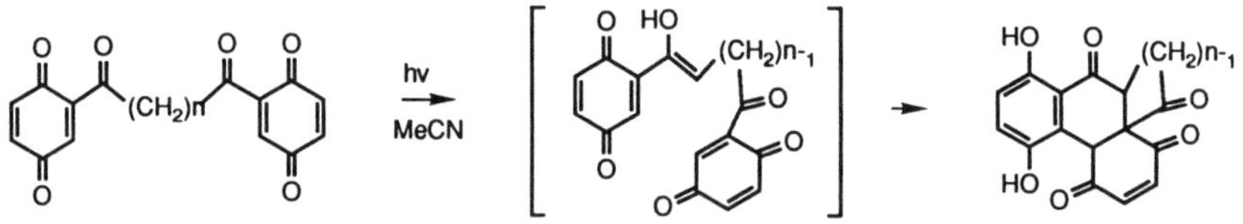

Thiele acid is the photolytic product of 2-chlorophenol in an alkaline solution [Guyon, 1982]. It is logical to assume that Diels–Alder dimerization would follow hydration of the ketene which results from Wolff rearrangement of the cyclic carbene.

hν
ROH
ROH
(dimerize)
COOR
ROOC

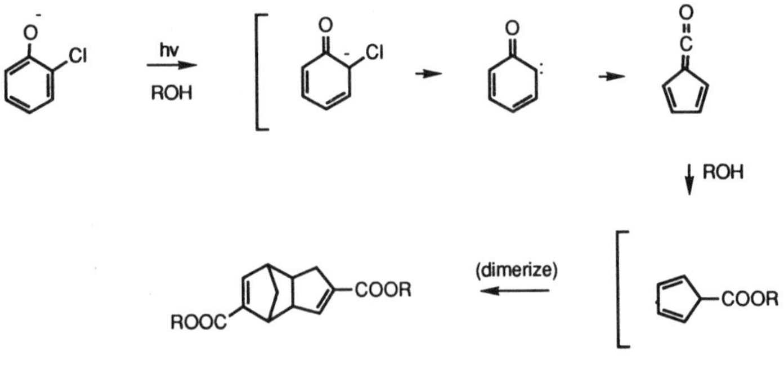

3,3-Dimethyloxindole loses carbon monoxide on irradiation to produce 2,2-bis(o-aminophenyl)propane [Döpp, 1979]. Decarbonylation and Diels–Alder dimerization of the imine intermediate are indicated. Fragmentation of the dimer should be favorable due to aromatization and relief of steric interactions.

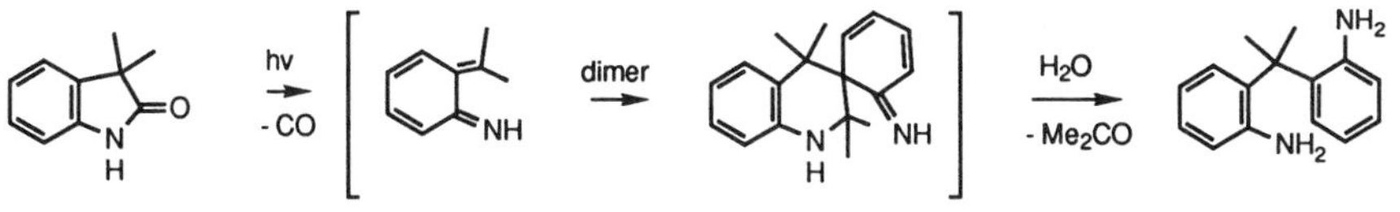

Generation of a conjugated dienol unit in an aliphatic chain containing a terminal vinyl group by pyrolytic isomerization of a 2-acyloxy-1,3-diene promoted an intramolecular Diels–Alder condensation [Shea, 1982].

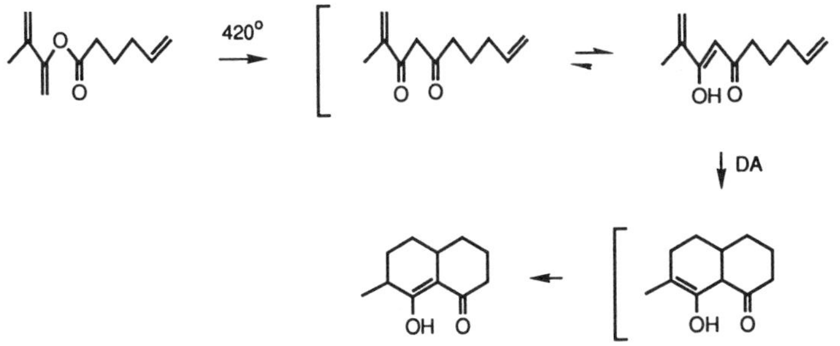

3-Trimethylsiloxy-1,3,7,9-decatetraene underwent spontaneous cycloaddition on treatment with potassium fluoride [Oppolzer, 1981b]. Release of the enone system by desilylation brought the molecule into a highly active state that the observed reaction was inevitable.

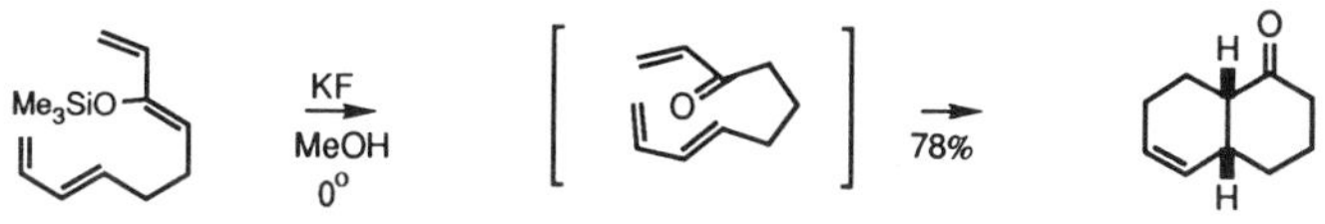

An intramolecular Diels–Alder reaction of a β-siloxyethyl ester was induced by desilylation in the presence of hydrofluoric acid [Roush, 1984]. Activation of the dienophilic double bond by the formation of a 1,3-dioxolenium ion was implicated, as the corresponding methyl ester did not undergo cycloaddition under the same conditions.

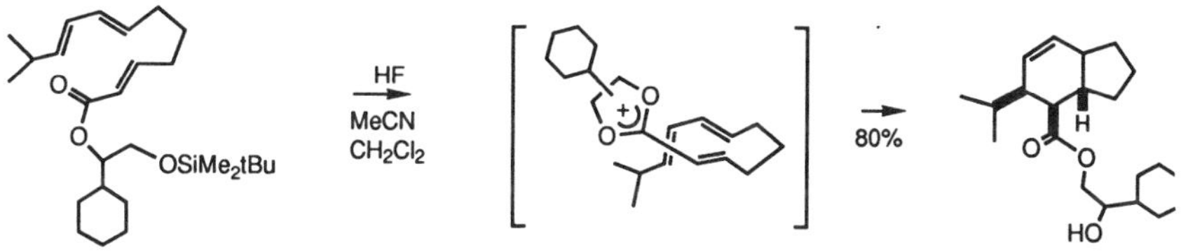

3-Alkenylindoles are rather unstable although they can be isolated. With respect to a carbazole synthesis based on Diels–Alder reaction of such molecules it is more convenient to intercept the nascent species with dienophiles [Noland, 1979].

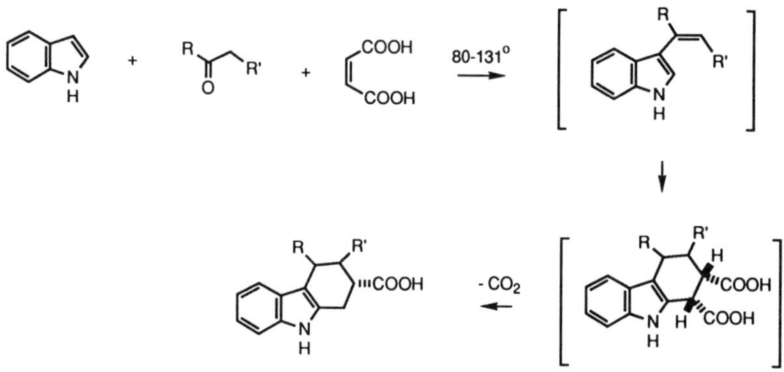

A tandem Wittig/Diels–Alder reaction sequence works well [Kozikowski, 1986]. With proper addends the adducts may be transformed into more difficultly accessible structures, for example medium-sized ring diketones by ozonolysis.

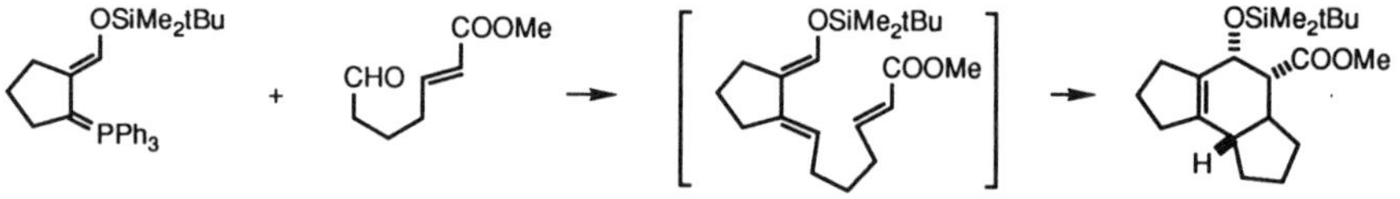

Another such tandem was employed effectively in the construction of a precursor of sinularene [Antczak, 1987].

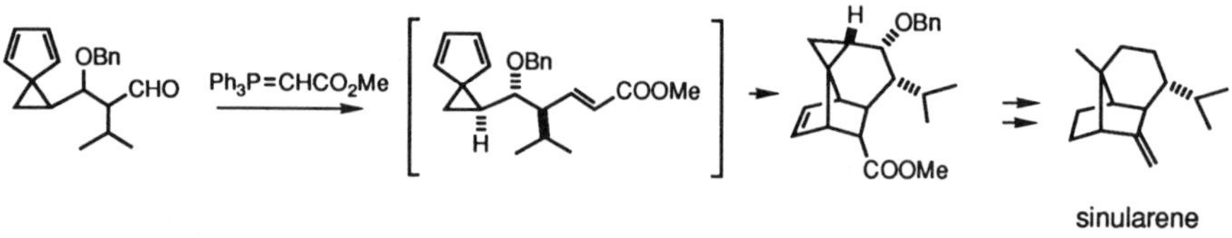

sinularene

The Horner–Emmons condensation product of 1-formyl[6]helicene undergoes spontaneous Diels–Alder reaction to relieve overcrowding [R.H. Martin, 1965a].

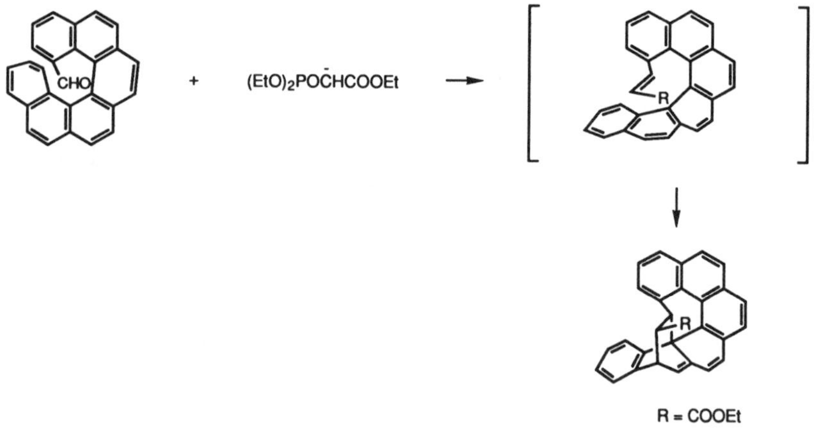

R = COOEt

A methyl γ-silycrotonate was shown to be in equilibrium with the conjugated ketene acetal as it was transformed into a dimethyl hydroxyphthalate on heating with dimethyl acetylene-dicarboxylate [Chan, 1982].

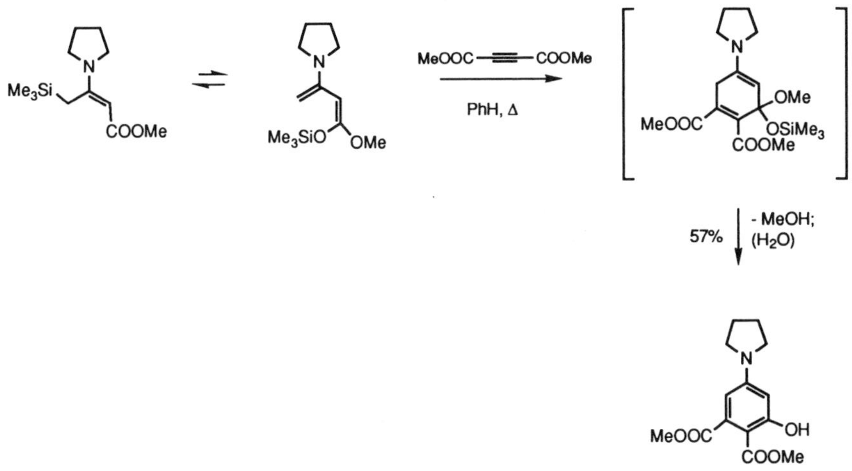

Deconjugation of a dienone in which the double bonds are endocyclic and exocyclic to a cyclopentane, resulting in a cyclopentadiene, can trigger an intramolecular Diels–Alder reaction when the α′ position of the ketone group bears a chain containing a dienophile. This strategy formed the basis of a scheme for synthesis of gascardic acid [Berubé, 1989].

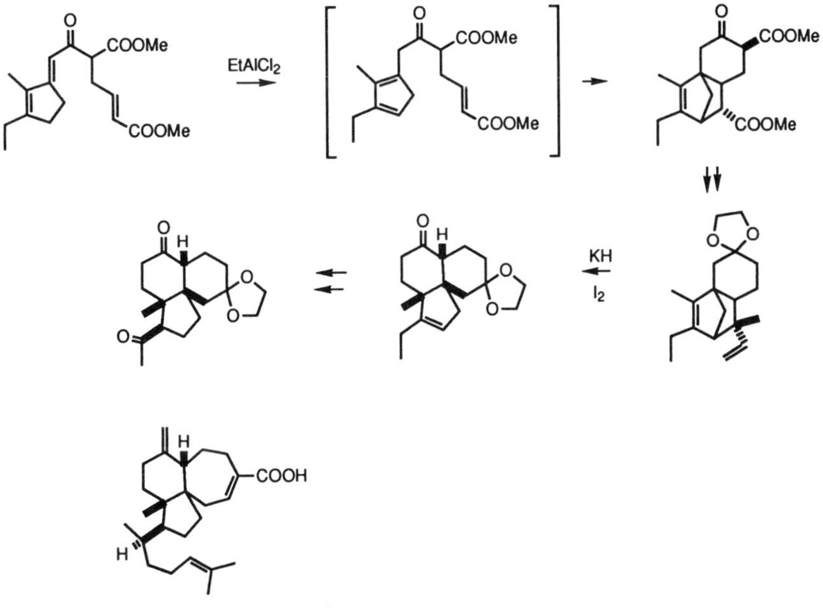

gascardic acid

On the other hand, enolsilylation of a cyclopentenone led to a concomitant intramolecular cycloaddition which served to construct a brexane precursor of sativene [Snowden, 1986]. The cyclopentanone subunit uncovered on hydrolysis of the adduct was readily expanded and converted into an isopropylcyclohexane ring.

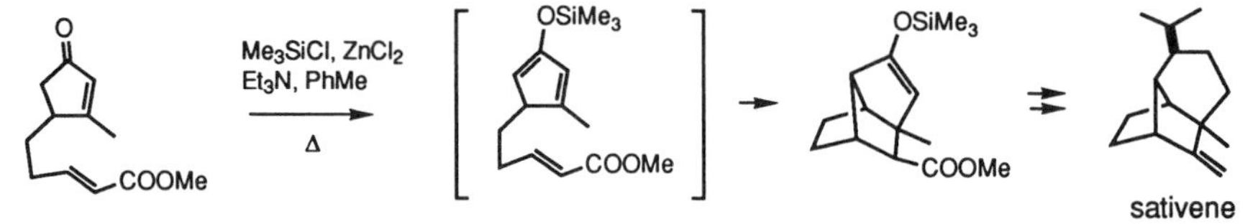

sativene

For an aliphatic system, slightly higher temperatures are needed to induce the cycloaddition. In an approach to $\Delta^{9(12)}$-capnellene [Ihara, 1991a], two enone systems were exposed to enolsilylation conditions, but only one of them could be converted into a diene system.

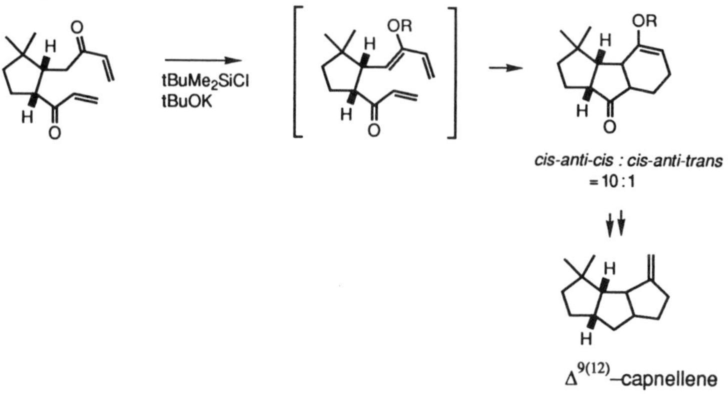

cis-anti-cis : cis-anti-trans
= 10 : 1

$\Delta^{9(12)}$–capnellene

A report describing the concomitant intramolecular Diels–Alder reaction with siloxydiene formation [Keck, 1981] is interesting in that the corresponding tetrahydropyranyl ether analog apparently did not cyclize.

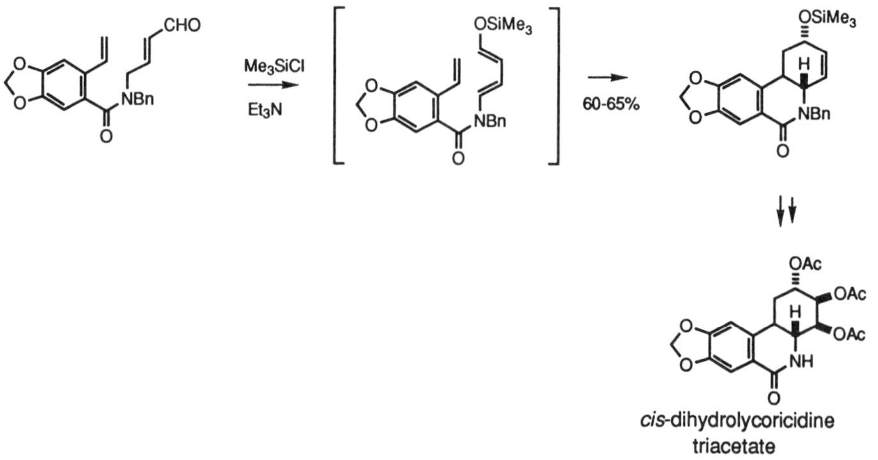

cis-dihydrolycoricidine
triacetate

In a previous paragraph it is described that a Prins reaction converts a skip diene into a conjugate isomer and the latter species undergoes a hetero-Diels–Alder reaction. There are reports on isomerization of propargyl group to an allenyl residue and the utilization of one of the allenyl double bond to participate in an intramolecular cycloaddition. This manipulation permitted the syntheses of platyphyllide [K. Hayakawa, 1986a] and (+)-4-oxo-5,6,9,10-tetradehydro-4,5-seco-furanoeremophilane-5,1-carbolactone [K. Hayakawa, 1988], a tricyclic intermediate of forskolin [Kanematsu, 1989], and (+)-sterpurene [Gibbs, 1989].

Besides the fact that an allene is more reactive than a propargyl triple bond in the Diels–Alder reaction, the concerted nature of the sulfoxide to sulfenate rearrangement also enabled the operation of a chirality transfer from a center to an axis and then to an emerging center [Gibbs, 1989].

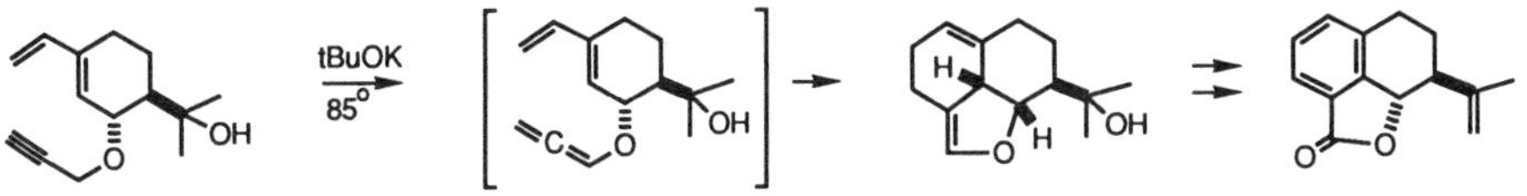

platyphyllide

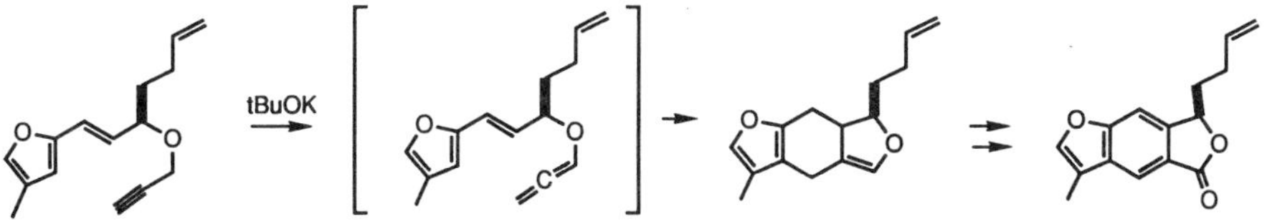

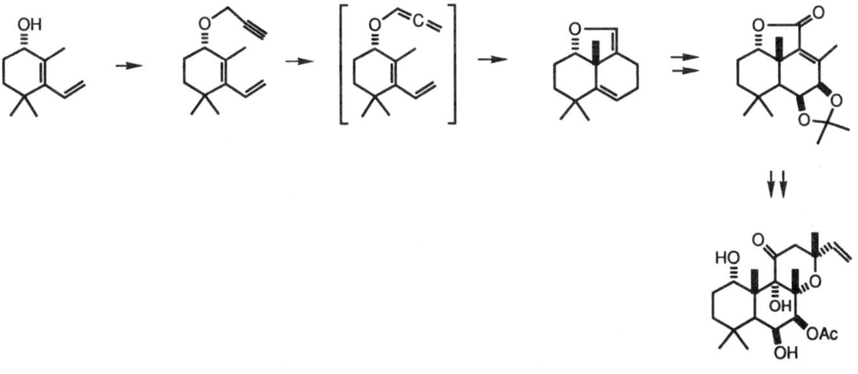

forskolin

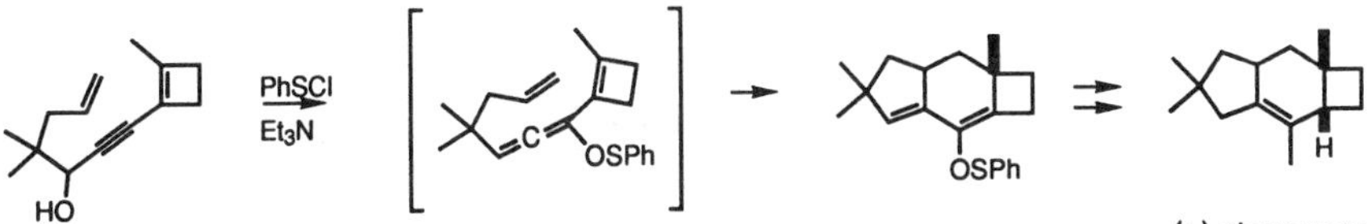

(+)-sterpurene

A particularly interesting "furan exchange" by this technique has allowed the preparation of peri-fused decalones [Y. Yamaguchi, 1987] which serves as models for halenaquinone, xestoquinone, viridin, etc. A more recent report describes the application of two such sequential furan ring transfers, in a synthesis of the sesquiterpene euryfuran [Kanematsu, 1991].

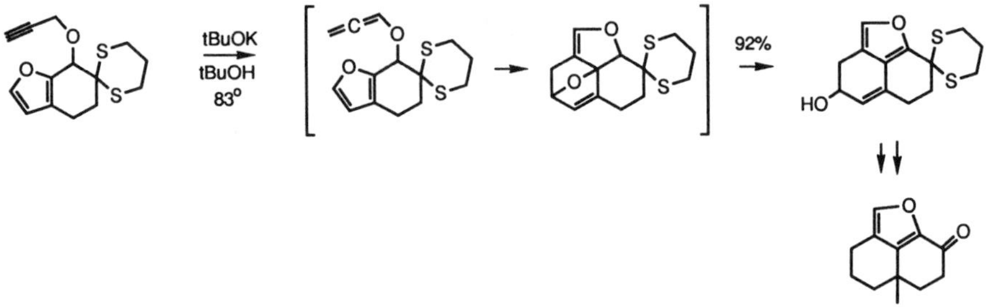

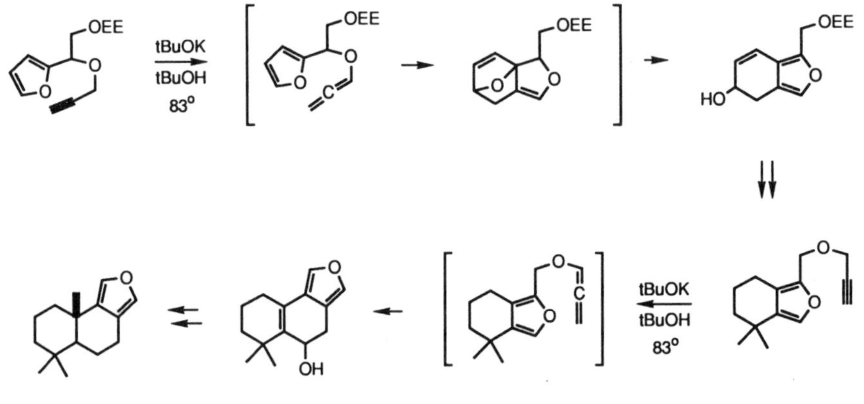

euryfuran

A very good plan for the elaboration of the *cis*-trikentrin skeleton [Yasukouchi, 1989] is via homologative synthesis of an allene from a propargylamine derivative (involving a redox process). Intramolecular cycloaddition led to a tetracyclic compound which is a partially hydrogenated indole. After aromatization the two benzylic methyl groups were readily crafted from the benzobicyclo[2.2.1]heptadiene system.

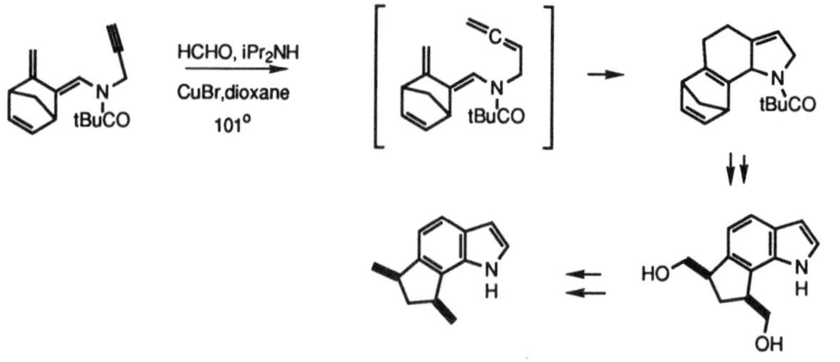

The key steps of the indole synthesis was incorporated previously in a total synthesis of hippadine [K. Hayakawa, 1987].

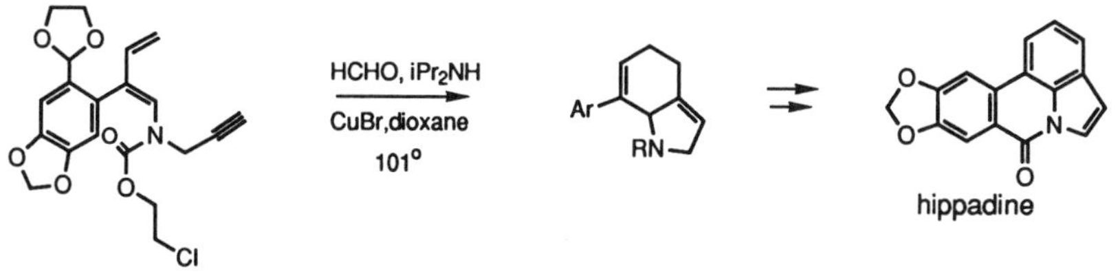

hippadine

The allenylcarbonyl dienophile may be generated by protodesilylation of a (trimethylsilylmethyl)propargyl derivative, and with a juxtaposed diene unit in the molecule a Diels–Alder reaction can occur in tandem with the desilylation process [Magnus, 1992].

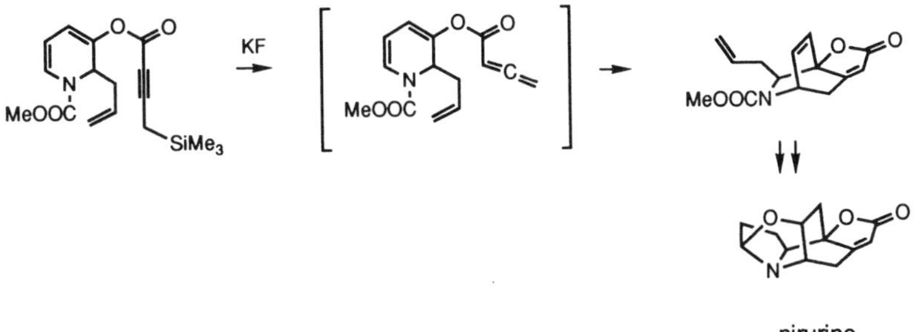

nirurine

A novel construction of the norsativane skeleton from a bicyclo[3.2.0]heptane precursor [Sigrist, 1986] also consists of a fragmentation to generate a molecule which possesses an allenyl ketone and a cyclopentadiene moiety. A facile cycloaddition between these structural subunits are expected.

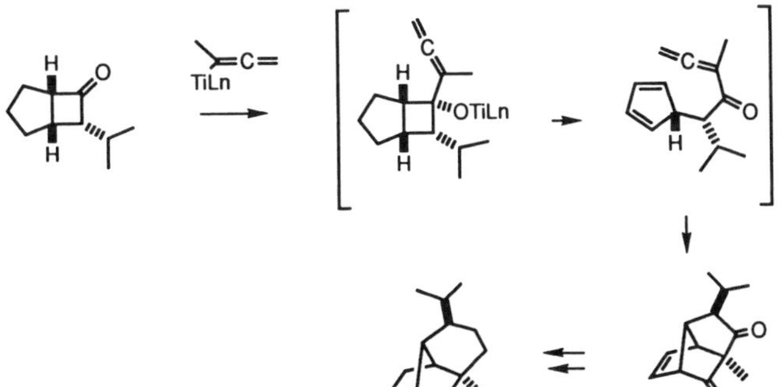

sativene

It should be noted that cyclopentadienes which bear an allyl ketone group generated by this fragmentation route are susceptible to [1.5]-hydrogen shift prior to the intramolecular Diels–Alder reaction [Wallquist, 1983], although the ratio of products from the isomeric cyclopentadienes may be affected by the substitution pattern of the dienophilic component.

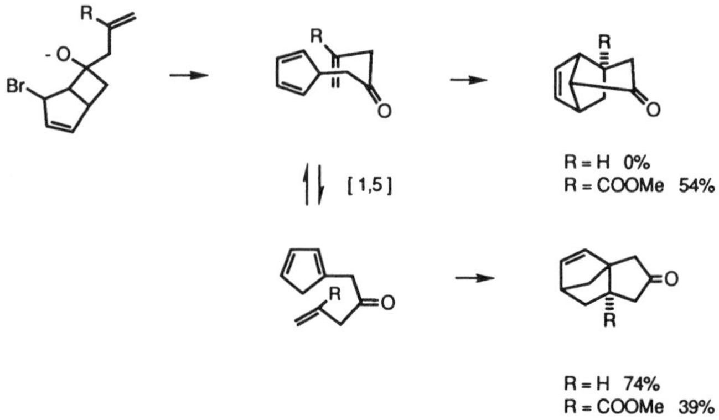

Allenylcarboxylic ester formation by palladium(0) displacement of propargyl carbonates and carbonylation can be coupled to an intramolecular Diels–Alder reaction to create polycyclic systems [Mandai, 1991b].

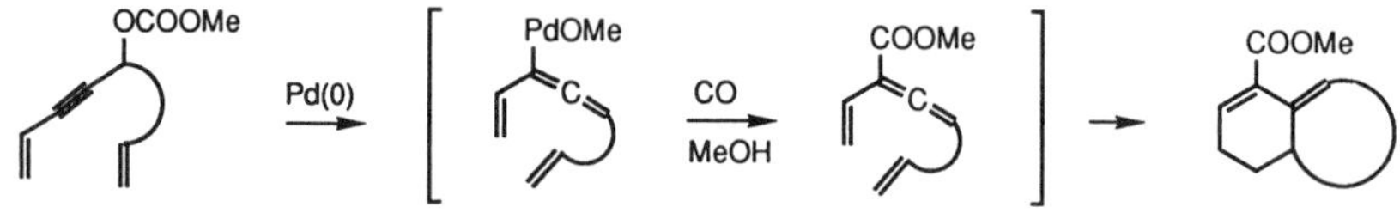

A simple isomerization (a formal 1,5-hydrogen migration) of an *o*-benzoquinone was implicated in the formation of a spirocyclic adduct when 4-methoxy-5-methyl-*o*-benzoquinone was heated with butadiene [Mazza, 1974]. The behavior of the quinone was unexpected. Later investigations [Danishefsky, 1977] showed a dramatic solvent effect in this reaction, the "normal" Diels–Alder reaction indeed proceeded in methanol.

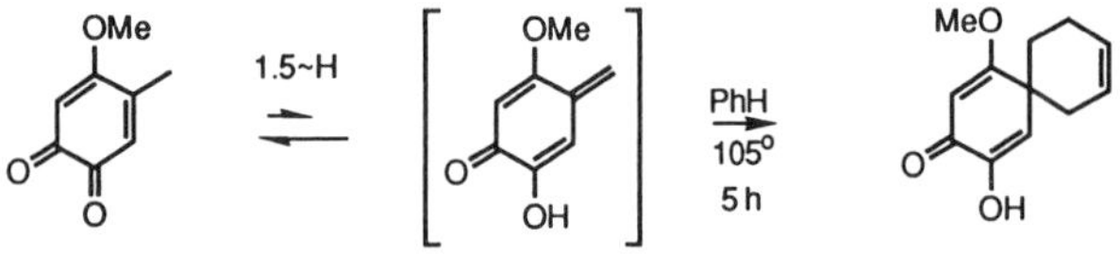

α-Keto dithio esters are active dienophiles. They may be generated via fragmentation and intercepted by dienes [Vedejs, 1983].

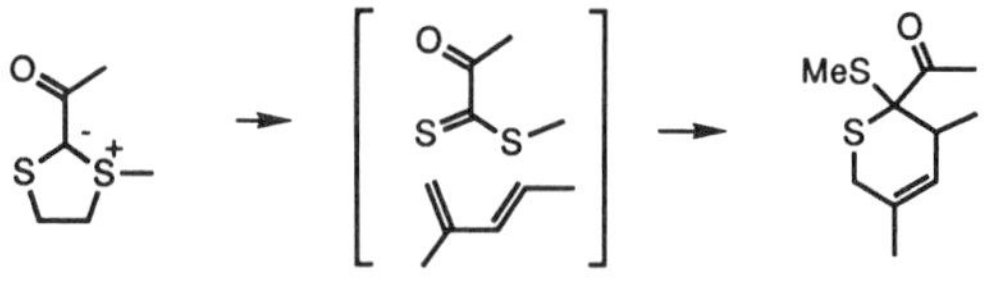

In a synthetic approach to quebrachamine [Takano, 1979a] the removal of a lactam carbonyl was essential. However, a fragmentation and intramolecular Diels–Alder reaction tandem intervened during treatment with phosphoryl chloride.

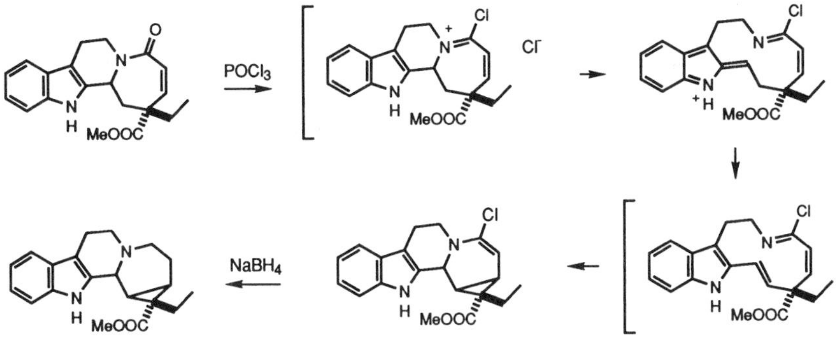

By dehydrosulfinylation of primary carbamates with thionyl chloride sulfinylimines are obtained. These unstable species may be trapped by a conjugate diene. The heterocycle derived from such an intramolecular reaction may be cleaved and caused to undergo a [2.3]-sigmatropic rearrangement to deliver an oxygen function to the β-carbon of the nitrogen atom. The whole process is characterized by excellent diastereocontrol that (E)-threo-sphingosine was synthesized readily from an (E,E)-diene, and (E)-erythro-sphingosine from the (E,Z)-isomer [Gqrigipati, 1984].

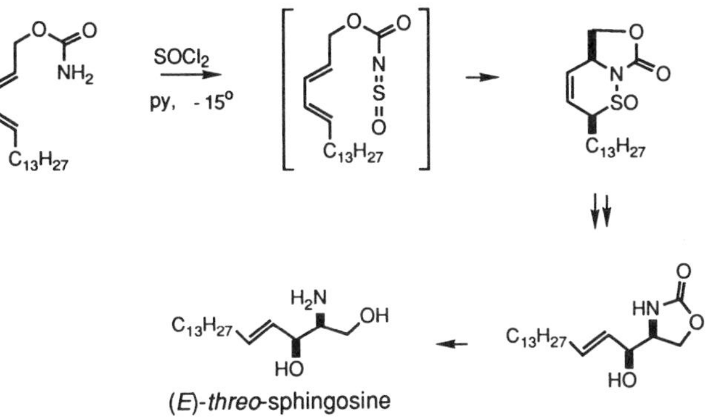

(E)-threo-sphingosine

The preferential formation of a cis-quinolizidinone by condensation of 2-acetonylpiperidine with an araldehyde has been explained by a mechanism involving retro-Michael fission of the iminium intermediate and a cycloaddition [Lantos, 1977].

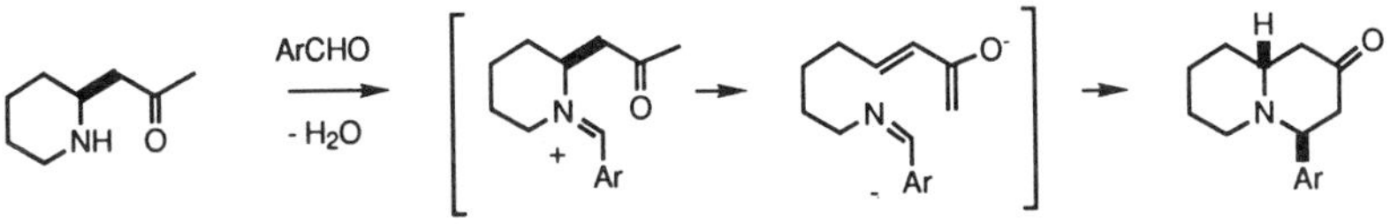

Biomimetic routes to various aspidosperma alkaloids have been investigated intensively, resulting in highly convergent syntheses. For example, a mixture of ibophyllidine and 20-epiibophyllidine emerged from such a reaction [Barsi, 1985].

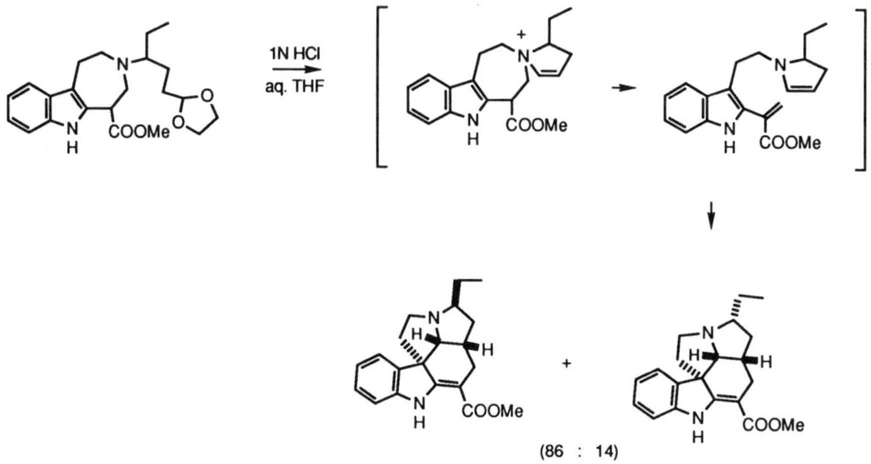

20-epiibophyllidine   ibophyllidine

A quaternary enamonium species could be the central intermediate of the process. This intermediate underwent a retro-Michael elimination and a Diels–Alder reaction, the latter highly favored by the presence of positionally matched polar substituents.

Formation of a secodine-type intermediate from an analogous, spiroannulated indoloazepine was rapidly followed by the re-formation of a pentacyclic product, vincadifformine [Kuehne, 1978; 1979b].

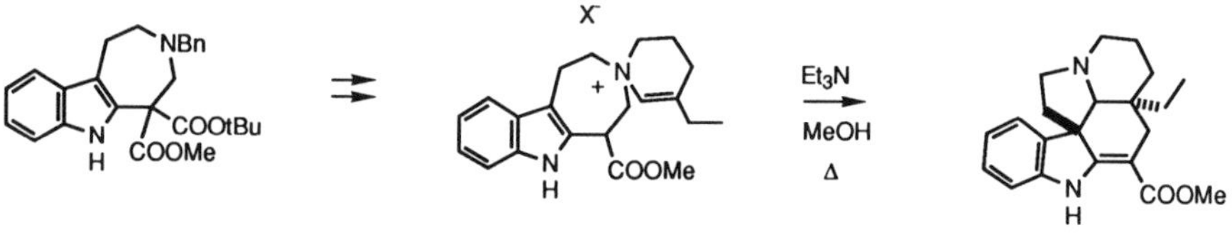

vincadifformine

By changing the alkylating species secodines of different substitution patterns are readily prepared in this manner. The judicious variation culminated in the synthesis of alkaloids such as pandoline [Kuehne, 1980].

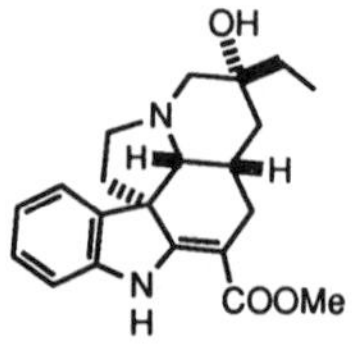

pandoline

More complex indoloazepines in which the diazaspirocyclic system characteristic of strychnos and aspidosperma alkaloids is present can also be induced to fragment to afford secodines via a retro-Mannich pathway [Kuehne, 1983]. Interestingly, the Diels–Adler reaction regenerates the spirocyclic portion but rearranges other parts of the molecule. This strategy was employed in a synthesis of minovincine [Kuehne, 1983], ibophyllidine [Kuehne, 1981]. tabersonine [Kuehne, 1982], among others.

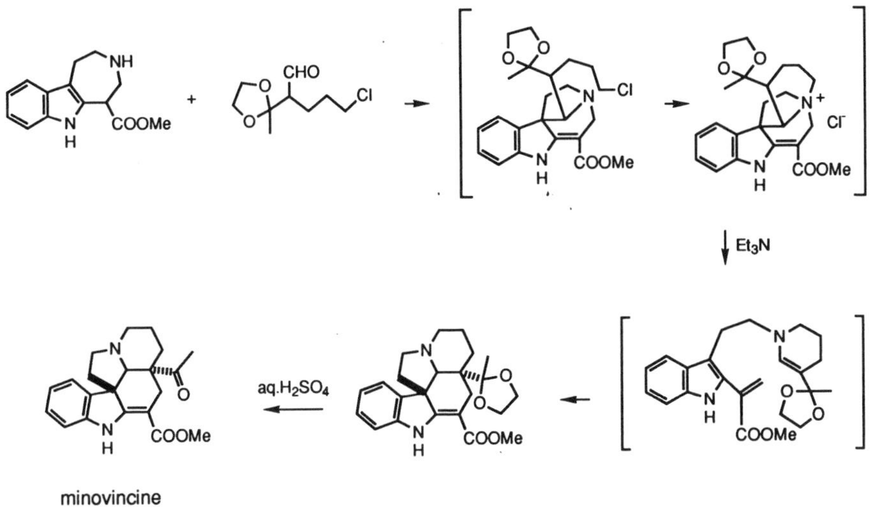

minovincine

Tetrahydro-β-carboline derivatives also provide suitable precursors of the secodine [Kuehne, 1979a].

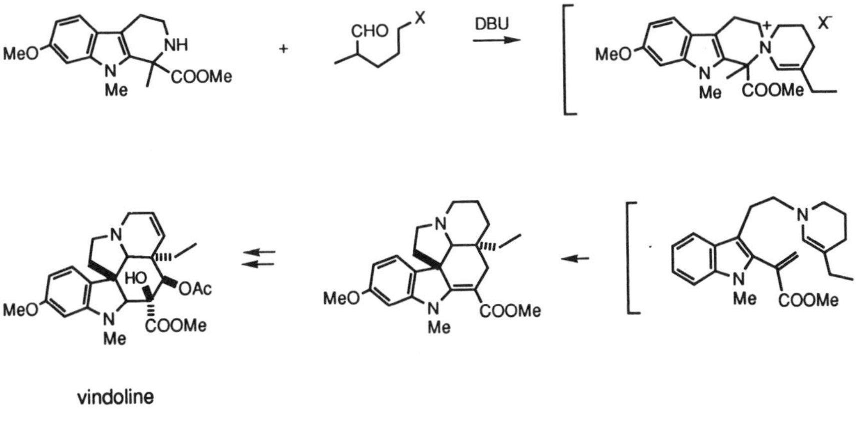

vindoline

An access to the tubotaiwine skeleton [Legseir, 1987] started from reaction of a tetrahydro-β-carboline with butyraldehyde. The imine underwent a tandem retro-Michael ring opening and a Diels–Alder reaction. Lactamization was then effected.

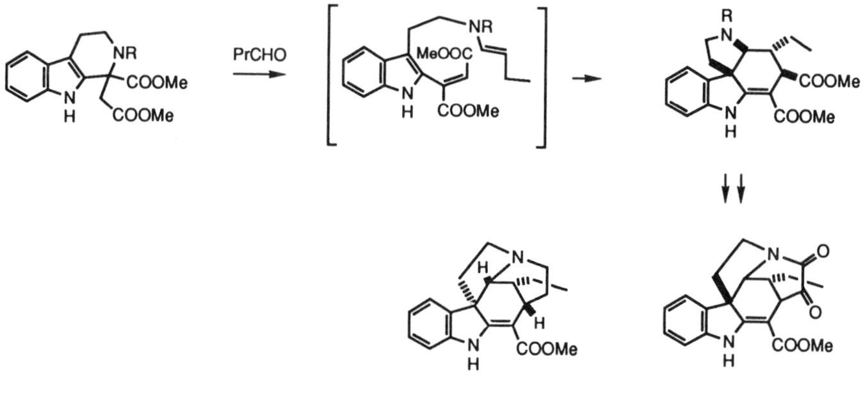

tubotaiwine

The thermal conversion of precondylocarpine to andranginine [Kan-Fan, 1974] is most likely the result of a tandem retro-Mannich, fragmentative deacetoxylation, and Diels–Alder reactions.

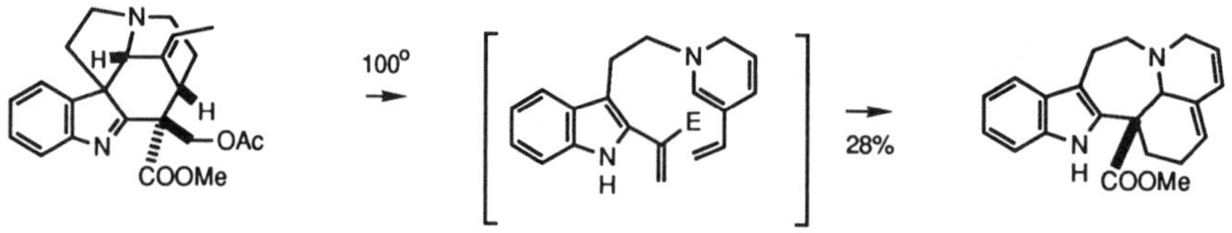

precondylocarpine
acetate

E = COOMe

andranginine

A Lewis acid-catalyzed tandem Michael addition/Diels–Alder reaction between an acryloylnaphthoquinone and (2,4-pentadien-1-yl)trimethyltin is a most delightful protocol for the construction of a tetracyclic intermediate of 11-deoxydaunomycinone [Naruta, 1988].

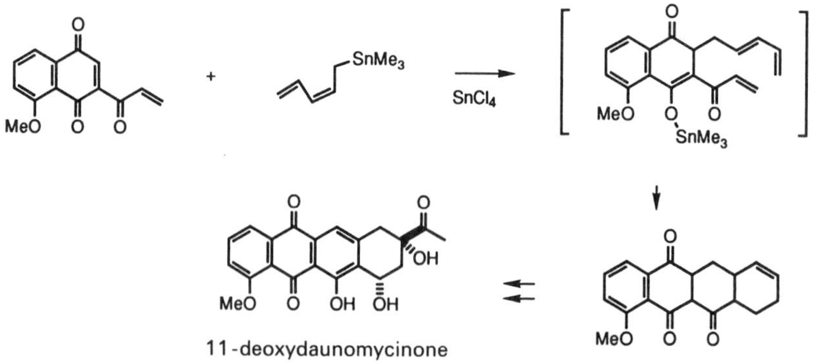

11-deoxydaunomycinone

An analogous condensation with an N-acryloyliminum ion as the electrophile led to isoquinolines [R. Yamaguchi, 1990].

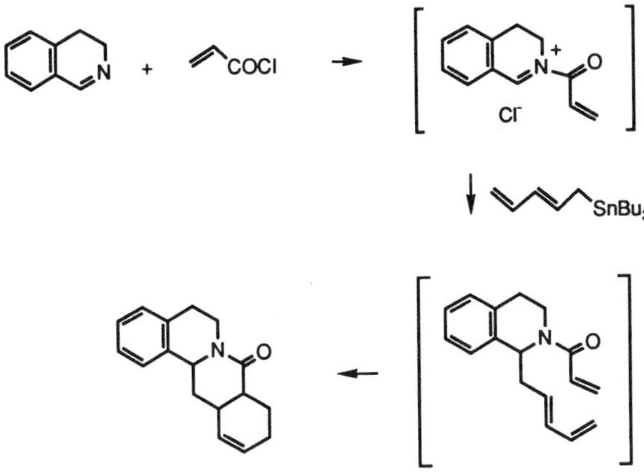
N
COCl
N
O
Cl⁻
SnBu3
N
O
N
O

8

OTHER CYCLOADDITIONS

8.1. [2 + 2]-CYCLOADDITIONS

The [2 + 2]-cycloaddition became a standard methodology for synthesis only quite recently. The photochemical method works best on conjugated carbonyl addends, whereas the thermal version is virtually limited to using ketenes and alkenes as coaddends. This latter cycloaddition, when proceeding by the $[_\pi 2_s + _\pi 2_a]$ mechanism, is orbital symmetry-allowed, and it should yield cyclobutanones with predictable stereochemistry. However, stepwise reactions occur in some cases.

Since ketenes are very unstable species, they must be generated in situ. A [2 + 2]-cycloaddition involving a ketene is necessarily part of a reaction tandem. The most frequently used methods for ketene formation are dehydrochlorination of acyl chlorides, Wolff rearrangement of α-diazo ketones, and zinc reduction of α-haloacyl halides. Additionally, pyrolysis of many compounds may lead to ketenes.

The success of intramolecular cycloaddition of ketenes is critically dependent on the tether. There must be a balance between entropy of activation and strain of the product. Consequently, three-atom tethers seem to offer the best compromise.

Thermolysis of cyclobutenones produces unsaturated ketenes which may be trapped intramolecularly [S. L. Xu, 1989].

Remarkably, vacuum distillation of β-ionylideneacetaldehyde gave a 50% yield of a tricyclic ketone [Smit, 1975]. It is quite certain that the reactive ketene was formed by [1.5]-hydrogen shift after isomerization of the α,β-unsaturated double bond.

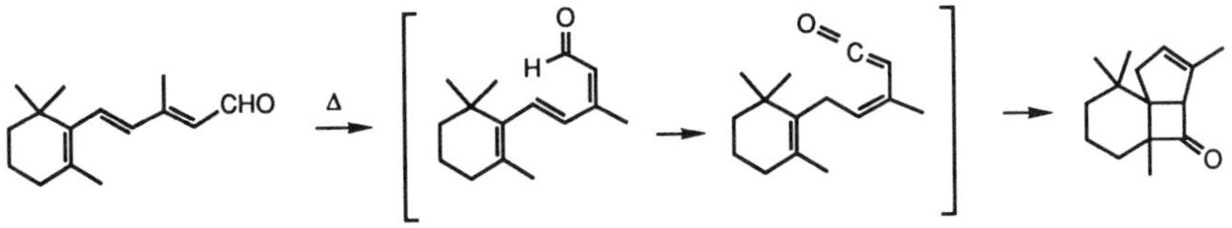

A similar mechanism must be operating in the conversion of a cycloheptatriene monoxide to a bicyclo[3.2.0]heptenone [Schiess, 1974; 1976].

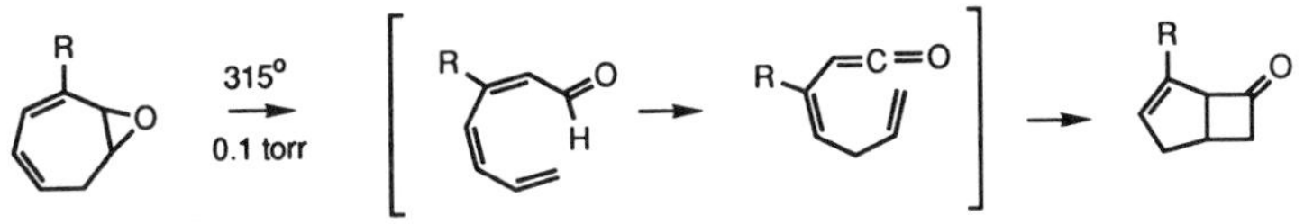

In the above reaction the alkene component was not present in the original molecule. In principle, any method of ketene formation which also generates the trapping component is valuable for creating new carbon frameworks. Thus, photoisomerization of a fused bicyclo[4.1.0]hept-2-en-4-one led to a rearranged product [Schultz, 1984].

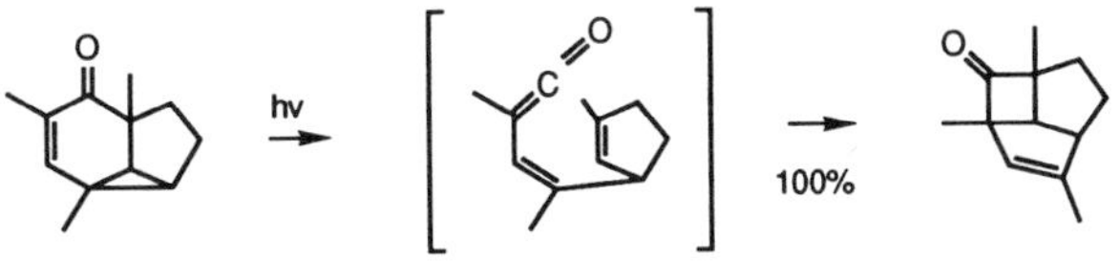

The photorearrangement of 3-chlorobicyclo[3.2.1]octa-3,6-dien-2-one [Goldschmidt, 1973] is readily accounted for on the basis of a cyclomutation/intramolecular [2 + 2]-cycloaddition sequence.

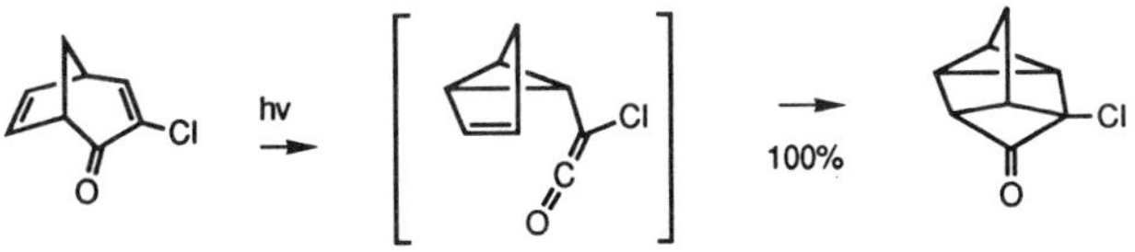

Photochemical or silver-catalyzed Wolff rearrangement of (2,3-diphenyl-cycloprop-2-enyl)methyl diazomethyl ketone led to a ketene, thence a strained cycloadduct [Masamune, 1965]. Although this adduct could be isolated, it underwent ring opening to afford an isomeric ketene.

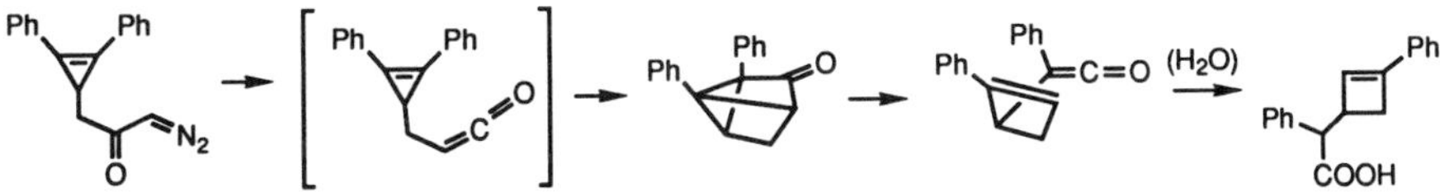

Diacrylamides undergo thermal rearrangements to furnish (acrylamido)acyl ketenes. Bicyclic lactams are obtained upon ensuing intramolecular cycloaddition [A. Alder, 1983]. An interesting substituent effect on the α-carbon atom of the acrylyl residue was observed.

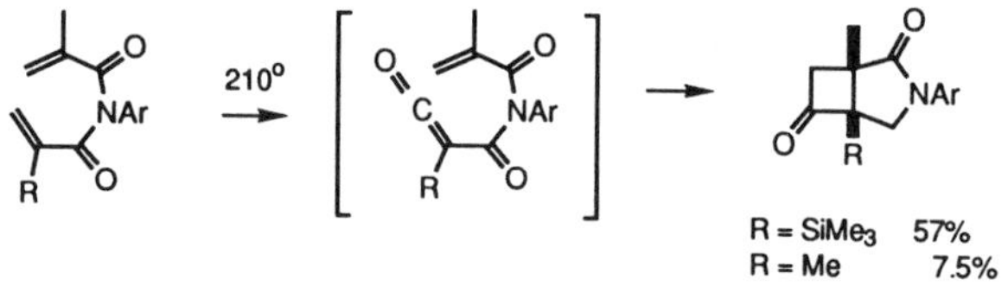

R = SiMe₃   57%
R = Me   7.5%

Metal-complexed ketenes are formed by the reaction of Fischer carbene complexes with alkynes. Intermediates of the ketenes are the chromacyclobutenes which can undergo electrocyclic opening and CO insertion. In the presence of a properly distanced olefin the ketene unit may be trapped [Wulff, 1985].

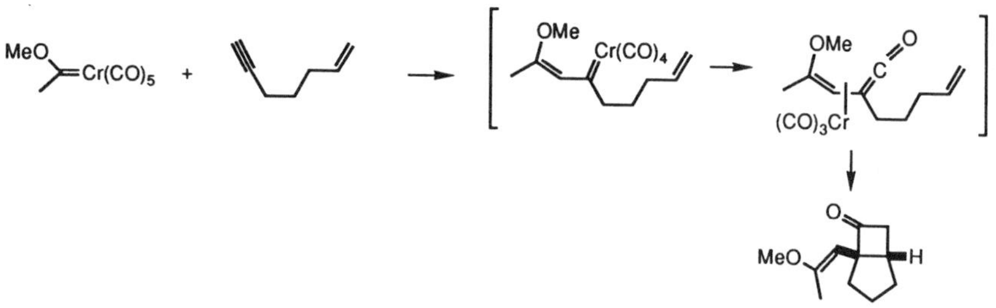

Intramolecular cycloadditions of α,β-unsaturated ketenes are even more general than those of saturated aldo- and keto-ketenes. Perhaps the conjugated double bond accelerates the reaction by lowering the HOMO energy or retarding dimerization and other side-reactions. Eminent examples of synthetic applications of these reactions are β-*trans*-bergamotene [Corey, 1985a; Kulkarni, 1985], 6-protoilludene [Oppolzer, 1986c], methyl dehydrojasmonate [S.Y. Lee, 1988], and retigeranic acid [Corey, 1985b].

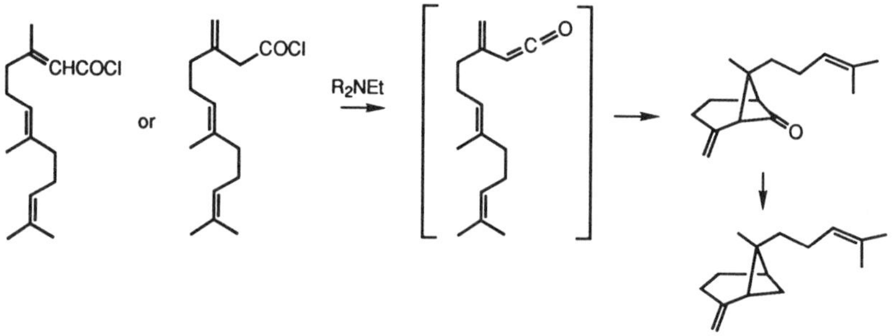

β-*trans*-bergamotene

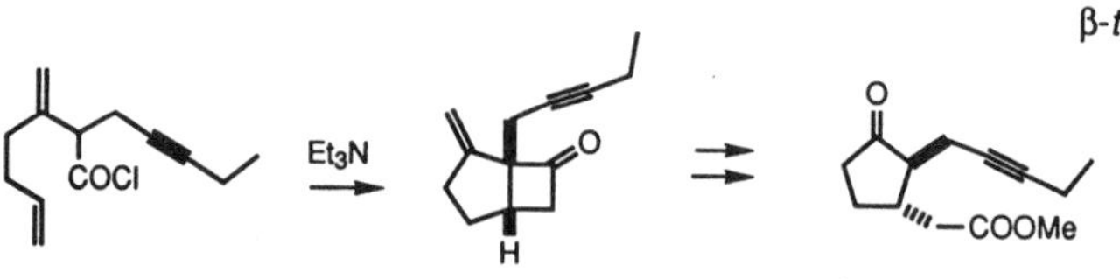

methyl dehydrojasmonate

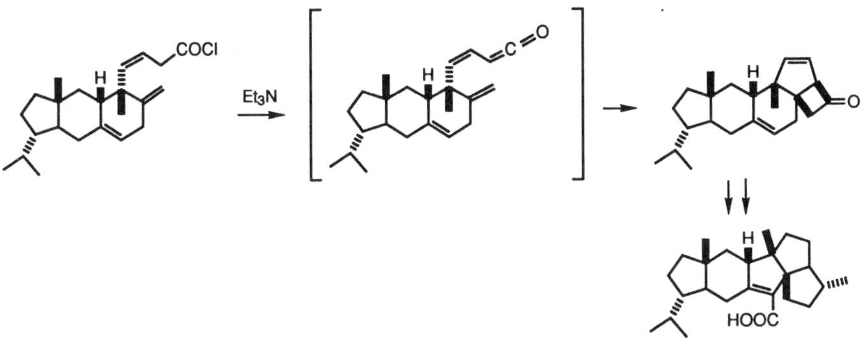

retigeranic acid

o-Allylbenzyl halides undergo palladium-catalyzed carbonylative cyclization to afford benzannulated enol lactones and/or indanocyclobutanones [G. Wu, 1991]. The ketone formation involves intramolecular [2 + 2]cycloaddition.

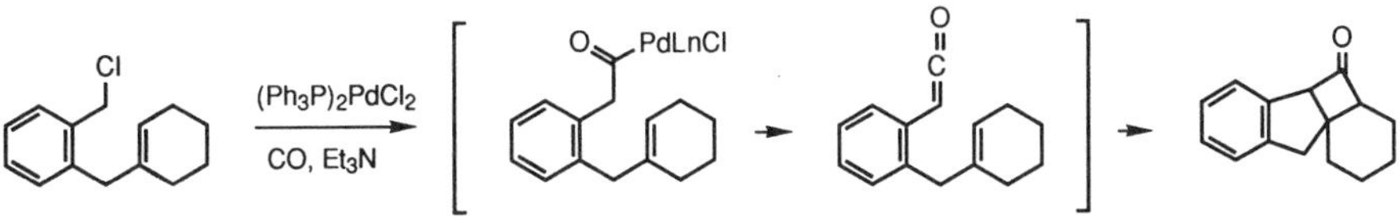

Two spatially juxtaposed alkenes readily undergo photocycloaddition. Two examples of tandem reactions which are terminated by such a process are the formation of homocubanone from a linear tricyclodienone [Cargill, 1971] which must be preceded by a [1.3]-sigmatropic rearrangement, and the conversion of basketene into an intriguing heptacyclic hydrocarbon [Allred, 1973], apparently via a retro-Diels–Alder fission, photochemically induced electrocyclo-reversion, [2 + 2 + 2]-cycloaddition, and the well-known [2 + 2]-cycloaddition.

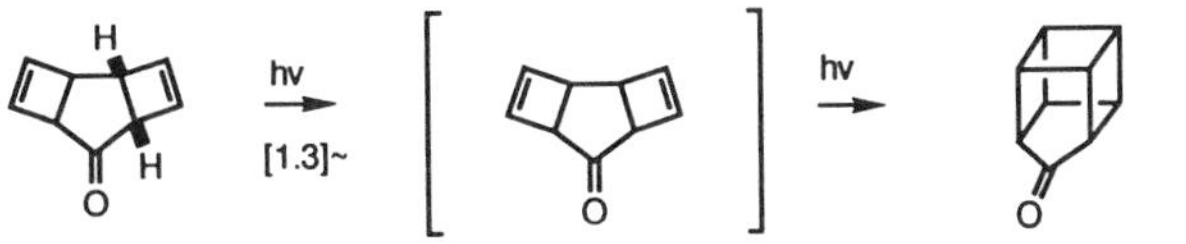

homocubanone

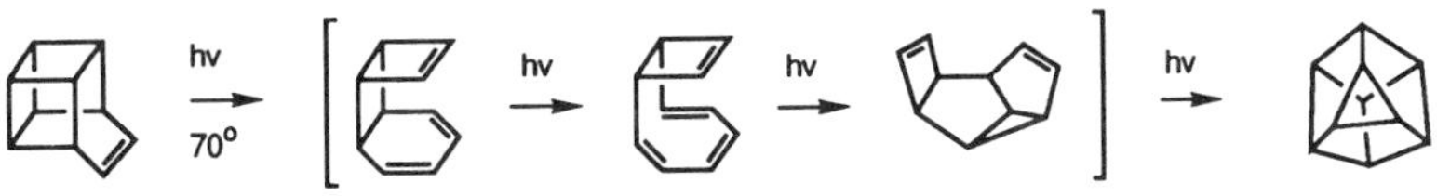

8.2. [2 + 2 + 2]-CYCLOADDITIONS

In 1866 Berthelot reported the thermal trimerization of acetylene to benzene at high temperatures. In more recent times many transition metal systems have been found to catalyze cyclotrimerization of alkynes under much milder conditions by the general mechanism:

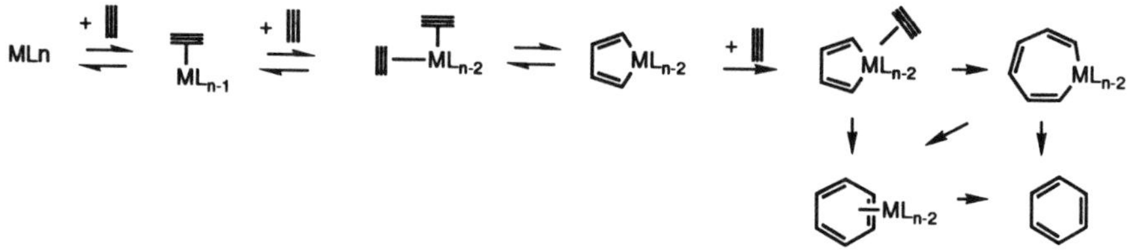

A particularly well-developed theme is the cocyclization of α,ω-diynes with alkynes in the presence of CpCo(CO)$_2$. This method for benzocycloalkene synthesis tolerates a great variety of substituents, including nitrogen-containing groups, as demonstrated by an approach to the protoberberines [Hillard, 1983].

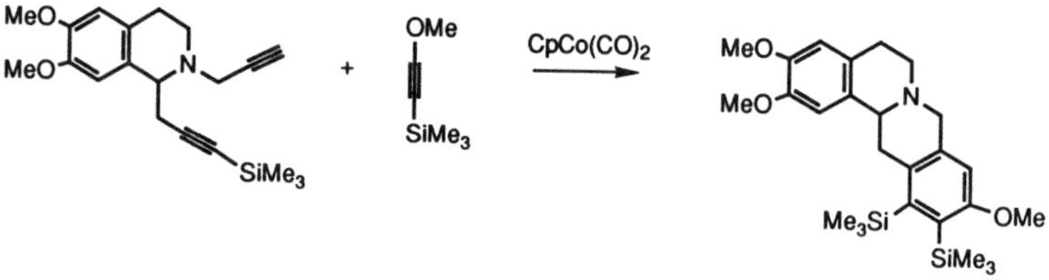

The catalytic cycloaddition to form three rings with complete stereocontrol over three centers greatly facilitated the synthesis of stemodin [Germanas, 1991]. Unfortunately, the (Z)-isomer failed to undergo a similar reaction, which would otherwise provide a convenient entry into the aphidicolane skeleton.

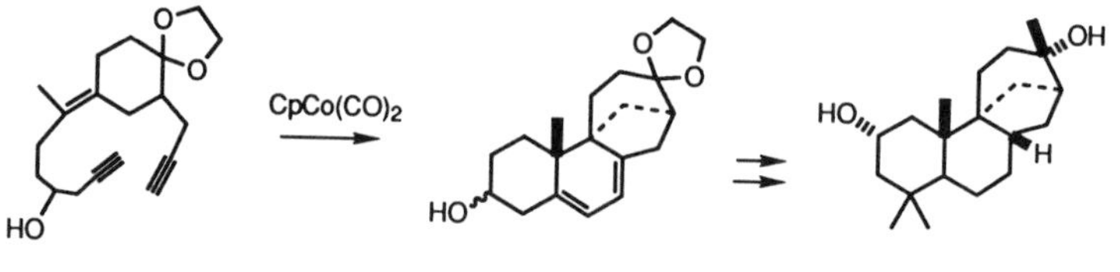

stemodin

Synthetically, the angularly fused cyclobutahydrindanone framework of illudol presents a considerable challenge, which has been met with unique efficiency by the intramolecular [2 + 2 + 2]-cycloaddition [E.P. Johnson, 1991]. A remarkable stereoselectivity at three critical centers was also observed.

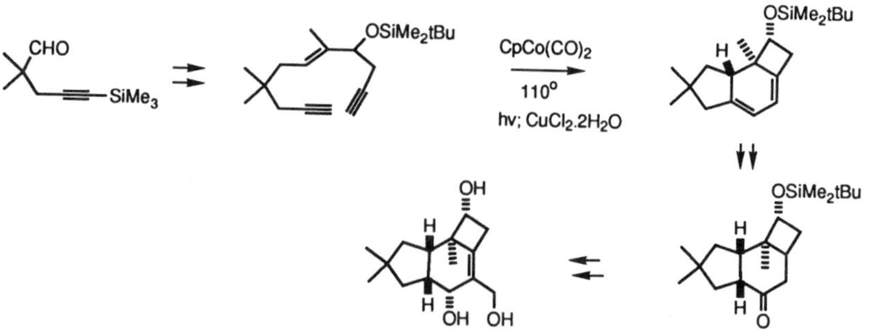

illudol

Severely strained benzocyclobutenes can be obtained rapidly by application of this method [Saward, 1975].

Since benzocyclobutenes are *o*-quinodimethane precursors, the tandem approach to A-aromatic steroids incorporating the co-oligomerization is very attractive. Accordingly, two routes to estrone have been accomplished: D → ABCD method [Funk, 1977], and A → ABCD method [Sternberg, 1982].

estrone

In the former route, bistrimethylsilylacetylene was one of the reactants. The silyl group at C-2 of the product was selectively removed by acidolysis and the required hydroxy group in the A-ring was introduced by oxidative desilylation.

It must be reemphasized that the cobalt-based catalyst is the cheapest but not the only effective compound to accomplish the transformation. An example is the construction of the indane nucleus of calomelanolactone [Neeson, 1988] via a rhodium(I)-mediated intramolecular reaction of a triyne.

calomelanolactone

Cocyclization of alkynes with nitriles affords pyridines. Steric factors are important in determining the ratio of regioisomers produced from unsymmetrical diynes. Pyridines with a sterically less encumbered substitution pattern are formed preferentially. Of course the synthesis of pyridoxine [Geiger, 1984] did not encounter any such problems.

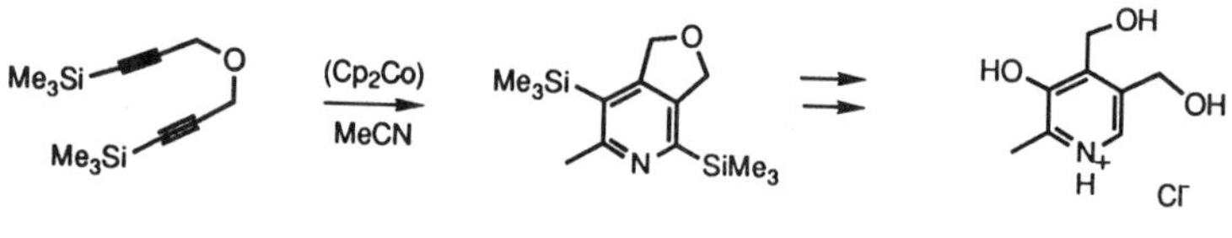

pyridoxine HCl

Replacing the alkene component in the cobalt-catalyzed [2 + 2 + 2]-cycloaddition with an isocyanate leads to pyridone. The reaction of ω-alkynyl isocyanates favors the placement of a bulky substituent of the alkyne coreactant at C-3 of the pyridone ring. Thus, a synthesis of camptothecin [Earl, 1983] exploited the trimethylsilyl group as a regiocontrol element.

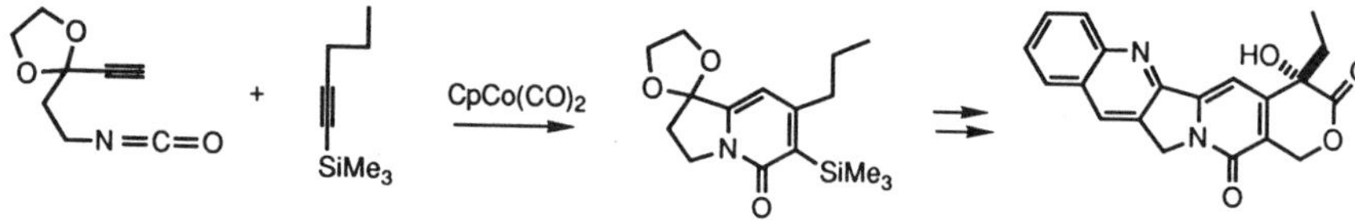

camptothecin

The discovery of a tandom Diels–Alder/double two-alkyne annulation of Fischer carbene complexes [Bao, 1991] greatly facilitates the construction of the steroid ring system.

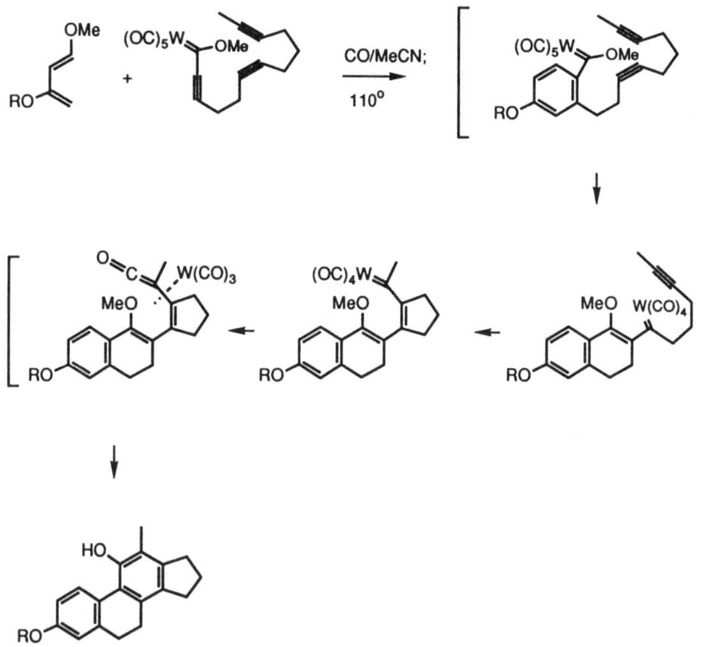

A very unusual cooperative [2 + 2 + 2]cycloaddition occurs in polyquadricyclanylidene[n]rotanes [Trah, 1987a] owing to excellent juxtaposition of the reacting components and relief of strain during the reaction.

8.3. FORMATION OF FIVE-MEMBERED RINGS BY CONCERTED CYCLOADDITIONS

The lack of a synthetic method comparable to the Diels–Alder reaction for the preparation of cyclopentanes was a great impediment to the development of their chemistry. Until recently, stereocontrolled formation of such rings frequently relied on ring contraction approaches. Fortunately, efforts to remedy this situation have borne fruitful rewards [Trost, 1986a].

There has been a tremendous progress in the synthetic application of arene–alkene photocycloadditions, especially the meta-cycloadditions [Wender, 1989b]. Despite the apparently concerted nature of these reactions which occur via exciplexes, the impressive formation of three CC bonds, new ring systems, and up to six new stereocenters makes their inclusion here quite appropriate.

As will be seen in the following examples, such a cycloadduct can be employed in synthesis of many structural types, for example cyclopentanes, cycloheptanes, bicyclo[3.2.1]octanes, and bicyclo[3.3.0]-octanes, by selective cleavage of one or two CC bonds.

While the *m*-cycloadduct of benzene and vinyl acetate has found a use in the elaboration of isoiridomyrmecin [Wender, 1983a], the propellane skeleton of modhephene was constructed from the indane–vinyl acetate adduct [Wender, 1982a].

isoiridomyrmecin

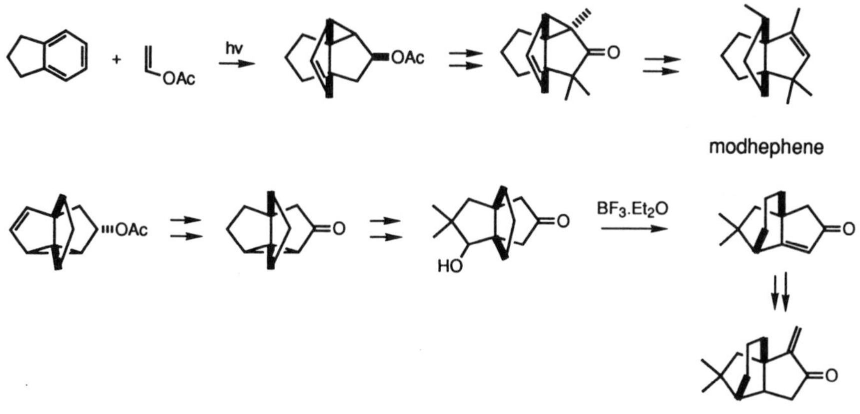

modhephene

decarboxyquadrone

The intramolecular arene–alkene *m*-cycloaddition is even more versatile. It has permitted completion of elegant and concise syntheses of cedrene [Wender, 1981b], hirsutene [Wender, 1982c], isocomene [Wender, 1981a], silphinene [Wender, 1985b], retigeranic acid [Wender, 1990b], laurenene [Wender, 1988c], rudmollin [Wender, 1986], and an approach to β-acoradiene [Wender, 1984] with excellent regio- and diastereoselectivities. This partial list of accomplishments manifests the accessibility to different bridged, condensed, and spiro ring systems.

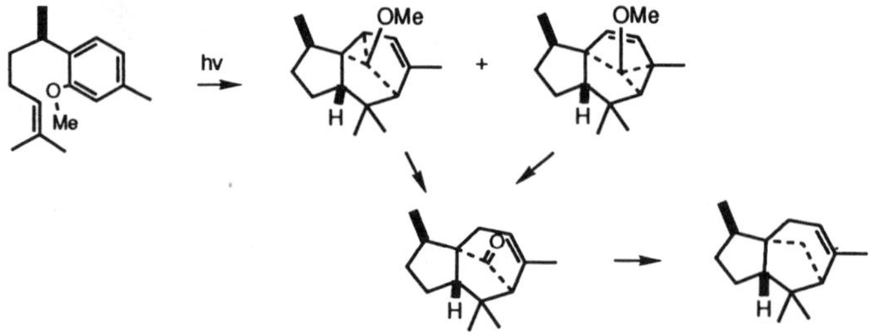

α-cedrene

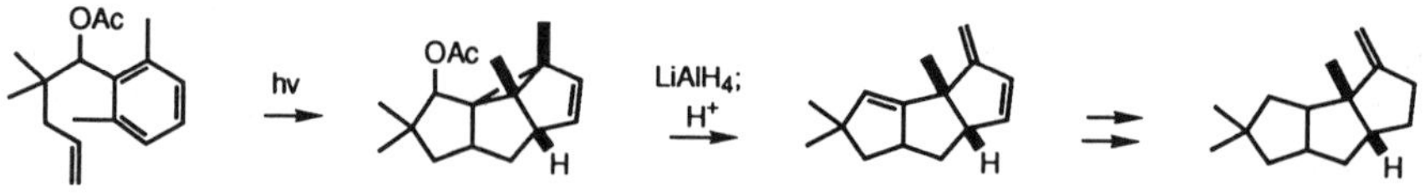

hirsutene

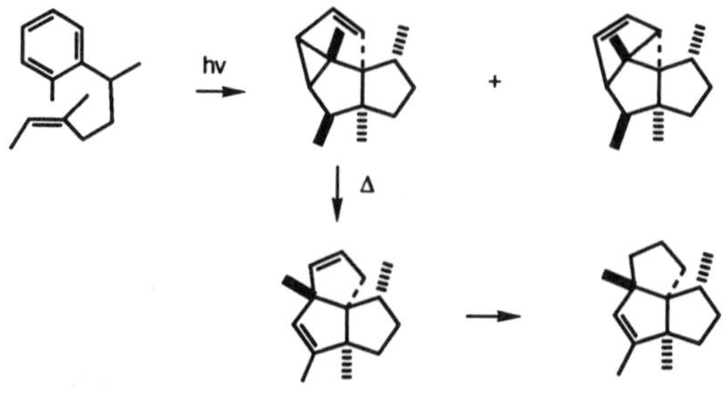

isocomene

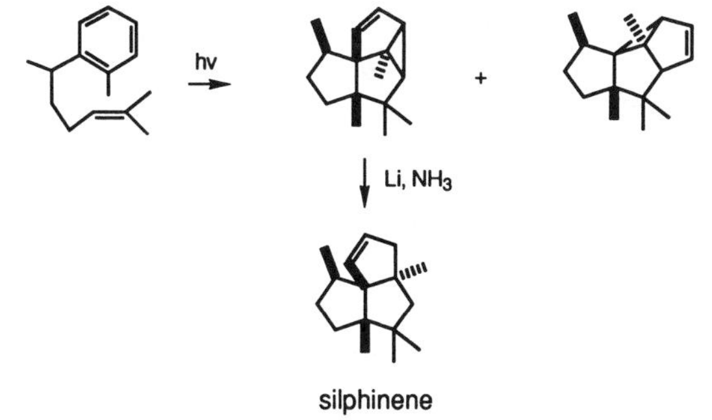
hv
+
Li, NH3
silphinene

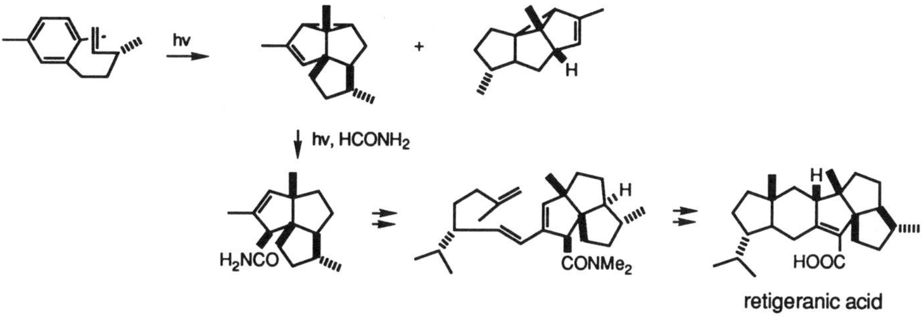
hv
+
H
hv, HCONH2
H2NCO
H
CONMe2
H
HOOC
retigeranic acid

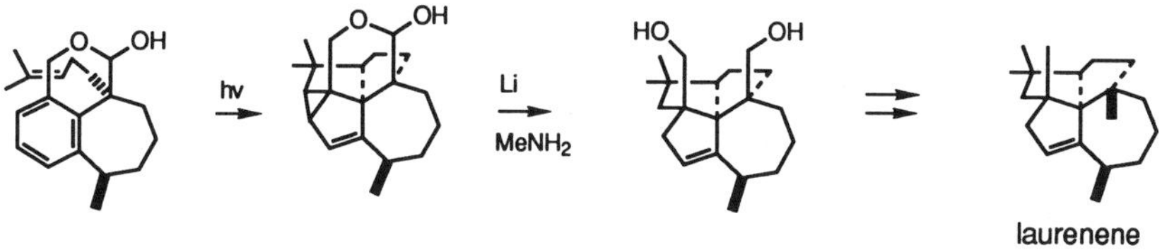
O
OH
hv
O
OH
Li
MeNH2
HO
OH
laurenene

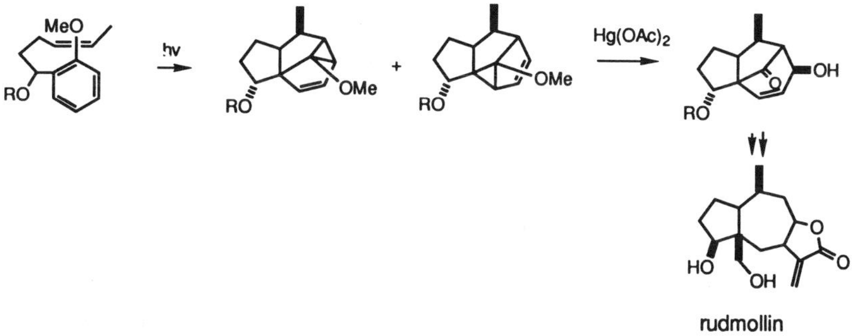
MeO
RO
hv
RO
OMe
+
RO
OMe
Hg(OAc)2
RO
O
OH
O
HO
OH
O
O
rudmollin

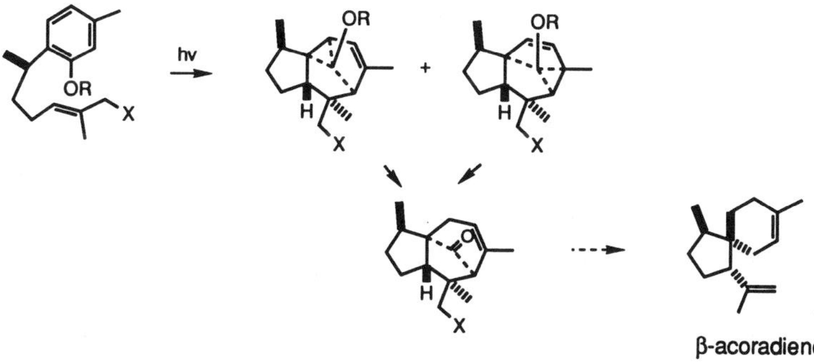
OR
X
hv
OR
H
X
+
OR
H
X
O
H
X
β-acoradiene

A particularly intriguing intramolecular arene–alkene *meta*-cycloaddition is that of a hydrophenanthrene [Palmer, 1980]. On further irradiation the cycloadduct undergoes elimination.

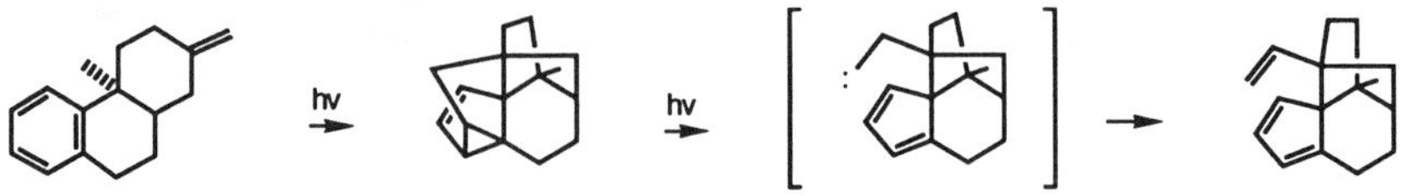

An appealing methodology of cyclopentane synthesis is the cycloaddition involving an all-carbon 1,3-dipole, a convenient synthetic equivalent of which being 2-(2-trimethylsilylmethyl)propenyl acetate. In the presence of Pd(O) catalysts its cycloaddition with electron-deficient alkenes is smooth. The exocyclic methylene group represents a useful handle for structural modifications.

Greatly simplified routes to many cyclopentanoid natural products have been developed. These include albene [Trost, 1982b], loganin [Trost, 1985], and brefeldin-A [Trost, 1986b].

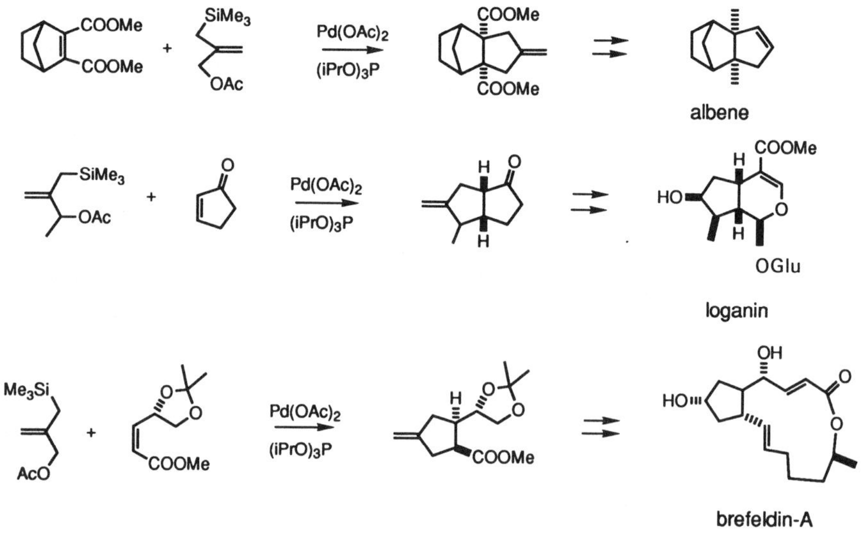

albene

loganin

brefeldin-A

The intramolecular insertion of the CC bond of a benzylidenecyclopropane [R.T. Lewis, 1988] is related to the above reaction. Also noteworthy is the nickel(O)-mediated cycloaddition of methylenecyclopropanes to 2-cyclopentenone to furnish diquinanes of a different substitution pattern from the Pd(O)-catalyzed reactions [Binger, 1988].

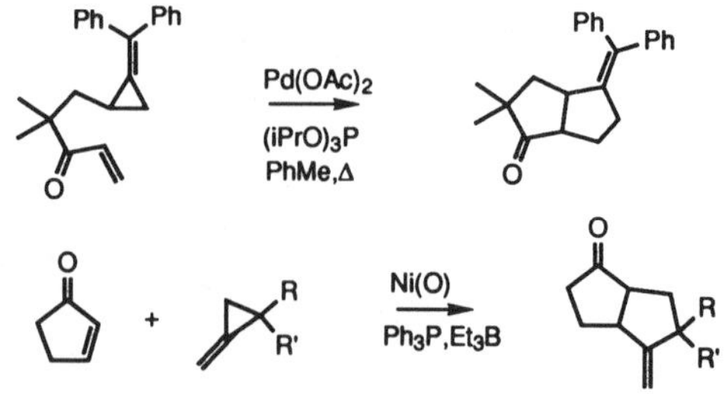

Instead of initiating the cycloaddition by a desilylative process the presence of a strongly acidifying substituent in the precursor of the 1,3-dipole helps the generation of the reactant via deprotonation [Shimizu, 1984].

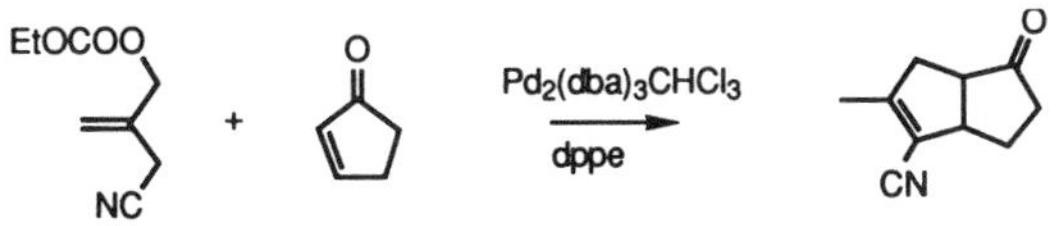

The functioning of allenylsilanes as latent 1,3-conjunctive species [Danheiser, 1981a] is due to the susceptibility of the vinyl cation intermediates to 1,2-migration. As indicated in the following reaction the products are silylcyclopentenes with a transposed silyl group.

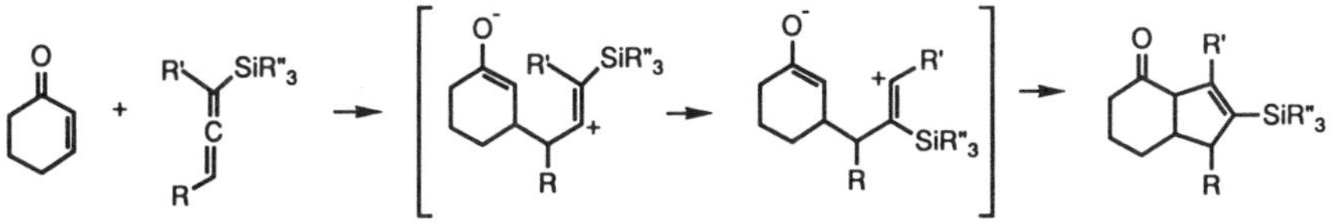

An azulene synthesis [D.A. Becker, 1989] represents an extension of this methodology.

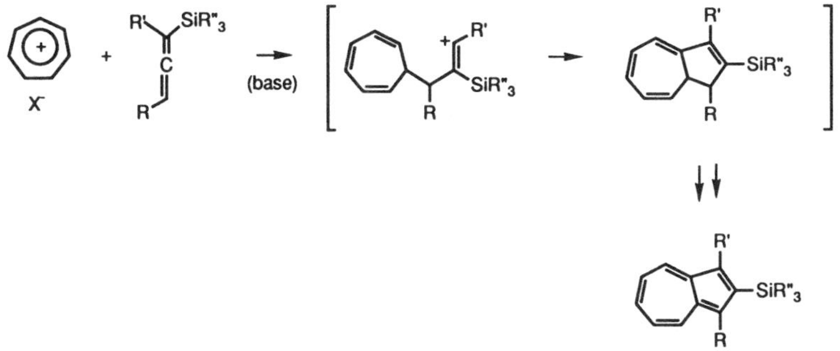

Allyl- and propargyliron compounds such as $(\eta^1\text{-allyl})$dicarbonyl$(\eta^3\text{-cyclopentadienyl})$iron and the $(\eta^1\text{-propargyl})$ analog react with Michael acceptors [Abraham, 1982; Bucheister, 1982] very similarly to allenylsilanes, with respect to the migration of the iron atom. The mechanisms are different in that the intermediates for the latter reactions are π-complexes.

Interestingly, oxytropeniumiron tricarbonyls are readily attacked by $(\eta^1\text{-allyl})$iron complexes [Watkins, 1984] resulting in the formation of hydrazulenes.

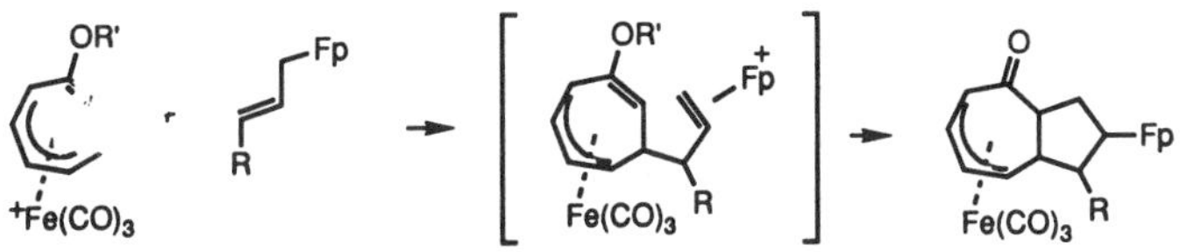

It is interesting to note that among other allylmetal complexes allylstannanes react with α,β-unsaturated acyliron complexes in the presence of aluminum chloride to produce cyclopentane derivatives [Herndon, 1991]. Allylsilanes tend to submit the allyl group to the acyliron complexes, while allylgermanes give mixtures of cyclic and open-chain products. Allyllead and allylmercury compounds decompose under the reaction conditions.

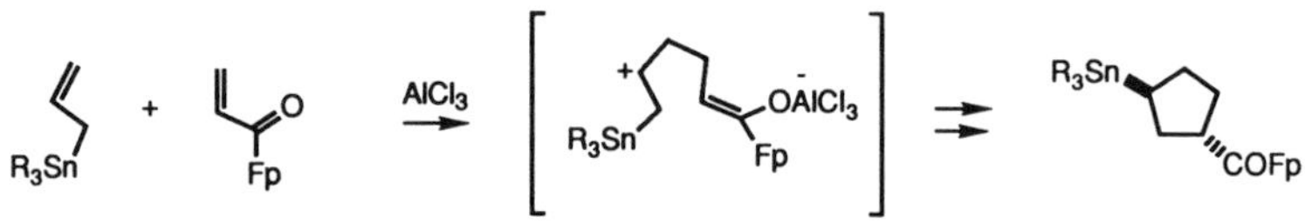

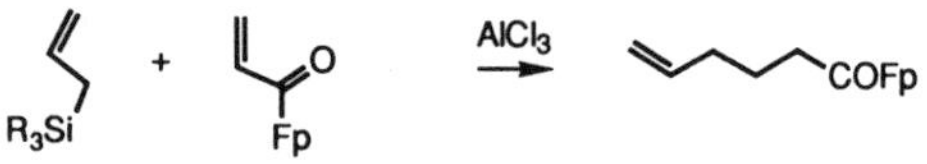

2-Oxyallyl cations are another series of 1,3-conjunctive reactants. α,α'-Dibromoketones are convenient sources of these species [Noyori, 1979a].

Generally, alkene trapping of 2-oxyallyl cations generated by debromination of α,α'-dibromoketones using an iron carbonyl unfortunately is not very efficient, owing to the nonconcerted nature of four-electron cycloadditions. Despite such defects the reaction has been employed in synthesis of α-cuparenone [Y. Hayakawa, 1978], camphor, and campherenone [Noyori, 1979b], an intramolecular cycloaddition operating in the latter two cases.

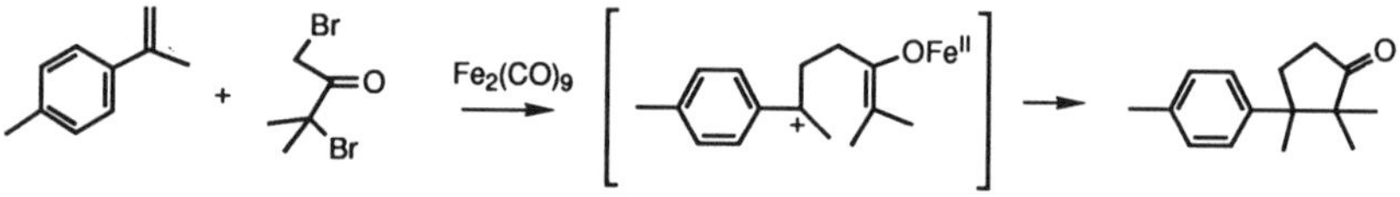

α-cuparenone

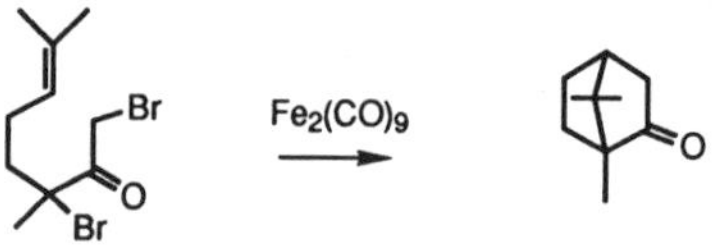

camphor

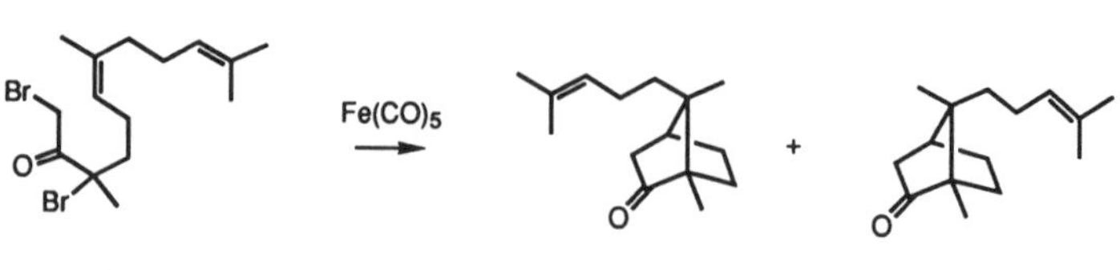

campherenone

In the α-cuparenone synthesis a high regioselectivity for the cycloaddition was observed. Perhaps the procyclic intermediates were formed reversibly, and only the more stable one collapsed in the forward direction. Consequently, it is instructive to compare this debrominative route with the ionization approach [Sakurai, 1979].

Zwitterionic 2-oxyallyl ions may be generated by fluoride ion promoted β-elimination of 1-chloromethyl-1-trimethylsilyloxiranes [Chan, 1974; Ong, 1976]. The initially formed methyleneoxiranes, which are tautomers of the cyclopropanones and the 2-oxyallyl ions, may react according to reagents presented to them.

8.4. HIGHER-ORDER CYCLOADDITIONS

Ionic [4 + 2]-cycloadditions are orbital symmetry-allowed. The entropy demand engenders the four-electron component of a pentadienyl cation to maintain a U-shape, for example by bridging the termini. When such conditions are met, cycloadditions may be induced.

Neolignans such as guianin which contain a bicyclo[3.2.1]octane subunit are excellent targets for testing synthetic potentials of the ionic [4 + 2]-cycloadditions. The oxygen substituent attaching to the one-carbon bridge of guianin must be present in the 4e addend, therefore an oxonium ion derived from a *p*-benzoquinone seems to be the best candidate [Büchi, 1977b]. Indeed, the cycloaddition proved successful, but depending on the reaction conditions the initially formed bicyclic intermediate could undergo rearrangement to give quinoid isomers possessing the benzofuran skeleton.

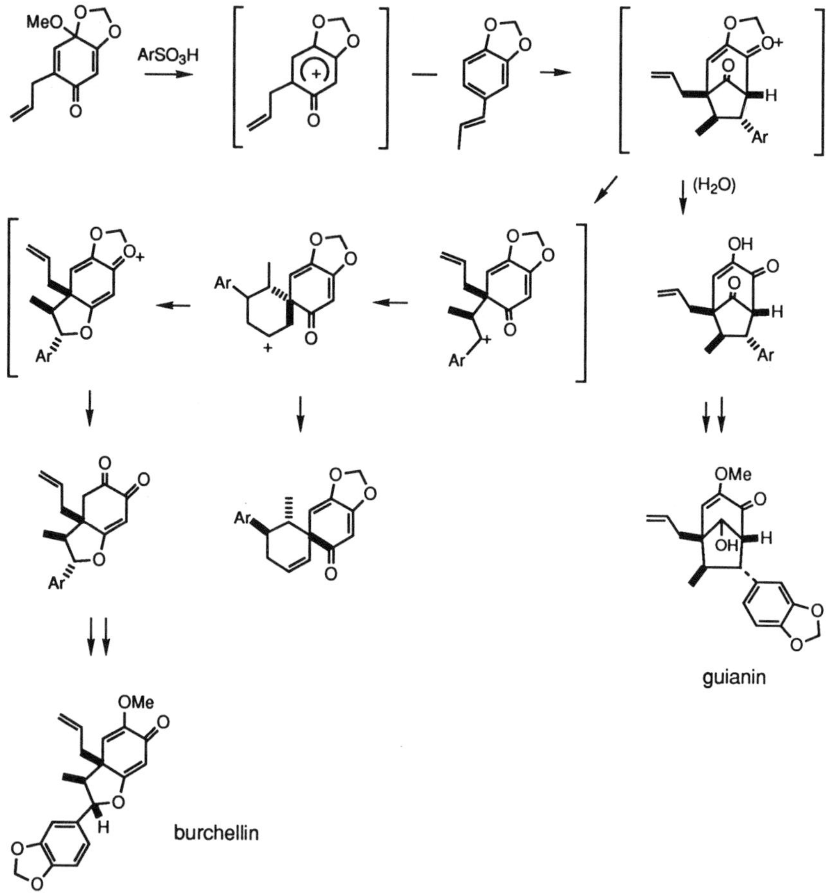

guianin

burchellin

Condensed [2 + 2]- and [3 + 2]-cycloadducts from titanium(IV)-promoted reaction of 1,4-benzoquinones and styrenes may be derived by cleavage of the bridged ring intermediates followed by CC or CO bond formation [T.A. Engler, 1988]. The nature of the Ti(IV) catalyst has a dramatic effect on the type of products obtained. Generally, TiCl$_4$ gives mainly the dihydrofurans, while TiCl$_4$-(iPrO)$_4$Ti mixtures tend to afford more benzocyclobutenes.

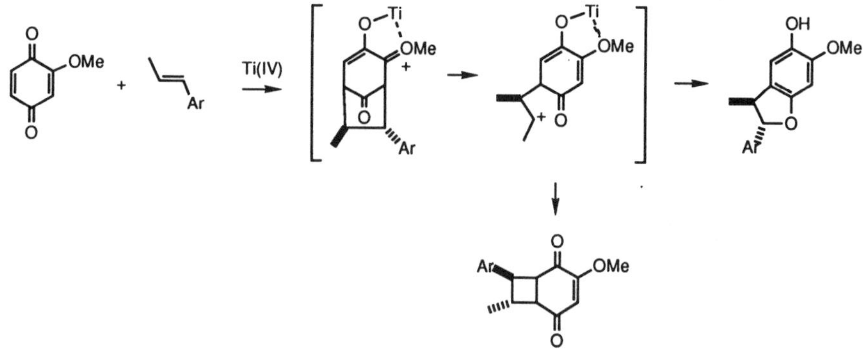

Chemically oxidized species of phenols have also been trapped by styrenes to furnish dihydrobenzofuran neolignans [S. Wang, 1991].

As expected, the intramolecular ionic [4 + 2]-cycloaddition is much more favorable. The conversion of perezone into the pipitzols [Joseph-Nathan, 1977] actually served as an inspiration to the synthetic approach to guianin indicated above.

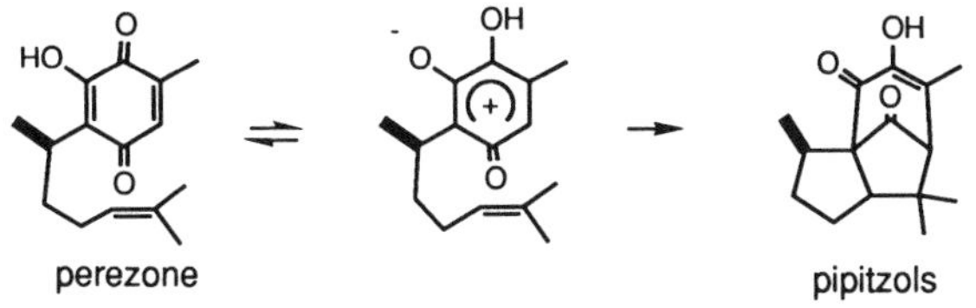

perezone   pipitzols

Electrochemical generation of the O-alkylquinones appears to be more efficient. The bicyclic precursor of helminthosporal [Shizuri, 1986b] has been obtained in an excellent yield. Subsequent syntheses of 8,14-cedranoxide [Shizuri, 1987], silphinene [Shizuri, 1989], and pentalenene [Shizuri, 1990] exploited the intramolecular version of the cycloaddition.

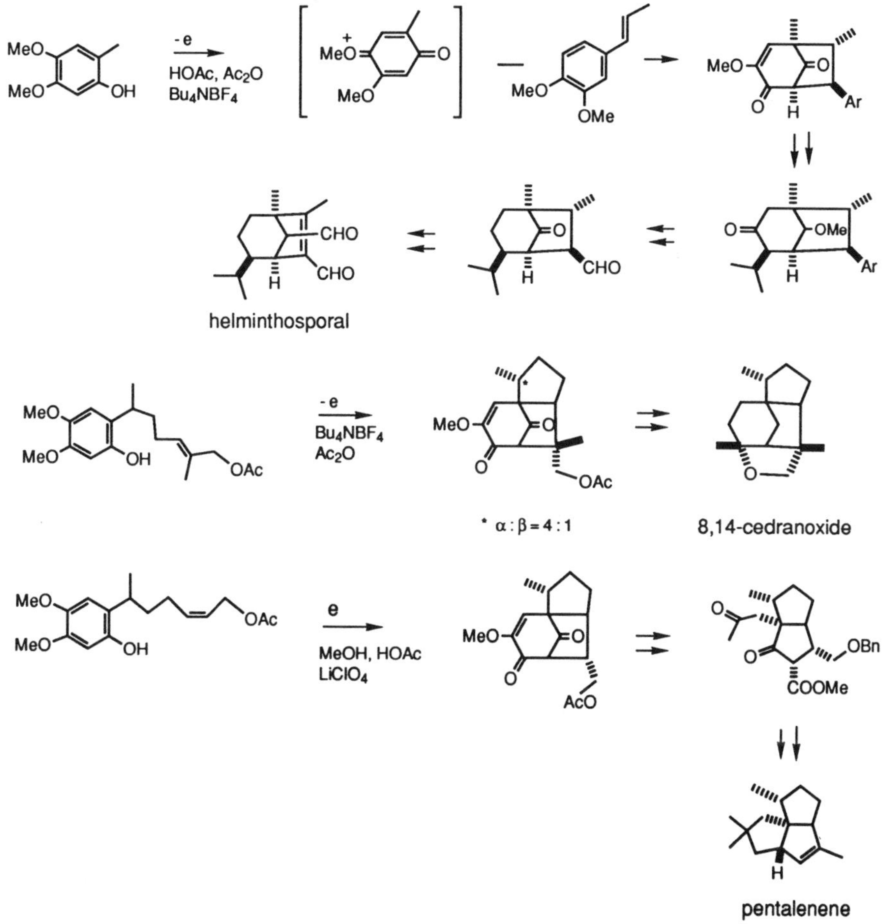

helminthosporal

* α : β = 4 : 1   8,14-cedranoxide

pentalenene

The cycloaddition of 2-oxyallyl species, generated in situ from α,α'-dibromoketones, with furan and pyrrole derivatives has been studied [Noyori, 1979]. While the reaction involving α,α'-dibromoacetone failed to give any adducts, $\alpha,\alpha,\alpha',\alpha'$-tetrabromoacetone proves to be an adequate surrogate.

This cycloaddition has contributed to the development of new and convergent approaches to tropane alkaloids [Noyori, 1978a] and C-nucleosides including showdomycin [Noyori, 1978b].

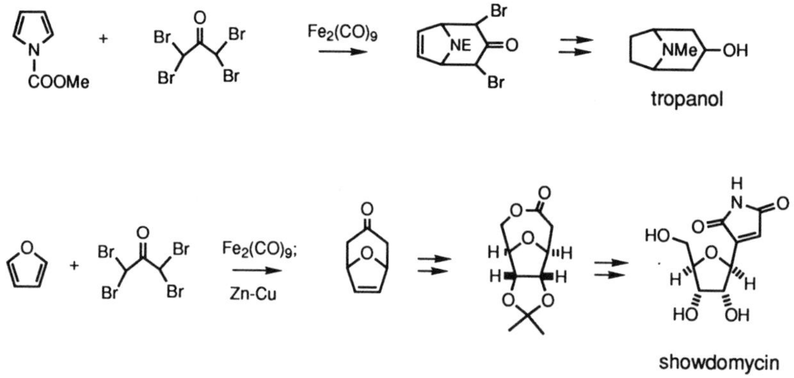

tropanol

showdomycin

A rather unusual cycloaddition was observed [Schultz, 1985] when 4-(2-furylmethoxy)-2,4,6-trimethyl-2-5-cyclohexadienone was irradiated. Photo-rearrangement of the dienone gave a zwitterionic species which was trapped by the furan.

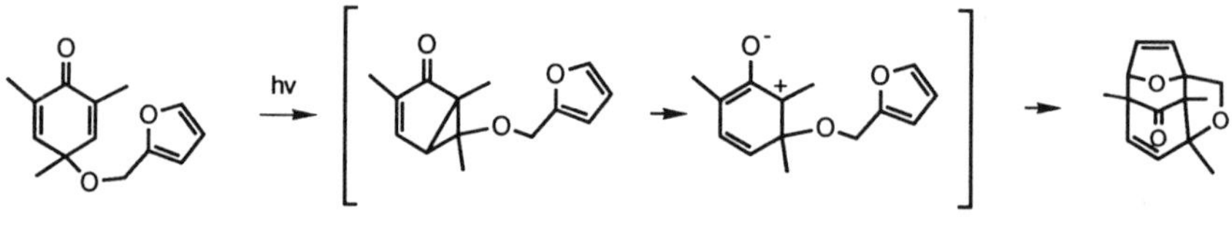

Another interesting method for generating the 3C,2e-addend is by Lewis acid coordination of dienals or cleavage of activated vinylcyclopropanes [Ohno, 1990].

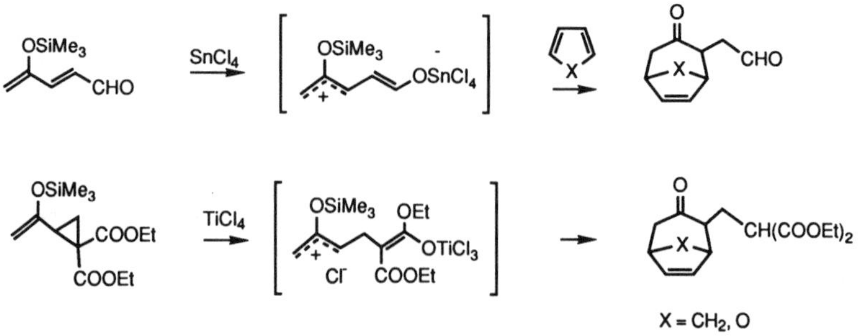

$X = CH_2, O$

Note that dimethyl 2-vinylcyclopropane-1,1-dicarboxylate undergoes ring cleavage to afford a formal 1,3-zwitterionic species in which the two ions are not directly related but individually stabilized [Shimizu, 1985]. The zwitterion has been trapped by methyl acrylate.

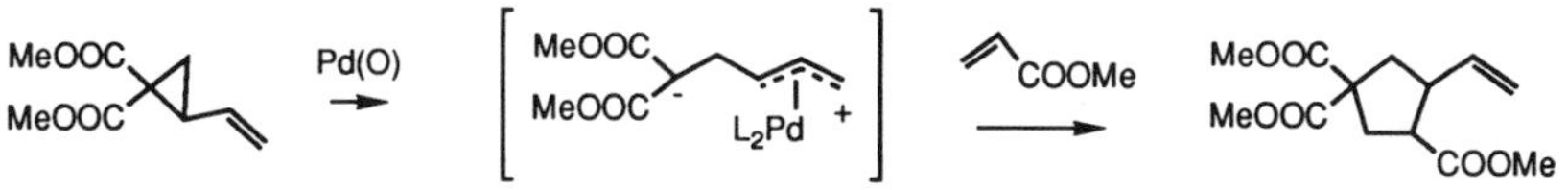

The cycloaddition of pyrylium 3-oxide zwitterions with alkenes to furnish 9-oxabicyclo[3.2.1]non-3-en-2-ones is isoelectronic with the Diels–Alder reaction. The pyrylium oxides are conveniently formed by thermal elimination of acetic acid from the 6-acetoxy dihydro-3-pyrones. Consequently, the elimination and the cycloaddition steps can be achieved in tandem.

The intramolecular version of such tandem reactions is particularly appealing, as attested by very successful synthetic approaches to several polycyclic terpenes including cryptofauronol and valeranone [Sammes, 1983], the preparation of a potential precursor of the tiglane, daphnane, and ingenane diterpenes [Wender, 1989a], and a useful intermediate for the synthesis of colchicine [Lupi, 1990].

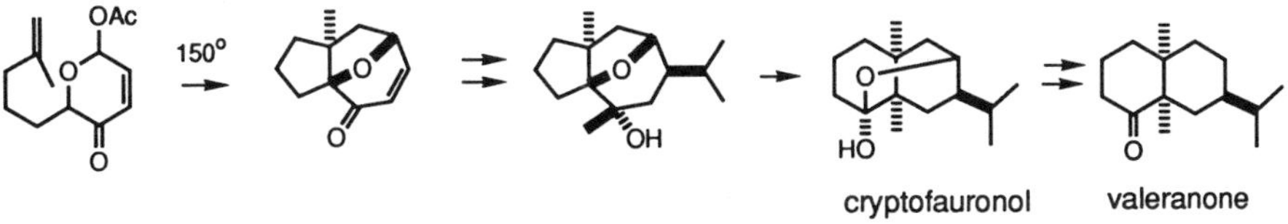

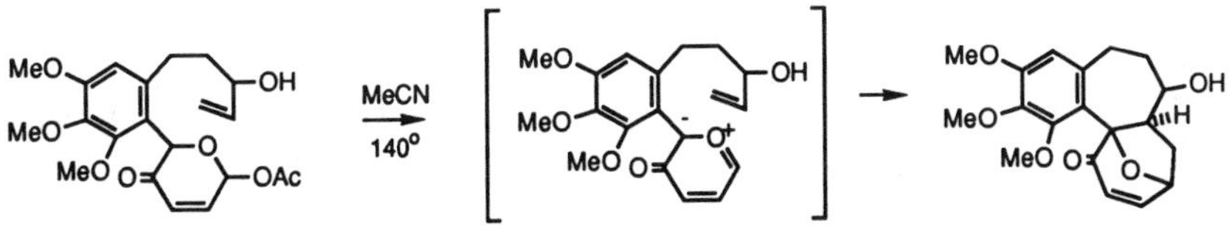

cryptofauronol   valeranone

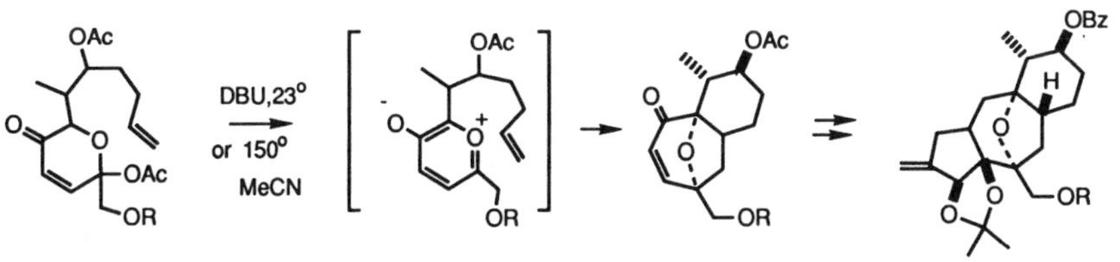

The intramolecular tropone–alkene photocycloaddition is a $6\pi + 2\pi$ process, an application of which is in the formation of a precursor of dactylol [Feldman, 1989].

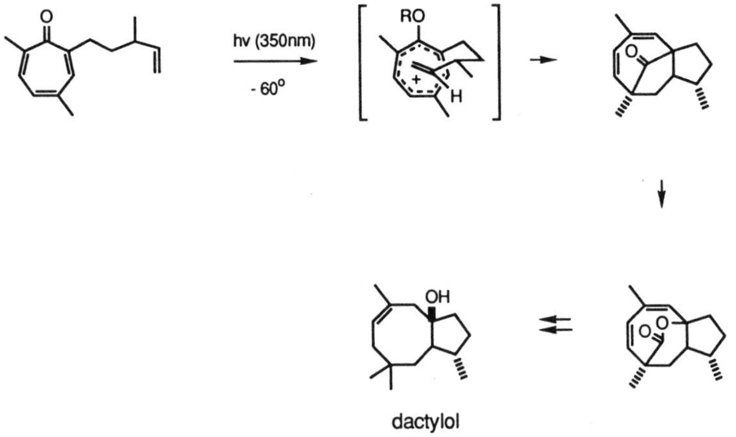

dactylol

Intramolecular [6 + 4]-cycloadditions in which the six-electron component is a fulvene [T.-C. Wu, 1983] is significant because it contrasts with the intermolecular mode which prefers the formation of Diels–Alder adducts. The facility of some of such reactions was shown by the tandem sequence initiated by enamination.

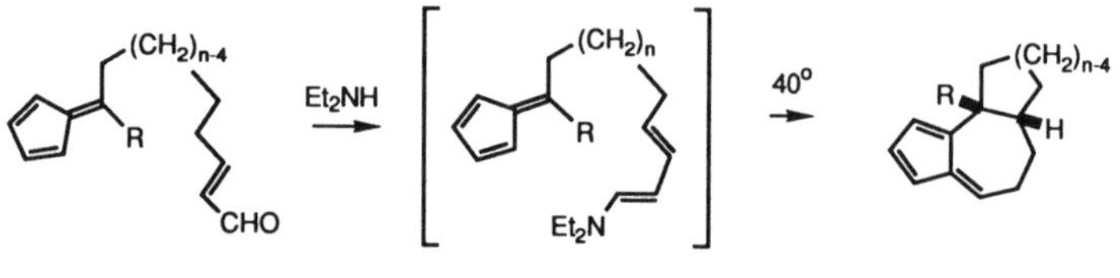

An o-quinodimethane generated in the presence of a fulvene in the same molecule was trapped. The cycloadduct arose from [6 + 4]-cycloaddition in the endo mode which is different from the exclusive exo-stereoselectivity exhibited by other such cycloadditions of fulvenes.

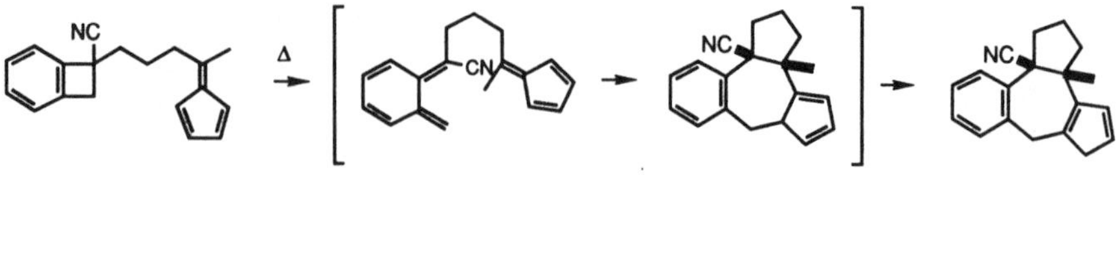

8.5. 1,3-DIPOLAR CYCLOADDITIONS

Contrary to the scarcity of methods for cyclopentane formation, five-membered heterocycles are relatively easily accessible, thanks to the 1,3-dipolar cyclo-additions. In such a process 6π-electrons are involved, four from the 1,3-dipole and two from the dipolarophile. Three types of 1,3-dipolar cycloadditions are known [Sustmann, 1974]: type I dipoles have relatively high-lying HOMOs and LUMOs, and are generally referred to as HOMO-controlled, or nucleophilic dipoles; type II dipoles possess very low-lying frontier orbitals, they are electrophilic, undergoing LUMO-controlled reactions; the intermediate type II dipoles are ambiphilic, their cycloadditions are accelerated by either a donor or acceptor substituent on the dipolarophile. While there are numerous 1,3-dipoles, and consequently many dipolar cycloadditions, only the few which have contributed enormously to carbogenic synthesis are mentioned here. Since the dipolar species are unstable, they must be generated in situ. Necessarily, most 1,3-dipolar cycloadditions are part of tandem reactions.

8.5.1. Tandem Dehydration–Cycloaddition

Nitroalkanes undergo dehydration on exposure to phenyl isocyanate to give the unstable nitrile oxides. These nitrile oxides may be trapped by alkenes to form isoxazolines. Since the reaction effects CC and OC bond formation, and cleaving N—O and C=N bond is rather easy, its value lies not only in the heterocycle preparation. In other words, the methodology provides an entry to 1,3-amino alcohols; and β-hydroxy ketones may be assembled in a manner different from the aldol condensation pathway. From the numerous applications of the method to complex synthesis the routes to chanoclavine-I (Kozikowski, 1980] and biotin [Confalone, 1980] may be cited.

chanoclavine-I

biotin

The formation of a dimeric product from a nitroalkyl acrylate [Asaoka, 1981] is interesting because it furnished a synthetic intermediate of pyrenophorin in one step.

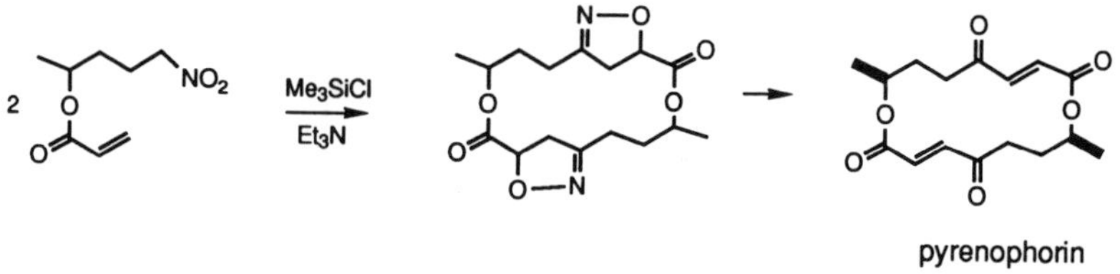

pyrenophorin

N-Acyl-N-alkyl-α-amino acids give oxazolone ylides very readily, for example by treatment with acetic anhydride. Such ylides react with acetylenic dipolarophiles to afford pyrrole derivatives upon decarboxylation of the cycloadducts. A route to the mitosane ring system [Rebek, 1984] has been developed on the basis of this process.

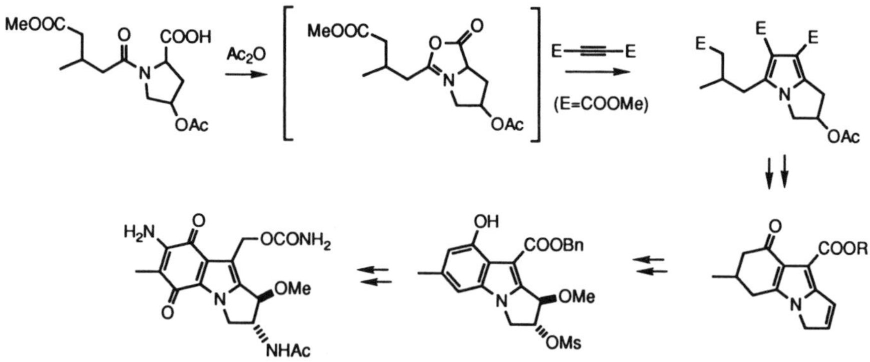

8.5.2. Elimination–Cycloaddition

1,3-Dipoles may be generated by elimination reactions. For example, reaction of N-(trimethylsilylmethyl)imines with an acyl chloride in the presence of an alkynoic ester leads to N-acyl-3-pyrroline-3-carboxylic esters [Achiwa, 1981]. The dipolar species emerged from either desilylation of the N-acyliminium ions or decomposition of the N-(α-chloroalkyl) amides.

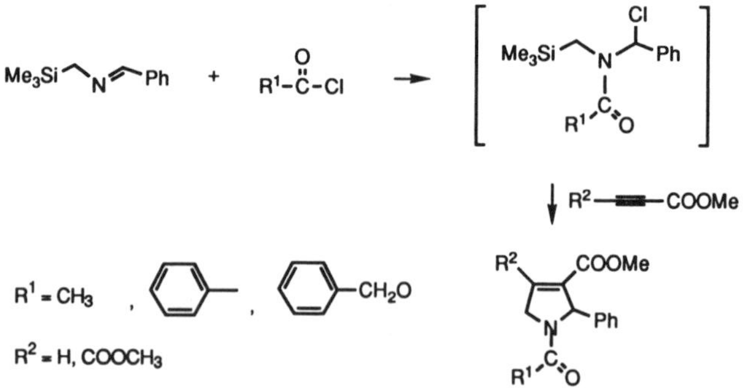

Azomethinium ylides are derivable from N-Alkylation of such imines followed by treatment with fluoride ion. A convenient assembly of eserethole [R. Smith,

1983] demonstrated such a protocol and the intramolecular trapping of a formamidine methylide.

eserethole

O-Alkylation of *N*-(trimethylsilyl)methyl amides facilitates desilyation. The 1,3-dipoles generated by this method cycloadd to various dipolarophiles [Vedejs, 1980].

Cleavage of the O—C(2) bond of the oxazolium ring by reaction with cyanotrimethylsilane and subsequent desilylation constitute another variant of ylide formation. An indoloquinone can be acquired via an intramolecular reaction of such an ylide with an alkyne group in a sidechain [Vedejs, 1989].

indoloquinones

In a synthesis of cocaine [Tufariello, 1979] the cyclic nitrone was generated from an isoxazolidine by thermolysis (elimination of methyl acrylate). A tandem intramolecular cycloaddition that ensued furnished a tricyclic product containing all the potential functionalities of the alkaloid.

cocaine

Simultaneous generation of nitrone and alkene moieties by fragmentation of *N*-hydroxy-*cis,cis*-5-(tosyloxy)decahydroquinoline [LeBel, 1985] also caused an intramolecular cycloaddition in the absence of a more reactive dipolarophile. The product is a tricyclic isoxazolidine.

Finally, a tandem cheletropic elimination of benzene and intramolecular 1,3-dipolar cycloaddition [Hoffmann, 1985] may be mentioned.

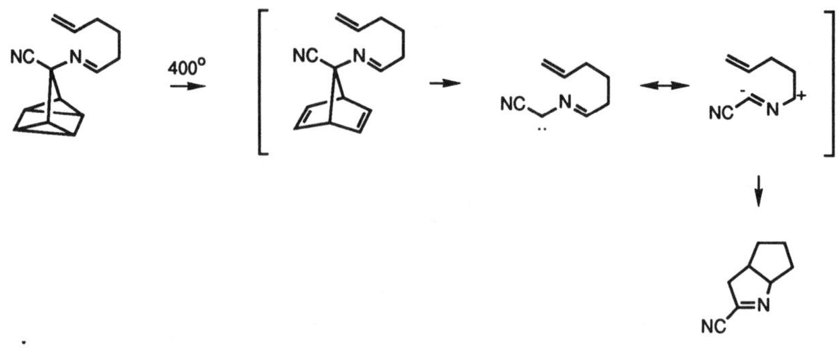

8.5.3. Condensation–Cycloaddition

Nitrones are another class of 1,3-dipoles which have witnessed extensive use in synthesis. The work of chanoclavine-I, paliclavine [Oppolzer, 1983a], and (+)-luciduline [Oppolzer, 1978c] attest to the versatility of the cycloaddition with which intriguing molecular skeletons can be constructed.

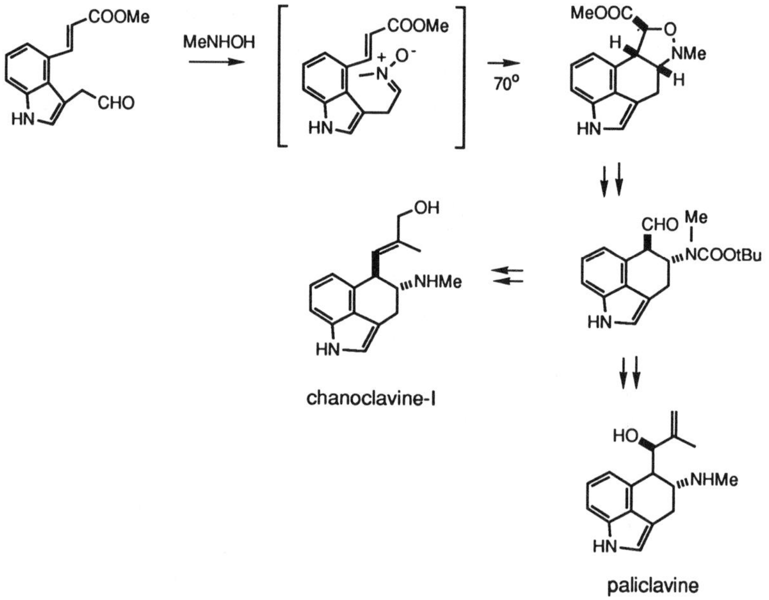

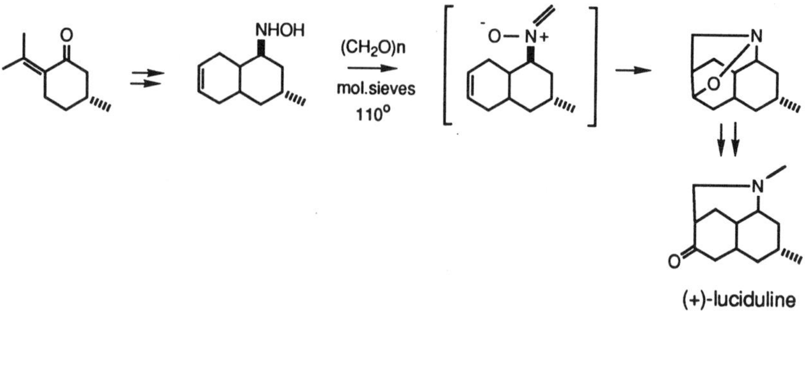

chanoclavine-I

paliclavine

(+)-luciduline

Iminium ylides can be generated by reaction of carbonyl compounds with secondary amines which contain an anion-stabilizing substituent at the α-position of the amino nitrogen. Consequently proline esters are readily accessed. A synthesis of Sceletium alkaloid-A$_4$ [Confalone, 1984] illustrated the formation of hydrindole system by a 1,3-dipolar cycloaddition.

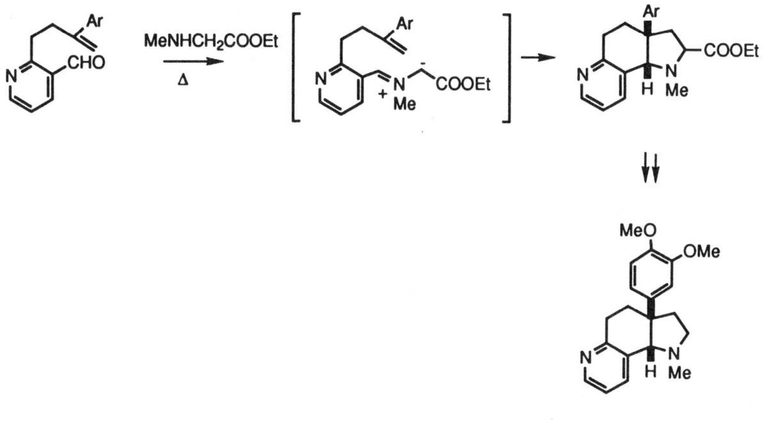

sceletium alkaloid A$_4$

The decomposition of α-diazoketones with intramolecular participation of a remote carbonyl group serves as a convenient route to oxonium ylides. Rhodium(II) salts are the preferred catalyst to promote the acylcarbenoid formation. In the presence of an appropriate dipolarophile an oxabicyclic structure may be formed, such as the molecular framework of the brevicomins [Padwa, 1990a].

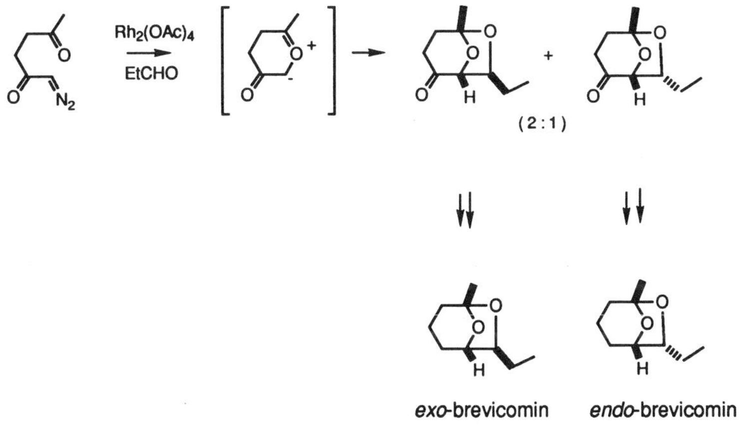

exo-brevicomin   endo-brevicomin

An intriguing polycyclization from an acyclic bis(α-diazoketone) involves formation of the n-donor by an initial cyclopropanation [Gillon, 1982].

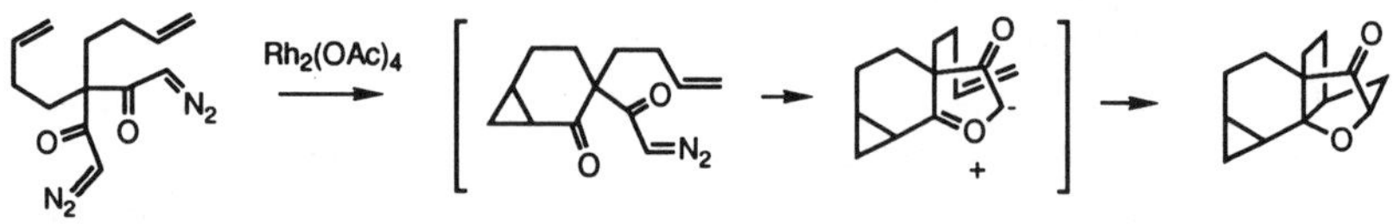

An oxime undergoes *N*-Alkylation to give a nitrone. In the presence of an added or intramolecular dipolarophile a tandem cycloaddition would take place [Grigg, 1989b]. An epoxide is an adequate alkylation agent and it may also be present in the same molecule as the oxime or introduced separately.

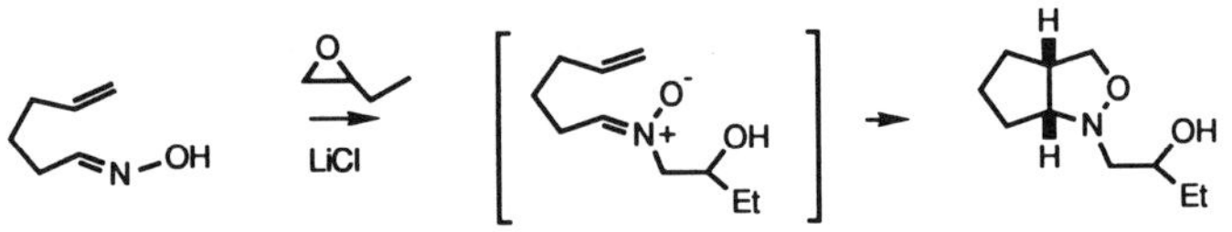

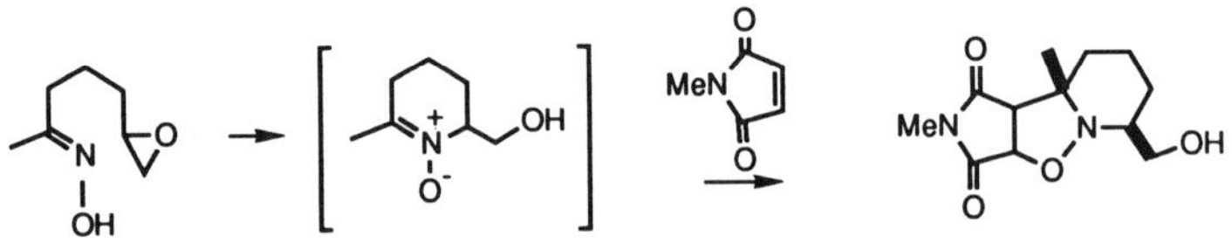

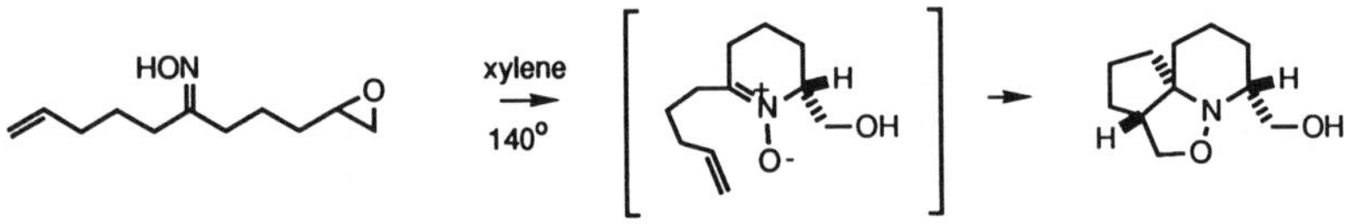

Nitrones obtained by reaction of alkenyl oximes with Michael acceptors may be induced to form isoxazolidines [Grigg, 1989a]. In the following example the formation of three rings is illustrated.

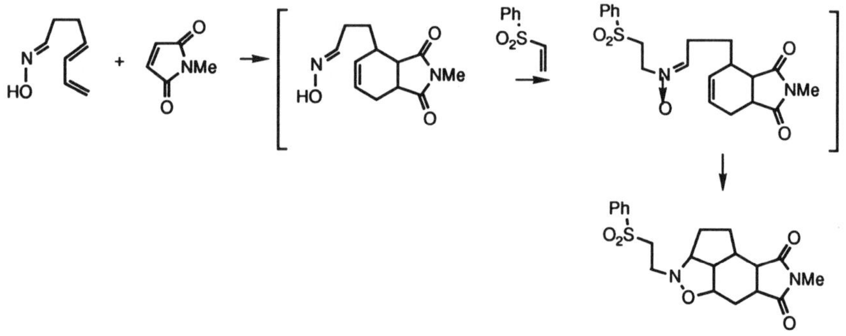

Two other versions of the same tandem reactions are those involving oximation of a carbonyl compound which contains a Michael acceptor and a dipolarophile locating at proper distances [Armstrong, 1987a], and the reaction of a ketoxime with a bifunctional Michael acceptor [Armstrong, 1987b; Normal, 1991].

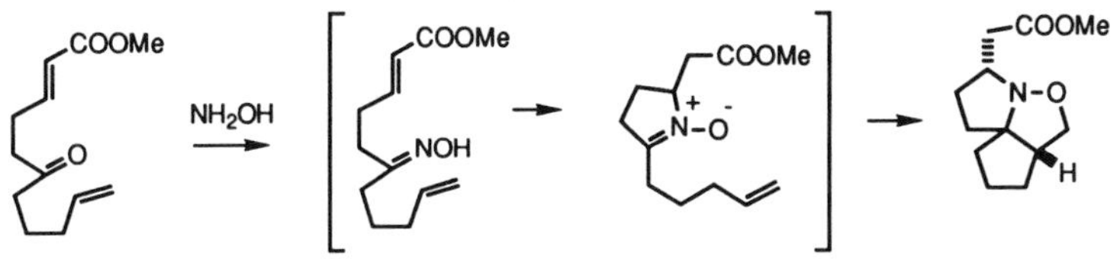

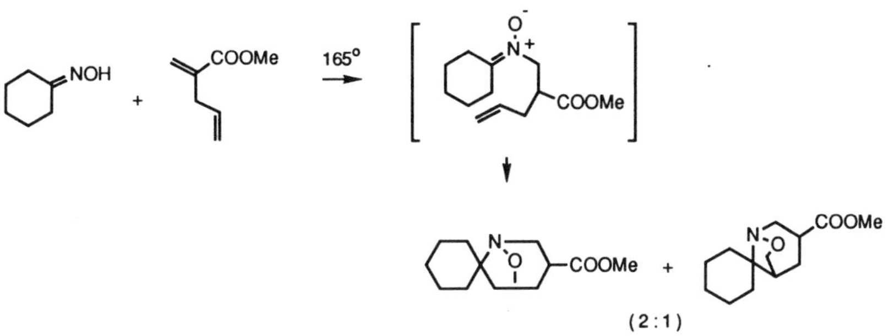

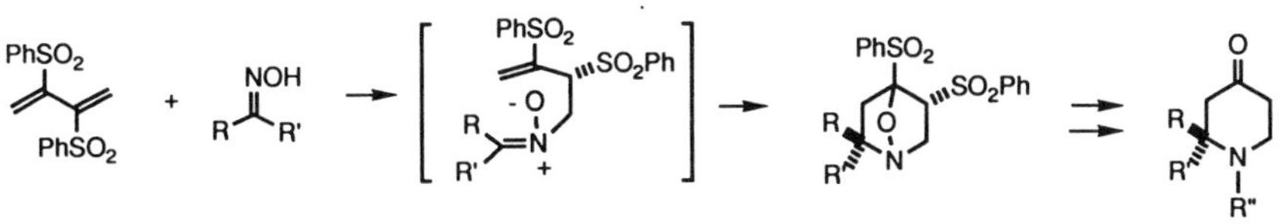

(2:1)

The reaction of *o*-allyl-β-nitrostyrene with *t*-butylisonitrile gave rise to a heterocyclic intermediate which cycloadded to the isolated double bond spontaneously [Knight, 1987]. A tricyclic isoxazoline was isolated as the product.

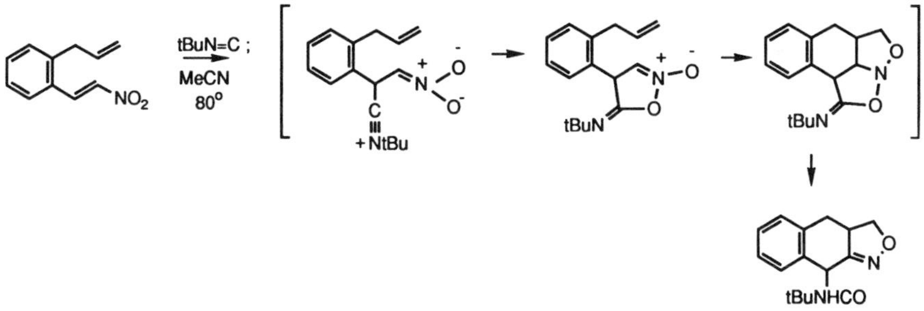

Allyl transfer from allyltrimethylstannane to a nitroalkadiene also promoted ring formation [Uno, 1989]. The very high *threo*-selectivity is noteworthy.

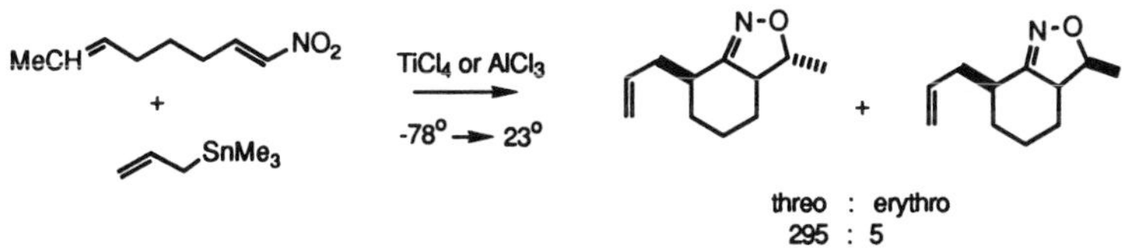

threo : erythro
295 : 5

Tandem [4 + 2]- and [3 + 2]-cycloadditions have been effected [Denmark, 1990]. Under Lewis acid catalysis the intermolecular [4 + 2]-cycloaddition of

unsaturated nitrones with alkenes proceeds at low temperatures, the intramolecular [3 + 2]-cycloaddition then follows.

If the ene reaction is considered as a condensation the following transformation
also belongs to the same category. The hetero-ene reaction between an oxime
and alkene generates a nitrone which may be trapped inter- or intramolecularly
[Grigg, 1990].

The zwitterionic intermediate from reaction of dehydronuciferine with
dimethyl acetylenedicarboxylate is easily transformed into a 1,3-dipole, due to
spatial proximity of the *N*-methyl group to the carbanion. Instead of ring closure
to form a cyclobutene, the adduct prefers isomerization and further reaction
(1,3-dipolar cycloaddition) [Menachery, 1983].

dehydronuciferine

E = COOMe

8.5.4. Oxidation–Cycloaddition

Although oximes are convertible into nitrones by methods indicated above,
the more popular dipoles derived from oximes are nitrile oxides which can be
generated by oxidation with positive halogen reagents. The syntheses of
hernandulcin [Zheng, 1989] and (+)-[6]-gingerol [Le Gall, 1989] are two
recent applications of the method in the construction of the β-ketol system, by
intra- and intermolecular cycloadditions, respectively. The asymmetric induction
by using a dipolarophile containing a chiral tricarbonyliron- complexed triene
is novel.

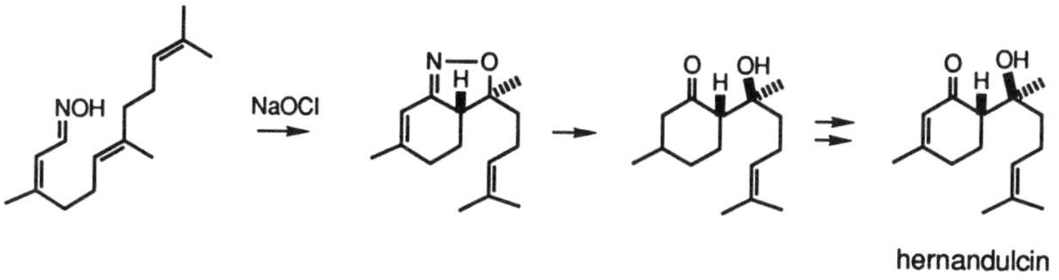

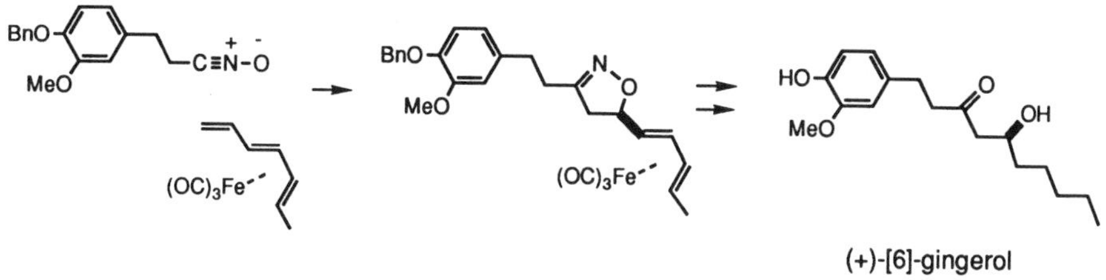

hernandulcin

(+)-[6]-gingerol

Stereospecific formation of unsaturated nitrile oxides by oxidative fragmentation of epimeric β-stannyl cycloalkanone oximes is indicated by the isolation of diastereomeric isoxazolines [Nishiyama, 1985].

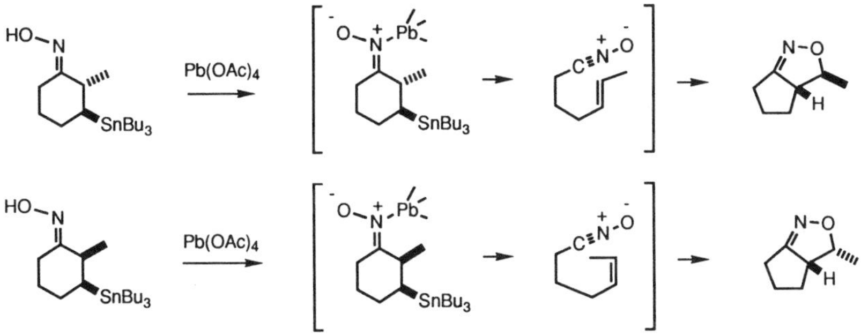

Dihydroberberines are quite unstable as they undergo oxidation readily in air. A pentenyl derivative was found to give a hexacyclic base upon attempted purification [Cushman, 1989]. An intramolecular 1,3-dipolar cycloaddition took place.

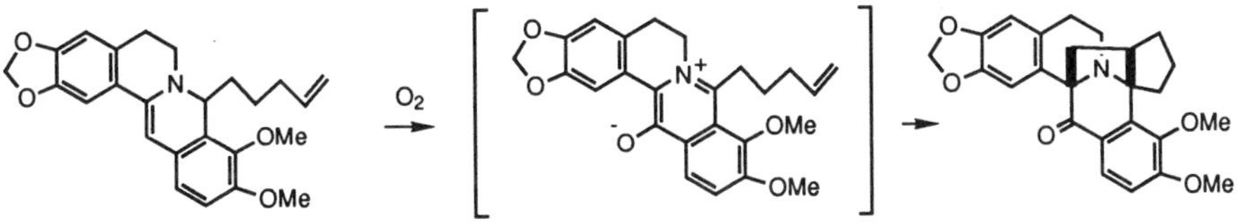

8.5.5. Cycloaddition of Photochemically Generated 1,3-Dipoles

Only an example of intramolecular 1,3-dipolar cycloaddition [Dittami, 1991] from a species generated photochemically is shown here. The increase in structural complexity of such tandem reactions deserves further study.

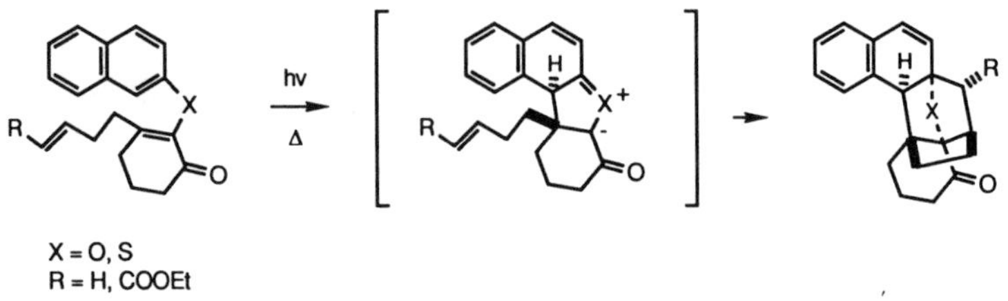

X = O, S
R = H, COOEt

2,7-Cyclooctadienone undergoes cycloaddition with enol ethers under irradiation [Matlin, 1989]. It has been proposed that the (E,Z)-dienone would cyclize conrotatorily to give a bicyclic oxyallyl species which can be trapped by the enol ethers.

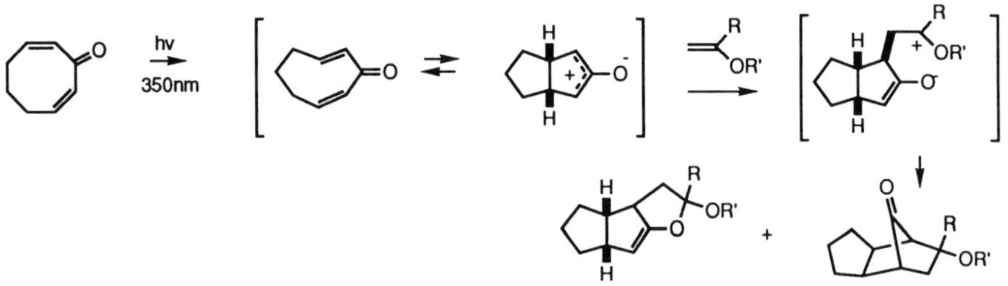

In polar aprotic solvents p-benzoquinone and cyclooctatetraene form an exciplex which rearranges spontaneously to furnish a zwitterion [R.M. Wilson, 1985]. The latter species has been trapped by acetonitrile.

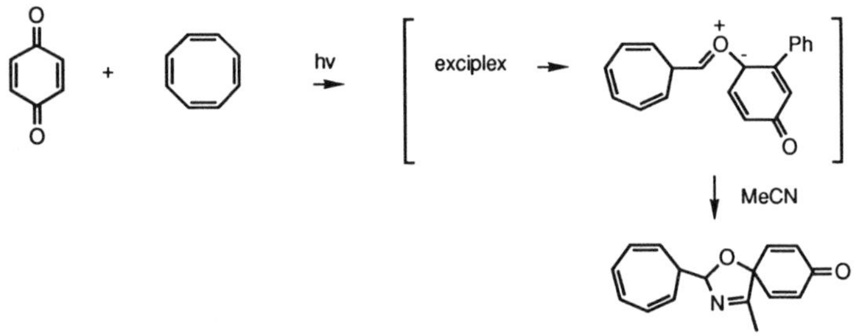

Allyl-substituted (2H)-azirines undergo photorearrangement to give 2-azabicyclo[3.1.0]hex-2-enes [Padwa, 1978] via an intramolecular 1,3-dipolar cycloaddition.

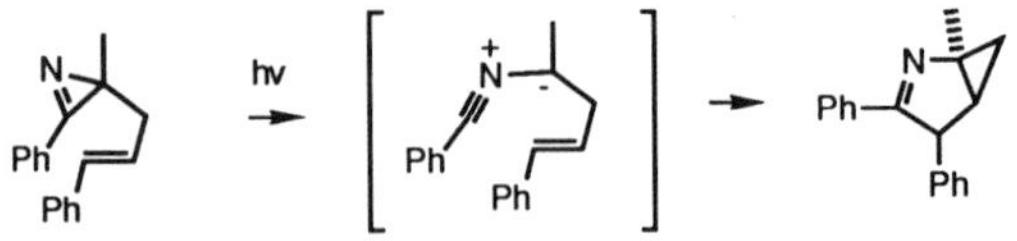

9

RETRO-DIELS–ALDER AND CHELETROPIC REACTIONS

The retro-Diels–Alder reaction denotes a thermal decomposition of a six-membered ring containing at least one endocyclic double bond into a diene and an unsaturated fragment. It is naturally subject to the same orbital symmetry requirements.

The use of retro-Diels–Alder reactions in synthesis [Ripoll, 1978; Ichihara, 1987] is in the context of regeneration of conjugate dienes or "dienophiles" from their masked forms after modification of other parts of the molecular architecture, and the construction of homoannular dienes. In the former exercise an unsaturation is present in the starting material or intermediate which is protected in the form of a Diels–Alder adduct; exactly the same atoms are involved in the bond formation and bond cleavage steps. On the other hand, in the homoannular diene synthesis invariably the alternative set of bonds in the Diels–Alder adduct is severed. Such mode of cleavage generally requires a lower energy of activation, as for example in the extrusion of some small molecules such as carbon dioxide and dinitrogen.

Since the retro-Diels–Alder fission is a thermal process, its most versatile employment would be to succeed another thermally allowed reaction. In such cases a tandem reaction sequence results.

9.1. DIELS–ALDER/RETRO-DIELS–ALDER REACTION TANDEMS

The majority of synthetically significant retro-Diels–Alder reactions makes up the Diels–Alder/retro-Diels–Alder cascade. Perhaps the best known examples are those dealing with the preparation of 1,3-cyclohexadienes from α-pyrones. The retro-Diels–Alder reaction step is highly favorable because a stable small molecule (CO_2) is eliminated.

α-Pyrone is converted into (E)-cinnamic acid on heating [White, 1972]. It is conceivable that decarboxylation of the Diels–Alder dimer would be followed by tautomerization of the dihydrocoumarin and configuration change of the double bond to give a thermodynamically much more stable compound.

The homoannular diene subunit in occidentalol is a distinct structural feature well-suited for elaboration by the method. The attractiveness of such a synthetic approach is even higher in view of the presence of a *cis* ring juncture. Thus, a Diels–Alder reaction of an α-pyrone with a methylcyclohexene derivative would give rise to the *cis*-hexalin [Watt, 1972].

Because of the undesirability of diastereoisomer formation in the first step, but the need for a sidechain at C-7 (terpene numbering), 4-methyl-3-cyclohexenone was used as the dienophile. The carbonyl group of the adduct was reserved as the foundation for the chain attachment. It is rather remarkable that the Diels–Alder reaction was regioselective to give the desired isomer. This observation may be rationalized in terms of electronic effect exerted by the ketone group (polarity alternation rule [Ho, 1991]).

α-copaene   occidentalol

Regarding synthesis of colchicine, the problem has always centered around construction of the tropolone system. A method that succeeded in accomplishing this goal was conceived on the premise that condensation of an α-pyrone with chloromethylmaleic anhydride would lead to an angularly chloromethylated 1,2-dihydrophthalic anhydride (in situ decarboxylation expected); an intramolecular alkylation (cyclopropanation) of the proper derivative would generate a norcaradiene. Emergence of a cycloheptatriene tautomer of the norcaradiene is a natural consequence [J. Schreiber, 1961].

colchicine

From time to time chemists encounter difficulties in the synthesis of benzenoid compounds because of special substitution patterns in the rings and structural requirement in sidechains which make conventional routes starting from preexisting aromatic substances less attractive. Such is the case of lasalocid-A. The α-pyrone route provided a neat solution [Ireland, 1980].

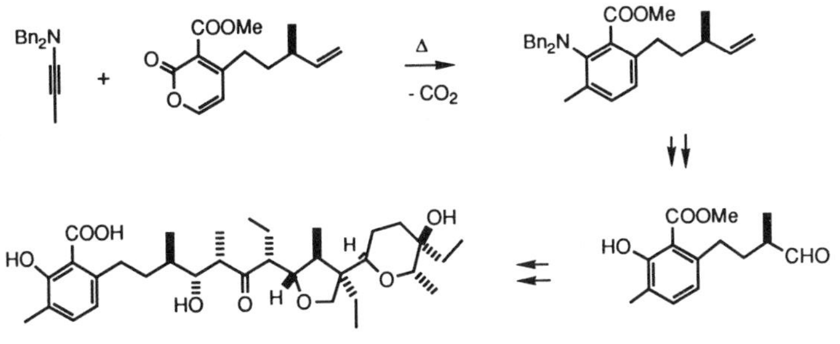

lasalocid

The few azafluoanthene alkaloids known to date share a common substitution pattern in the ABC-ring moiety. In terms of synthetic economy the use of a common intermediate to elaborate many members of the family is highly commendable. A tetracyclic α-pyrone precursor, with its expected reactivity toward keteneacetals, has proved to be of great service in the synthesis of rufescine and imelutine [Boger, 1984b].

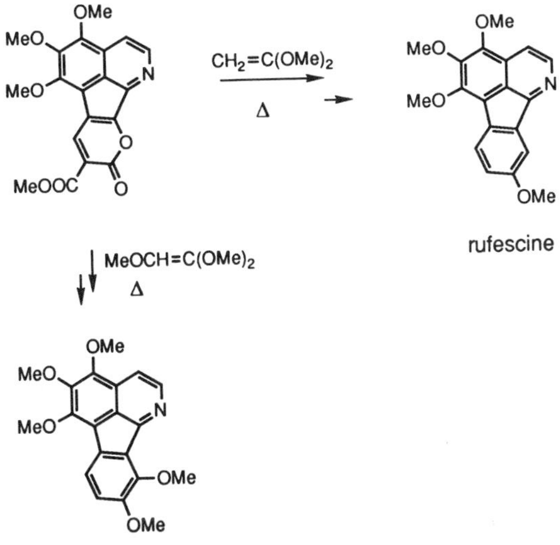

rufescine

imeluteine

Formation of the pyridocarbazole ring system from indolo-α-pyrones by condensation with 3,4-pyridyne was successful, although regiochemical control could not be exercised [May, 1984].

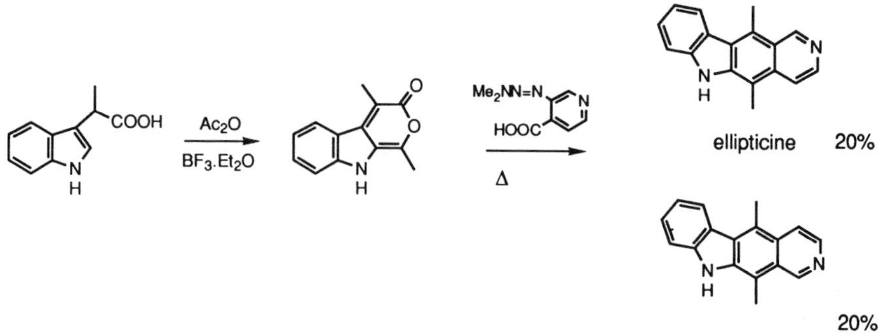

ellipticine   20%

20%

Better regioselectivity was observed in the condensation with α-chloroacrylo-nitrile [Narasimhan, 1985]. The primary product underwent decarboxylation and also a dehydrochlorination, giving a useful intermediate for the synthesis of olivacine.

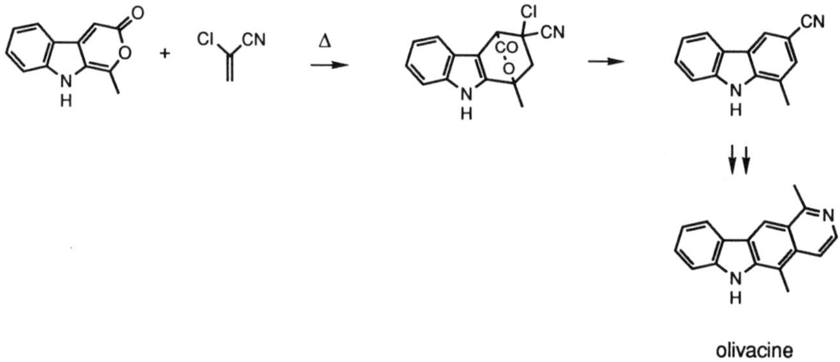

olivacine

It is interesting to note that the enolates of indologlutaric anhydrides also behave as dienes [Tamura, 1984a], as shown by the following reaction. This is an extension of the base-catalyzed cycloaddition of homophthalic anhydrides [Tamura, 1984b].

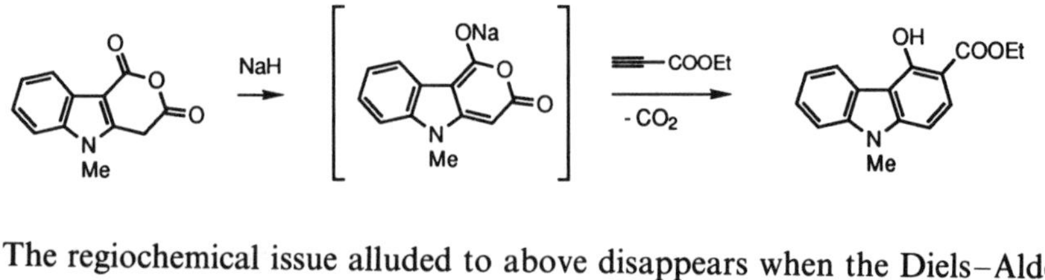

The regiochemical issue alluded to above disappears when the Diels–Alder reaction becomes intramolecular. Thus, 1-({ω − 1}-alkynyl)pyrano[3,4-b]indol-3-ones were directly transformed into [a]-annulated carbazoles [Moody, 1988]. A particularly interesting application of the Diels–Alder/retro-Diels–Alder reaction sequence is found in the assembly of the staurosporine aglycone [Moody, 1990].

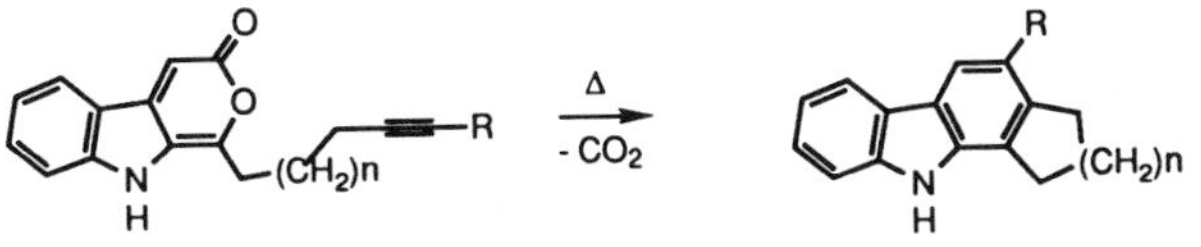

n = 1, 2, 3
R = H, COOMe, SiMe₃

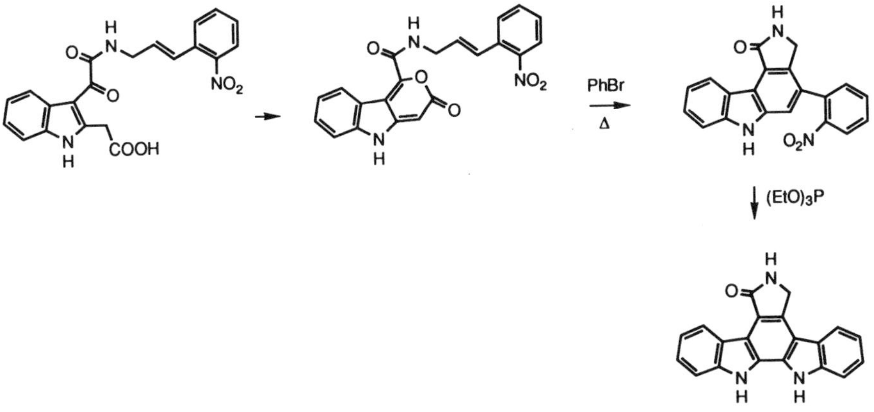

staurosporine aglycon

1,3-Cyclohexadienes obtained from the Diels–Alder/retro-Diels–Alder reaction tandem are very versatile compounds amenable to regio- and stereocontrolled structural modifications, as validated by the excellent service of a hydroisoquinolones as CD-ring synthon for α-yohimbine [S. F. Martin, 1985] and reserpine [Martin, 1987].

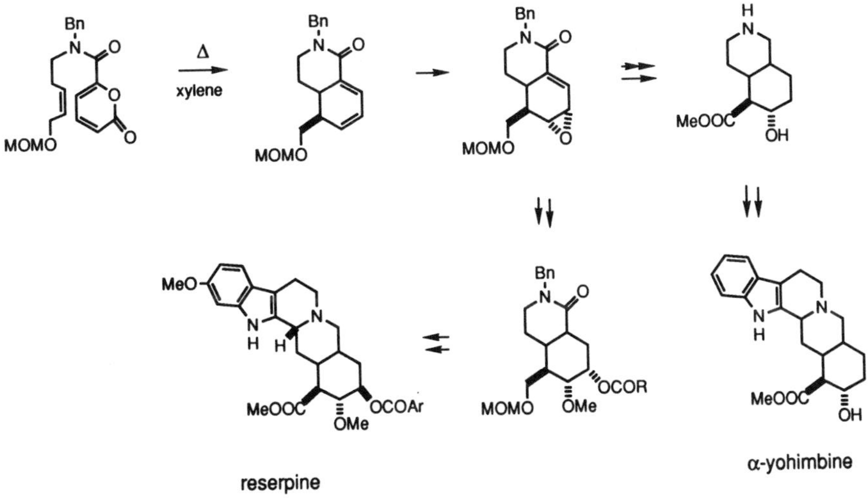

reserpine

α-yohimbine

Perhaps the most widely employed strategy involving α-pyrones to furnish a diene subunit to the products is in the area of anthraquinone synthesis. Thus, pachybasin, chrysophanol, and helminthosporin have been elaborated in a few steps [Jung, 1978].

pachybasin

Based on a similar scheme of cycloaddition–decarboxylation a convenient passage into podophyllotoxin [Jones, 1989b] has been developed. There was a rapid aromatization of the *o*-quinodimethane species generated by the decarboxylation step. A suprafacial 1,5-hydrogen migration established the two *cis* substituents in the dihydronaphthalene nucleus.

podophyllotoxin

The condensation of 3-hydroxy-α-pyrone with 5,8-diacetoxy-7-chloro-2-methyl-1,4-naphthoquinone afforded digitopurpone diacetate [Cano, 1983] as a result of reaction of the pyrone with the more reactive naphthoquinone in equilibrium and decarboxylation of the adduct. The equilibrium is a formal internal redox transacetylation.

digitopurpone
diacetate

Benzocyclopropene is an avid dienophile as indicated by its reaction with 5-carbomethoxy-α-pyrone [Neidlein, 1989]. The isolated product is a 1,6-methanol[10]annulene which resulted from electrocycloreversion and retro-

Diels–Alder decarboxylation, although the timing of these two reactions cannot be determined. The reaction of benzocyclopropene with tetracyclones also provided cyclodecapentaenes. There was a decarbonylation (cheletropic reaction) instead of decarboxylation.

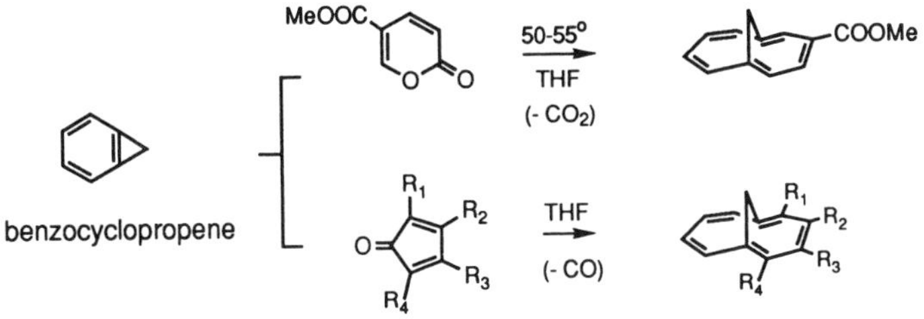

An intriguing synthetic route to the naphthalide system is via pyrolysis of propargyl coumarin-3-carboxylate [Kraus, 1979].

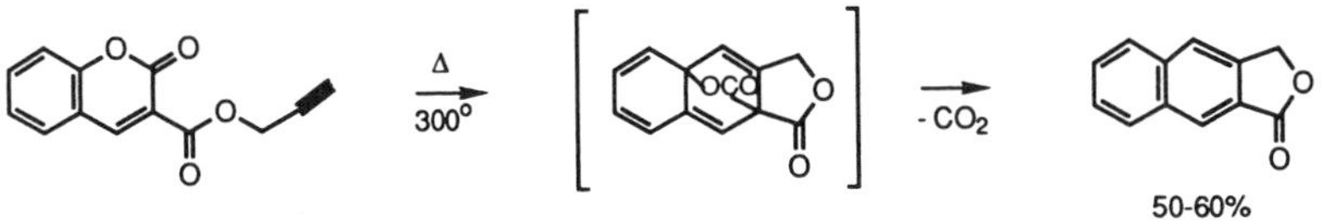

The unexpected result from the reaction of α-pyrone with bis(trimethylsilyl)-acylene is the predominant formation of m-bis(trimethylsilyl)benzene [Seyferth, 1967]. Decarboxylation of the Diels–Alder adduct may have led to a prismane or a benzvalene precursor. The cause for such an anomalous pathway must be the unfavorable interaction of the two bulky substituents in a vicinal relation.

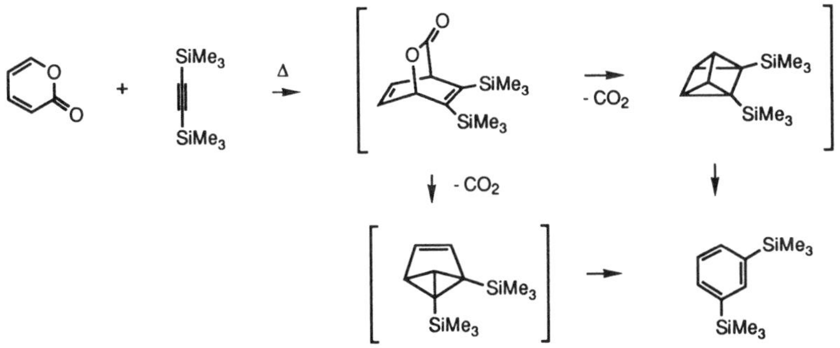

Replacement of the carbonyl oxygen of the α-pyrone with a pentacarbonyl-chromium group increases the diene reactivity as well as the tendency of the adducts for the retro-Diels–Alder reaction in which chromium hexacarbonyl is expelled [S. Wang, 1990]. Both reactions take place at room temperature.

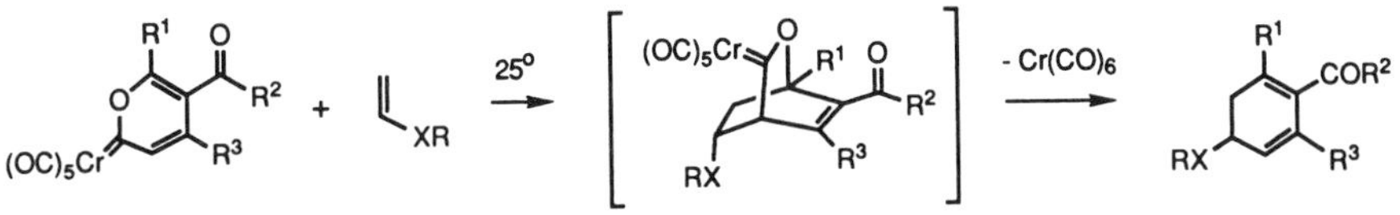

Two of the three products obtained from singlet oxygen reaction of a tetrasilylated phenylene derivative are diketones with an alkyne unit [Mestdagh, 1986]. The unusual cleavage is readily rationalized by a retro-Diels–Alder reaction of an endoperoxide intermediate.

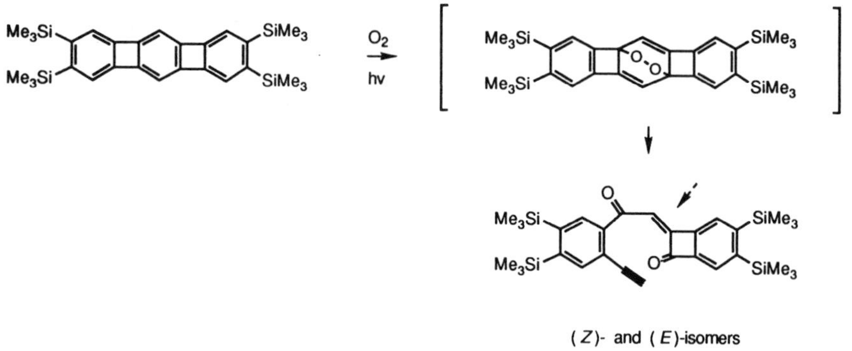

(Z)- and (E)-isomers

It might be convenient to include the few Diels–Alder/cheletropic tandem reactions which are of synthetic significance in this section. A formal Diels–Alder cycloaddition involving a benzyne as dienophile, followed by in situ decarbonylation constituted the crucial step in a synthesis of corydaline [Saá, 1986].

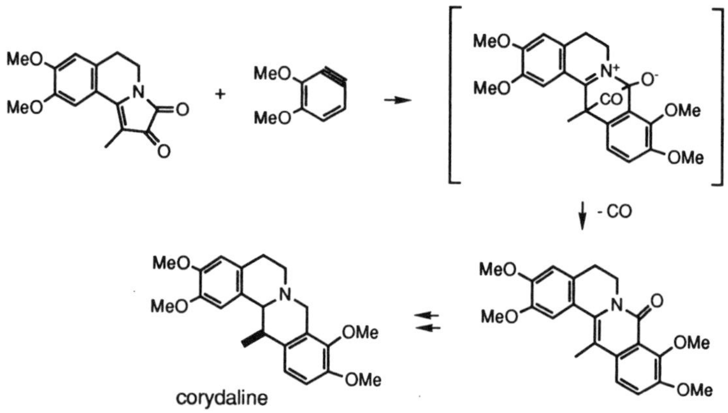

corydaline

Decarbonylation from the Diels–Alder dimer of cyclopentadienone also preceded the formation of 1-indanone [DePuy, 1964].

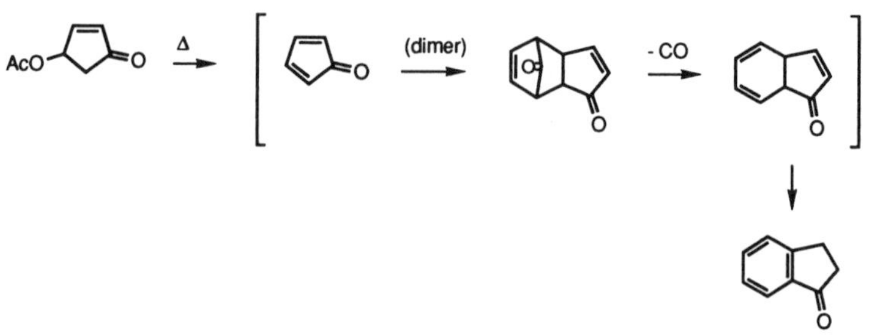

A most unusual decarbonylative dimerization of dicyclobuteno-*p*-benzo-quinone [Oda, 1990] occurs at > 50°C, possibly via cheletropic reaction of the "homo" Diels–Alder adduct.

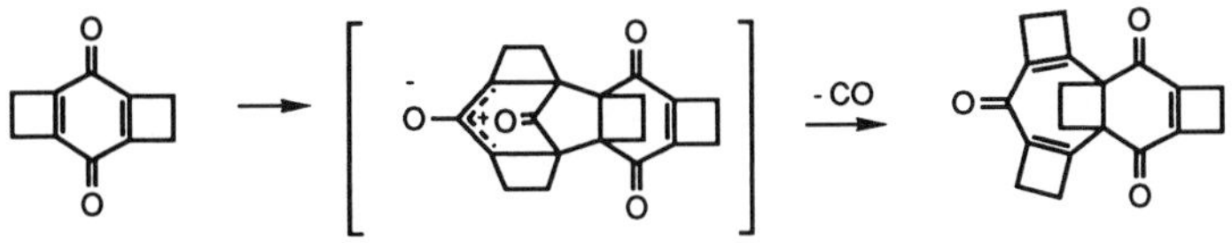

N-Aminopyrroles have found some use as dienes. Their reaction with electron-deficient acetylenes provides directly benzenoid compounds [Schultz, 1979]. Sometimes the carbamates behave better, as shown in a synthesis of juncusol [Schultz, 1981].

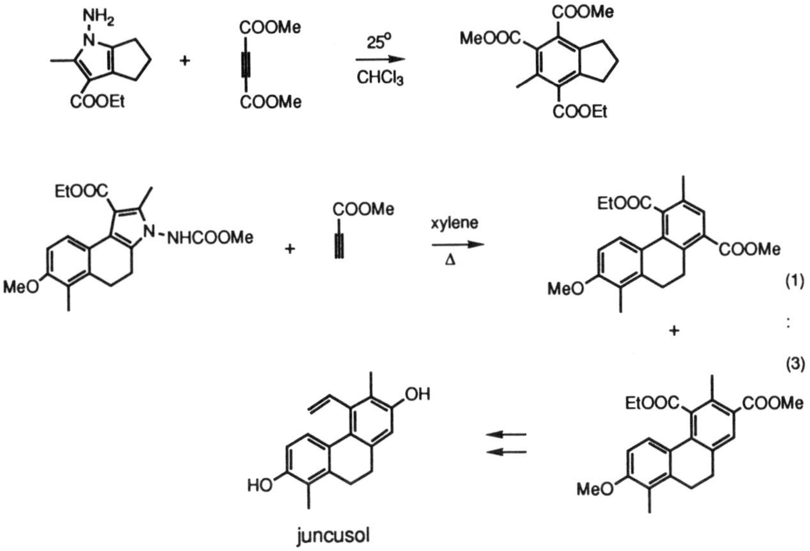

By far the most widely employed extrusible subunit is sulfur dioxide. Consequently the preparation of 1,3-dienes [Bloch, 1982] and 1,2-4-trienes [Bloch, 1983] from tricyclic sulfones is a very efficient process.

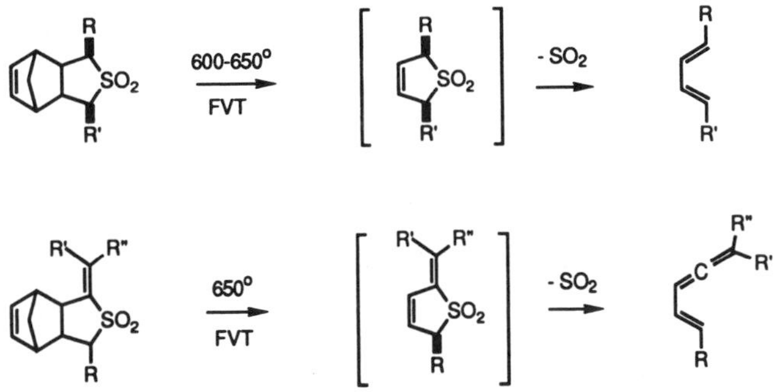

In view of steric crowding *o*-di-*t*-butylbenzene and its derivatives are quite difficult to obtain. It is therefore significant that the readily available 3,4-di-*t*-butylthiophene 1,1-dioxide undergoes Diels–Alder reactions with various alkynes (benzyne, 1-hexyne, diphenylacetylene, methyl propynoate, etc.) to give directly the aromatic compounds [Nakayama, 1988]. The reaction with phenyl vinyl sulfoxide in refluxing 2-chlorotoluene led to *o*-di-*t*-butylbenzene which might have involved thermal elimination of phenylsulfenic acid from the adduct prior to the cheletropic reaction.

The same reaction scheme is applicable to the synthesis of *o*-bisadamantyl-benzenes [Nakayama, 1990].

The dimerization of 3-halo-2,5-dialkylthiophene 1,1-dioxide is also accompanied by cheletropic reaction and elimination [Gronowitz, 1988].

An interesting approach to [*c*]-fused furans involves intramolecular Diels–Alder reaction with the aid of an enone to manipulate the double bonds of the furan ring [K. Ando, 1991]. Thus, a Diels–Alder reaction enables cheletropic elimination of sulfur dioxide to unveil a diene system for a subsequent Diels–Alder reaction. Retro-Diels–Alder reaction of the adduct splits off the enone to give the fused furan.

Extrusion of other stable smaller molecules such as dinitrogen and organonitriles from Diels–Alder adducts of heterocyclic compounds are also favorable. By this process many pyridine derivatives have been prepared from 1,3-diazines (pyrimidines). An intramolecular cycloaddition followed by elimination of cyanic acid was the basis of a synthesis of actinidine [Davies, 1981].

actinidine

Pyridazines are popular Diels–Alder addends for the generation of benzene rings, and 1,2,x-triazines for pyridines. Thus the thermal reaction of alkynyl pyridazines is an excellent method for affording benzocyclo structures and particularly substituted indolines, in view of the ready availability of such substrates from reaction of 3-chloropyridazines. There is hardly a more convenient way to prepare the highly substituted intermediate for synthesis of the cAMP phosphodiesterase inhibitor PDE-II [Boger, 1984c].

R = NH₂ PDE - I
R = Me   PDE - II

A very effective synthesis of polysubstituted aromatic rings is by the reaction of 1,2,4,5-tetrazines with dienophiles, followed by twofold extrusion of dinitrogen. This is the basis of several syntheses of complicated aromatic and heteroaromatic substances, for example streptonigrin [Boger, 1985], PDE-II [Boger, 1986b], and prodigiosin [Boger, 1987].

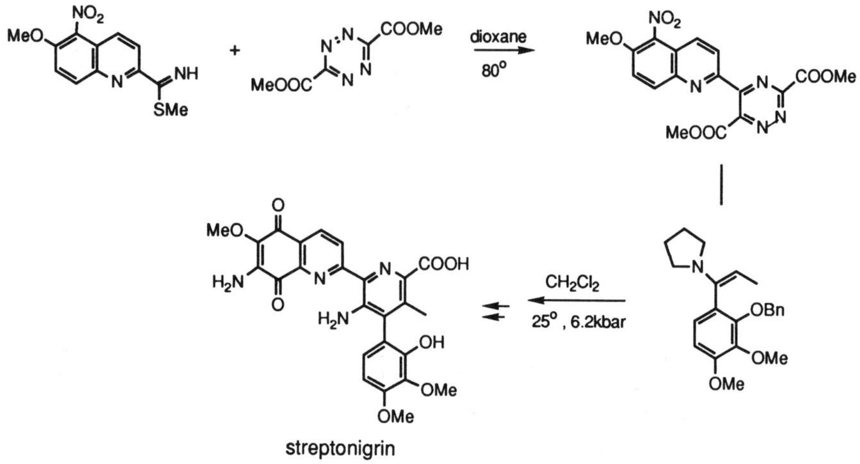

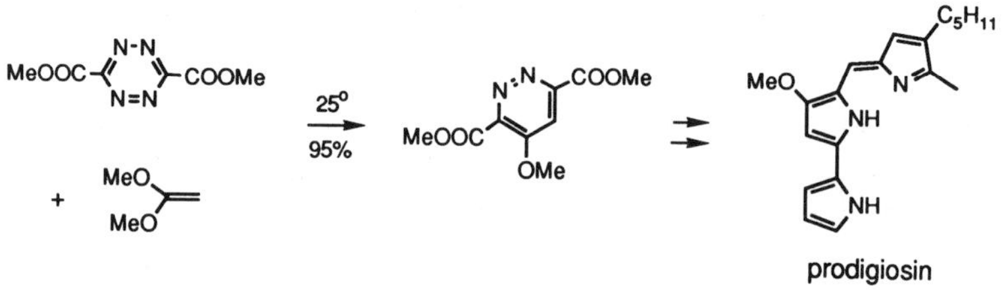

streptonigrin

prodigiosin

The extensive use of the Diels–Alder/retro-Diels–Alder tandem makes the synthesis of *cis*- and *trans*-trikentrin [Boger, 1991] very efficient. The tricyclic skeleton of these compounds was constructed by an intramolecular version (with concomitant elimination of methanesulfinic acid) on a substrate assembled from a 1,2,4,5-tetrazine.

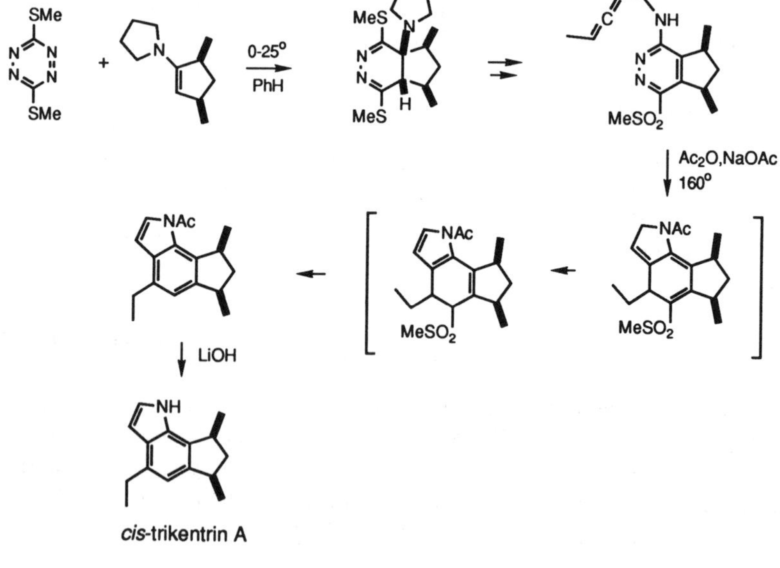

cis-trikentrin A

Homo-retro-Diels–Alder reactions are very facile for diazanorcarenes. A concise synthesis of semibullvalenes [Schuster, 1983] from 1,2,4,5-tetrazines and 3,3-dicyclopropenyls took advantage of a sequence of alternating Diels–Alder and retro-Diels–Alder reactions, the terminating extrusion of dinitrogen being a homo-retro-Diels–Alder reaction.

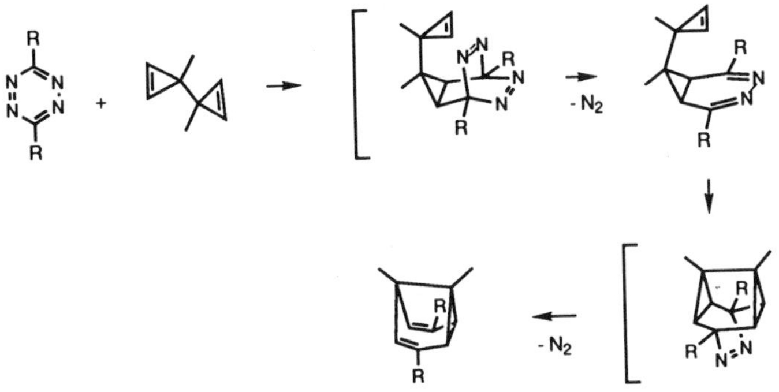

An approach to unusual heterocycles such as thieno[3,4-*b*]furan [Moursounidis, 1986] is based on a Diels–Alder/retro-Diels–Alder reaction tandem.

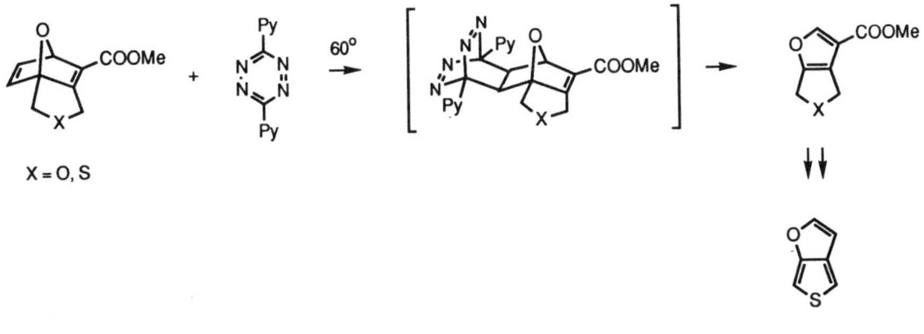

X = O, S

thieno[3,4-b]furan

The cycloaddition of indole with tetramethyl 1,2-diazine-3,4,5,6-tetracarboxylate gave, after expulsion of dinitrogen, eliminative aromatization, and lactonization, a phenanthridone [S.C. Benson, 1990].

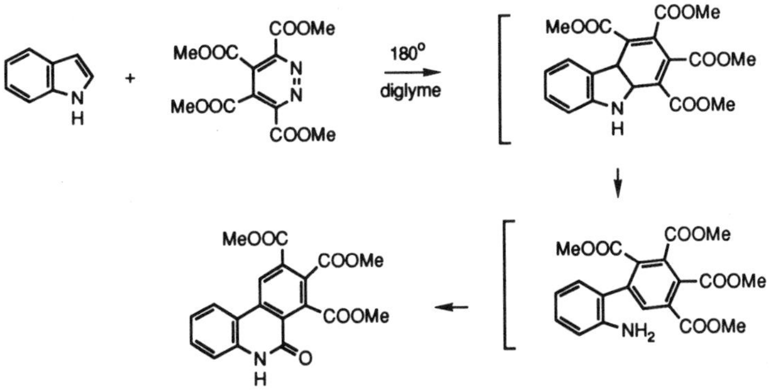

In an approach to fabianine [Sugita, 1986], condensation of a pulegone enamine with 4-methyl-1,2,3-triazine furnished a hydroquinoline derivative. Hydration of the product completed the synthesis.

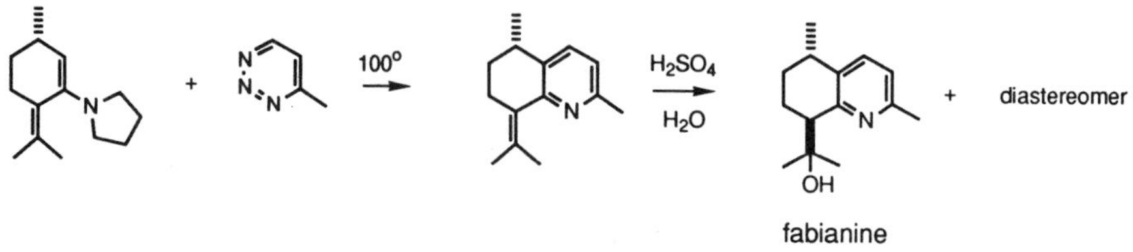

fabianine

A convenient and mild procedure for the synthesis of isoindoles is by reaction of napthalene-1,4-imines with 3,6-dipyridyl-1,2,4,5-tetrazine [Priestley, 1972]. The two retro-Diels–Alder reactions following the cycloaddition are so facile that the intermediates were not detectable even at below 0°C.

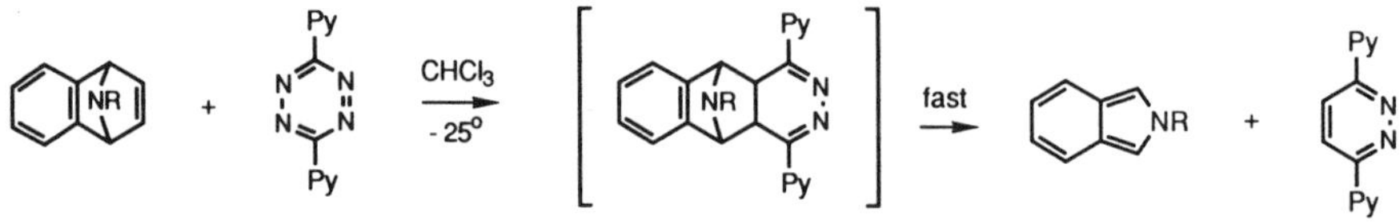

Extrusion of dinitrogen appears to be a more favorable process than decarboxylation [Hegmann, 1987] in the reaction shown below.

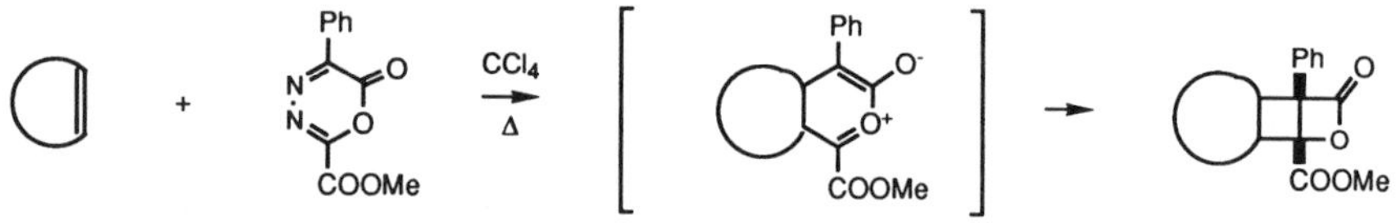

β-Acylaminoacrylates afford 1,3-oxazin-6-ones on heating. Prolonged reaction at slightly higher temperatures gives 2,4,6-trisubstituted pyridines via a Diels–Alder condensation, decarboxylation, and nitrile elimination.

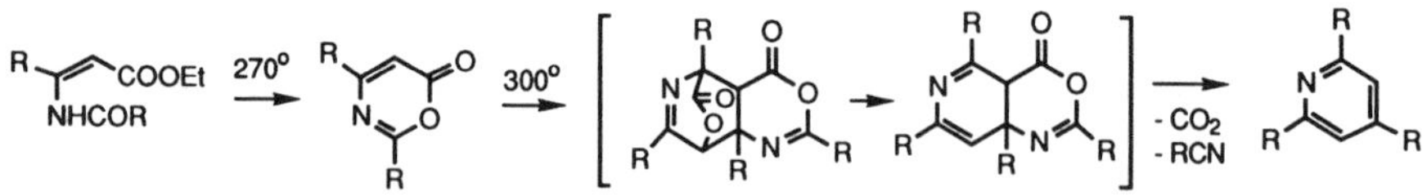

It is remarkable that N-methyl-N-propargyl isoquinoline-1-carboxamide was transformed into a naphthalene derivative [Wuonola, 1991] on heating in xylene. Of the two stages of the reaction the elimination of hydrogen cyanide from the cycloadduct requires much lower energy of activation.

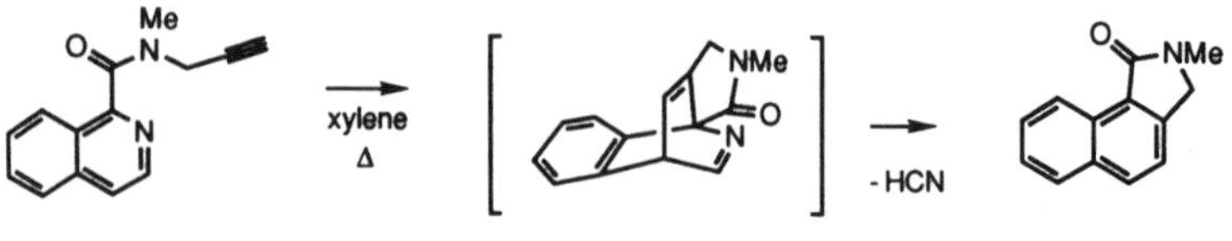

Oxazoles condense with alkynes in the Diels–Alder reaction fashion. The adducts tend to lose hydrogen cyanide or an organonitrile to generate furan derivatives as the final products. Particularly valuable to synthesis is the intramolecular version of the tandem reaction sequence and it has been exploited in the synthesis of natural products: ligularone [Jacobi, 1984], paniculide-A [Jacobi, 1987], norsecurinine [Jacobi, 1989], and gnididione [Jacobi, 1990].

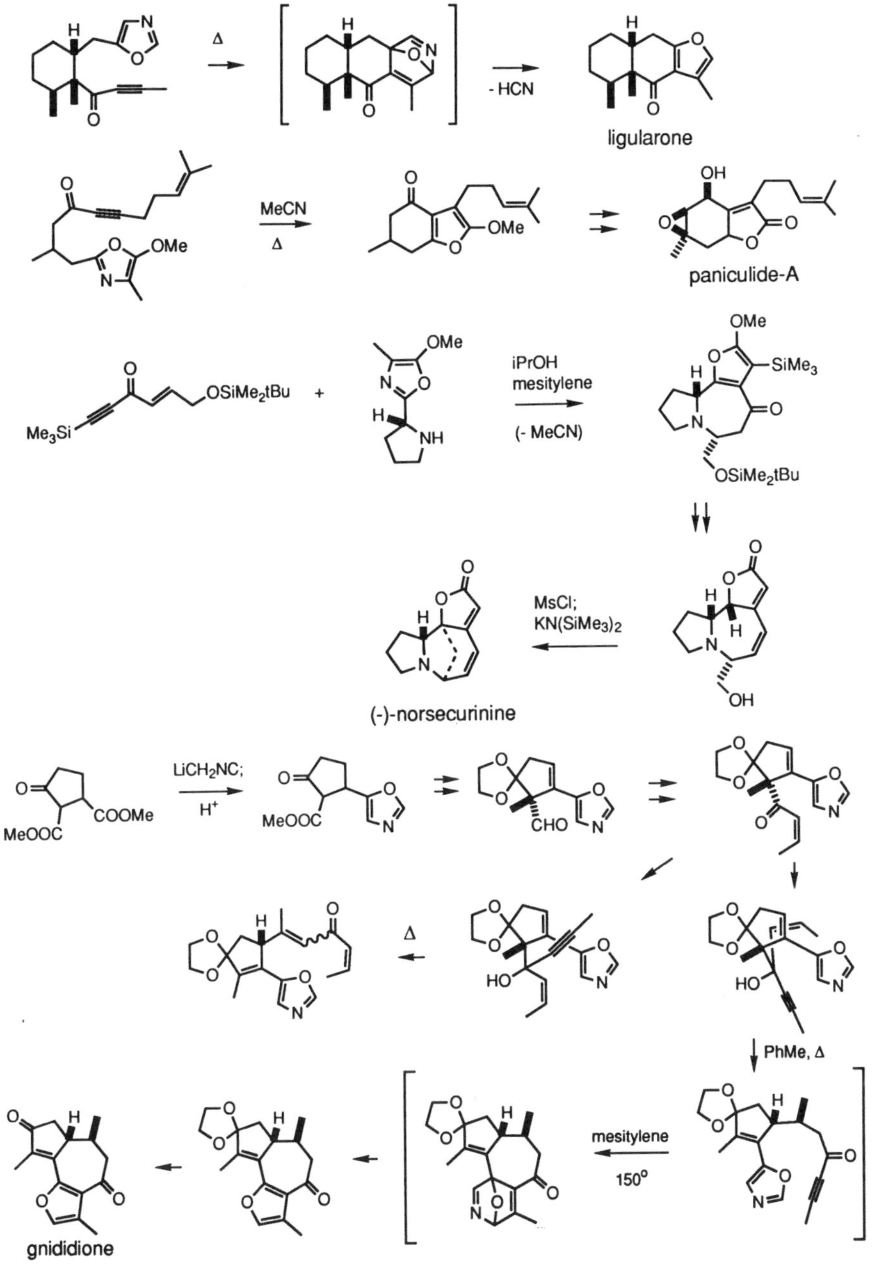

ligularone

paniculide-A

(-)-norsecurinine

gnididione

In the gnididione synthesis the Diels–Alder reaction was set up by an oxy-Cope rearrangement. At lower temperatures the subsequent Diels–Alder reaction may be prevented, but in refluxing mesitylene, three consecutive thermal reactions occurred.

6-Phenyl-5-azaazulene participates in Diels–Alder reactions across the 2-azadiene system [Hafner, 1970]. It if followed by regeneration of the aromatic system.

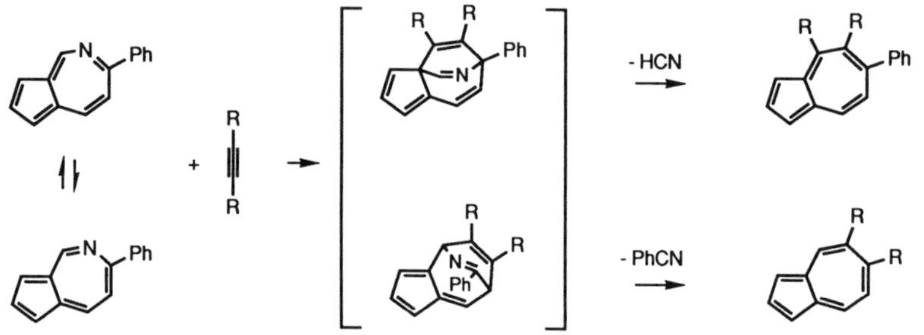

Retro-Diels–Alder elimination of hydrogen cyanide appears to be the final step in the formation of benzaldehyde by a thermal reaction of 7,8-diazabicyclo[4.2.2]deca-2,4,7,9-tetraene N-oxide [Olsen, 1982]. The prior steps are an intramolecular Diels–Alder reaction, retro-Diels–Alder isomerization, and proton shift.

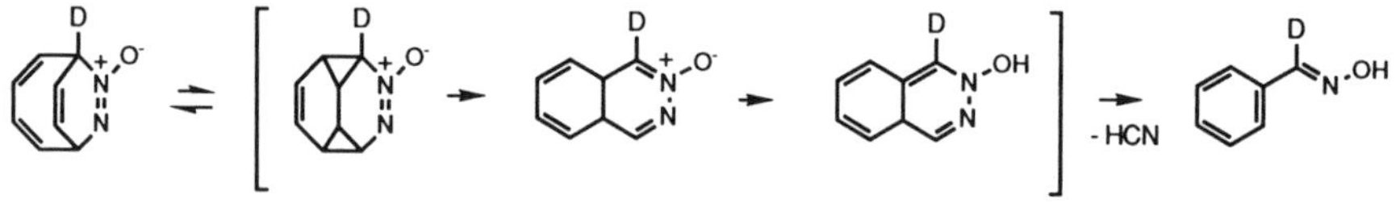

The 2H-pyran tautomers of acyclic dienones cycloadd to alkyne dienophiles, and the adducts aromatize on losing an aldehyde [Salomon, 1976].

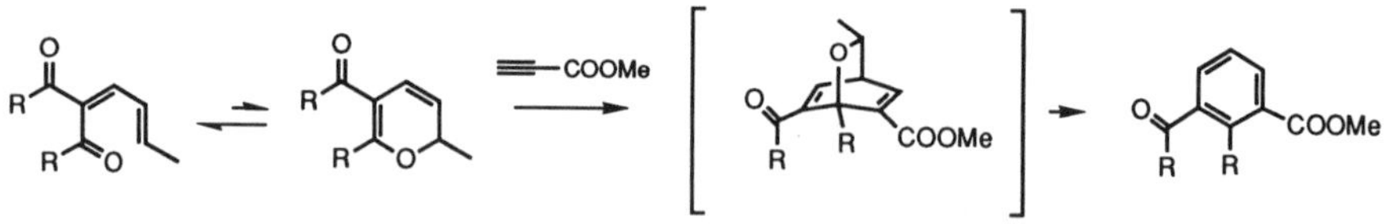

The formation of o-bistrifluoromethylbenzene by the reaction of 2,2-dimethyl-1-oxa-2-sila-3,5-cyclohexadiene with hexafluoro-2-butyne [Hussmann, 1983] is similar. "Dimethylsilicone" was eliminated.

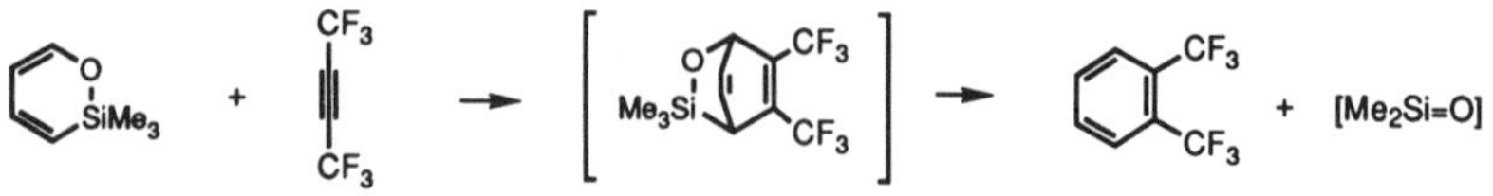

An indirect 4-amination of quinolinium salts [Fuks, 1969] is by reaction with ynamines. Formally it involves a Diels–Alder reaction and elimination of acetylene from the adducts.

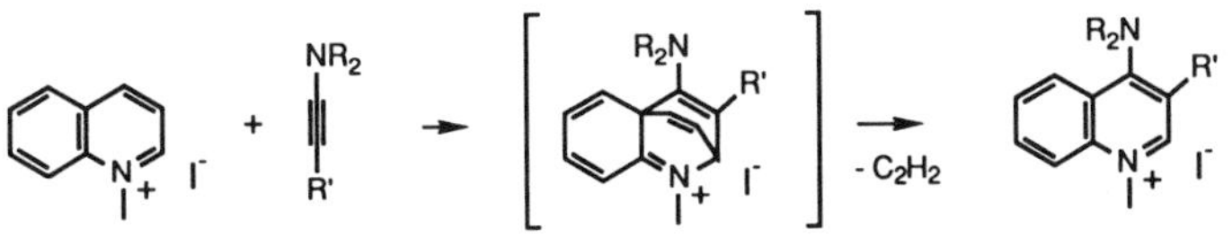

At sufficiently high temperatures bicyclo[2.2.2]octadienes can lose ethylene. Thus, benzene derivatives are available from 1,3-cyclohexadienes and alkynes via the Diels–Alder/retro-Diels–Alder reaction sequence. A route to macrolide aromatic polyketides which possess an alkyl-β-resorcylate skeleton is based on this method [Birch, 1990]. It also constitutes a reliable access to many anthraquinones, including emodin [Krohn, 1980].

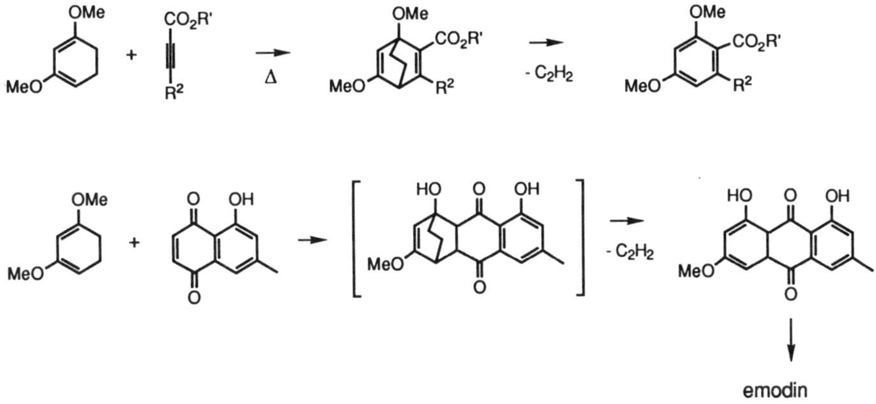

emodin

Substituted phenols are readily elaborated from 2-siloxy-6,6-dimethyl-1,3-cyclohexadienes and alkynes [Rubottom, 1977]. The cycloadducts lose isobutene.

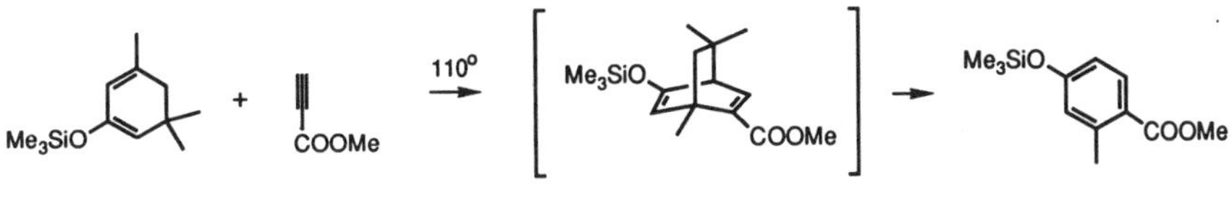

Diels–Alder adducts that aromatize by elimination of 1,4-cyclohexadiene were found to decompose at 60°C [Papies, 1980].

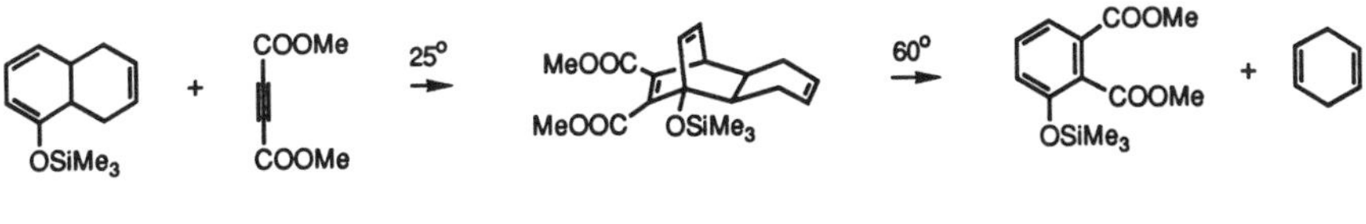

An apparent diene exchange in a Diels–Alder adduct [Farina, 1986] may actually have involved a Diels–Alder reaction followed by a retro-Diels–Alder reaction.

The transformation of a hydroanthracene into a propellane [Gilbert, 1976] can be rationalized by a reaction pathway through a cycloadduct.

A most intriguing observation is the generation of [4,4]metacyclophane and [4,4]paracyclophane upon thermolysis a bridged Dewar benzene in a KOH-conditioned ampoule at 200°C [Kostermans, 1987].

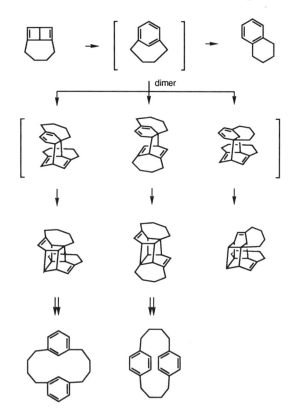

The [4,4]cyclophanes were generated from the [4]metacyclophane isomer by two consecutive Diels–Alder reactions (inter- and intramolecular), followed by retro-Diels–Alder process at the two cyclohexene moieties. The mechanism was supported by results of specifically deuterated substrates.

Two (*i, o*)-bicyclo[n.2.2]alkadienes have been shown to combine with two molecules of hexafluoro-2-butyne by a sequence of ene reaction, Diels–Alder cycloaddition, and retro-Diels–Alder reaction to give 4-substituted 1,2-bistrifluoromethylbenzene [Gassman, 1981].

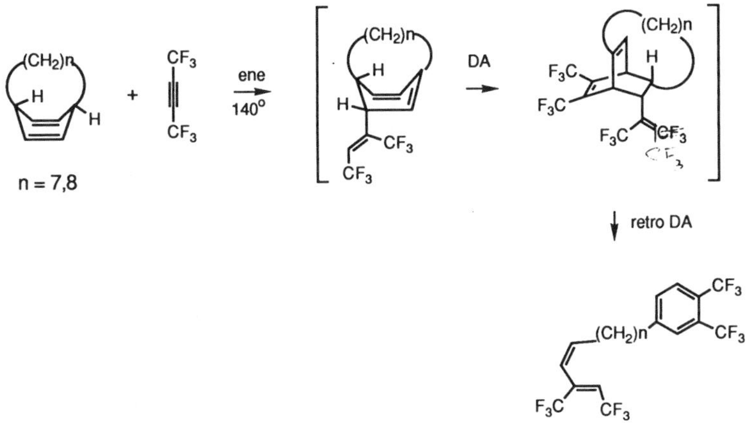

The reaction of activated alkynes to barrelene furnishes 1,2-disubstituted naphthalenes [Zimmerman, 1964]. Apparently the [2 + 2 + 2]-cycloadducts are susceptible to retro-Diels–Alder reaction to give 9,10-dihydronaphthalenes which dehydrogenate readily.

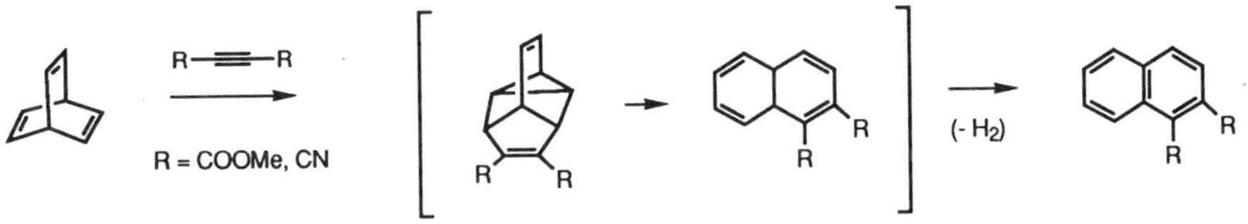

A pentalene was synthesized from 1-cyanoazulene [Hafner, 1982] via a series of cycloadditions with an ynamine. Retro-Diels–Alder decomposition of the 1:2 adduct terminated the reaction.

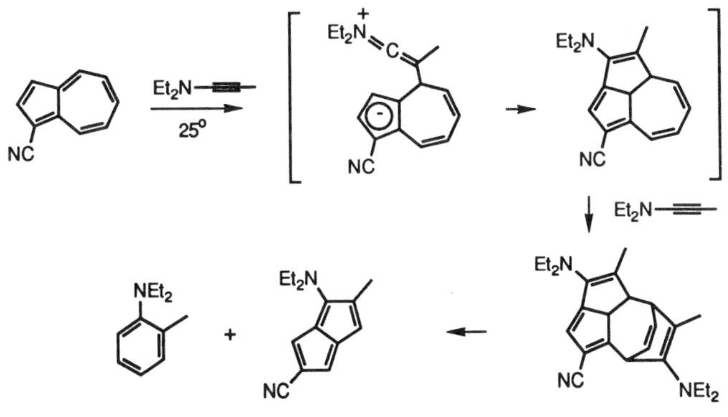

A retro-Diels–Alder scission following an intramolecular 1,3-dipolar cyclo-addition has been observed [Brokatzky, 1981].

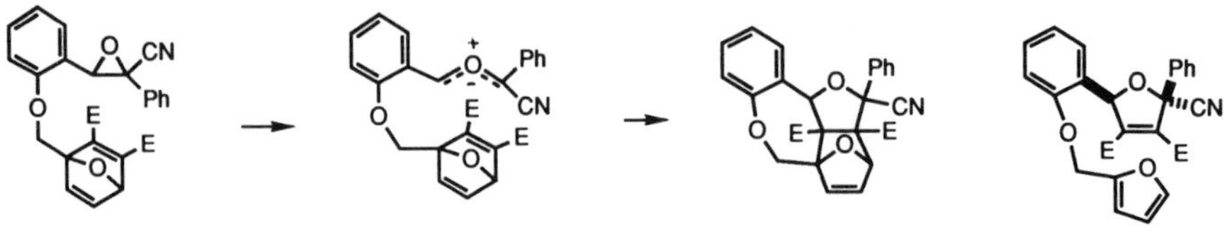

E = COOMe

A novel synthesis of actinidine [Nitta, 1981] involved pyrolysis of 3,5-dimethyl-4-propargyloxypyridine and subsequent Wolff–Kishner reduction. The thermal reaction likely consisted of isomerization of the propynyl group to an allene, intramolecular Diels–Alder reaction, retro-Diels–Alder reaction, and cyclization of the ketene.

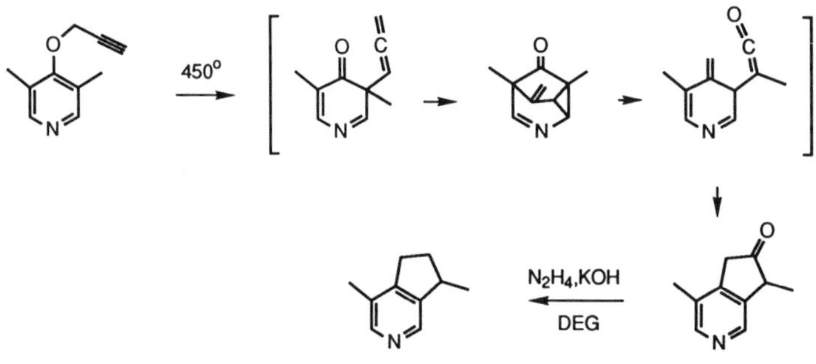

actinidine

9.2. RETRO-DIELS–ALDER REACTIONS FOLLOWING OTHER PROCESSES

The C_{2v}-homobasketene decomposes thermally to give benzene and cyclopenta-diene [Grimme, 1985] by way of two consecutive retro-Diels–Alder reactions. The triene intermediate can be trapped by 4-phenyl-1,2,4-triazoline-3,5-dione.

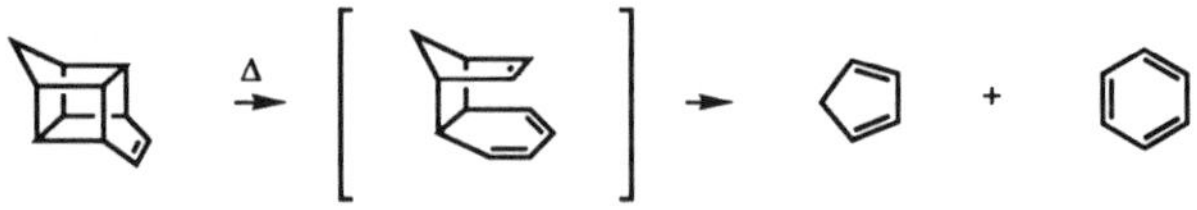

C_{2v}-homobasketene

The photoisomer of *cis*-4a,10a-dihydrophenanthrene is thermally labile [Vedejs, 1974], and is susceptible to Cope rearrangement. However, the dearomatization provides a driving force for the tandem retro-Diels–Alder cycloreversion.

The major photosensitized reaction products of tetracyclo$[4.3.0.0^{2,4}.0^{3,7}]$-non-8-ene derivatives are the bishomobenzenes, apparently formed by a [1.3]-sigmatropic rearrangement process. These bishomobenzenes are very susceptible to cycloreversion [Prinzbach, 1971]. Similar intermediates have also been postulated for the photolysis of a pyrAzoline [Tsuruta, 1968].

A pericyclic reaction cascade, probably terminated by a retro-Diels–Alder reaction, was found to provide azocines [Elix, 1970]. In this route cyclo-octatetraene contributes four skeletal carbon atoms to the final products, while 1,2,4-triazines furnish the rest of them.

This one-step synthesis uses the tricyclic adduct of cyclooctatetraene and dimethyl acetylenedicarboxylate. Condensation of this adduct with a triazine continues with elimination of dinitrogen, electrocyclic opening of the azabicyclo-[4.2.0]octadiene, and secession of dimethyl phthalate. However, it is possible that the order of the last two reactions is reverse.

An excellent route to corannulene [Scott, 1991] is via the formation of a fluoranthene diester in one step. This process involves aldol condensation, Diels–Alder reaction, cheletropic elimination, and retro-Diels–Alder reaction.

corannulene

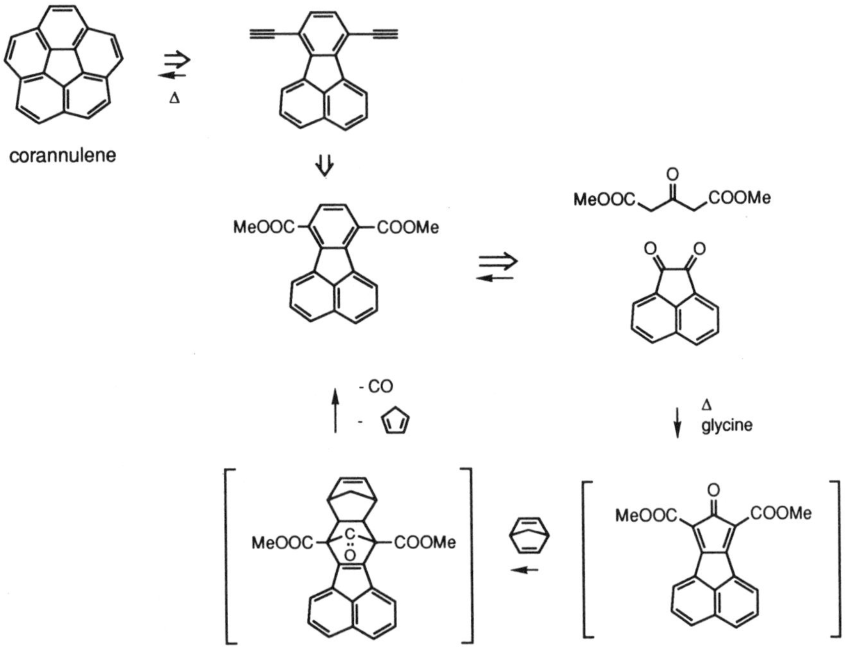

Oxidation of 1,6-methano[10]annulen-11-ol led to naphthalene only [Vogel, 1966]. Decarbonylation of the tricyclic ketone tautomer must be extremely facile.

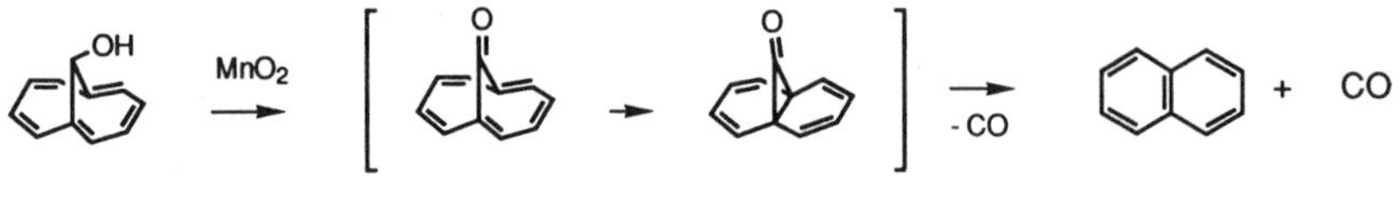

A double retro-Diels–Alder reaction occurred spontaneously on oxidation of a tetracyclic dihydrazine [Shen, 1971]. *cis*-9,10-Dihydronaphthalene was produced in high yield.

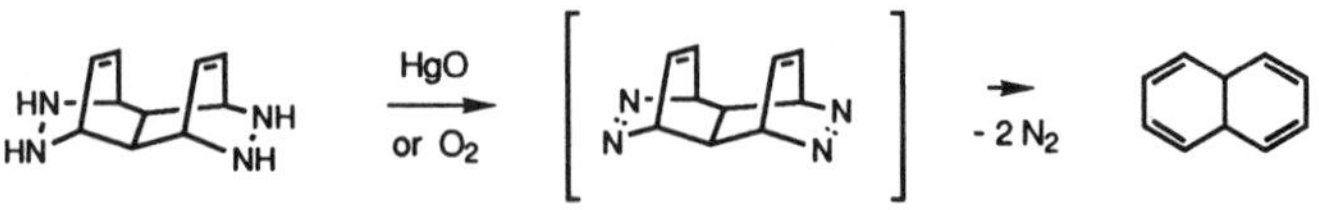

Creation of an activated (e.g. strained) cyclohexene usually leads to a retro-Diels–Alder reaction. Thus, decarboxylative silyl group transfer of trimethylsilyl 2-oxobicyclo[3.3.1]nonane-1-carboxylate occurred on heating [Bloch, 1976a]. The anti-Bredt olefin thus generated decomposed immediately by the retro-Diels–Alder fission.

Bridgehead olefins may also be formed by rearrangement of carbenes [Wolf, 1973]. As expected, trienes were found to be the final products.

Among the acetolysis products of (7-methylnorbornadienyl-7-)methyl brosylate is a cyclopentadiene [Bly, 1983]. Its formation has the effect of strain relieving.

The reaction of 7-chloronorbornadiene with sodium cyanide also was followed by a retro-Diels–Alder reaction of the tricyclic product [Tanida, 1965].

Treatment of bicyclo[2.2.2]oct-5-en-2-one with an alkoxide base at high temperatures led to the 2-phenyl derivative [Cheng, 1977]. The aldol condensation product underwent deconjugation and elimination of ethylene. This experiment was part of a homoenolization study.

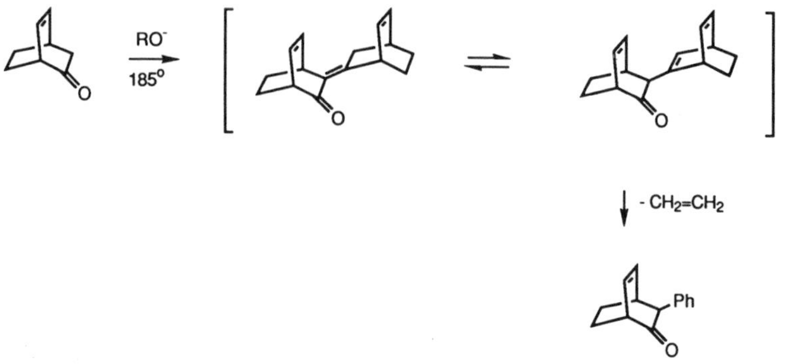

Retro-Diels–Alder reactions attended by aromatization derive much driving force from such action. It is easy to rationalize the behavior of a hexacyclic diketone upon tosylhydrazone formation in that the major product turned out to be pentacyclic inspite of a new heterocycle being made [Coxon, 1990]. The transformation was mediated by alcoholysis of the monotosylhydrazone and transannular azo group participation to give the retro-Diels–Alder reaction substrate.

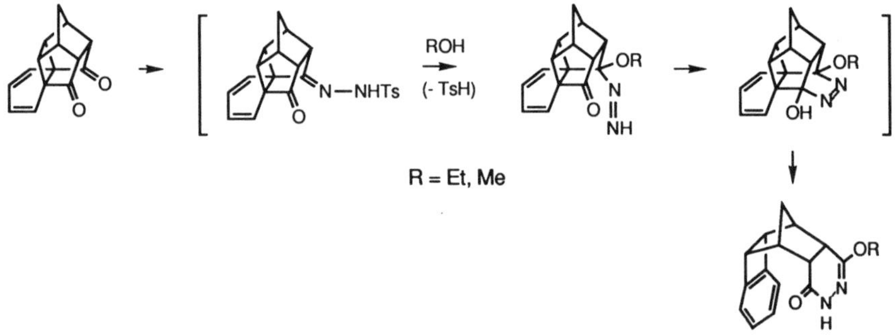

A rather unusual retro-Diels–Alder reaction of a 7-azabicyclo[2.2.1]hept-5-en-2-one system was implicated in the photochemical extrusion of a tricyclic triazoline [Schultz, 1980].

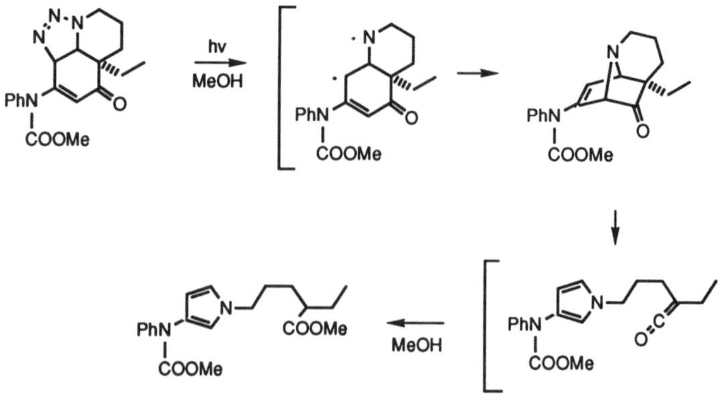

7-Phenylbicyclo[2.2.1]norborn-2-en-7-ide was found to undergo a retro-Diels–Alder reaction [Bowman, 1976]. The generation of the aromatic phenylcyclopentadienide anion and ethylene supplied the driving force.

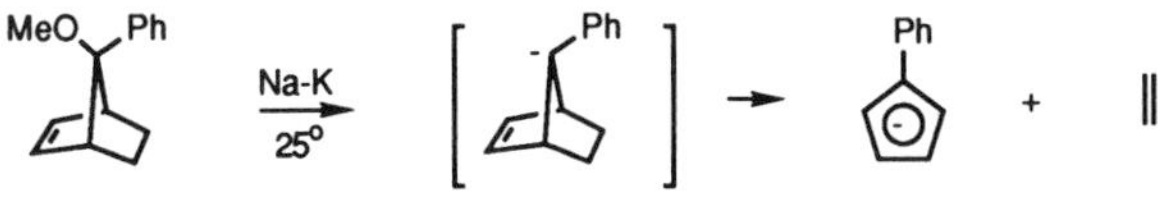

1-Bromo-3a,7a-dihydroindan-2-one does not undergo Favorskii rearrangement due to prohibitively high strain of both intermediate and product. The preferred reaction pathway is dehydrobromination to form a transient tetracyclic enone which is decomposed via a retro-Diels–Alder reaction [Paquette, 1972a].

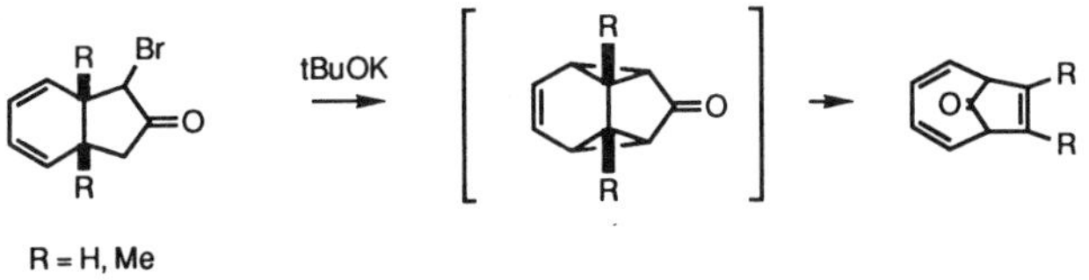

R = H, Me

A Cope elimination was conducted in an attempt at generation of a 2-azabarrelene precursor [Sliwa, 1977]. However, under the experimental conditions a retro-Diels–Alder reaction was favored as it liberated benzene.

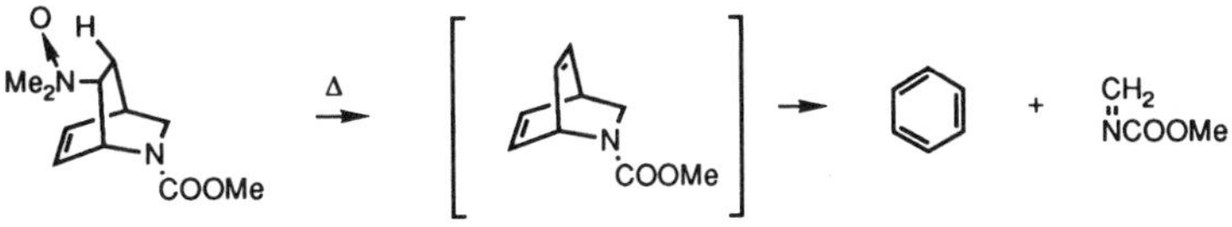

Flash vacuum pyrolysis of phenyl propargyl ether gave benzocyclobutene and 2-indanone [Trahanovsky, 1972; Riemann, 1977]. After the aromatic Claisen rearrangement there were a Diels–Alder/retro-Diels–Alder reaction tandem and the cyclization reaction.

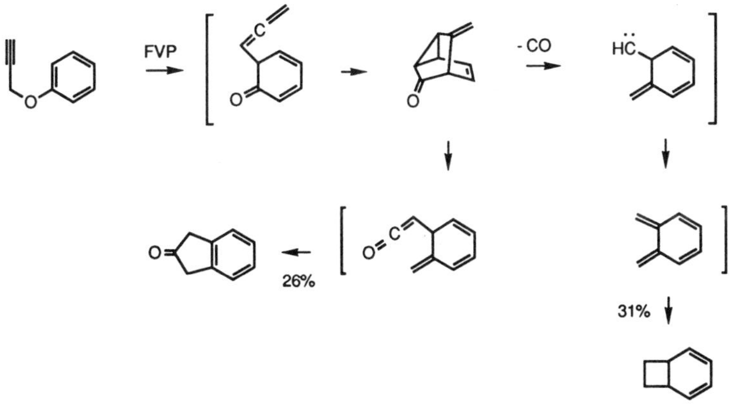

An extensive skeletal reorganization of a tetracyclic ketone to give a benzocyclooctadienone [Miyashi, 1977] was initiated by the scission of two cyclopropane rings while forming a new one. However, this ring suffered cleavage (at an alternative internal bond) by a retro-Diels–Alder process.

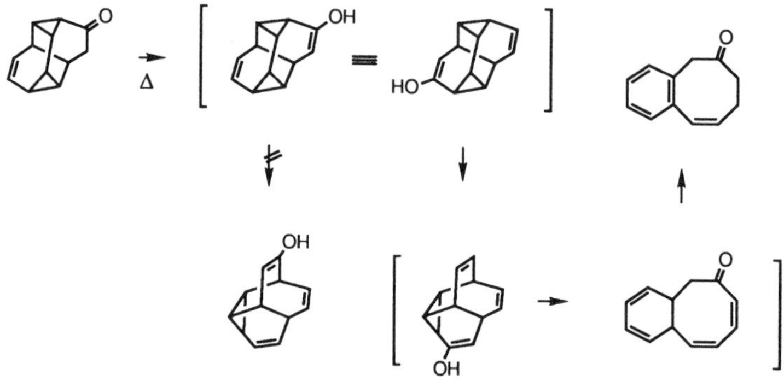

Activation of a substrate toward retro-Diels–Alder reaction by oxidation is witnessed in the following examples. The air oxidation of the Diels–Alder adducts (regioisomers) gave rise to the unstable diquinones which were stabilized by elimination of ethylene [Krohn, 1979]. Note that the concentration of the diquinone species need not be high to derive the complete conversion of the substrate to the anthracyclinone products.

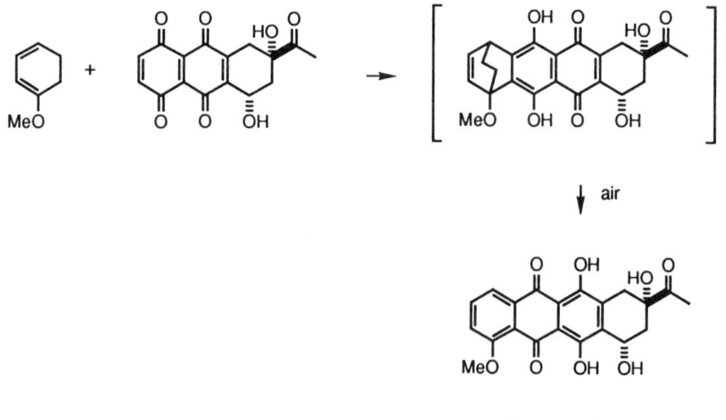

daunomycinone

Oxidation of a diazasnoutane also triggered a retro-Diels–Alder reaction to generate semibullvalene [Paquette, 1970].

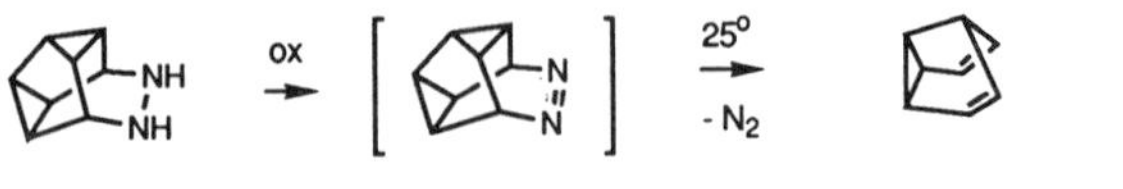

semibullvalene

9.3. SOME OTHER CHELETROPIC REACTIONS

An efficient azulene synthesis is based on [6 + 4]-cycloaddition of 6-dimethylaminofulvene with thiophene dioxides [Mukherjee, 1979]. The cyclo-adducts promptly eliminate dimethylamine and sulfur dioxide.

The Ramberg–Bäcklund olefin synthesis [Paquette, 1977] is a versatile method for oxidative coupling of alkyl halides. It is applicable to the preparation of some very strained olefins such cyclobutenes [Paquette, 1971b].

Mechanistically, the initial stage of a Ramberg–Bäcklund reaction is similar to that of the Favorskii rearrangement of unstrained α-haloketones. However, the episulfone intermediates undergo cheletropic reaction rather than ring opening by the attack of a nucleophile. (Note: in the absence of nucleophiles cyclopropanones and cyclopropenones lose carbon monoxide.) A very convenient one-pot reaction [Vedejs, 1978] consists of treating the sulfone with hexachloroethane in the presence of sodium hydride. More harsh conditions involving KOH and carbon tetrachloride are effective for more robust sulfones.

The ability of α-sulfonyl sulfones to undergo electrophilation and the Ramberg–Bäcklund reaction simplifies the preparation of certain alkenes [Ranasinghe, 1989].

Conjugate polyene circuits are conveniently assembled by a Michael addition-promoted Ramberg–Bäcklund reaction [Burger, 1981]. For example, an isomeric mixture of β, γ; δ,ε; ζ,η-trienyl sulfone has been obtained from readily available substances in a few steps. The sulfone can be used further in Julia olefination.

Allenyl chloromethyl sulfone is a useful building block. Ramberg–Bäcklund reaction of its Diels–Alder adducts generates 4,5-dimethylenecyclohexenes [Block, 1990] which may be submitted to the same reaction sequence to build up molecules with a linear array of 1,4-cyclohexadiene cells.

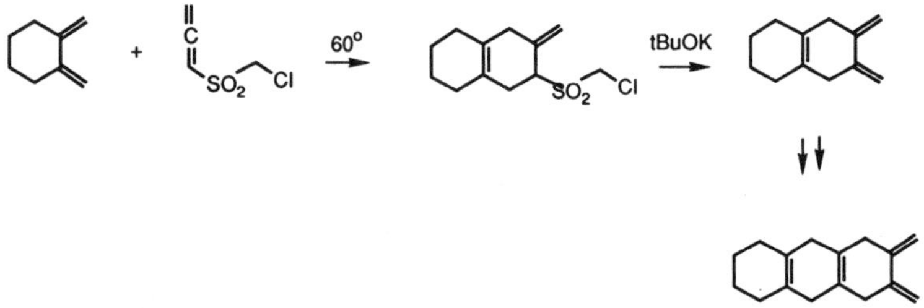

A very convenient synthesis of cyclopentenones [Hendrickson, 1985] is based on the assembly of α'-allylcarbinyl triflyl sulfones and the oxidation of the allylic alcohol group. Because of the high acidity of the C-H which is flanked by sulfonyl and triflyl groups, an intramolecular Michael reaction is an inevitable consequence. Elimination of the triflyl group and cheletropic reaction follow.

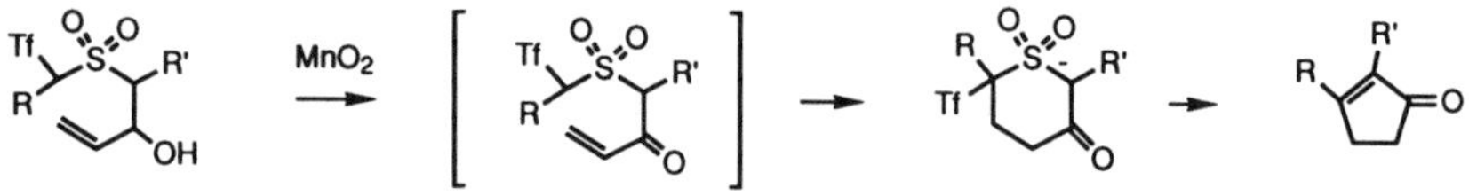

A version of the Eschenmoser fragmentation consists of heating an aziridinoimine derivative of an α,β-epoxy ketone [Felix, 1972]. The decomposition step involves a cheletropic reaction.

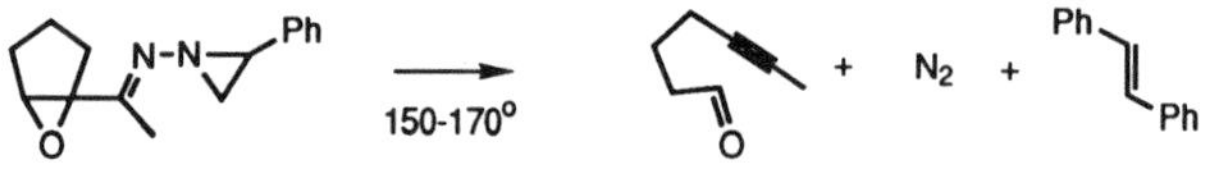

10

ELECTROCYCLIC REACTIONS

Electrocyclic reactions are intramolecular reactions dealing with the formation of a single bond between the termini of a linear conjugate π-electron system, and the reverse process. The substrates may be neutral or charged species. Depending on the number of π-electrons involved, and the conditions (thermal or photochemical), a conrotatory or disrotatory mode of cyclization or cycloreversion may take place, with the result that the configurations of the substrates the the products are correlated.

These modes are strictly governed by conservation of molecular orbital symmetry. The general rules are: thermal cyclization reaction of a (4n) π-electron array adopts a conrotatory action, and that of a (4n + 2) π-electron system adopts a disrotatory mode; in the first excited state (photoreaction) reverse relationships hold.

10.1. ELECTROCYCLIC REACTION CASCADES

A combination of two or more electrocyclic reactions, be they thermally or photochemically induced, or of mixed origins, is within this category of reactions. Given the structural features with which an electrocyclic reaction may be induced, and the primary product is capable of undergoing further electrocyclic reactions, a compound may be transformed into a novel skeleton in a single operation. Needless to emphasize, such cascade reactions, with proper control, are of enormous value to synthesis.

An early attempt at gaining access to [10]annulene [Masamune, 1967] was by thermolysis of bicyclo[6.2.0]deca-2,4,6,9-triene. However, due to the intrinsic instability of the medium-size ring compounds, trans-9,10-dihydronaphthalene was obtained. This was apparently the result of two electrocyclic reactions:

conrotatory ring opening and disrotatory ring closure of the (E,Z,Z,Z,Z)-isomer of cyclodecapentaene.

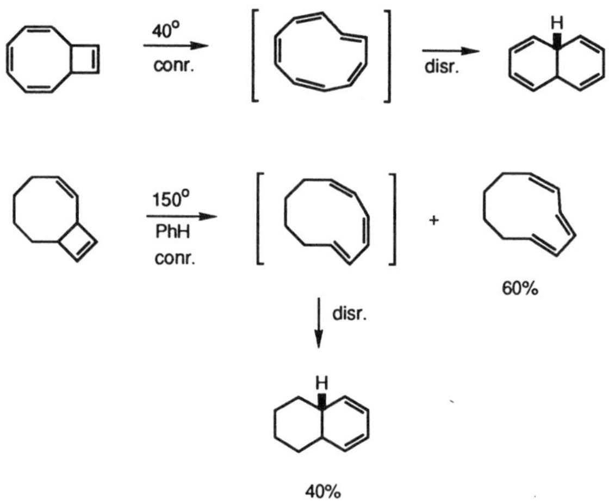

The two modes of conrotatory opening of bicyclic dienes are possible, and this aspect has been investigated [Dauben, 1981].

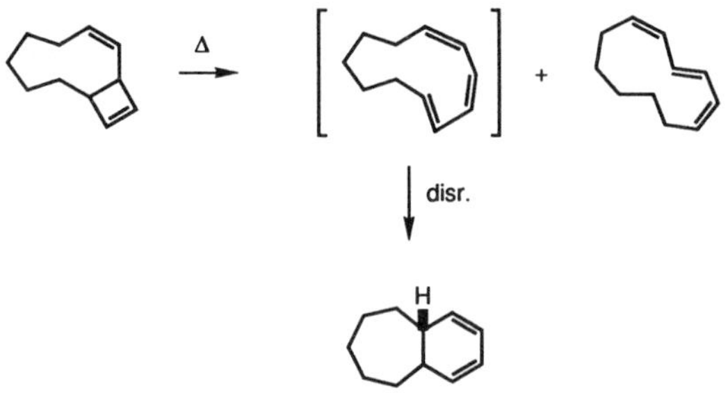

The conversion of bicyclo[6.1.0]nonatriene to the *cis*-bicyclo[4.3.0]nona-triene isomer by rhodium(I) catalysis [Powell, 1971] passes through the monocyclic tetaene which cyclizes immediately.

N-Carbethoxy-1,2-dihydroazocine reverts to an acyclic azatetraene at 130°C and under the same conditions the latter species undergoes a six-electron cyclization to give 2-vinyl-*N*-carbethoxy-1,2-dihydropyridine [Okamura, 1969].

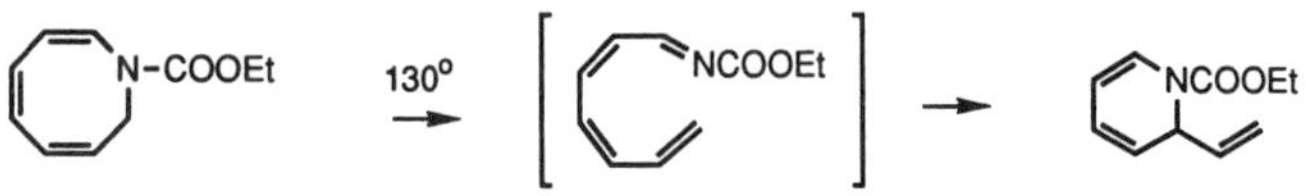

In the field of natural products, an application of consecutive electrocyclic reactions is best represented by a synthesis of (+)-occidentalol [Hortmann, 1973]. A *trans*-fused homoannular hexalincarboxylic ester was irradiated at low temperature to induce a conrotatory ring opening, and on warming to −20°C, the monocyclic triene was transformed into two *cis*-fused hexalins, because there are two modes of rotatory action. On reaction with methyllithium one isomer gave (+)-occidentalol, whereas the other diastereoisomer yielded (−)-7-epi-occidentalol.

7-epi-(-)-occidentalol   (+)-occidentalol

A very efficient phenol synthesis from a combination of vinylketenes and acetylenes is based on a pericyclic cascade of three reactions. A [2 + 2]-cycloaddition of the two substrates leads to a cyclobutenone which is susceptible to ring opening under the reaction conditions. Reclosure of the resulting dienyl ketene is thermodynamically driven, as it forms a six-membered ring capable of aromatization.

A synthesis of maesanin [Danheiser, 1990] shows the overall process in some detail. Several other natural products containing a resorcinal portion have also been acquired, including grifolin and the antibiotic DB-2073 [Danheiser, 1984].

maesanin

The scope of this transformation has been broadened by the development of a convenient procedure for the preparation of siloxyacetylenes from esters [Kowalski, 1988].

Hydroquinone and *p*-quinones are now readily available by thermal manipulation of 4-oxycyclobut-2-enones which carry an unsaturated substituent at C-2 [Liebeskind, 1986; Perri, 1986]. These substances undergo thermolysis to give dienyl or enynyl ketenes which cyclize spontaneously. Interestingly, an O- to C- allyl migration observed in certain substrates was very useful for a synthesis of nanaomycin-D [Foland, 1989].

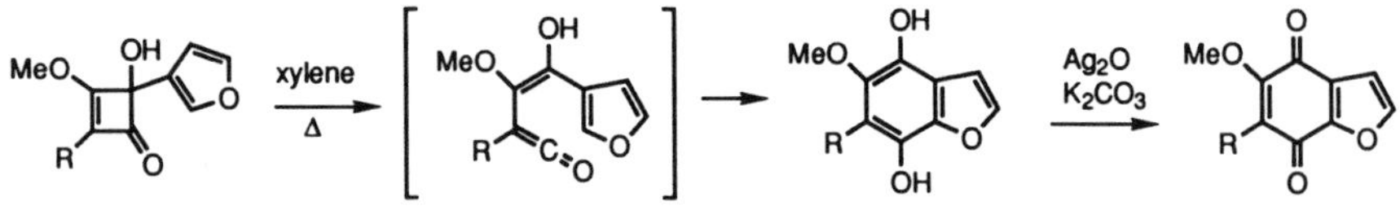

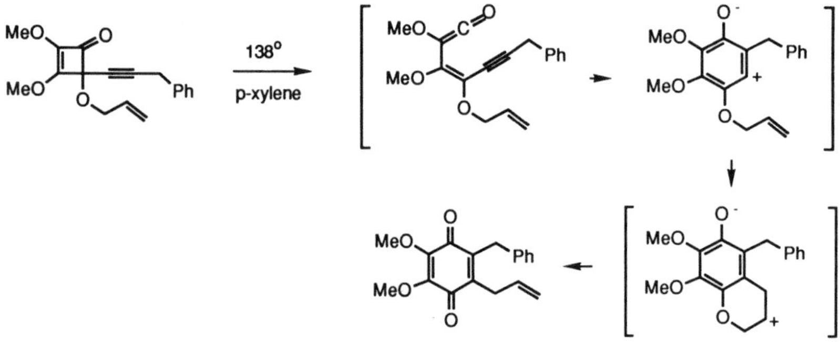

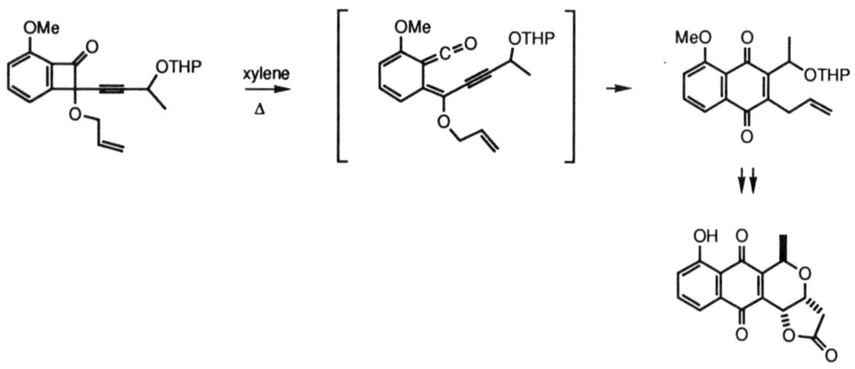

nanaomycin-D

The more difficultly accessible hydroxybenzoquinones may now be synthesized from 3-*t*-butoxy-3-cyclobutene-1,2-diones [Enhsen, 1990]. Of course the *t*-butyl group of the final products is removable by treatment with acid.

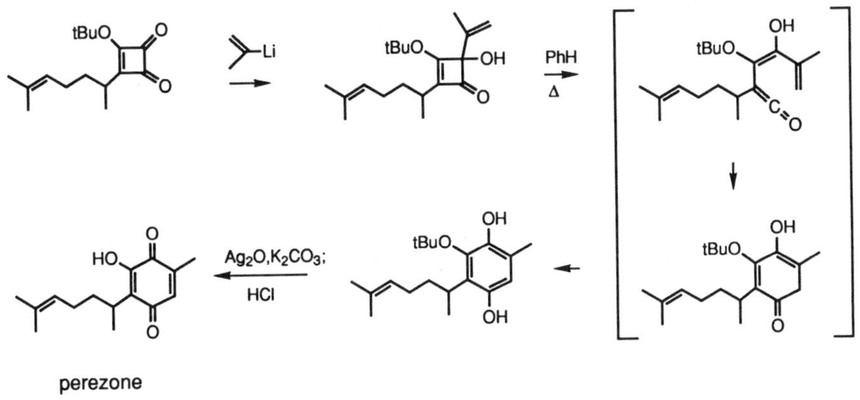

perezone

A 4-vinyl-2-cyclobutenone, obtained by an unexpected rearrangement of the primary [2 + 2]-cycloadduct of 2,4-dimethyl-2-cyclobutenone and 1-(*N*,*N*-diethylamino)propyne, was transformed into a phenol on heating at 140°C [Ficini, 1977].

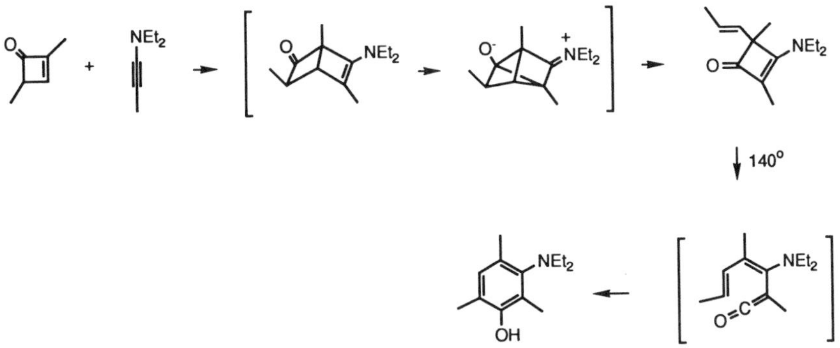

Thermolysis of 1-vinylbenzocyclobutenes gives rise to *o*-quinodimethane intermediates to which a six-electron cyclization pathway is open. When the substrates are heated in the presence of manganese dioxide, naphthalenes can be obtained directly [Shishido, 1990], although compounds of a more sensitive nature are probably better prepared by carrying out the dehydrogenation at a separate stage [Shishido, 1989d].

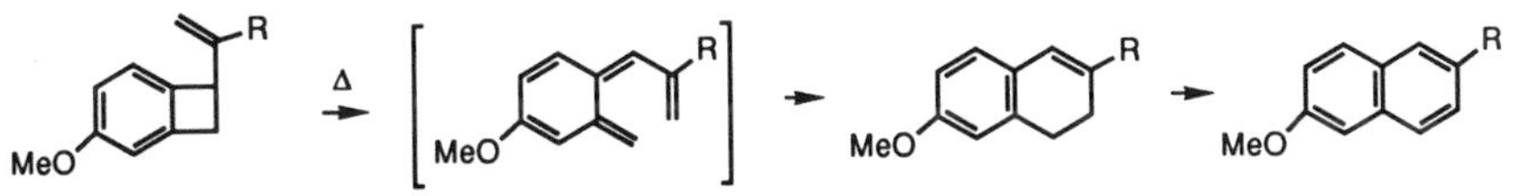

The product of Bamford–Stevens reaction of a benzocyclobutenone tosyl-hydrazone in the presence of benzene may have arisen from two electrocyclic reactions of the spirocyclic adduct [O'Leary, 1978].

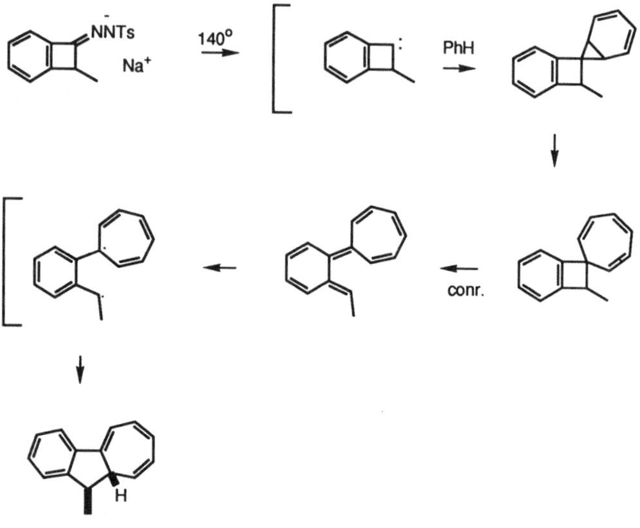

The photoisomerization of xanthene to 6H-dibenzo[b,d]pyran in aqueous acetonitrile [Huang, 1991] involves O–C$_{Ar}$ bond homolysis and subsequent formation of a spirobenzocyclobutene. The logical fate of the latter species is electrocyclic opening of the four-membered ring followed by electrocyclization.

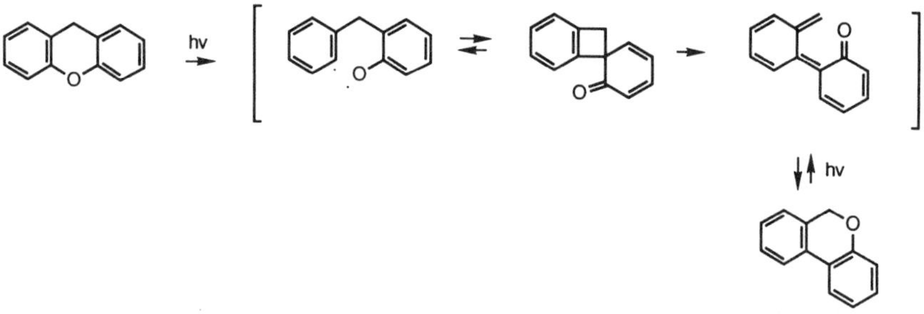

The facile electrocyclic opening of vinylcyclobutenolone and reclosure of the ketene intermediate to a naphthoquinone formed the basis of a convergent approach to α-citromycinone [Swenton, 1984].

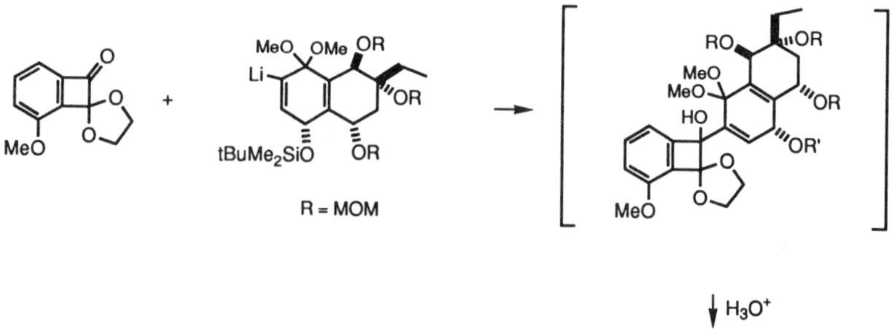

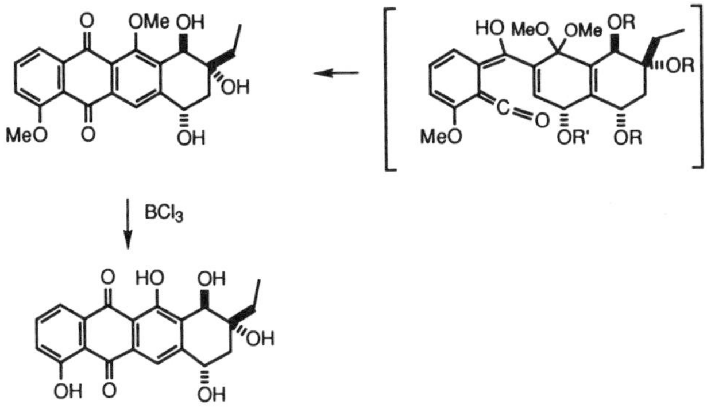

α-citromycinone

Chromium tricarbonyl complexes of unsaturated ketenes are intermediates in the naphthol synthesis from arylalkoxycarbene complexes and alkynes. Very extensive investigations have been devoted to the application of this synthetic method (cf. deoxyfrenolicin synthesis [Semmelhack, 1985a]).

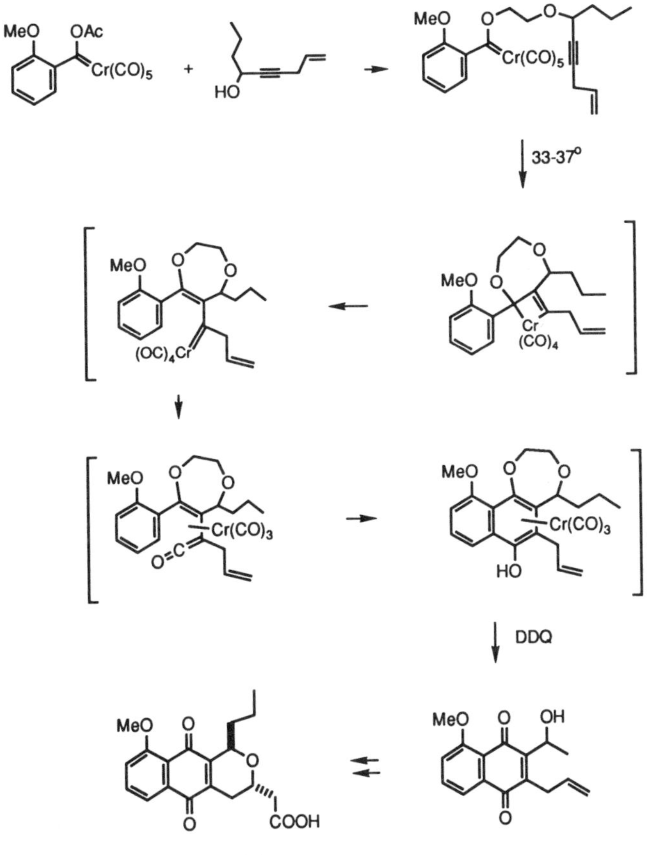

deoxyfrenolicin

The enol carbonate of a β-tetralone synthesized by this method was used in a formal synthesis of atisine [Shishido, 1989b]. The enol carbonate was cleaved and transformed into a triene for elaboration of a hydrophenanthrene derivative by an intramolecular Diels–Alder reaction.

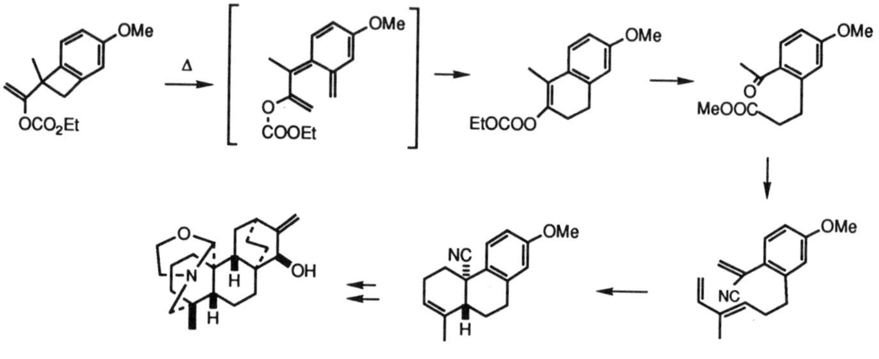

atisine

α-Tetralones are also accessible in a similar manner [Hickman, 1991].

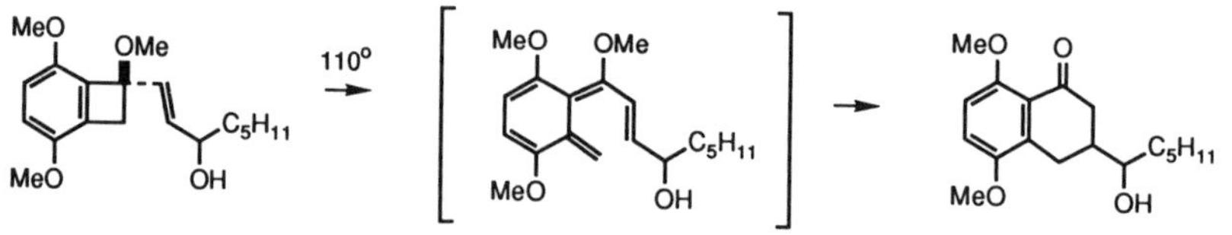

As dihydroisoquinolines are the thermolysis products from the corresponding iminobenzocyclobutenes, the method has great value in alkaloid synthesis. Thus, the access to xylopinine [Kametani, 1973] and yohimbone [Kametani, 1974b] represents only two examples of the application.

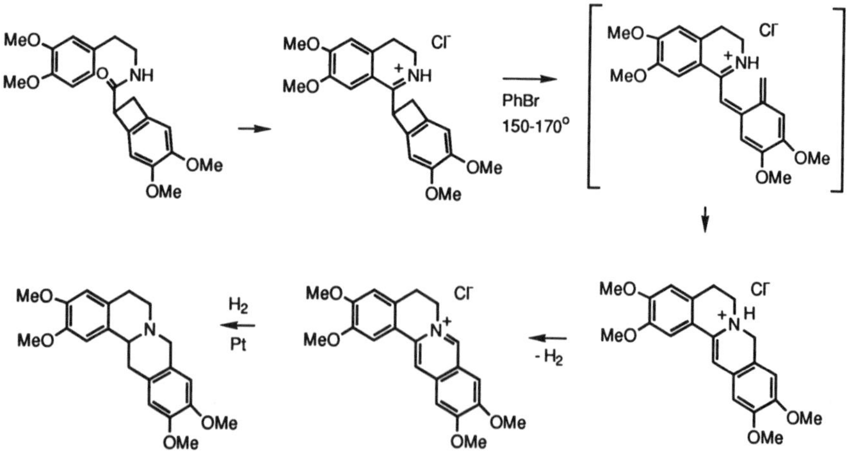

xylopinine

protoberberine

Several benzocyclobutenyl ketoximes have been converted into isoquinolines [Shishido, 1989a]. The electrocyclization was apparently followed by dehydration of the *N*-hydroxy-1,2-dihydroisoquinolines.

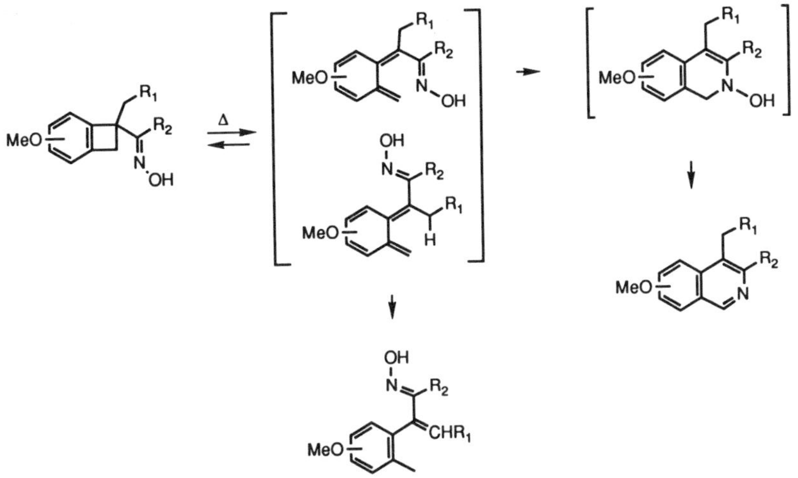

The pyrolytic transformation of a benzocyclobutenecarboxylic acid into an isochroman-3-one suggested a simple route to sendaverine [Kametani, 1979a]. The electrocyclic reaction must be preceded by a geometrical isomerization of the double bond conjugated to the carboxyl group of the initially formed *o*-quinodimethane. This isomerization was definitely facilitated by the methoxy substituent.

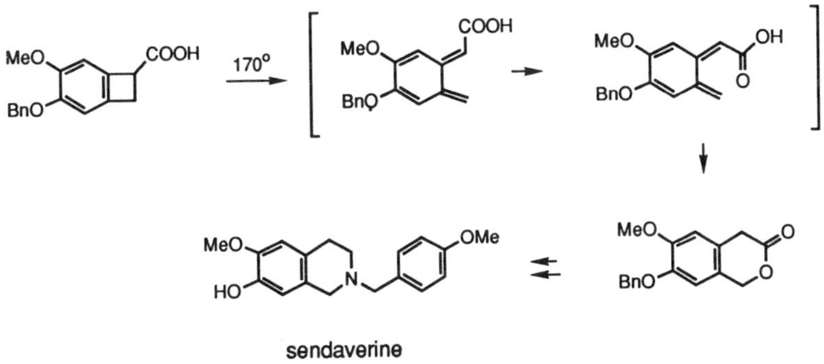

sendaverine

Related to the above is the (*Z*) to (*E*) isomerization of the styrenic double bond of 5-phenyl-2,4-pentadienealdehydes [Schiess, 1970]. Both the reaction activation parameters and trapping experiments showed the operation of an electrocyclic mechanism.

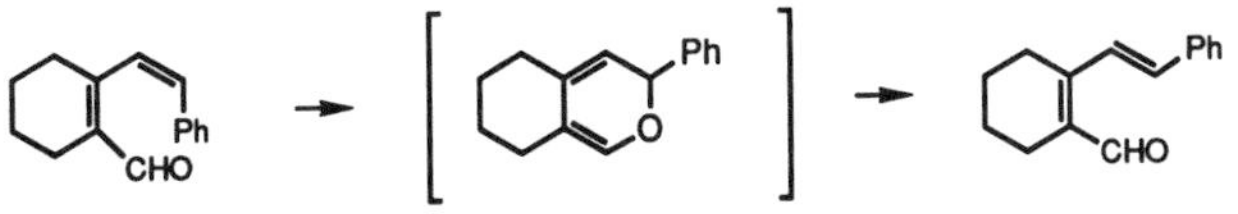

The endiandric acids are racemic natural products which are likely derived from acyclic polyenes by a series of electrocyclic and Diels–Alder reactions. This hypothesis has received strong support by the access of these compounds from thermal cyclization of the proper polyenes. Thus, in a stepwise stereocontrol synthesis of endiandric acid-A and -B [Nicolaou, 1982c] a conjugated (E,Z,Z,E)-tetraene prepared as the precursor actually underwent consecutive electro-cyclizations in situ, resulting in a bicyclo[4.2.0]-octadiene. This ring skeleton is the same as that of endiandric acids E, F, and G. Accordingly, these latter compounds were also obtained by proper modification of the sidechains [Nicolaou, 1982a].

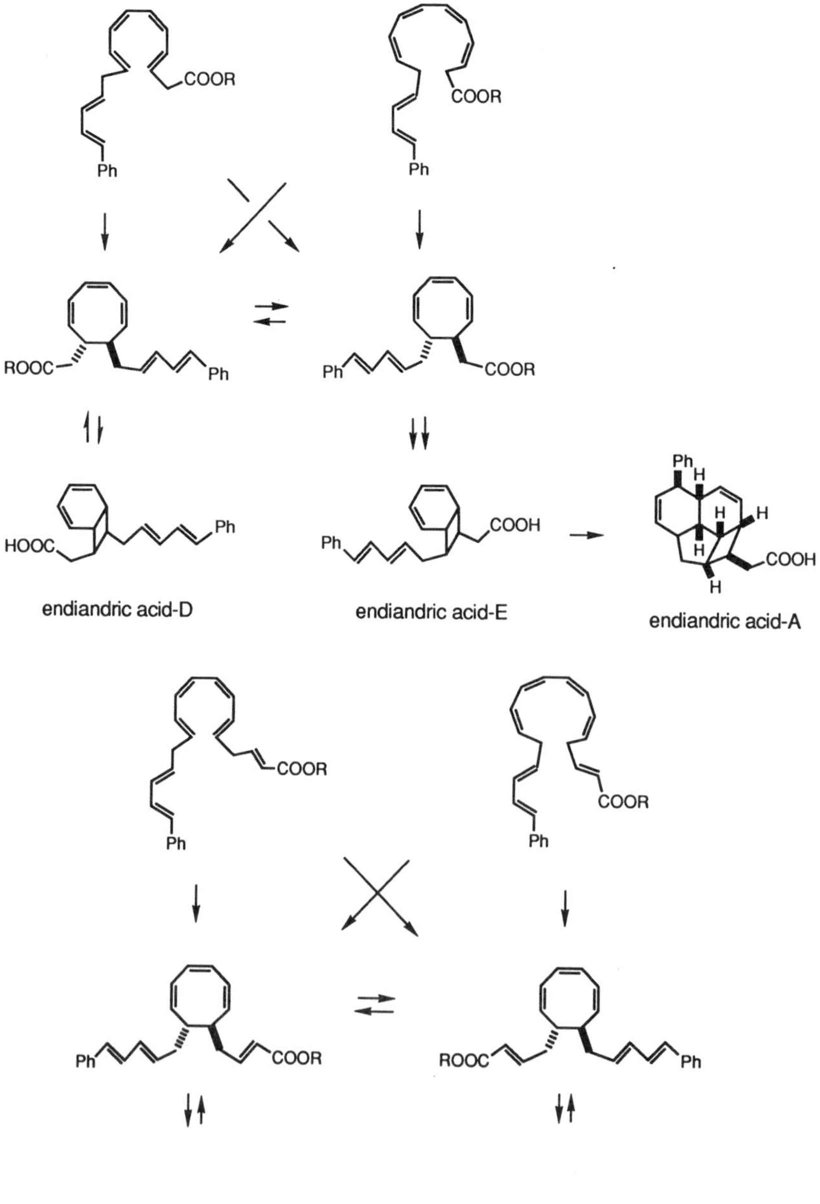

endiandric acid-D   endiandric acid-E   endiandric acid-A

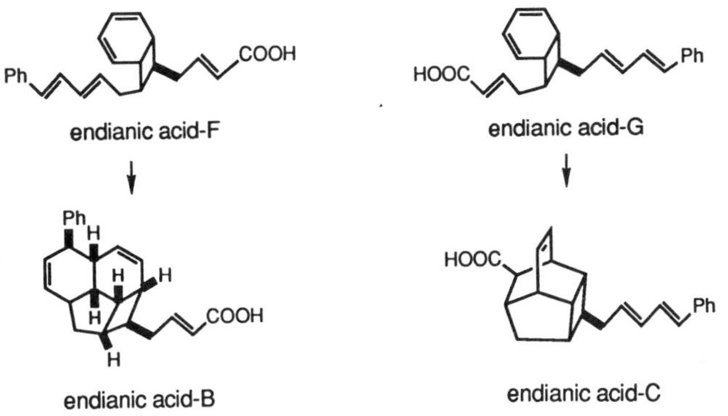

endianic acid-F

endianic acid-G

endianic acid-B

endianic acid-C

10.2. CYCLOADDITION OR CYCLOREVERSION/ELECTROCYCLIC REACTION TANDEMS

Tandem reactions terminated by an electrocyclic reaction may also be initiated by a cycloaddition or retro-cycloaddition. A very useful two-carbon ring expansion involving an enamine and an acetylenecarboxylic ester is such a process. The cyclobutene derivatives are rarely isolated as they generally undergo ring opening in situ.

While the Woodward–Hoffmann rules specify conrotatory ring opening, the products (medium-sized rings) are (Z,Z)-dienes. The inconsistency between theory and experimental results have been resolved [Visser, 1982]. Indeed, an (E,Z)-diene have been characterized by x-ray crystallography, and it was shown to isomerize to the (Z,Z)-isomer.

Applications of the ring expansion procedure are numerous. It constituted the key step for developing the hydrazulene skeleton of velleral [Froborg, 1978].

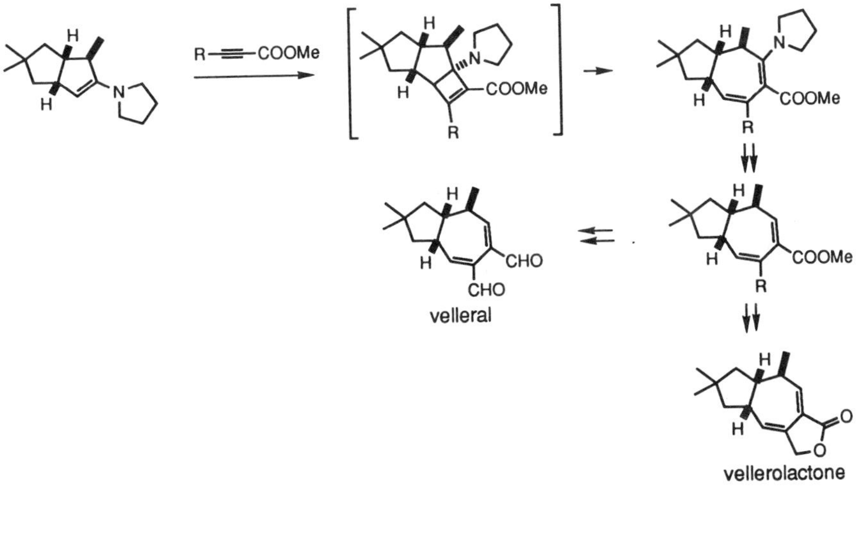

velleral

vellerolactone

The photoreaction of p-benzoquinone with 5-phenyl-4-pentynoic acid proceeded by a [2 + 2]-cycloaddition which was followed by ring opening and lactonization [R.M. Wilson, 1980].

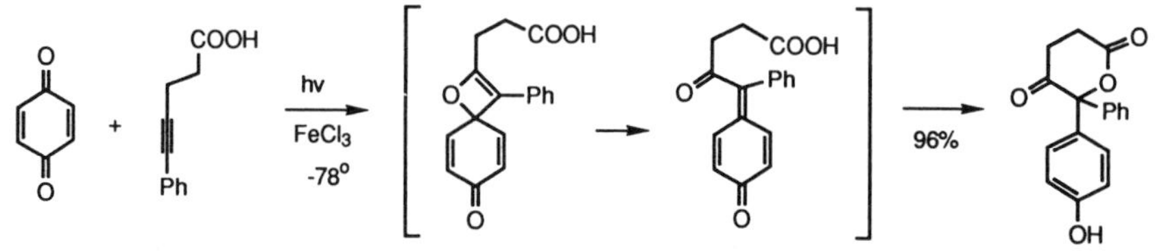

Tetrathiafulvalene undergoes [2 + 2]-cycloaddition with certain highly electron-deficient alkynes with its central double bond. However, the cyclo-adducts are susceptible to electrocycloreversion [Hopf, 1991].

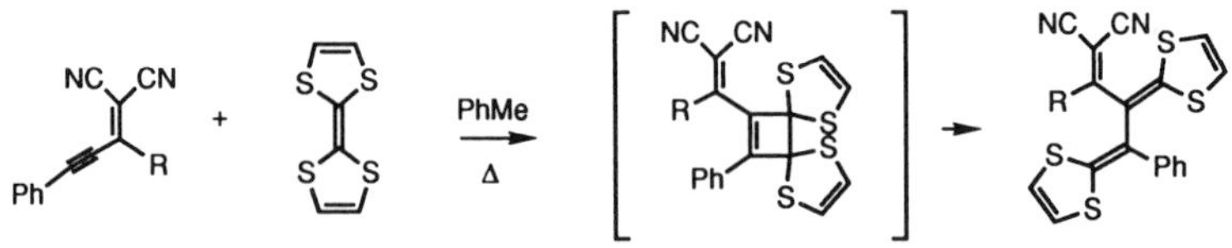

A pyrolytic product of 2-ethynylbiphenyl is benzazulene [Brown, 1972]. The carbene derived from rearrangement of the acetylene is readily intercepted by the distal benzene ring; that subsequent electrocyclic opening is a logical consequence.

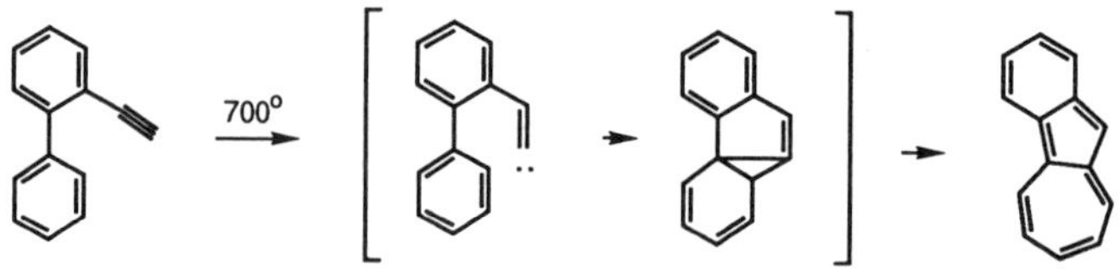

The thermal decomposition of pivaloyldiazomethyl t-butyldiphenyl silane proceeds via a 1,3-$C \rightarrow O$ migration and nitrogen loss. Intramolecular carbene cycloaddition to one of the benzene rings attaching to the silicon is followed by an electrocyclic opening [Brückmann, 1986].

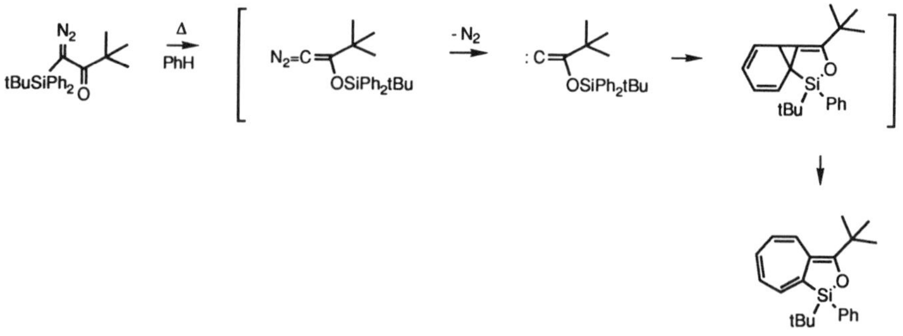

Lead tetraacetate oxidation of 4,4-bis(p-dimethylaminophenyl)phthalazin-1-one gave rise to a benzoazulenone [Kuzuya, 1980]. The result can be explained by invoking the generation and intramolecular interception of an o-quinodimethaneketene. The tetracyclic ketone is a norcaradiene derivative whose tautomerization (electrocyclic opening) is expected.

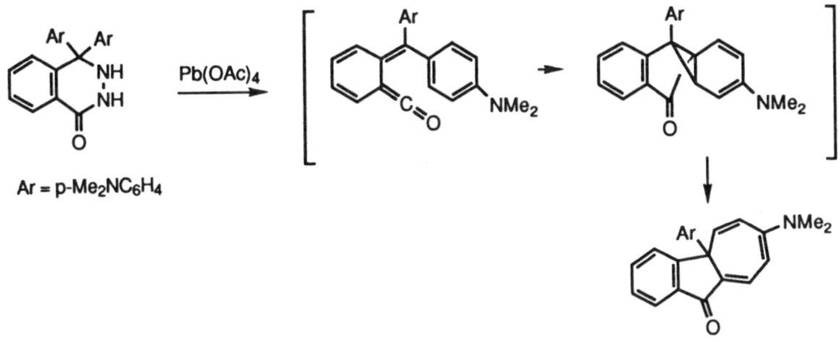

At high pressure the condensation of 3-carbomethoxy-α-pyrone with cyclopropenone propyleneacetal yields the acetal of tropone-3-carboxylic ester [Boger, 1986a]. The Diels–Alder/retro-Diels–Alder pathway is followed, and the norcaradiene tautomerizes subsequently.

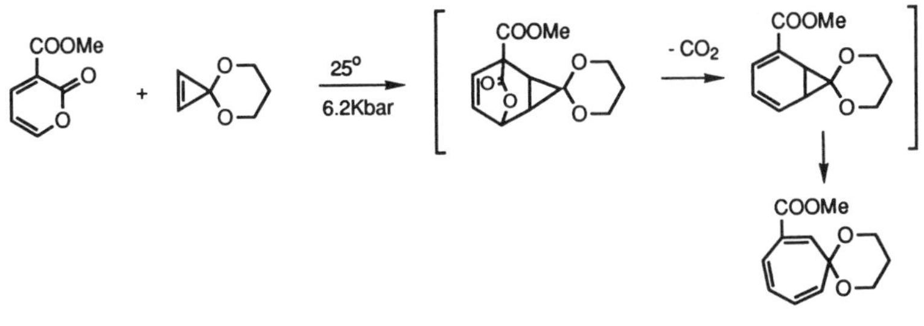

Pyrano[3,4-b]indol-3-ones undergo cycloaddition with mesoxalic esters with loss of carbon dioxide [Pindur, 1989]. The pyran ring of the products is prone to open on account of the higher stability of the indole derivatives.

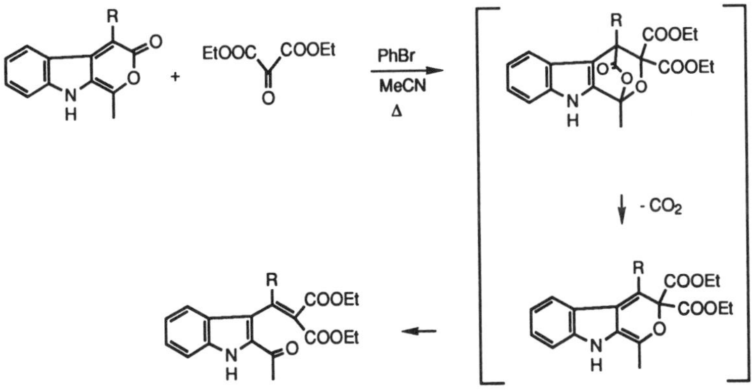

A study of the 1,4-elimination reaction from cyclohexenes required specifically deuterated substrates. These compounds were obtained from Diels–Alder reactions of the butadienes, the latter species were derived from thermally labile cyclobutanes via the corresponding cyclobutenes [Hill, 1978].

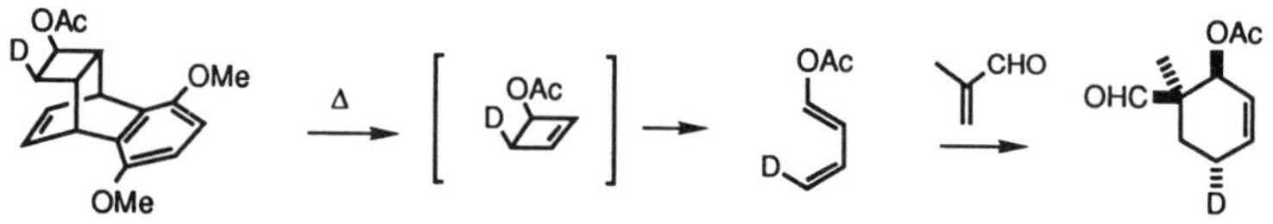

Other cyclobutane derivatives that are suitable for the generation of butadienes include the cyclopentadiene adducts [Trost, 1978].

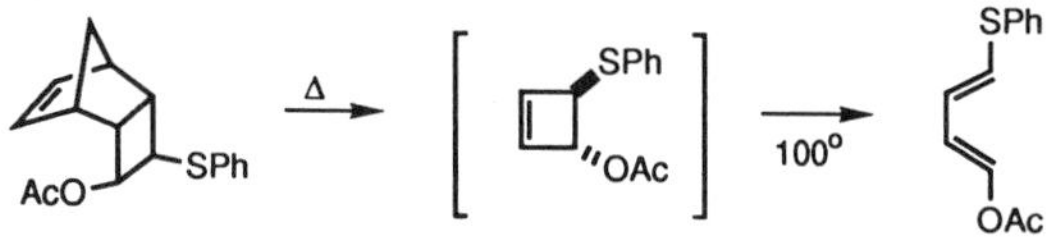

A series of symmetry-allowed changes, that is three electrocyclic reactions interconnected by a Diels–Alder/retro-Diels–Alder reaction tandem can be adduced for the reaction of a 1,8-disubstituted (Z,Z,Z,Z)-octatetraene with 1,4-naphthoquinone to give anthraquinone and an (E,E)-1,3-butadiene [Meister, 1963].

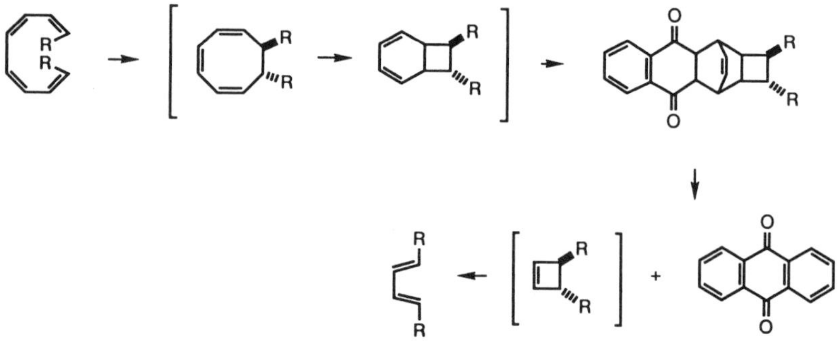

A dilactone which is the formal carbon dioxide adduct of *o*-formylphenylacetic acid lactone gave isocoumarin on pyrolysis [Bleasdale, 1983]. Undoubtedly decarboxylation was the first step, but the product underwent electrocyclic opening and 1,5-hydrogen shift (a redox exchange between the aldehyde and the ketene moieties) before reclosure of the heterocycle.

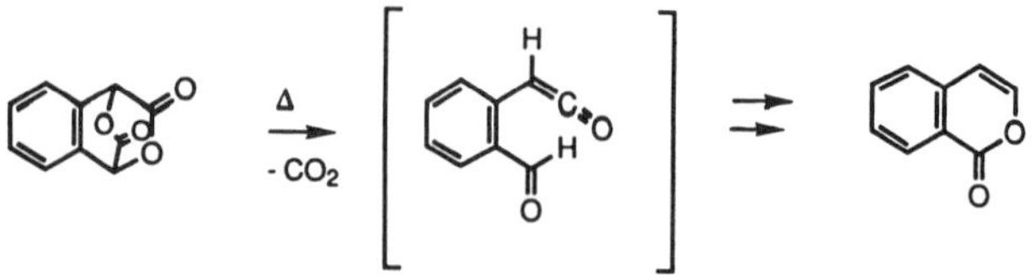

The thermal reorganization of annulated bicyclo[4.2.0]octatrienes are known to afford cyclooctatetraenes [Paquette, 1974b]. An intramolecular Diels–Alder cycloaddition followed by retro-Diels–Alder fission and disrotatory electrocyclic opening of the cyclohexadiene moiety accounts for the results.

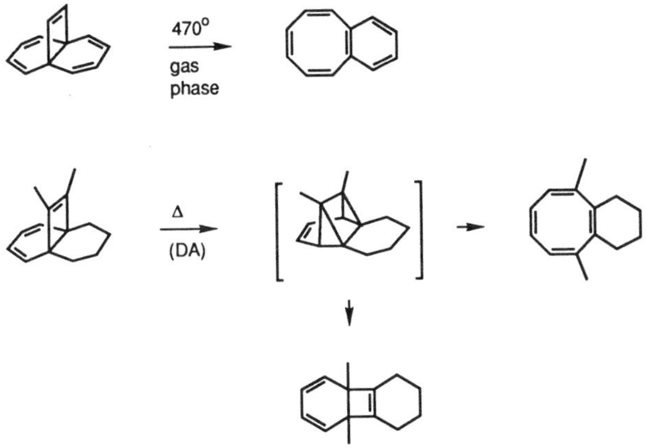

The reaction of benzocyclopropene with 1,2,4-triazines provides an entry to 3,8-methano-1-aza[10]annulenes [J.C. Martin, 1984]. The reaction tandem of Diels–Alder/retro-Diels–Alder reactions prior to electrocyclic tautomerization is evident.

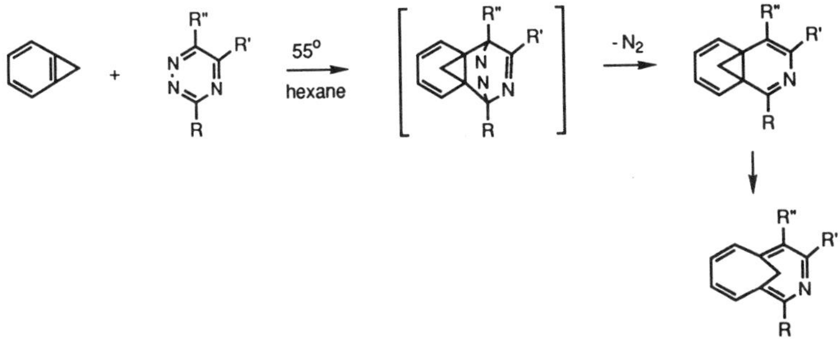

A diazabasketene was found to undergo thermal elimination of hydrogen cyanide in an intriguing fashion [McNeil, 1971]. The reaction may be rationalized as involving a retro-Diels–Alder reaction, a 1,3-alkyl shift (via diradical?), another retro-Diels–Alder fission, and electrocycloreversion.

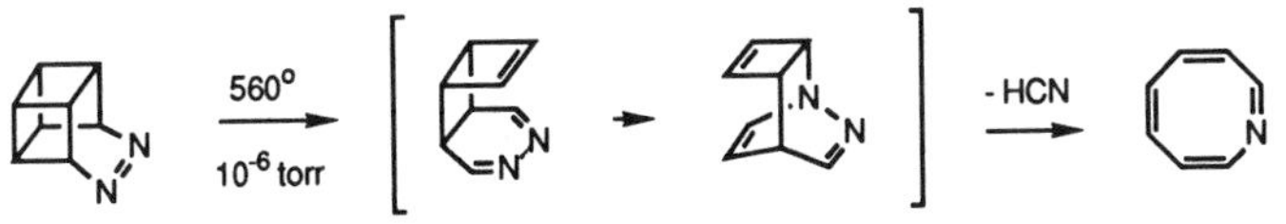

A fascinating intramolecular enyne metathesis mediated by a palladacyclo-pentadienetetracarboxylic ester leads to the formation of bicyclic dienes in which one of the double bonds is at a bridgehead position [Trost, 1991c]. Electrocyclic opening is the terminating step of such reactions.

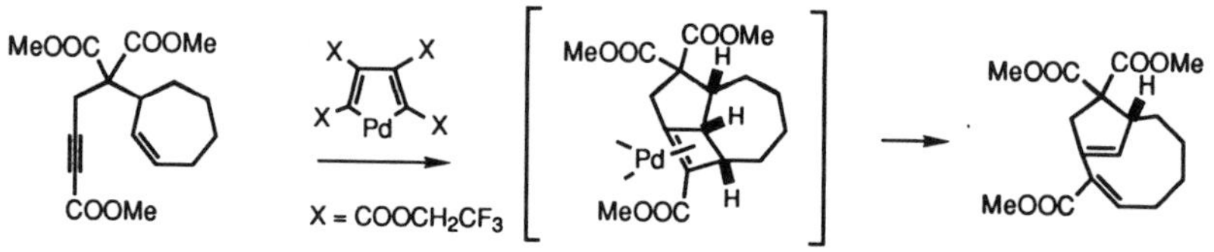

A bridged semibullvalene was converted into a fused cyclooctatetraene. The intricate mechanism as elucidated by deuterium labeling [Paquette, 1973] depicts an initial cyclopropane cleavage to a diradical, and subsequently a 1,3-bridged cyclooctatetraene. Because of the strain of such a ring system, tandem electrocyclization, intramolecular Diels–Alder reaction, retro-Diels–Alder reaction, and electrocyclic opening ensued.

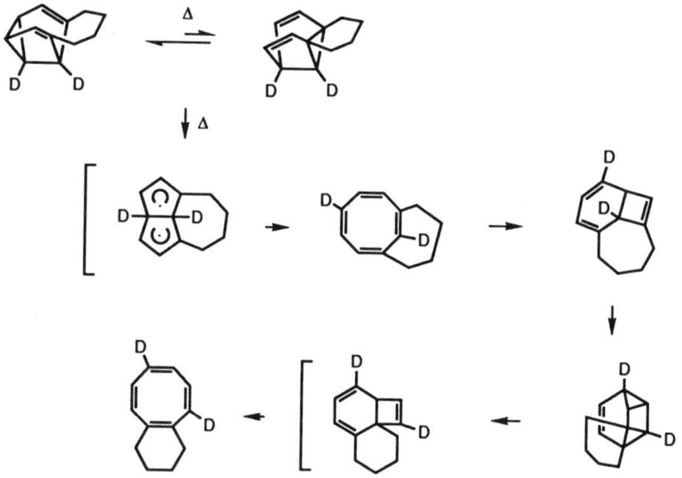

4,5-Epoxy-2-cyclopentenones liberated from the cyclopentadiene adducts may undergo $[_4\pi_a + {_2\pi_a}]$ cycloreversion and electrocyclization at proper temperatures to furnish α-pyrones [Klunder, 1981].

10.3. REARRANGEMENT–ELECTROCYCLIC REACTION TANDEMS

It is natural to expect that an electrocyclic reaction may form a tandem with sigmatropic rearrangement because both these processes belong to the pericyclic reaction group, and they can be often induced under the same set of conditions.

The formation of (4E)-4-methyl-3-methylene-1,4-hexadiene from the thermal reaction of 1-ethylidene-3-methyl-2-methylenecyclobutane is readily explained by the tandem rearrangement–ring opening sequence [Gajewski, 1972].

Cyclodeca-1,2,4-triene is prone to undergo thermal [1.5]-sigmatropic rearrangement to give a conjugate triene. however, the isomer could not be isolated as it underwent a 6-electron electrocyclization [Minter, 1979]. Interestingly, in the thermolysis of bicyclo[7.1.0]deca-2,3-diene the cyclopropane ring was cleaved and re-formed.

trans-bicyclo[4.4.0]deca-2,4-diene

The mechanism for the thermal conversion of 5- into 3-substituted pyrones proceeding via electrocyclic opening, [1.5]-hydrogen migration, and electrocyclization is supported by ^{18}O-labeling experiments and the blocking of the reaction by a 6-methyl group [Pirkle, 1975].

It is possible that the annulative formation of 2-naphthols by pyrolysis of o-methylbenzylidene derivatives of Meldlrum's acid proceeds via [1.5]-hydrogen shift prior to elimination of acetone and carbon dioxide to give the ketene intermediates which then undergo electrocyclization and enolization [Brown, 1974]. The generality of this transformation has been shown by the preparation of hydroxy-indoles, carbazoles, benzothiophenes, benzofurans, and dibenzofurans [G.J. Baxter, 1974].

1-Alkenylnaphthalenes are pyrolytic products of cyclobutenodihydronaphthalenes [Criegee, 1970]. From the substitution pattern it was concluded that the reaction involves rearrangement. A reasonable mechanism prescribes the occurrence of two [1.5]-sigmatropic shifts, an allylic rearrangement, and electrocyclic opening.

The generation of a di(o-tolyl)ketene at 250°C initiated a subsequent [1.5]-hydrogen shift and electrocyclization [Hug, 1972a]. Furthermore, a sequence of [1.5]- and [1.7]-hydrogen migrations prior to electrocyclization has been authenticated in the formation of 2-alkyl-3-chromenes from pyrolysis of o-(1E,3-alkadienyl)phenols [Hug, 1969]. The reversible [1.5]-shifts are a prerequisite to isomerize the styrenic double bond to permit the progression of the [1.7]-sigmatropic shift. This [1.5]-shift was replaced by a homo-[3.3]-sigmatropy in the thermal conversion of cyclopropa[c]-chromenes into 2-alkyl-3-chromenes [Hug, 1971].

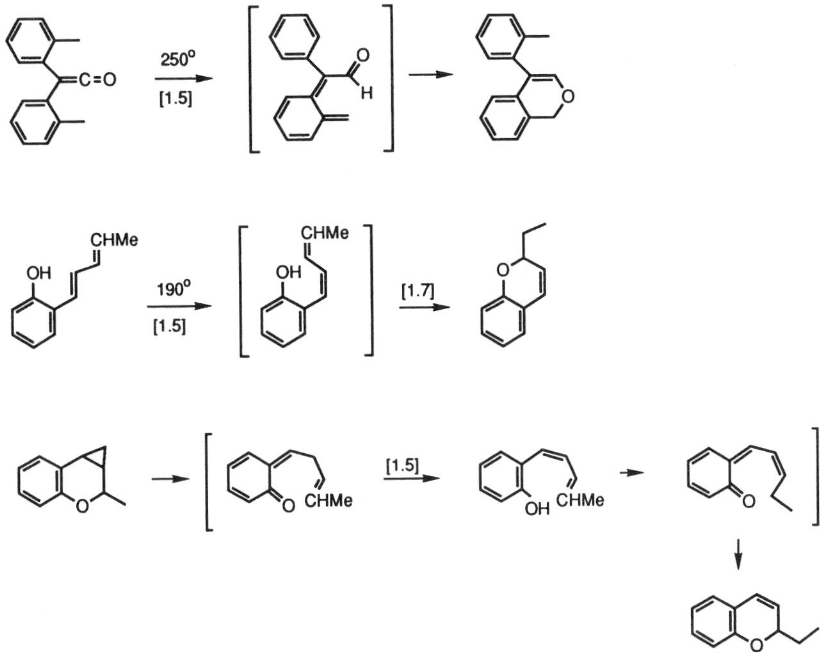

Thermal conversion of 5-substituted 5-methyl-1,3-cyclohexadienes to *m*-substituted toluenes [Schiess, 1975] operates in two pathways. When the substituent is a phenyl group, electrocyclic opening is preferred; however, [1.5]-sigmatropy is more rapid for substrates containing an acyl group.

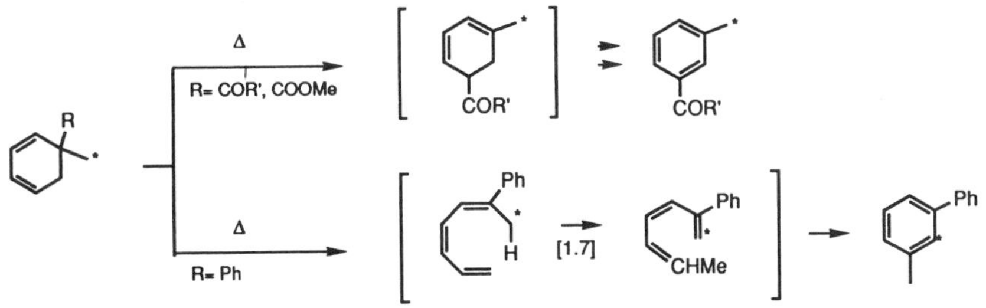

The photochemical reaction of 1,1-dimethylindene affords 2,2-dimethyliso-indene, probably via electrocyclization, [1.3]-sigmatropic rearrangement, and electrocyclic opening [Palensky, 1977]. Note that the last step should be disrotatory.

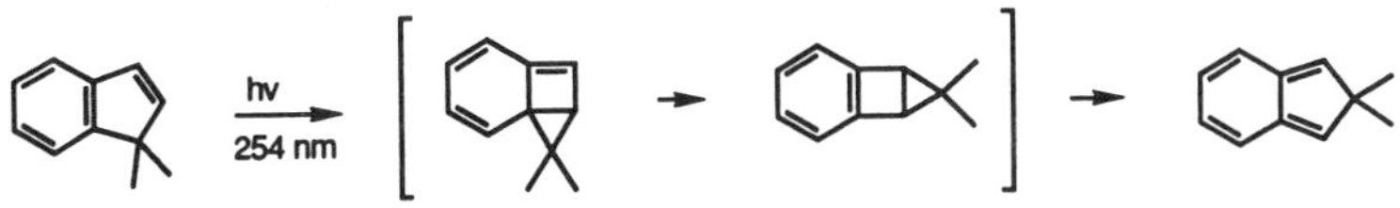

The photoinduced electrocyclic reaction of a spirocyclic diene [Oren, 1984] is favorably followed by a [1.7]-hydrogen shift. The reaction conditions also promoted further electrocyclization of the terminal diene unit.

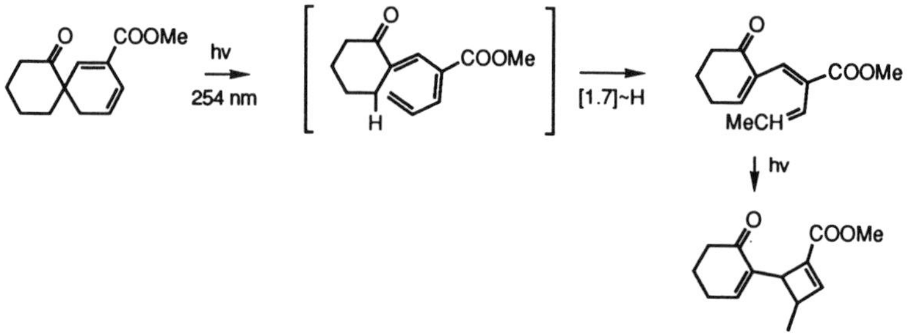

A very expedient approach to the alkaloids ellipticine and olivacine is based on the electrocyclization of pyridine-3,4-quinodimethane intermediates [Kano, 1982]. The reactive species were in turn generated by a thermally induced 1,5-hydrogen shift.

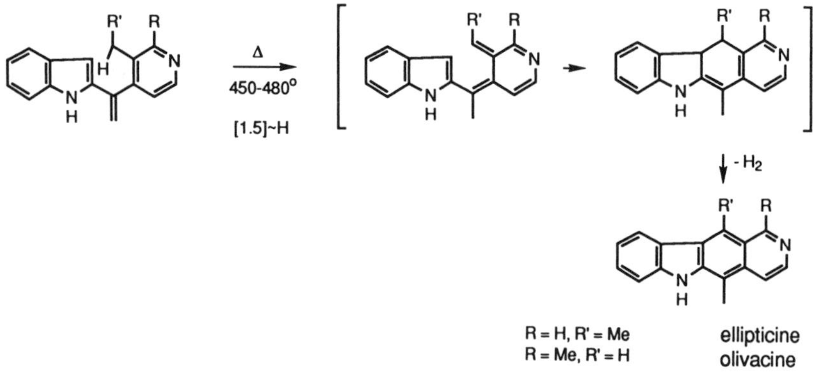

R = H, R' = Me   ellipticine
R = Me, R' = H   olivacine

Dehydrocycloguanandin has been acquired, albeit in a 9% yield, by etherification of a dihydroxyxanthone with 3-bromo-3-methylpropyne [Quillinan, 1972]. Three reactions were involved after the O-propargylation: Claisen rearrangement, prototropic shift and electrocyclization.

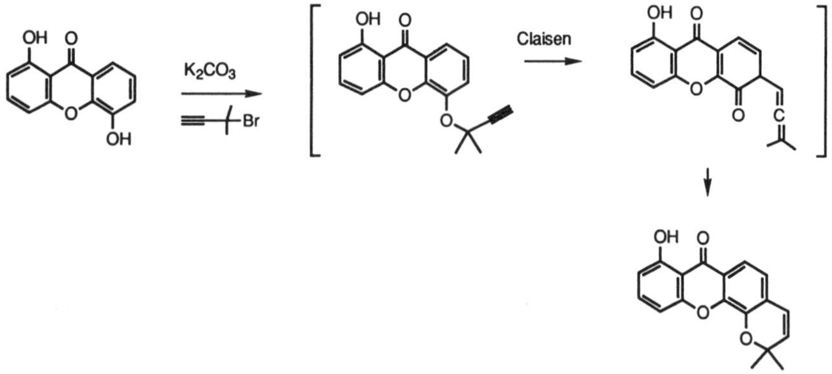

Benzopyran and benzofuran derivatives were produced from thermolysis of a pyridyl substituted propargyl phenyl ether [Attwood, 1991]. The phenolic product from a Claisen rearrangement can undergo proton transfer to the pyridine site or the allene moiety (sigmatropy). In the latter pathway an electrocyclization leads to a benzopyran.

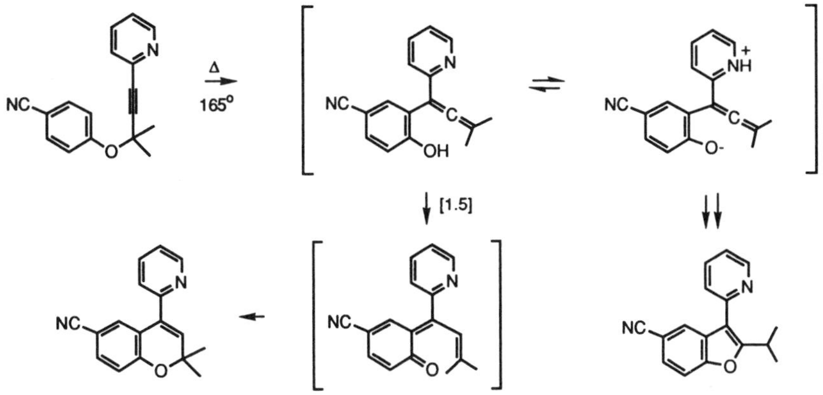

The photochemical product of a benzonorcaradiene is a indanocyclobutene [Kato, 1976], formed by [1.5]-sigmatropic rearrangement and two subsequent electrocyclic reactions.

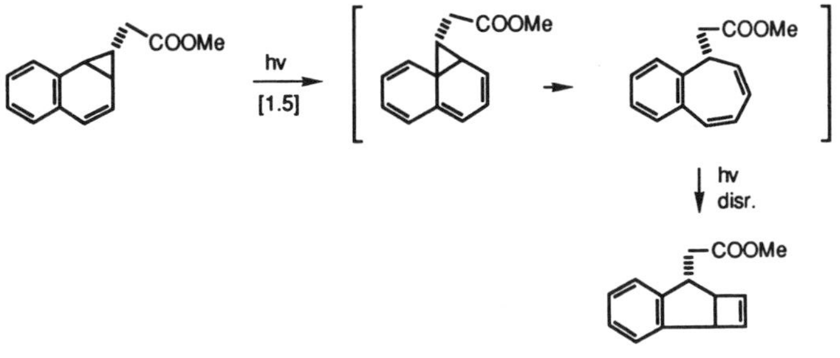

The formation of an aryl acetate from cross-conjugated enynone on treatment with tosic acid and isopropenyl acetate at high temperature [Jacobi, 1988] could involve a [1.7]-hydrogen migration after the enolacetylation and before the electrocyclization.

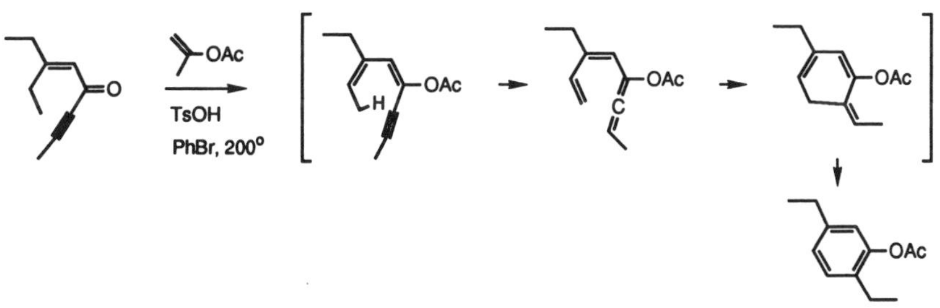

It has been suggested that the intriguing thermolysis of fidecene [Beck, 1982] is initiated by a 18-electron conrotatory ring closure. This symmetry-forbidden process is followed by [1.9]- and [1.5]-hydrogen shifts before termination by another electrocyclization involving 14-electrons.

A [2.3]-sigmatropic rearrangement that converts an allylic sulfenate into a 1,3-transposed sulfoxide has found many uses in synthesis. Particularly interesting is the rearrangement involving propargyl sulfenates from which allenyl sulfoxides are obtained. With chiral propargyl alcohols, the process results in the production of chiral allenes.

A system that is conducive to electrocyclization of the allenyl sulfoxide has demonstrated a ring formation reaction with chirality transfer from a center through an axis to another center [Okamura, 1985].

Very mild conditions are required to effect the formation of benzene derivatives from propargylic phosphate esters of relatively unstrained enediynes [Nagata, 1989]. A [2.3]-sigmatropic rearrangement to the allenyl phosphonate isomers initiated the cyclization to diyl intermediates. (The central double bond of enediyne systems might not play a role in the Bergman cyclization, but it is crucial to constraint such molecules to reactive conformations. In this context, the cyclizations are not electrocyclic.)

In an approach to seychellene [Snider, 1988] a projected vinylcyclobutanol ring expansion led to an anionic oxy-Cope rearrangement instead. Such a rearrangement also superseded the desired transformation of the corresponding acetylenic alcohol, and in this latter case the terminating reaction was an electrocyclic opening.

(17%)   (25%)

Pericyclic reactions can take place across an extended π-system such as poly(methenopyrrolenes). A spectacular example is the photochemical antarafacial [1.16]-hydrogen shift with a tandem, thermally induced conrotatory electrocyclization that resulted in the formation of a corrin skeleton [Yamada, 1969].

Beckmann rearrangement in the absence of external nucleophiles may induce intramolecular trapping of the nitrilium ion intermediate by an olefin linkage. One such reaction of a phenone oxime derivative was further succeeded by an electrocyclization [Sakane, 1983].

10.4. ELECTROCYCLIC REACTIONS INDUCED BY ENOLIZATION, ELIMINATION, CONDENSATION, OR OTHER PROCESSES

A mechanism involving retro-ene elimination of propene from diallyldimethylsilane and electrocyclization of the silene accounts for the formation of 3-dimethylsilacyclobutene [Block, 1978].

Vinylogous aminomethylene Meldrum's acid derivatives afford 1*H*-azepin-3(2*H*)-ones on thermolysis [McNab, 1987]. It is likely that the transformation involves elimination of acetone and carbon dioxide, hydrogen migration, and electrocyclization.

(major)

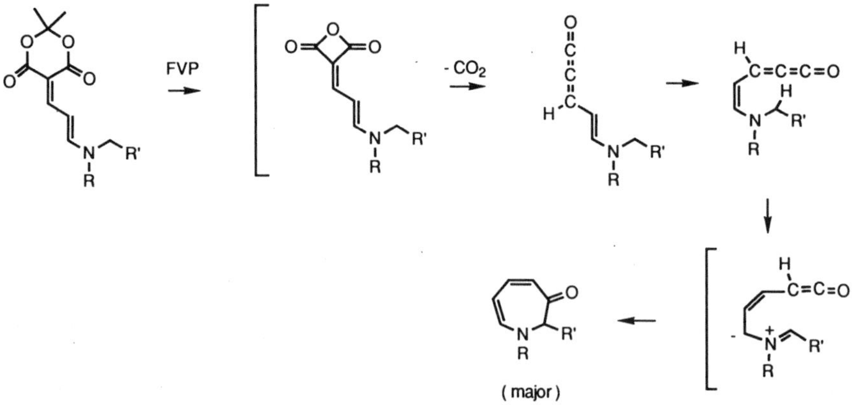

The tendency for electrocyclic opening of a bicyclo[3.1.0]hexenide anion [Radlick, 1967] is considerably enhanced by 1,5-bridging to a conjugate diene because an aromatic species would be created. Thus, deprotonation of the tricyclic triene led to cleavage of the CC bond which is common to all three rings.

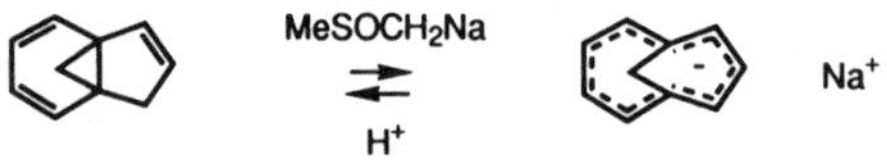

Two reaction pathways are possible to account for the phenol formation from a butynoyloctalone: intramolecular Michael addition, and electrocyclization of an ynedienolate ion. The product was a key intermediate for pseudopterosin-A [Corey, 1989].

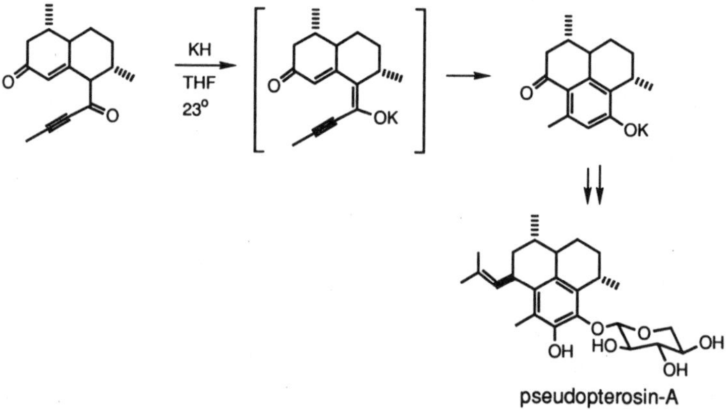

pseudopterosin-A

A synthesis of juncusol [Jacobi, 1991] is based on *de novo* formation of the more highly substituted phenolic ring. Thus, electrocyclization is set up by enolacetylation and a subsequent prototropic shift; aromatization of the product follows.

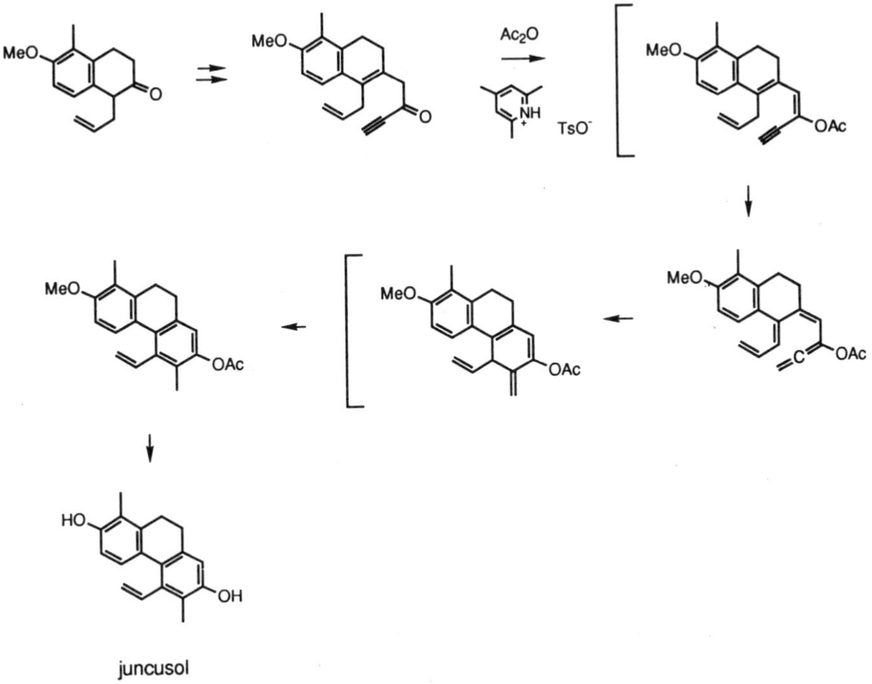

juncusol

The photochemical reaction in basic solution of a β-keto ester derived from β-ionone gave a product with the drimane skeleton [J.D. White, 1978]. The key to the transformation is an (E) to (Z) isomerization.

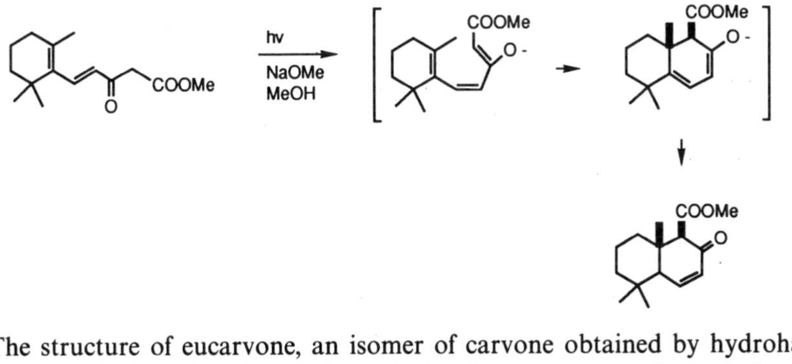

The structure of eucarvone, an isomer of carvone obtained by hydrohalogenation and dehydrohalogenation of the latter compound, caused some confusion as some of its derivatives are bicyclic. Its formation is now easily understood in terms of a reaction sequence involving 1,3-elimination, enolization, and disrotatory opening of the cyclohexadiene moiety.

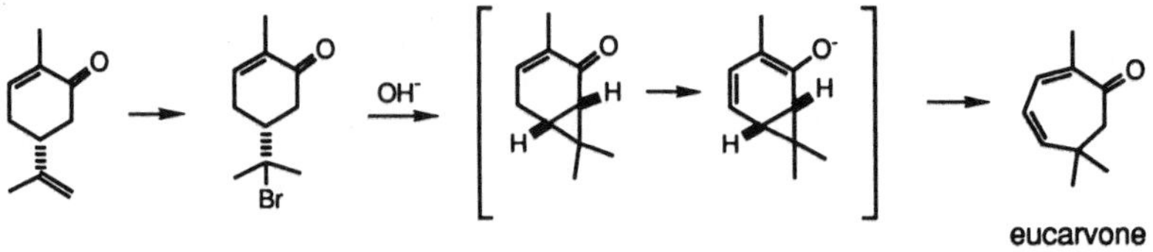

eucarvone

While tautomerization of 3-caren-2-one is favored by a relief of strain and the extension of the conjugate system, the isomeric bicyclo[4.1.0]hept-2-en-4-one is more resistant to such changes. However, the equilibrium may be shifted by trapping the enolate, as shown during a synthesis of spiniferin-I [Marshall, 1983].

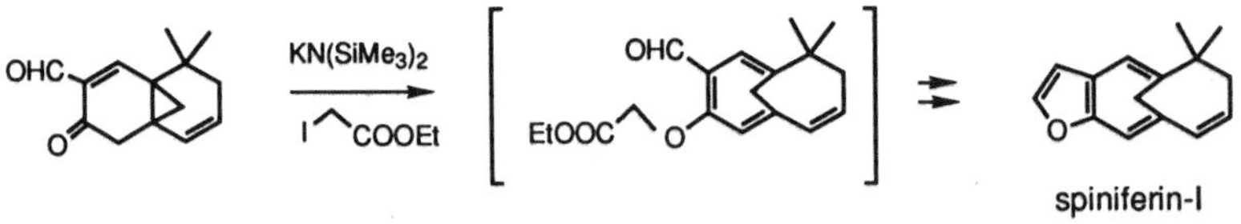

spiniferin-I

It must be emphasized that the enolization–electrocycloreversion process may be driven by some irreversible reactions that follow it. Such is the case in the formation of 4,5-cyclopentanotropone from ferricyanide oxidation of 4-(4-nitrobutyl)phenol [Kende, 1986]. The formation of a spiro compound was succeeded by an intramolecular Michael addition, enolization, electrocyclic reaction, and terminated by the expulsion of a nitrite ion.

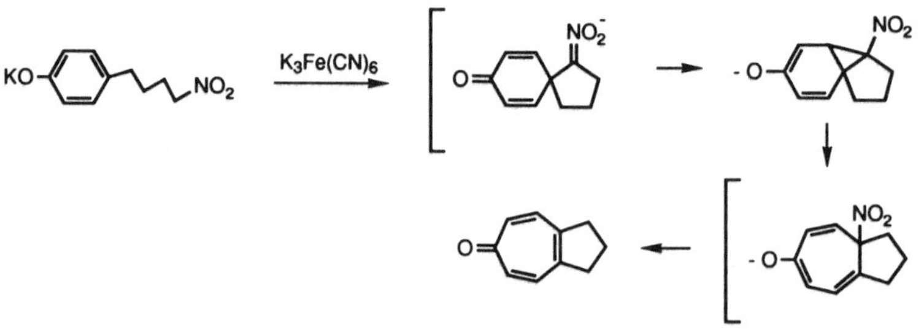

A more recent synthesis of colchicine [Banwell, 1990] was also based on electrocyclic opening and elimination reaction to establish the tropolone ring system (cf. [Banwell, 1985]).

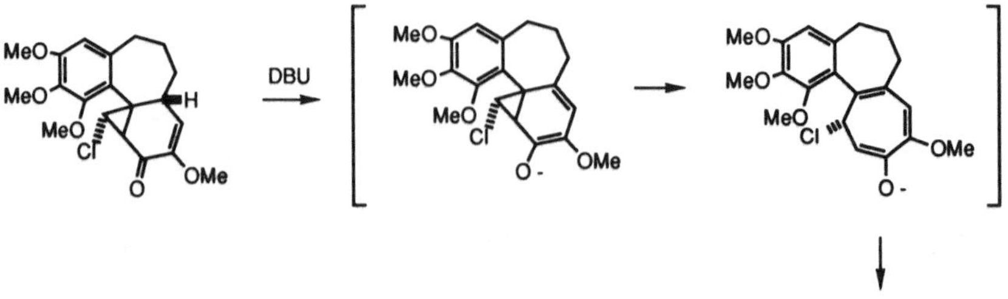

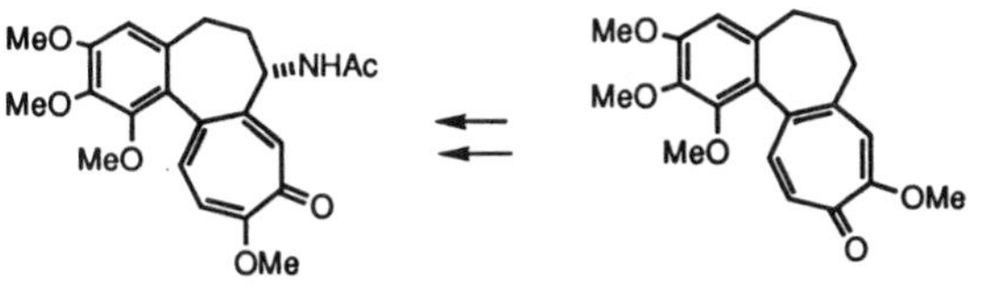

colchicine

An alternative mode of access to 7-halonocaradienolate ion is by closure of the three-membered ring, such as the method used in a synthesis of lettucenin-A [Monde, 1990].

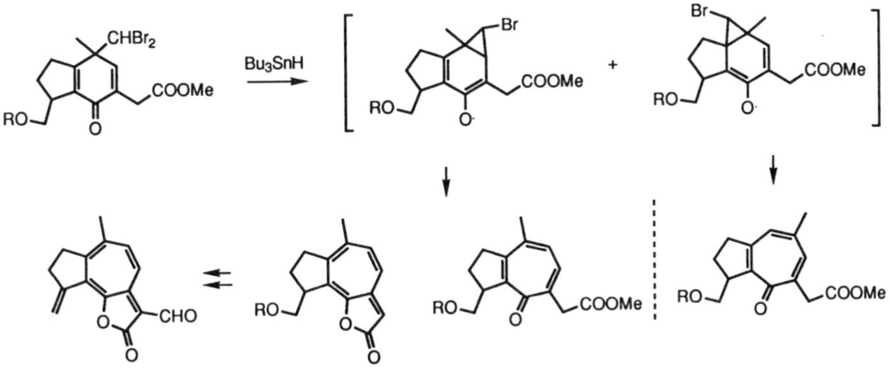

lettucenin-A

There are many more instances in which an electrocyclic reaction occurs after an elimination. One example is the photochemical transformation of the o-(dicarbethoxymethyl)biphenyl carbanion into ethyl 10-hydroxyphenanthrene-9-carboxylate [Yang, 1969]. The ketene intermediate cyclized and then enolized.

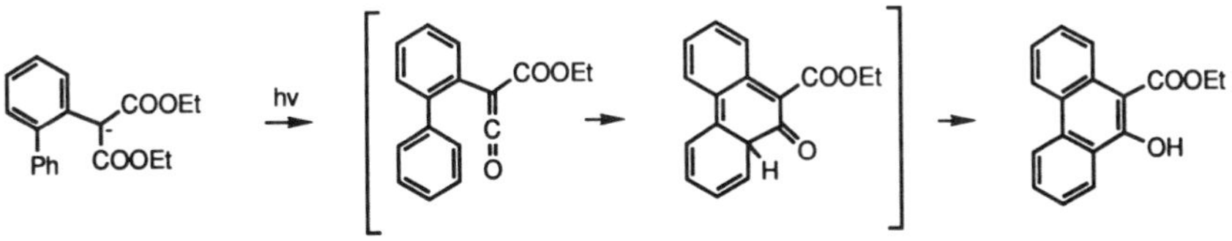

The disrotatory electrocyclic opening of trienes to give [1,3]cyclooctatetraenophanes upon dehydration of tricyclic dienols is subject to steric hindrance to the movement of the interior vinylic hydrogen by the polymethylene loop. Indeed tricyclic trienes with a shorter bridge enjoy a greater kinetic stability [Paquette, 1990b].

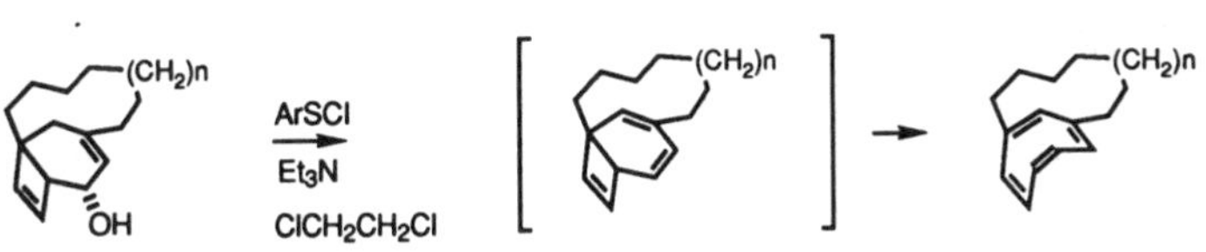

The major reaction product from tetracyclone and cyclopropenone is 3,4,5,6-tetraphenyltropone [Oda, 1972]. Its generation is adequately accounted for by a reaction sequence of cheletropic decarbonylation, of the Diels–Alder adduct, followed by electrocyclic opening.

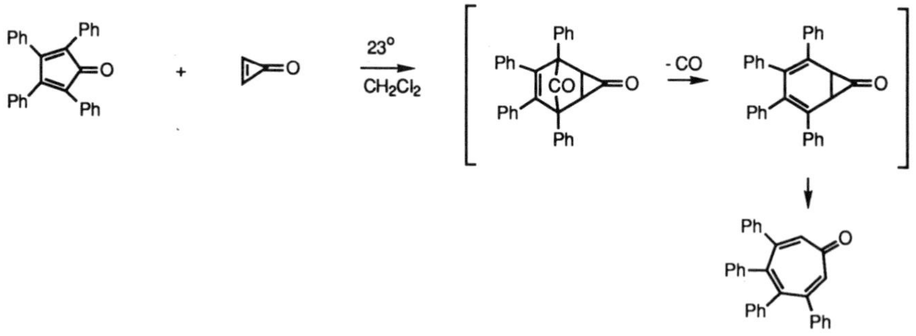

Pyrolysis of 2-furfuryl benzoate led to 2-methylene-3-cyclobutenone [Trahanovsky, 1973]. An allenyl ketene was implicated as the intermediate.

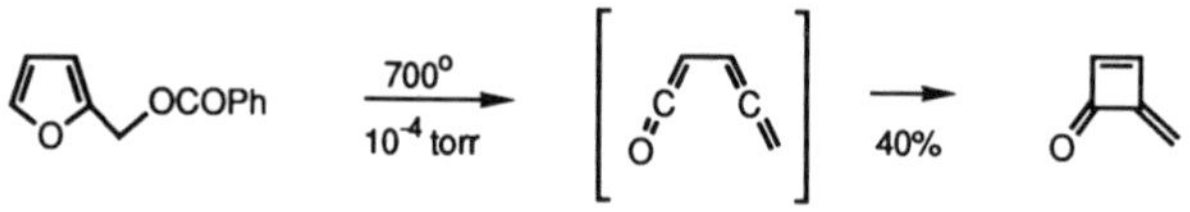

Flash vacuum pyrolysis of N-substituted aminomethylene derivatives of Meldrum's acid generates unsaturated ketenes. Different pathways are available to stabilize these cumulene-type species. When the N-substituent is a phenyl group, a [1,3]-prototropy and electrocyclization constitute the major mode of transformation, leading to 4-quinolinone [Gordon, 1983].

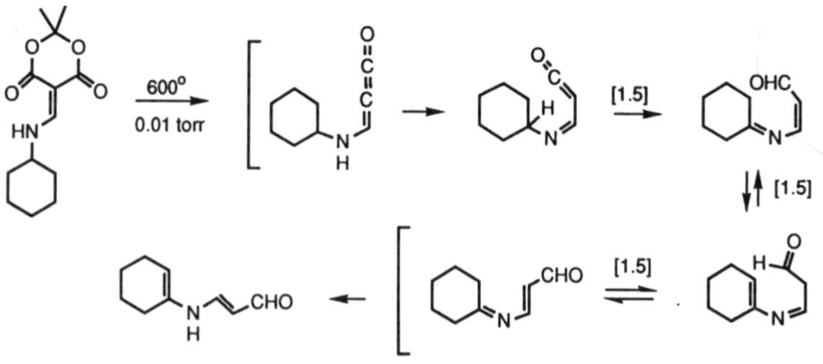

2-Azidotropone transmutates into o-hydroxybenzonitrile on heating [Hobson, 1967]. Upon loss of dinitrogen, ring cleavage gives a ketene which undergoes electrocyclization readily [Moore, 1979].

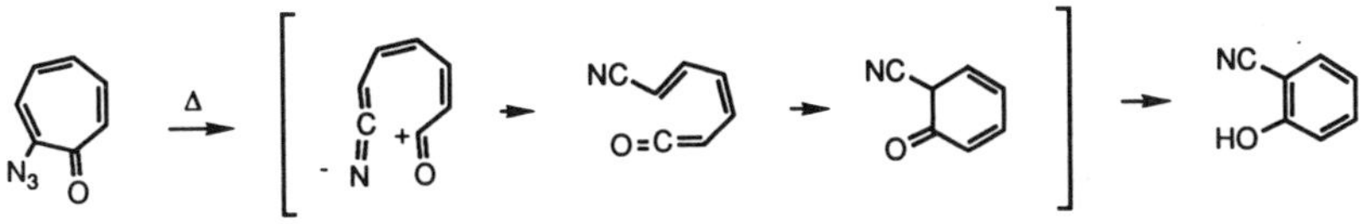

Photolysis of 9-azido-9-demethoxyisocolchicine in dioxane effected an analogous contraction of the tropone ring [Staretz, 1991]. However, a quite different behavior of 10-azido-10-demethoxycolchicine was observed; ketene trapping by the acetamido group to form a cyclic imide was clearly preferred even when methanol was used as solvent.

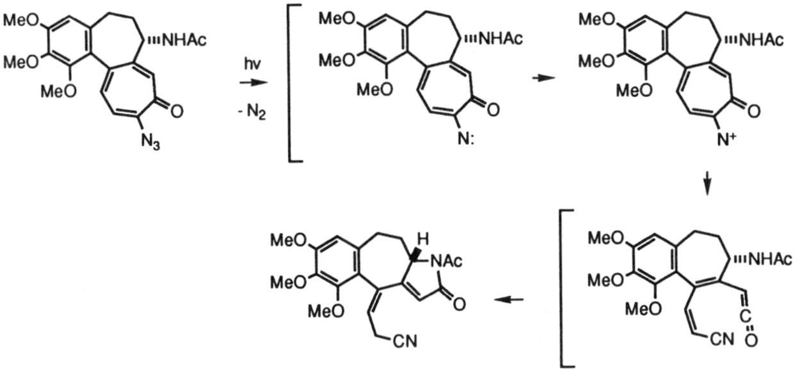

Photoextrusion of dinitrogen from an azotetracyclic compound accompanied by cyclopropane opening was followed by a rearrangement to give a bicyclo-[6.1.0]nonatriene [Paquette, 1971a]. Conrotatory opening of the bicyclic nitrogen-free intermediate apparently was rather favorable even at $-25°C$ and the resulting (Z,Z,E,Z)-cyclononatetraene reclosed by a disrotatory mode would be an excited state reaction.

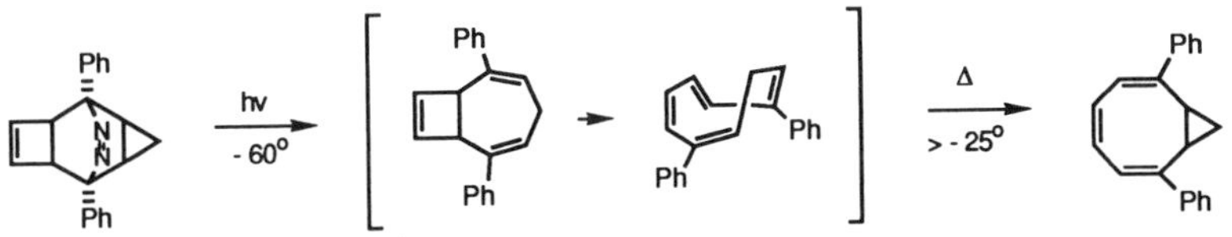

Pyrolysis of tricyclo[4.2.0.0^{1,4}]octane yielded two hydrocarbons which could be accounted for by a mechanism involving homolysis to a 1,4-diradical and ring opening [Wiberg, 1980]. The cyclobutene suffered electrocyclic cleavage.

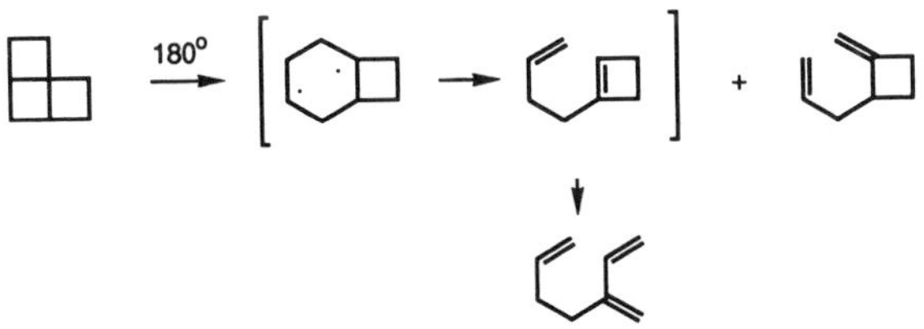

A diradical intermediate has also been implicated in the reaction of azulene with dimethyl acetylenedicarboxylate [Klärner, 1982] which gives rise to the 1,2-heptalenedicarboxylic ester. It appears that the highly strained Diels–Alder adduct readily forms a diradical, which can be transformed into a norcaradiene by breaking of a CC bond. Electrocyclic opening of the norcaradiene results in the heptalene derivative.

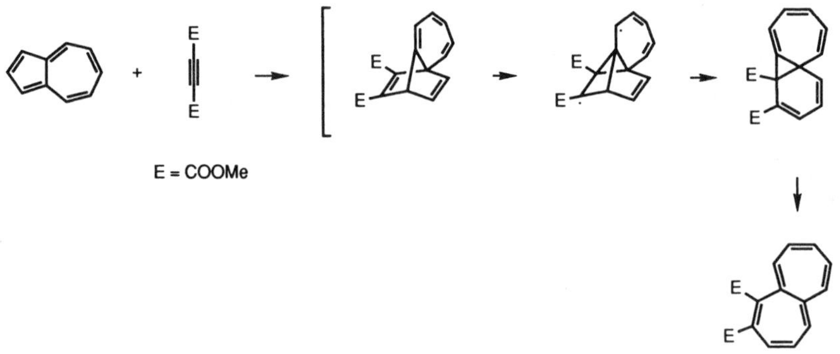

E = COOMe

Propargylic pseudoureas undergo thermolysis to give 2-pyridones [Overman, 1980b]. Involved in this transformation are a [3.3]-sigmatropic rearrangement, double bond shift, and elimination from the urea an amine, before an electrocyclization of the dienyl isocyanate.

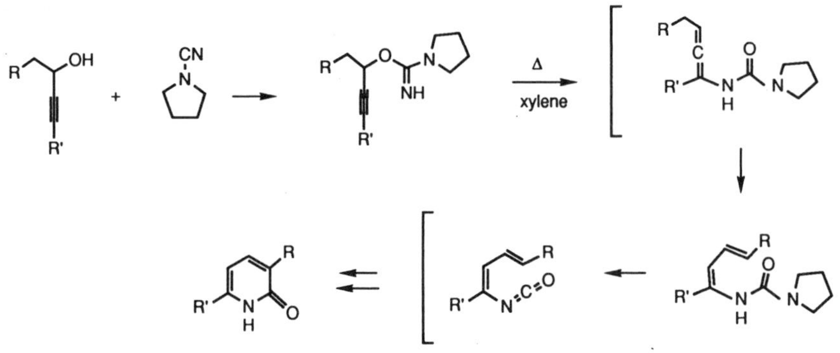

It is also evident that the pyrolytic formation of olivacine [Bergman, 1978] from a 3-(bisindol-3-yl)methylpyridine proceeded through elimination of an indole moiety to generate the electrocyclization-prone indole-2,3-quinodimethane species.

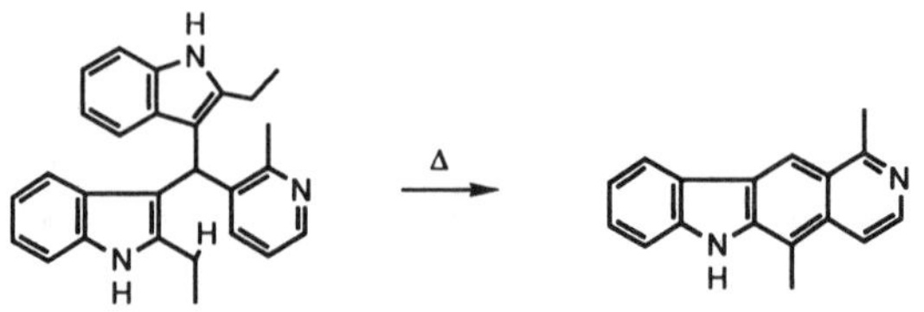

A spectacular thermal transformation of a chloride salt of nickel(II) oxasecorrinate into the neutral nickel(II) D-pyrrolocorrinate was observed during model studies of vitamin B$_{12}$ synthesis [Eschenmoser, 1976]. A dehydration was apparently followed by a conrotatory electrocyclization of the 16-electron system.

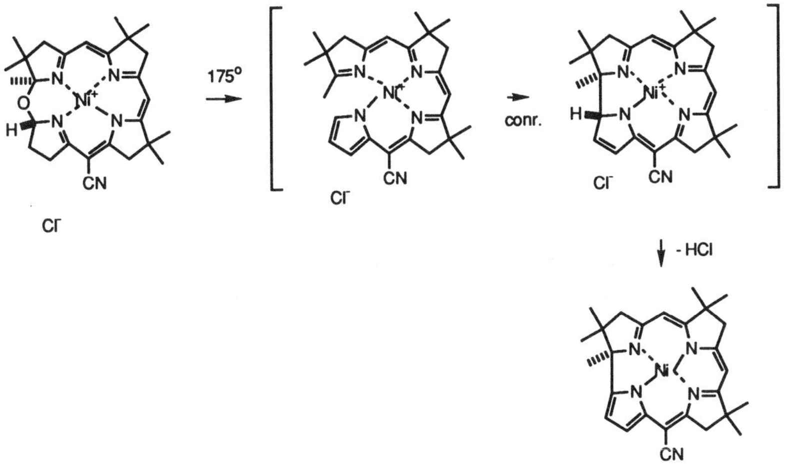

A bicyclic product has been detected in the methylation of octalene dianion. Instead, 1,8-dimethyl[14]annulene was isolated [Vogel, 1982]. The electrocyclic opening pathway for the primary product is exceptionally favorable.

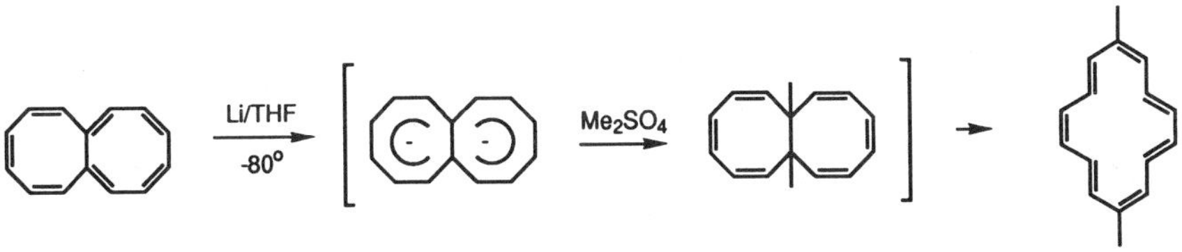

Treatment of 8-bromo-8-cyanoheptafulvene with copper led directly to a tricyclic dihydroheptafulvalene [Kuroda, 1976]. The coupled product is a 16-electron system and it underwent conrotatory closure spontaneously.

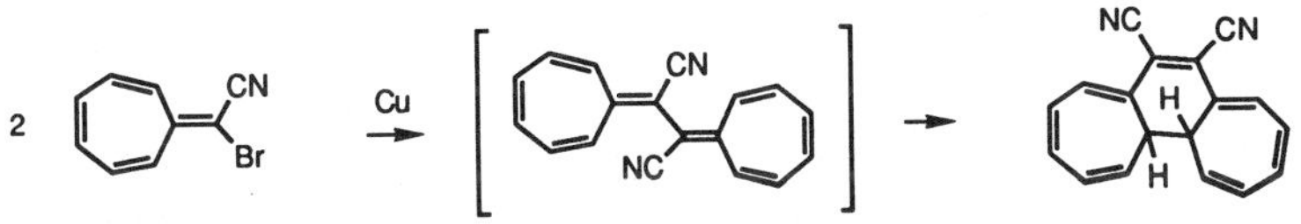

gem-Dibromocyclopropanes are transformed into allenes on treatment with methyllithium. The debrominative ring opening of the bisdibromocarbene adduct of *o*-divinylbenzene is followed by electrocyclization [Brinker, 1985]. The unstable *o*-quinodimethane derivative reacts readily with oxygen.

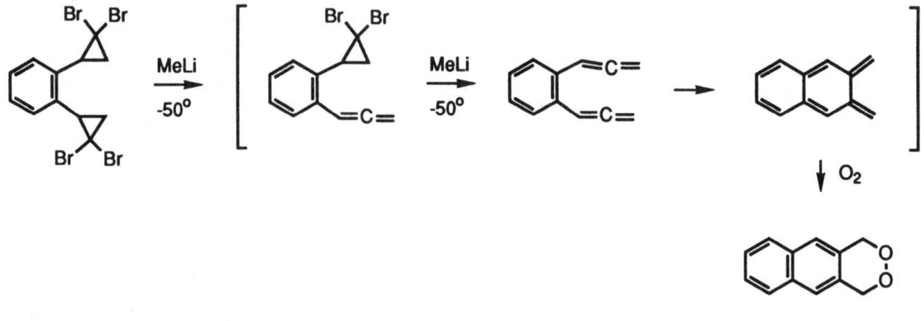

The Nazarov cyclization of cross-conjugated dienones is an important method for the synthesis of cyclopentenones. This cyclization is subject to stereocontrol by the Woodward–Hoffmann rules. A clever application of the Nazarov cyclization to synthesis of nootkatone [Hiyama, 1979] consisted of treatment of a 2-yne-1,4-diol with sulfuric acid. There were a hydration of the triple bond and double hydration to unfold the dienone. Conrotatory ring closure of the dienone under acid catalysis led to the bicyclic compound with the *vis-cis* dimethyl pattern.

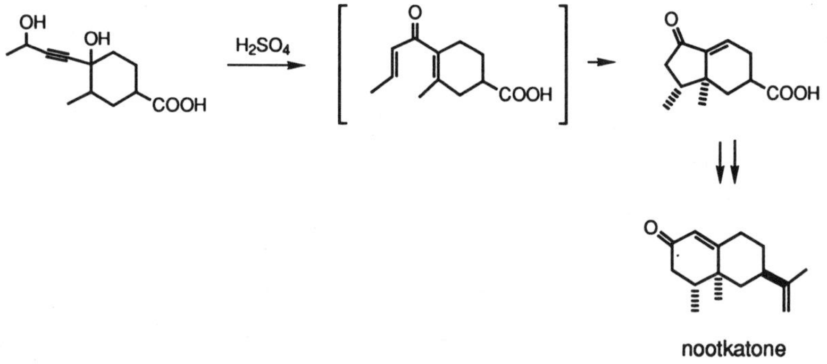

nootkatone

Isomeric oxypentadienyl cations have been generated by epoxidation of 1,2,4-trienes [Doutheau, 1983]. The ability of such ions to undergo spontaneous cyclization enables the synthesis of cyclopentenones from very different precursors.

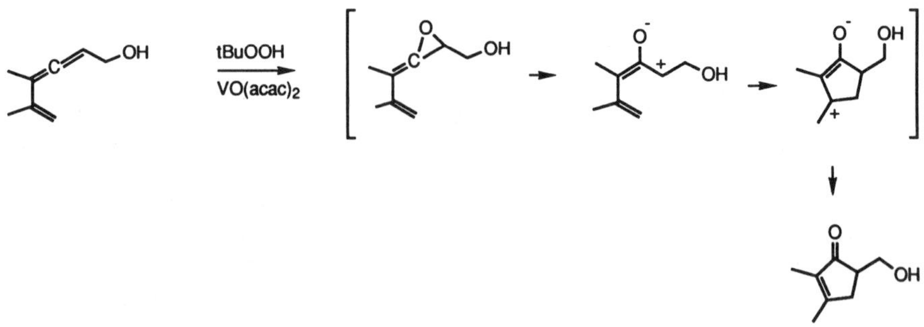

Certain cyclic enediynes are sterically disposed to cyclization. As the electrocyclization would generate a very strained cyclohexatetraene subunit, the cyclization is probably not electrocyclic. The products from cyclization of such enediynes exist better as benzene-1,4-diyl species.

Very intensive research activities have been devoted to these structures since the discovery of the very potent DNA-cleaving antibiotics of the calicheamicin-type. A model study [Nicolaou, 1988] has concentrated on the synthesis of cyclic enediynes from α,α'-dialkynyl sulfones by the Ramberg–Bäcklund reaction.

Condensation reactions giving rise to electrocyclizable systems also induce further reactions of the primary products. Thus, electrocyclization of an alkylidenation product from 5-pentyl-1,3-cyclohexanedione and citral was crucial to a synthesis of cannabichromene [Tietze, 1982].

cannabichromene

Pyridines have been obtained from thermal decomposition of 4-allyloxazolin-5-ones [Götze, 1976]. An azatriene system, generated by an aza-Cope rearrangement, decarboxylation, and prototropic shift, is ready to undergo electrocyclization. The resulting dihydropyridine is stabilized by dehydrogenation.

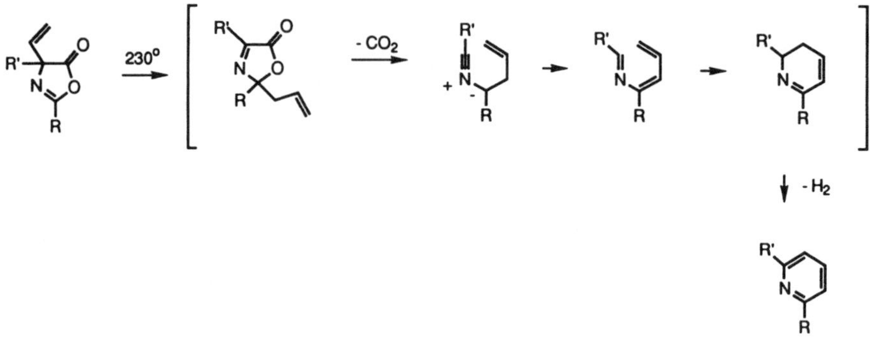

A route to benzoic acid esters involves sequential admixture and heating of three reactants [Minami, 1982]. A Michael adduct is formed and participates in a Horner–Emmons reaction. On heating the product loses methanesulfenic acid to give a conjugate triene; electrocyclization of the latter compound is followed by elimination of methanethiol.

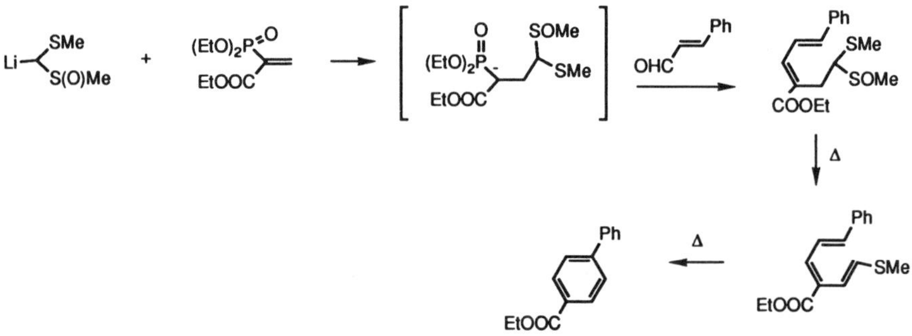

Formation of aromatic ring via condensation and electrocyclization (and subsequent elimination) is illustrated by the following examples [Jutz, 1972].

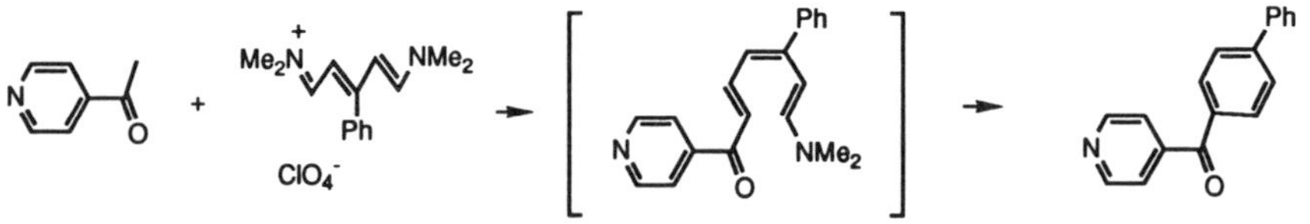

Elimination of dimethylamine from the [2.3]-sigmatropic rearrangement products of allyl phenylpropargyl dimethylammonium salts prepare such molecules for cyclization to afford biphenyls [Jemison, 1972].

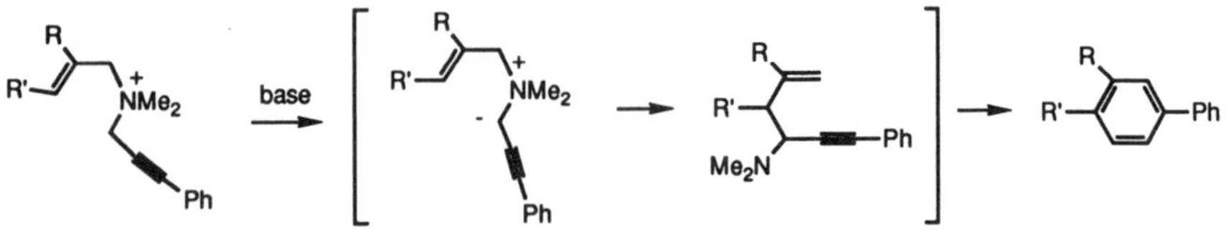

Condensation of diphenyldiazomethane with 3-(*p*-nitrophenyl)-3-cyclobutene-1,2-dione [Ried, 1972] followed an unusual course in which a Michael addition was succeeded by decarbonylation and expulsion of dinitrogen. An electrocyclization process for the resulting ketene led to an α-naphthol derivative.

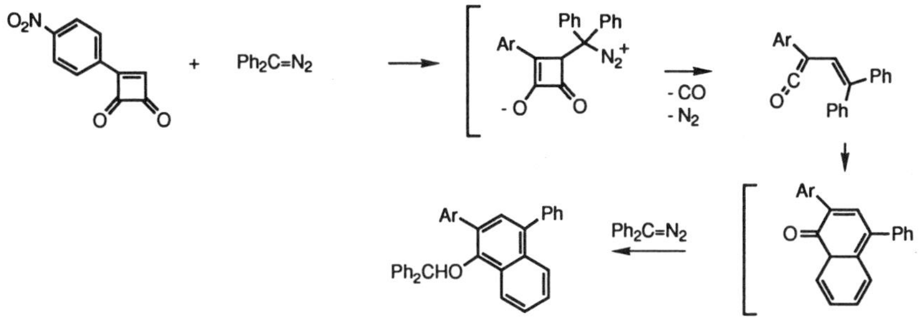

The intramolecular Friedel–Crafts acylation process employed in the elaboration of the tetracyclic skeleton of aklavinone [Confalone, 1981] was thwarted by the formation of a spirocyclic ketone. Fragmentation of the latter compound followed by electrocyclization indeed gave an anthracycline derivative, but the substitution pattern changed completely as to render this approach to synthesis of related antiobiotics useless.

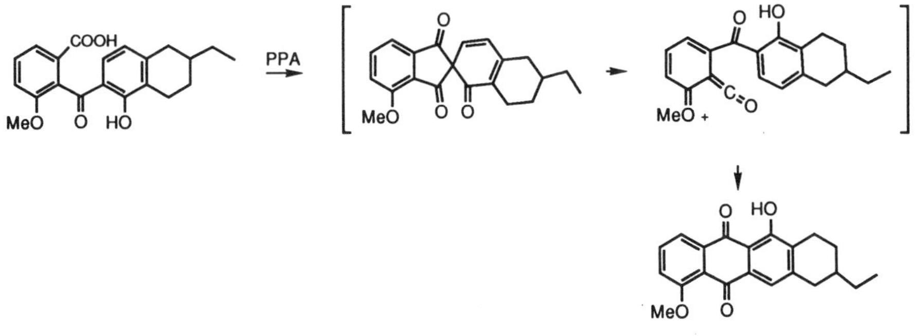

A mechanism for the formation of a *trans-vic* dimethyloctalone from reaction of 2-methylcyclohexanone with 3-penten-2-one in dimethyl sulfoxide advocated the intermediacy of a trienolate anion and its disrotatory ring closure [Scanio, 1971]. It is intriguing that the initial condensation would not follow the common pattern of Michael addition.

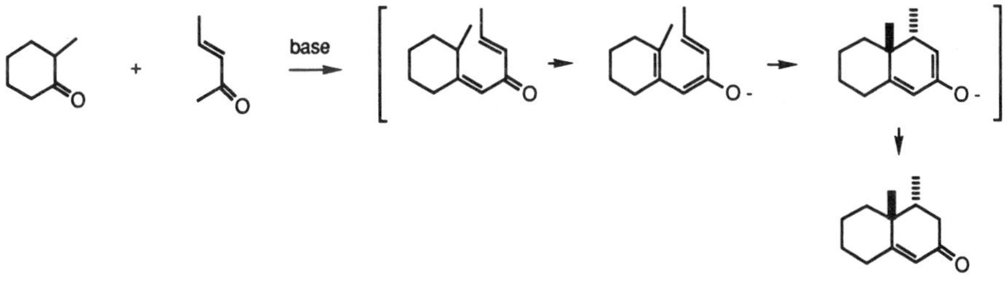

2-Methyl-3-(1Z,3Z-pentadienyl)-2-cyclohexenone did not undergo a concerted disrotatory electrocyclic closure [Ramage, 1970], presumably due to severe steric interactions in the transition state. The product has a *trans-vic* dimethyl substitution pattern.

It should also be noted that when the trienone was generated in situ by pyrolytic decomposition of a carbonate strong dependence of the product on the pyrolysis time was observed. At a longer period the product is a bicyclic dienone containing an ethyl sidechain. Apparently, a thermal equilibration via cycloreversion to the monocyclic trienone allowed isomerization to an isomer via a [1.7]-hydrogen shift whose electrocyclization is sterically more favorable.

A synthesis of [n](2,4)pyridinophanes where n = 6–9 has been accomplished by the reaction of N-(1-phenylethenyl)iminotriphenylphosphorane with 2-cycloalkenones [Kanomata, 1989]. The resulting Schiff bases apparently electrocyclized readily. Aromatization (dehydrogenation) of the dihydropyridine products led to the observed products.

Disrotatory electrocyclization of the conjugate triene portion produced from an intramolecular McMurry reaction of a bis-cinnamaldehyde occurred in tandem with the coupling [Tanner, 1983].

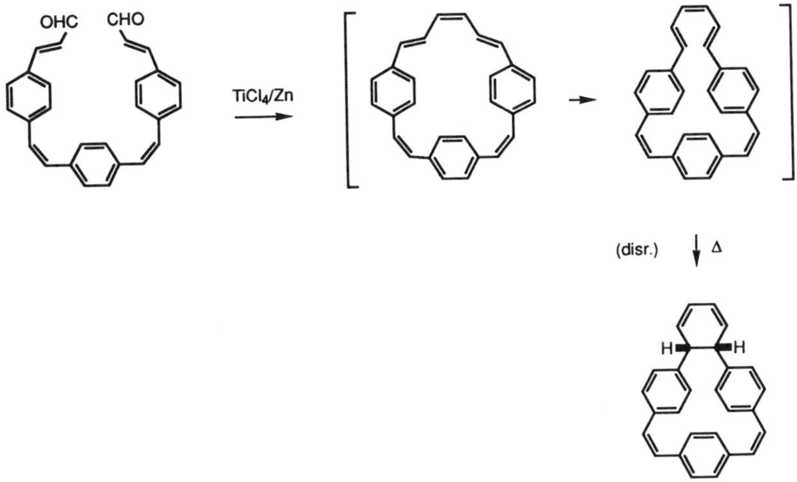

(disr.)   ↓ Δ

The bioactivity of dynemicin owes to an arene-1,4-diyl, generated through cyclization of the enediyne chromophore. An acid-catalyzed hydrolysis of an epoxide in a model compound [Nicolaou, 1990] activated the cyclization. In this case the polyunsaturated system was only locked by a steric constraint.

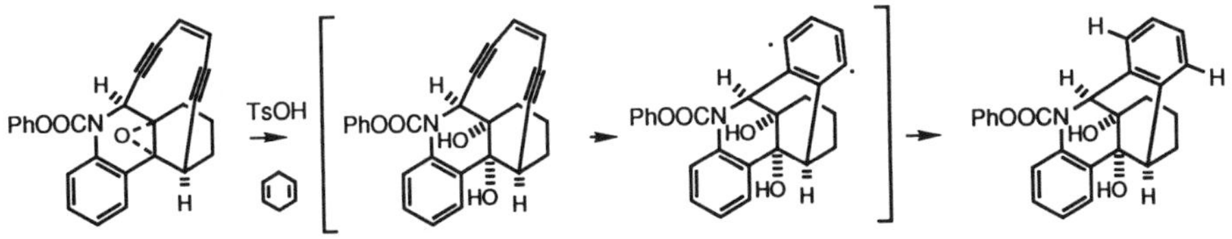

Electrocyclic reactions often follow *N*-acryloylation of 1-alkyl-3,4-dihydro-β-carbolines (also in the dihydroisoquinoline series). This reaction sequence provided a convenient access to a tetracyclic enamine intermediate for the synthesis of *eburnamona* alkaloids such as vincamine [Danieli, 1980].

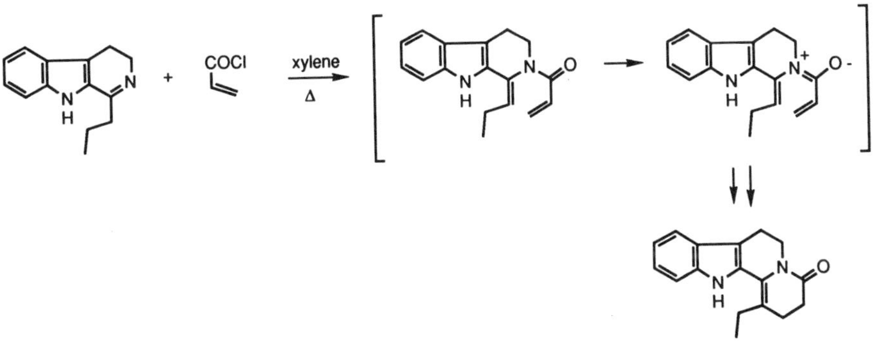

Generally, cycloheptatrienes are much more stable than their norcaradiene tautomers, the generation of the latter species in reactions almost assures the isolation of the trienes. It was therefore quite unexpected that a tricyclic ketone containing a norcaradiene moiety was the predominant product from an intramolecular carbenoid insertion into an arene [Kennedy, 1988]. The stabilizing influence came from the carbonyl group, as tautomerization attented the reduction of that functionality.

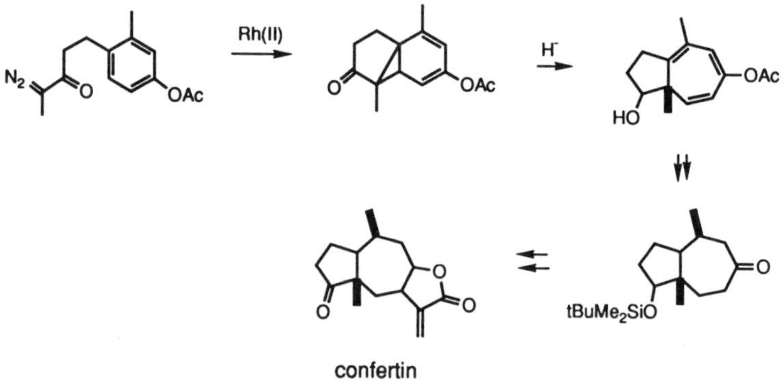

confertin

Note that the electrocyclic opening may defer to other processes such as the retro-Michael fission shown in the following equation [Iwata, 1981].

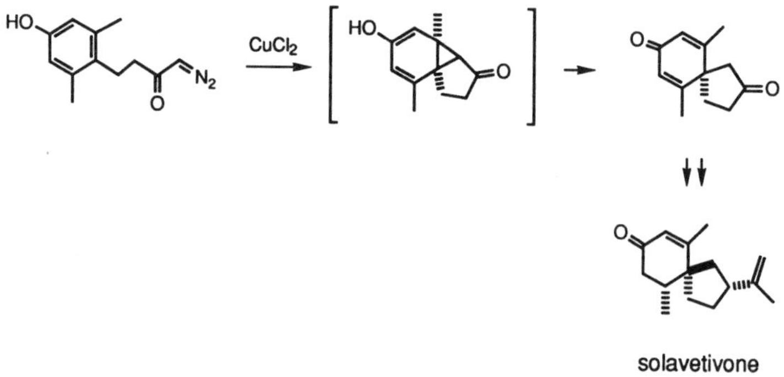

solavetivone

11

ENE, RETRO-ENE, AND SOME OTHER THERMAL REACTIONS

11.1. ENE REACTIONS

The ene reaction [Alder, 1943; Hoffmann, 1969] is a thermal reaction of an olefin containing an allylic hydrogen atom (*ene*) with an electron-deficient unsaturated compound (*enophile*) to form a 1:1 adduct via a cyclic six-electron transition state. Generally, it is a concerted suprafacial process.

An intramolecular ene reaction is more facile and the enophile may be an unactivated double or triple bond. Thus when pinane derivatives (e.g. pinanol [Strickler, 1967]) undergo homolytic cleavage of the four-membered ring substantial amounts of iridenes are also produced, as a result of the ene reaction that follows. The ene reaction requires less energy of activation than the ring cleavage step.

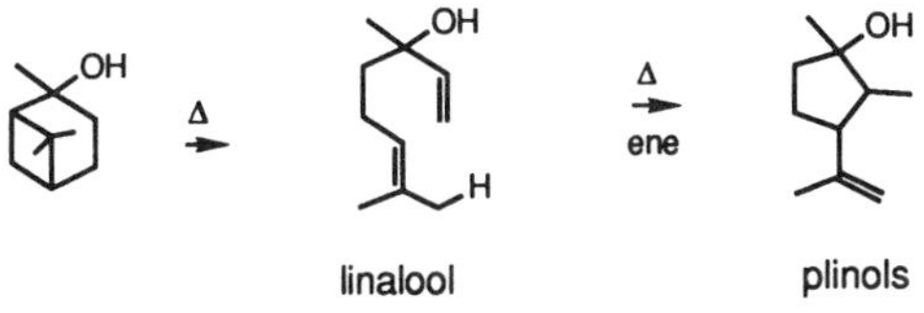

linalool   plinols

Sometimes a carbonyl group may act as the enophile, consequently, thermolysis of certain cyclobutyl ketones leads to homoallylic alcohols. This opportunity for tandem transformation has been exploited in the establishment of the octalol system with an angular hydroxyl group [Wender, 1980b,c, 1982b; J.R. Williams, 1980].

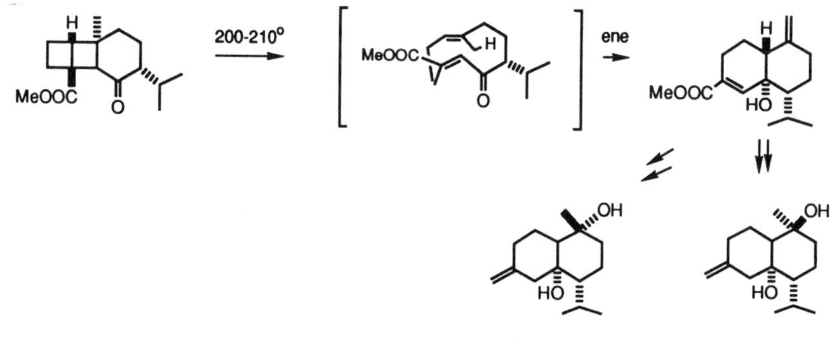

calameone   isocalamendiol

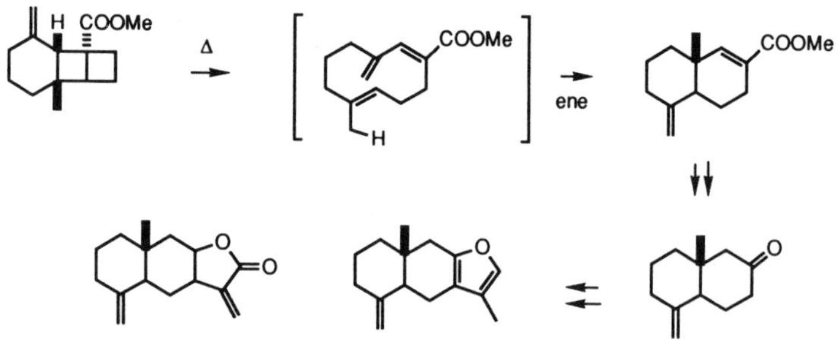

isoalantolactone   atractylon

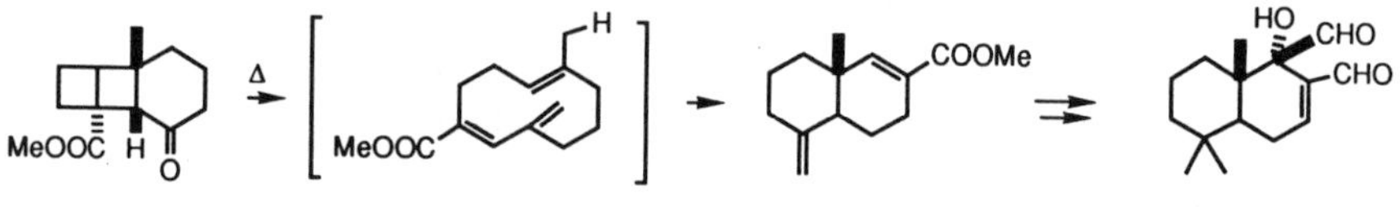

warburganal

In the course of a verrucarol synthesis [Trost, 1982a] that featured a Diels–Alder reaction an intramolecular ene reaction occurred. This event turned to be a blessing as it provided protection of one of the carbonyl groups and rigidified the molecule so that stereoselective reactions could be achieved more readily. At a later stage, the functional groups were resurrected by a retro-ene reaction.

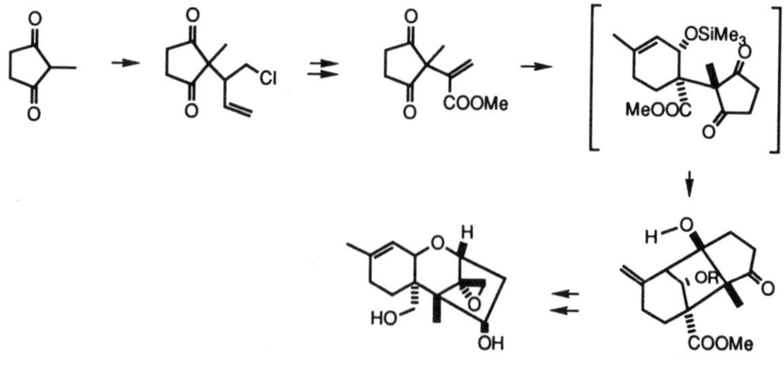

verrucarol

The ring strain of the norbornane which is fused to two isopropylidenecyclopropane rings is a major factor that increases the propensity of *exo–endo* isomerization via 1,3-diradical intermediates. The *endo–endo* isomer is sterically predisposed to an ene reaction [Bloch, 1976b], and since the reaction is irreversible in the sense that even more energy is supplied to the pentacyclic product the newly formed CC bond would not be broken in preference to those constituting the cyclopropanes.

Pyrolytic generation of a γ,δ-insaturated acylimine from the acylcarbamate was found to give a lactam [Koch, 1983]. The ene reaction stabilized the reactive acylimine.

The formation of a bicyclo[3.3.0]octane system in one step [Leyendecker, 1974] from a 6,9-dien-1-one is very useful. Tandem Conia cyclization, a special kind of ene reaction, is indicated. Two consecutive Conia reactions of 3,3-di-(3-butynyl)cyclopentanone represents the most convenient approach to a propellane system [Drouin, 1975].

The remarkable catalytic effect of Lewis acids on the ene reaction [Snider, 1980] lowers the reaction temperature from the 200–300°C range to ambient temperature. Alkylaluminum halides are particularly effective because the alkyl group functions as a proton scavenger.

Sequential ene reactions of alkylidenecycloalkanes with β-unsubstituted α,β-unsaturated carbonyl compounds have been observed [Snider, 1983a]. The formation of condensed bicyclic alcohols in one step is very appealing as synthetic method in view of its efficiency.

The tandem Claisen rearrangement/ene reaction was investigated [Ziegler, 1984] with the aim of constructing functionalized bicyclo[3.3.0]octanes. Valuable information concerning the nature of intermediates and temperature dependence of the reaction sequence was gained. Reversible oxa-ene reaction after the Claisen rearrangement was established.

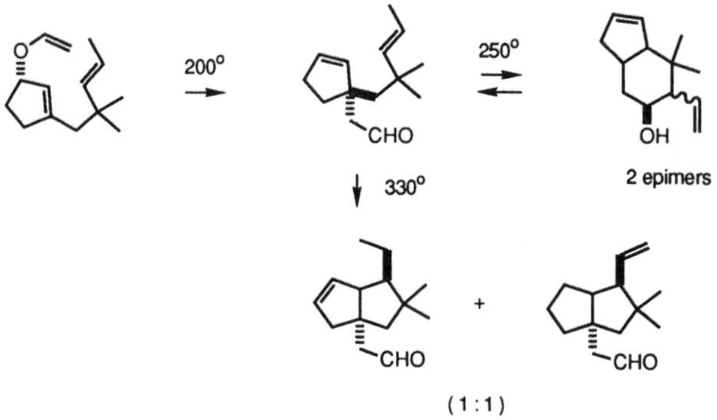

(1:1)

The stereoselective formation of a five-membered ring has been applied to the acquisition of a (+)-estrone precursor that contains the CD-ring component [Mikami, 1990]. In the Claisen rearrangement the (S) and (Z)-character of the allylic alcohol was transmitted to the (14S)-chirality and a high 8,14-syn selectivity via a chairlike transition state. The following ene reaction adopted an endo transition state in which the large AB-ring assumed a pseudoaxial orientation.

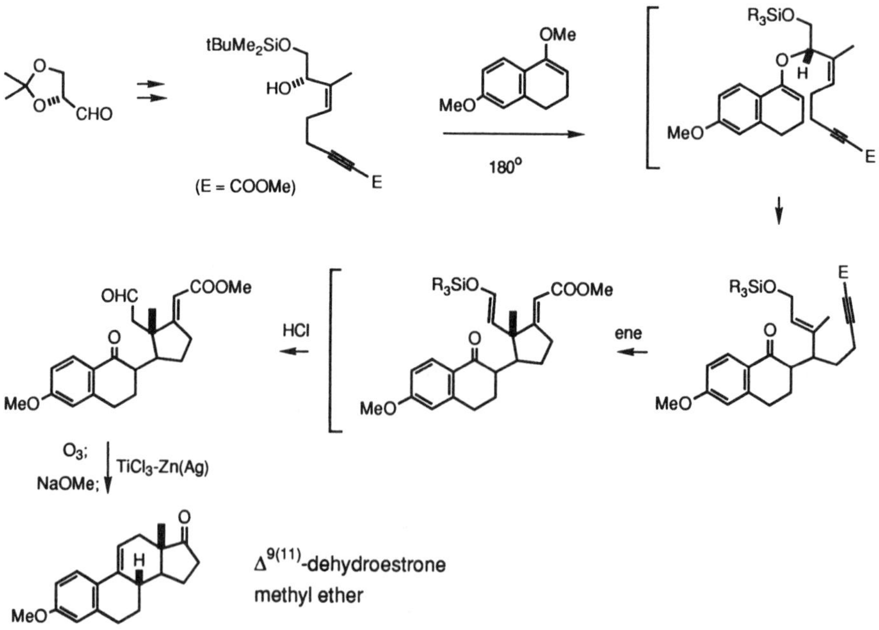

$\Delta^{9(11)}$-dehydroestrone
methyl ether

From a simpler substrate a CD-ring synthon was obtained [Takahashi, 1988]. However, the tandem reaction sequence was seriously undermined, both in terms of yields and stereoselectivity, by other types of ene reactions.

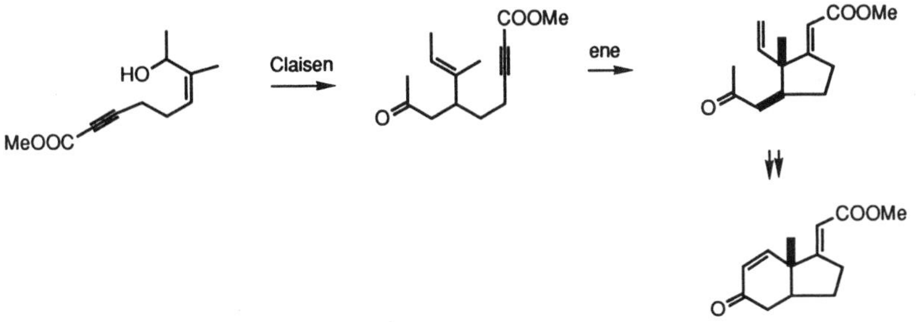

Pyrolysis of glauconic acid acetate gave a diquinane derivative [Barton, 1965]. The structure of the product suggests its formation via tandem Cope rearrangement and ene reaction.

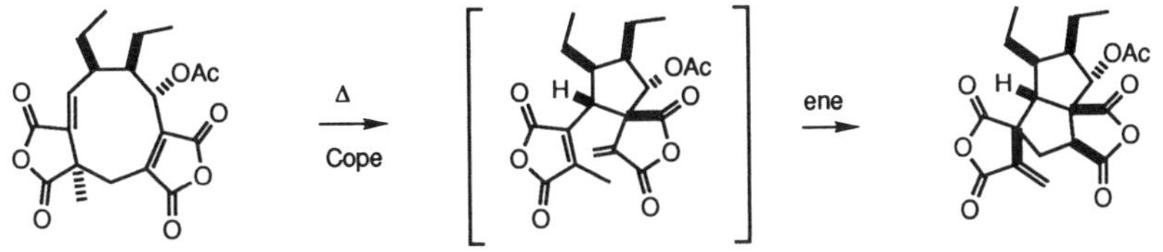

glauconic acid acetate

The key step in a synthesis of isocarbacyclin [Mandai, 1990] is the thermolysis of a (β-sulfinylethoxy)cyclopentene. The thermal elimination generated a vinyl ether which participated in a Claisen rearrangement. The presence of an alkenyl *cis* to the nascent acetaldehyde chain invited the latter to undergo an ene reaction.

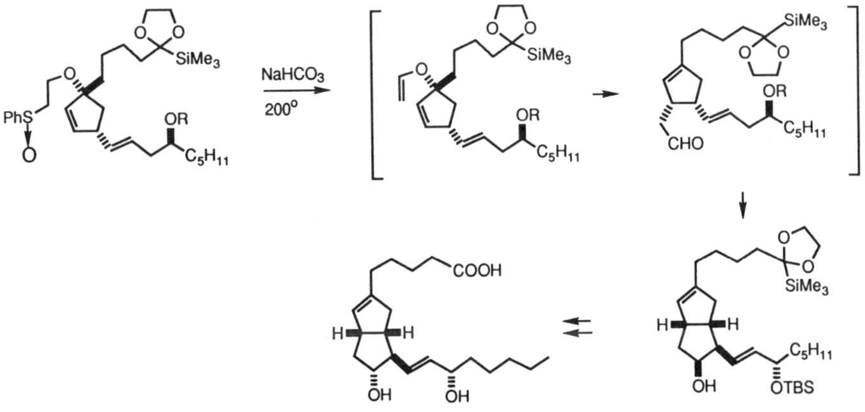

An intermolecular tandem Claisen rearrangement/ene reaction is that in which one of the reactive components (ene or enophile) is generated in the presence of the other. 2-(Phenoxymethyl)acrylic acid apparently underwent

Claisen rearrangement and lactonization prior to the occurrence of an ene reaction with *N*-phenylmaleimide [Sunitha, 1985].

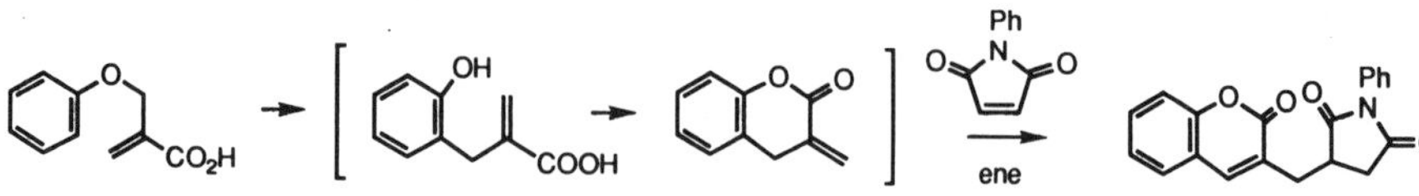

When 7-propagyloxycycloheptatriene was heated a series of 1,5-hydrogen shift and Claisen rearrangement occurred, and culminated in the formation of a tricyclic trienone by an ene reaction [Pryde, 1974].

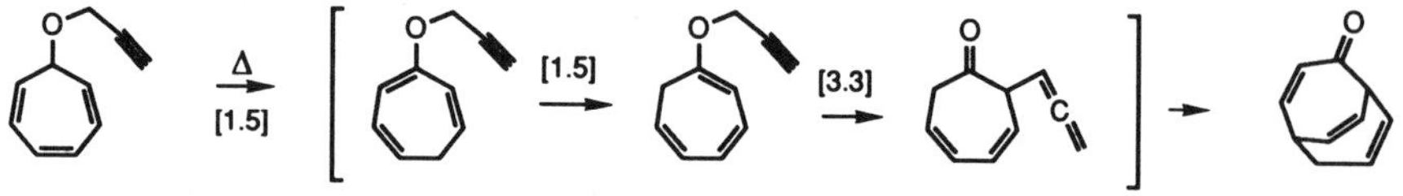

It is not surprising that an ene reaction in tandem with cycloaddition or cycloreversion can be effected. Thus, the 1:2 adducts of furfural and isoprene have tetracyclic structures [Wenkert, 1988] because in the *cis-anti-cis*-octahydrodibenzofurans the angular aldehyde group is spatially proximal to the allyl system of the distal ring.

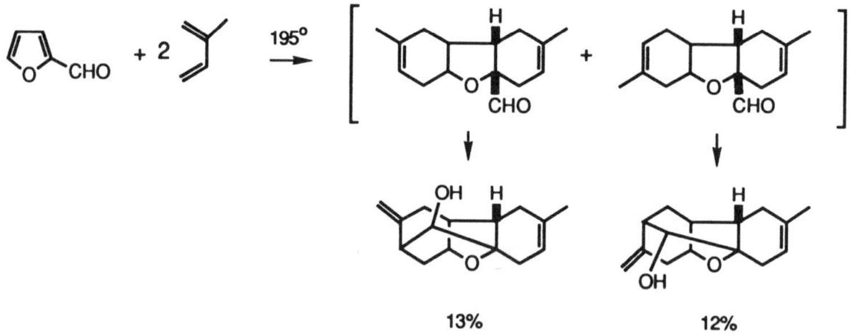

An intramolecular ene reaction of a proper acyl nitroso olefin led to a cyclic hydroxamic acid with the double bond allylic to the nitrogen atom. This reaction was the key step to some alkaloid syntheses including mesembrine [Keck, 1982].

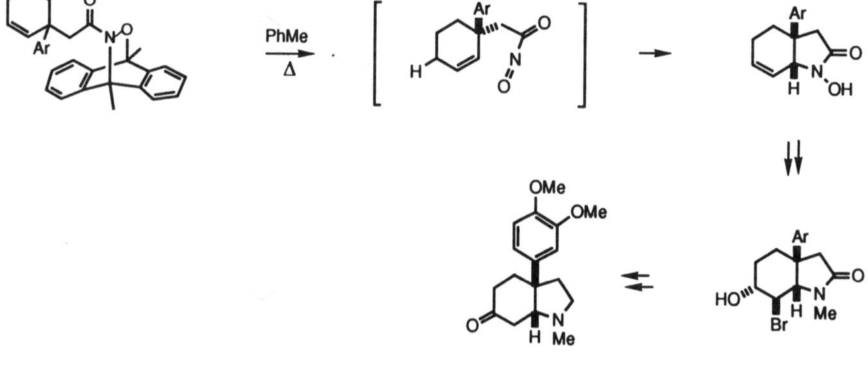

mesembrine

One of the products from the reaction of benzyne with toluene is 2-benzylbiphenyl [Brinkley, 1972], apparently arising from two consecutive ene reactions.

The reaction of 2-vinylnaphthalene with benzyne also goes beyond a [4 + 2]-cycloaddition as benzyne is also an excellent enophile [Ittah, 1977].

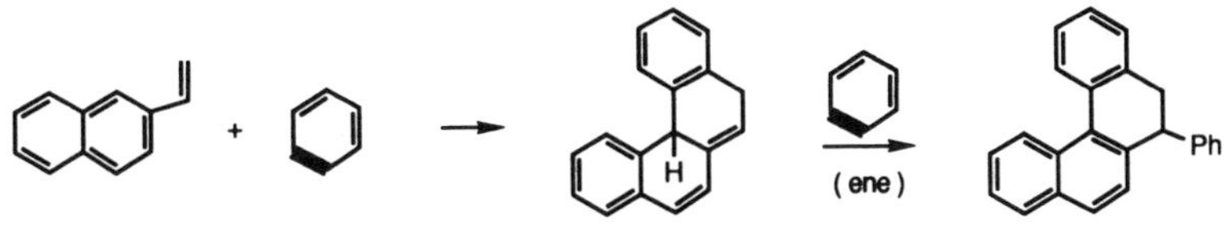

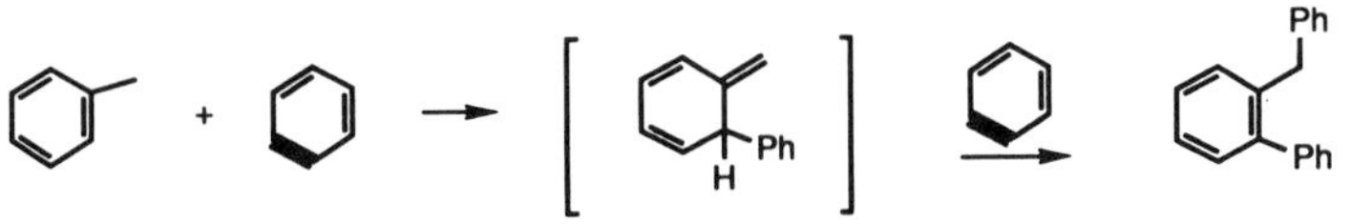

The Conia cyclization is a special type of intramolecular ene reaction in which the enol form of a carbonyl group acts as the ene component. An example of tandem Cope rearrangement/Conia cyclization [Conia, 1966] is shown now.

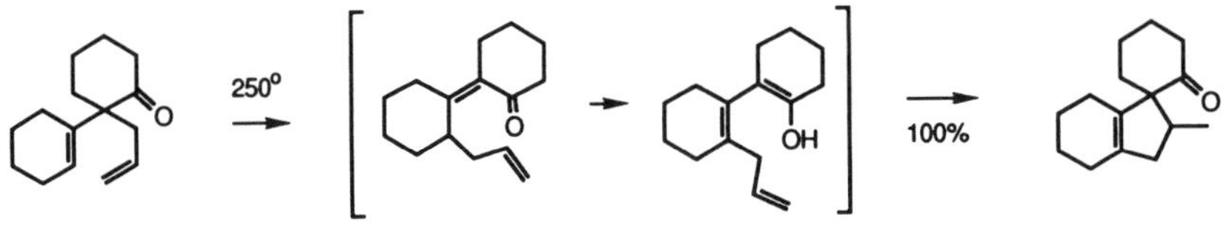

While the metallo–ene reaction in tandem with functionalization maneuvers is discussed elsewhere, an annulation initiated by Pd(O)-catalyzed alkylation has been observed [Trost, 1991a]. The cyclization occurred as a result of adventitious presence of oxygen which oxidized the Pd(O) species to the Pd(II) state, which is effective in causing an intramolecular ene reaction.

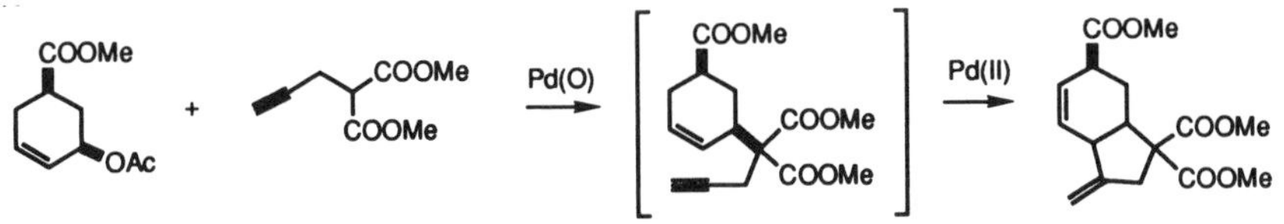

11.2. RETRO-ENE REACTIONS

Homoallylic alcohols are fragmentable via a retro-ene reaction. However, it is possible to manipulate the timing of this fragmentation so that it occurs in tandem with other thermal processes. In the following reaction an ene reaction took precedence to the retro-ene reaction even though the homallylic system was originally present [Onishi, 1980].

Pyrolysis of methyl 7-methyl-3-oxo-6-octenecarboxylate was thought to afford a keto ketene. Intramolecular [2 + 2]-cycloaddition and a retro-ene reaction then followed to afford 4-isopropenyl-1,3-cyclohexanedione [Leyendecker, 1976].

The generation of a de-O-allyl compound from 3-(3,3-dimethylallyl)-4-(3,3-dimethylallyloxy)quinoline-2-one [Grundon, 1985] on thermolysis is due to elimination of a C_5 unit after a Claisen rearrangement by a retro-ene reaction. (Note: this latter process is a retro-Conia reaction of intermolecular version. Its consequence is similar to the McLafferty rearrangement encountered in mass spectrometry, the difference being in the electronic states – homopolar vs cation radical.)

The retro-ene reaction was found to be the major reaction pathway to relieve strain of certain bridgehead dienes which were generated by thermal decomposition of acetates [Tobe, 1983].

A tandem intramolecular ene reaction and retro-hetero-ene fragmentation pathway accounts for the thermolytic transformation of a dienylalkyl propargyl ether into an enyne aldehyde [Shea, 1988].

o-Allyltoluene generated by thermolysis of a norbornen-7-one [Gajewski, 1990] was the consequence of a tandem cheletropic – retro-ene reactions. The competing reaction is the formation of tetralin which requires a higher activation energy, as shown by product ratio measurement at various temperatures.

Tetralin : o-allyltoluene - temp.dependent:

61°	0.9 : 1
96°	1.3 : 1
240° / 0.01torr	5.6 : 1

Aromatization via the retro-ene elimination of an acetaldehyde unit provided the driving force for the thermolytic formation of pyridines from 4a,7a-dihydrocyclopenta[*e*][1,2]oxazines [Faragher, 1977]. Preceding the retro-ene reaction are a prototropic shift which enables a retro-Diels–Alder cycloreversion and a Michael reaction to form the dihydropyridine ring.

Pyrolysis of 1,2-diallyloxyanthraquinone led to 2-allyl-1-hydroxyanthra-quinone [Wong, 1979], as a result of Claisen rearrangement and retro-ene elimination of acrolein.

11.3. VINYLCYCLOPROPANE-TO-CYCLOPENTENE REARRANGEMENTS

These rearrangements occur because much less strained molecules are produced. The reaction is quite valuable in view of the ready availability of vinylcyclo-propanes by reaction of conjugate dienes with carbenoid reagents.

There are a few interesting systems arising from other thermal reactions. For example, a tricyclic dimer of cyclooctatetraene undergoes electrocyclic and

Diels–Alder reactions on heating, and at higher temperatures, the cage structure was transformed into another diene via the vinylcyclopropane rearrangement [Moore, 1964].

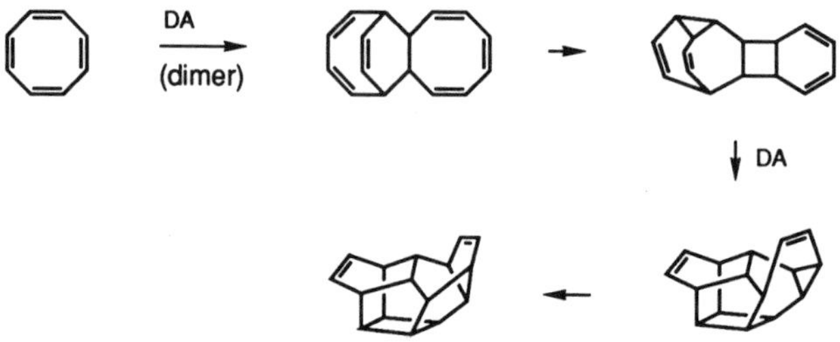

Two products were detected from the thermolysis of a cycloheptadiene which is fused to a benzocyclobutene nucleus [Kato, 1971]. The cyclopentenodihydro-naphthalene was shown to be derived from the [1.5]-sigmatropic rearrangement product which contains a vinylcyclopropane moiety.

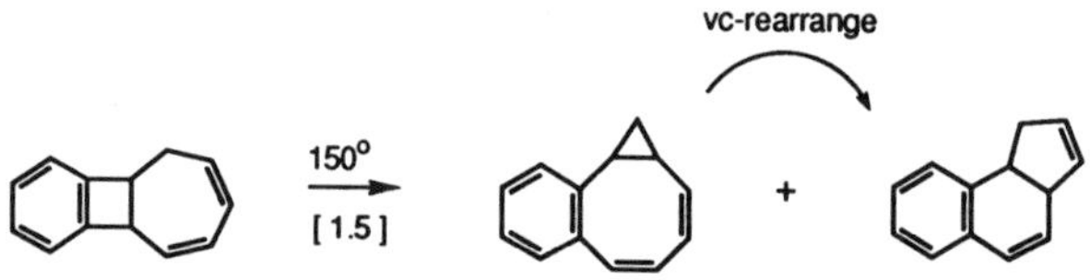

A fascinating thermal reorganization that interconverts bridged and condensed ring structures by tandem Cope–vinylcyclopropane to cyclopentene rearrangements has been documented [Miyashi, 1979].

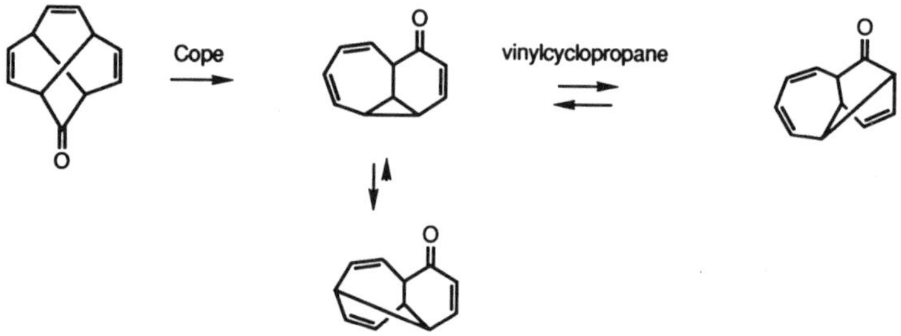

The thermal behavior of the cyclobutenobullvalene [Labows, 1967] is different from that of dihydrobullvalene in that the four-membered ring undergoes cleavage by a Cope-type reorganization which is followed by the vinylcyclopropane rearrangement.

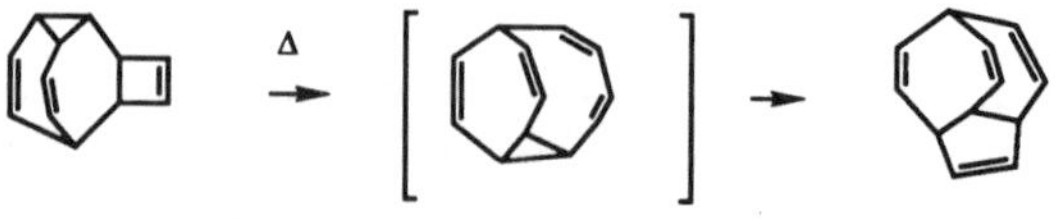

Finally, it is of interest to note that a seemingly contrathermodynamic conversion of a cyclopentene moiety into a vinylcyclopropane took place in tricyclo[3.3.0.0^{2,6}]octa-3,7-diene upon its formation by dehydrochlorination [Meinwald, 1969b, Zimmerman, 1969b]. The rearrangement product is semibullvalene.

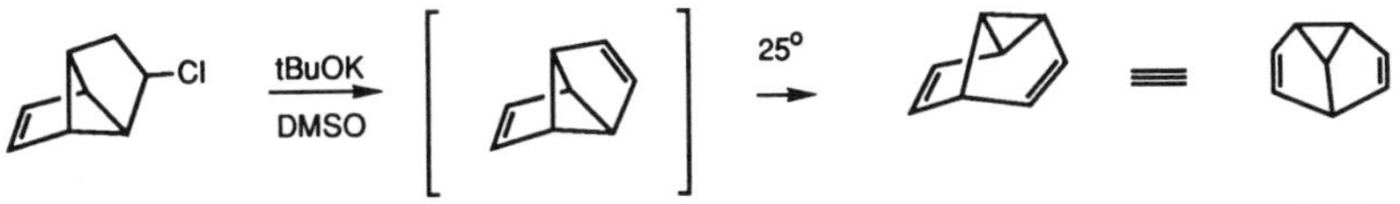

semibullvalene

12

SIGMATROPIC REARRANGEMENTS

In a sigmatropic rearrangement there is migration of a σ-bond along a conjugated π-electron circuit to a new position. The change is defined by the order $[i, j]$ when the termini of the resulting system are $i - 1$ and $j - 1$ atoms away from the original bonded loci.

A σ-bonded group can migrate to the same surface of the π-electron system (suprafacial process) or to the opposite face (antarafacial process). For those rearrangements in which both i and j are greater than 1, the topographical selection rules are as shown in Table 1.

TABLE 1

$i + j$	Thermal Reactions	Photochemical Reactions
$4n$	antara-retention	supra-retention
	supra-inversion	antara-inversion
$4n + 2$	supra-retention	antara-retention
	antara-inversion	supra-inversion

Generally, the migrating group retains its configuration after the sigmatropic shift; however, if the migrating group possesses and uses a p-orbital to form the new bond with the migrating terminus, an inversion of configuration at the migrating group results.

12.1. [1,n]-SIGMATROPIC REARRANGEMENTS

These are the simplest class of sigmatropic rearrangements [Spangler, 1976]. While tandem reaction sequences involving the [1.3]-sigmatropic rearrangement

308

are rare, a very interesting series of such reactions for sterically interactive ene/enedione systems has been reported [Mehta, 1991]. This sequence can be described as involving metathesis via [2 + 2]photocycloaddition and reversion, followed by [1.3]-sigmatropy.

Circumambulation of the single carbon bridge over the five-membered ring in methylated bicyclo[3.1.0]hexenyl cations via [1.4]-sigmatropic process has been studied by nmr at various temperatures [Childs, 1968]. In the case of a heptamethyl derivative, averaging of signals due to the basal methyl groups started as temperature was raised, while the gem-dimethyl group remained unchanged.

[1.4]-Sigmatropic rearrangement of such suprafacially constrained molecules must proceed with inversion of the migrating group. Evidence has been adduced by examining a Favorskii reaction of such a system [Brennan, 1968; Zimmerman, 1968].

The thermally induced suprafacial 1,5-hydrogen (also other substituents) shift is a very well-known phenomenon, exhibited especially readily by cyclic dienes. Many other reactions can set off the hydrogen migration.

4-Alken-1-yn-3-ols can be induced to undergo [2.3]-sigmatropic rearrangements by treatment with a sulfenyl chloride or a phosphinyl chloride. However, the resulting 1,2,4-trienyl derivatives are less stable than the conjugate trienes, with which they are related by the [1.5]-sigmatropy. Consequently, only the latter compounds have been isolated [K.-M. Wu, 1990a].

6,6-Dimethyl-6*H*-dibenzo[*b,d*]pyran is photolabile, and it is readily transformed into a mixture of 4-hydroxy-9-methyl-9,10-dihydrophenanthrene and 9-methylphenanthrene [Bowd, 1980]. The phenol formation pathway involves electrocyclic opening, [1.7]-hydrogen shift, electrocyclization, and a thermally induced [1.5]-hydrogen migration. 9-Methylphenanthrene arises from a divergent electrocyclization product.

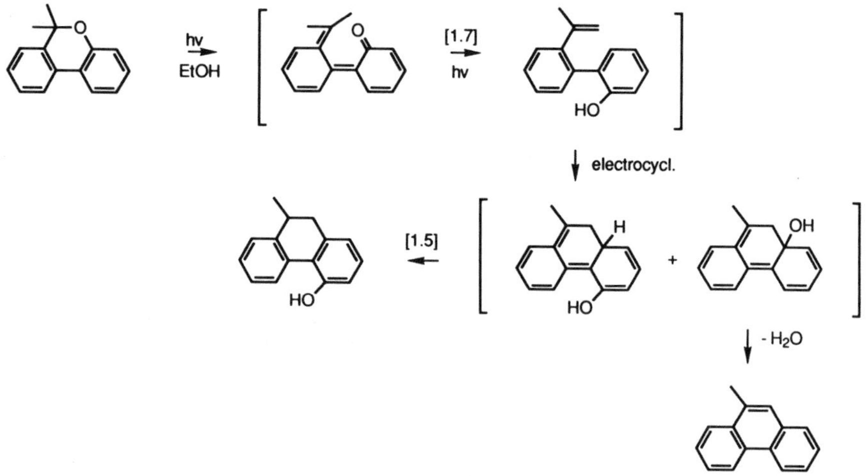

A convergent synthesis of 6-methylpretetramide [D.H.R. Barton, 1970] demonstrated the usefulness of electrocyclization in ring formation, which in this case the crucial step was apparently followed by [1.5]-hydrogen shift and rearomatization of the A-ring.

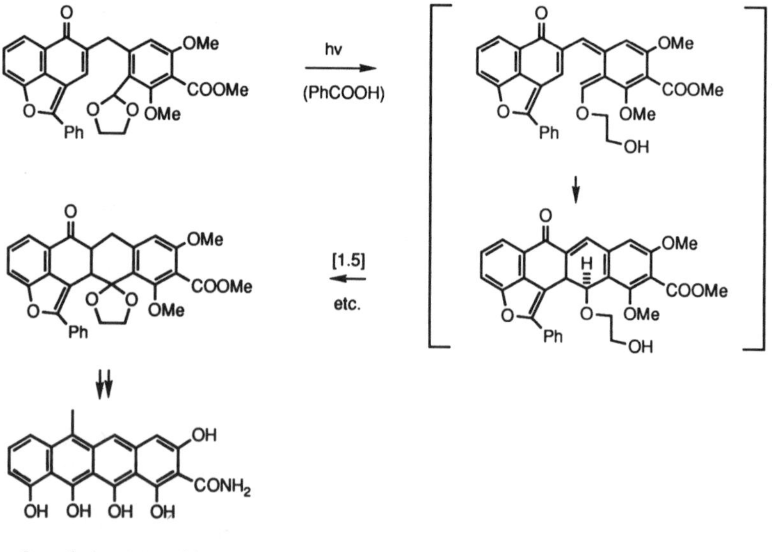

6-methylpretetramid

This pericyclic reaction tandem also represents the pathway for the transformation of (*E,E*)-*o*-dipropenylbenzene into 2,3-dimethyl-1,2-dihydro-naphthalene [Heimgartner, 1970].

The photochemical transformation of 2-phenyl-1,2-dihydronaphthalene into a dibenzocyclooctatriene [Lamberts, 1984] has been shown to be the combined operation of an electrocyclic opening/closure and a thermally induced [1.5]-hydrogen migration in the aromatization step. Interestingly, 1-phenyl-1,2-dihydronaphthalene behaves somewhat differently [Laarhoven, 1985] in that the final step is the ring closure to give *cis*-dibenzobicyclo[3.3.0]octa-2,7-diene instead of the sigmatropic rearrangement.

Irradiation of 1,2-bis(diphenylmethylene)cyclobutane at various temperatures furnished evidence for the existence of an electrocyclized product [Kaupp, 1978] and the requirement of thermal energies for the [1.5]-hydrogen migration. Thus, at lower temperatures increasing amounts of starting material was recovered.

[1.5]-Hydrogen translocation from an imino ketone was observed when the *N*-cyclohexylaminomethylene derivative of Meldrum's acid was subjected to flash vacuum pyrolysis [Gordon, 1983]. The product was an enaminoenaminone.

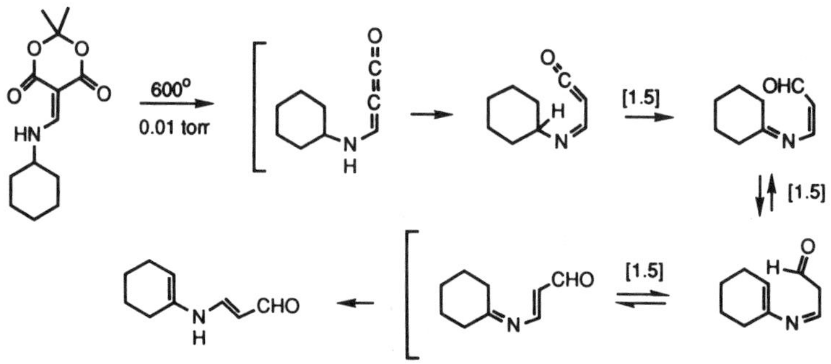

Two [1.5]-oxygen migrations together with prior electrocyclization of 8,16-oxa[2.2]metacyclophane-1,9-diene constituted the pericyclic tandem which was the key to its thermal transformation into 1-hydroxypyrene [B.A. Hess, 1969].

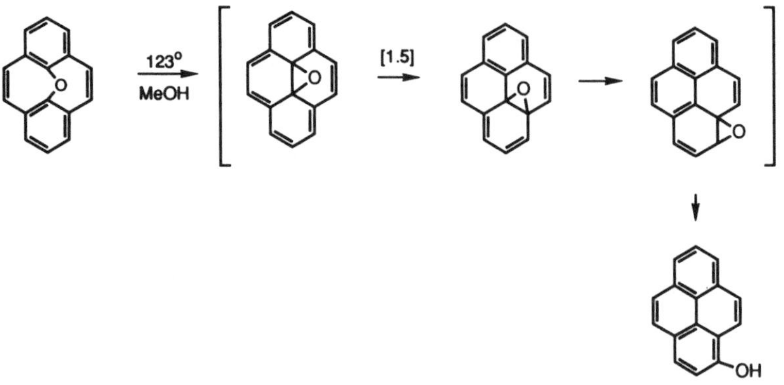

3-(N-Benzoylamino)-1,2-dihydronaphthalenes are photorearranged to give tetrahydrophenanthridinones with a *trans* ring juncture [Ninomiya, 1973]. The reaction consists of conrotatory electrocyclization and [1.5]-sigmatropic hydrogen shift to reestablish the aroyl moiety.

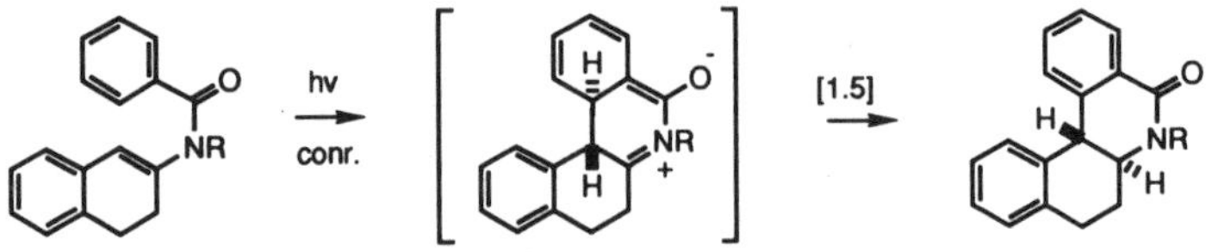

The tricyclic products from intramolecular [8 + 2]-cycloaddition of alkenyl-heptafulvenes [C.-Y. Liu, 1983] are subject to [1.5]-hydrogen shifts.

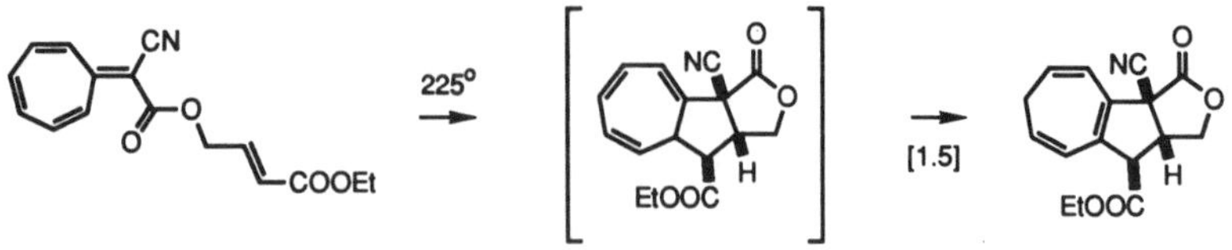

Carbon–carbon bond migrations of 5,5-disubstituted cyclopentadienes are also common. A particularly interesting case is the following thermolytic rearrangement of 1,2-dehydroaspidospermidine [Hugel, 1991], leading to novel indole and indolenine products. All these products are expected on the basis of a tandem [1.5]-sigmatropic manifold.

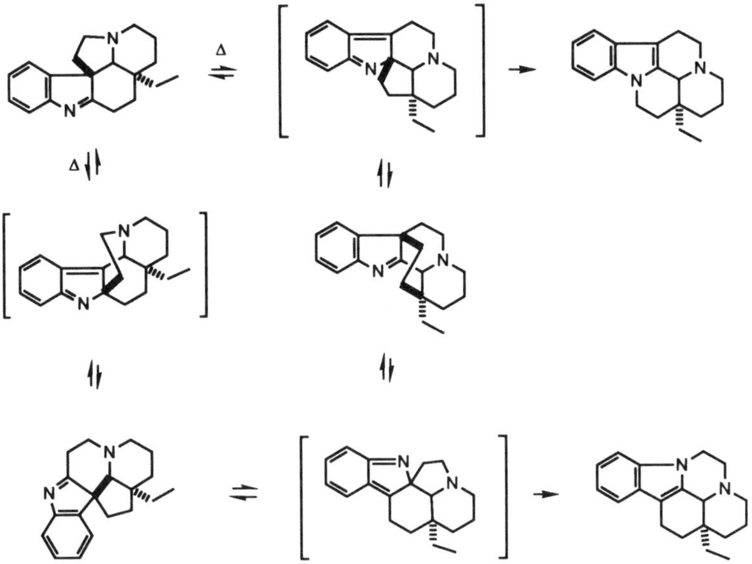

Decarboxylation of a Diels–Alder adduct of indenone and dehydrohomophthalide induced a [1.5]-benzoyl migration [Vanderzands, 1983] which transformed an *o*-quinodimethane into a dihydronaphthalene.

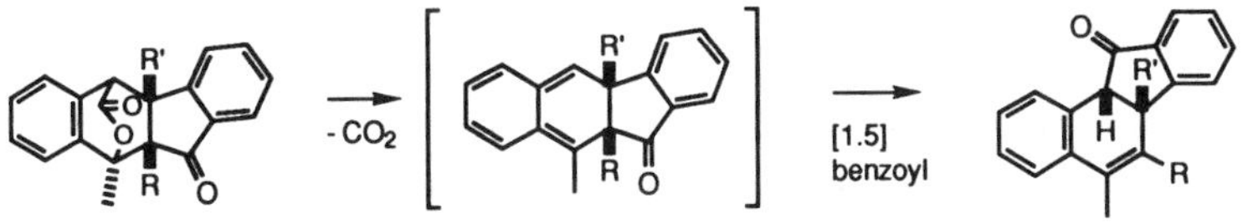

Similarly, aromatization of an *o*-quinodimethane by [1.5]-silyl group transfer accounts for the last stage of a pyrolytic transformation of *o*-tolyl trimethylsilyl ketone into *o*-trimethylsilylmethylbenzaldehyde [Shih, 1982].

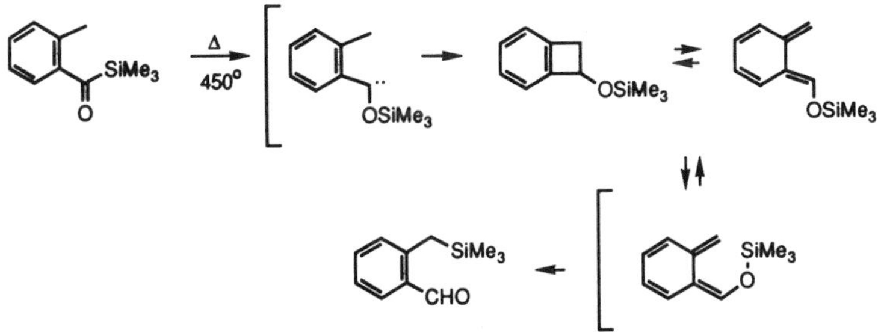

A [1.5]-transfer of an acetyl group to return the cyclohexadienone intermediate from a Claisen rearrangement of an allyl aryl ether to the aromatic state accounts for the appearance of an abnormal product [Falshaw, 1973].

The generation of abnormal Claisen rearrangement products has been known for some time [Marvell, 1962]. These compounds arise via a homo-[1.5] O → C hydrogen migration from the primary products which usually bear alkyl groups at the sp³ center of the allyl residue, and a reversed process comprising the alternative bond sequence around the alkylcyclopropane.

By virtue of both ring strain and the high p-character of endocyclic bonds a cyclopropane is an excellent surrogate of a C=C bond in many reactions. Thus, an analogy to the latter steps of the abnormal Claisen rearrangement has been identified for the behavior of β-tosyloxy ketoximes toward base treatment when they are fully substituted at the α-position [Clark, 1978].

The Conia rearrangement of β-(trimethylsilyl)alkynyl ketones leads to ring formation. When the α-position of such a β,γ-unsaturated ketone is not blocked, conjugation would follow and the isomer avails itself to silyl transfer [Kende, 1988].

The thermal stabilities of two meta photocycloadducts of benzene and (E)-2-butene differ. At 250°C, only the minor cycloadduct underwent [1.5]-

hydrogen migration; however, at 310°C the major cycloadduct also rearranged to the same diene [Srinivasan, 1971]. Direct [1.5]-hydrogen shift of the major adduct was not possible, but at the higher temperature its isomerization to the minor adduct enabled the sigmatropic rearrangement to occur.

A novel photochemical transformation of a *cis*-1-(9-anthryl)2-benzoylethene into a furano-annulated 5*H*-dibenzo[*a,d*]cycloheptene has been rationalized [Becker, 1985] in terms of a [4 + 2]-cycloaddition, electrocyclic reaction, and [1.5]-sigmatropic shift.

Thermolysis of acyloxybenzocyclobutenes leads to *o*-formylbenzyl ketones [Schiess, 1985] by way of electrocyclic opening and [1.5]-shift of the acyl group.

[1.5]-Sigmatropy of both carbon and hydrogen is the crucial feature for the isomerization of 3,4-benzotrophilidene into the 1,2-benzo isomer. However, other intricate processes also occur prior to the inception of the reaction pathway [Gruber, 1970], as evidenced by deuterium scrambling from the seven-membered ring into the aromatic portion. The results are consistent with a mechanism involving formation of nonaromatic intermediates via [1.5]-hydrogen shifts and electrocyclic reactions revolving around a bisnorcaradiene intermediate.

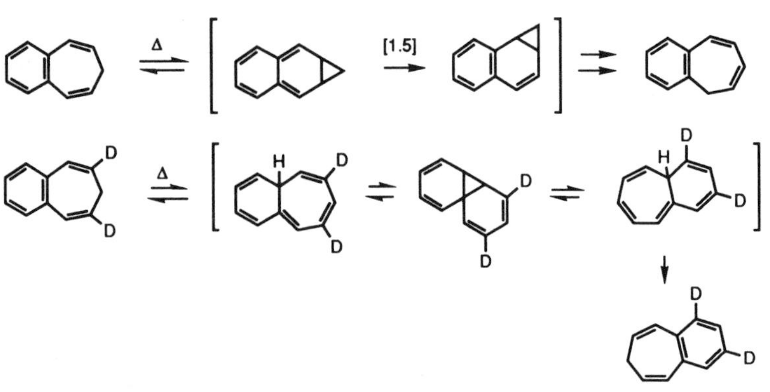

Supporting evidence for the mechanism also came from a study of a related bicyclo[5.4.0]undecapentaene [Bradbury, 1981].

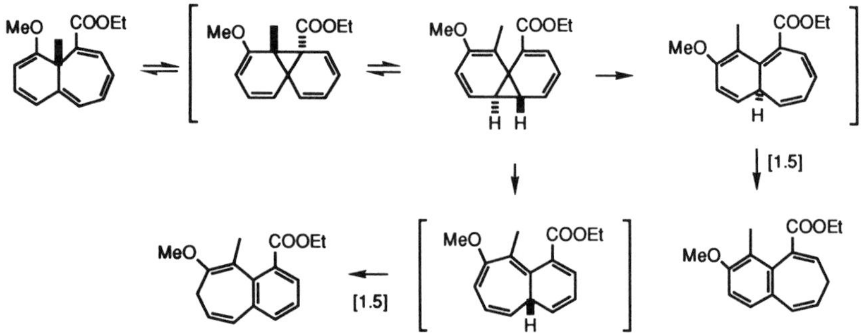

The following reaction is described here, even though it does not terminate in a sigmatropic rearrangement, because of its apparent significance and selectivity. The conversion of dimethyl tricyclo[4.2.2.0^{2,5}]deca-3,7,9-triene-7,8-dicarboxylate into the 2,6-naphthalenedicarboxylic ester [Avram, 1957] is likely the result of tandem Cope rearrangement, Diels–Alder reaction, retro-Diels–Alder reaction, [1.5]-sigmatropic rearrangement, cycloreversion, and dehydrogenation. The interesting feature of this mechanism is the [1.5]-sigmatropy prior to opening of the four-membered ring. Note that there are two possible modes, but the alternative rearrangement would disrupt one of the conjugated ester systems.

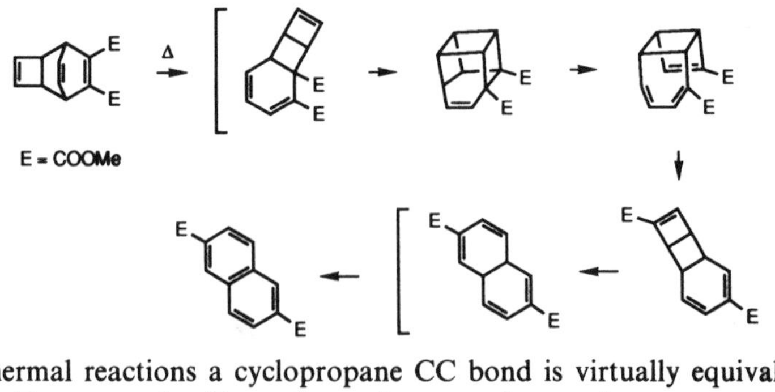

E = COOMe

In thermal reactions a cyclopropane CC bond is virtually equivalent to a π-bond. Consequently, "homo" reactions are frequently encountered. Thus, *cis*-2-vinylcyclopropylalkanes undergo isomerization readily to provide 1,4-dienes, as exemplified by the generation of a bicyclo[3.3.1]nonadiene system from 7,7-dimethylbicyclo[4.1.1]octa-2,4-diene [Young, 1980]. Three other orbital symmetry-allowed thermal reactions ([1.5]-alkyl shift, retro-ene reaction, and intramolecular Diels–Alder cycloaddition) precede the homo[1.5]-hydrogen transfer.

By a homo[1.5]-hydrogen shift a tetracyclic arene-alkene *m*-cycloadduct is transformed into dehydroisocomene [Wender, 1981a]. The other cycloadduct is less reactive because it must undergo a vinylcyclopropane rearrangement first.

isocomene

2-Vinylindene is the product from thermal decomposition of 5,6-benzotricyclo[3.2.0.0²,⁷]-hept-5-ene [Adam, 1980]. It is reasonable to consider its derivation from a tandem retro-Diels–Alder reaction/[1.5]-hydrogen migration.

A most remarkable pericyclic reaction sequence attends the thermal conversion of heptahendecafulvadiene to a pentacyclic product [Beck, 1984]. The sequence consists of disrotatory 20e-electrocyclization, [1.9]-sigmatropy, 10e-, and 6e-electrocyclizations, cycloreversion, and [1.5]-hydrogen shift.

The [1,7]-shift is less frequently observed due to the much less favorable entropic factors. Furthermore, at least two double bonds of a triene system must have an (Z)-configuration.

In the light of such steric restrictions it is of interest to witness the conversion of a vinylallene isomer of vitamin-D into two conjugate trienes [Hammond, 1978]. These minor products arose from a series of [1.5]- and [1.7]-hydrogen migrations. The [1.7]-shift must be antarafacial.

It is important to note that an allylic hydroxyl group attaching to one end of the triene system has a profound effect on the [1.7]-hydrogen migration. Migration to the *syn* face of the hydroxyl is preferred [K.-M. Wu, 1990b].

Of particular interest is the thermolytic behavior of (E)- and (Z)-propenylbenzocyclobutenes [De Camp, 1974]. The steric disposition of the terminal methyl group in the o-quinodimethane intermediates dictates the product formation pathway, for example via either electrocyclization or [1.7]-hydrogen migration.

A laser flash photolysis study of diphenylamines indicates the generation of zwitterionic transients with lifetimes of about a millisecond at room temperature in methylcyclohexane [Grellmann, 1982]. The zwitterionic species are formed by electrocyclization, and their charges disappear upon sigmatropic shift (e.g. [1.8]-hydrogen shift).

Photochemical conversion of 9,10-dihydronaphthalene to bullvalene proceeds via $[_\pi 4_s + _\pi 4_a]$-cycloaddition, a retro-Diels–Alder reaction, and a $[_\pi 4_a + _\pi 2_a + _\sigma 2_s]$ rearrangement [Doering, 1967].

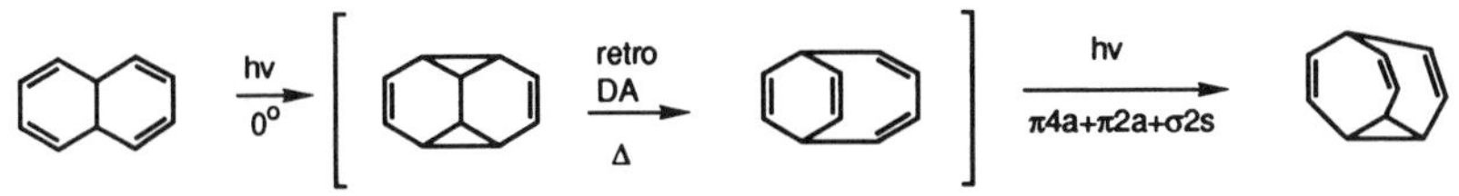

A method for the synthesis of syn-1,6:8,13-bismethano[14]annulene [Vogel, 1983] is via a tandem 14e-electrocyclization to a strained allene and [1.13]-hydrogen shift.

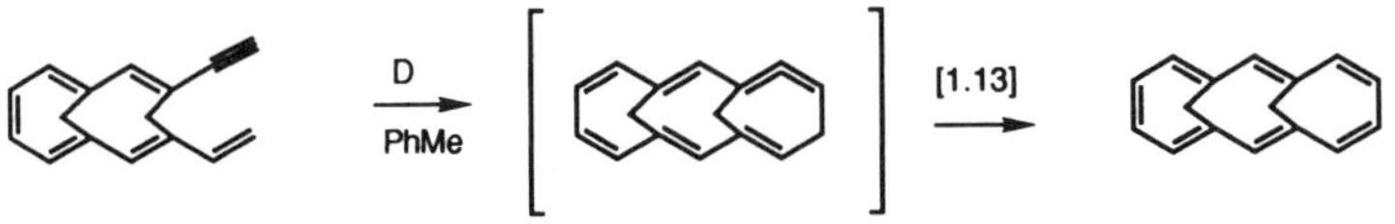

[1.17]-Sigmatropic rearrangements within the porphyrin framework have been witnessed [Callot, 1974].

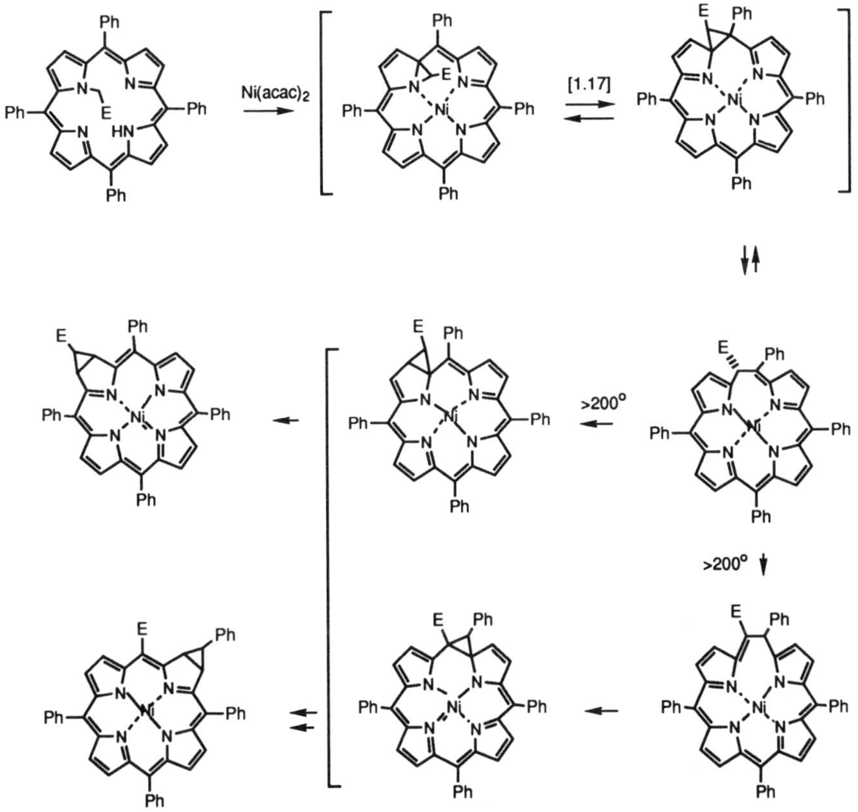

12.2. [3.3]-SIGMATROPIC REARRANGEMENTS

The all-carbon [3.3]-sigmatropic rearrangement is commonly known as the Cope rearrangement. The active orbitals involved are the σ-orbital of the single bond to be broken and the π_g orgital. Generally, a chair-like transition state is favored because there is a destabilizing nonbonding interaction between the central carbon atoms in the boat-like transition state. However, steric constraints could force the substrate to undergo rearrangement via the boat-like transition state.

The concerted nature of the [3.3]-sigmatropic rearrangements makes them very useful in synthesis. For example, the stereospecificity and stereoselectivity of the Cope rearrangement can be exploited in the establishment of two adjacent stereogenic centers and configurations of two double bonds.

Several transition metal species (e.g. Pd(II)) catalyze Cope rearrangements [Lutz, 1984]. However, the oxy-Cope rearrangement, and particularly the anionic version, is most significant with respect to applications [Paquette, 1990a]. The product of such a rearrangement is an enol or enolate ion and the immediate tautomerization to give a carbonyl compound is a strong driving force for the reaction. In terms of structural changes the (anionic) oxy-Cope rearrangement leads to δ,ε-unsaturated compounds which are homologs of the aliphatic Claisen rearrangement products.

The anionic oxy-Cope rearrangement is accelerated by factors of up to 10^{15}. Sometimes the rearrangement products are obtained directly from the reaction of β,γ-unsaturated carbonyl compounds with alkenylmetals [Paquette, 1983a].

The aliphatic Claisen rearrangement of allyl vinyl ethers [Ziegler, 1988] is completely analogous to the Cope rearrangement. Of course the conversion of the substrates into carbonyl compounds render the reaction irreversible. It also proceeds preferentially via the chair-like transition state.

Many aliphatic Claisen rearrangements involve in situ generation of the allyl vinyl ethers; therefore the rearrangement itself is tandem to the formation of the substrate.

Due to the intrinsic instability of *simple* aliphatic enols, allyl vinyl ethers are usually prepared from allyl alcohols. Besides the procedure involving mercuric acetate–catalyzed exchange with alkyl vinyl ethers (cf. reaction with (E)-(carboxyvinyl)trimethylammonium betaine [Büchi, 1983]), other variants developed since the 1960s have expanded the versatility and scope of the Claisen rearrangement. Thus, chain extension to yield methyl ketones has become feasible by using 2-methoxypropene or 2,2-dimethoxypropane to form ethers with the allylic alcohols, including the acid-sensitive tertiary alcohols such as linalool [Marbet, 1967]. Acid catalyzed transacetalization of simple ortho esters with allylic alcohols permitted the synthesis of γ,δ-unsaturated esters [Johnson, 1970]. The exchange with amide acetals leads to the analogous amides [Wick, 1964].

By far the most significant development is the rearrangement of ketene silylacetals [Ireland, 1972] which occurs at ambient temperature or below. The possibility of controlling the enolate stereochemistry is of great importance in obtaining products of defined configurations.

12.2.1. Cope Rearrangements

12.2.1.1. Pericyclic Reaction-Induced Cope Rearrangements

The most fascinating combination of pericyclic reactions is the degenerate Cope rearrangement. This rearrangement has a very low energy of activation and it generally occurs with 1,5-hexadienes embedded in rigid frameworks such that the entropy favors the molecular reorganization greatly. Since the degenerate rearrangement converts one structure into an identical one, it can be identified only by isotopic labels or by NMR methods. The most remarkable molecule that exhibits such a continuous tandem Cope rearrangements is bullvalene [Doering, 1963]. All C–H units in this $(CH)_{10}$ system become equivalent at $>100°C$, as evidenced by a single proton absorption band in the NMR [Schröder, 1963]. In other words, all $10!/3 (= 1,239,600)$ structurally identical valence tautomers are interconverting and no two carbon atoms are permanently bonded to each other in this molecule.

bullvalene

Another fluxional molecule that undergoes degenerate Cope rearrangement is hypostrophene [McKennis, 1971].

hypostrophene

The two sterically constrained cyclobutadiene moieties liberated from the metal complex of a superphane [Gleiter, 1988] underwent spontaneous intramolecular Diels–Alder reaction. Degenerate Cope rearrangements of the product has been detected.

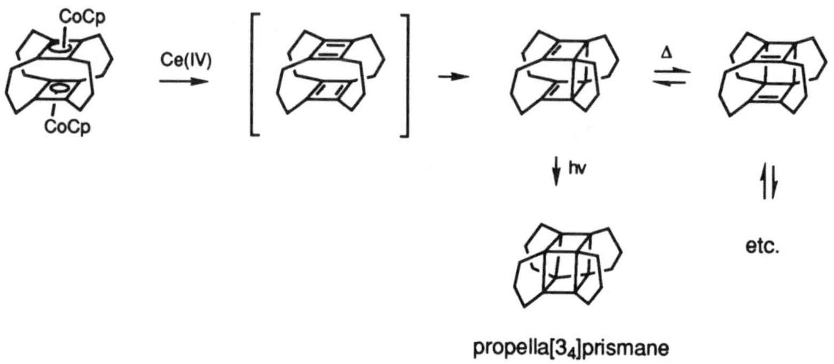

propella[3₄]prismane

The mechanism for the very intriguing conversion of cyclododeca-1,5,9-triene to [6]-radialene has been deciphered [Dower, 1986] by isotopic labeling. A tandem Cope rearrangements is indicated.

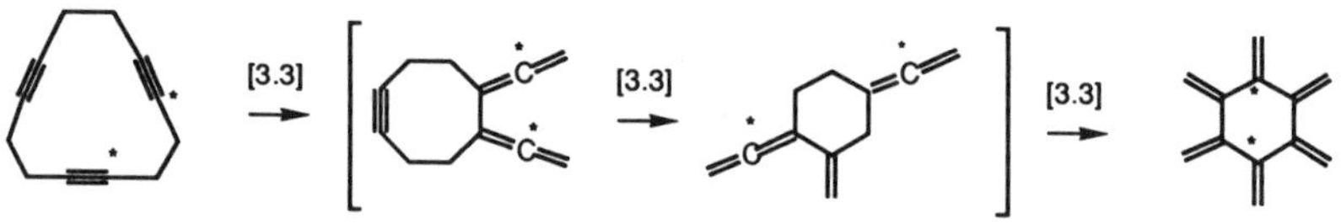

Tandem thermal rearrangements to afford 2,3-disubstituted 1,3-butadienes have been described [Hopf, 1985].

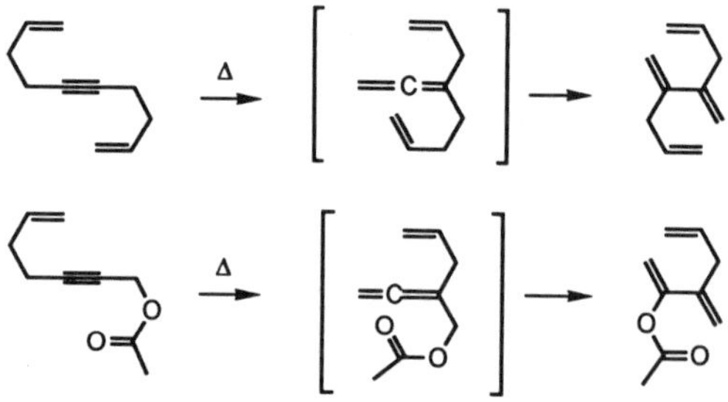

Two consecutive Cope rearrangements constitute the reaction pathway for the conversion of caryophyllene into isocaryophyllene [Ohloff, 1967].

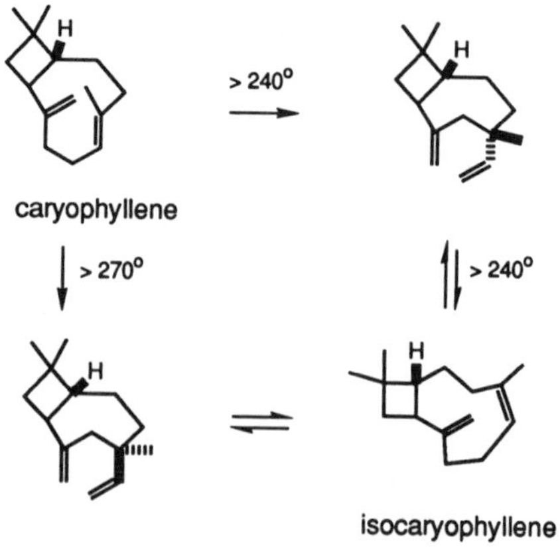

caryophyllene

> 240°

> 270°

> 240°

isocaryophyllene

The (E)-to-(Z) isomerization of (E,Z)-1,5-cyclooctadiene has been shown with methyl labels to proceed via a tandem Cope rearrangement process [Berson, 1972b], with chairlike and boatlike transition states for the first reaction forming a *cis*-1,2-dipropenylcyclobutane, and the ring expansion phase, respectively.

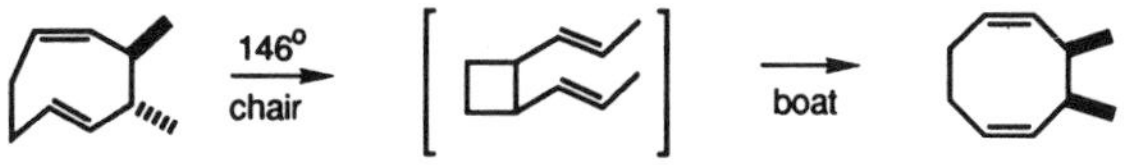

The *syn-anti* isomerization of bicyclo[6.1.0]nonatrienes [Lewis, 1975] involves two Cope rearrangements which are interposed by two electrocyclic reactions.

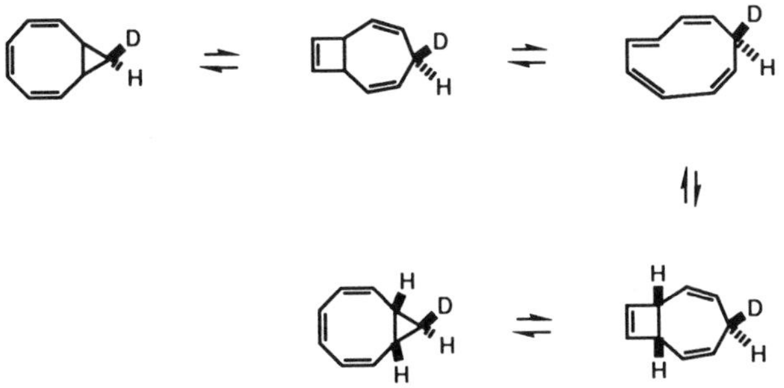

The scope of the Cope or oxy-Cope rearrangement-mediated ring expansion has been extended to the addition of eight carbon atoms in one step. Such a process greatly facilitated the synthesis of $(3Z)$-cembrene-A [Wender, 1985a] and muscone [Wender, 1983b], and it appears that two tandem Cope rearrangements were involved [Wender, 1985b]. There was a remarkable acceleration of the second rearrangement by the enolate.

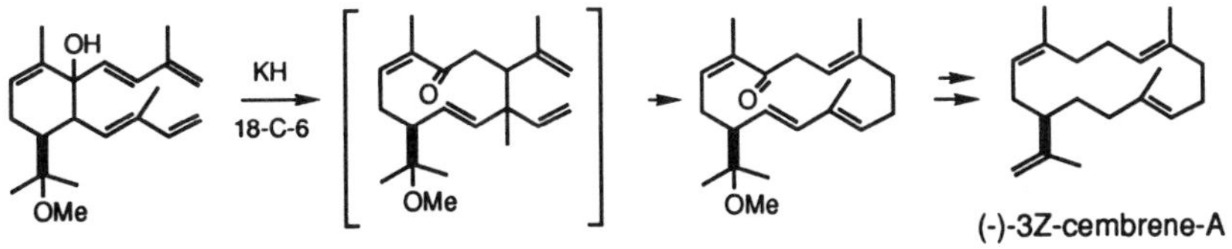

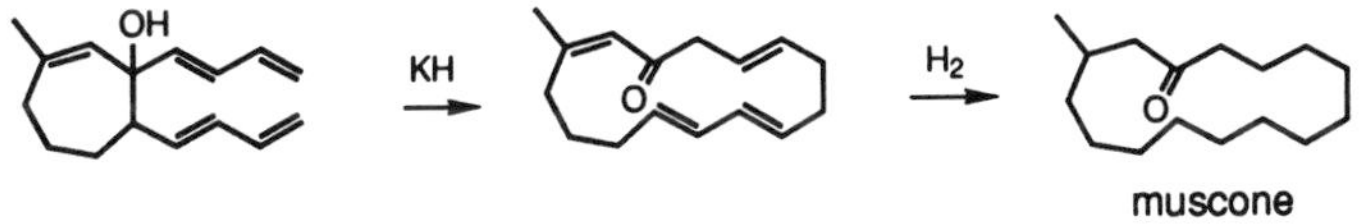

(-)-3Z-cembrene-A

muscone

The Cope rearrangement played an important role in the synthesis of linderalactone and related substances [Gopalan, 1980]. The thermal behavior of the various isomers has simplified the task enormously.

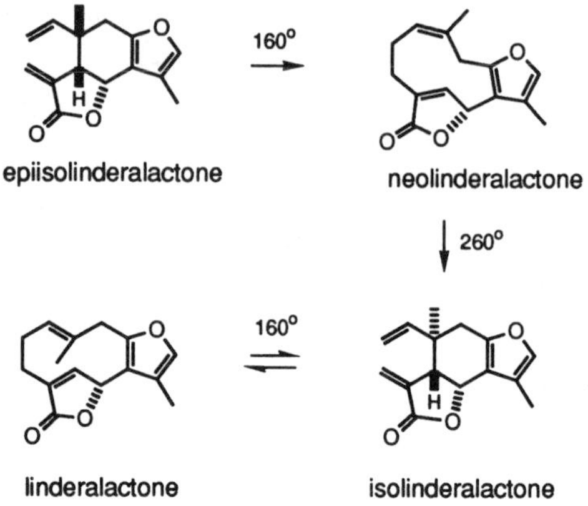

epiisolinderalactone   neolinderalactone

linderalactone   isolinderalactone

An equilibrium of isolinderalactone and linderalactone is established at 160°C via a Cope rearrangement. More interestingly, epiisolinderalactone is converted into neolinderalactone, and at higher temperature, into isolinderalactone.

A closely related process is the conversion of curzerenone to pyrocurzerenone [Hikino, 1968]. Consecutive Cope and oxy-Cope rearrangements serve to transform the elemane-type molecule into a furanocadinane skeleton. Dehydration with attendant aromatization is expected under the reaction conditions.

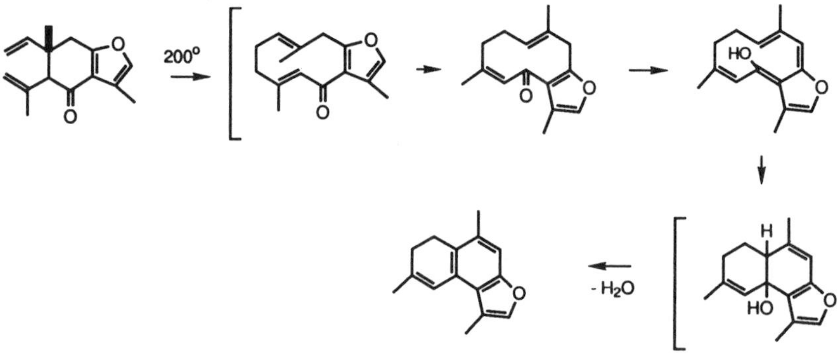

pyrocurzerenone

Since many terpenes retain the 1,5-diene remnants from their acyclic precursors (e.g. geranyl, farnesyl derivatives) the Cope rearrangement can be exploited to advantage in their synthesis. Naturally consecutive sigmatropic rearrangements of a well-designed intermediate would be most effective for achieving the synthetic goal, in view of the irreversibility rendered by the succeeding reaction(s).

A siloxy-Cope–Cope rearrangement sequence served well in the construction of bishomocitral [Mikami, 1981].

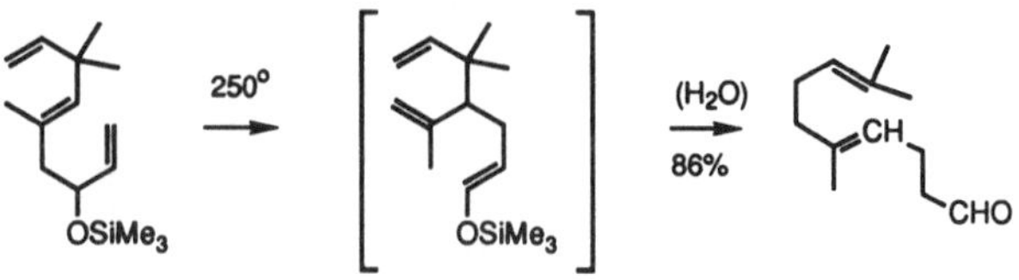

The combination of Claisen and Cope rearrangements is an even more powerful synthetic method because the very versatile carbonyl group is established in the final products. An eminent example is the one-step synthesis of β-sinensal [Thomas, 1969]. The same process can be used to prepare torreyal [Thomas, 1970].

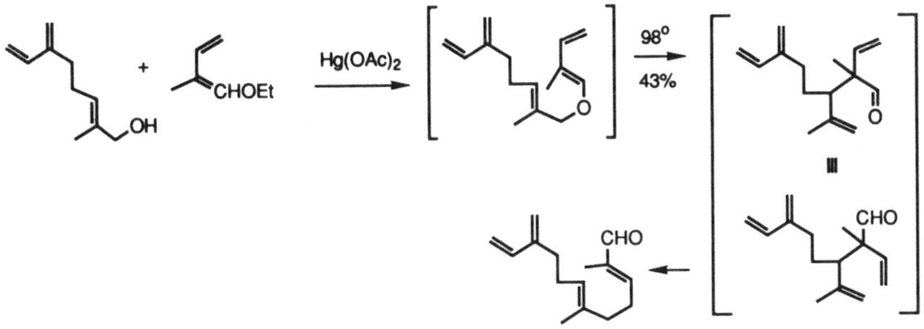

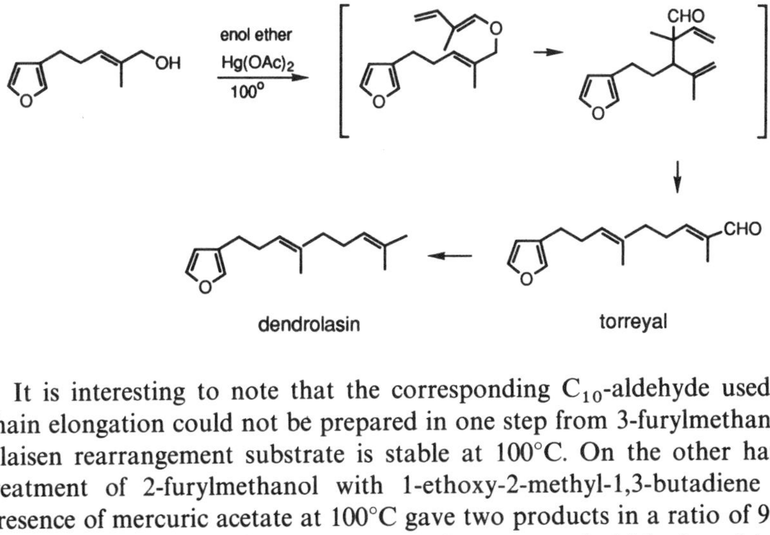

dendrolasin   torreyal

It is interesting to note that the corresponding C_{10}-aldehyde used in the chain elongation could not be prepared in one step from 3-furylmethanol. The Claisen rearrangement substrate is stable at 100°C. On the other hand, the treatment of 2-furylmethanol with 1-ethoxy-2-methyl-1,3-butadiene in the presence of mercuric acetate at 100°C gave two products in a ratio of 9:1. The major compound was shown to be an α,β-unsaturated aldehyde, arising from a Claisen–Cope rearrangement tandem.

A convenient synthesis of citral [Leimgruber, 1974] consists of reaction between dimethylallyl alcohol and an acetal of the corresponding aldehyde.

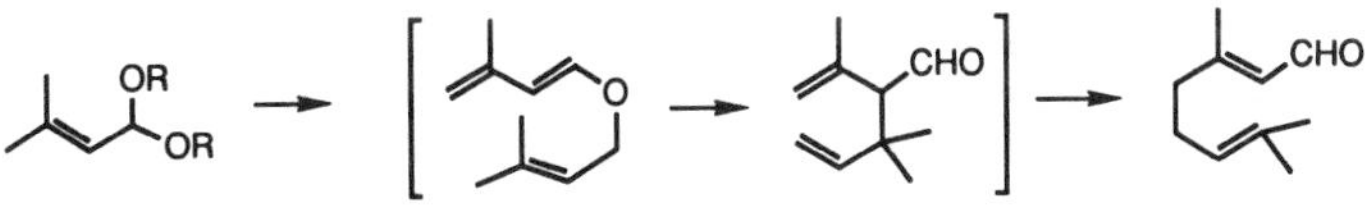

Alternatively, diprenyl ether may be monochlorinated and dehydrochlorinated. At 150°C citral was produced [Suzuki, 1983].

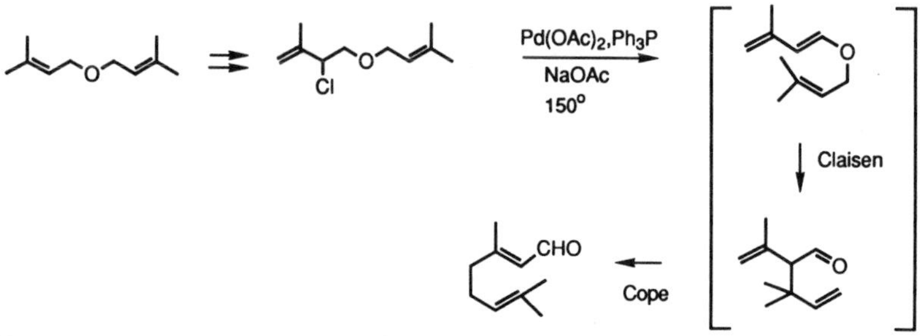

The preparation of β,γ; ζ,η-unsaturated esters may also enlist the service of consecutive Claisen and Cope rearrangements [Fujita, 1978]. The Cope rearrangement can be arrested by conducting the reaction at lower temperatures (145–155°C vs 180–200°C).

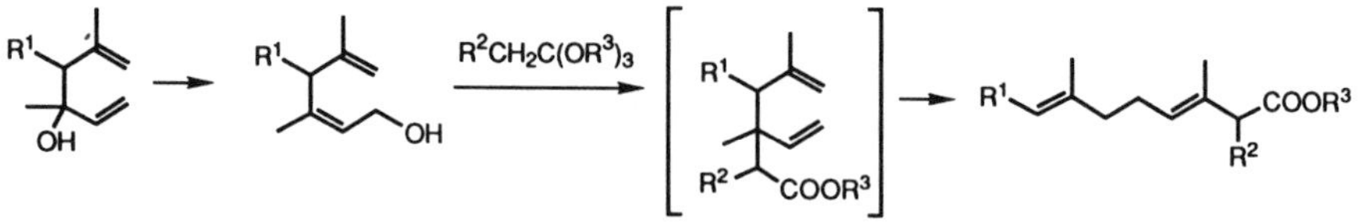

It is of course easier to separate the two processes by employing the much milder version (e.g. Ireland protocol) of the Claisen rearrangement, as shown in a synthesis of the cecropia juvenile hormone [Frater, 1975].

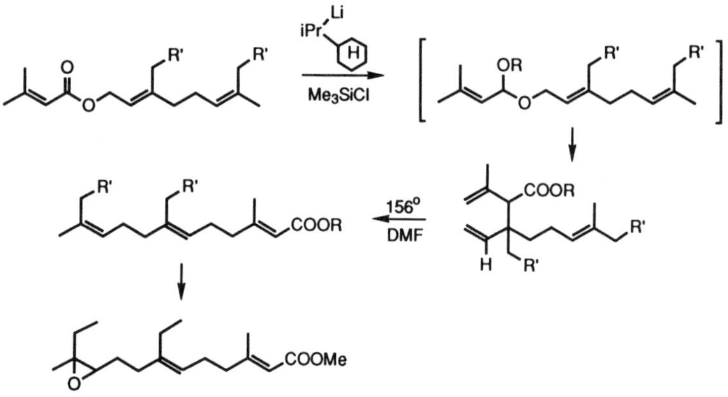

The Claisen rearrangement of a proparyl vinyl ether may be as facile as that of the corresponding allyl vinyl ether. The resulting allenyl ketone may undergo a Cope rearrangement when another double bond is present in the active position [Bowden, 1973].

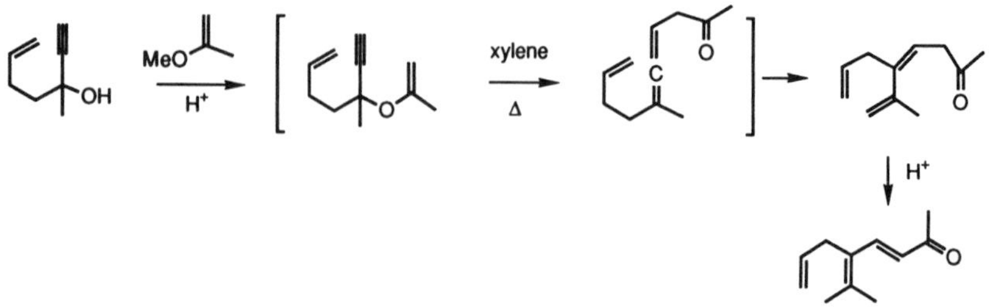

Two Cope rearrangements succeeding a Claisen process are indicated in the following transformation [Cookson, 1973].

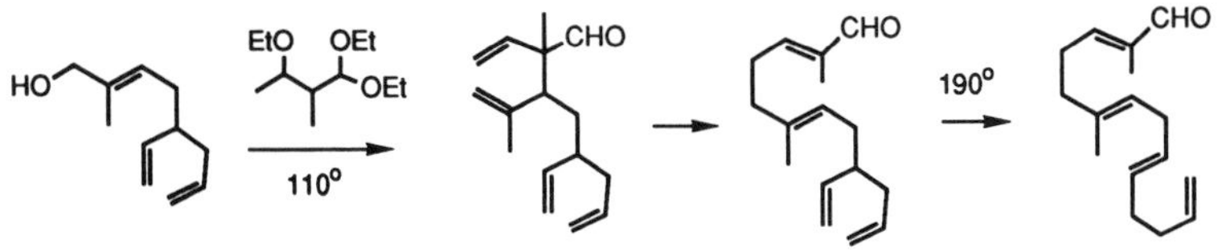

The thio-Claisen rearrangement of ketene S,N-acetals [Tamaru, 1980] generally proceeds in the temperature range of 25–85°C. A Cope rearrangement of the products can be effected without isolation of the initial products.

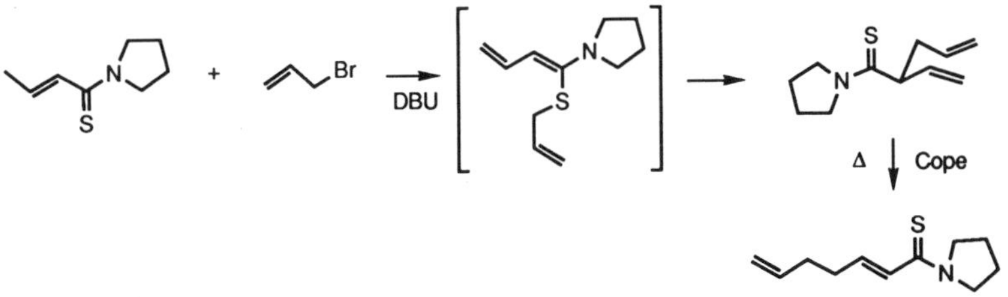

Normally, aromatic Claisen rearrangement terminates at an o-position when it is unsubstituted. A recent finding [Maruoka, 1990] concerning the ability of certain bulky organoaluminum compounds to direct the rearrangement to the p-position of a dimethylallyl 2,4-dialkylphenyl ether is of synthetic and mechanistic significance. Probably because the coordinated aluminum species which undergo rearrangement tend to avoid steric interactions with an ortho substituent, and therefore are forced to dispose the allyl chain to lie proximally to the o-substituent.

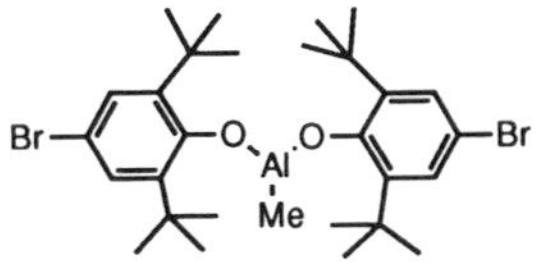

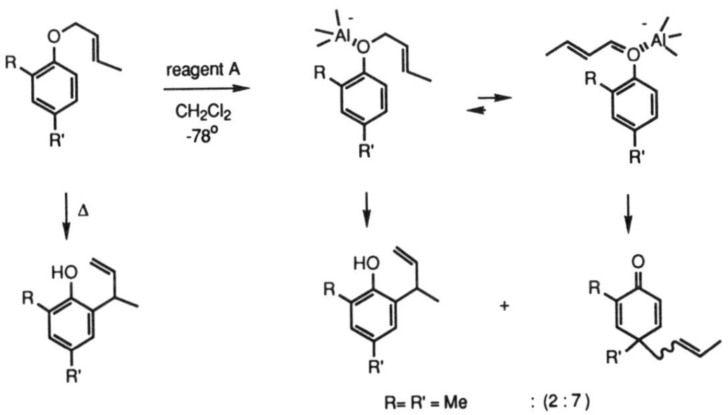

R = R' = Me   : (2 : 7)
R = t-Bu; R' = Me : (0 : 1)
R = Me; R' = Cl   : (1 : 5)

The traverse of an allyl group from the aryl ether to the central carbon atom of an *o*-propenyl sidechain as first reported in 1926 [Claisen, 1926] is the consequence of a tandem Claisen–Cope rearrangements [Lauer, 1956; Schmid, 1956]. An allyl residue can even insert into a para sidechain via a three-staged rearrangement process [Nickon, 1964]. The out-of-ring migration is the second Cope rearrangement.

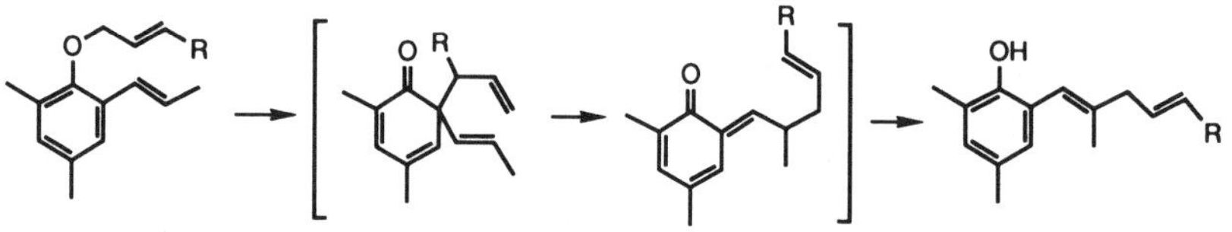

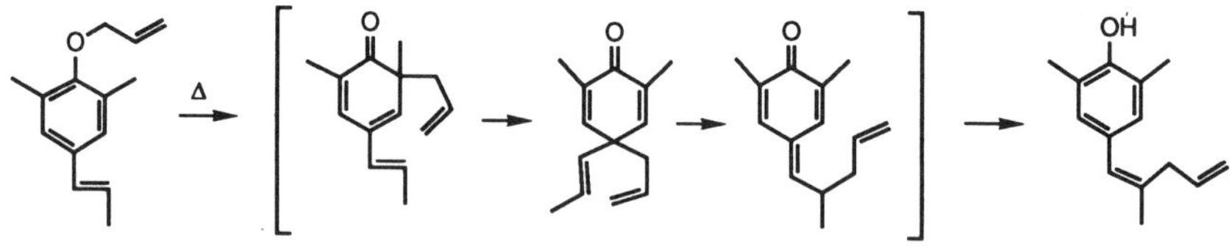

In a synthesis of gravelliferone [N. Cairns, 1987; Massanet, 1987] the introduction of the methylbutenyl substituent at C-3 of the coumarin skeleton was accomplished by thermolysis of the 7-dimethylallyloxy derivative. The methylbutenyl group migrated to an angular site by tandem aromatic Claisen–Cope rearrangements which are generally considered as *para*-Claisen rearrangement, before lodging in the heterocycle via another Cope rearrangement. The last step permitted rearomatization of the ring system.

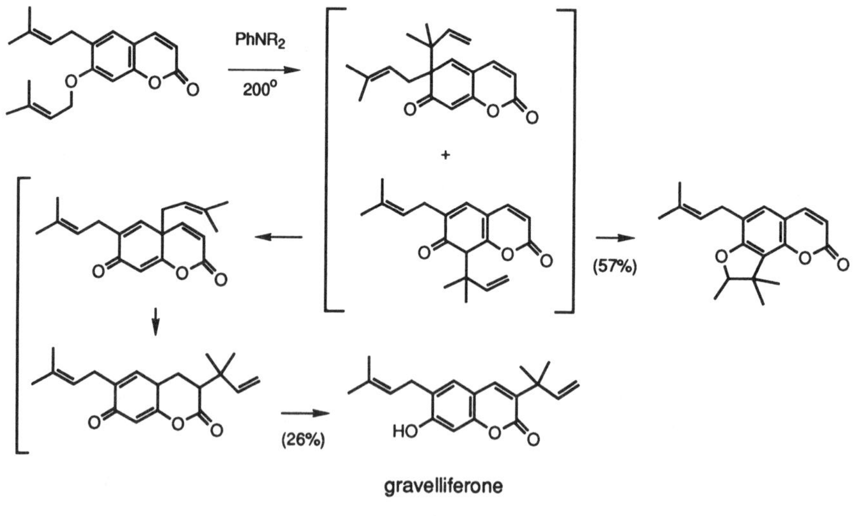

gravelliferone

As expected, thermolysis of the bis-2,6-xylyl ether of (*E*)-2-butene-1,4-diol led to the formation of the symmetrical 1,4-diaryl-2-butene [Thyagarajan, 1967].

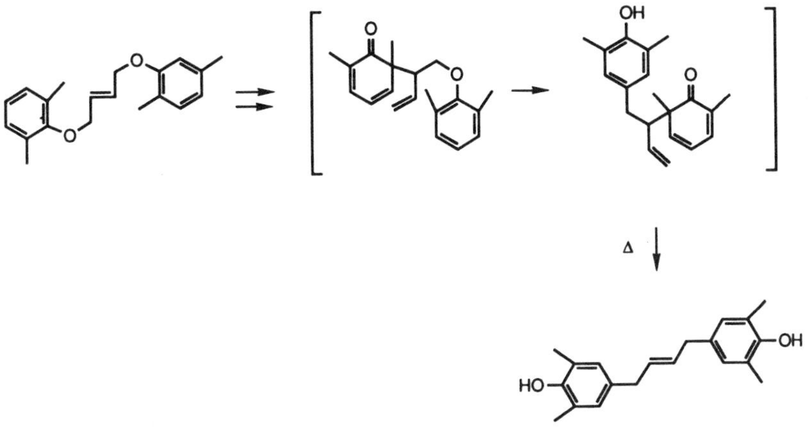

The aromatic Claisen rearrangement of 2-allyloxy-3-hydroxybenzaldehyde gave several abnormal products due to the presence of a free hydroxyl group which directed [1.2]- and [2.3]-sigmatropic rearrangements [Kilenyi, 1991]. Different pathways were followed when the corresponding phenolic ether was submitted to the same reaction conditions.

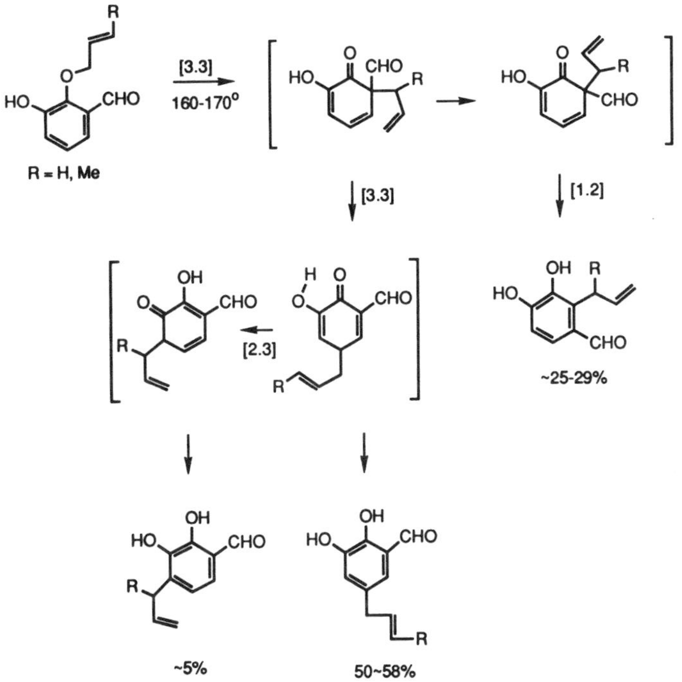

The Claisen–Cope rearrangement tandem occurring in a purine framework are represented by the following examples [Leonard, 1974; B.N. Holmes, 1976].

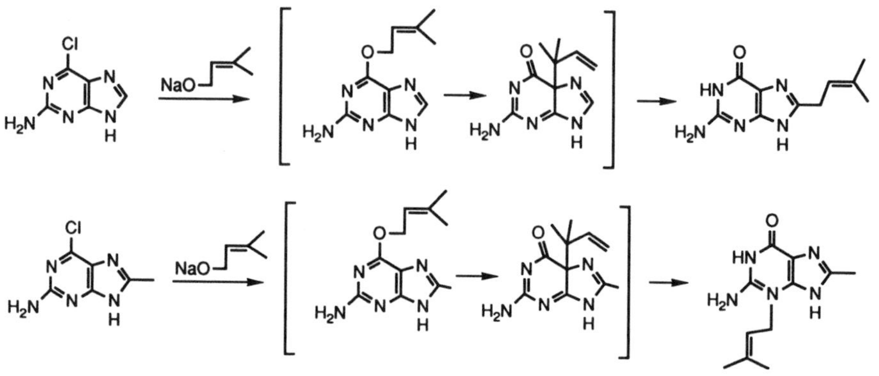

Thermolysis of tropolone allyl ethers was the basis for synthesis of nootkatin [Y. Kitahara, 1958] and procerin [Y. Kitahara, 1964]. It is apparent that tandem Claisen–Cope–Cope rearrangements were involved, and most interestingly two isomeric ethers in each series were converted into the same terpene.

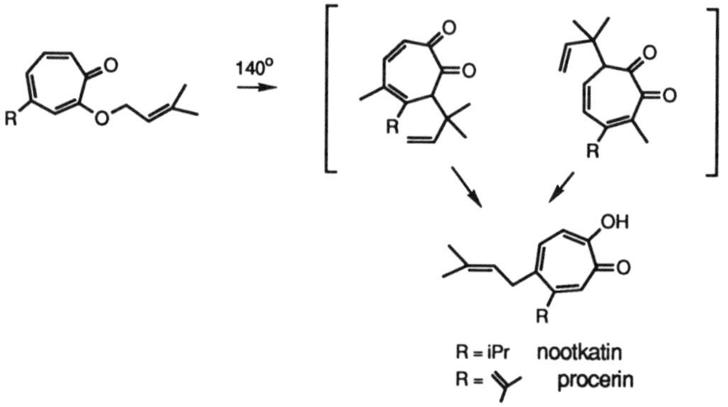

R = iPr   nootkatin
R = ⅄   procerin

[2.3]-Sigmatropic rearrangement of allyl vinyl sulfonium ylides may be accompanied by a Cope rearrangement [Labuschagne, 1975]. Similarly, it has been found that while the [2.3]-sigmatropic rearrangement of dicinnamyl dimethylammonium ylide in dimethyl sulfoxide is the major reaction course at room temperature, its allylic amine product diminished as reaction temperature was raised (in methanol solvent) [Jemison, 1980]. Thus, after 15 h at 85°C it was completely replaced by an enamine, due to a tandem Cope rearrangement.

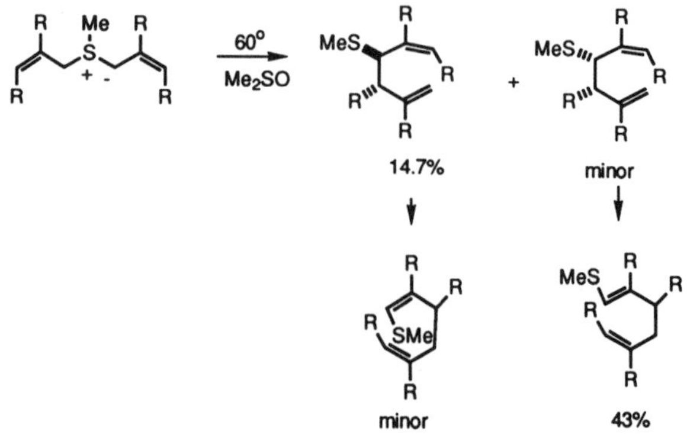

14.7%   minor

minor   43%

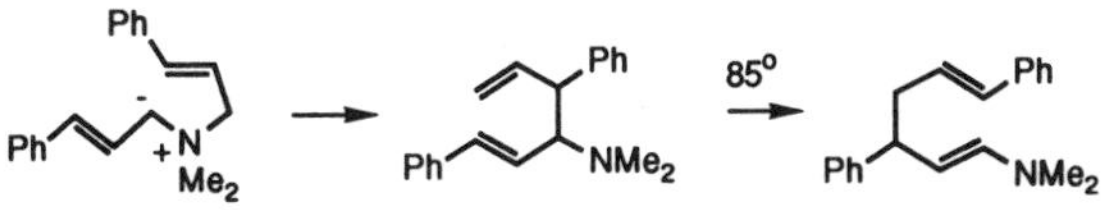

In the presence of alkali the monogeranyl ether of hydroquinone favors [2.3]-rearrangement which is followed by a Cope rearrangement [Harwood, 1991]. The product has been converted into alliodorin.

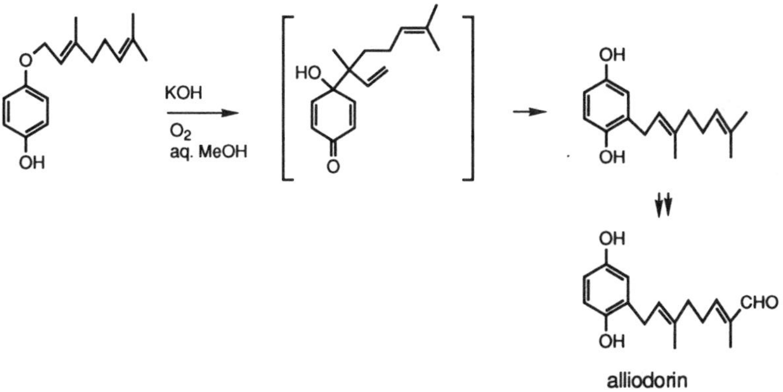

alliodorin

On base treatment 1,5-dien-3-ols undergo allylic transposition via fragmentation of the C-3/C-4 bond. When the allylic alcohol moiety is incorporated into a carbon framework such as the bicyclo[3.2.0]hept-2-ene system the rearranged compound is liable to further transformations, in this case, an anionic oxy-Cope rearrangement [Jung, 1983].

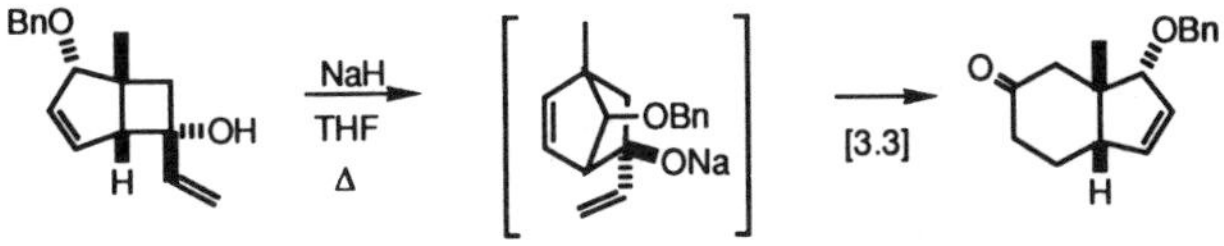

It has been established that treatment of a 9-vinylbenzobicyclo[4.2.1]nonatrien-9-ol with potassium hydride at low temperatures led to a rapid equilibrium with three other allylic alcohols [Miyashi, 1982]. Eventually all of the alcohols were depleted by an anionic oxy-Cope rearrangement.

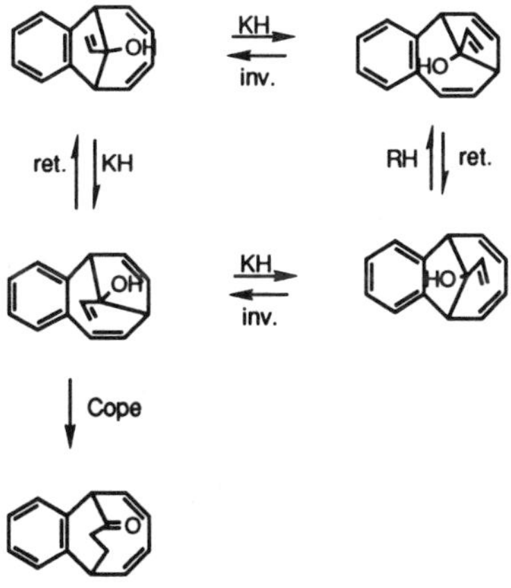

Tandem [1.3]-/[3.3]-sigmatropic rearrangements of medium-sized ring bis-*O*-silyl enolates derived from keto lactones have been observed [Khan, 1991]. The siloxy enol ether (ex ketone) appears to exert a crucial effect favoring the [1.3]-sigmatropy over the Claisen rearrangement mode.

A series of [1.3]-, [1.5]- and [3.3]-sigmatropic rearrangements was responsible for the transformation of a vinyl carbinol to a ring-expanded tricyclic triene [Miyashi, 1978].

The similar behavior of three-membered ring bond to a π-bond is well-known. In thermolysis of 2,2-dialkyl-1-alkynylcyclopropanes and epoxides, the [1.5]-hydrogen transfer with ring opening is usually followed by the [3.3]-sigmatropic rearrangement [Dalacker, 1974; Karpf, 1977].

$X = O, CH_2$

Tandem [2.3]-Wittig/oxy-Cope rearrangements have been designed and examined [Nakai, 1986]. The sequence provides a versatile method for the synthesis of δ,ε-unsaturated carbonyl compounds. It is noted that the (E/Z) stereochemistry of the oxy-Cope rearrangement is not dependent on the erythro–threo ratio of the [2.3]-Wittig rearrangement product. While anionic oxy-Cope and siloxy-Cope rearrangements give rise to a modest degree of (E)-stereoselection, this selectivity is greatly enhanced (from the 70% range to the 90% range) at higher temperatures (ca. 170°C in decane). Examples of application are syntheses of brevicomin, oxocrinol [Mikami, 1982] and 5-trimethylsilyl-5-alkenals [Mikami, 1983].

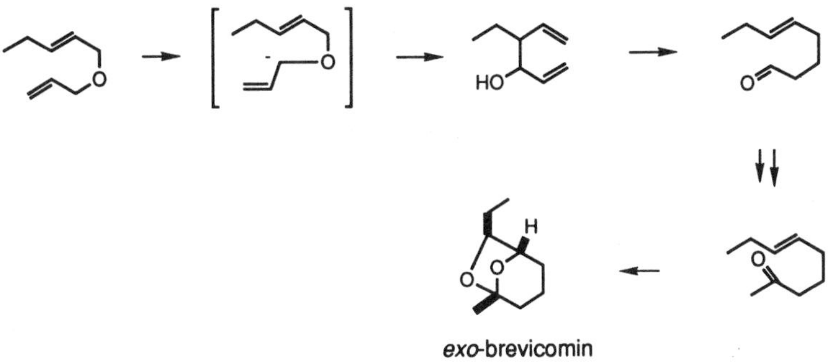

exo-brevicomin

The Cope rearrangement of *cis*-1,2-divinylcyclopropanes are extremely facile, owing to the relief of strain associated with the change. As a result of a remarkable *cis*-stereoselectivity for cyclopropanation of oxygenated dienes, and the chemoselectivity in the carbenoid reaction of 1-substituted dienes in which the least hindered double bond is attacked, it is possible to develop a convenient route to pseudoguaiane precursors based on the cyclopropanation–Cope rearrangement protocol [Cantrell, 1991].

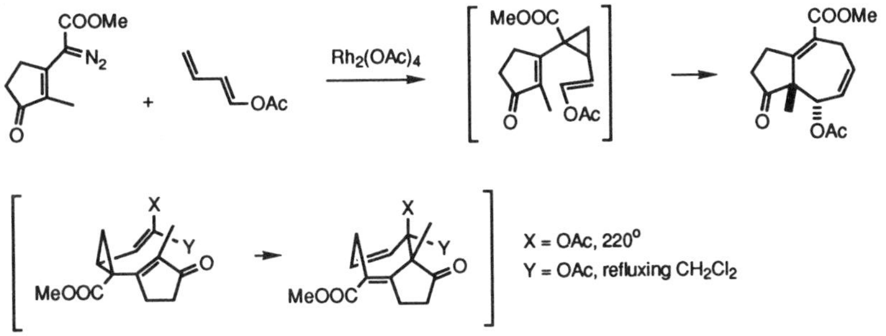

Formation of three new rings in one step was observed on treatment of an α-diazoketone containing well-juxtaposed alkyne and diene subunits with rhodium(II) acetate [Padwa, 1991]. The last reaction is a Cope rearrangement.

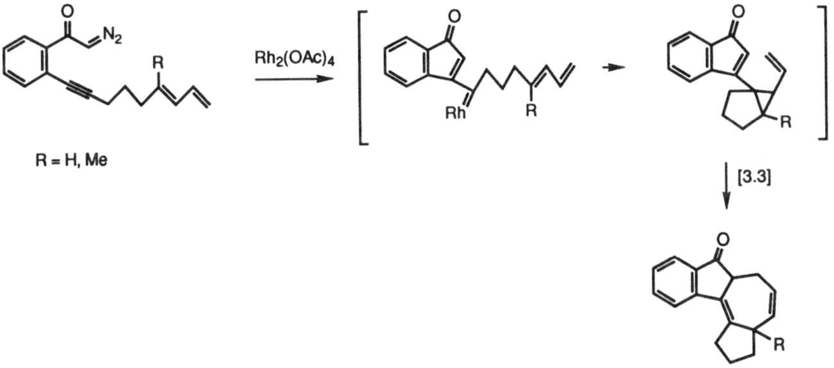

The Cope rearrangement occurs via a boatlike transition state, and therefore the rate is greatly dependent on the geometry of the double bond.

An apparent intention of achieving an intramolecular Diels–Alder reaction between a diene and a propargyl ether was thwarted [Hayakawa, 1986b]. Instead, isomerization of the triple bond to an allene induced a [2 + 2]-cycloaddition which generated a 1,5-diene with a stereochemical predisposition for the Cope rearrangement.

A thermally induced tandem reaction involving electrocyclic opening of a cyclobutenone, [2 + 2]-cycloaddition of the ketene with a conjugate diene, and subsequent Cope rearrangement [Danheiser, 1982] is a valuable method for synthesis of cyclooctadienones.

The thermal instability of 3-ethoxypropenylidenecyclopropane is due to its dimerization ([2 + 2]cycloaddition); the dimer can undergo a tandem Cope rearrangement at appropriate temperatures [Kienzle, 1991].

cis-9,10-Dihydronaphthalene is found in the pyrolysate of snoutene [Paquette, 1971]. The deannulation can be rationalized in terms of a mechanism involving retro-Diels–Alder reaction, electrocycloreversion, Diels–Alder reaction, and Cope rearrangement.

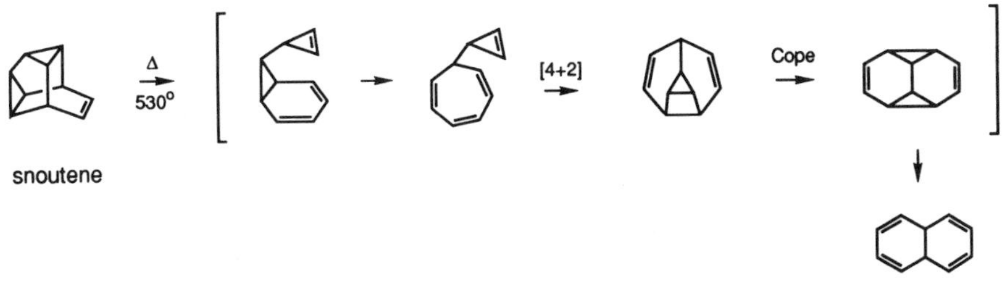

snoutene

The [4 + 2]-cycloadduct from tropone and cyclopentadiene did not arise directly as originally thought. Confirmation of the mechanism involving a Cope rearrangement of the primary adduct came from an unambiguous synthesis of the latter compound by liberation from its tricarbonyliron complex and its thermal behavior [Franck-Neumann, 1977].

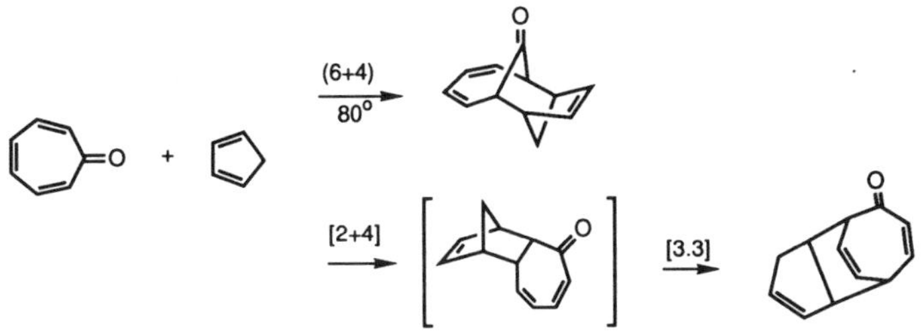

The cycloadducts of isobenzofulvenes with tropone are quite unstable, and they undergo a Cope rearrangement at room temperature [Paddon-Row, 1974; Tegmo-Larsson, 1978]. In the transition states the substituent(s) assumed quasi-axial conformations.

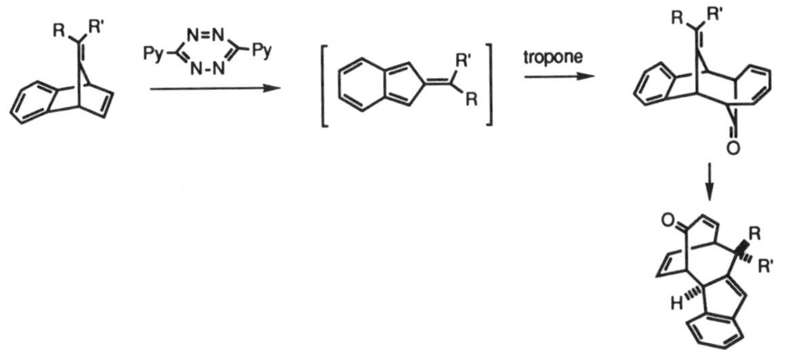

The spatial relationship between the exocyclic double bond and the diene moiety in the initial cycloadducts of the above reactions is the same as that in the formal adduct from the isobenzofulvene and benzene which has been acquired from decarbonylation of the bridged ketone [Warrener, 1978]. It is not surprising that the facile Cope rearrangement also followed immediately the generation of the hydrocarbon.

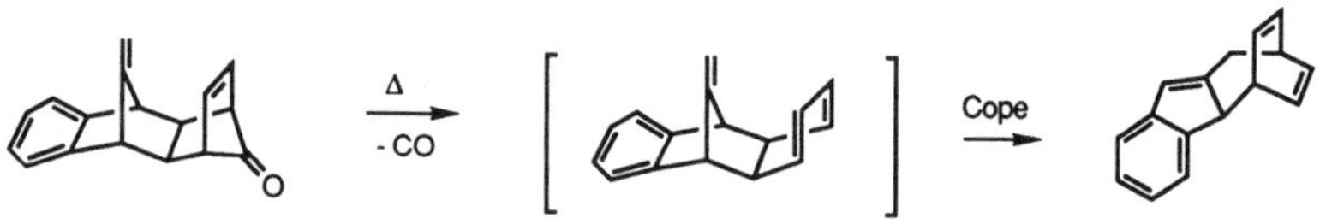

Cyclopentadiene forms a 1:2 adduct with cyclooctatetraene. In fact the dimer of cyclooctatetraene that reacts with cyclopentadiene is pentacyclic, containing a fused dihydrobullvalene substructure. It was considered that this dimer acted as a dienophile toward cyclopentadiene in the Diels–Alder reaction, but the adduct underwent a Cope rearrangement [Bratby, 1977].

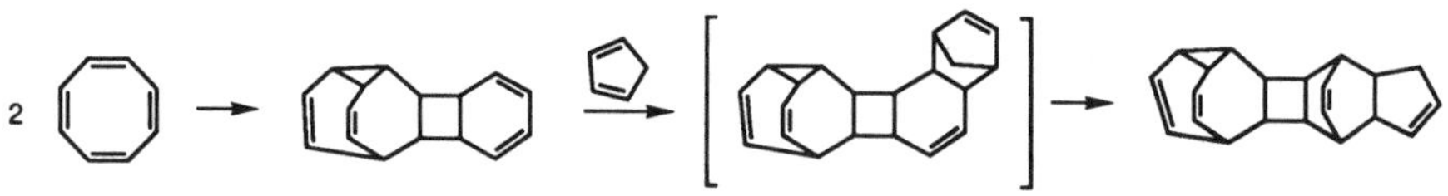

Five consecutive thermal reactions took place when a pentacyclic triene was heated with tetrachlorothiophene dioxide in chloroform [Gravett, 1991]. These reactions are: Diels–Alder reaction, cheletropic elimination, retro-Diels–Alder reaction, Diels–Alder reaction, and Cope rearrangement.

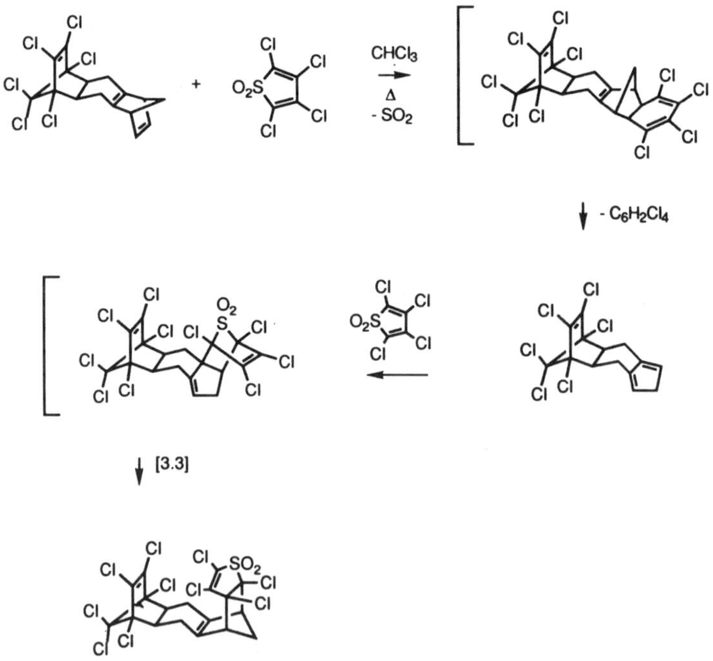

An intriguing series of reactions took place between dimethyl acetylene-dicarboxylate and cycloheptatriene [Goldstein, 1965]. The product arose from tandem ene reaction, electrocyclization, and Cope rearrangement.

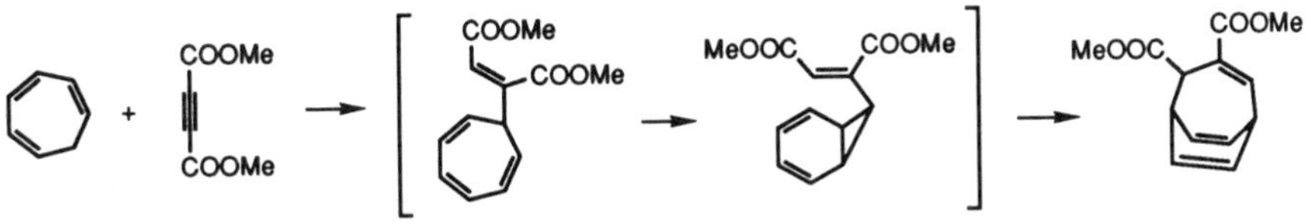

A reaction pathway comprising of ene reaction and Cope rearrangement appears to be more favorable than the intramolecular Diels–Alder reaction of a 14-membered cyclic dienyne [Y.-C. Xu, 1990].

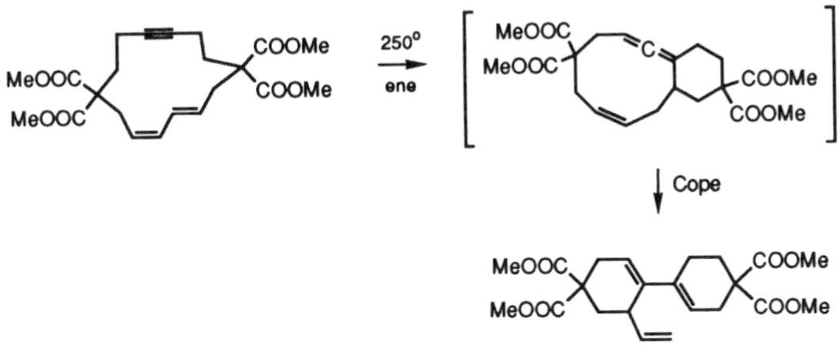

1,2,3,4,5-Pentamethyl-5-vinyl-1,3-cyclopentadiene is photoisomerized to the bicyclo[3.2.0]-hepta-2,6-diene derivative [U. Burger, 1984]. Direct irradiation at 254 nm effected electrocyclization with the resultant compound undergoing a rapid Cope rearrangement. On the other hand, sensitized photolysis (at 300 nm) was shown by deuterium labeling experiments to proceed via a di-π-methane rearrangement, leading to the same product.

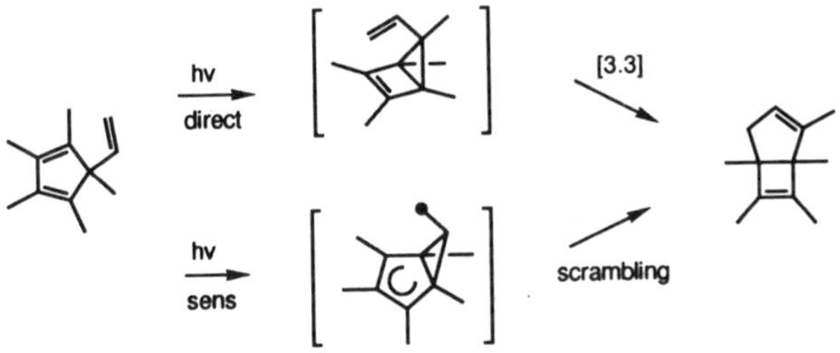

Although the structure of the thermolysis product of basketene suggests a simple cleavage of a cyclobutane, its formation actually involves the retro-Diels–Alder reaction–Cope rearrangement tandem [Westberg, 1969; Vedejs, 1971]. This reaction mechanism is also supported by a related study on the behavior of diazabasketene [McNeil, 1971]. In the latter case the Cope rearrangement product was converted into azacyclooctatetraene by two more reactions (i.e. retro-Diels–Alder reaction and electrocycloreversion).

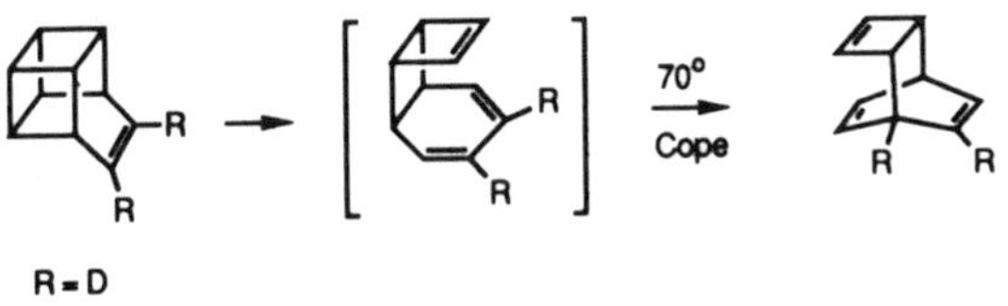

R = D

The reaction pathway leading to methyl 4-(3-pyridyl)butanoate by pyrolysis of cocaine [Novak, 1991] has been elucidated. After elimination of benzoic acid, the azabicyclic system breaks down via a retro-Diels–Alder reaction, imino-ene reaction, aza-Cope rearrangement, hydrogen shifts, and aromatization.

Among the reaction pathways for tropylideneketene in the presence of 2-dimethylaminotropone is a fascinating combination of four pericyclic reactions with a posterior elimination of dimethylamine [Morita, 1974]. These pericyclic reactions are [2 + 2]-cycloaddition, [1.7]-oxy shift, electrocyclization, and Cope rearrangement.

12.2.1.2. *Cope Rearrangement Induced by Other Reactions*

The union of a vinyl group with an allyl residue is expected to induce Cope rearrangement of the product on thermolysis. When structural features are introduced to the molecule such that the rearrangement is highly favored, the initial coupling product might not be isolable. It is thus possible to take advantage of the tandem process to accomplish synthetic goals in a more efficient manner. *cis*-Divinylcyclopropanes are species that undergo Cope rearrangement without much activation energies, thus 1,4-cycloheptadienes may be prepared in one step by generating such cyclopropanes [Marino, 1976; Piers, 1978].

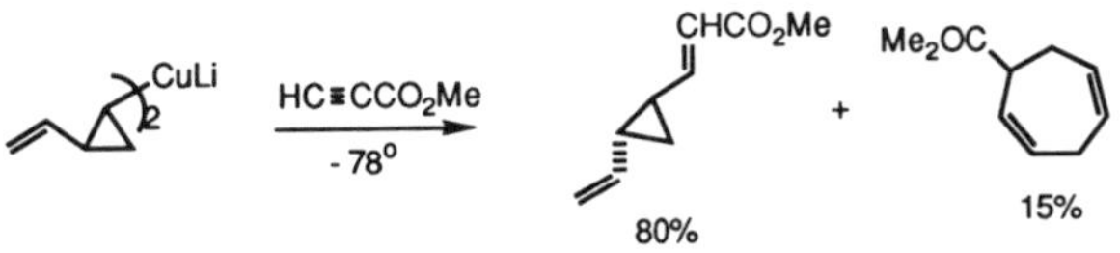

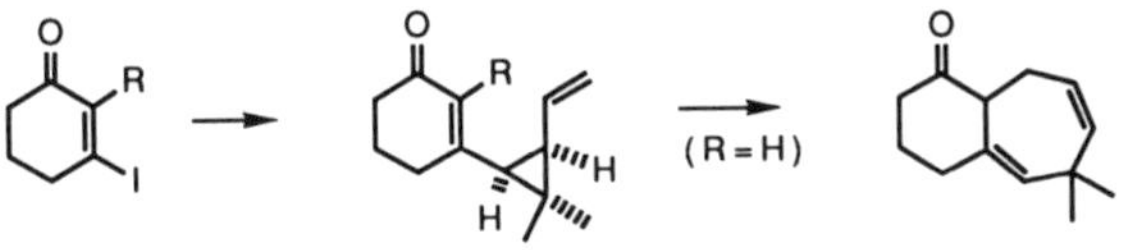

Since the coupling reaction gives rise to both *cis* and *trans*-divinylcyclopropanes, it is expedient to heat the reaction product mixture directly so as to achieve isomerization and rearrangement of the *trans* isomer.

A method that guarantees exclusive formation of a *cis*-divinylcyclopropane derivative, most easily adapted to fused ring systems, is the halogen replacement of a *cis*-2-vinylcyclopropyl bromide with configuration retention by an alkenyl residue. The Cope rearrangement product can be obtained in high yield under milder conditions [Wender, 1980a].

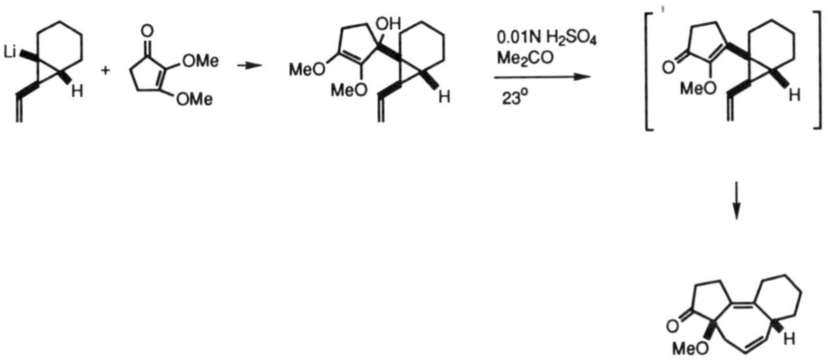

A Cope rearrangement induced by enol ether formation was the basis of a model synthesis of phorbol [Wender, 1988a].

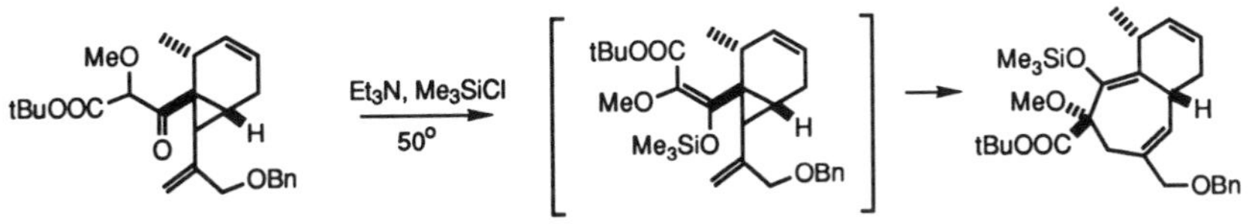

The stereoselective formation of *cis*-disubstituted cyclopropanes by an addition–trapping tandem on cyclopropenone ketals can be extended to the synthesis of 4,5-disubstituted 2,6-cycloheptadienones [Nakamura, 1988]. Thus, addition with an alkenylcuprate reagent and trapping by an iodoalkene in the presence of a Pd(O) catalyst results in a *cis*-dialkenylcyclopropane which is susceptible to Cope rearrangement.

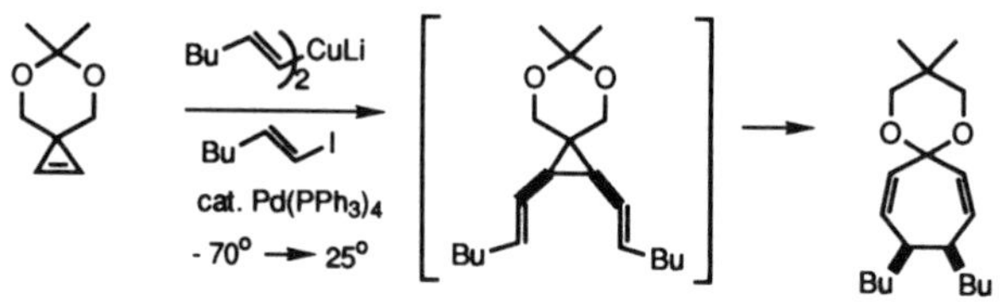

Another method for the construction of 1,4-cycloheptadienes is by intramolecular addition of a vinylic carbene to a conjugate diene, the adduct being disposed to rearrange [H.M.L. Davies, 1988].

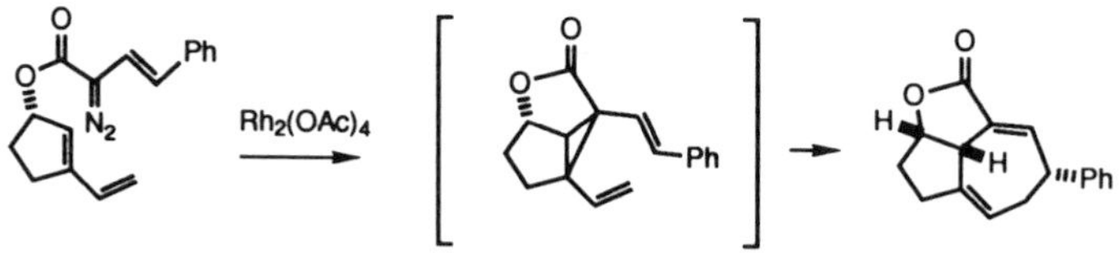

Rhodium(II) acetate-catalyzed decomposition of vinyldiazomethanes in the presence of *N*-(alkoxycarbonyl)pyrroles leads to 8-azabicyclo[3.2.1]octa-2,6-dienes [H.W. Davies, 1991]. This very efficient synthesis of the tropane skeleton has been adopted to an elaboration of ferruginine.

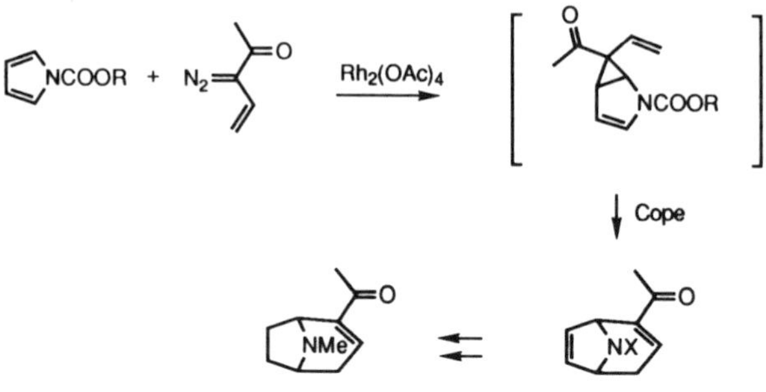

ferruginine

A desulfurative dimerization of a conjugated sulfoxonium ylide on heating [Tamura, 1973] was initiated by a [2.3]-sigmatropic rearrangement. The very reactive Michael acceptor generated by elimination was trapped by another ylide. Cyclopropanation and Cope rearrangement terminated the reaction.

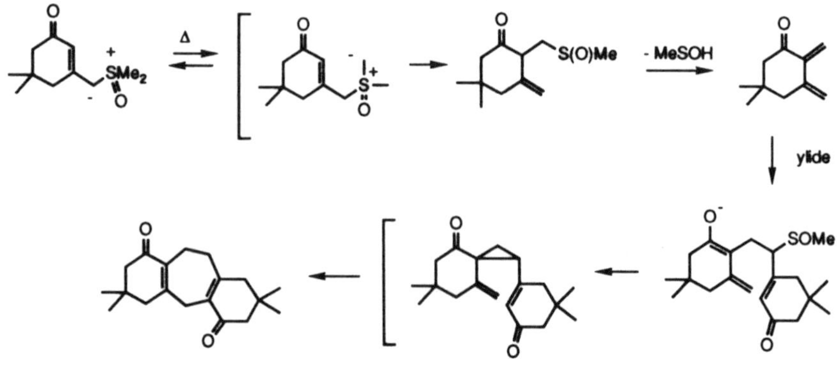

As anticipated, a locked *cis*-divinylcyclopropane underwent immediate Cope rearrangement upon its generation by dehydrobromination [Warner, 1983].

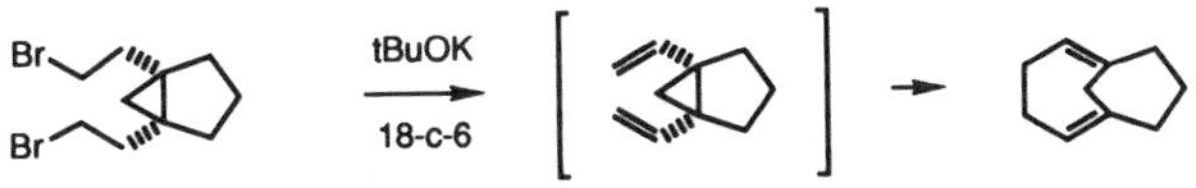

The spatial proximity of the two double bonds of (*E,E*)-1,5-cyclodecadienes is the cause of the heightened reactivities related to such molelcules. In the absence of external reagents the Cope rearrangement is a strain-relieving process for them.

The conversion of germacrane sesquiterpenes to the elemanes is well documented. It is therefore not surprising that the only isolable compounds from intramolecular coupling of (*E,E*)-1,10-dibromo-2,8-decadienes are 1,2-divinylcyclohexane derivatives [Corey, 1969a].

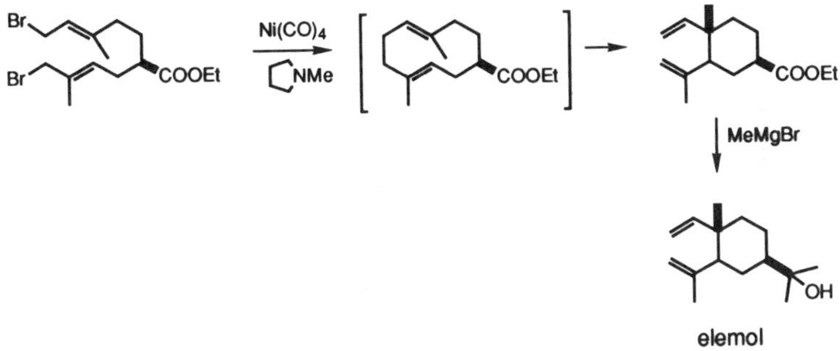

elemol

Hofmann or Cope elimination of 1,6-bisdimethylaminocyclodecane produces substantial amounts of *trans*-1,2-divinylcyclohexane in addition to an isomeric mixture of 1,6-cyclo-decadienes [Grob, 1958]. Apparently, (*E,E*)-1,5-cyclodecadiene is the origin of the divinylcyclohexane.

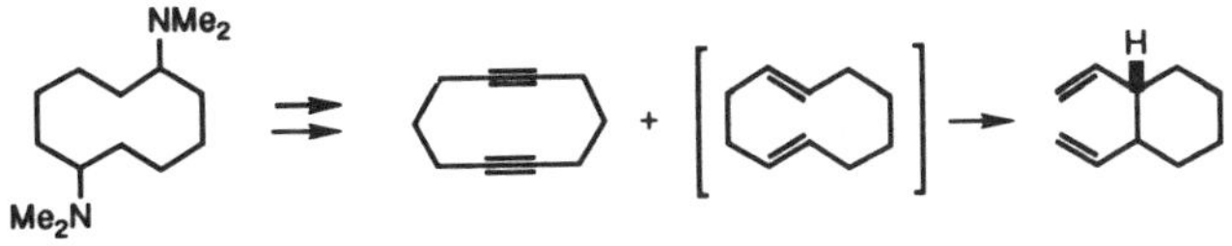

Related to the above reactions is the solvolytic behavior of γ-amino sulfonates which are incorporated in a decahydroquinoline system which yields *N*-alkyl-2-vinylcyclohexylamines. Interestingly, the products actually arise from fragmentation of the intercyclic CC bond but the intermediates undergo cationic aza-Cope rearrangement [Grob, 1967; Marshall, 1969]. Such tandem reactions have been applied to a synthesis of isochanoclavine-I [Kiguchi, 1989].

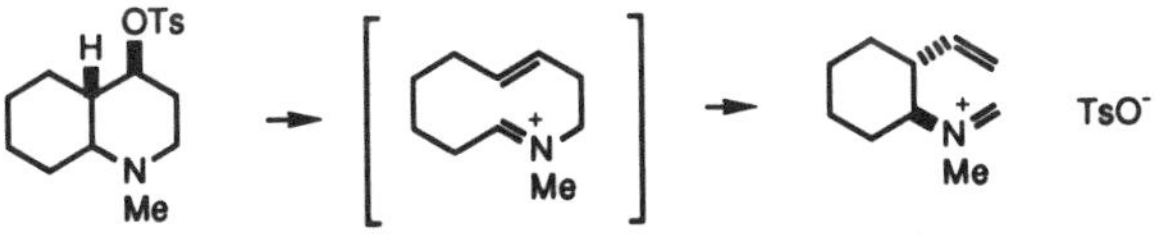

The deoxygenative coupling of dicarbonyl compounds with low-valent titanium reagents has a very broad scope. However, an attempt at synthesizing the orthogonal diene from a hydroheptalenedione led only to a dimethylene-[3.3.2]propellane [McMurry, 1987]. A spontaneous Cope rearrangement of the expected product is implicated.

The driving force for the facile [3.3]-sigmatropic rearrangement of the keto ester derivative of *endo*-dicyclopentadiene upon reduction of the conjugated ketone group [Suri, 1988] is the placement of the ester group into a conjugate system.

The sole product from pyrolysis of bicyclo[5.2.2]undec-1(9)-en-7-yl acetate is 2,8-dimethylenebicyclo[5.2.0]nonane [Tobe, 1983], as a result of acetic acid elimination and a rapid [3.3]-sigmatropy. This behavior is in contrast with the generation multiple products from the lower and higher homologs, suggesting the unique favorableness of this particular bridgehead diene to undergo Cope rearrangement.

An anionic oxy-Cope rearrangement must be responsible for the production of a bicyclo[5.2.1]decenedione from a vinylpentalenolone [Tice, 1981]. While steric constraints prevented a direct rearrangement, configurational inversion at the carbinol center by a retro-aldol–aldol reaction sequence followed by a rapid oxy-Cope rearrangement could channel all the material into the observed product. Interestingly, a different reaction pathway prevailed when the substrate was exposed to potassium hydride in an aprotic solvent.

The occurrence of a Cope rearrangement from the enol form of a methyl ketone [Bessiere-Chretien, 1973] was rather surprising, although the process broke the four-membered ring and thereby transforming the molecule into a less strained one.

At least part of the product from the Knoevenagel reaction of 5-norbornene-2-*endo*-carboxaldehyde is a tetrahydroindenecarboxylic acid [Miyano, 1982], due to the intervention of a Cope rearrangement.

It is interesting to note that the elimination *vic*-hydroxysilane by base treatment can be coupled to an anionic oxy-Cope rearrangement [Hudrlik, 1981].

Perhaps the enone generated from an α-cyanohydrin trimethylsilyl ether suffered unfavorable electronic interactions with the methoxy substituent at an adjacent carbon atom; a succeeding Cope rearrangement and subsequent elimination of methanol remedied the situation [Evans, 1976].

lapachol methyl ether

Thiophenol addition to alkylidenecyclopropanes places the heteroatom at the central carbon of the allene linkage. When a *cis*-vinyl group is present the vinyl sulfide will participate in a Cope rearrangement immediately to give a 1,4-cycloheptadiene derivative [P.M. Cairns, 1982].

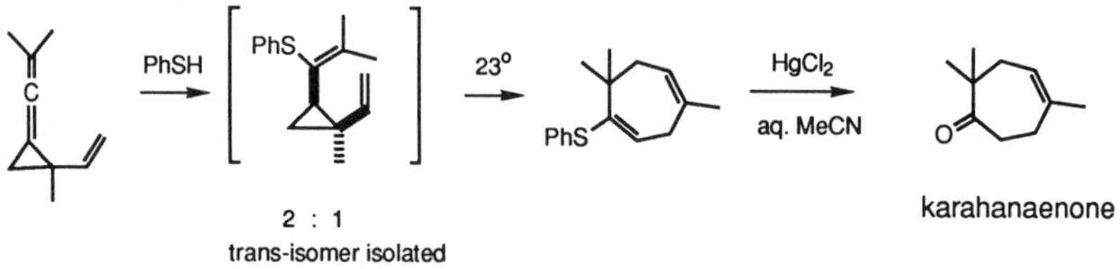

2 : 1
trans-isomer isolated

karahanaenone

A convenient method for the conversion of allyl azides into α-allylated nitriles consists of Staudinger reaction and treatment of iminophosphoranes directly with ketenes [Molina, 1991]. The last step involves an aza-Wittig reaction and a very special aza-Cope rearrangement.

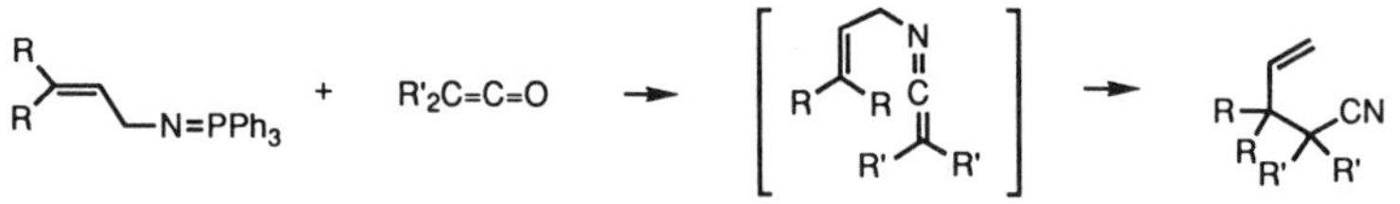

Ring expansion of 2-vinylazacyclic compounds has been achieved when they were exposed to ketenes [Edstrom, 1991]. *N*-Acylation and aza-Cope rearrangement of the zwitterionic adducts were involved.

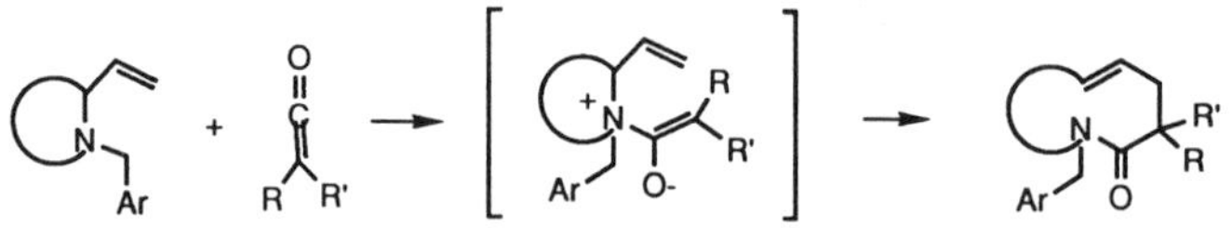

Zwitterions of another type, formed by *N*-alkylation of 2-azabicyclo[2.2.2]-oct-5-enes with propynoic esters, are susceptible to the aza-Cope rearrangement [E.W. Baxter, 1989]. This reaction sequence provides an excellent entry into the hydroisoquinoline ring system, and its potential for the synthesis of several yohimbe alkaloids is evident.

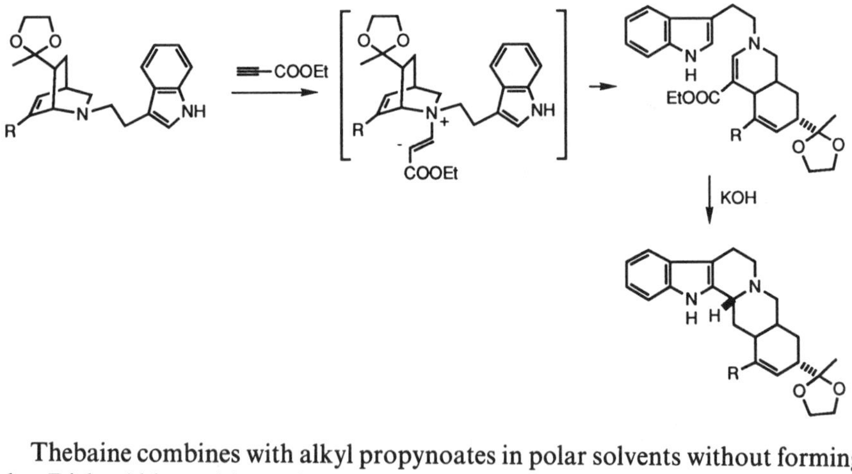

Thebaine combines with alkyl propynoates in polar solvents without forming the Diels–Alder adducts in any appreciable quantities. Formation of the products is accompanied by the destruction of the piperidine ring [K. Hayakawa,

1981] which can be rationalized in terms of an analogous mechanism as the above reaction.

thebaine

There is strong evidence for a tandem mechanism for the fluoride ion-promoted intramolecular Michael addition to form an eight-membered ring [Majetich, 1988]. The *vis-cis* divinylcyclobutane intermediates generated by the desilyative Michael reaction undergo an enolate-accelerated Cope rearrangement.

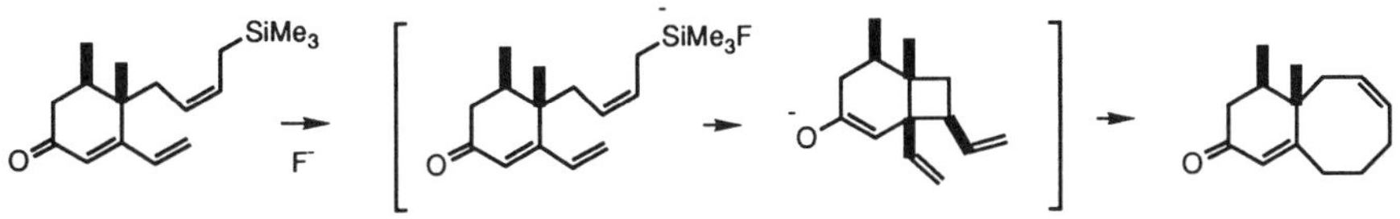

Pyrolysis of 3-thiabicyclo[3.2.0]heptane *S,S*-dioxide generates 1,5-hexadienes with the extrusion of sulfur dioxide. When the primary products carry conjugationable substituents at the allylic positions a Cope rearrangement would be favored [J.R. Williams, 1981].

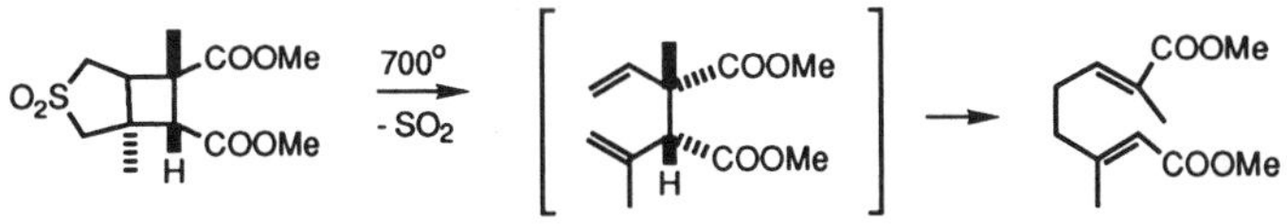

δ,ε-Unsaturated ketenes are a special kind of 1,5-dienes, and it is natural that in the absence of nucleophiles these substances would undergo a Cope rearrangement. Photochemical generation of such ketenes indeed is followed by the expected change [Freeman, 1965; Chapman, 1968].

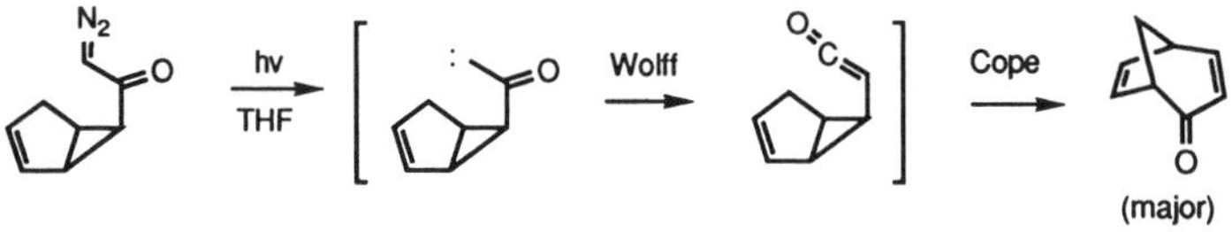

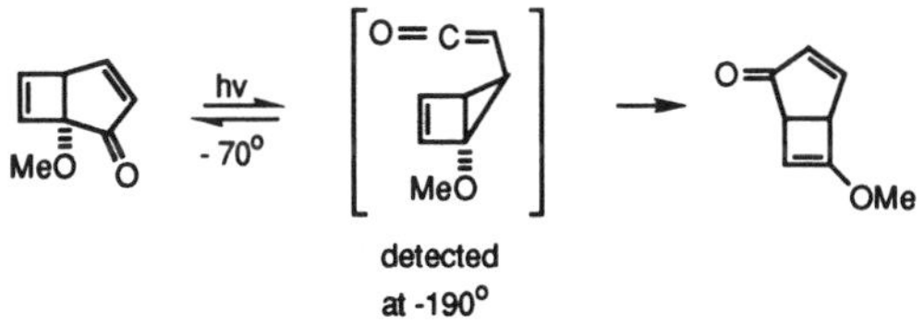

(major)

detected
at -190°

The analogous reaction is also known for δ,ε-unsaturated isocyanates [Vögtle, 1965].

A useful method for the synthesis of certain cycloheptadienes entails the reaction of dienynes with Fischer carbene complexes [Harvey, 1991b]. The mechanism may consist of metallacyclobutenes, electrocyclic opening to generate new metal carbene species which add to one of the double bonds, and a Cope rearrangement.

12.2.2. Claisen and Other Hetero-Cope Rearrangements

12.2.2.1. Claisen Rearrangements from Atypically Generated Substrates

The scope of the Claisen rearrangement has been greatly broadened as a result of development of methods to acquire its substrates besides the conventional procedure. Since such substrates are susceptible to the rearrangement thermally, it is very convenient to carry out the subsequent reaction without isolation of the often sensitive vinyl ethers.

A synthesis of phoracantholide-J [Petrzilka, 1978] via the Claisen rearrangement is significant because it also established the double bond in the correct position. The enol ether, generated by phenylselenoetherification followed by oxidative elimination of the selenyl substituent, underwent the rearrangement in situ. Apparently very mild conditions were required for the rearrangement of the keteneacetal.

phoracantholide-J

Another tandem thermal elimination–Claisen rearrangement process was employed in the elaboration of the eight-membered ring of a potential precursor of acetoxycrenulide [Ezquerra, 1990].

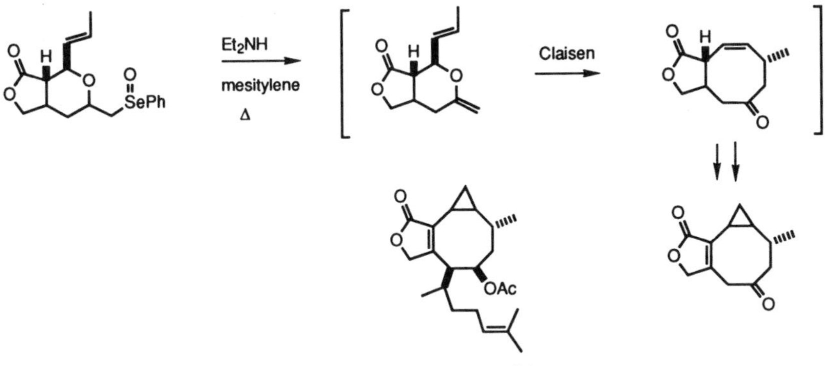

acetoxycrenulide

(In connection with these reactions is the facile [3.3]-sigmatropic ring expansion of cyclic thiocarbonates [Harusawa, 1991]. The ring size of the cyclic thiocarbonate determines the double bond geometry of the product.)

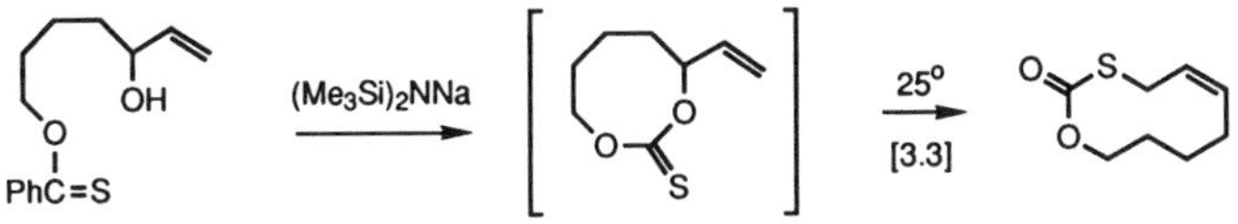

Dehydrohalogenation of 2-(1-bromoalkyl)-5-vinyltetrahydrofurans at high temperatures gives 4-cycloheptenones [Demole, 1969]. Many natural products containing a seven-membered ring have been synthesized with the aid of this process. It should be mentioned that the reaction profile can be changed by transition metal catalysts [Trost, 1981], and substrates such as these are directed to [1.3]-rearrangement to give 3-vinyl cyclopentanones.

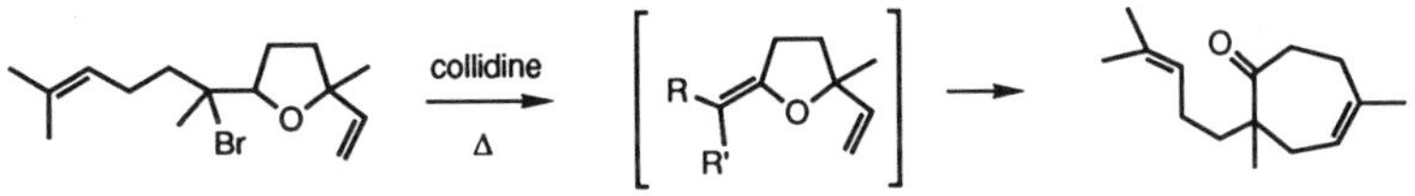

The minor product from reaction of 4-methoxy-5-(p-methoxyphenoxy)-o-benzoquinone with 1-octen-3-ol is the symmetrical bisether. This compound is prone to rearrangement at room temperature [Reinaud, 1988].

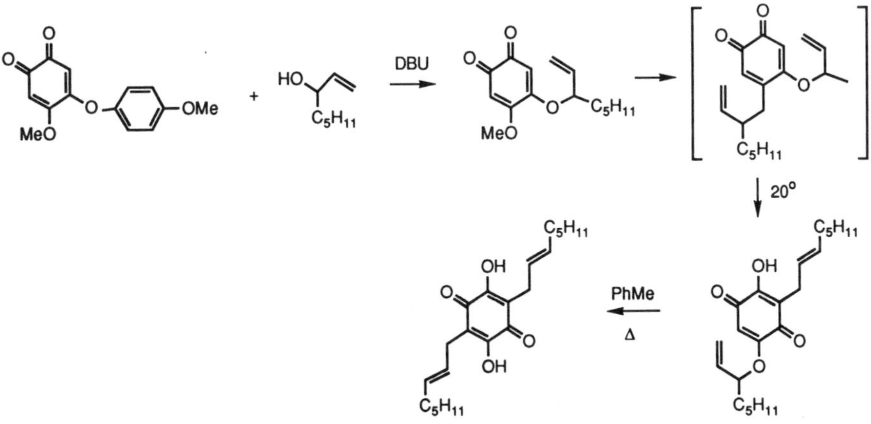

The more general alkoxy exchange is limited to ortho esters, and the Claisen rearrangement products are necessarily in the higher oxidation state. However, the protocol is very popular because of its generality and simplicity, without sacrificing stereocontrol. For example, it proved superior to other methods for stereoselective crotylation of a γ-lactone that constituted an important step of a compactin synthesis [Kozikowski, 1987].

compactin

Generation of allyl vinyl ether from an intramolecular addition of an allyl alcohol to an alkyne linkage could lead to the Claisen rearrangement product directly [Marvell, 1986].

A tricyclic indole derivative was formed by an analogous reaction of a phenylhydroxylamine [Coates, 1979].

The abnormal Nef reaction of nitronobornenes has been explained in terms of heterolysis of the bicyclic system [Ranganathan, 1976]. However, this author considers that the [3.3]-sigmatropic pathway is equally convincing.

Transient adducts are formed from 2-allyl cyclic ethers and dichloroketene. As these zwitterionic species are set up for a Claisen rearrangement [Malherbe,

1978], ring-enlarging lactone formation is readily achieved. A logical application of this tandem reaction is a synthesis of phoracantholide-J from 2-methyl-6-vinyltetrahydropyran.

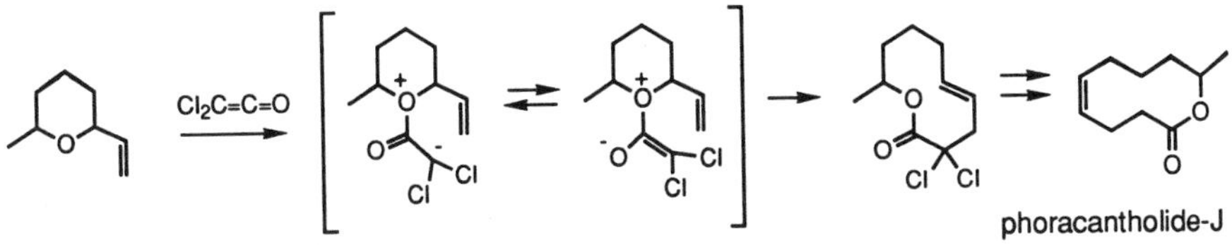

A Michael addition-induced Claisen rearrangement of the allyl ether of 1,2-cyclohexanedione [Koreeda, 1985] proceeded under exceptionally mild conditions. Furthermore, the group just introduced served as an effective regulator of the steric course.

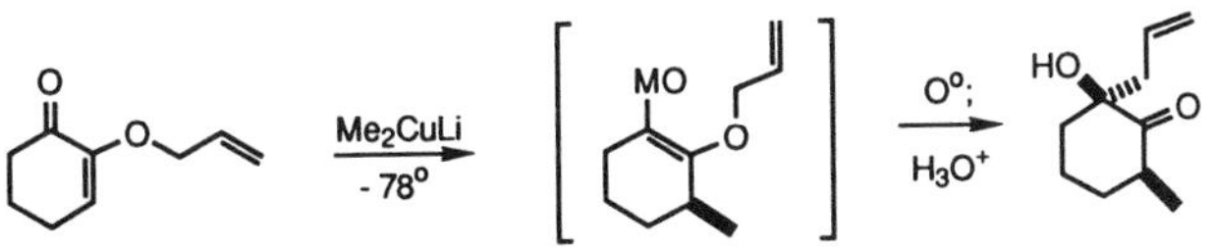

The aluminum chloride-catalyzed reaction of allylic sulfides with methyl propiolate begins with a Michael reaction and ends by a [3.3]-sigmatropic rearrangement [Hayakawa, 1982].

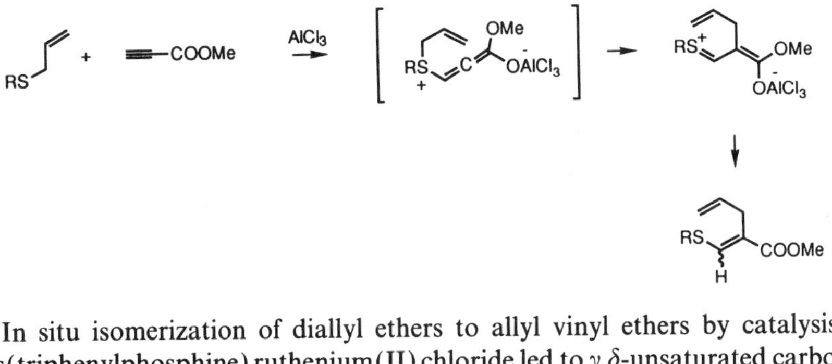

In situ isomerization of diallyl ethers to allyl vinyl ethers by catalysis of tris(triphenylphosphine)ruthenium(II) chloride led to γ,δ-unsaturated carbonyl compounds [Reuter, 1977]. The reaction is the basis of a very simple synthesis of dihydrojasmone [Tsuji, 1979] and the acylic precursor of a steroid CD-ring synthon [Stork, 1982a].

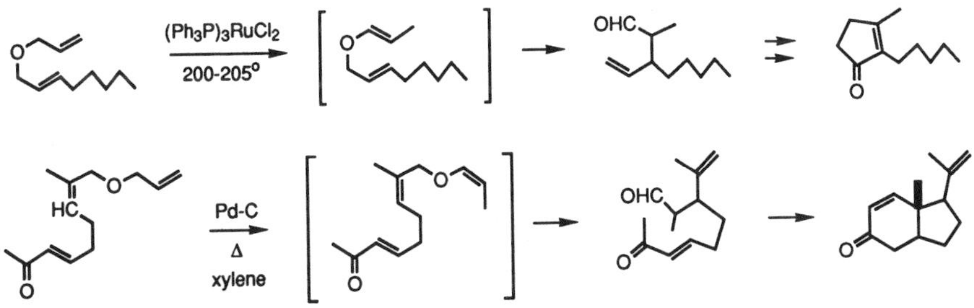

The conversion of dihydropyranyl ketimines to the aminocyclohexenes [Lipkowitz, 1979] must be due to isomerization of the imines to the thermally active enamines.

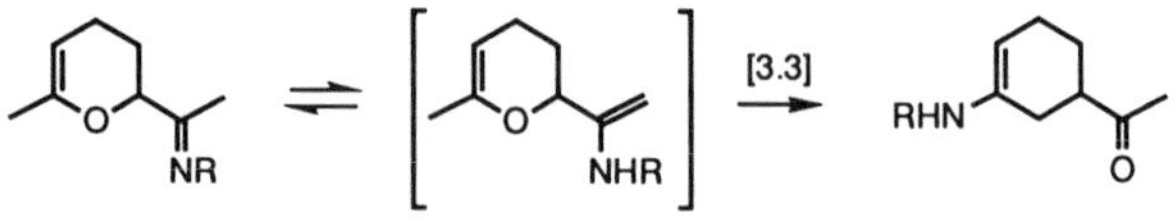

Epoxidation of norbornadiene also causes a rearrangement via π-participation [Meinwald, 1963]. The bicyclo[3.1.0]hexenealdehyde is sterically predisposed to undergoing further transformation (Claisen rearrangement) to give a bridged vinyl ether. A prostacyclin precursor was obtained when the properly substituted norbornadiene was epoxidized [A.D. Baxter, 1983].

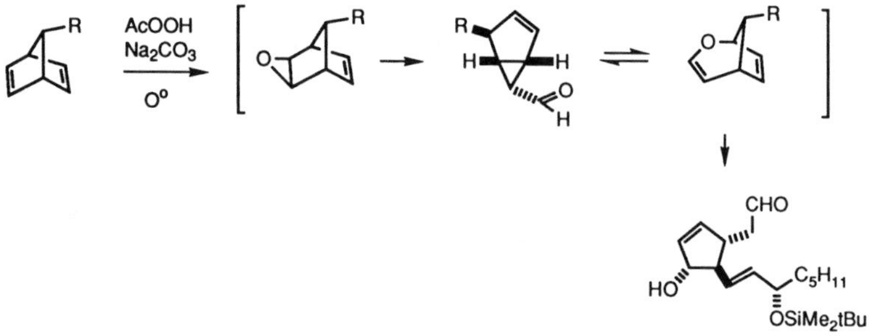

prostacyclin precuror

In the course of a synthesis of (+)-ophiobolin-C [Rowley, 1989], establishment of the sidechain in the C-ring was initiated by a tandem Brook and Claisen rearrangements. Thus, transformation of the α-silyl ester into a silyl keteneacetal triggered the stereospecific CC bond formation.

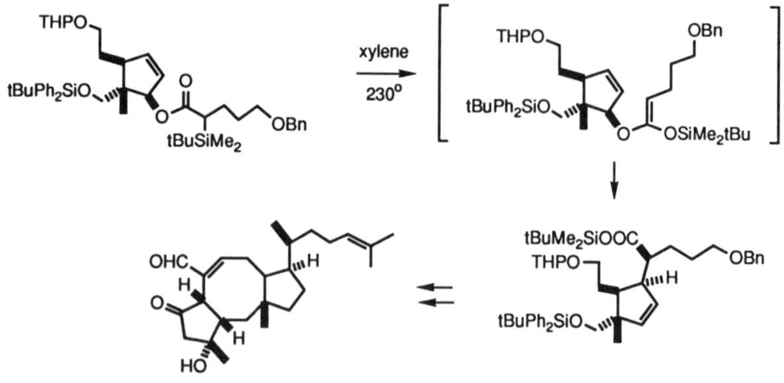

(+)-ophiobolin-C

The reaction of methyl 6-oxo-5-phenyl-1,3,4-oxadiazine-2-carboxylate with 2,3-dimethyl-1,3-butadiene is very complicated as it consists of more than four steps [Hegmann, 1988]. Between the Diels–Alder cycloaddition and the Claisen rearrangement are a retro-Diels–Alder and retro-Michael reactions. The retro-Diels–Alder reaction led to a ketene which was trapped by the double bond. Participation of the α-keto ester subunit at this stage pushed the zwitterionic intermediate toward the retro-Michael fission–Claisen rearrangement cascade.

12.2.2.2. Tandem Sigmatropy Ended by a Claisen Rearrangement

Contrary to the conventional Cope rearrangement which is reversible, the Claisen rearrangement does not suffer from such a defect as a useful synthetic reaction. Since both rearrangements are inducible by heat, the Cope rearrangement product can be channeled away by a tandem Claisen rearrangement. Most frequently the Claisen rearrangement requires a lower energy of activation; therefore intermediates from the tandem Cope–Claisen rearrangements cannot be isolated.

A Cope–Claisen rearrangement tandem transformed cis-2-ethynyl-3-vinyl-oxirane into cis-2-ethynylcyclopropylcarboxaldehyde [Manisse, 1977]. Interestingly, the corresponding aziridine gave rise to the azepine, prototropic shift was more favorable than the aza-Cope rearrangement.

A detailed study of the thermal behavior of the isomers of a 2,3-divinylcyclohexyl vinyl ether [Ziegler, 1982b] indicated that the Cope–Claisen rearrangement product is more accessible from the cis-divinyl isomers. Under certain conditions the trans isomer underwent equilibration without entering the irreversible pathway. This particular process is characterized by the unusual property of having $\Delta G^{\ddagger}$ (Claisen) $\gg \Delta G^{\ddagger}$ (Cope).

With careful design the tandem Cope–Claisen rearrangements are a powerful tool for building molecules with three contiguous asymmetric centers in one step. For example, incorporation of the diene and vinyl ether substructures in the periphery of a five-membered ring enabled the preparation of intermediates for synthesis of pseudoguaianolides. A high stereoselectivity manifested the preference for a chairlike transition state in the Cope rearrangement and the occurrence of the Claisen rearrangement on the face further removed from the newly established allyl sidechain.

R = H, Me

78   22
major   minor

Further exploitation of the tandem reactions resulted in a synthesis of estrone methyl ether [Ziegler, 1982a]. The stereoselectivity is lower in this case.

(2 : 1)

estrone methyl ether

Another outstanding application of the same process is in a synthesis of heliangolide and dihydrocostunolide [Raucher, 1986]. The $\Delta G^{\ddagger}$ for both rearrangement steps are about the same. The substrate was elaborated from carvone.

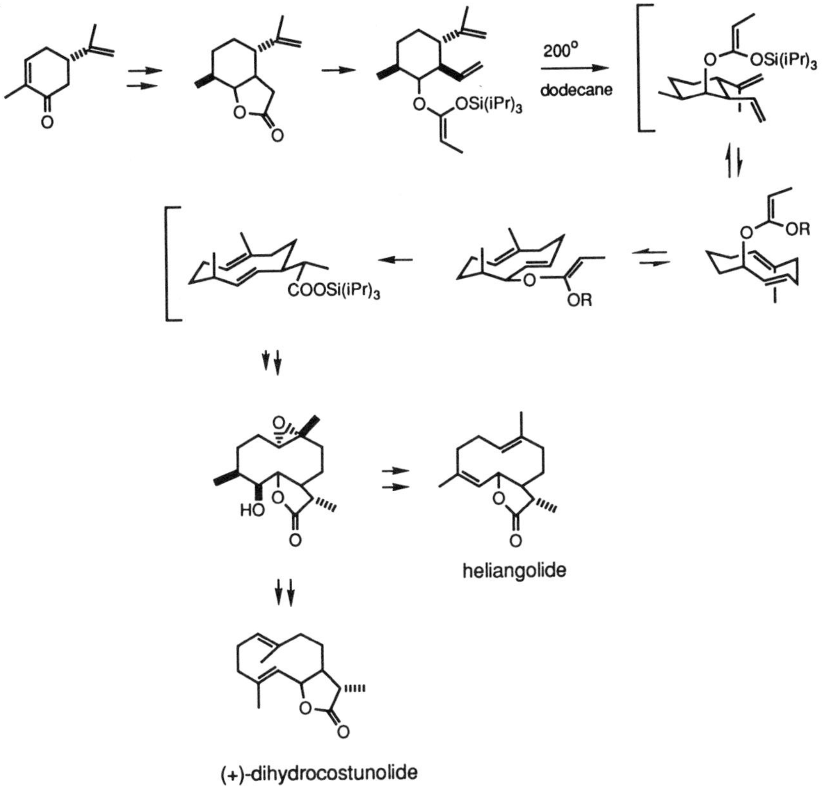

The condensation of oxepin with dimethyl 3,4-diphenylcyclopentadienone-2,5-dicarboxylate led to both [6 + 4]- and [4 + 2]-cycloadducts [T. Ban, 1980]. While the former compound was found to be quite stable, the Diels–Alder adduct underwent a sequence of Cope–Claisen rearrangements in refluxing benzene.

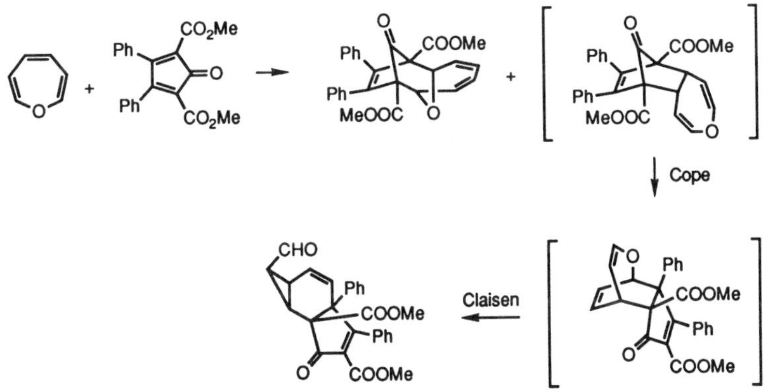

The noncatalyzed Fisher indole cyclization of a system from which aromatization was blocked by a methyl and a dimethylallyl group gave

N-dimethylallylskatole in 9% yield [Baldwin, 1977]. Formaation of this indole derivative is readily rationalized by invoking the conventional mechanism with deviation at the end (aza-Cope rearrangement instead of prototropic shift).

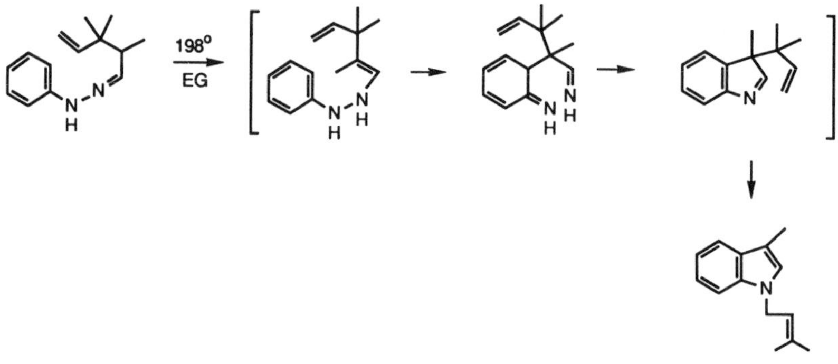

By virtue of hetero-Claisen rearrangements, dihydrobenzothiophenes and indoles have been obtained [Majumdar, 1972; Thyagarajan, 1974]. The substrates for the rearrangements were the aryl propargyl sulfoxides and propargylaniline *N*-oxides, respectively. A [2.3]-sigmatropy transformed these compounds into allenyl ethers which are susceptible to reaction analogous to the aromatic Claisen rearrangement.

The rearrangement of the *N*-oxides are very facile. Oxidation of the anilines with a peracid at room temperature delivered the indoles directly.

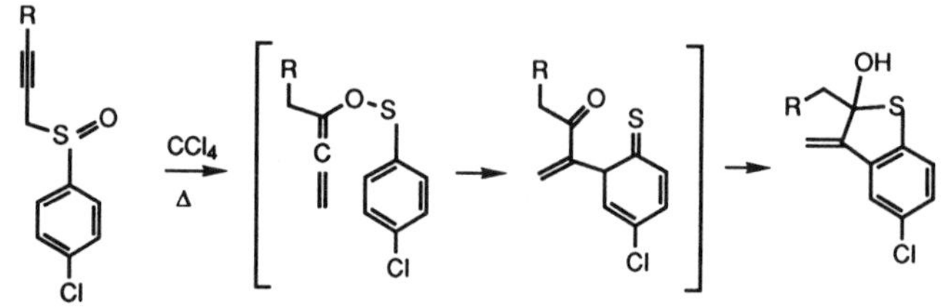

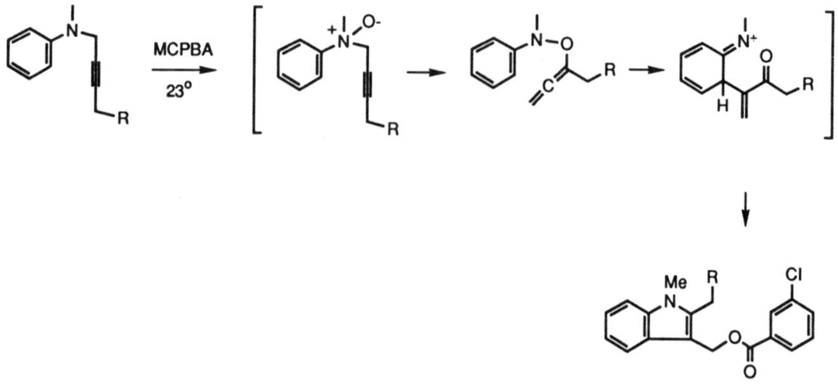

As mentioned previously, a small ring, particularly three-membered, intervening between an alkyl group and an unsaturation, is capable of rendering the molecular segment active toward [1.5]-sigmatropic shift of a hydrogen atom from the alkyl substituent to the unsaturated moiety. A sophisticated application of this reaction is that of a molecule containing a 2-ethynyl-3-methyloxirane unit [Karpf, 1977], the sigmatropic product being an allenyl vinyl ether. The thermal reaction resulted in the generation of a γ,δ-alkynyl ketone.

A pericyclic reaction sequence of some synthetic importance is the thermal conversion of allyl benzocyclobutene-1-carboxylates into 4-allylisochroman-3-ones. Because of the torquoselectivity in the electrocyclic opening of the four-membered ring to form (E)-o-quinodimethanes, the α-position of the ester carbonyl group must be fully substituted. The isomers which may participate in a six-electron electrocyclization would undergo ring formation, and the resulting species are perfect substrates for the Claisen rearrangement.

Synthesis of calabar bean alkaloids such as physovenine, physostigmine [Shishido, 1986] and geneserine [Shishido, 1987], and the more complex aspidosperma alkaloid vindoline [Shishido, 1989c] has taken full advantage of the structural attributes of these tandem pericyclic reaction products.

physostigmine

geneserine

A Claisen rearrangement following the event of electrocyclic opening–closure and aromatization [Perri, 1990] delivered the allyl hydroquinone. Isolation of the p-hydroquinone from the reaction is due to in situ autoxidation.

The Diels–Alder adducts of N-acylnitroso compounds are susceptible to [3.3]-sigmatropic rearrangement and this behavior has been exploited in a synthesis of cis-1,2-diols [Backenstrass, 1990].

12.2.3. [2.3]-Sigmatropic Rearrangements

Besides the above sigmatropic rearrangements, those belonging to the [2.3] order are most thoroughly investigated and they have the best potential of synthetic application. The [2.3]-sigmatropic rearrangements are isoelectronic with the [3.3] counterparts, although they involve five centers only. Both anionic and zwitterionic [2.3]-sigmatropic rearrangements are well represented.

Allylic oxidation of alkenes by selenium dioxide is a useful reaction which proceeds via tandem ene reaction/[2.3]-sigmatropy [Sharpless, 1972].

α'-Tosylamination of enol silyl ethers have been reported [Magnus, 1991]. The reaction involves reaction with TsN=Se=NTs and it proceeds via C—Se bond formation with double bond transposition (i.e. an ene reaction) and a [2.3]-sigmatropic rearrangement.

2,3-Disulfinyl and diphosphonyl 1,3-butadienes are potentially very useful building blocks for synthesis. They are readily available by reaction of 2-butyne-1,4-diol with sulfenyl chlorides [Okamura, 1982] or with phosphinyl chlorides, via two [2.3]-sigmatropic rearrangements of the diesters. Spontaneous rearrangement of 2-alkynyl esters of phosphorus-based acids to give allenyl-phosphoryl compounds is well documented [Mark, 1969].

The (Z) to (E) isomerization of the Δ^{13}-double bond which was required during a synthesis of the prostaglandins [J.G. Miller, 1974] was effected by treatment of the allylic alcohol with an arenesulfenyl chloride. By virtue of the reversible [2.3]-sigmatropic rearrangement of the sulfenate ester and the conformational mobility of the various intermediates the isomer with a double bond in the more stable configuration accumulated.

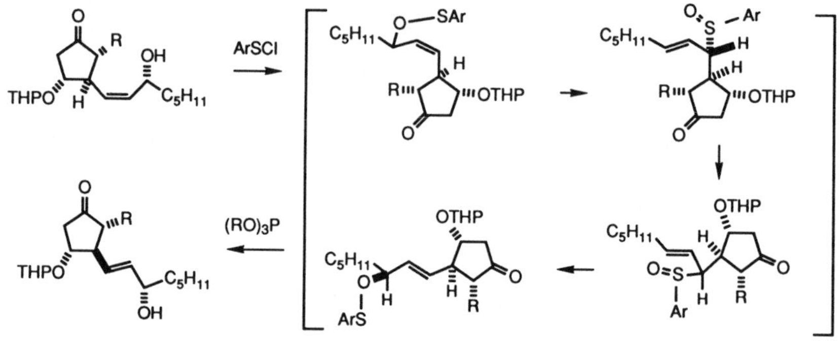

The interception of silylenes by allyl ethers is followed by immediate rearrangement of the ylides [D. Wang, 1984]. This reaction is the basis of a transpositional silylation method.

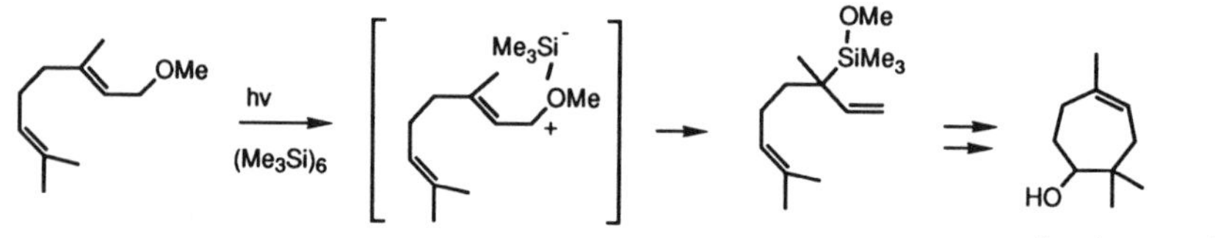

karahanaenol

The soft divalent sulfur atom is a good nucleophile toward soft acids such as alkyl halides. The intramolecular S-alkylation with an allylic halide may engender a [2.3]-sigmatropic rearrangement if the α-position is sufficiently activated (e.g. by a carbonyl group) [Vedejs, 1988b].

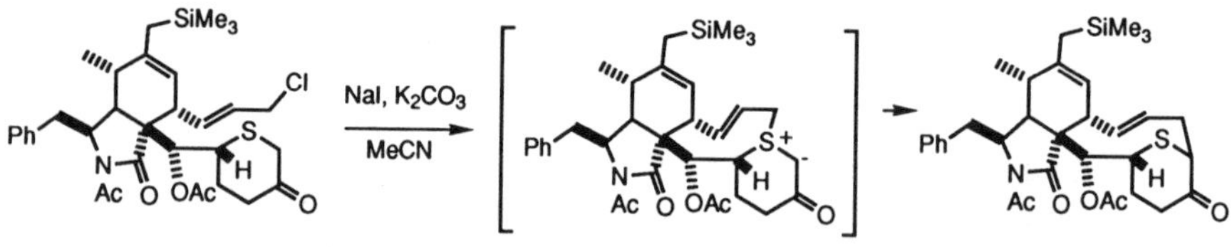

It is not surprising that a [2.3]-Sigmatropic rearrangement followed an intramolecular interception of a carbene by an allylic sulfide [Kondo, 1972; Kido, 1991].

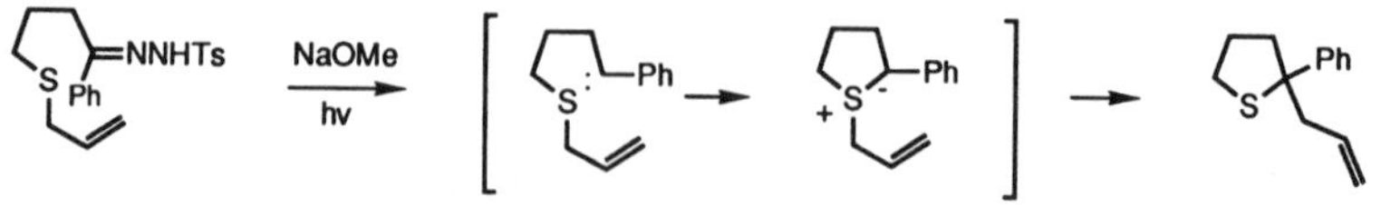

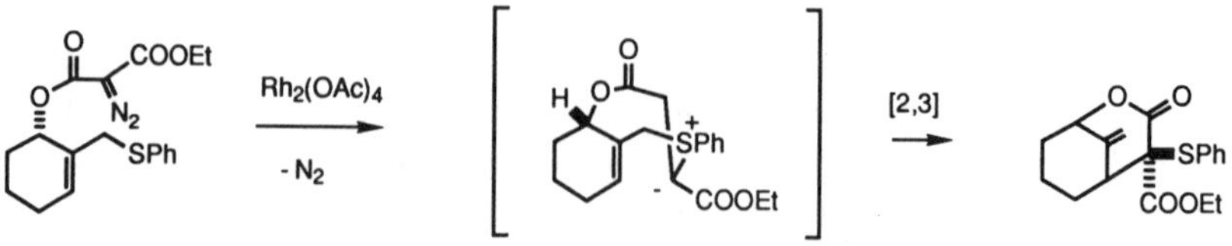

Carbanions of allylic thioacetals are susceptible to [2.3]-sigmatropic rearrangement. A unique version of this reaction involves an anion flanked by a tosylazo group, as it is accompanied by concurrent fragmentation [Evans, 1977].

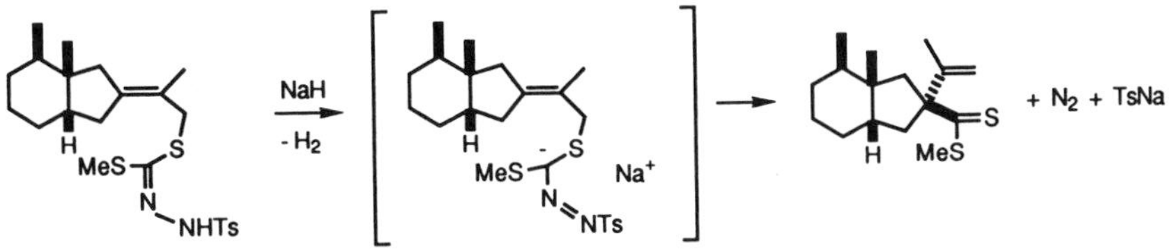

There is enormous synthetic potential of this rearrangement as a ring-enlarging protocol. Very interesting developments in the rearrangement of oxonium ylides [Pirrung, 1986] should be noted for the different types of precursors and the relationships of various reactive sites. Complementary strategies for synthesis of certain compounds may be considered on the basis of the two approaches.

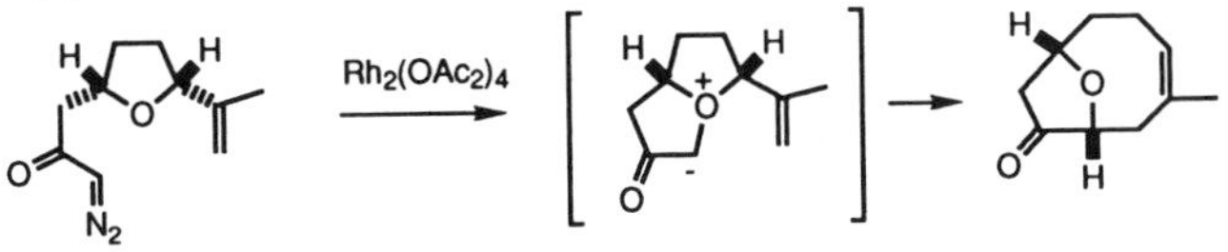

The deletion of the unsaturation in the oxonium ylides (i.e. their precursors) forces such species to undergo the Stevens rearrangement ([1.2]-sigmatropy) [Eberlein, 1991].

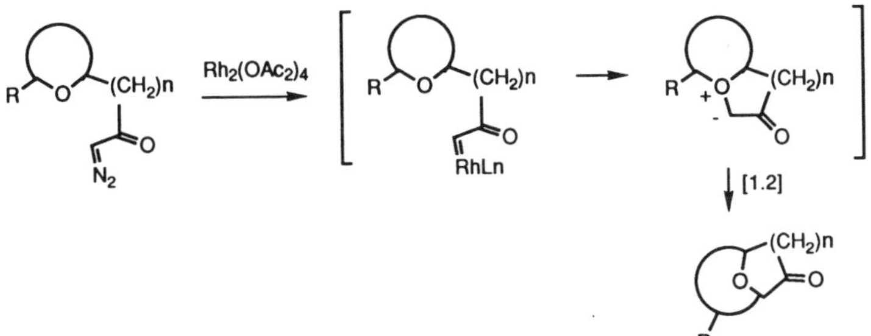

Certain α-diazomethyl alkynyl ketones undergo Rh(II)-catalyzed tandem alkyne insertion-ylide formation-[2.3]-sigmatropic rearrangement in the presence of allyl sulfides to afford γ-allylthio cyclic enones [Hoye, 1992].

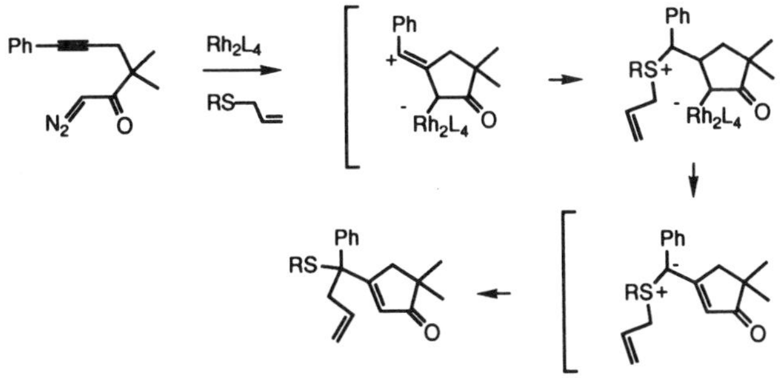

A rare [2.5]-sigmatropy of a foiled carbene resulted in an acetylene [Jones, 1989a].

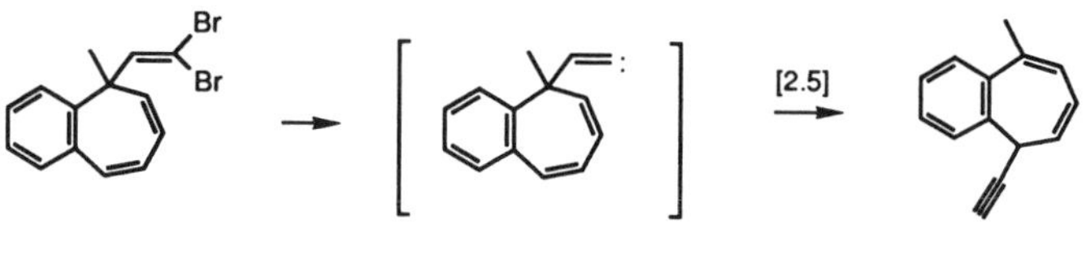

12.2.4. Sigmatropic Rearrangements of Higher Orders

Much fewer examples of tandem reactions are known that consist of a higher-ordered sigmatropic rearrangement as the terminating process. A significant example is the reaction of tropothione with diazomethane from which a tricyclic dithiepane was shown [Huisgen, 1985] to be the product.

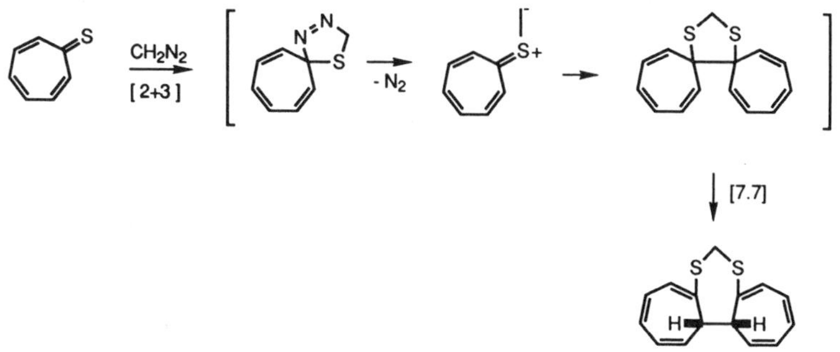

A 1,3-dipolar cycloaddition of the sulfonium methylide derived from tropothione with the thione forms a dispiro dithiolane which can be isolated at low temperatures. However, this dithiolane is thermally unstable and it undergoes ring expansion by way of a [7.7]-sigmatropic rearrangement very readily.

13

REARRANGEMENTS AND FRAGMENTATIONS

13.1. REARRANGEMENTS

Besides sigmatropic rearrangements which are described in Chapter 12, some tandem reactions pertaining to other rearrangements will be discussed in the following paragraphs. In view of the huge numer of such reactions, only very few representative cases can be included, and most of these cases involving cationic intermediates. Thus, many interesting multistep rearrangements (tandem!) are taken into consideration.

13.1.1. Skeletal Rearrangement of Carbocations

Carbocations are generated by ionization of certain compounds (e.g. alcohols, ethers, halides,...), by protonation of alkenes, or by hydride abstraction. The fate of the cations is critically dependent of their structures.

First we should mention the isomerization of tetrahydrodicyclopentadienes into adamantane under Lewis acid catalysis which is initiated by hydride transfer. Adamantane has a diamond lattice and is essentially strain-free.

From *exo*-tetrahydrodicyclopentadiene at least 2897 pathways are available for the isomerization even if highly strained and primary cations and those highly strained species are excluded as intermediates [Whitlock, 1968]. The picture becomes more complex if all other tricyclodecane isomers are considered (cf. the following scheme [E.M. Engler, 1973]). However, it is evident that, from the calculated heats of formation, the most stable isomer is adamantane.

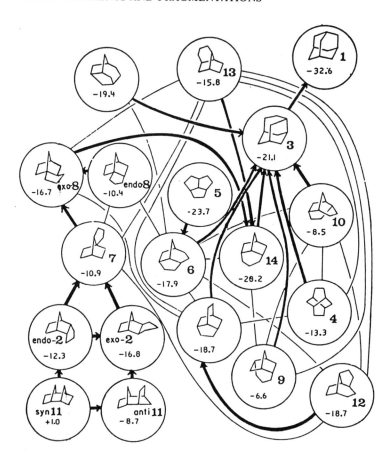

The conversion of binorbornylidene into sesquibicyclooctene [Nelson, 1985] is favorable because strain is relieved from each CC bond migration.

sesquibicyclooctene

The biomimetic formation of pentalenene from a bicyclic diene [Pattenden, 1987] also led to another product through a more complex 1,2-rearrangements.

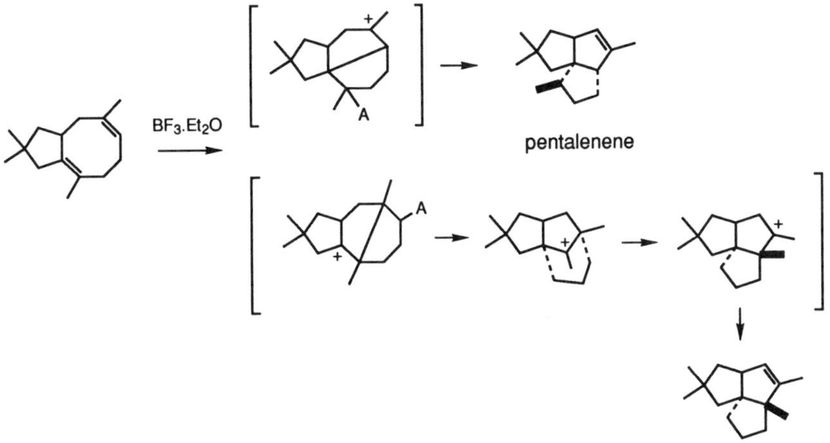
BF3.Et2O
A
pentalenene
A

One of the solvolytic products from 1,1′-bishomocubyl mesylate is a symmetrical alcohol [Dauben, 1966b]. Its emergence may be indicated by the following mechanism.

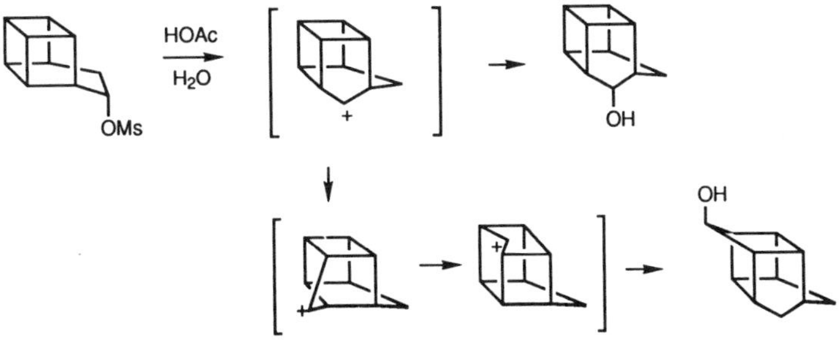
HOAc
H2O
OMs
OH
OH

Expansion of small rings is thermodynamically favorable. Consequently, cyclopropyl- and cyclobutylcarbinyl cations are prone to rearrange. A tandem series of such rearrangements can lead to fascinating skeletons [Fitjer, 1988a].

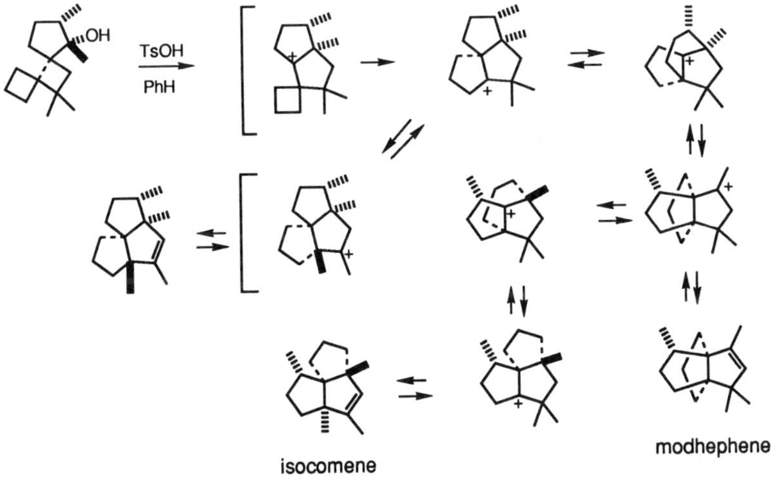
OH
TsOH
PhH
isocomene
modhephene

Most interestingly, the synthesis of modhephene and isocomene has been executed by application of the rearrangement cascade of a cation derived from the same dispiro[3.0.4.2]undecane, under conditions of kinetic control and thermodynamic control, respectively.

An analogous reaction of a polyspiroannulated carbocyclic alcohol resulted in a coronane intermediate [Wehle, 1987].

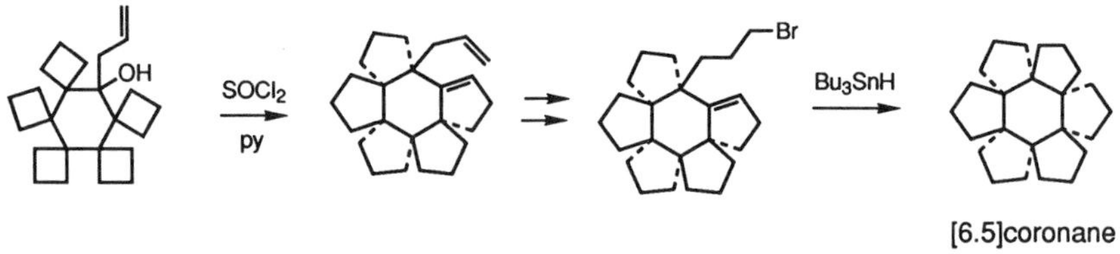

[6.5]coronane

Ring expansion of spiroannulated vinyl epoxide on exposure to Pd(O) catalyst [Kim, 1991] is also thermodynamically favorable.

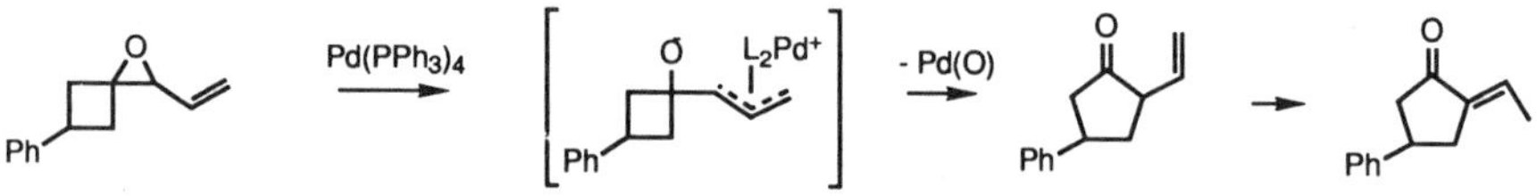

The transformation of a propellane into a fused tricycle [Paquette, 1986] was mediated by a series of 1,2-bond migrations after the initial ionization. Perhaps the generation of a benzene ring also contributed to the facility of the rearrangement.

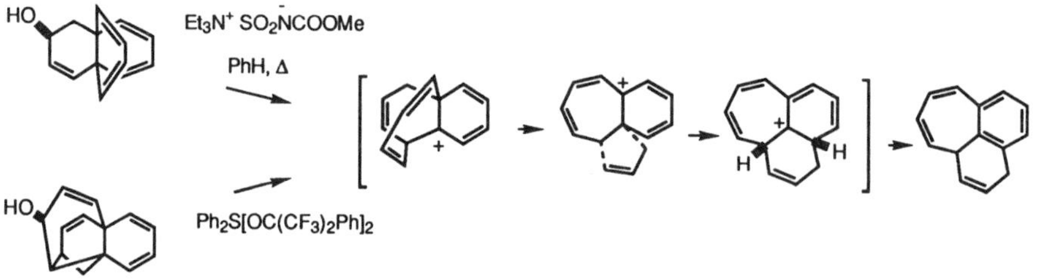

Herbertene is formed as the major product upon treatment of a drimane derivative with acid [Frater, 1982]. A possible mechanism for this unexpected transformation indicates a series of carbocationic rearrangements after an assisted ionization of the homoallylic alcohol.

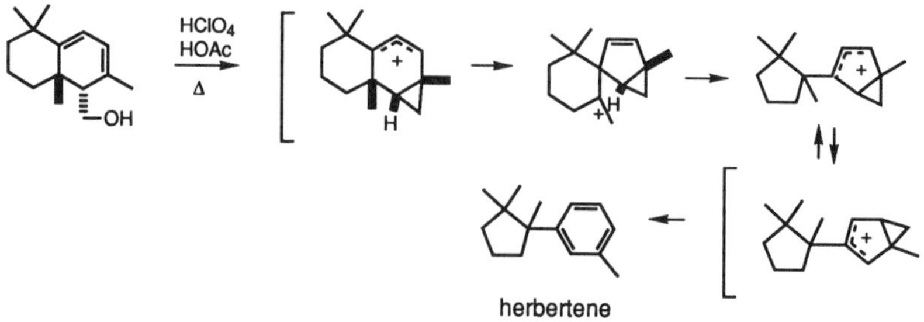

herbertene

Carbocationic rearrangements involving the bicyclo[2.2.1]heptyl system is one of the most extensively studied. Due to the possible intervention of Wagner–Meerwein rearrangement, Nametkin rearrangement, *endo*-2,6-hydrogen shift, and other processes, very intriguing results have emerged. To illustrate the intricacy of such processes the partial racemization of camphor and functionalization of its unactivated positions (e.g. C-8, 9, 10) may be understood in terms of the following mechanism [Finch, 1969].

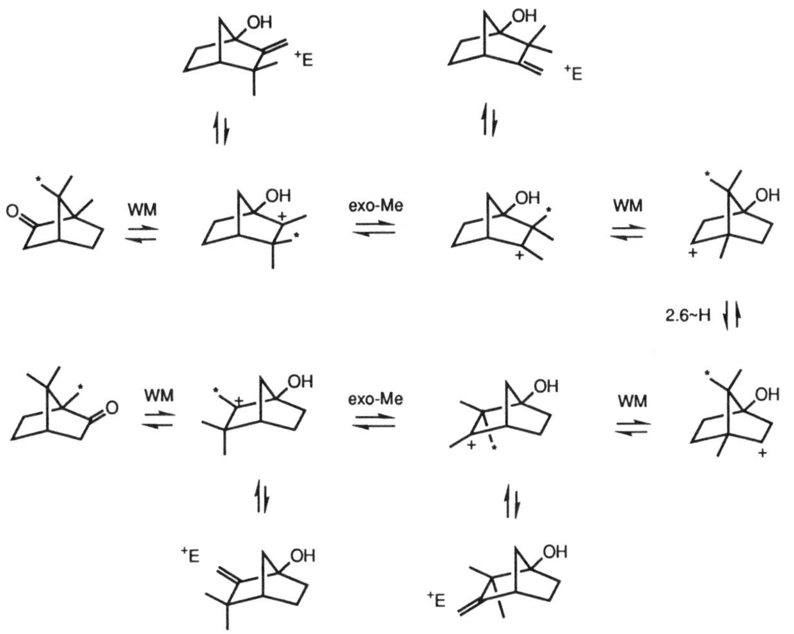

Sometimes results can be deceptive, as revealed by deuterium scrambling in the 8-deltacyclyl system [P.K. Freeman, 1970].

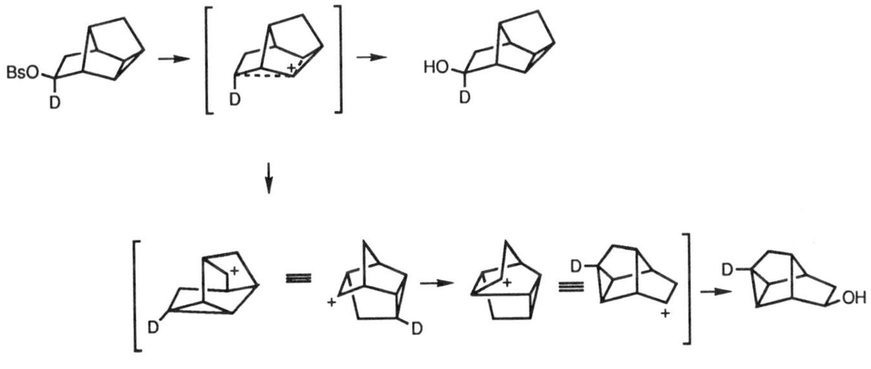

The rearrangement of a tricyclic diol into a new skeleton has been exploited in an approach to copacamphor and ylangocamphor [Kasturi, 1988].

ylangocamphor   copacamphor

The rather unexpected formation of a bridged ring system by intramolecular capture of a cyclopentenyl cation with an acetylenic sidechain was exploited in a synthesis of longifolene [Volkmann, 1975].

longifolene

The favorable ionization of benzylic alcohols is further facilitated by the removal of steric interactions from 1-hydroxymethyl[6]helicene [R.H. Martin, 1975b]. Thus, exposure of the alcohol to acid led to a spiroindene. Ring contraction followed the intramolecular Friedel–Crafts alkylation.

Temporary destruction of aromaticity also attended the Fischer indole synthesis of an *N*-iminotetrahydroquinoline [Fusco, 1975]. Two consecutive 1,2-shifts of methyl groups rather than a 1,3-shift have been demonstrated.

The pinacol rearrangement is a very common reaction. Epoxides are even more reactive substrates owing to strain relief during C—O bond cleavage. As expected, further enhancement of the reactivity is witnessed in oxaspiropentanes and, in fact, epoxidation of cyclopropylidene derivatives in unbuffered solutions has led to isolation of cyclobutanone products [McCullough, 1988], due to catalysis by the coproduced carboxylic acid.

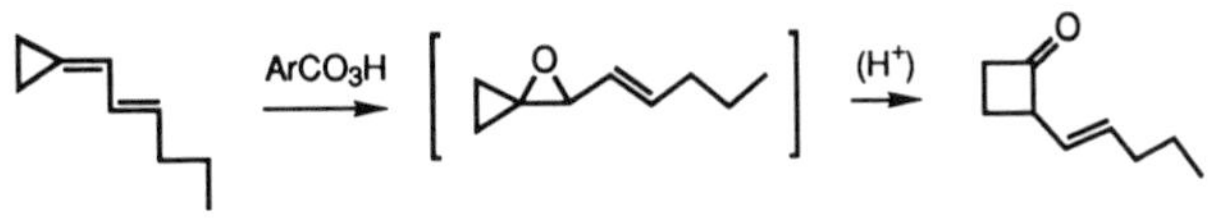

Another method of generating 1-oxycyclopropylcarbinyl cations is by electrophillic attack on the vinylcyclopropanes. An intramolecular alkylation of such a system aaffords spirocyclic ketones [Trost, 1988].

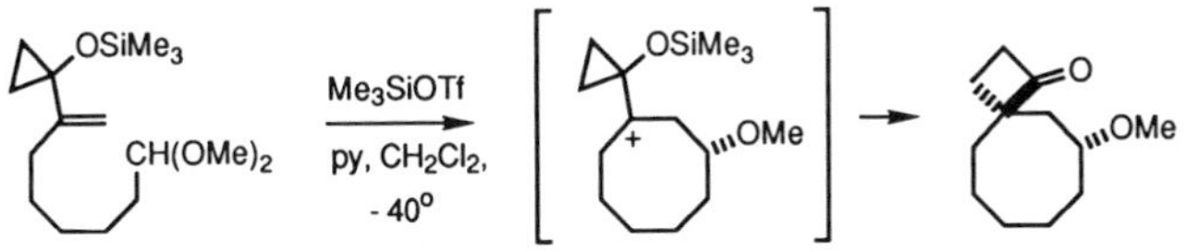

A process of multiple 1,2-shifts initiated by epoxide opening reorganized a 5:4-fused carbocyclic system to a ketone containing six- and three-membered rings. Marasmic acid was synthesized on the basis of this crucial step [Tobe, 1990]. The rearrangement facilitates the synthetic process enormously as the precursor is very readily available via photocycloaddition and simple transformations.

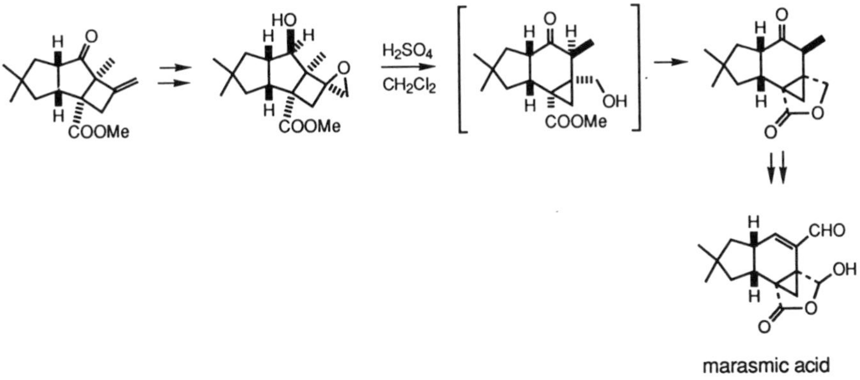

marasmic acid

A synthesis of frontalin [R.M. Wilson, 1984] consisting of laser (364 nm) irradiation of a bicyclic azoalkane to generate a 1,3-diradical which was trapped by oxygen. The dioxabicycloalkane underwent a pair of 1,2-rearrangement to provide the pine beetle pheromone.

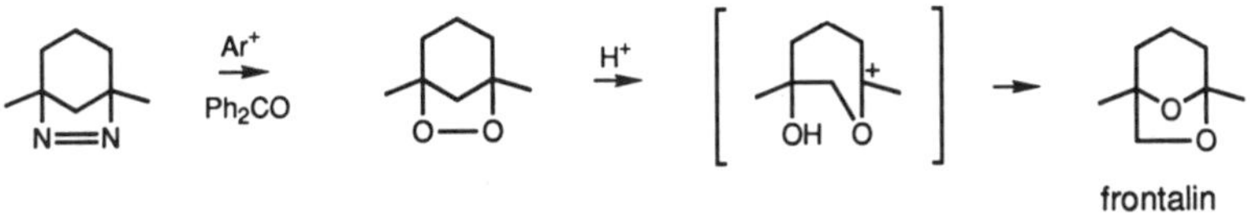

frontalin

Cyclic acetals of 1-vinyl-1,2-cycloalkanediols undergo ring-enlarging tetra-hydrofuran annulation [Herrinton, 1987]. Apparently three tandem reactions are involved in this useful transformation: O—C bond heterolysis, Prins cyclization, and pinacol rearrangement.

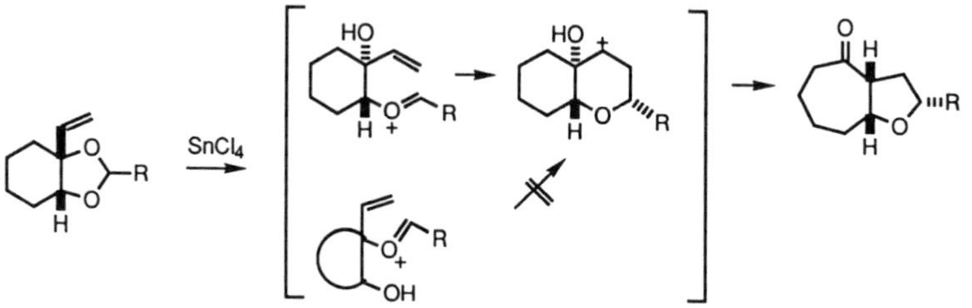

Carbocycles are also available by using modified substrates [Hirst, 1989].

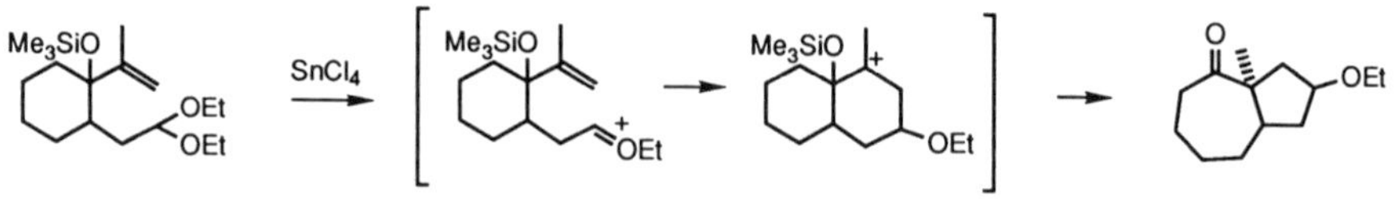

An example of α-hydroxycarbenium ion generation by protonation of specially substituted allylic alcohols and their rearrangement is the following [Cargill, 1970].

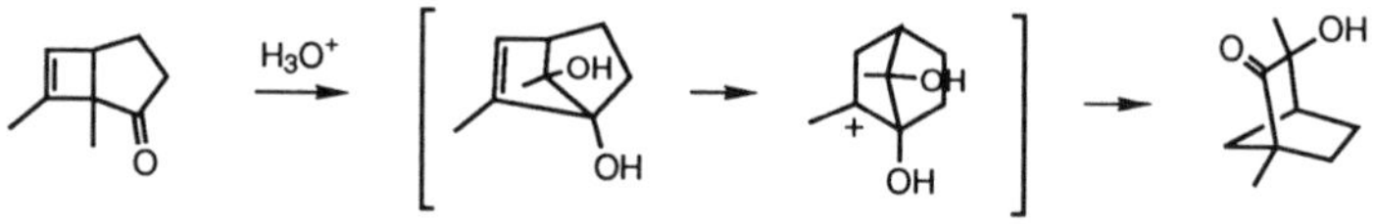

The ring contraction in tandem with cyclization of α-sulfonyl-α-(2-trimethyl-silylmethyl-2-alkenyl)cycloalkanones [Trost, 1983; 1990] in the presence of a Lewis acid constitutes a useful structural modification leading to spirocyclic compounds. The ring contraction step is a typical 1,2-rearrangement.

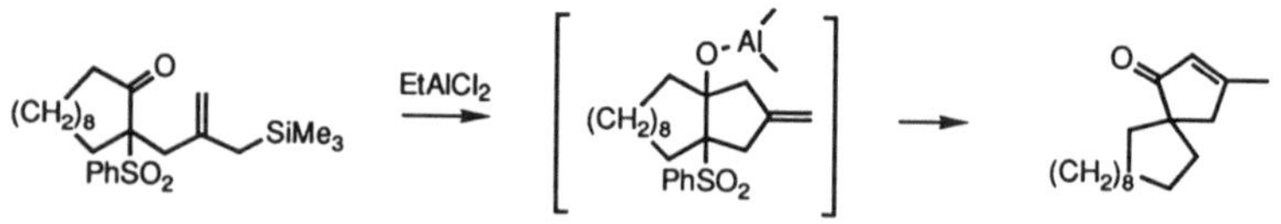

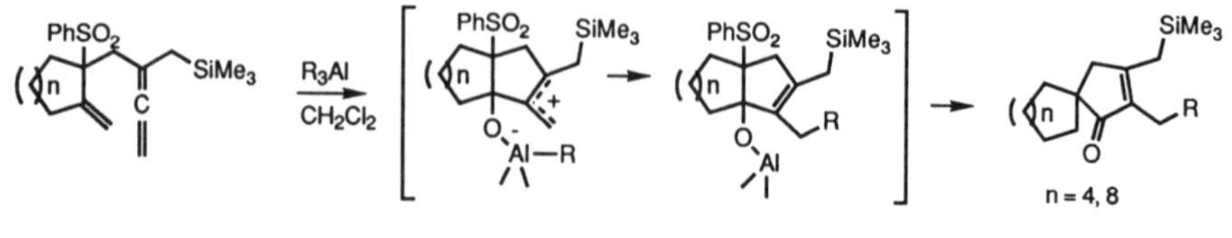

The carbonyl group is activated by protonation or coordination with a Lewis acid. The incipient oxycarbenium ion undergoes various transformations depending on its structure. For example, quantitative yield of a [3.3.3]propellanone has been obtained from acid treatment of a dispiro[3.0.4.2]undecanone [Fitjer, 1988b].

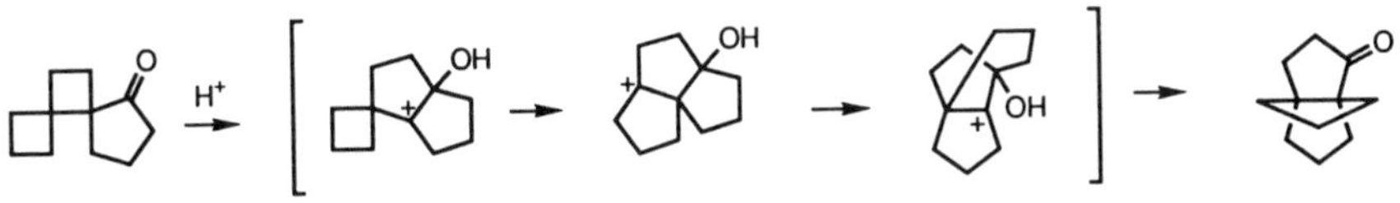

Strained ketones of simpler constitution that are converted into new skeletons are exemplified by the following [Stork, 1969; Hantawong, 1985].

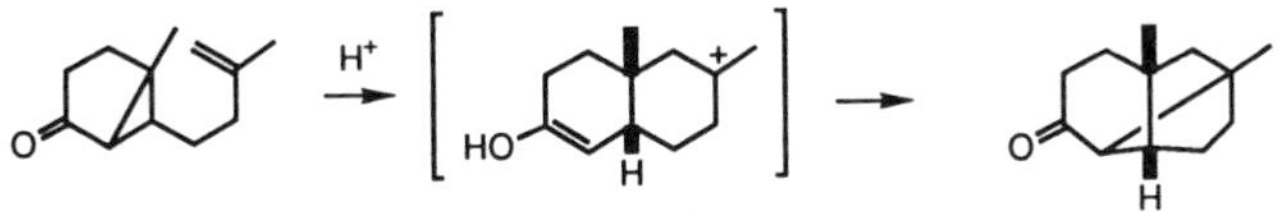

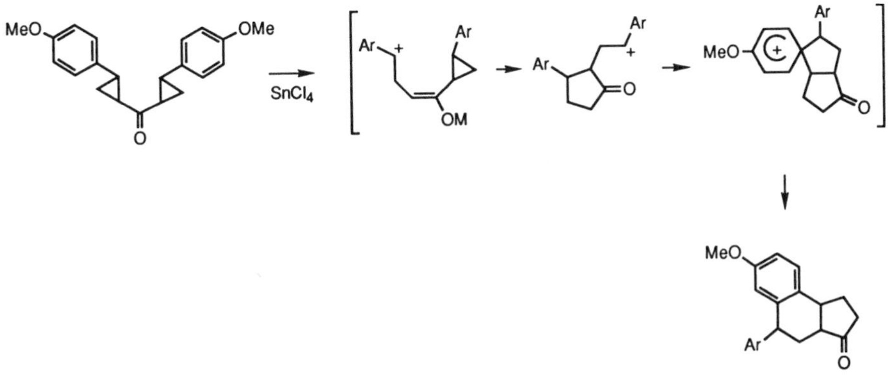

By a two-staged rearrangement a methylenebicyclo[4.2.0]octanone was converted into the isomeric bicyclo[3.2.1]octenone [Duc, 1975], via a hydroxybicyclo[3.3.0]octyl cation. This reaction is the basis for an isophyllocladene synthesis.

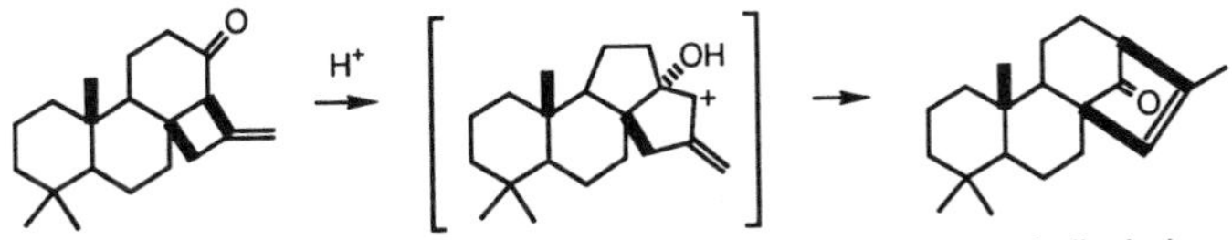

isophyllocladene

The emerging importance of triquinanes has furnished a tremendous stimulus to develop their chemistry, for example synthesis, etc. A method for the preparation of angular triquinanes is by rearrangement of cyclobutano-fused hydrindanones, and a route to silphiperfol-6-ene [Kakiuchi, 1989] based on this rearrangement has been devised.

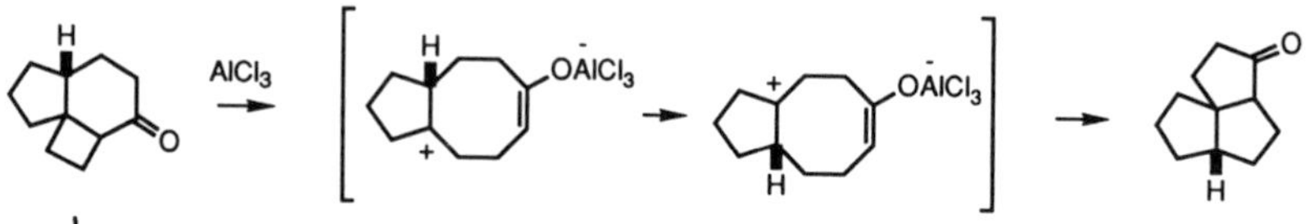

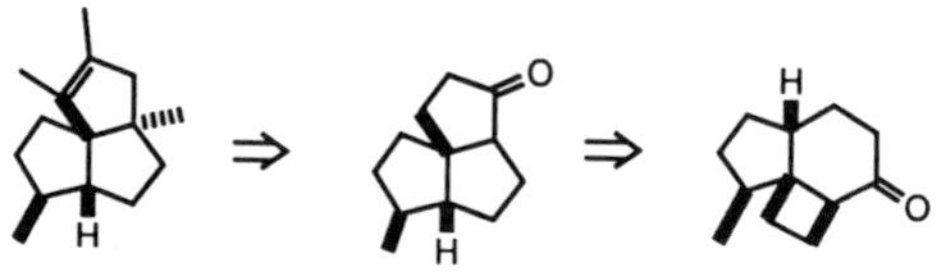

silphiperfol-6-ene

A cage diketone which the six-membered rings are constrained in the boat shape rearranged to an isomer on exposure to trifluoroacetic acid [Hirao, 1979]. The rearrangement expanded the two cyclobutane rings and established two new cyclohexanones in the chair conformation.

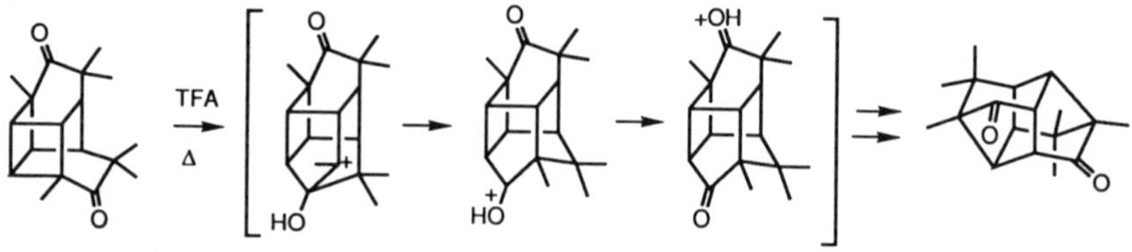

1,3-Bishomocubanones also undergoes a novel skeletal reorganization [Ogino, 1990] on treatment with boron trifluoride etherate or silver salt. Although the product contains two cyclopropane rings in place of the two cyclobutane rings originally present in the molecule, calculations showed that it has a lower energy.

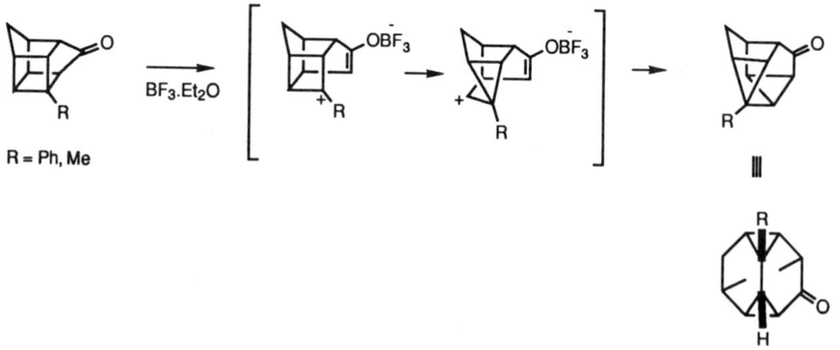

R = Ph, Me

Steroid synthesis is often closely related to the construction of *trans*-hydrindenones. A convenient route to these substances is via a Lewis acid-catalyzed cyclization–rearrangement [Snider, 1983b].

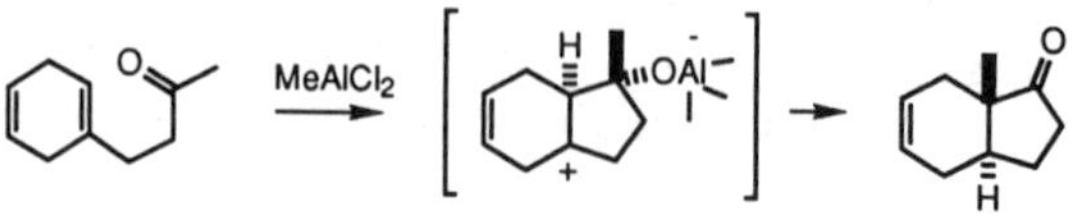

The silicon-directed Nazarov cyclization allows the synthesis of thermo-dynamically less stable cyclopentenones. However, the reaction is subject to subtle structural as well as catalyst variations [Denmark, 1988], as shown by the rearrangement below.

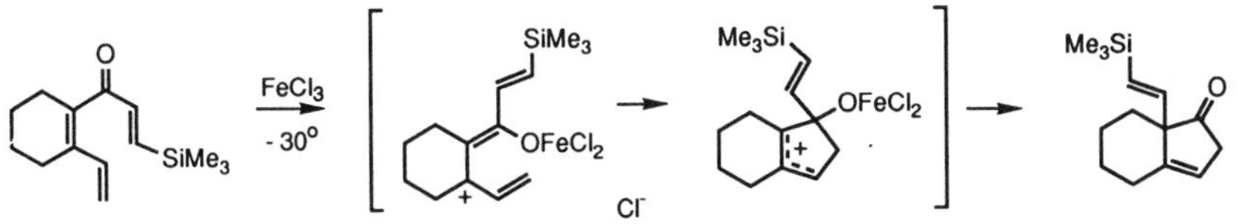

Intramolecular O-acylation of a cyclobutanone appears to be the key feature to a ring contraction–lactonization [R.D. Miller, 1991].

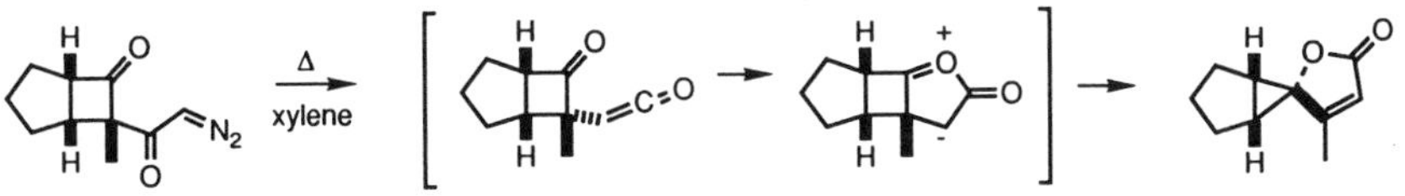

A tetracyclic synthon for the aspidosperma alkaloids has been produced by a rearrangement route [Takano, 1978b] involving CC bond migration from the α- to β-position of the indole nucleus. The reactive intermediate was formed by an intramolecular alkylation of an α-diazoketone.

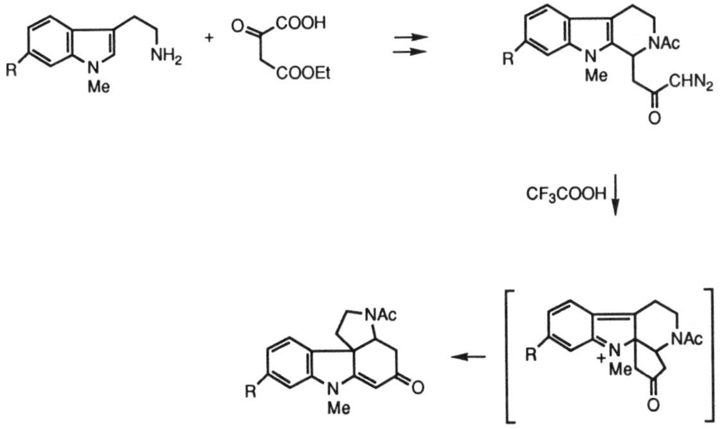

Such alkylation–rearrangement tandems have been exploited for synthetic purposes in the tetralin series [Mander, 1991]. Intermediates of analogous structures are implicated in the oxidative coupling of 1-benzyltetrahydro-isoquinoline alkaloids such as laudanosine [Kupchan, 1973]. Accordingly, these intermediates undergo a series of 1,2-alkyl shifts.

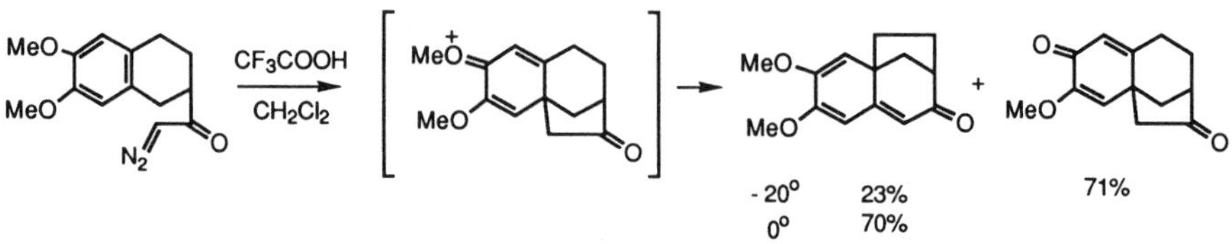

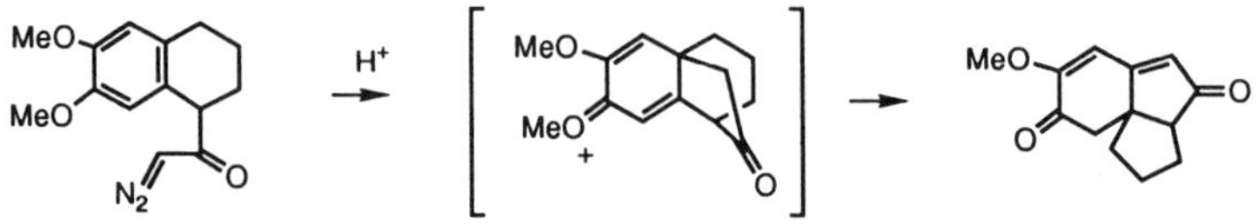

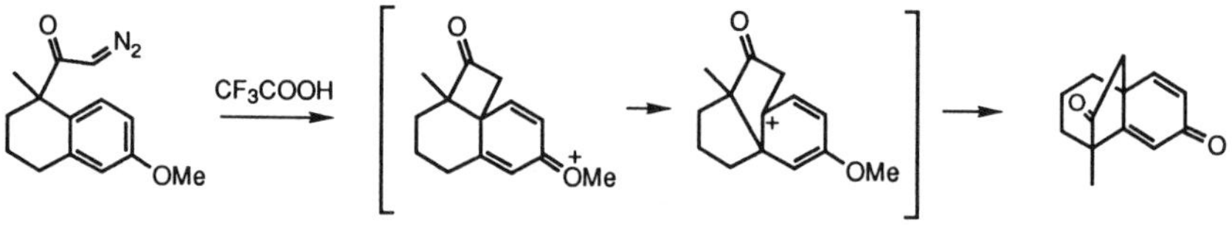

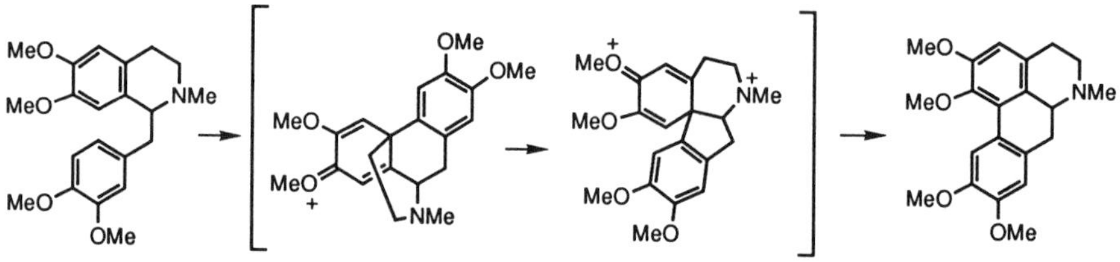

laudanosine

glaucine

Aryl migration after an intramolecular Friedel–Crafts alkylation established the tricyclic skeleton of colchicine [D.A. Evans, 1978]. It is remarkable that among the two interconvertible spiran intermediates only one favors the aryl migration (stereoelectronic effects).

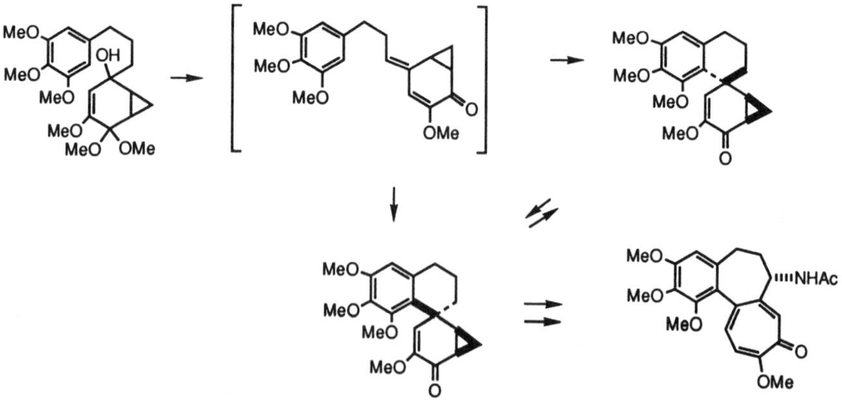

colchicine

A formal reductive 1,2-migration observed in the Clemmensen reduction of 1,3-diketones [Vincent, 1969] proceeded via the 1,2-cyclopropanediols. Ketonization with cleavage of the small ring led to the products.

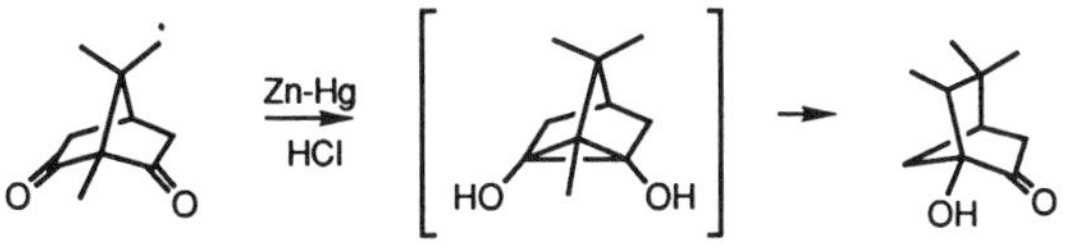

13.1.2. Anionic Rearrangements

In a synthesis of hirsutene [Dawson, 1983] the transformation of a bridged ring moiety into a condensed ring system formally involves a 1,2-CC bond migration. In fact, homoenolization attended this rearrangement.

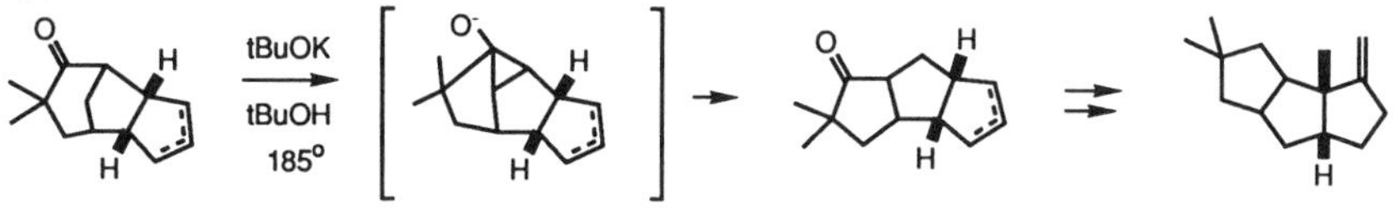

hirsutene

Homoenolization accounts for racemization of (+)-camphilenone and deuterium incorporation into the molecule [Nickon, 1962]. By the same mechanism 2-brexanone is transformed into 2-bredanone [Nickon, 1965].

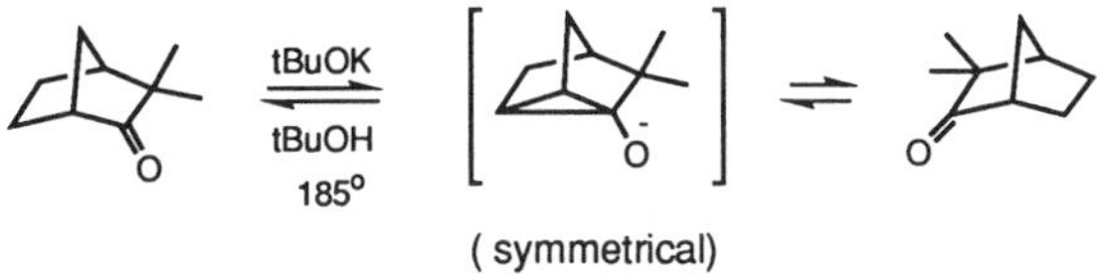

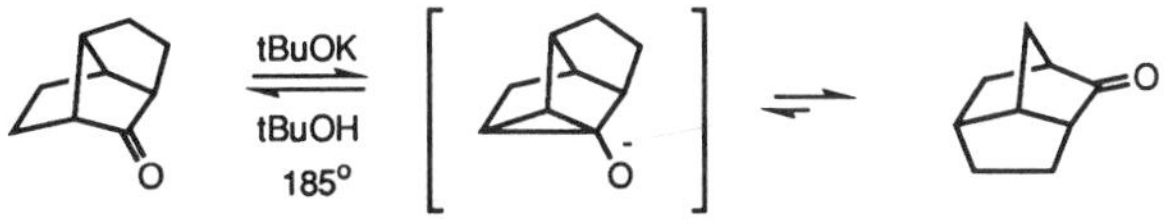

(symmetrical)

brexan-2-one   brendan-2-one

60%

Ketonization of a homoenolate ion and reflexive intramolecular trapping constitute an excellent method for the synthesis of 3-cyclopentenols [Danheiser, 1981b]. The precursors of such vinylsubstituted homoenolate ions are available from alkoxycarbene addition to conjugated dienes.

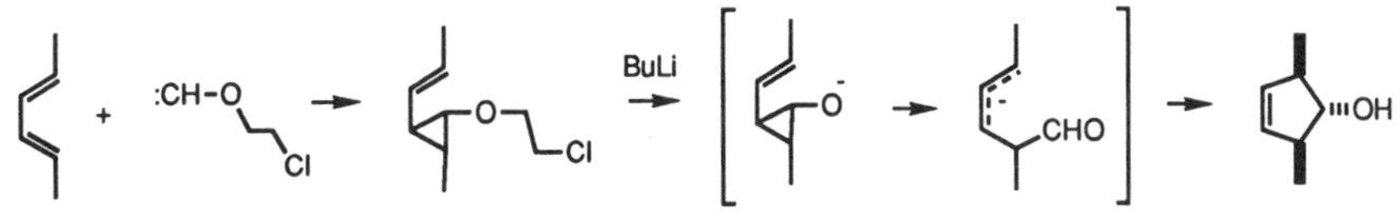

Treatment of 4,5-dihydroxyhomocubanone with sodium methoxide led to a brendanedione [R.D. Miller, 1973] by way of homoketonization, α-ketol exchange (rearrangement), and another homoketonization.

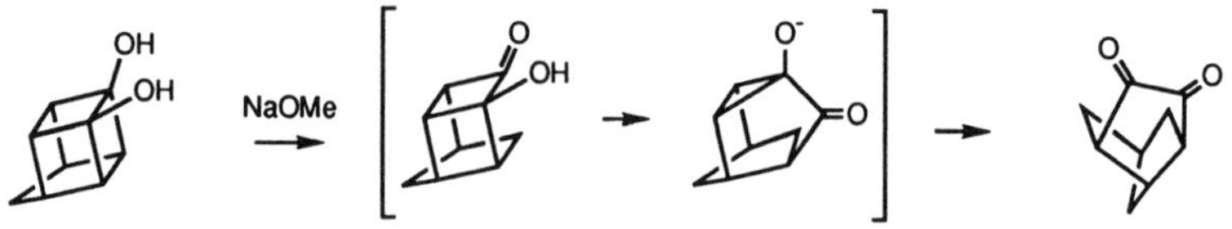

A tandem aldol condensation–pinacol rearrangement has been devised to construct the tricarbocyclic framework of quadrone [Monti, 1982]. Interestingly, the skeleton of the aldol precursor was derived from a rearrangement of an α-alkoxy ketone.

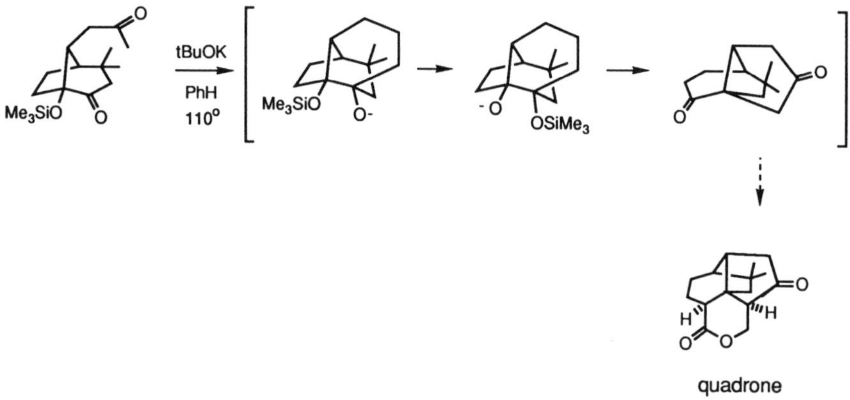

quadrone

The reaction course of benzyne trapping with substituted acetonitriles is dependent on the substituent of the acetonitrile. Normal alkylation occurs when the substituent is an alkyl group. However, the major products from reactions of arylacetonitriles come from a tandem addition–rearrangement process [Khanapure, 1988].

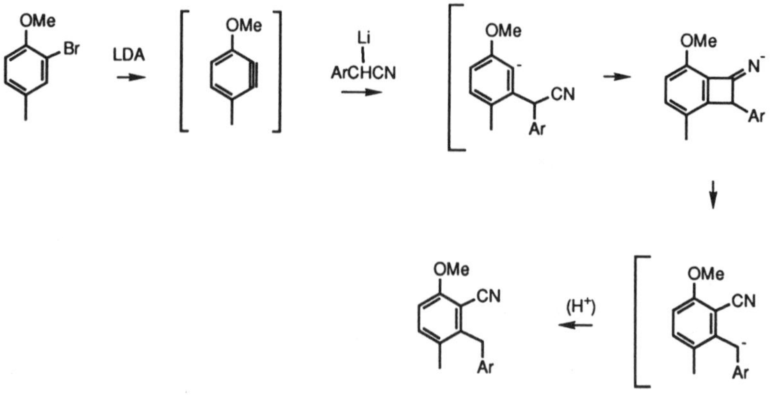

13.1.3. Other Rearrangements

Thallation of alkenes often leads to rearranged skeletons because the thallium atom in the solvatothallation products is still trivalent and it is very keen to undergo reduction. The most readily reaction pathway is by its expulsion via a rearrangement process, as indicated in the following examples.

[Taylor, 1976]

[Abley, 1973]

[Murakami, 1981]

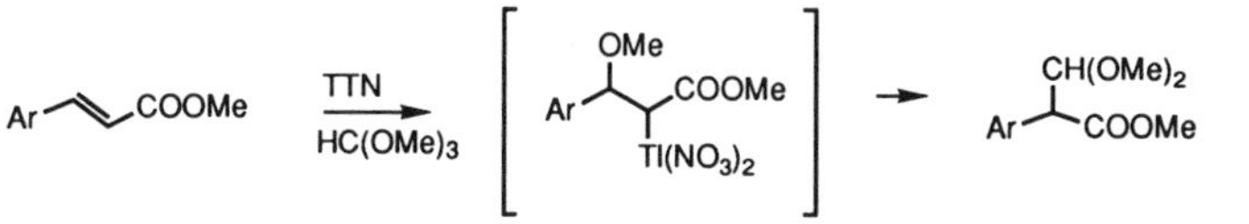

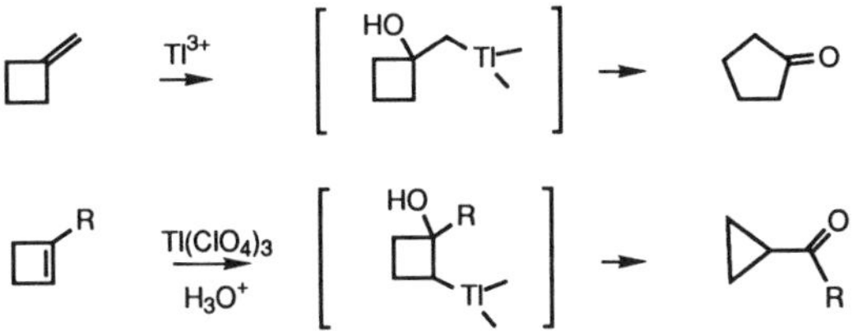

Manipulation of organoboranes is an integral part of alkene functionalization via hydroboration. Usually a reagent used to attack the boron atom of the borane is capable of inducing a 1,2-rearrangement. Alkaline hydrogen peroxide is a typical reagent.

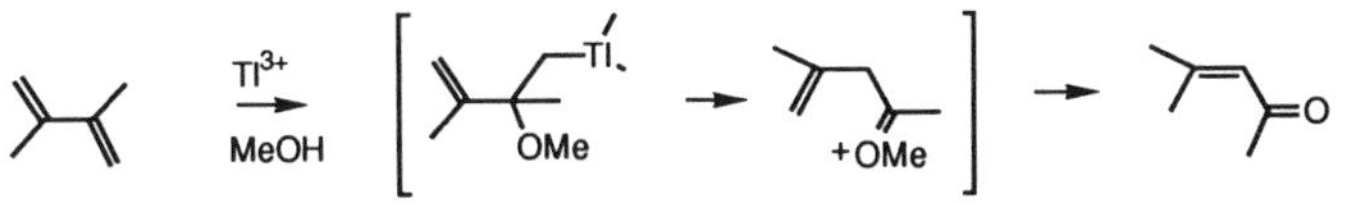

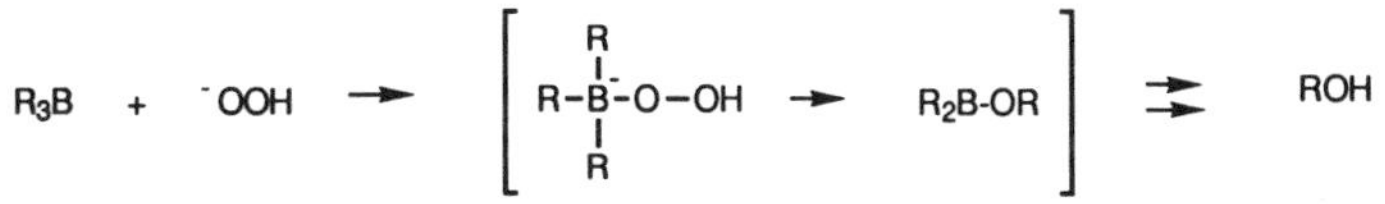

Of both synthetic and mechanistic interest is the participation of a homoallylic azido group at the end of hydroboration, which resulted in the formation of a pyrrolidine derivative [Evans, 1987]. Apparently an azaboracycle was formed, and rearrangement with expulsion of dinitrogen led to ring contraction.

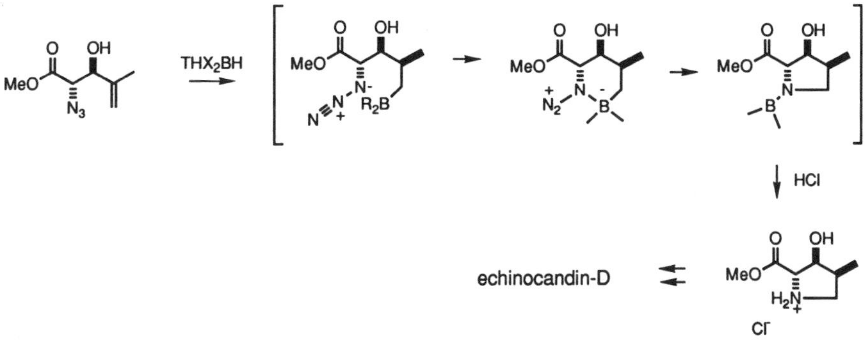

Carbon atom migration from boron to carbon can also occur. Thus rearrangement is induced upon intramolecular alkylation of 2-indolylboronates at C-3 [Ishikura, 1991]. Because of aromatic stabilization the rearrangement products eliminate borane with facility.

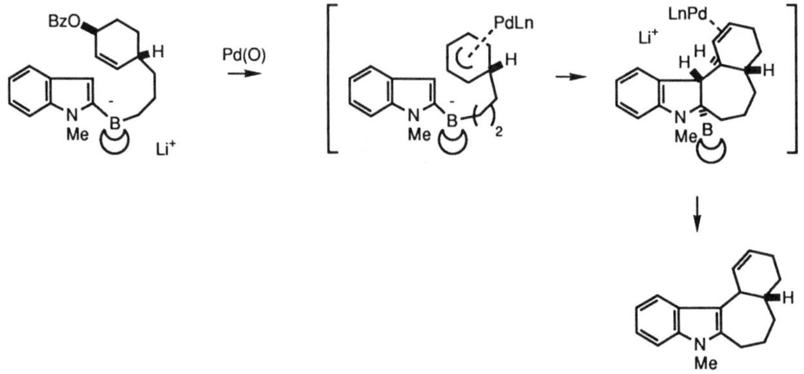

Thermolysis of 1-(but-1-en-3-ynyl)-2-phenylcycloalkene oxides affords mainly furan derivatives [Eberbach, 1992], as a result of intramolecular 1,7-cyclization of the 1,3-dipolar tautomers, followed by rearrangement.

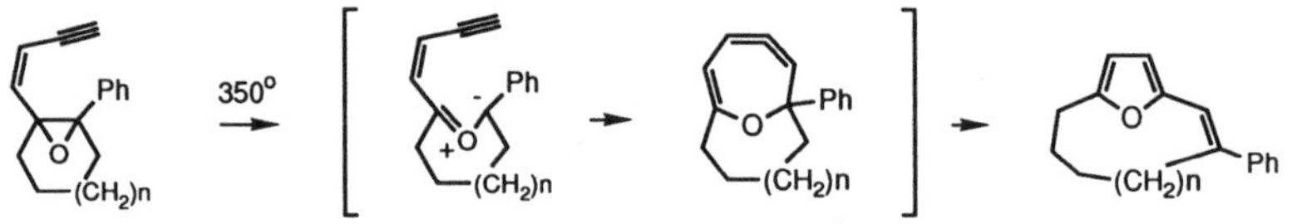

n = 1,2,3

Dyotropic rearrangement designates an intramolecular exchange of two σ-bonded groups at two adjacent sites. Relevant to our discussion is one such process in tandem with aldolization and lactonization upon reaction of α-amino aldehydes with ketene acetals [Reetz, 1989].

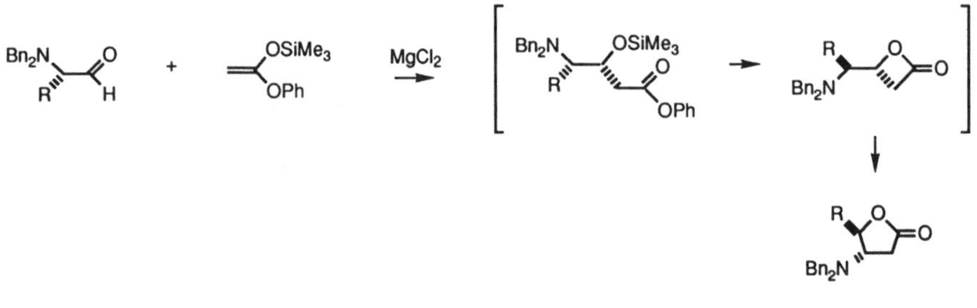

The photoracemization of a dihydropyrazine which is laterally fused to two norbornyl moieties is a fascinating rearrangement involving a sequence of bond migrations.

The oxazole ring has been used as a latent, activated carboxyl group [Wasserman, 1986]. Although the detailed mechanism for the demasking–activation by dye-sensitized photooxygenation is unknown, evidence for 2,5-endoperoxide intermediates is irrefutable.

The decomposition of the endoperoxide to give a triamide may be visualized to proceed via a Baeyer–Villiger type rearrangement to give an imino anhydride which then undergoes an O- to N-acyl shift.

To conclude this section an example of photochemical rearrangement of macrocyclic 2-phenylcycloalkanones [Lei, 1986] is given. The favorable reaction pathway for such compounds is an α-cleavage; interestingly, the long chain separating the acyl and the benzylic radicals allows reclosure of a larger ring at the para position.

13.2. FRAGMENTATION REACTIONS

A surprisingly large number of fragmentation reactions are known in organic chemistry. Many common reactions have their retrograde counterparts; thus, aldol, Claisen, Michael, Mannich, Diels–Alder, and ene–Prins reactions can be reversed under proper conditions. Only a few classes of fragmentation reactions are presented here, mainly because these are amenable to combine with other types of reactions in a tandem manner.

13.2.1. Tandem Reactions Terminated in a Retro-Aldol or Retro-Claisen Fission

The retro-aldol reaction can be accomplished by thermolysis without added reagents, although it is strongly catalyzed by acids and bases. Of course, structural characteristics of the substrate contribute to the ease of such a reaction. For example, 2-acylcyclopropanols and cyclobutanols fragment virtually spontaneously upon their formation. Ring strain is relieved.

The intricate rearrangement of santonic acid to give parasantonide has been formulated [Woodward, 1963] to proceed via an intramolecular aldol condensation followed by a retro-aldol cleavage.

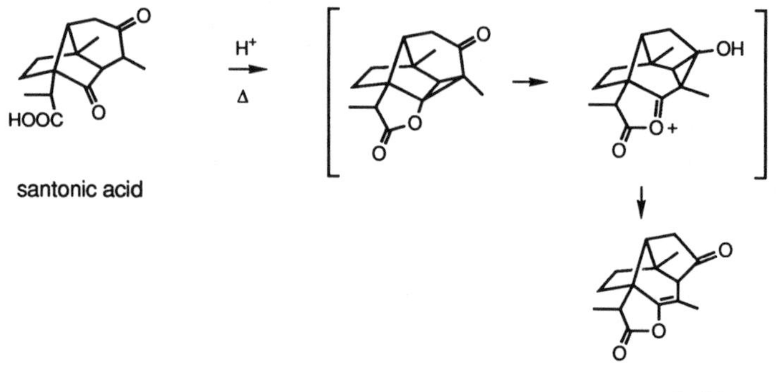

santonic acid

parasantonide

Solvolysis of 5-(bromomethylethylidene)-2,2-dimethyl-1,3-dioxane-4,6-dione under mildly alkaline conditions afforded the 2-acetonyl derivative of Meldrum's acid [Hunter, 1980]. The envisaged mechanism consists of several steps: hydration, 1,3-elimination, and retro-aldol fission of the acylcyclopropanol.

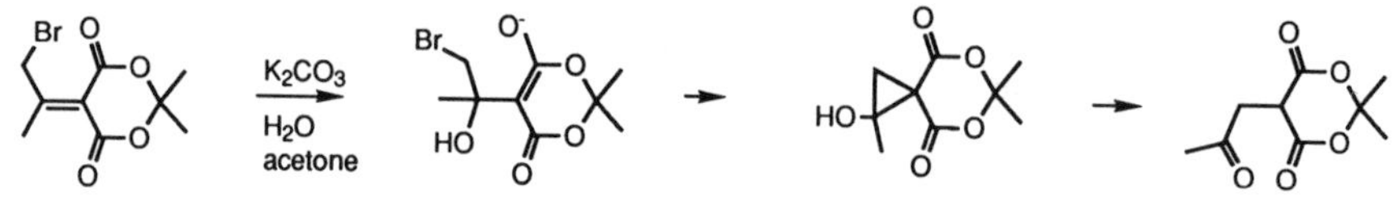

On reaction with Grignard reagents, cyclopropenone furnishes 2-substituted resorcinols [Oda, 1972]. This reaction appears to proceed via a double Michael–aldol reaction tandem to give tricyclic ketol intermediates which undergo an ethanologous retro-aldol fission.

The retro-aldol fission or, more properly, retro-Knoevenagel reaction, of β-hydroxy esters formed by an intramolecular Knoevenagel reaction of 2-oxocycloalkylacetic esters completes a ring enlargement protocol [Xie, 1989; Bit, 1991]. Interestingly, 2-acetonyl-2-carbalkoxycyclo-pentanones behave differently than their homologs on base treatment [Xie, 1988]. Instead of forming a bicyclic enone, ring expansion results.

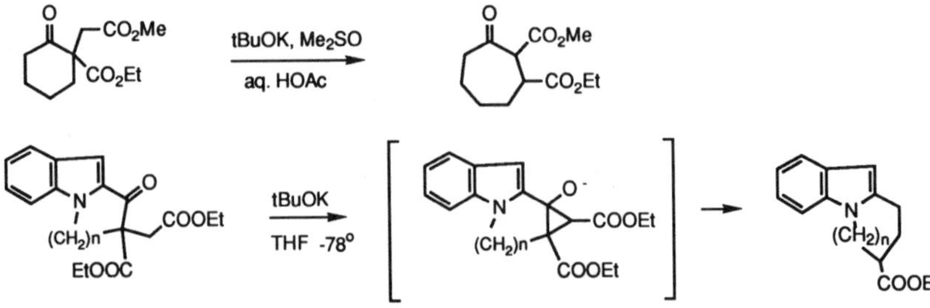

n = 1, 2

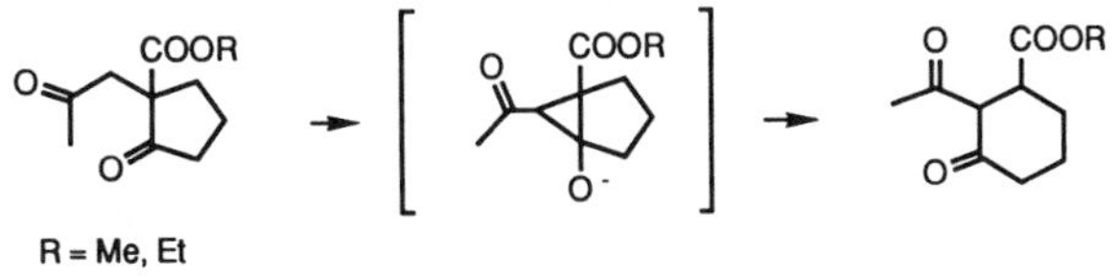

R = Me, Et

2,2'-Diphenyl-2,2'-biindan-1,1',3,3'-tetrone gives 2,3-diphenyl-1,4-naphthoquinone on treatment with alkali [Beringer, 1963]. A mechanism consisting of retro-Claisen, aldol, and retro-aldol reaction steps accounts for the skeletal change.

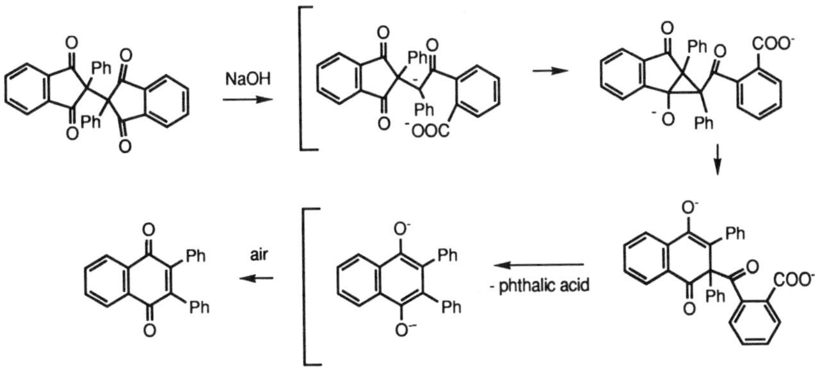

Furan ring formation during demethylation of an anthraquinone methyl ether [Ahmed, 1983] may have involved an intramolecular alkylation and a retroaldol reaction for the cleavage of the cyclopropane intermediate. This process might have some significance in the biogenesis of certain benzofuran natural products.

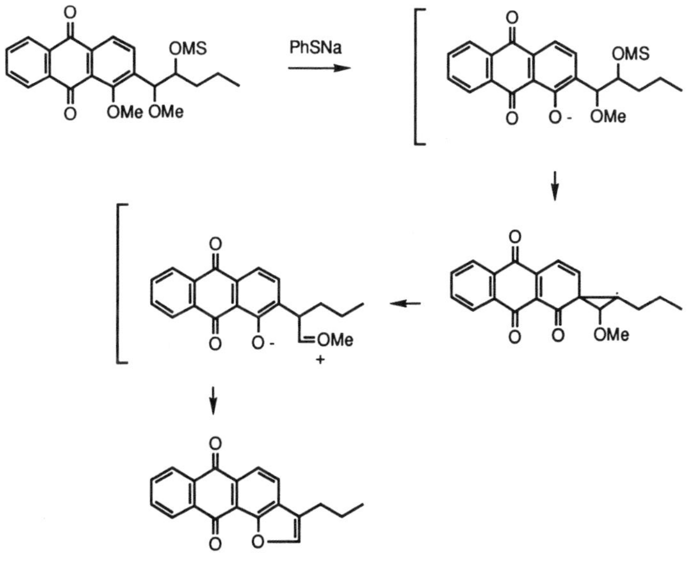

Rhodium(II)-catalyzed reaction of α-diazoketones in the presence of alkenes has become a method of choice for the preparation of acylcyclopropanes. When a furan is used as the alkene acceptor the cycloadduct is prone to fragment by a vinylogous retro-aldol reaction to give a 1,4-diacyl diene. Both inter- and intramolecular versions of the tandem cycloaddition–fragmentation reactions have been examined, revealing excellent synthetic potentials [Wenkert, 1989; 1990].

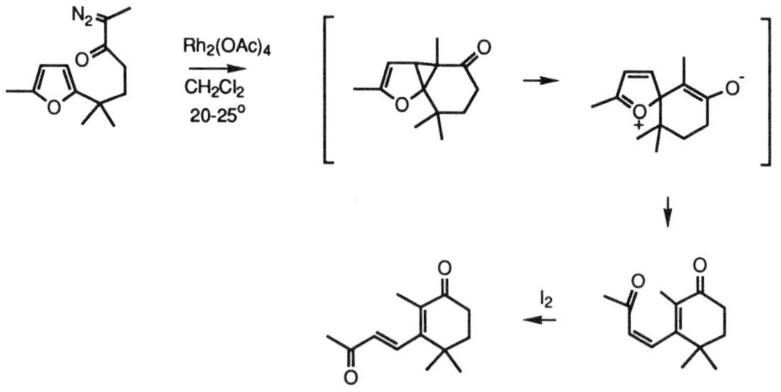

4-oxo-β-ionone

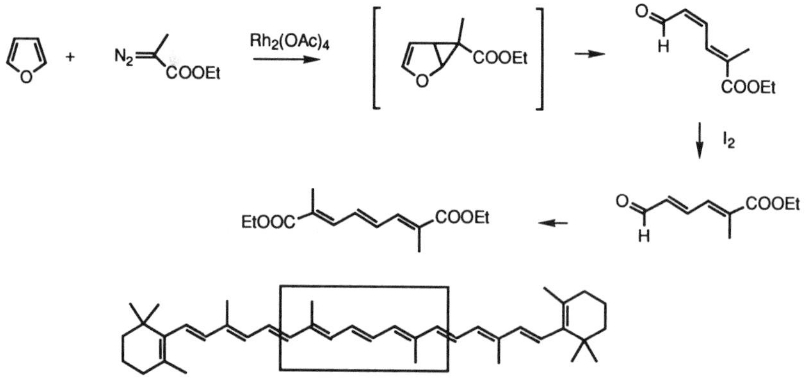

β-carotene

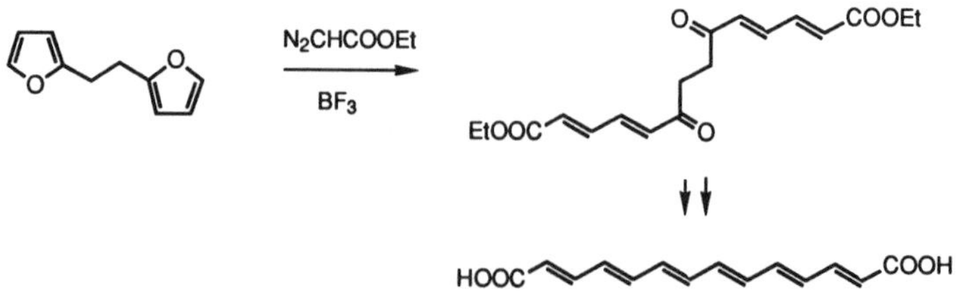

corticrocin

The formation of a stilbene from alkali treatment of a 2,2-diarylethyl aryl ether [Gierer, 1977] most likely proceeded via a spirocyclic cyclohexadienone intermediate. The decomposition of the latter species is a formal retro-aldol reaction (benzenologous, ethanologous, and vinylogous).

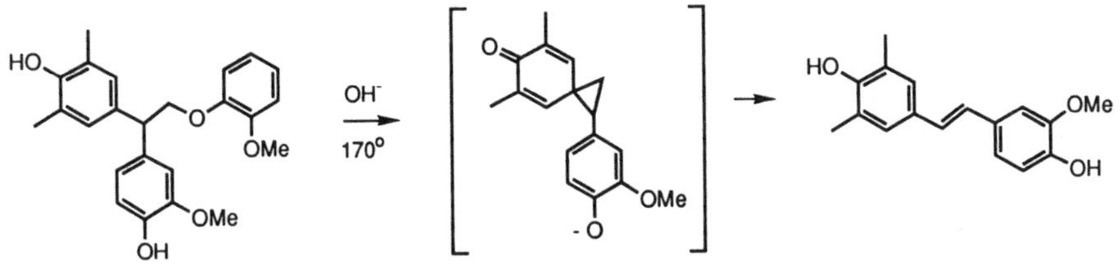

The conversion of 1,3,3-triphenyl-1,4-pentanedione to the 1,4,4-triphenyl isomer [Yates, 1971] probably involved formation and fission of cyclopropanol intermediates.

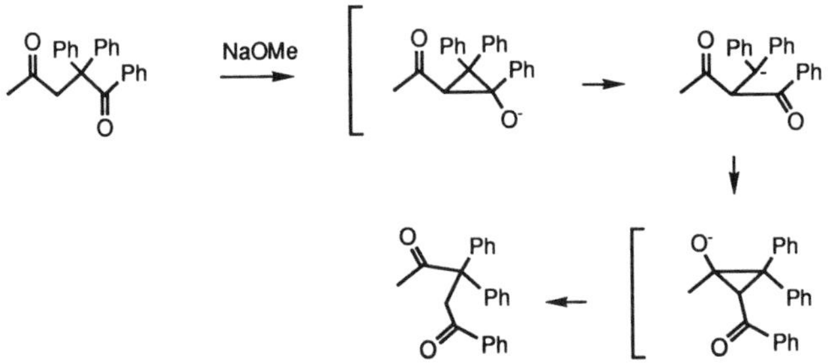

$4(\alpha/\beta)$-Hydroxysantonenes undergo oxa-di-π-methane rearrangement on uv irradiation [East, 1974]. However, the photoproducts contain a 2-acylcyclopropanol moiety which is susceptible to retro-aldol cleavage.

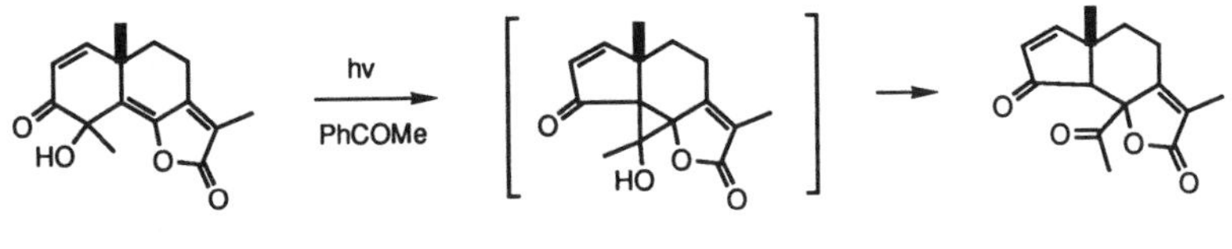

The formation of a cycloheptanedicarboxylic acid during an apparent attempt at hydrolysis of an α-hydroxy aldehyde dithioacetal [Ranu, 1988] is due to intervention of a 1,2-rearrangement, retro-aldol fission, and subsequent oxidation.

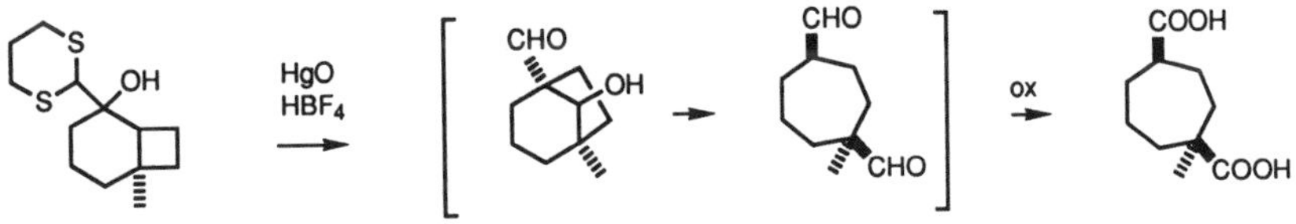

A more straightforward ring cleavage is represented by the reductive fragmentation of 2-carbalkoxycyclobutanones [S. Goldstein, 1981].

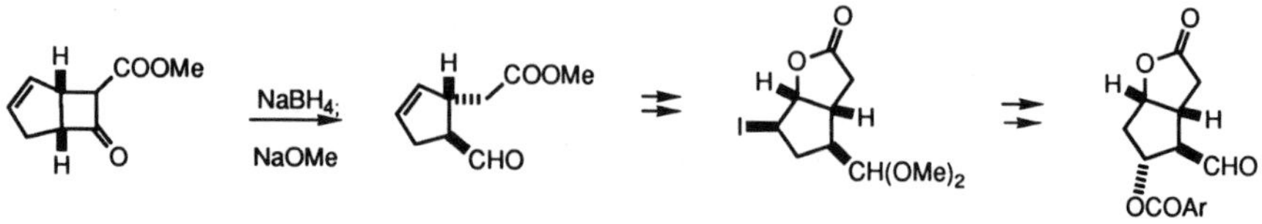

Corey's lactone
for prostanoids

Enolizable β-dicarbonyl compounds form [2 + 2]photocycloadducts with various olefins to give 1,5-dicarbonyl substances (de Mayo reaction), as a result of spontaneous, strain-relieving retro-aldol fragmentation of the 2-acylcyclo-butanol adducts. This reaction has been exploited extensively, and an early application is in a synthesis of loganin [Büchi, 1973; Partridge, 1973].

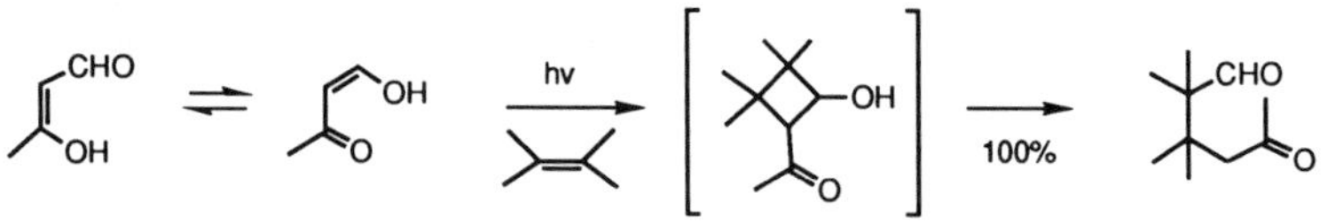

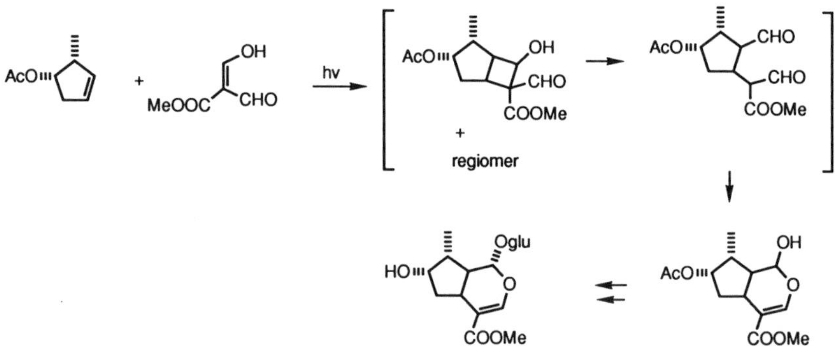

loganin

The usefulness of this reaction for preparation of cyclohexenones is also evident.

A variant of the de Mayo reaction employs (5H)-5,5-dimethyl-3-furanone as a photoaddend. Alternative methods for the ring cleavage step have been developed. For example, the photoadducts may be submitted to reaction with a methylmetal and photolysis of the corresponding nitrite esters [S.W. Baldwin, 1982].

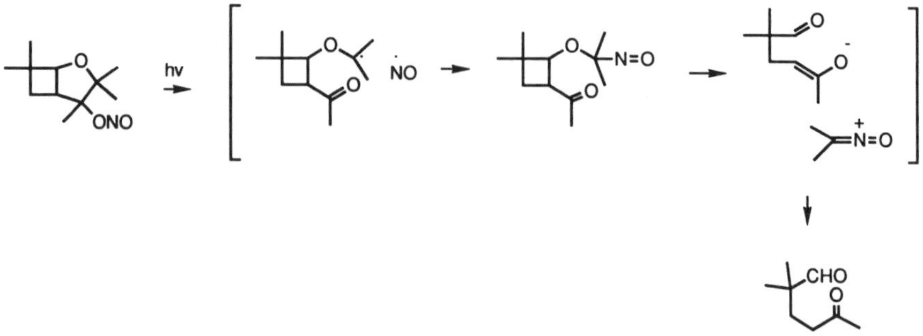

The diol derived from mutilin liberates glycollic aldehyde on base treatment [Arigoni, 1962]. Transannular hydride shift prior to and after the retro-aldol reaction must be involved.

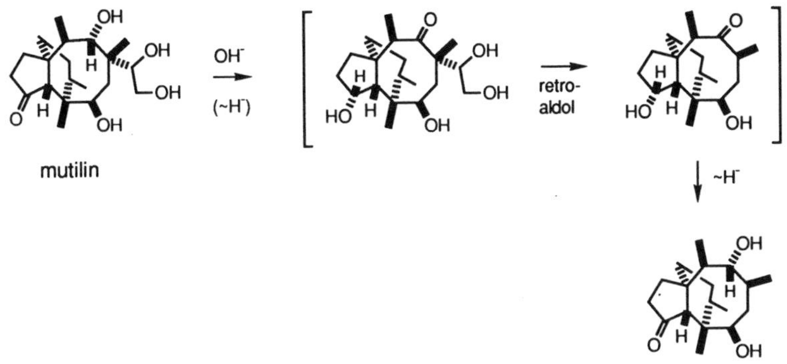

mutilin

The degradation of ishwarone included hydroboration of isoishwarone [Govindachari, 1969]. The production of a keto aldehyde with the eremophilane skeleton was the result of a retro-aldol cleavage, and it afforded evidence for the carbon framework and particularly the bridging positions.

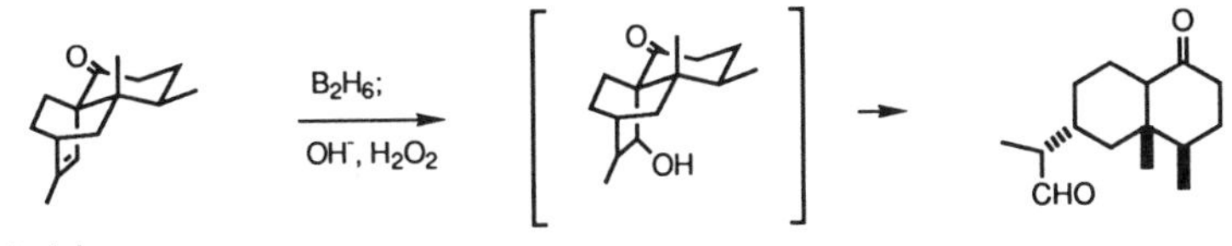

isoishwarone

3-Acetyl-6-styryl-α-pyrone is transformed into 4-acetyl-3,5-diphenylcyclo-hexanone [Takeuchi, 1991] on alkali treatment. The decarboxylation product generated after saponification of the heterocycle can dimerize via a reflexive Michael reaction to form the cyclohexanone ring. Furthermore, under the reaction conditions the two 2-ene-1,5-dione subunits are susceptible to hydration and retro-aldol fission.

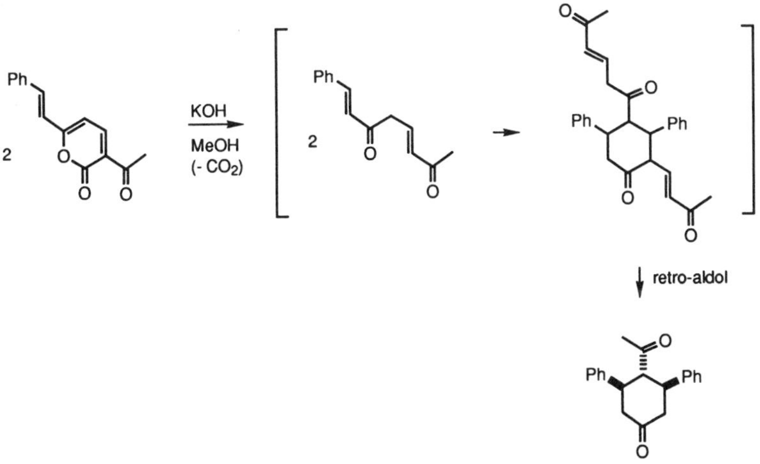

Coupling of 1,8-dihydroxyanthraquinone with methyl vinyl ketone under reducing conditions [Krohn, 1985; Koerner, 1991] perhaps proceeded via tandem Diels–Alder/retro-aldol reactions. The modest yield of the product was due to low regioselectivity of the Diels–Alder reaction step.

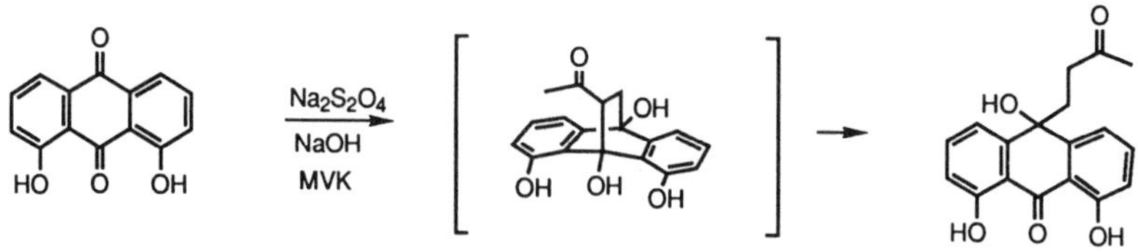

An interesting transformation of an AB-seco epoxide derivative of totarol occurred on acid treatment [Gambie, 1969]. Participation of the exocyclic methylene group in the epoxide ring opening was followed by a retro-aldol fission.

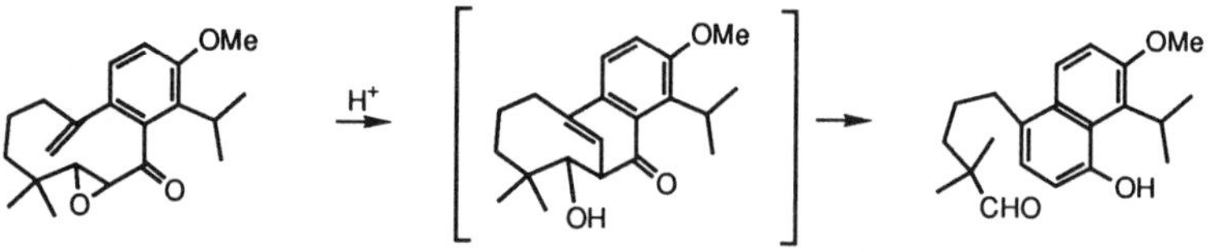

An indole derivative is formed by peracid oxidation of N-(4-phenoxy-2-butyn-1-yl)tetrahydroquinoline [Majumdar, 1987]. To account for this unusual transformation a mechanism involving [2.3]- and [3.3]-sigmatropic rearrangements, cyclization, and dimerization, has been proposed. The dimer is a β-hydroxy iminium species which is prone to eliminate formaldehyde in a process analogous to the retro-aldol fragmentation.

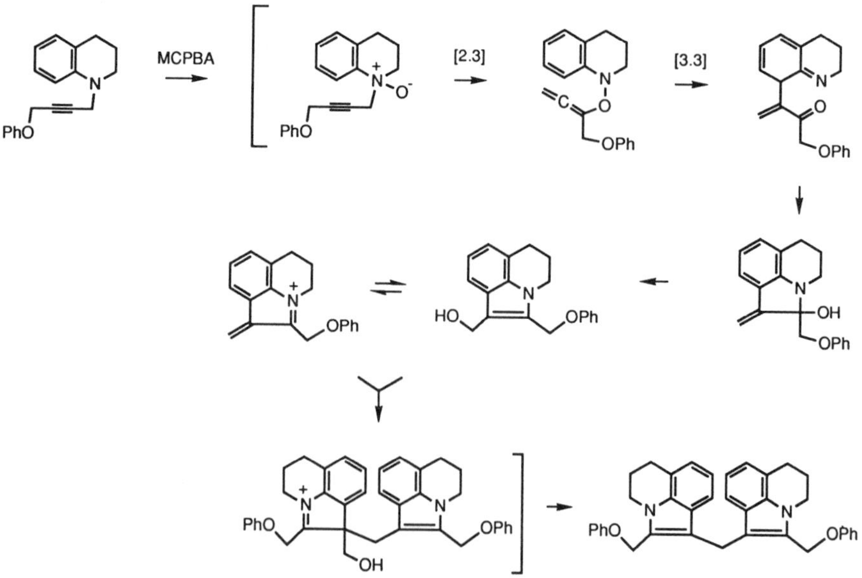

Constitutional change and redistribution of functionalities have been observed in the intramolecular condensation of certain nitroalkanediones [Lorenzi-Riatsch, 1981]. Apparently, such molecules tend to undergo aldol condensation followed by a retro-Henry reaction.

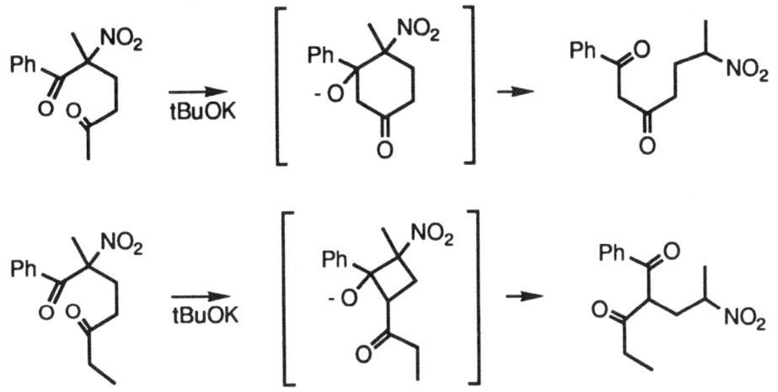

Alkylation of 2-nitrocycloalkanones with *p*-benzoquinone gives lactones directly [Stach, 1986].

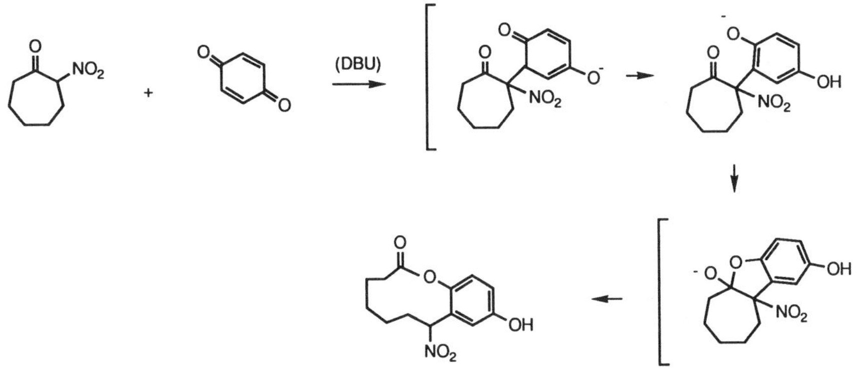

The fragmentation step of retro-Claisen reaction of 1,3-diketones is actually a retro-aldol cleavage. A useful strategy to gain access to certain nitrogenous heterocycles depends on the intramolecular formation of γ-oxocarbinolamines followed by fragmentation [Ohnuma, 1983].

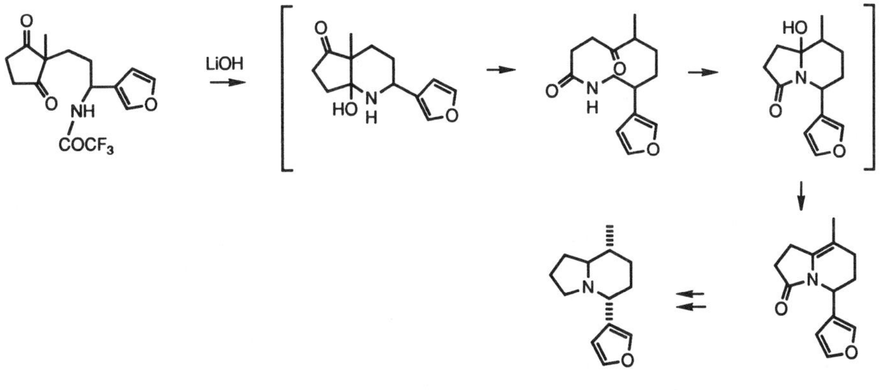

Nuphar indolizidine

The facility of retro-Claisen cleavage of small ring compounds is understandable. Exploitation of molecular strain to induce such a reaction for synthesis has been very successful. The small ring ketones are not necessarily present in the exposed form; frequently during unveilment of the active functionality by hydrolysis the retro-Claisen cleavage takes place. Thus, the tandem processes were involved in a synthesis of brefeldin-A [LeDrian, 1982] in which the cleavable cyclobutanone unit was incorporated in the photoisomer of α-tropolone methyl ether. In a synthetic approach to *erythro*-juvabione [Ficini, 1974] the retro-Claisen fission substrate was obtained by ynamine cycloaddition. For an access to methyl dehydrojasmonate [S.Y. Lee, 1988] the release of the acetic ester chain followed an oxidative cleavage of an exocyclic methylene group.

brefeldin-A

erythro-juvabione

Further examples include establishment of the bridged carbocyclic systems of atisine [Guthrie, 1966] and related substances from the [2 + 2]-photocycloadducts of cycloalkenones and allene. The key reactions consist of oxidative cleavage of the exocyclic methylene group and retro-Claisen cleavage prior to accessing the carbon framework of the synthetic targets. Particularly interesting is a reaction sequence leaving to a tricarbocyclic intermediate of stemodin [Piers, 1982] in which three tandem reactions were shown to occur.

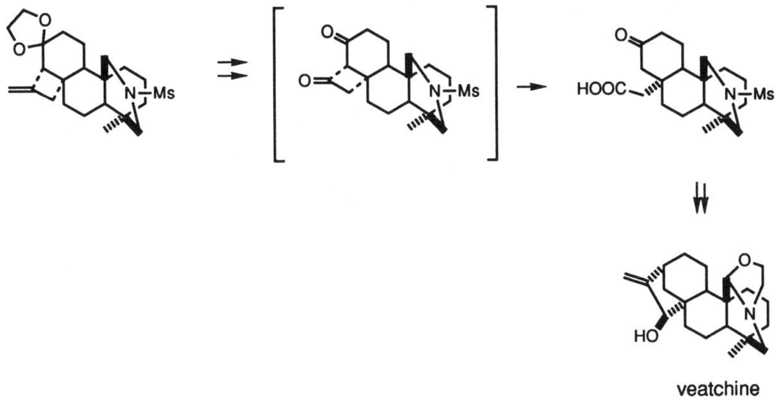

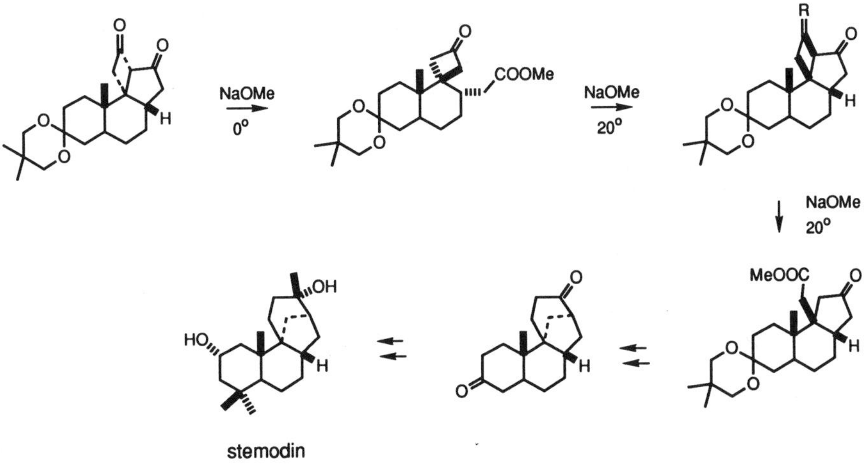

veatchine

stemodin

ent-Kaurene was correlated with phyllocladene by its conversion into an enantiomeric tricyclic keto ester [Cross, 1963] by virtue of a Claisen–retro-Claisen tandem on a diastereomer.

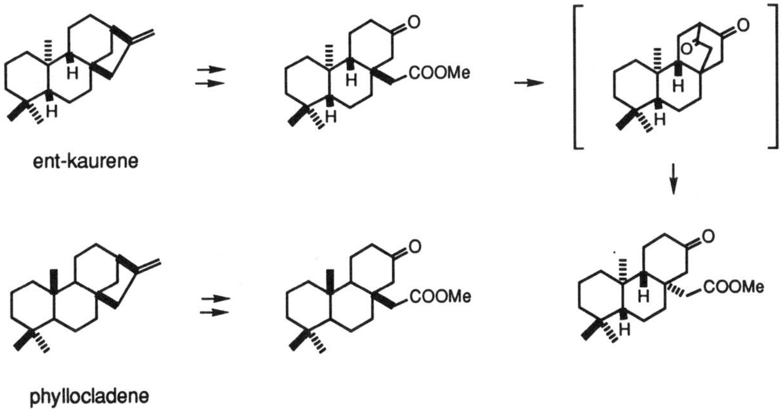

ent-kaurene

phyllocladene

The reaction of 3-methyl-2-cyclohexenone with methyl sodiocyanoacetate has been found to proceed via a Michael addition and a Claisen–retro-Claisen reaction sequence [Hill, 1975].

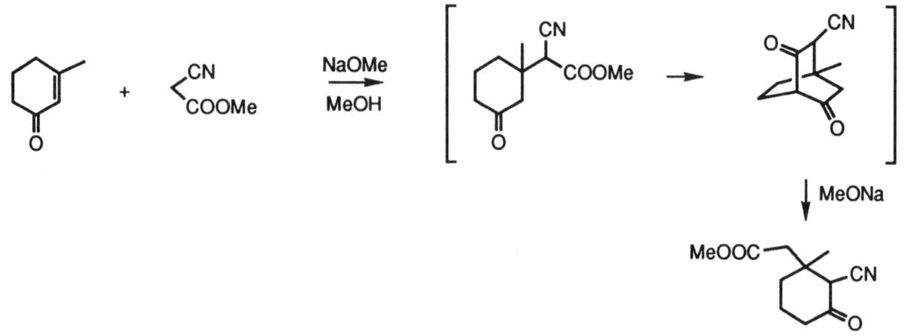

The Japp–Klingemann reaction of β-keto esters involves formation of diazo intermediates which undergo deacylation very readily. Its service in the synthesis of 2-indolecarboxylic esters is unique, and a recent application is found in an approach to methoxatin [Corey, 1981].

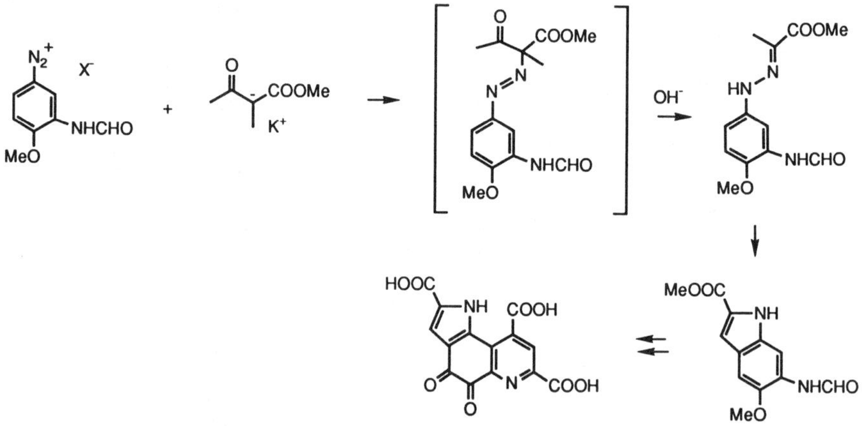

methoxatin

In the classic synthesis of quinine [Woodward, 1945] the alicyclic portion of a *cis*-decahydroisoquinolone intermediate was split to furnish two carbon chains by reaction with ethyl nitrite.

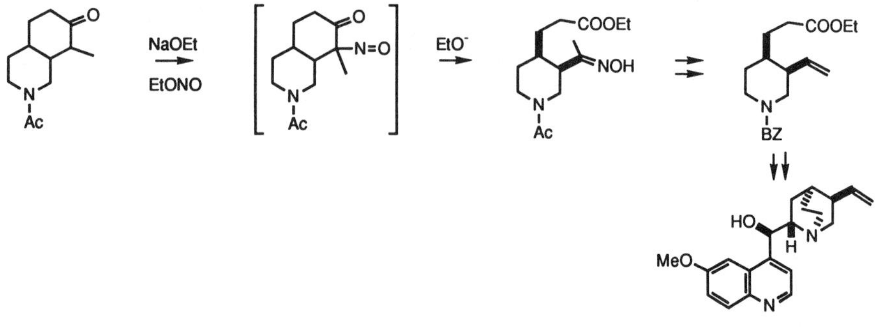

quinine

A very unusual route to methyl geranate started with fragmentation of a bicyclo[3.3.0]octanone epoxide [Tsuzuki, 1977]. The two rings were destroyed by isomerization to a monocyclic diketone and a subsequent retro-Claisen fission.

The reaction of β-diketones such as 1,3-cyclohexanedione with 1,3-propanedithiol ditosylate in the presence of a base leads to 2-acyl-1,3-dithianes [Bryant, 1975]. Diacyldithianes are the intermediates which suffer deacylation.

β-Iminoketones are also subject to analogous cleavage reactions. Such a reaction constitutes part of a synthetically most useful transformation in which a 3-acylindolenine generated from photoisomerization of an *N*-acylindole is attacked by a primary amino group intramolecularly [Ban, 1983]. The mesocyclic lactam has served as intermediate for eburna, aspidosperma, and strychnos alkaloids.

quebrachamine

tubifolidine

ebumamine

strempeliopine

When dehydration is unfavorable a β-ketol system may undergo two other pathways of CC bond fission, involving either a retro-aldol process or deacylative elimination. A combination of intramolecular aldol condensation and deacylative fragmentation has been developed into a powerful methodology for assembling new skeletons, as demonstrated by its applications to the synthesis of buslnesol [M. Tanaka, 1988], acorenone-B [Nagumo, 1990], trichodiene [M. Tanaka, 1991].

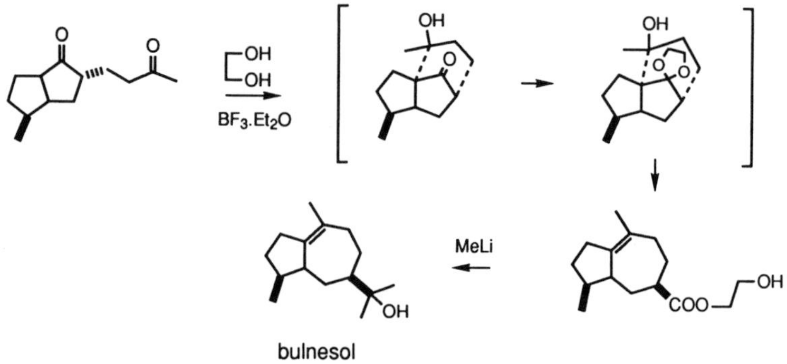

bulnesol

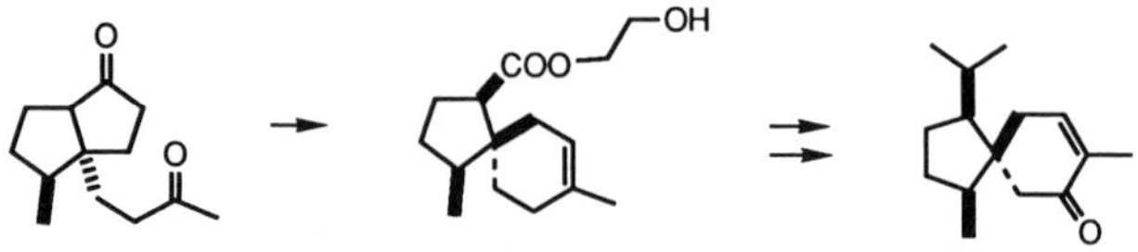

acorenone-B

A method for reductive succinoylation of ketones [Nakamura, 1977] consists of reaction of the corresponding acetals with 1,2-bis(trimethylsiloxy)cyclobutene in the presence of Lewis acids such as $AlCl_3$, $TiCl_4$, $SnCl_4$, and $SbCl_5$. Under these conditions the aldol–Mukaiyama condensation products undergo ring-opening deacylative elimination to give the enol silyl ethers.

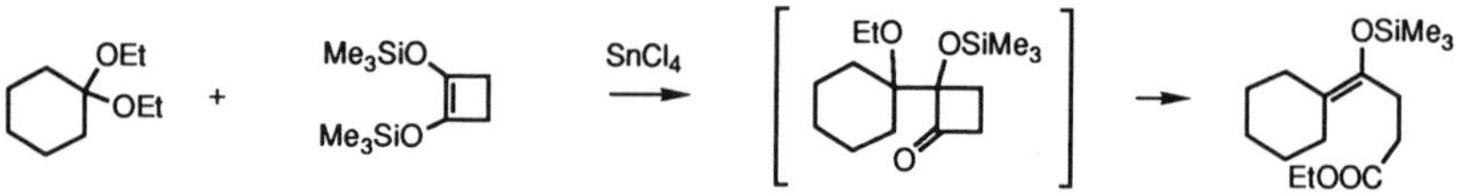

13.2.2. Retro-Mannich Reactions

Deacetoxyneoberberine acetone splits off the three-carbon bridge on acid treatment [S. Naruto, 1975] as a result of a double retro-Mannich reaction.

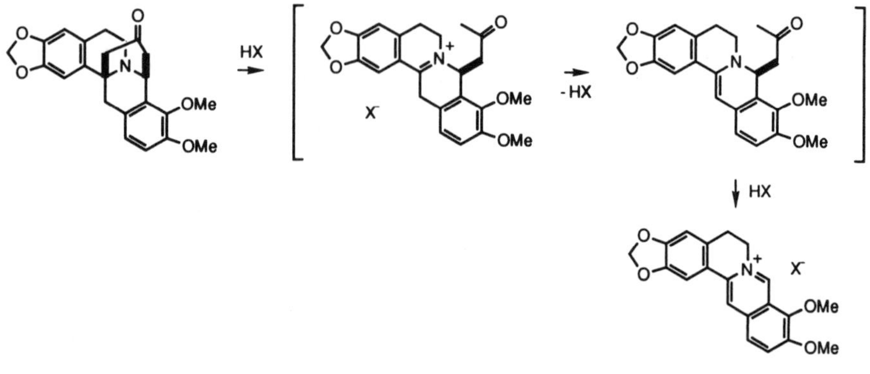

The transformation of a 4,4-dichloro-1-azaadamantanedione to the 5-chlororescorcinol on treatment with triethylamine in aqueous ethanol [Risch, 1991] follows the pathway of a Grob fragmentation and two retro-Mannich reactions.

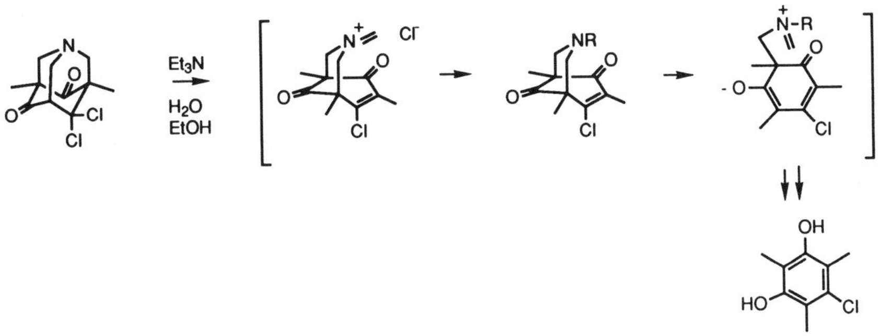

The Mannich–retro-Mannich reaction couple in either direction is a common occurrence in indole alkaloid chemistry. For example, its intervention in the conversion of a formal hydroxylation product of vincadifformine to vincamine by acid [Hugel, 1981] is rather convincing.

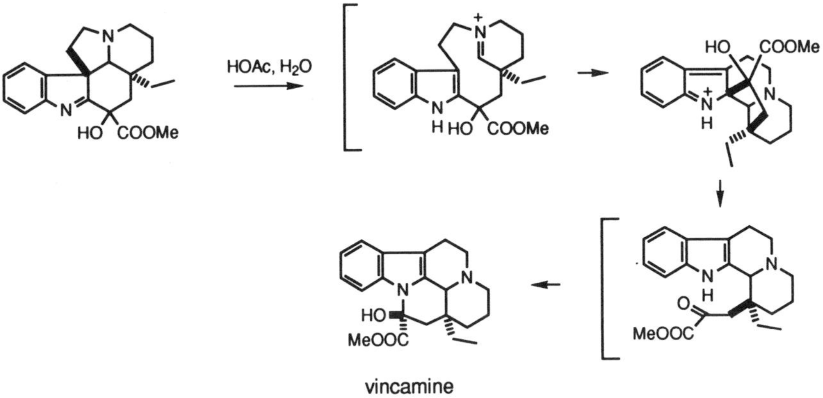

vincamine

A retro-Mannich reaction following an intramolecular alkylation of a diacetylpiperidinylindole permitted the formation of a substituted carbazole which is a convenient precursor of olivacine [Naito, 1986].

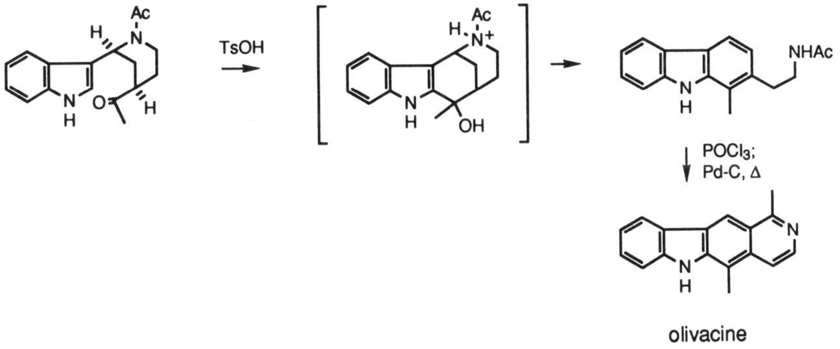

olivacine

The thermal rearrangement of an azepinedicarboxylic ester to an aminofulvene derivative [Childs, 1967] has been considered proceed via an intramolecular Diels–Alder cycloaddition and subsequent retro-Mannich reaction.

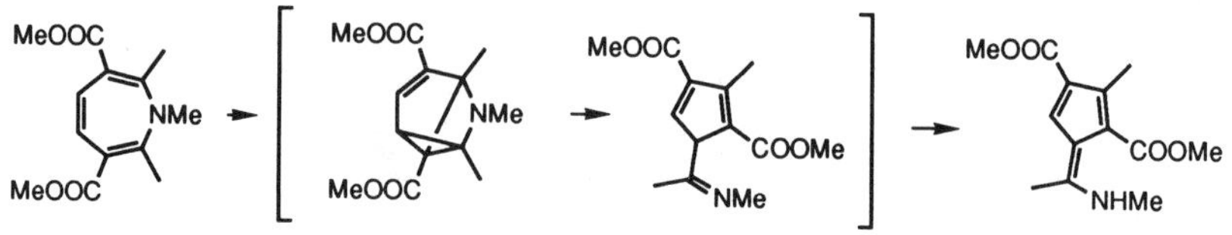

Photochemical rearrangement of thebaine under reducing conditions led to a mixture of two dibenzo[d,f]azonine alkaloids, neodihydrothebaine and bractazonine [Theuns, 1984] in a 4:1 ratio. The rearrangement set up fragmentable systems.

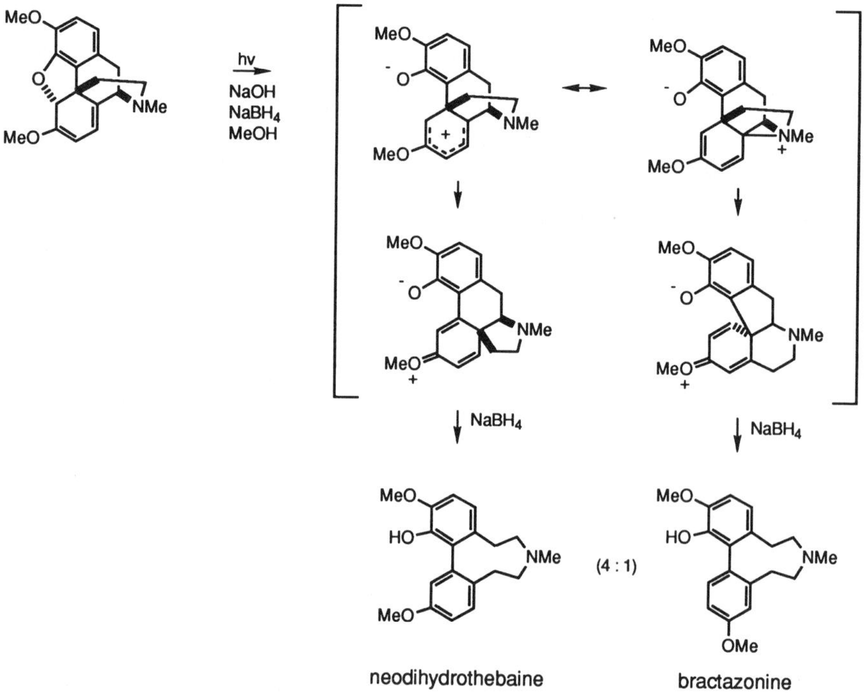

(4 : 1)

neodihydrothebaine   bractazonine

13.2.3. Grob Fragmentation

The Grob fragmentation reaction which generates a carbonyl group and an alkene is very well studied mechanistically. Its synthetic value has been enhanced by coupling to some other reactions. In other words, these other reactions are used to set up the fragmentable species to achieve particular skeletal transformations.

A case in which peripheral fragmentation of a decalin derivative as a key step toward synthesis of saussurea lactone [M. Ando, 1978] was hinged on an isomerization of an epoxy mesylate. Note that the unsaturated aldehyde obtained from the fragmentation was reduced in situ.

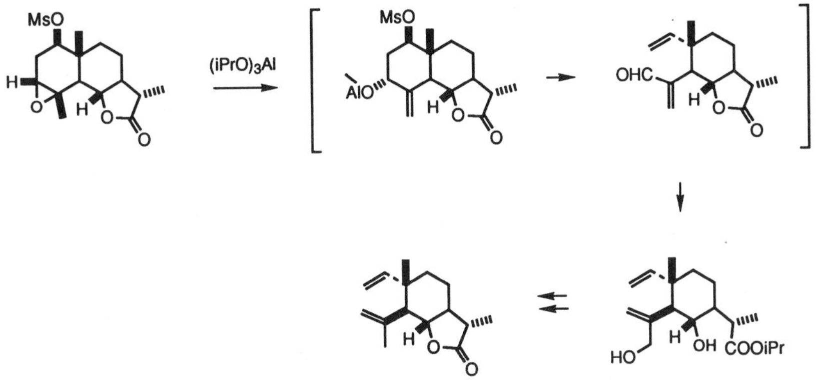

saussurea lactone

A synthetically significant fragmentation concerns with the preparation of cyclodecadienes from octalin sulfonates in which the leaving group is homoallylic and subangular. Thus, hydroboration followed by treatment with alkali accomplishes the desired transformation which constituted the crucial step in a synthesis of hedycaryol [Wharton, 1972]. A similar intermediate was used in an approach to globulol [Marshall, 1974].

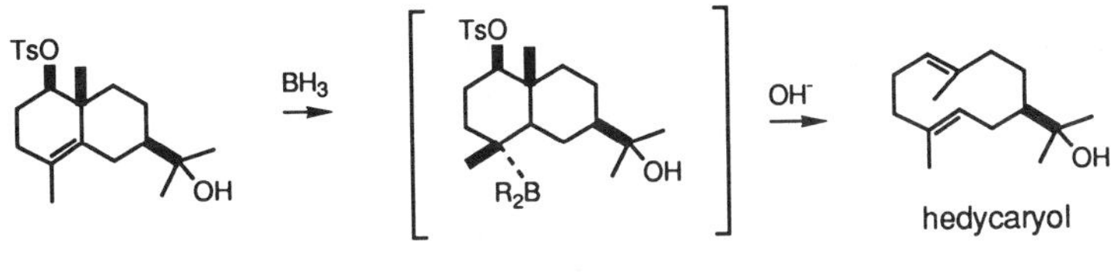

hedycaryol

Certain β,γ-epoxy tosylates are liable to hydrate and become fragmentable [Gray, 1977].

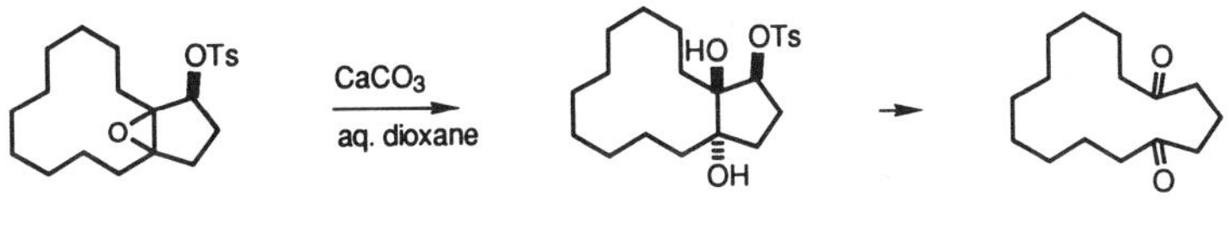

A synthesis of (+)-hinesol [Chass, 1978] conceived the closure of the six-membered ring by a Dieckmann-type condensation simultaneously with the generation of the three-carbon sidechain in the cyclopentane moiety. Thus, a β-tosylaoxy ketone was set up for the tandem transformation by elaboration of β-pinene.

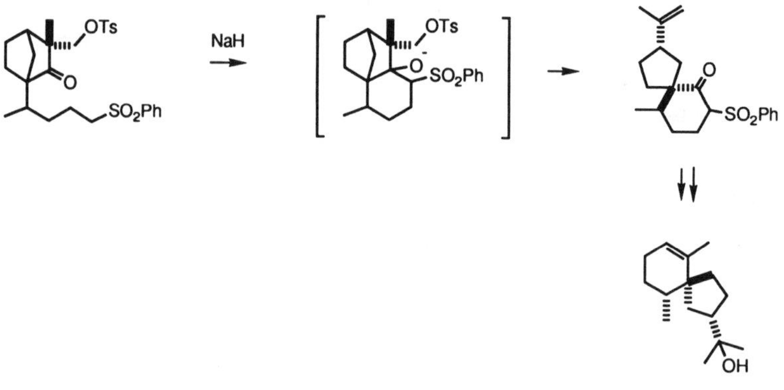

hinesol

Protonation of the methylenecyclopropene system is favored by its formation of an aromatic carbocation. The accessibility of the hydrates, even in minute amounts in equilibrium, of methylenebromocyclopropenes avails a fragmentation pathway for strain relief [Billups, 1976].

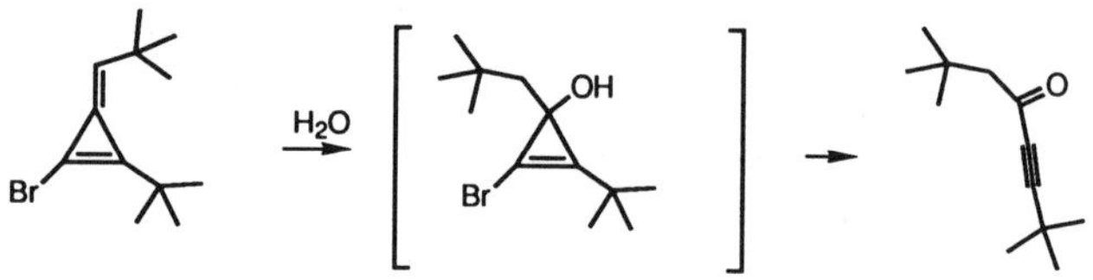

A Grob fragmentation occurred upon hydride reduction of bischloromethyl-tricyclo[$2.1.0.0^{2,5}$]pentanone [Dowd, 1982].

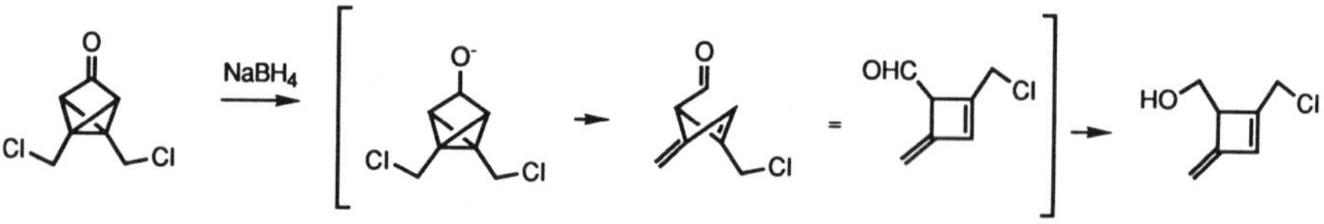

Cleavage of the C-9/C-10 bond of $9\alpha,11\alpha$-epoxy-1,4-cholestadien-3-one accompanied its reduction with lithium aluminum hydride [Lythgoe, 1980]. The fragmentation is favored by aromatization of the A-ring.

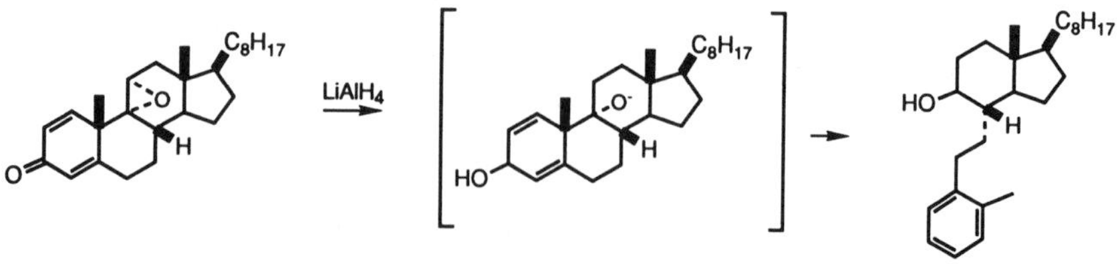

An interesting alkylation of acetoacetic esters with substituted α-chloro-α'-haloacetones gives 2-acrylylacetoacetates [Sakai, 1985]. Under the reaction conditions a Favorskii rearrangement of the dihaloacetone takes place, and the ensuing 2-chlorocyclopropanone is intercepted by the β-keto ester. Fragmentation represents the best course presented to the adduct.

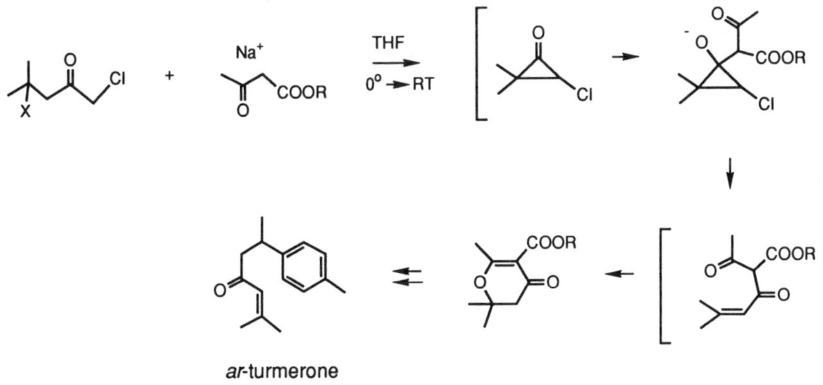

ar-turmerone

The reaction of 17-halo-16α-methyl-20-oxosteroids with dimethyloxosulfonium methylide affords products with a cyclobutanone spiroannulated to C-17 [Wiechert, 1970]. This transformation must be initiated by a Favorskii rearrangement, but the cyclopropanone intermediates are subject to attack by the ylide. Ring expansion with expulsion of dimethyl sulfoxide terminates the reaction.

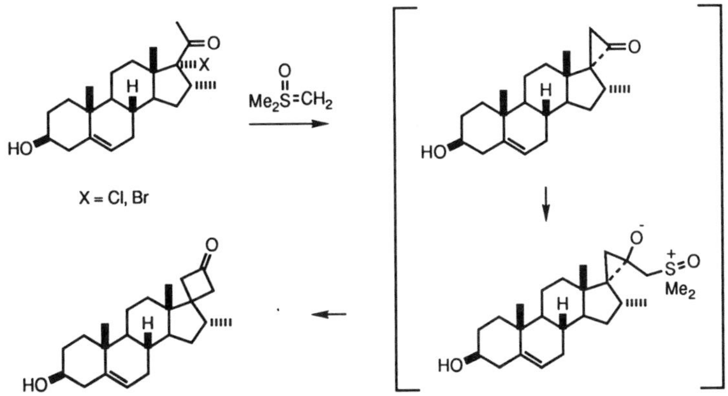

X = Cl, Br

The mechanism for the transformation of the dichloroketene adduct of cyclopentadiene to tropolone under acetolytic conditions has been elucidated [Bartlett, 1970]. Enolization and S_N2' displacement of a chlorine atom precede a Grob-type fragmentation to sever the intercyclic bond.

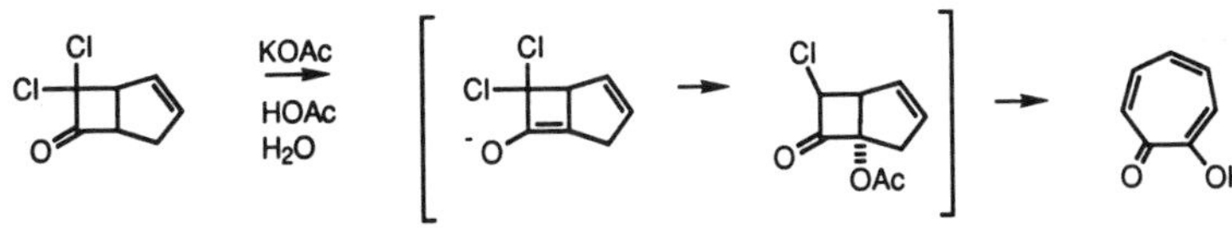

Facile fragmentation of the Paterno–Büchi reaction products of 3′,5′-dimethoxybenzoin esters is the crucial feature of a method for protection of carboxylic acids as these esters [Sheehan, 1971]. Thus the mild and selective deprotection procedure of photolysis depends on the generation of fused oxetanes from which the expulsion of carboxylate ions is assisted by the methoxy groups. This process is closely related to the Grob fragmentation.

13.2.4. Other Fragmentations

Two retro-Michael reactions ensue upon the reaction of pentacyclic diketones, which are derived from cyclopentadiene and 2,5-dihalo-1,4-benzoquinones, with excess ethyl diazoacetate in the presence of boron trifluoride etherate [Marchand, 1988]. The diketone becomes fragmentable when a carbon atom is inserted within the disjointed circuit.

To conclude this chapter on fragmentation and rearrangement it is perhaps appropriate to mention the behavior of dipyrromethanes toward acids. These substances tend to fragment and recombine to afford mixtures of compounds when the two pyrrole rings are not identically substituted [Jackson, 1989]. This reactivity has critical significance in the biosynthesis of porphyrins.

A = acetic
P = propionic

14

FREE RADICAL REACTIONS

The synthetic perspectives of free radical reactions have undergone an enormous change in recent years. Previously, such reactions were ignored or avoided by synthetic chemists except those working in the polymer field. In other words, chain reactions which are valued by the polymer chemists are for the most time detrimental to others who deal with monomeric compounds. Renewed appreciation and new knowledge of free radicals, and particularly the design of proper substrates to control reaction patterns, have contributed to accelerate the integration of free radical reactions into the repertoire of synthetic methodology. As numerous free radical reactions effect the formation of more than one bond, and these processes are nonsynchronous, they are naturally tandem reactions.

The chemoselectivity of free radical reactions is due to the limiting factor specifying only those functional groups that have bimolecular rate coefficients $k > 10^2$ (1/mol.s) are attacked. Consequently, carbon-centered radicals display substantial chemoselectivities, allowing their use in reactions with complex molecules. As an example, the common practice or necessity of OH and NH protection is exempted.

The ambident reactivity pattern of conjugated carbonyl compounds toward nucleophiles is suppressed in radical reactions. The carbonyl group is untouched, as orbital interactions are the dominant factor.

The different transition states favored by radical and ionic reactions are reflected in the behavior of 5-hexen-1-yl radical and the cationic counterpart toward cyclization. The preferred unsymmetrical transition state for the radical cyclization leading to a five-membered ring is faster, owing to ring strain effects (cf. Baldwin rules). Contrarily, the cyclohexyl cation is formed as a result of higher electron density at the terminal carbon of the olefin. The complementarity of ionic and radical processes to provide solutions to synthetic problems is evident.

The very fast radical reactions commonly used in synthesis have the implications that racemization of adjacent or remote centers of chirality and radical rearrangements are relatively rare.

14.1. VICINAL CC BOND FORMATIONS

Pioneering work in the formation of polycyclic compounds by free radical reactions was laid down [M. Julia, 1971; 1974] with emphasis in stereochemical and synthetic aspects. A synthesis of the drimane skeleton [Breslow, 1968] using such a method is remarkable with respect to the high degree of stereoselectivity.

A popular method of free radical generation is by dehalogenation of halocarbons with organotin hydrides. When applied to halo compounds containing a sterically accessible CC π-bond, cyclization with radical transfer to a new position is expected. The new radical may then react with a trapping agent. With limitation of the tin hydride (initiator) concentration the cyclization is uncomplicated since the intramolecular CC bond formation is entropically favored. The great synthetic potential of the tandem reactions (cyclization–trapping) can be appreciated by noting the simple and stereocontrolled assembly of the steroid CD-ring synthons [Stork, 1983b] and an elegant approach to prostaglandin F_{2_α} [Stork, 1986].

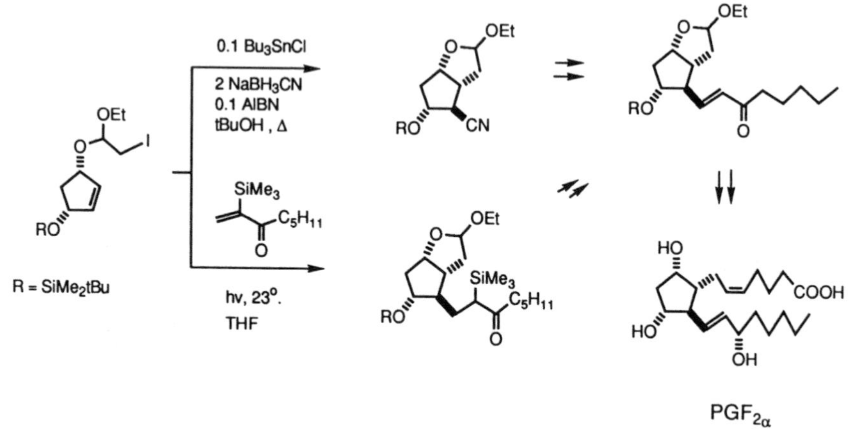

From the above examples the synthetic utility of β-bromoacetals is manifested. The advantage is the retention of residual functionality near the initiating site. Another interesting feature of the reaction is the generation of quaternary carbon centers without complications despite the sensitivity of free radical additions to steric effects.

The involvement of alkyne linkages in the role of free radical transmitter or intermediary trapping agent is shown in the following synthetic applications to a model of cardiac aglycones [Stork, 1983a], prostaglandin F_{2_α} [Stork, 1990a] and isoiridomyrmecin [Kilburn, 1990].

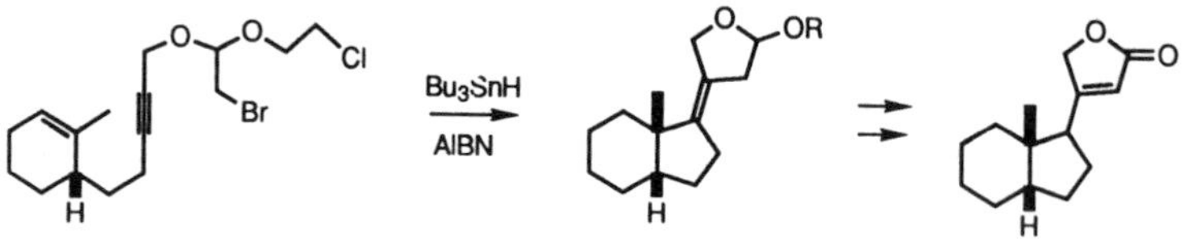

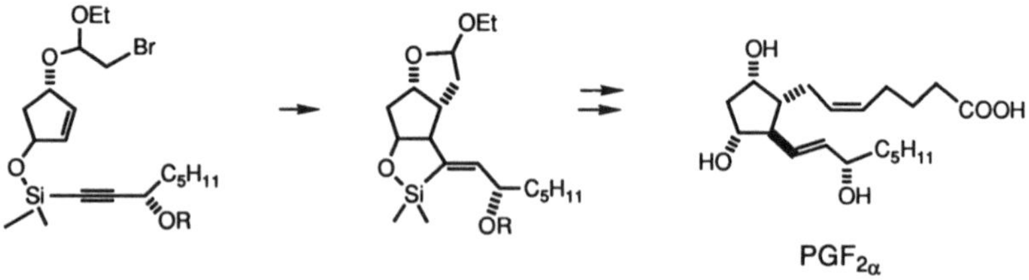

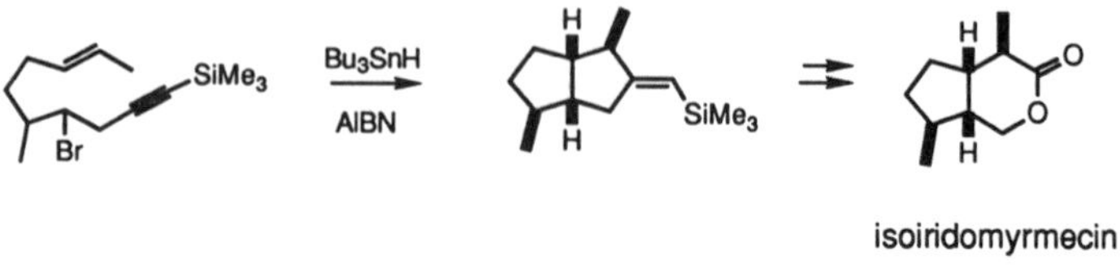

isoiridomyrmecin

Aryl and vinyl radicals are also readily generated by treatment of the corresponding halides with a tin hydride. Thus, the oxidohydrophenanthrene skeleton of the morphine alkaloids has been acquired by an aryl radical initiated intramolecular cycloaddition across the cyclohexane double bond and trapping with an oxime ether [Parker, 1988].

α-NHOMe 40%
β-NHOMe 31%

A vinyl radical transmission via addition to a triple bond resulted in a conjugate diene system which is susceptible to attack by tributyltin hydride [Parsons, 1988]. The product contains an allyltin residue.

The lysergane skeleton has been assembled very efficiently by a twofold vicinal addition [Cladingboel, 1990].

X = SPh

X = OMPM

Acyl radicals seem best to be generated from *Se*-phenylseleno esters, also by treatment with a tin hydride. These radicals participate in CC bond formation with an intramolecular double bond at rates greater than hydrogen abstraction from tributyltin hydride (reduction) and decarbonylation, therefore they are very useful for synthesis of cyclic systems.

Tandem 6-*endo-trig*/5-*exo-dig*, 6-*endo-trig*/6-*exo-dig* or 6-*endo-trig*/6-*exo-trig* free radical cyclizations have been demonstrated, as well as cyclization–intermolecular additions and addition–cyclization reactions [Boger, 1990a].

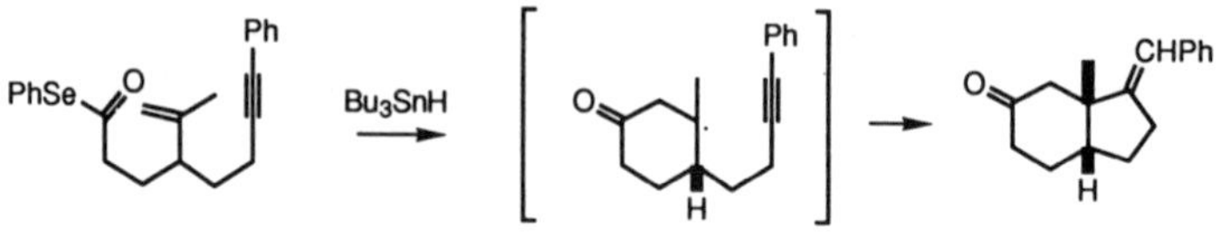

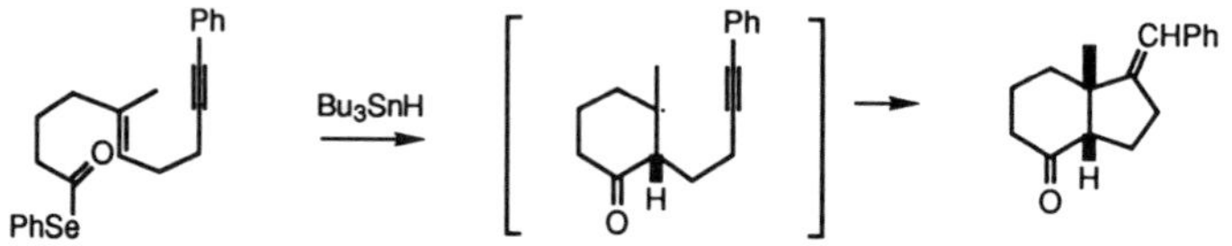

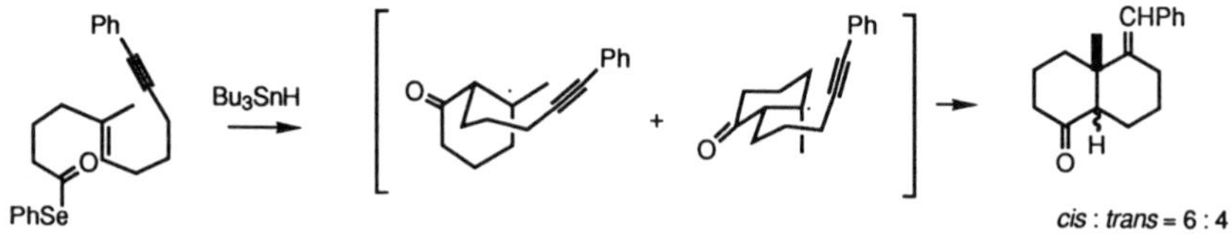

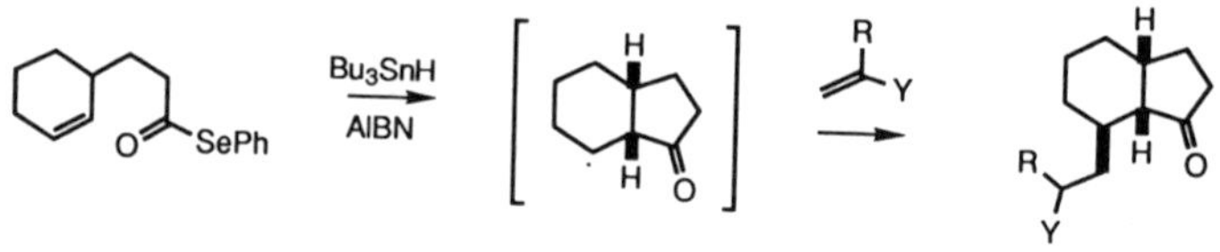

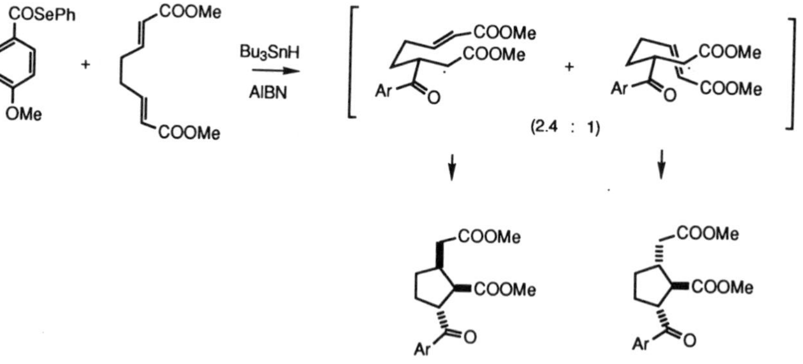

cis : trans = 6 : 4

(2.4 : 1)

The diastereomeric products of the intermolecular addition arose from predominantly 1,5-*cis*- and exclusively 1,2-*trans*- diastereoselective cyclization of the hex-5-enyl radicals. The intramolecular 5-*exo-trig* cyclization apparently proceeded faster than C-1/C-2 bond rotation.

A 5-hexenyl radical generated by additive transfer of a homoallylic radical to an acrylic ester can undergo cyclization to form a five-membered ring. Further reaction with another molecule of the acrylic ester leads to a 1 : 2 adduct [Barton, 1988].

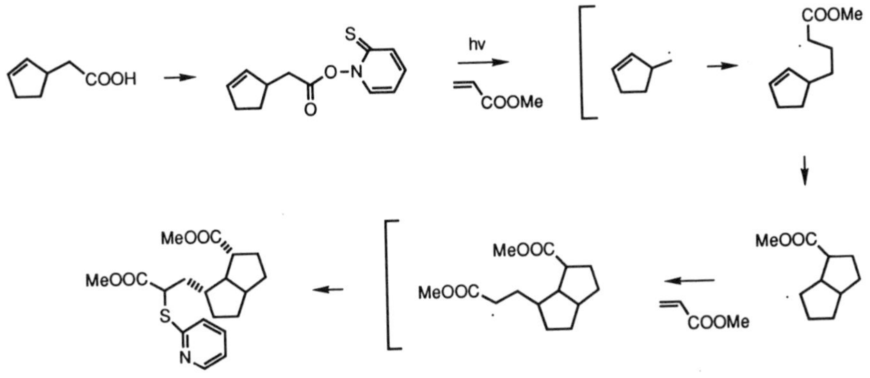

Imidoyl radicals derived from selenoimidates behave similarly to the corresponding acyl radicals. The formation of a chromanodihydroquinoline system [Bachi, 1989] is readily understood.

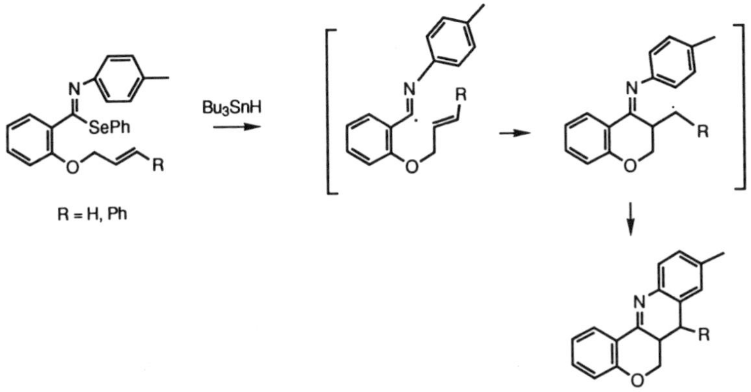

The addition of a 5-alkyn-1-yl radical to phenyl isocyanide leads to cyclopentano[b]quinoline derivative [Curran, 1991a].

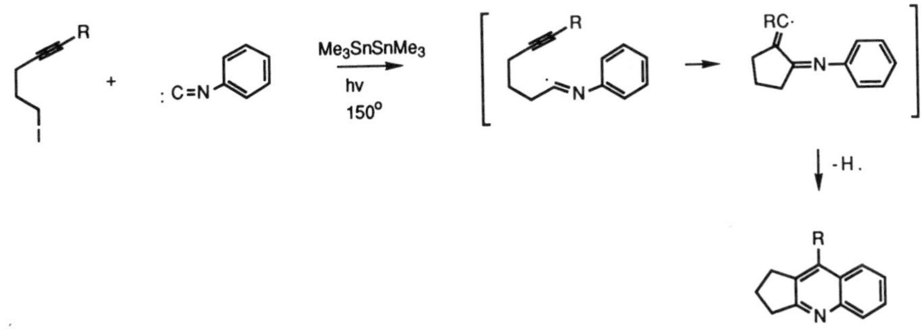

Establishment of the heterocycle (D-ring) and stereoselective introduction to the angular substituents of quassinoids have been accomplished in one operation [Kim, 1991] by a tandem intramolecular cyclization process initiated by an alkoxymethyl radical.

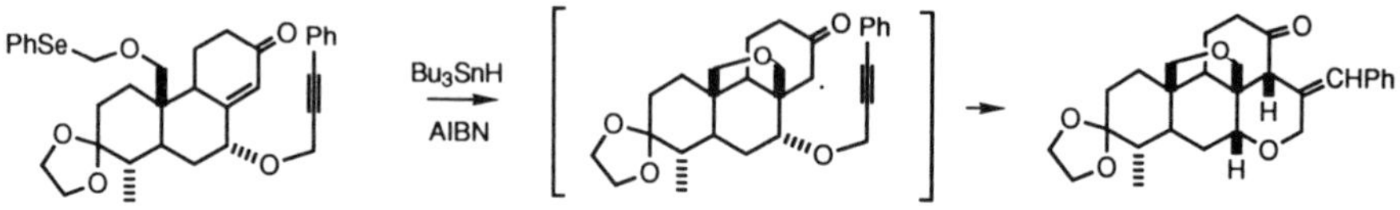

Homoallylic xanthates containing another allyl or propargyl group cyclize on treatment with tributyltin hydride to give thionolactones annulated to a cyclopentane [Iwasa, 1991].

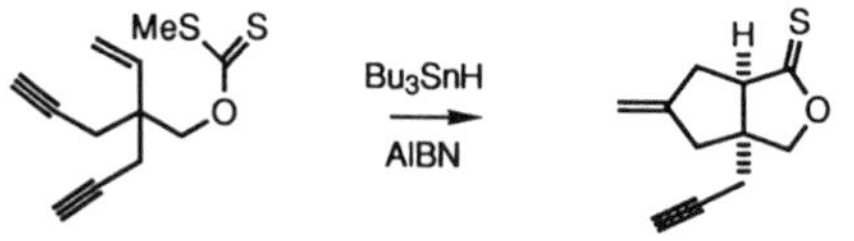

The diquinane nucleus of (−)-α-pipitzol has been constructed by the free radical cyclization [Pak, 1990] of a cyano enyne which is embedded in a pyranose template. The cyano group acted as the trapping agent.

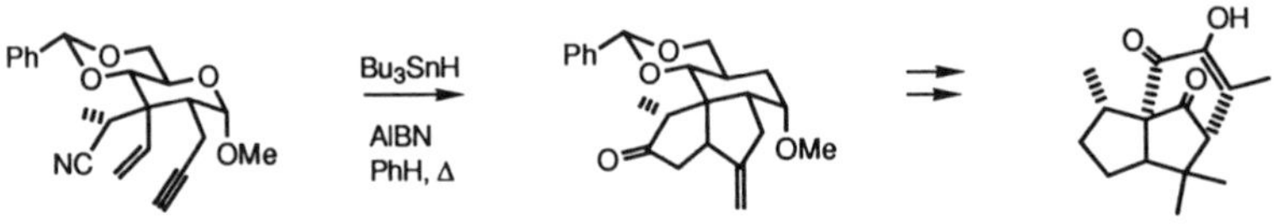

(-)-α-pipitzol

The predilection of radical cyclization to form five-membered rings has invoked a unified strategy for synthesis of triquinane sesquiterpenes [Curran, 1990]. To construct linear and angular triquinanes a cyclopentene ring serves as the switching post to receive and dispatch the carbon radical. However, propellanes cannot be formed by analogous double cyclization.

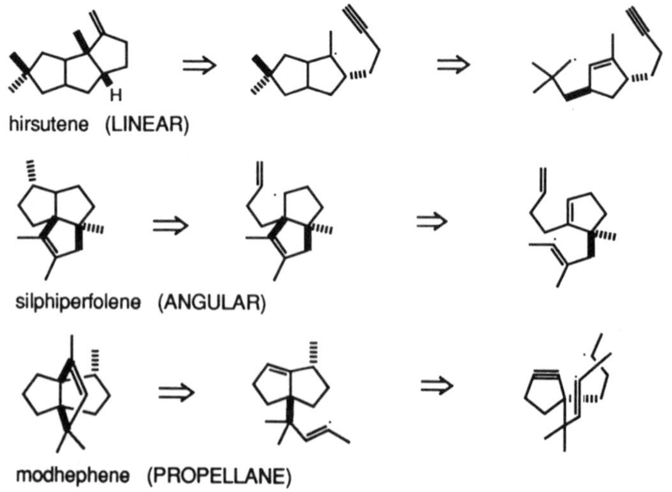

hirsutene (LINEAR)

silphiperfolene (ANGULAR)

modhephene (PROPELLANE)

There are at least three approaches to the linear triquinanes. With respect to natural product synthesis the presence of an exocyclic methylene group in hirsutene and $\Delta^{9(12)}$-capnellene favors the "forward" route of tandem cyclization.

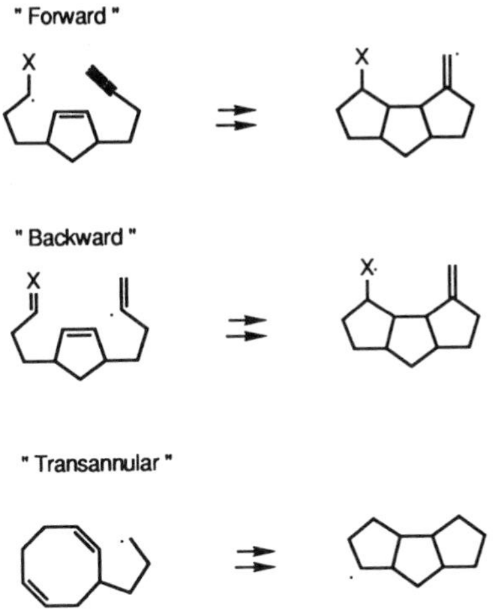

" Forward "

" Backward "

" Transannular "

As expected, the cyclization steps are stereospecific, the stereochemistry of the ring junctures being determined at earlier stages when the substituents were introduced into the cyclopentene nucleus [Curran, 1985a; b].

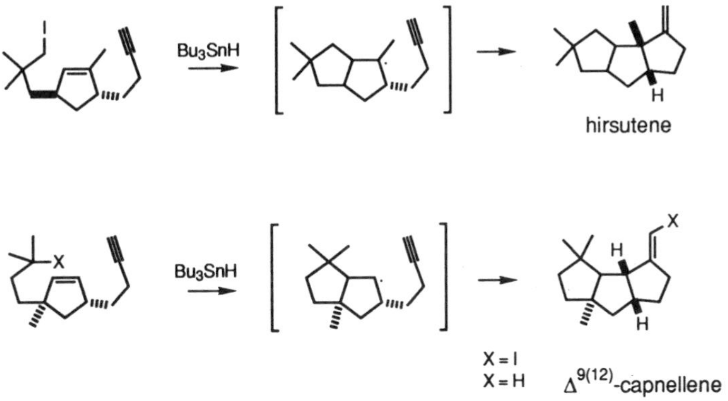

Bu₃SnH

hirsutene

Bu₃SnH

X = I
X = H $\Delta^{9(12)}$-capnellene

With catalytic amounts of the tin hydride, iodine atom transfer occurs. This variant of radical cyclization is valuable because there is much less tin-containing co-products, and more highly functionalized compounds are obtained.

There are also three modes of radical-mediated double cyclization leading to an angular triquinane. In fact, two of these would be different only if the existing cyclopentene ring or the two sidechains of the substrates carry additional substituents. The other strategy involves radical transfer through a conjugate diene unit.

" Forward " Cyclopentene Strategy

" Backward " Cyclopentene Strategy

Cyclopentadiene Strategy

A synthesis of silphiperfolene [Curran, 1987] was accomplished by means of a vinyl radical initiated cyclization. The addition was terminated by a vinyl group, the undesirable steric course that led to the preponderant production of the *endo*-methyl epimer could be reversed by ketalization of the cyclopentenone carbonyl.

3 : 1 α-Me : β-Me isomers

2.5 : 1
β-Me : α-Me isomers

silphiperfolene

In the above synthesis the carbonyl group was present only because it helped the preparation of the substrate. In the event it accorded an unexpected advantage to the synthesis by modifying the orientation of the vinyl acceptor after ketalization.

Oxygenated linear triquinanes such as hypnophillin are in principle approachable by the "backward" route (scheme on p. II) employing an aldehyde as the radical trap at the end of the double cyclization. However, only hydrostannation of the triple bond was observed.

The problem was solved by reverting to the forward scheme in which the α-oxy radical was generated by one-electron reduction of the aldehyde with samarium(II) iodide. Tandem cyclization furnished the triquinane precursor of hypnophilin.

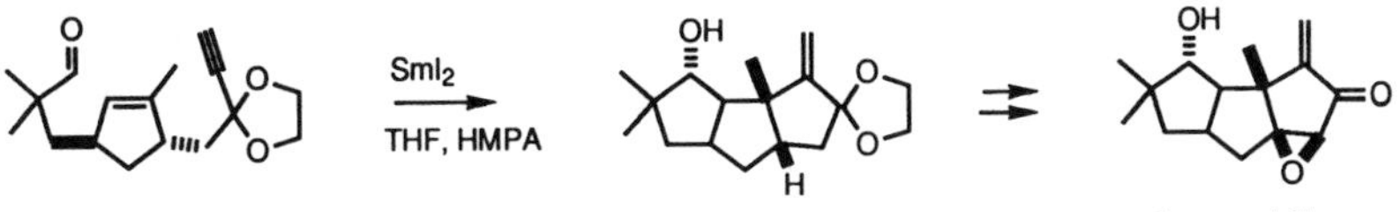

hypnophilin

An example of cathodic reduction of allyl pentenyl ketones [Kariv-Miller, 1989] which resulted in the formation of some bicyclo[3.3.0]octanols should be mentioned.

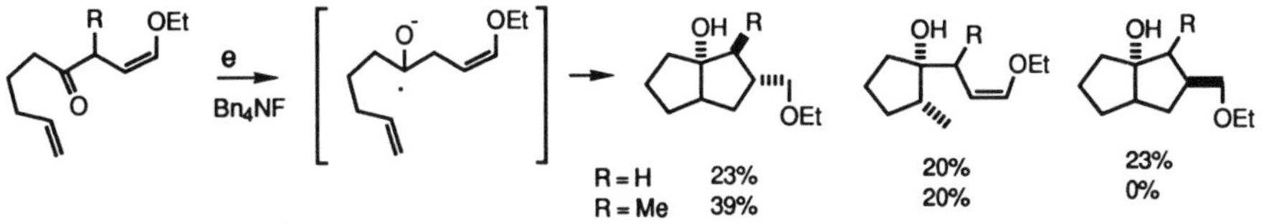

	R = H	R = Me
	23%	39%
	20%	20%
	23%	0%

Radicals can be generated by oxidation of active C—H compounds such as β-diketones, β-keto esters, etc. with manganese(III) ion. Certain polycyclic skeletons have been created by this method [Dombroski, 1990; Snider, 1991] via tandem vicinal CC bond formation.

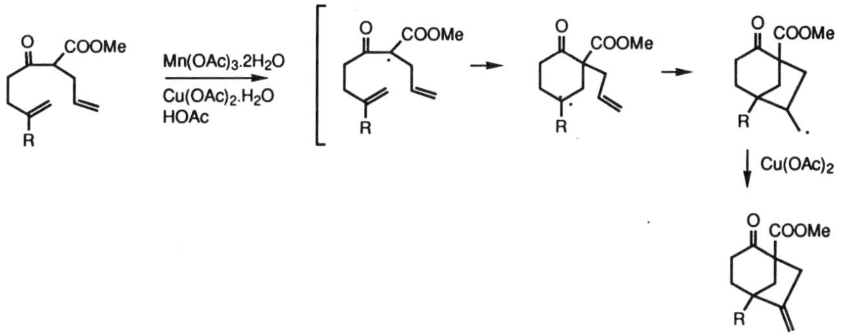

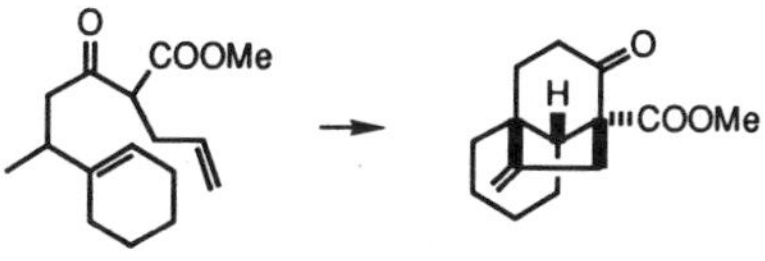

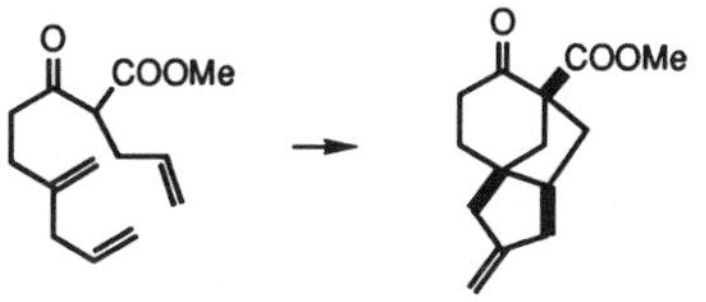

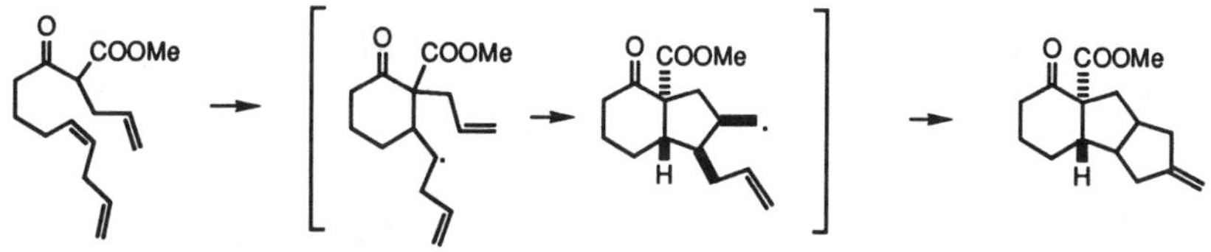

Oxidative cyclization of δ,ε- and ε,ζ-unsaturated enol silyl ethers derived from phenones can lead to tricyclic ketones [Snider, 1990]. The aromatic ring serves as the radical trap.

14.2. 1,3-DIYL CYCLOADDITIONS

Singlet trimethylenemethanes (2-methylene-1,3-propanediyls) embedded in a five-membered ring are readily captured by electron-deficient alkenes. The diylophilicity of such alkenes toward the diradicals roughly parallels their dienophilicity; and beneficial to synthetic ventures aiming at application of diyl capture, some poor dienophiles (e.g. cyclopentenones) behave remarkably well. Intramolecular diyl trapping generally proves useful in quinane synthesis.

With part of the diyl system in a cyclopentane ring the cycloaddition yields diquinane derivatives. It is natural to extend the utility in the area of triquinane synthesis, in view of the discovery of biologically active substances with triquinane skeletons.

Convenient diyl precursors are bicyclic diazenes which are available from fulvenes and azodicarboxylic esters. Accordingly, selective hydrogenation of the strained double bond of the Diels-Alder adducts, hydrolysis, and oxidation furnish the diazenes. With respect to synthesis of natural triquinane sesquiterpenes the regiochemistry and stereochemistry of the ring junctures must be resolved. The intramolecular version seems to provide the solution.

Experimental results indicate that conformational and nonbonded interactions are important factors in the diyl trapping reactions while electronic effects play a relatively insignificant role. The intramolecular diyl trapping proceeds without complication of dimerization even if the diylophile is unactivated, when the two reactants are separated by three bonds. Successful syntheses of hirsutene [Little, 1981], $\Delta^{9(12)}$-capnellene [Little, 1983], and coriolin [Van Hijfte, 1985] attest to the efficacy of this tool. Furthermore, variation of the tether length enables the elaboration of ring systems other than the linear triquinanes [Little, 1986], and the use of C=X compounds as diylophiles lead to heterocycles.

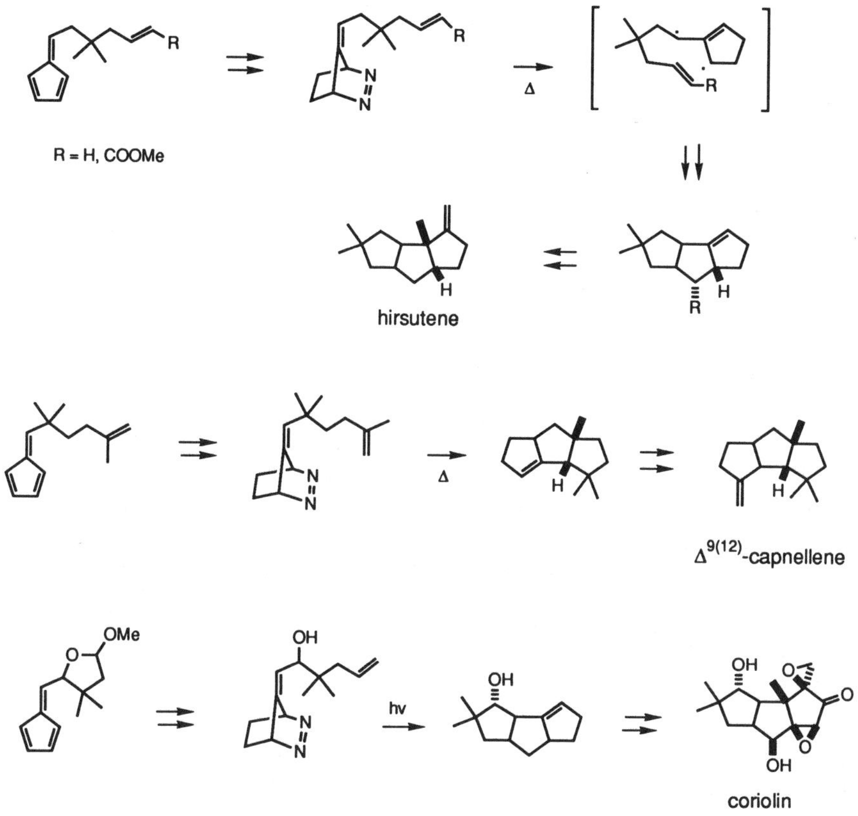

R = H, COOMe

hirsutene

$\Delta^{9(12)}$-capnellene

coriolin

At least in some cases the diyl trapping reaction is nonconcerted. This notion was suggested by an analysis of the products from the work related to the capnellene synthesis. Two newly formed rings made up a bicyclo[3.2.1]octane nucleus.

14.3. ADDITION–FRAGMENTATION TANDEMS

A formal insertion of a two-carbon unit to dimethyl 2-isopropylidenecyclopropane-1,1-dicarboxylate by means of free radical reaction with electron-rich alkenes [Singleton, 1991] involves addition of a thiyl radical to the double bond of the small ring compound, fragmentation, and reflexive reaction with the alkenes which is culminated in cyclization and regeneration of the thiyl radical.

Intramolecular radical addition to a carbonyl group may cause the formation of rearranged products. Thus, on treatment with tributyltin hydride, several 1-haloalkyl-2-oxocycloalkanecarboxylic esters have been transformed into cyclic keto esters in which the two functional groups are separated by the same number of carbon atoms as in the haloalkyl chain [Dowd, 1989]. The ring expansion proceeded via intramolecular alkyl radical attack on the ketone and fragmentation of the CC bond to give an ester-stabilized radical which is capable of hydrogen abstraction from the tin hydride reagent to renew the chain reaction.

Analogous bicyclic oxy radicals derived from β-stannylcycloalkanones undergo fragmentation to afford expanded cycloalkenones [J.E. Baldwin, 1989].

Further transformation may follow the ring expansion via the radical addition–fragmentation tandem. For example, the nascent cyclic carbon radical arising from homolysis of the intercyclic CC bond of such an adduct may add to a triple bond in a sidechain [Boger, 1990b].

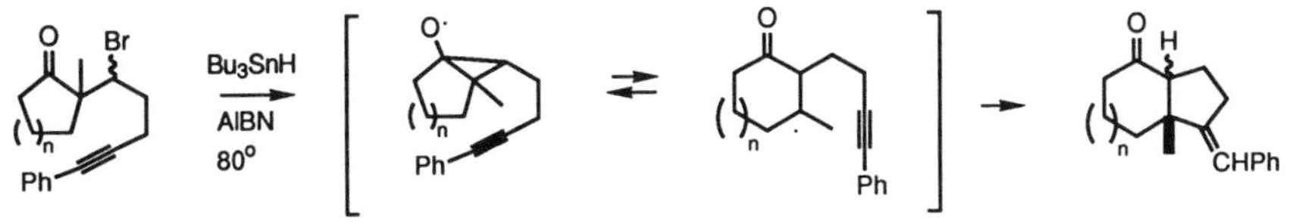

2-([ω-1]-alkynyl)cycloalkanones undergo a series of interesting reactions with tributyltin hydride [Nishida, 1990]. It is reasonable to assume that addition of the tin radical to the triple bond would be followed by an intramolecular attack of the alkenyl radical on the ketone. Fragmentation and readdition of the transposed radical to the vinylstannane effects an overall 1,2-rearrangement of the intermediate.

Structural comparison indicates the insertion of the internal sp-hybridized carbon of the alkyne group into the CC bond between the carbonyl and the proximal α-carbon atom, when the alkynyl chain is linked to the α-carbon via a heteroatom. On the other hand, the intermediate derived from a substrate with a homotypical chain may insert into the same position with both its sp-hybridized carbon atoms.

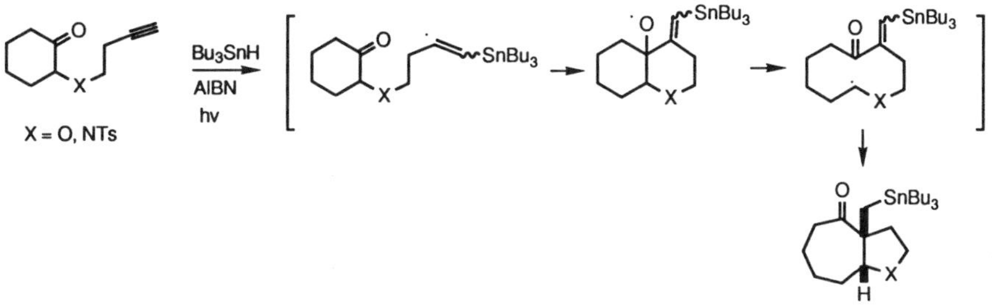

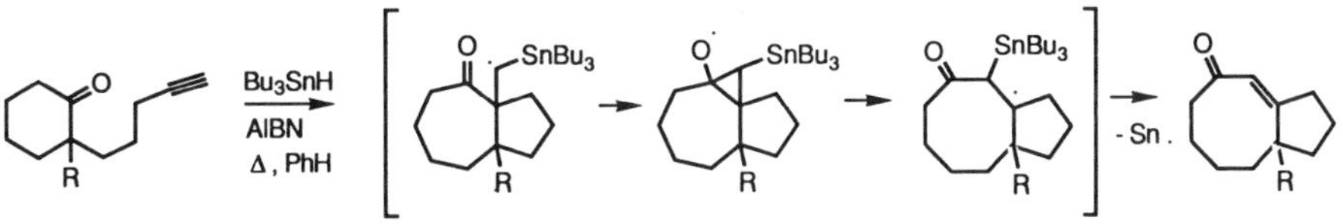

trans-α-Decalone is formed by the slow addition of tributyltin hydride to 7-(3-bromopropyl)bicyclo[3.2.0]-heptan-6-one [Down, 1991], when the oxohydrazulenyl radical is given an opportunity to undergo fragmentation.

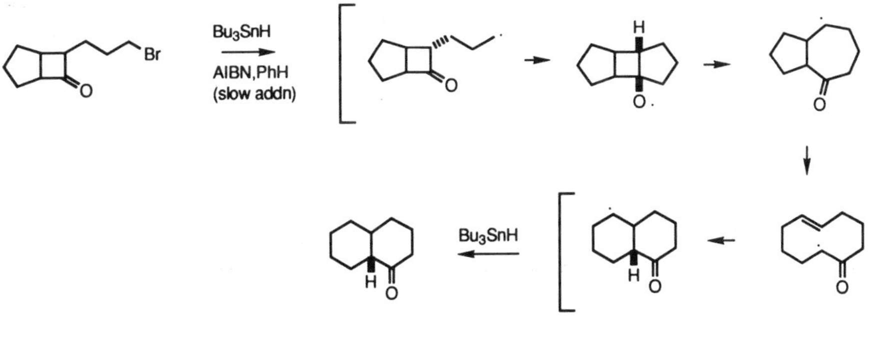

The oxidation of angular hydrindenols with mercuric oxide-iodine mixture may follow different pathways according to the initial CC bond homolysis [O'Dell, 1988]. Certain products may arise from re-interception of the carbon radical with the ketone.

14.4. REACTIONS OF CARBINYL RADICALS ADJACENT TO A SMALL RING

As a rule carbinyl radicals adjacent to a small ring fragment immediately upon their formation to relieve angular strains. What would take place next depends on the structural characteristics of the substrates. For example, conformationally rigid cyclopropylcarbinyl radicals may be directed to undergo homolysis to give the less stable primary radicals, and in the presence of a juxtaposed radical trap in a sidechain, very rapid ring formation would ensue. Stereospecific creation of spirocyclic systems is feasible [Harling, 1988].

A radical next to an oxirane ring behaves similarly in the generation of an allyloxy radical. However, hydrogen abstraction from a sidechain delivers a 5-hexen-1-yl radical which is prone to cyclization [Rawal, 1990]. The overall transformation of 1,2-reductive cyclization should find some applications in synthesis of complex ring systems.

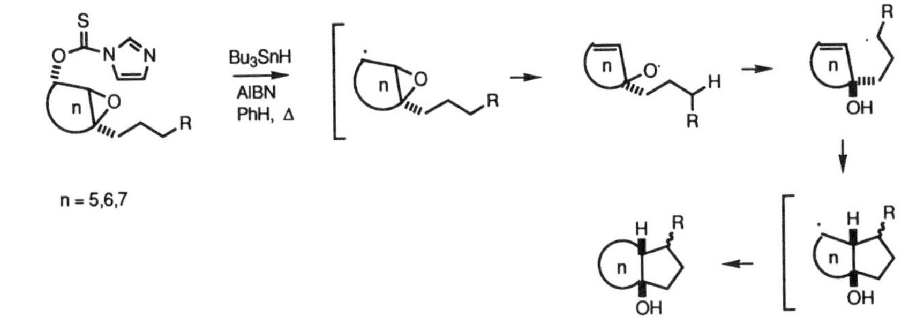

n = 5,6,7

14.5. MISCELLANEOUS RADICAL REACTION TANDEMS

Cyclopentadienes behave as radical transmitters, especially when both donor and acceptor are placed within the same molecule and individually accessible to one of the conjugate diene termini. This conjugate addition has been studied in the context of a crinipellin-A synthesis [Schwartz, 1990].

The tandem cyclization has been a problem, the requirement of obtaining a tricyclic intermediate with an *endo* isopropyl group is not easy.

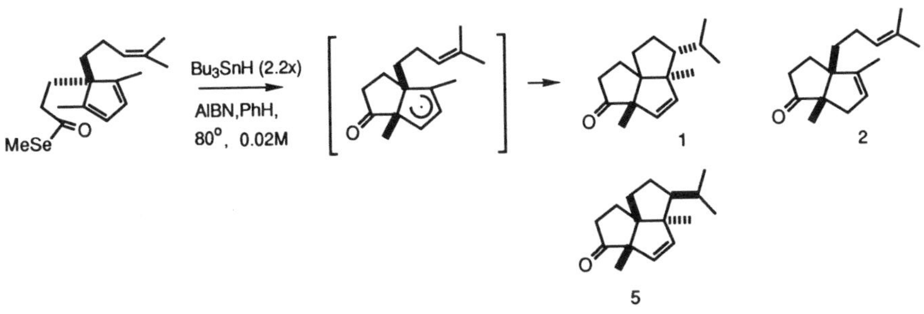

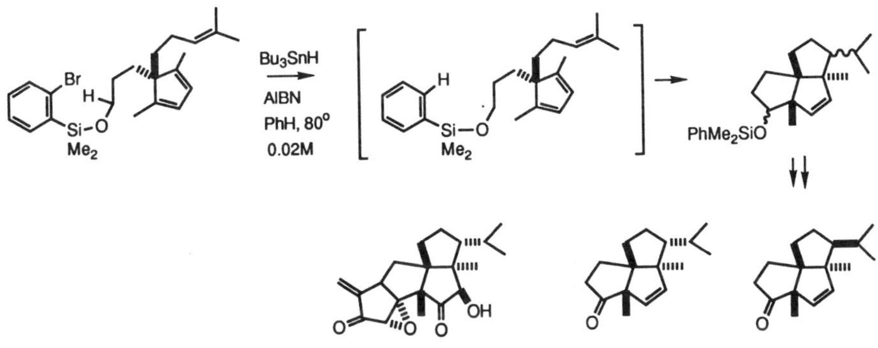

crinipellin-A

1,1-Diradicals (not carbenes) must be generated in stages, and the behavior of such diradicals is interesting to observe. Formation of bicyclo[3.3.0]octanes from such species has been demonstrated [Nagai, 1990].

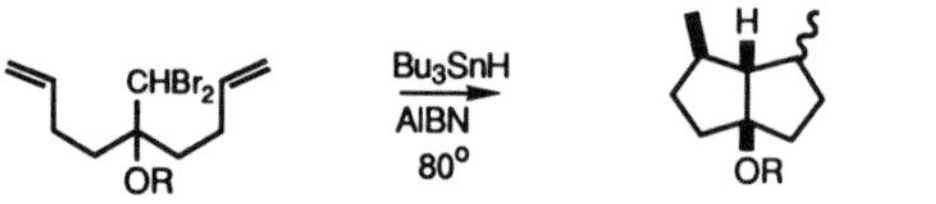

αMe : βMe = 8 : 2

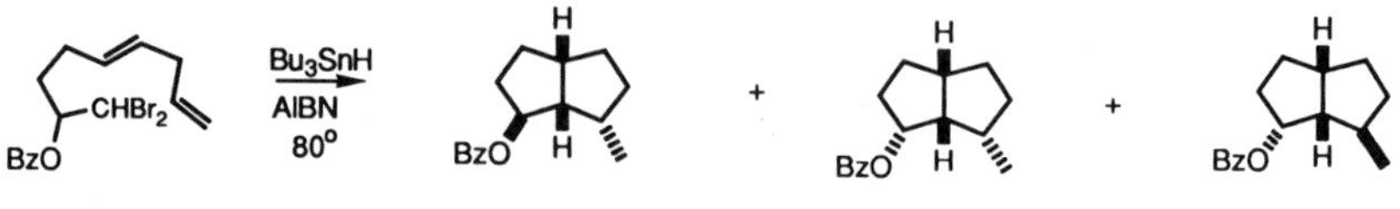

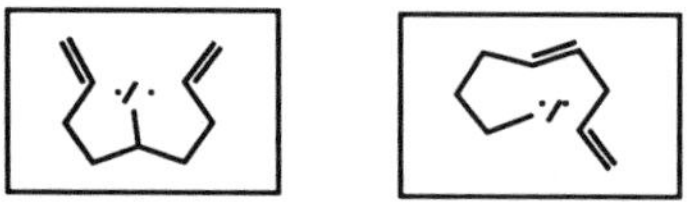

The thermal reaction of 4-(o-tolyl)ethynylcyclobutenones gives benzo-chromanes via electrocyclic opening and recyclization of the ketene intermediates. However, the recyclization can only give phenol O,C-diradicals which must seek a low-energy pathway which consists of hydrogen abstraction and O,C-coupling [K. Chow, 1990b].

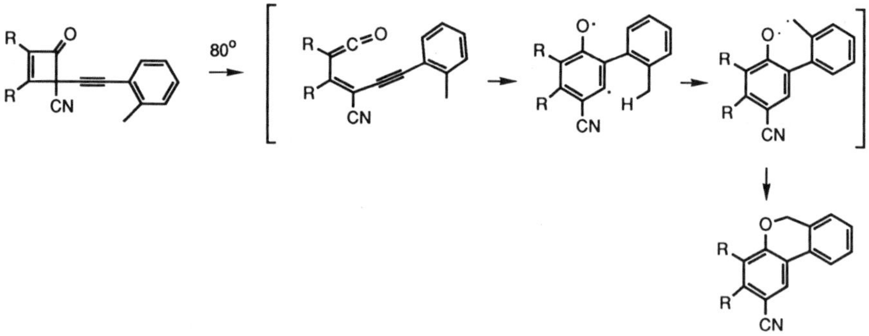

Acyl silanes are excellent radicophiles. Cycloalkyl silyl ethers have been acquired by an intramolecular reaction of ω-bromoacyl silanes with tributyltin hydride [Y.-M. Tsai, 1991]. A radical Brook rearrangement following the cyclization renders the process irreversible.

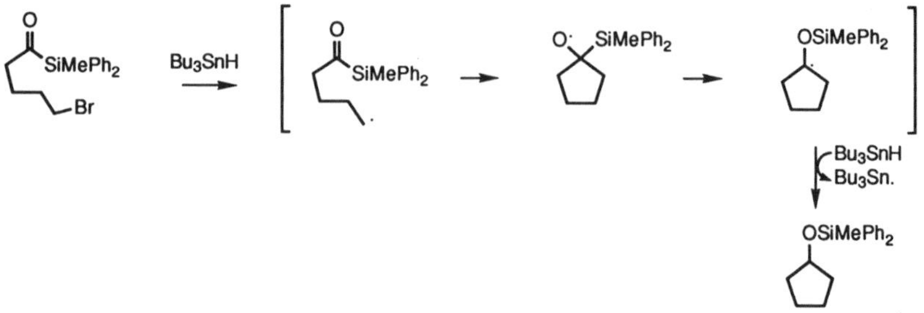

An access to the indolizidine skeleton of gephyrotoxin-223AB [Cordero, 1990] consists of thermal reorganization of a spiroannulated isoxazolidine. It is likely that generation of a ketone group and an aminyl radical accompanies the cyclopropane ring homolysis. Subsequent N–C bond formation delivers the azabicyclic system.

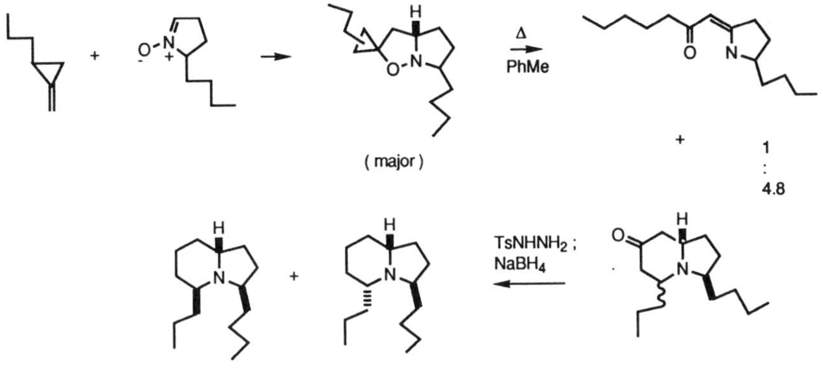

(major)

+

1
:
4.8

TsNHNH$_2$;
NaBH$_4$

GTX-223AB

The Barton nitrosation at spatially accessible carbon atoms has been used as a relay device. The presence of a radicophile in the vicinity of carbon radical can scavenge the latter species in competition with its combination with the nitroso radical. Thus, a 18-cyanocorticoid has been prepared by photolysis of a steroidal 11β-nitrite containing a 20-cyano group [Kalvoda, 1972]. Oxidation of steroidal 20-cyanohydrins with lead(IV) acetate–iodine also introduces the cyano group into C-14 [Kalvoda, 1968]. In this case the alkoxy radical derived from the cyanohydrin function serves as the hydrogen abstractor.

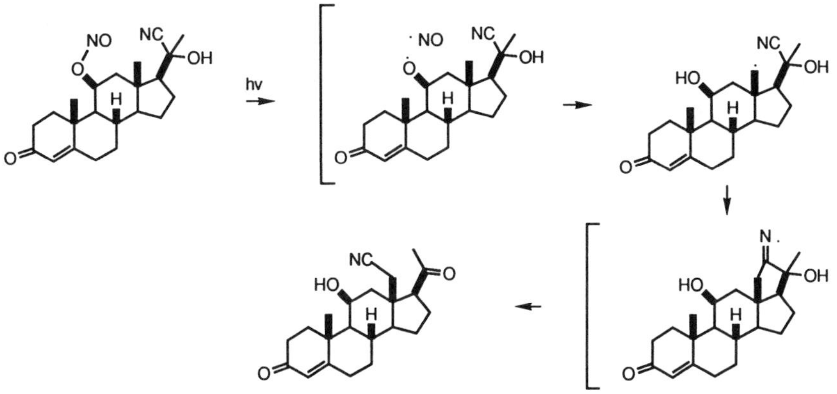

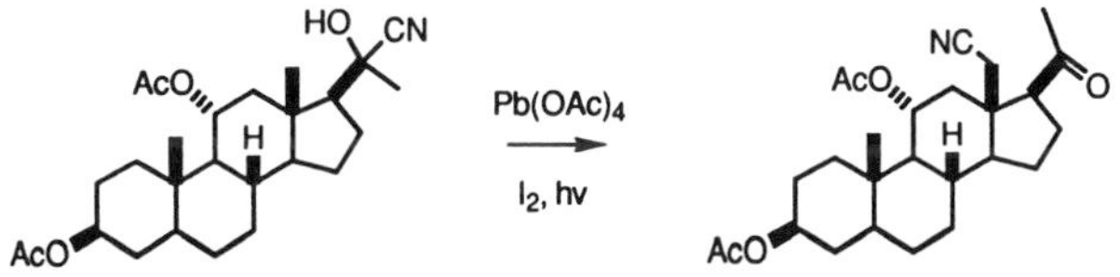

Finally, it must be mentioned that the Hofmann–Löffler–Freytag reaction and remote functionalization (Breslow) are tandem reactions in the realm of free radicals.

15

MISCELLANEOUS TANDEM REACTIONS

15.1. WITTIG AND RELATED REACTIONS

15.1.1. Michael–Wittig Reaction Tandems

Olefin synthesis scored an enormous advance when the Wittig reaction [Maryanoff, 1989] was discovered. Most tandem process that are anchored by a Wittig olefination are initiated by a Michael reaction, and the most useful of them are those leading to ring formation, although reactions exemplified by the following [Minami, 1982] are also of synthetic value.

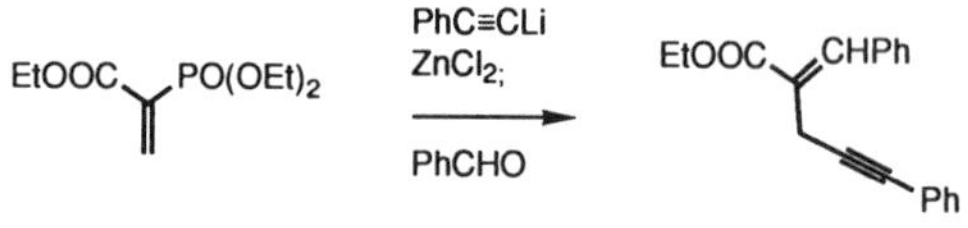

$(E:Z) = 3:1$

The Wittig reagents may assume the role of a donor or an acceptor in the Michael reaction step, although the popularization of vinyltriphenylphosphonium salts seems to have greatly unbalanced the relative importance of the two modes of annulation.

Syntheses of occidol [Dauben, 1973a] and β-damascenone [Büchi, 1971; cf. S.F. Martin, 1977] exemplify the donor capacity of conjugated phosphoranes toward enones. It should be remarked that the Michael reaction does not involve directly the electronegative carbon atom of the ylide. Consequently the overall process is a 3 + 3 annulation.

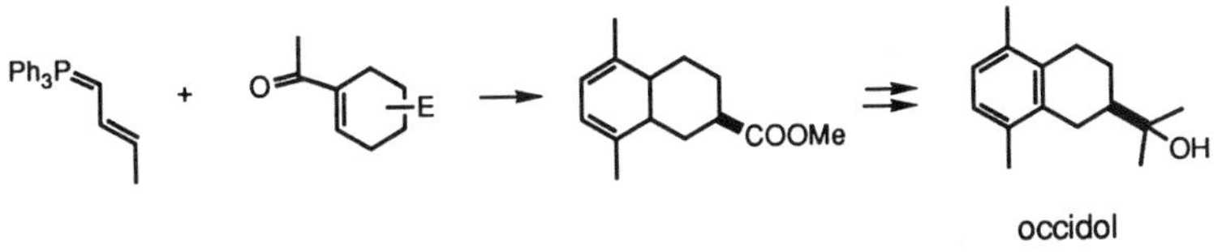

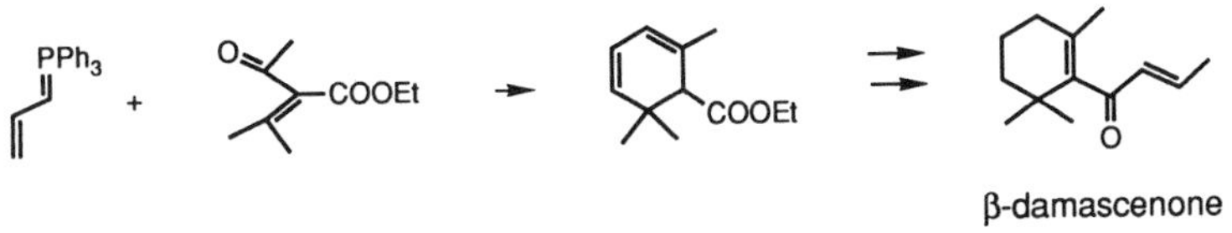

occidol

β-damascenone

The effectiveness of this tandem reaction is reflected in the creation of 1,3-cyclohexadienes in which the Wittig reaction step gives rise to a bridgehead double bond [Dauben, 1973b; Masamune, 1976].

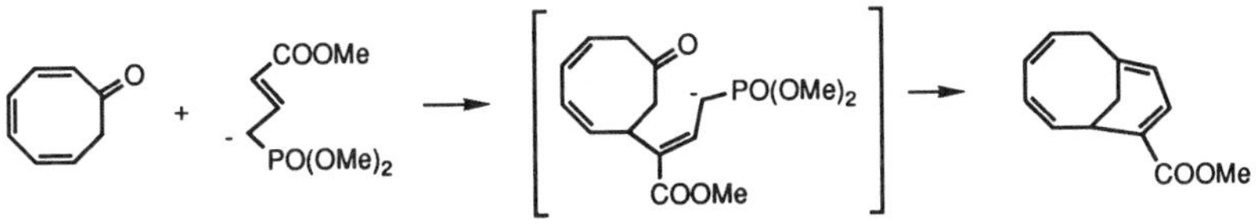

The acetone α,α'-dianion, stabilized in each side by an ester group and a phosphonium ion, condenses with α,β-unsaturated aldehydes readily to afford cyclohexenones [Pietrusiewicz, 1983].

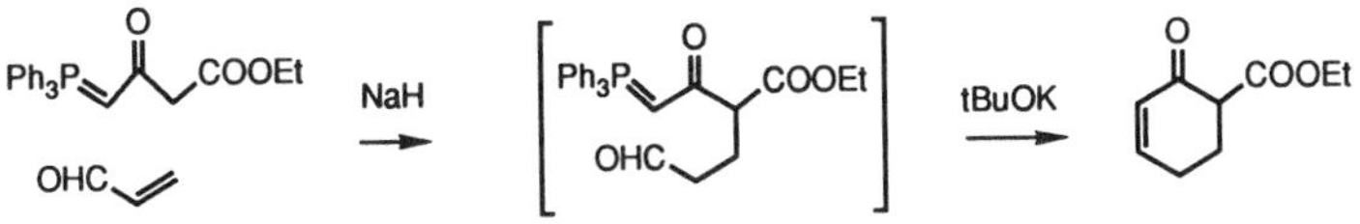

Functionalized cycloalkenes have been obtained by tandem Michael–Wittig reactions involving vinylphosphonium salts. The donors must be bi- or polyfunctional compounds in which the aldehyde or ketone unit remains unenolized throughout, so as not to interfere with either of the two stages of the reaction.

A cyclohexenecarboxylic ester was considered as an intermediate for the synthesis of ibogamine [Kuehne, 1985a]. Since its substitution pattern is neither agreeable to a Diels–Alder approach, nor amenable to reductive elaboration from an aromatic compound, a new method must be implemented for its acquisition. The Michael–Wittig tandem served well for the purpose.

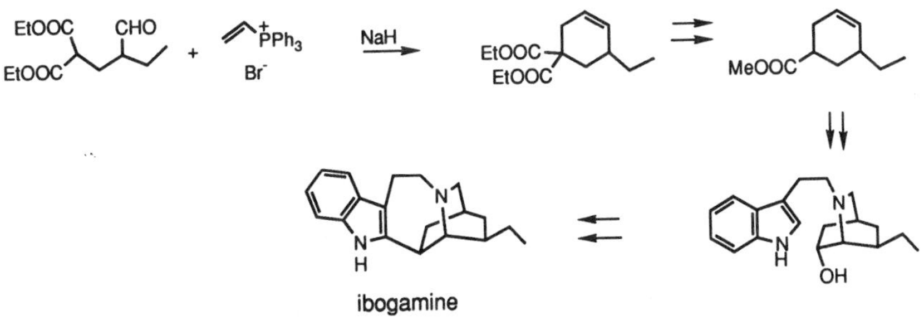

ibogamine

Certain structural variations in the vinyl group of the vinylphosphonium salts allow assembly of more complex cyclic olefins. One example is the synthesis of a bicyclo[3.2.0]heptene from a cyclobutenyltriphenylphosphonium salt [Okada, 1989].

When the α-carbon of the vinylphosphonium ion bears a hetero substituent, the annulation results in functionalized cycloalkenes. Products from such variants (e.g. with α-thio substituents) are more valuable synthetic intermediates, as demonstrated by the conversion of a diquinane derivative, thereby obtained, into chrysomelidial, loganin, and hirsutene [Hewson, 1985].

hirsutene   loganin   chrysomelidial

Suredly, the cyclic frameworks and functionalities in compounds assembled via the present process depend on the donor structures. Proper alteration has enabled extension of the method to synthesis of other cyclopentanoids including dihydrojasmones and the methylenomycins [Cameron, 1984].

methylenomycin B   dihydrojasmone

Reaction of a γ-keto ester α-anion with an α-thiovinylphosphonium salt provides a latent cyclopentanone-3-carboxylic ester. While it is difficult to generate the donor species by a deprotonation procedure, such a nascent enolate can be elaborated via desilylative fragmentation of a 2-siloxycyclopropane-carboxylate. Under aprotic conditions and in the presence of the vinylphosphonium salt, the annulation product emerges [Marino, 1985]. Effective entries into polyquinane sesquiterpenes are indicated [Marino, 1987].

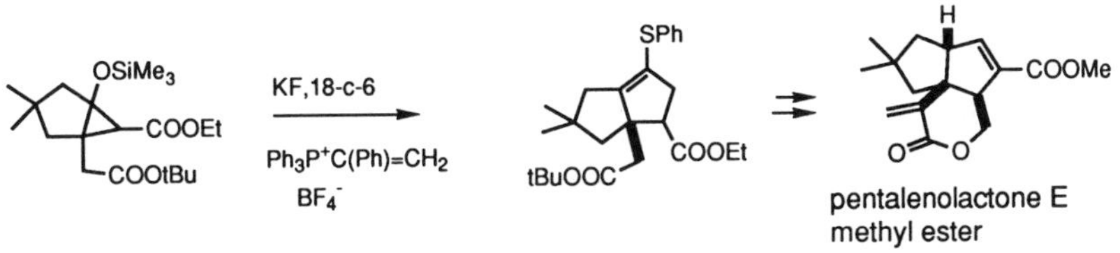

pentalenolactone E
methyl ester

As anticipated, vinylphosphonates are able to function in the same way as the corresponding phosphonium species [Minami, 1986b].

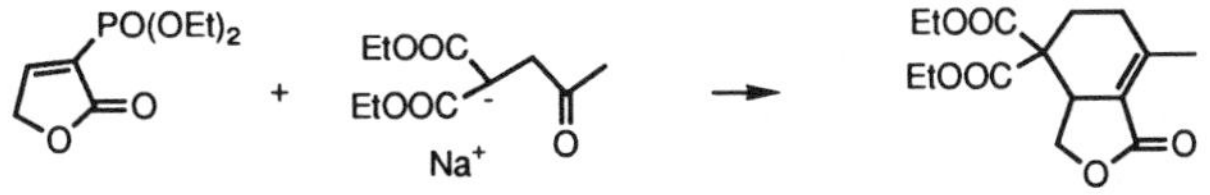

Heterocyclic alkenes are readily available by the same method, when proper donors are selected [Schweizer, 1964; Brandänge, 1971; McIntosh, 1978; Klose, 1980; Muchowski, 1980].

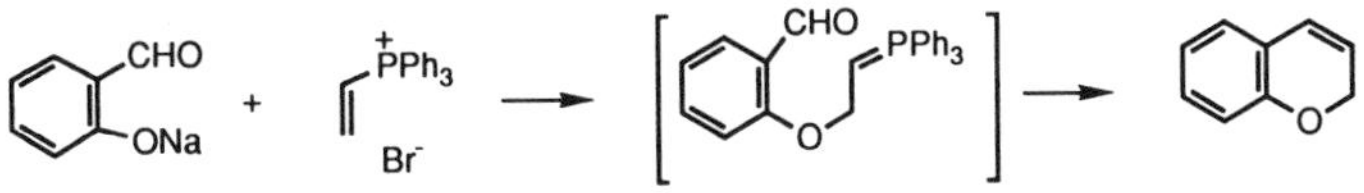

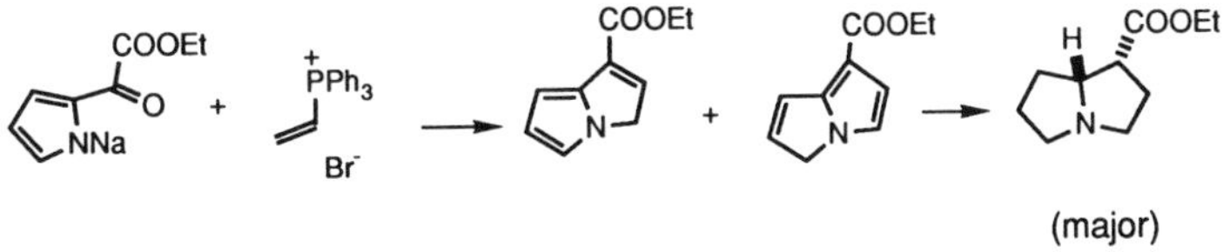

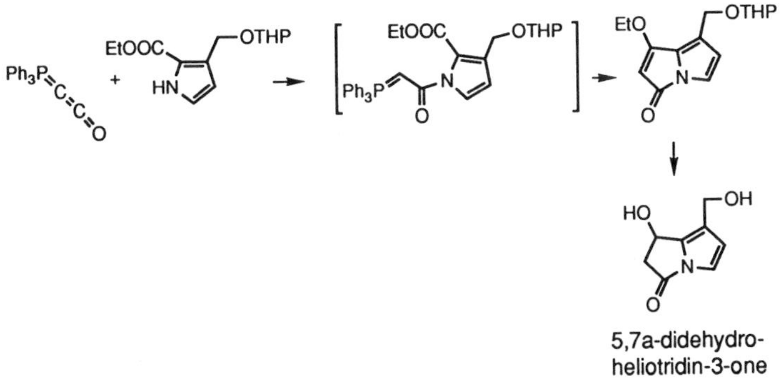

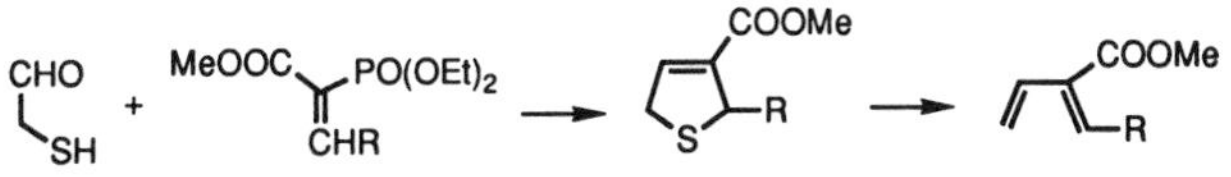

(major)

5,7a-didehydro-
heliotridin-3-one

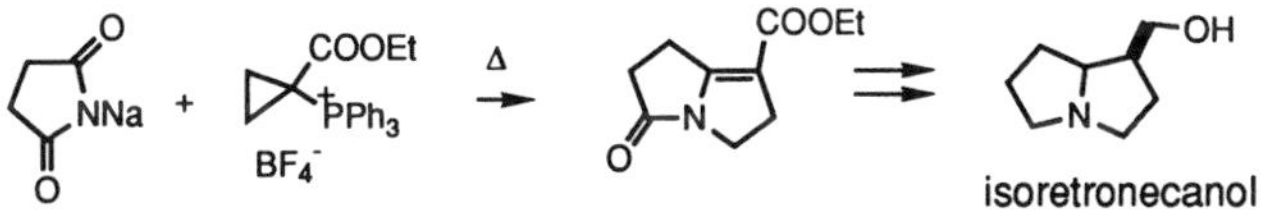

isoretronecanol

Enolates add to butadienylphosphonium salts in a 1,4-manner. The resulting ylides then undergo intramolecular Wittig reactions. When such a phosphonium salt bears a hetero-substituent at C-2, the final product would be a latent cyclohexenone [S.F. Martin, 1978].

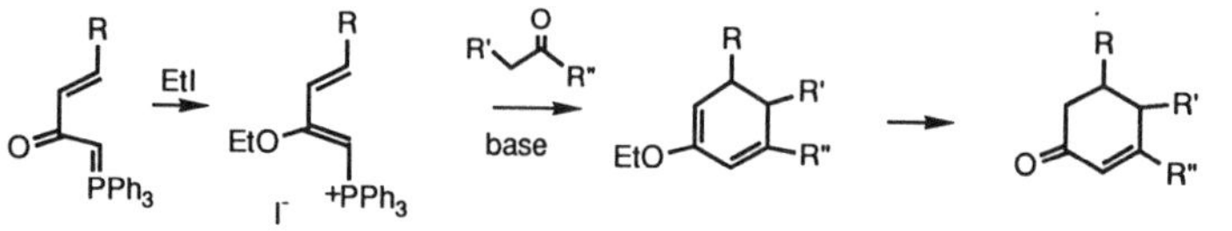

3-Hetero-substituted butadienylphosphonium species are formed by elimination of triphenylphosphonium chloride from the 1,4-bistriphenylphosphonium salts of 3-substituted 2-butenes. Accordingly, the tandem Michael–Wittig reactions generate isomeric cyclohexdienyl derivatives [Pariza, 1983]. Note that the elimination is regioselective because 1,3-disubstituted dienes are more stable than the 1,2-isomers.

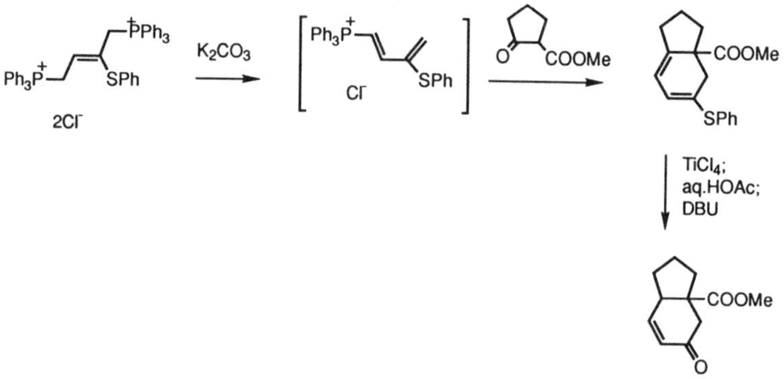

A three-component annulation involving a ketone enolate and two molecules of vinyltriphenylphosphonium bromide is achievable [Posner, 1985]. Two Michael reactions precede the intramolecular Wittig reaction.

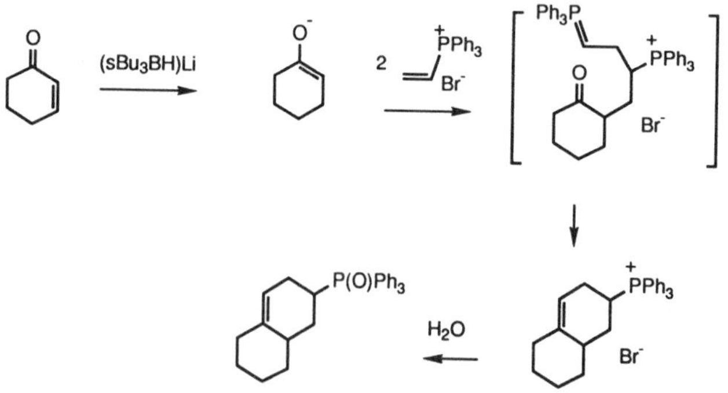

The Michael–Michael–Wittig reaction cycle among an enolate, an enone and vinylphosphonium salt is even more interesting, as it delineated a convergent assembly of 9,11-dehydro-8-epiestrone methyl ether [Posner, 1981]. Hydrogen chloride in refluxing methanol readily isomerizes the product into known precursors of estrone and estradiol.

(1 : 3)

As is well known, 1,1-disubstituted cyclopropanes in which both substituents are acceptor groups are susceptible to attack by nucleophiles. Exploitation of this reactivity has now included cyclopentene formation via the tandem Michael–Wittig reactions [Dauben, 1977b], using a (1-carboalkoxycyclopropyl)-triphenylphosphonium salt. A concise route to many spirovetivane sesquiterpenes was thereby developed.

β-vetispirene   β-vetivone   hinesol

15.1.2. Wittig Reactions Preceded by Other Base-catalyzed Processes

A few other reaction tandems, with their scope not yet explored, are terminated by the Wittig olefination. The combination of alkylation and the olefination has several examples: a synthesis of cyclopentadienes from α-haloketones [Hatanaka, 1990; 1991], and the insertive deoxygenation of a pyrylium salt to form a benzoic ester [Griffin, 1975] belong to this category. On the other hand, the annulation of dihydrocarvone [Büchi, 1975] by its reaction with (4-bromo-2-buten-1-yl)triphenylphosphonium bromide in the presence of a base may have actually involved elimination and the Michael–Wittig tandem.

2,2-Disubstituted 1,3-cyclohexanediones are converted into 2-cyclohexenones on reaction with dimethyl methylphosphonate anion and chlorotrimethylsilane [Y. Yamamoto, 1990]. Among other synthetic applications the tandem alkylation–retro-aldol–Wittig reactions have been used in an elaboration of α-acoradiene.

α-acoradiene

Pyrylium salts are very reactive toward nucleophiles. On exposure to alkylidenetriphenyl-phosphoranes they yield benzene derivatives [Märkl, 1962], apparently via acyclic dienenone intermediates which undergo ring closure (intramolecular Wittig reaction).

During a synthesis of (+)-norpatchoulenol [Liu, 1990] it has been found expedient to conduct a Claisen condensation and a Wittig reaction in one step.

The two ketone groups of the nonenolizable 1,3-diketone intermediate has such different steric hindrance that one of which is immune to attack by the Wittig reagent.

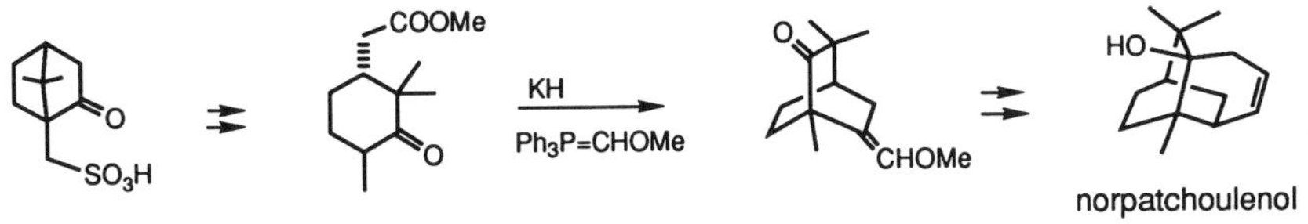

norpatchoulenol

Interestingly, a (2-oxocyclopentane)acetic ester underwent Claisen condensation also when it was exposed to a Wittig reagent [Brain, 1972]. However, the more strained bicyclic diketone apparently preferred fragmentation to yield a new keto ester which was promptly olefinated.

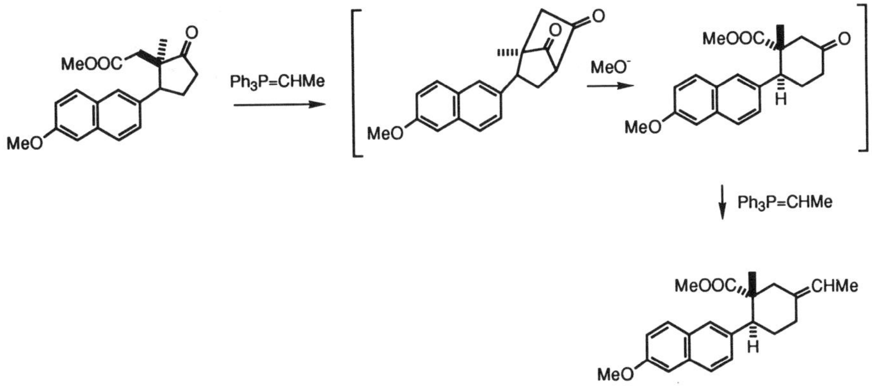

Enol lactones react with Wittig reagents or α-phosphonyl carbanions to give enones [Henrick, 1968]. This process represents an improvement over the method involving Grignard reaction.

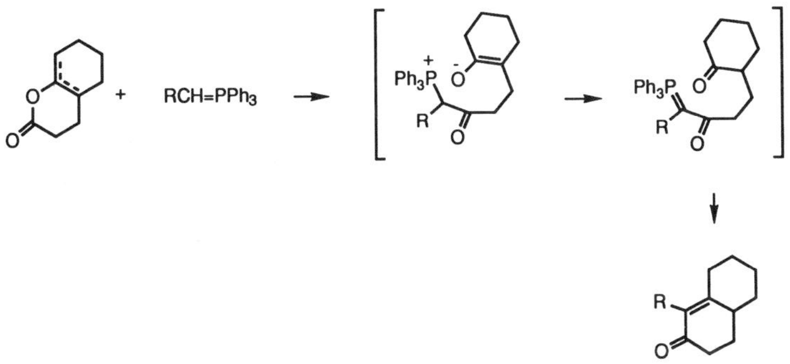

Macrocyclization to complete the synthesis of baccharin-B5 [Still, 1984] enlisted a two-step Wittig and Horner–Emmons reactions. It might be possible to modify the approach into a tandem mode, although the efficiency and stereoselectivity could be lower.

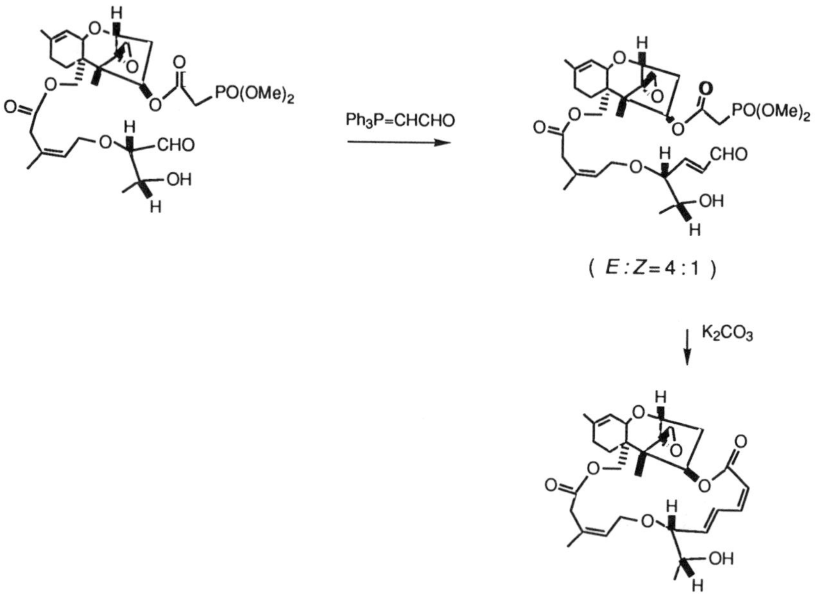

(*E*:*Z* = 4:1)

baccharin-B5

Tandem reactions from a quaternary phosphonium salt of cyclic phosphine with two aldehydes [I. Yamamoto, 1987] are interesting, despite that the second reaction involves a phosphonyl carbanion as donor, and it does not result in a diolefin.

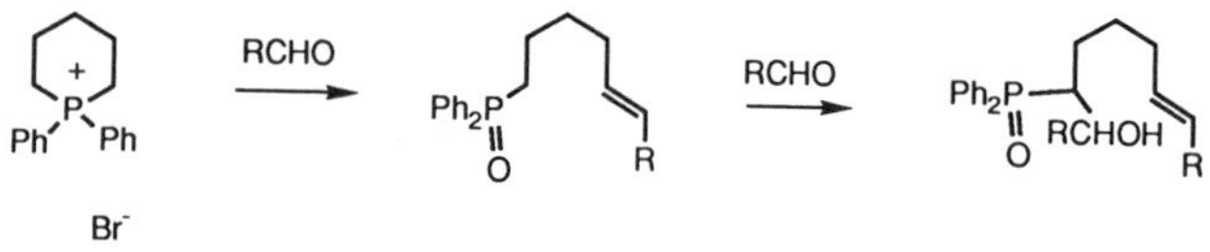

15.1.3. Some Other Olefination Reactions

Numerous modifications of the Wittig reaction are known, but their discussion is omitted here unless they are part of certain tandem processes.

The Tebbe reagent is a source of titanethylene which react with carbonyl compounds in an analogous manner to the Wittig reagent. Actually the Tebbe reagent is superior with respect to its reactivity toward esters and lactones to afford enol ethers. The equilibrium between titanacyclobutanes and the alkene–titanethylene components is also significant because of its role in olefin methathesis.

The combination of the two types of reactions in tandem constitutes a fascinating transformation of a 1,6-bridged 2-norbornene-5-carboxylic ester into a linearly fused ring system which is a useful precursor of $\Delta^{9(12)}$-capnellene [Stille, 1986].

$\Delta^{9(12)}$-capnellene

While the thioacetal function is recognized as a protecting device for the carbonyl group, its transformation into a methylene or a (trimethylsilyl)methylene group has been achieved by reaction with methylmagnesium iodide or (trimethylsilyl)methylmagnesium chloride in the presence of a nickel catalyst [Ni, 1990]. Apparently, the coordination with nickel polarizes the C—S bond of the thioacetal so that the carbon atom is rendered susceptible to attack by the Grignard reagent. The S-bound nickel may then insert into the remaining C—S bond of the original thioacetal function to complete the removal of the sulfur-containing portion of the molecule by the well-precedented β-hydride elimination of transition metal complexes.

The use of cyclopropyl Grignard reagents affords diene products [Ng, 1989]. Direct β-hydride elimination after the alkylation is now unfavorable on two accounts (angle strain of the product and torsional strain of the transition state), therefore ring opening to give the homoallylic η^1-nickel species predominates.

The Peterson olefination [Ager, 1990] complements the Wittig reaction in some way. A rather unusual generation of an α-silyl carbanion is by disilylation of a *gem*-disilylcyclopropene [Halton, 1990].

The tandem Wittig–Peterson olefination of benzophenone results in an allene [Gilman, 1962]. The vinylsilane intermediate could not be isolated.

$$Ph_2C=O \quad + \quad Ph_3P{=\!\!\!\diagup}_{SiMe_3} \quad \longrightarrow \quad \left[Ph_2C{=\!\!\!\diagup}_{SiMe_3} \right] \xrightarrow{Ph_2CO} \quad Ph_2C=C=CPh_2$$

15.2. ELECTROPHILATIONS

A requisite for the alkylation of carbonyl compounds is enolate formation. Of great interest is the decarboxylative allylation of allyl β-ketocarboxylates by treatment with Pd(O) or Mo(O) complexes [Tsuda, 1980; Tsuji, 1984]. The formation of η^3-allyl complexes from these allyl carboxylates on contact with the zero-valent metal species liberates β-keto carboxylate anions which are liable to decarboxylate, generating the enolates. Since the allylmetal species are electrophilic they react with the enolates readily. The overall tandem reactions proceed unver very mild conditions.

In a system in which the cleavage of an allyl ether linkage triggers a retro-aldol reaction the C-allylated diketone is the product [Kaczmare, 1986].

Generation of a mutually reactive ion-pair by Pd(O)-catalyzed allylic epoxide opening, and using the alkoxide ion to deprotonate a remote carbon site, could result in ring formation [Trost, 1982c]. This template-directed macrocyclization generally affords good yield of the products.

The same sort of mechanism apparently operates in the isomerization of 6-alkoxy-6H-pyran-3-ones to furnish 4-alkoxy-5-hydroxy-2-cyclopentenones [Mucha, 1989].

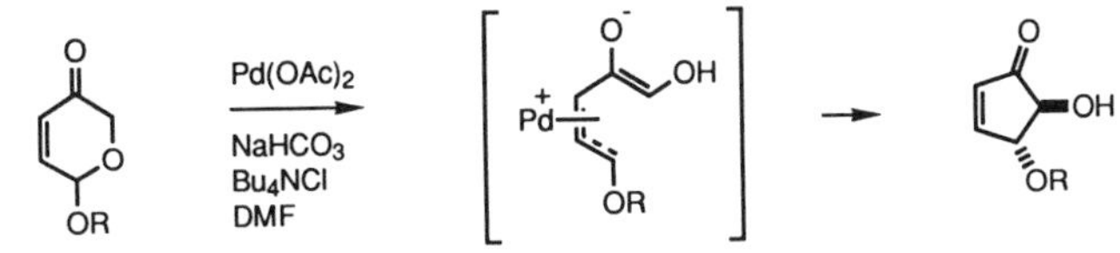

There exist certain tandem reactions which may be classified under *transmissive alkylation.* Hydroxyl compounds must be activated (e.g. sulfonylation) before they can be used as alkylating agents, yet specially constituted hydroxyl compounds may serve the purpose. In situ formation of epoxide fulfills the activation requirement [Ho, 1972; Gutzwiller, 1978].

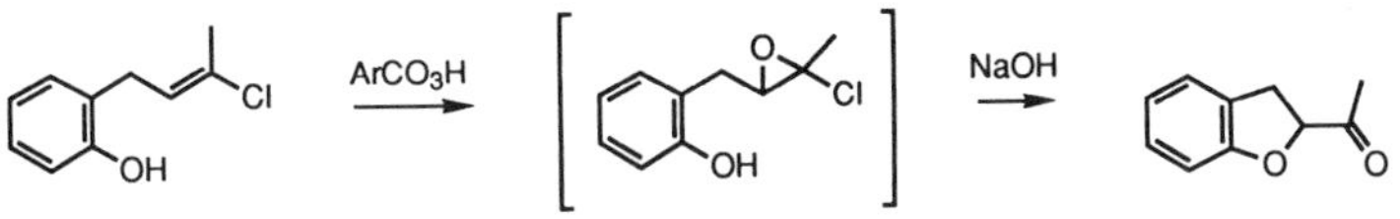

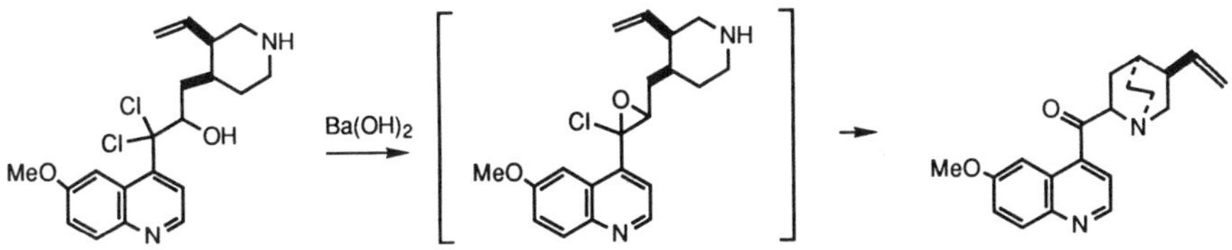

quininone
quinidinone

The other variant involving activation of a nucleophile, as exemplified by the conversion of fulvenes into spirocyclic products [Antczak, 1984].

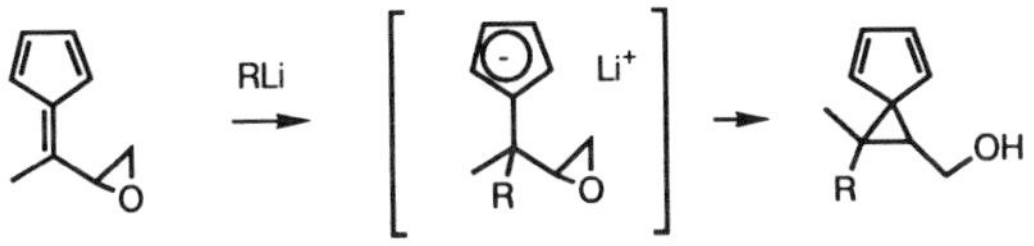

A Payne rearrangement was found to occur after an intramolecular aldol condensation [Swindell, 1990]. The rearrangement concerns with movement of epoxide positions.

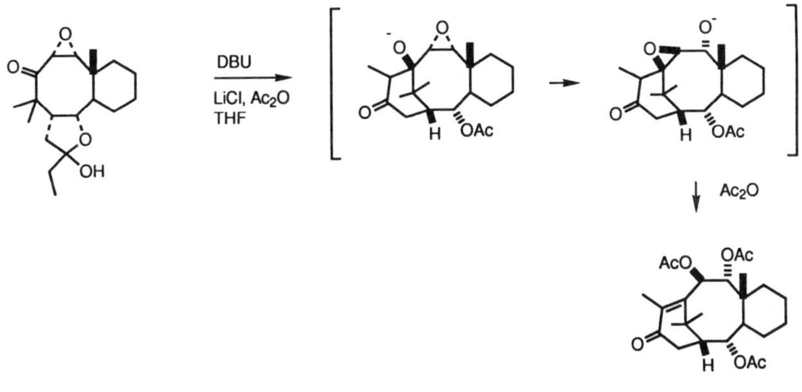

Related to this transmissive alkylation is the conversion of epoxides to episulfides, e.g. by reaction with triphenylphosphine sulfide [Chan, 1972].

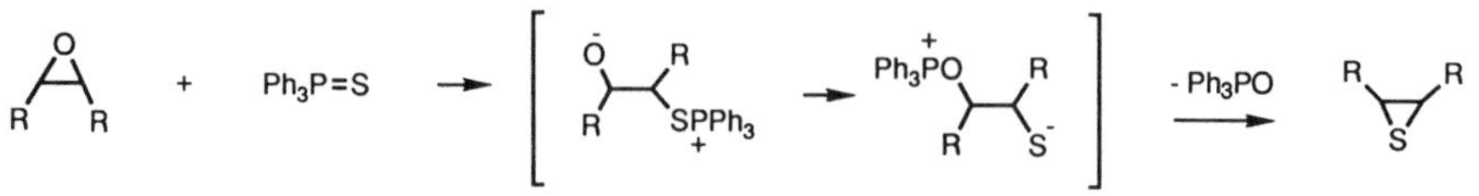

Prolonged reaction of allylic epoxides with alkynyl borates gives 1,5E-hexadien-3-ols and cyclohexenols [Mas, 1986]. The latter compounds arise from the (Z)-alkenyl boranes via tandem anionic oxy-Cope rearrangement and deboronative cyclization.

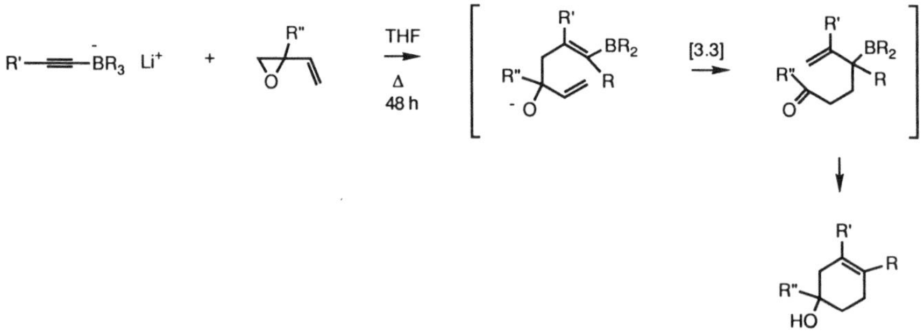

The conversion of a 2-vinylaziridinecarboxylic ester to the 3-pyrroline derivative by reaction with iodotrimethylsilane [Hudlicky, 1987] consists of analogous activation. Of particular importance is that this isomerization gives products of different substitution pattern from those obtained by pyrolysis, as shown.

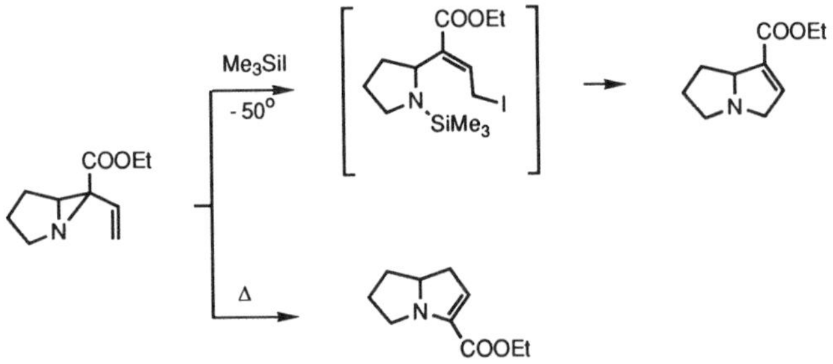

The direct synthesis of N-substituted carbazoles from amines and 2,5-dimethoxytetrahydrofuran [Kashima, 1986] is quite useful. Bisannulation of the N-substituted pyrroles which are the initial products is a Friedel–Crafts reaction.

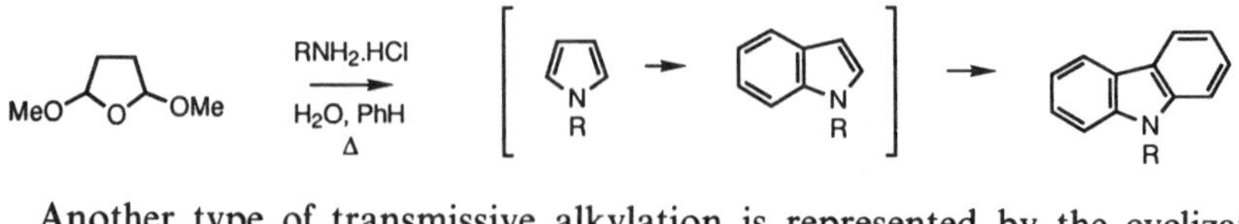

Another type of transmissive alkylation is represented by the cyclization leading to the pyrrolizidinone system [Kametani, 1986]. Here a carbenoid species is trapped by a divalent sulfur atom to form a ylide. Reorganization of ylide by a formal intramolecular displacement of the sulfur substituent from its original position completes the alkylation.

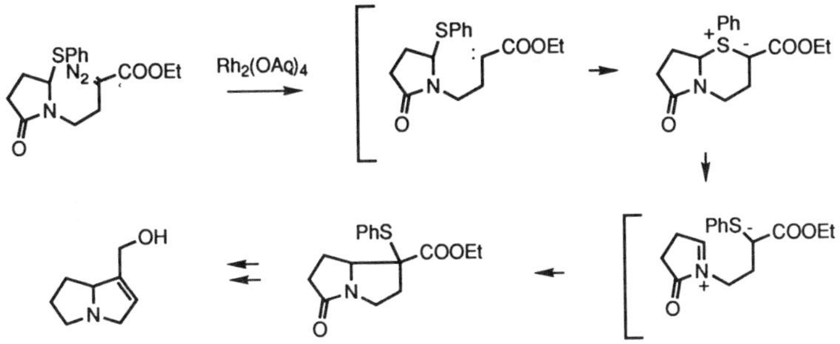

supinidine

Less efficient is the cylization which requires the generation of a less stable cationic site [Kametani, 1988].

The usefulness of controlled double alkylation in one operation is self-explanatory. The following examples merely serve to indicate a number of situations and structural characteristics which warrant the synthetic approaches.

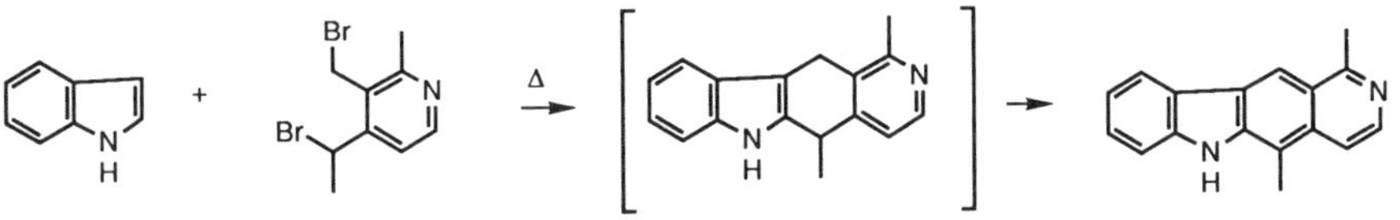

olivacine

[Kametani, 1975a]

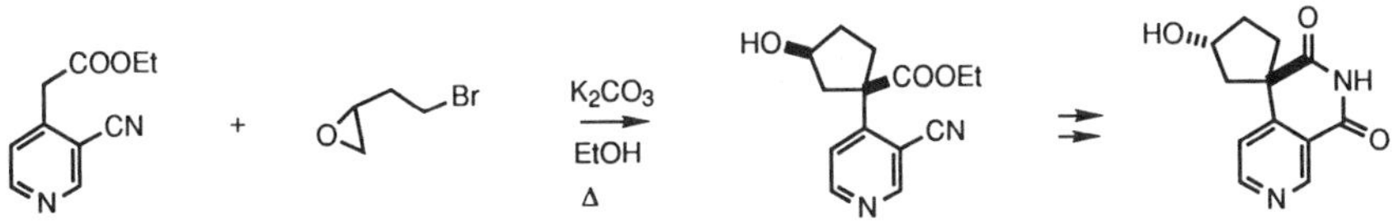

sesbanine

[Kende, 1980]

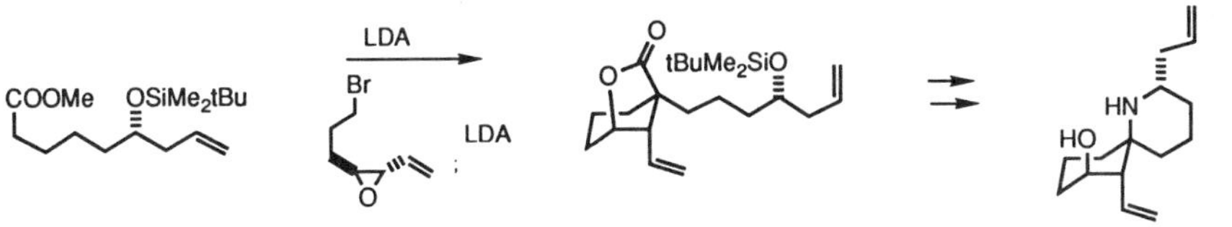

histrionicotoxin-235A

[Stork, 1990b]

An attempted intramolecular *O*-alkylation of a di-(7,7-norbornadienyl) derivative has resulted in the formation of a rearranged ketone [Trah, 1987b]. Fragmentation of the norbornadienol prior to a *C*-alkylation is indicated.

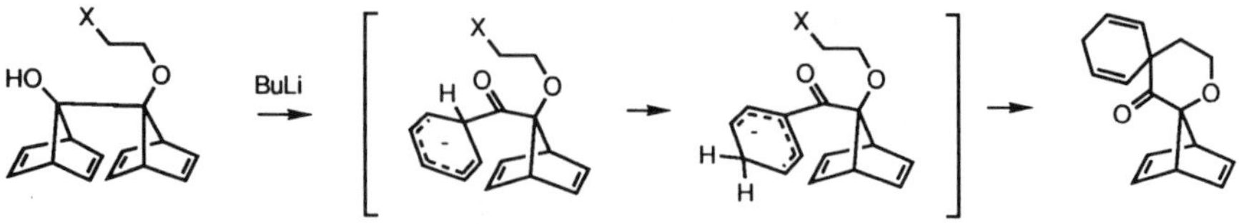

Ring formation by reaction of enol silyl ethers with dimethylacetals which contain an allylsilane pendant [T.V. Lee, 1989] has been reported.

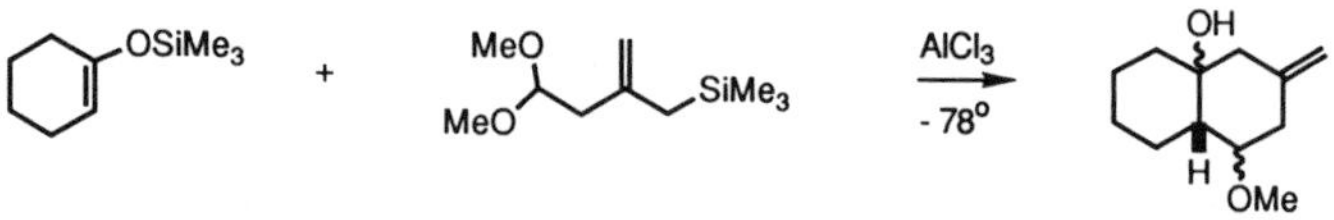

One-pot preparation of p-quinones from N,N-diethylbenzamides is possible. After such an o-lithiobenzamide is exposed to an araldehyde, re-lithiation at the ortho position of the original aldehyde group enables ring closure by displacing the diethylamide ion [Watanabe, 1980].

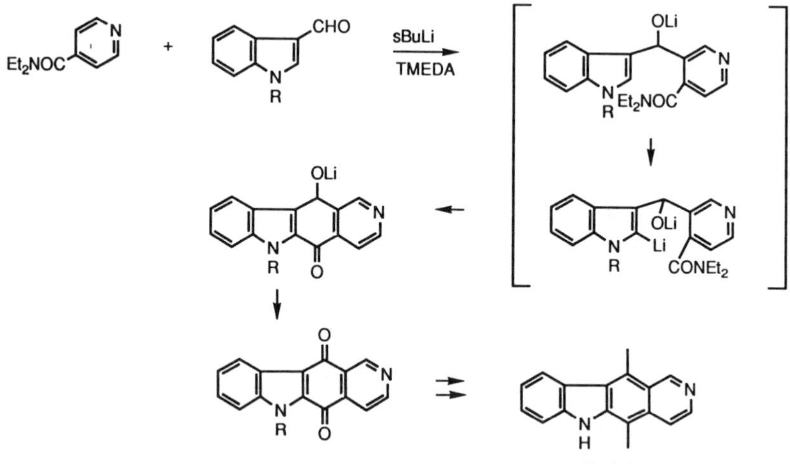

ellipticine

Heterocycle formation by the reaction of bifunctional reagents with imines is exemplified by an approach to erysodienone methyl ether [Chou, 1987].

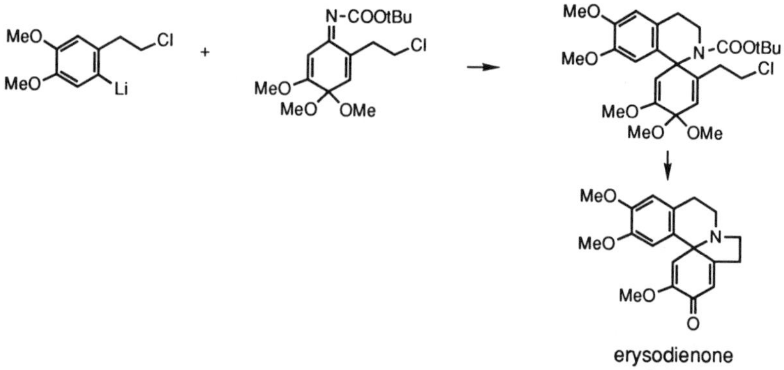

erysodienone
methyl ether

The treatment of *N*-(ω-chloroacyl-*o*-cyanoanilines with a strong base leads to azacyclically fused quinoline-2-ones [Vinick, 1989], apparently via a Ziegler–Thorpe–type cyclization and intramolecular *N*-alkylation of the resulting metalloimines.

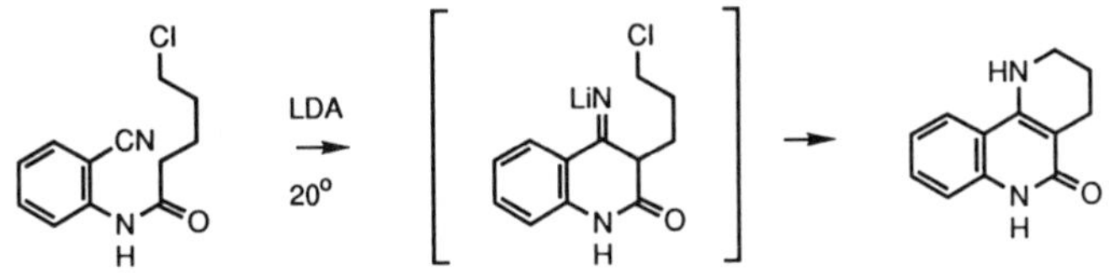

Intramolecular *N*-alkylation may be preceded by a Michael reaction. The tandem process formed the basis for a remarkable synthesis of lycopodine [Kraus, 1987] in which the Michael acceptor is a bridgedhed olefin.

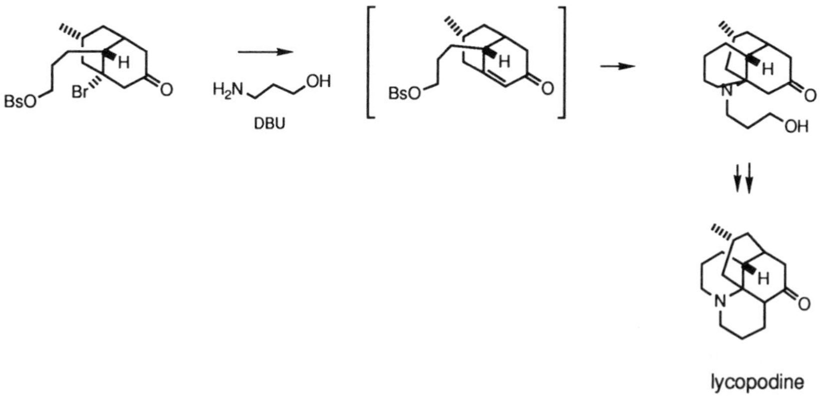

lycopodine

Cyclic anhydrides react with Grignard reagents to give hydroxy carboxylic acids due to the chemoselectivity of the keto carboxylate intermediates. The reaction of divalent Grignard reagents yield spirolactones [Canonne, 1980].

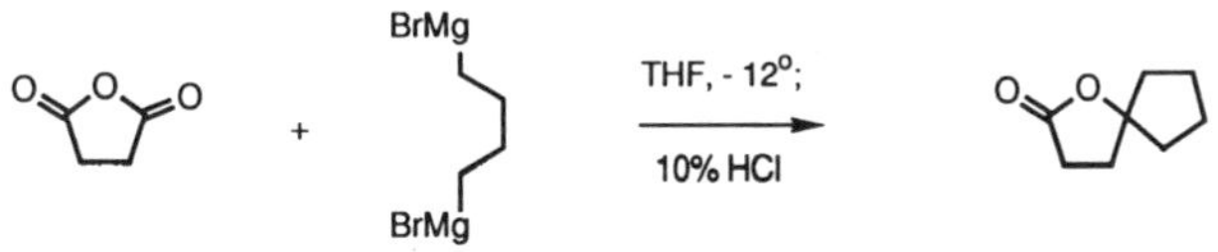

The synthesis of 2,6-disubstituted and 2,4,6-trisubstituted pyrylium salts [Balaban, 1959; Praill, 1959] by reaction of acid anhydrides with propene or 2-substituted propenes in the presence of a strong acid (e.g. perchloric acid) can be visualized as involving consecutive acylations followed by heterocyclization and dehydration.

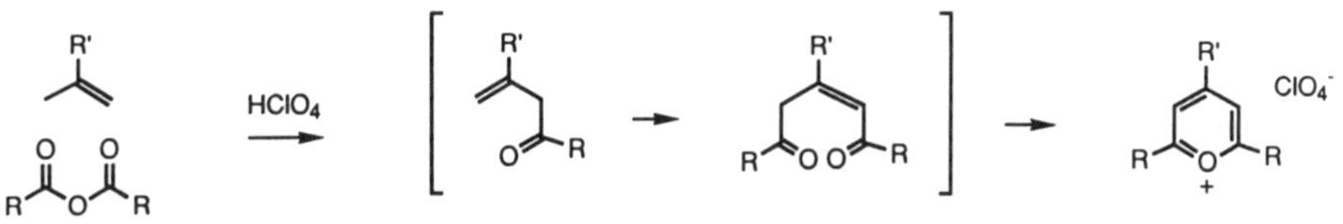

1-Acetyl-2-indanone and 3-acetyl-β-tetralone are obtained by treatment of 3-(o-chlorophenyl)-2,4-pentanedione and 3-(o-chlorobenzyl)-2,4-pentanedione, respectively, with excess potassium amide [Harris, 1964]. In each case a dianion is formed at the sidechain and it attacks the benzyne moiety to create the alicyclic ring.

An analogous cyclization was used in a synthesis of cephalotaxine [Semmelhack, 1973]. However, a photostimulated $S_{RN}1$ version proved much superior.

2-Acyl-3-hydroxyquinolines are availableby a Smiles rearrangement [Bayne, 1975]. The individual reactions made up of this transformation are Michael reaction, a fragmentation related to the retro-Claisen fission, decarboxylation, and cyclodehydration involving a nitroso group as the electrophile.

The selenonyl group behaves similarly to the sulfonyl group in certain ways. When phenyl vinyl selenone acts as a Michael acceptor toward 2-formyl-3-pentanone in the formation of a cyclopropane derivative [R. Ando, 1983] the adduct must undergo an intramolecular aldol–retro-aldol reaction tandem and displacement of the selenonyl group.

Electrophilation following rearrangement is not uncommon. As an example the acetylation of thujopsene [Daeniker, 1972] which gives rise to a product of different carbon framework may be mentioned.

thujopsene

During an examination of the α-(alkoxy)allyl stannane mediated carbo-cyclization to generate a ten-membered ring in analogy to the cembranolide synthesis it was found that the BF_3-catalyzed reaction is of the 1,3-transposed $S_{E'}$-type, presumably due to steric constraints on the formation of the smaller ring. The observed results seem to indicate that pentacoordinated stannanes serve as self-replicating catalytic transfer intermediates [Marshall, 1991b].

The Pummerer rearrangement produces sulfonium ions in which the α-carbon is highly electrophilic. Intramolecular trapping of such carbocations are readily achieved by moderately nucleophilic species including an aromatic rings and alkenes. Particularly useful to synthesis are those tandem reactions of β-keto sulfoxides leading to β-keto sulfides. The functional groups of the products offer multiple possibilities for manipulations. Typical applications of the method are included in syntheses of olivacine [Oikawa, 1976], 16-methoxytabersonine [Cardwell, 1988], cephalotaxine [Ishibashi, 1990], and that of a cyclopentanone [Ishibashi, 1986].

olivacine

16-methoxytabersonine

cephalotaxine

Intramolecular trapping of the sulfonium ion by a carbonyl group is shown in the following example of furan formation [de Groot, 1984].

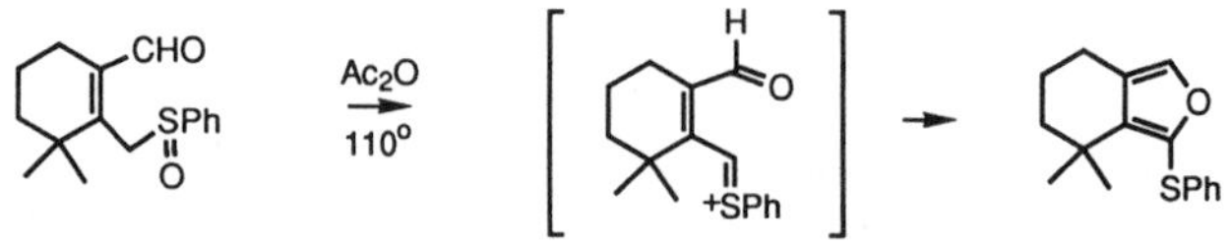

An excellent tactic for closure of the six-membered heterocycle of olivacine is by a tandem Beckmann rearrangement/Bischler–Napieralski cyclization of the properly substituted carbazole [Wenkert, 1962], since both reactions require essentially the same conditions. Note that the presence of a methyl group in the carbazole nucleus renders the cyclization regiospecific.

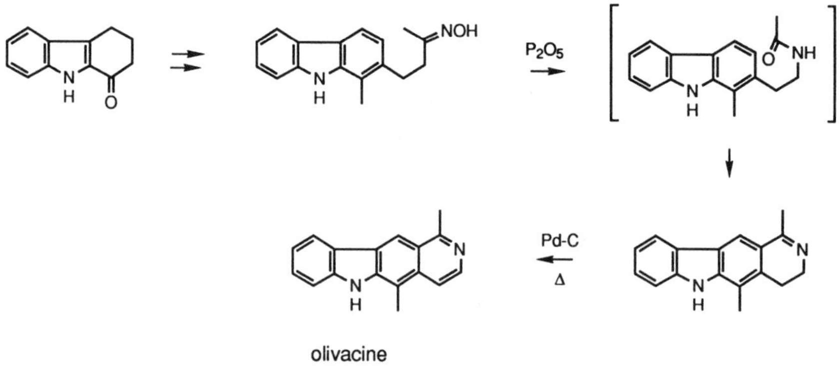

olivacine

Trialkylaluminums are Lewis acids which can induce Beckmann rearrangement and transferring an alkyl group to the incipient imine α-cation. Naturally an application of the tandem reactions to synthesis of solenopsin-A and -B was explored [Maruoka, 1983].

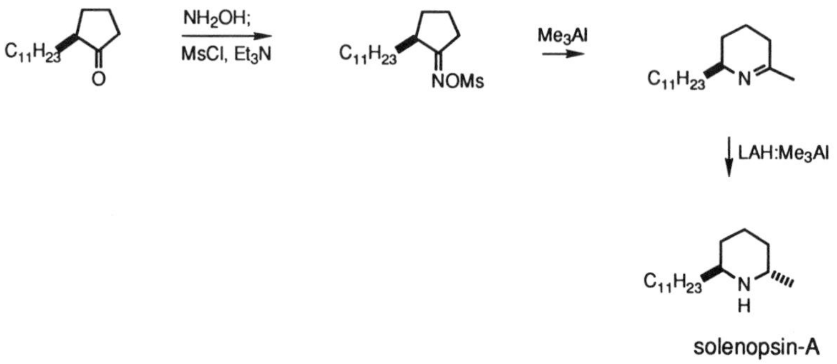

solenopsin-A

Other nucleophiles which do not interfere with the Beckmann rearrangement may be present as trapping agents. These include enol silyl ethers [Matsumura, 1983]. Intramolecular interception of the imine α-cation by an allylsilane is also possible [Schinzer, 1991]. However, the rearrangement takes precedence only in the (Z)-oxime derivatives, because the allylsilane residue prefers participation directly in the ionization of the (E)-oxime mesylates.

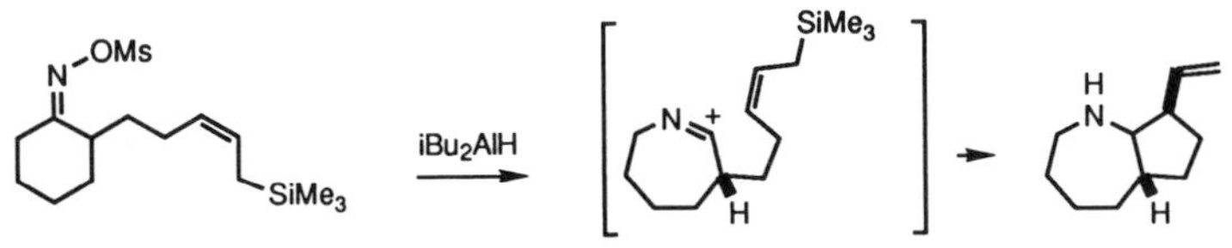

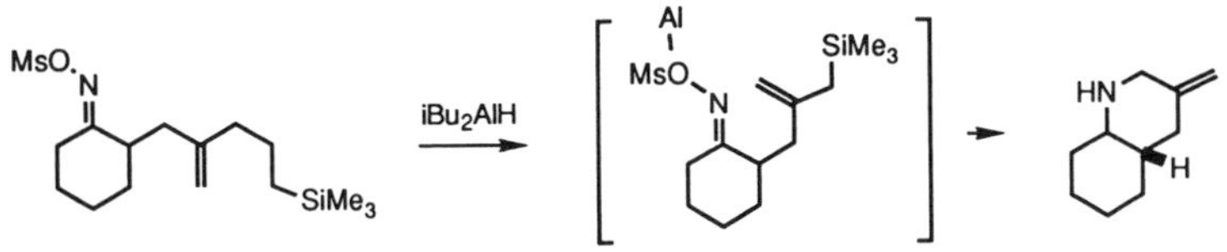

Fragmentation of α-alkoxycycloalkanone oximes and chain extension can be similarly performed in tandem under aprotic conditions and in the presence of nucleophiles such as enol silyl ethers and allylsilanes [Fujioka, 1990].

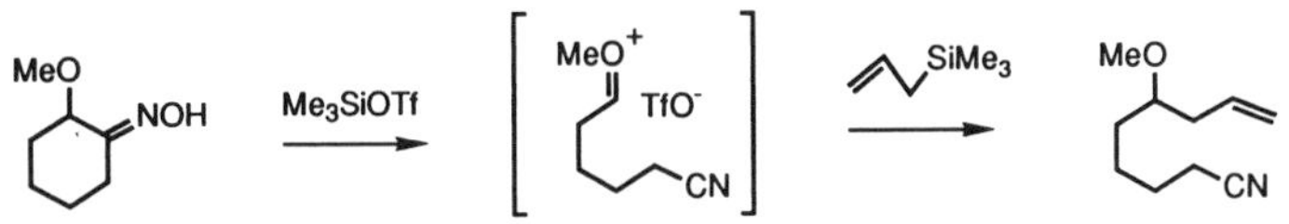

During synthesis of gymnomitrol and the barbatenes [Kodama, 1979] the removal of two primary hydroxyl groups of an intermediate was attempted by the reaction sequence of mesylation, displacement with sodium sulfide, and reductive desulfurization. However, one of the products from the displacement reaction is a ketone which does not contain sulfur [Ito, 1991]. Sodium sulfide apparently acted as a base to induce a tandem Grob fragmentation/intramolecular alkylation.

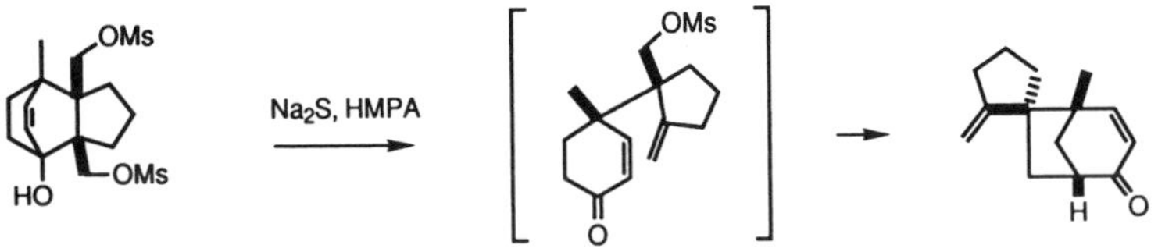

There are reports on intramolecular alkylation in which the enol/enolate ion is generated by an oxy-Cope rearrangement [Sworin, 1987; Paquette, 1989b, 1990b]. Necessarily the alkylation step is an S_N2' reaction.

Interestingly, the S_N2' displacement occurs with equal facility from either the front or the back [Paquette, 1990b].

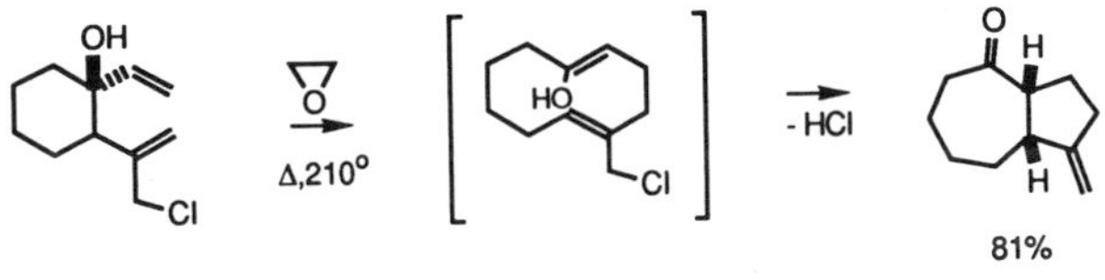

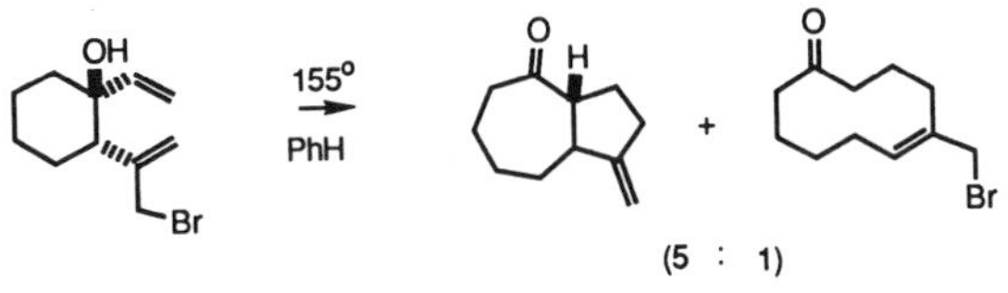

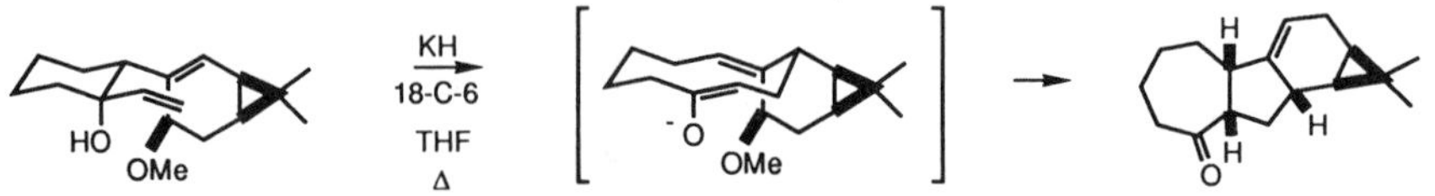

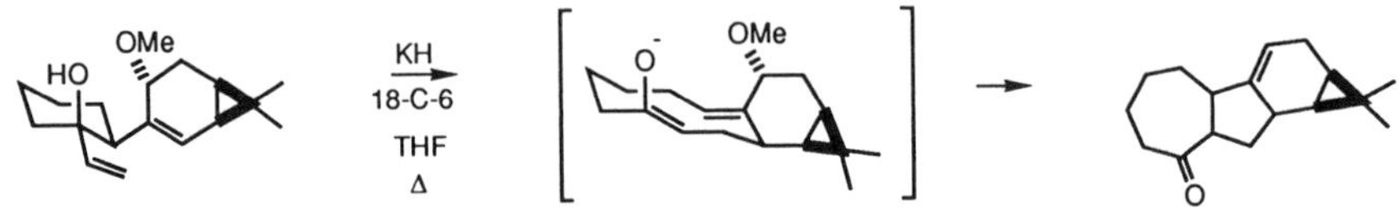

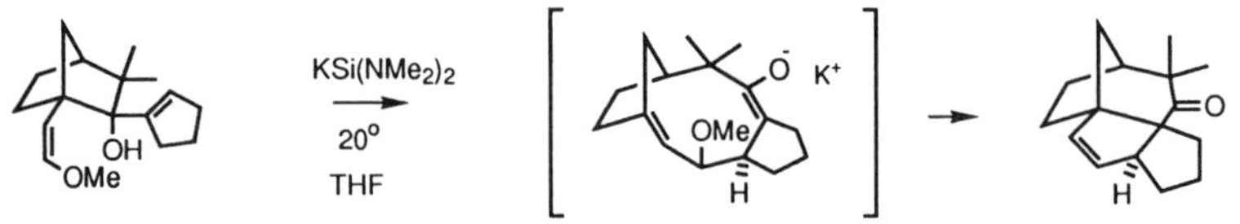

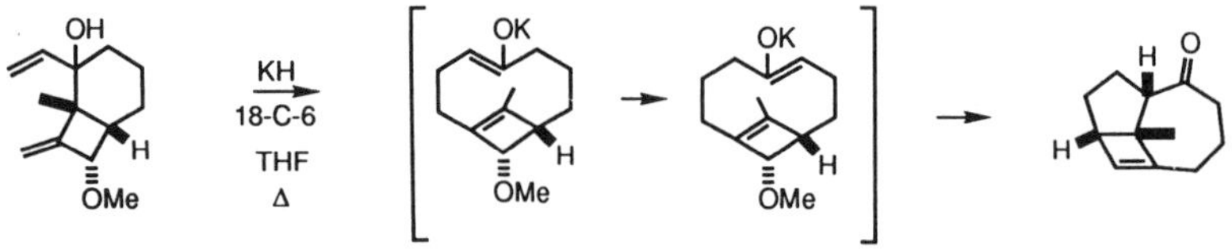

Cyclization via double bond participation during solvolysis is well known. A route for the synthesis of lubimin [Murai, 1982] based on such a reaction to form the five-membered ring is perhaps not as notable if the cyclization were not in tandem with a fragmentation scheme. It must be emphasized that the scheme accrues the advantage of control at two adjacent stereo-centers which were generated in the Diels–Alder reaction step at the beginning.

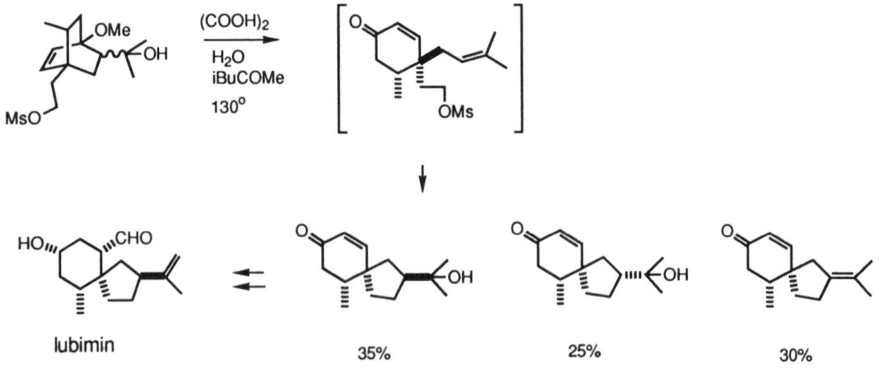

lubimin   35%   25%   30%

This synthesis was actually patterned after an approach to nootkatone [Dastur, 1974] with modification. In the nootkatone synthesis the ring closure after the fragmentation is a Michael-type reaction.

nootkatone

Spiroannulation of a three-membered ring at the α-carbon of cyclopentanone has been observed [Bravo, 1968] when 2-chlorocyclopentanone was treated with dimethyloxosulfonium methylide. It is reasonable to assume that 2-methylenecyclopentanone was the intermediate, which is susceptible to attack by the reagent to afford the spiro ketone. Formation of 2-methylenecyclopentanone must have involved displacement and elimination steps.

The last step of the above spiroannulation consists of a Michael reaction/1,3-elimination tandem. An analogous process is that involving addition to a vinylogous Michael acceptor and 1,5-elimination [Minami, 1983].

Alkylation and acylation of heteroatoms is commonplace, yet the following cases deserve mention due to their tandem nature. While the based catalyzed isomerization of propargyl epoxides leads to 1,2,3-trienes [Marshall, 1991a], these cumulenes are susceptible to intramolecular attack of an allylic alcohol which results in furan derivatives.

Terpinolene diepoxide has been converted into 2,2,6-trimethyl-6-vinyltetra-hydro-pyran-3-one [Ho, 1983] on heating with alumina in toluene. The reaction has been rationalized in terms of a fragmentation/O-alkylation mechanism. It is ethanologous to the reaction of tetramethylallene diepoxide [Crandall, 1968].

The desilylative functionalization of benzoquinones which contain a trimethylsilylmethyl group with concomitant reduction of the quinone system [Karabelas, 1990] is mechanistically interesting. o-Quinonemethide intermediates are likely involved, and such species add immediately available nucleophiles.

6,6-Disubstituted 2,4-cyclohexadienones are photolabile, as they undergo electrocyclic opening readily to furnish unsaturated ketenes. A very unusual synthesis of (+)-aspicilin [Quinkert, 1988] has been accomplished using an intramolecular trapping of the nascent ketene to form the macrolactone.

(+)-aspicilin

2-Bromobicyclo[3.2.0]heptan-6-ones undergo skeletal rearrangement on reaction with nucleophiles to give 7-substituted norcamphor derivatives [Mitch, 1960; S.M. Roberts, 1974]. In fact such products are generated via a 1,3-dehydrobromination and subsequent ring opening.

The insertion of a C=N moiety to a strained, allylic CC bond of the pinenes is the basis of the synthetic elaboration of several indole alkaloids, that is hobartine, makomakine, and aristoteline [Mirand, 1982; Stevens, 1983a]. The conversion of a cyclobutane into a piperideine ring was initiated by mercuration of the double bond which elicited ring cleavage to generate the tertiary carbocations. After trapping of the cations by the added nitrile molecule (Ritter reaction) the reclosure to a much less strained six-membered ring is energetically favorable, because, in the absence of a good nucleophile, the positive charge of the intermediate cannot be dissipated in any other way.

(+)-makomakine

Both α- and β-pinenes suffer insertion at the more highly substituted allylic cyclobutane bond when initiated by mercuration. While optically active compounds are obtained from β-pinene, racemic products result from the reaction of α-pinene. The reason is that in the ring closure step the allylic mercurials derived from α-pinene may react at the ipso carbon or the allylic carbon site, each yielding the same structure (but in an enantiomerically opposite sense).

The reaction of amines with carbonyl compounds is a typical alkylation, applying to both the direct Schiff condensation and the conjugate Michael addition. An interesting process is the coupling of 4-alkene-1,2,3-tricarbonyl compounds with primary amines which gives rise to 3-hydroxypyrrole-2-carbonyl derivatives. Application of this reaction to a synthesis of prodigiosin [Wasserman, 1989] is shown.

prodigiosin

The condensation of isatin with phloroglucinol started a very concise synthesis of acronycine [Hlubucek, 1970]. It is considered that an $N \rightarrow O$ transacylation followed by intramolecular C- and N-alkylations served to establish the acridine skeleton.

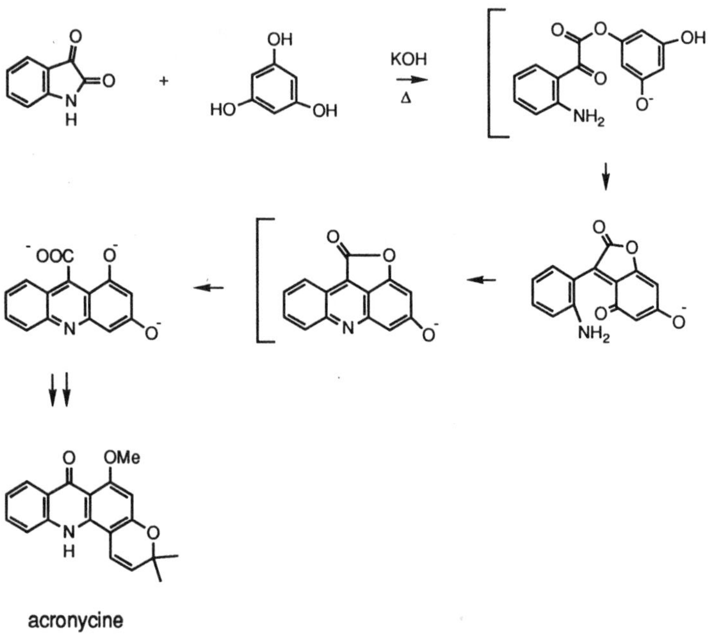

acronycine

2-Substituted indoles are available via sequential N-chlorination of anilines, treatment with α-sulfenyl ketones, and a tertiary amine [Gassman, 1973]. A [2.3]-sigmatropic rearrangement introduces the ketone sidechain to the ortho position of the regenerated anilines, and an intramolecular Schiff condensation–isomerization furnish the products.

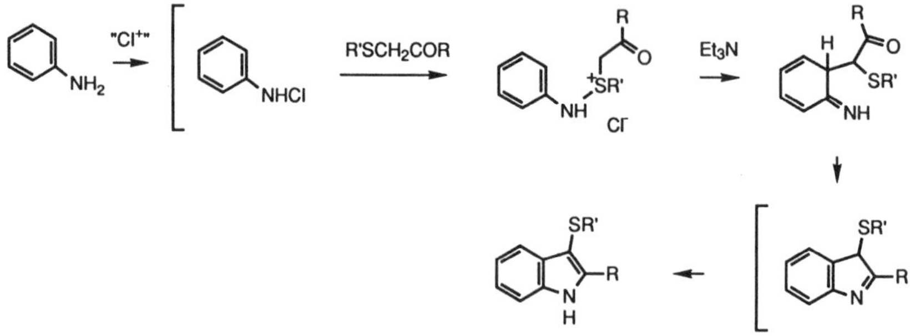

A convenient route to 2-vinylindoles [Wilkens, 1987] consists of a three-component reaction. Thus the nitrone formed from N-phenylhydroxyl-amine and an aldehyde is trapped by an allene, the resulting 5-methylene-isoxazolidine being susceptible to a series of transformations: Claisen-type rearrangement, retro-Michael ring opening, imine formation, and isomerization.

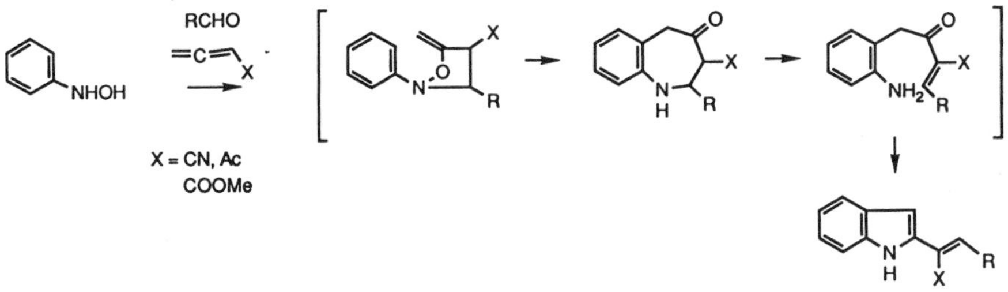

A similar intermediate is formed by condensation of an (*E,E*)-2,4-alkadienal with nitrosobenzene [Defoin, 1986]. The rearrangement step is preceded by a hetero-Diels–Alder reaction and enolization. The final product is tricyclic.

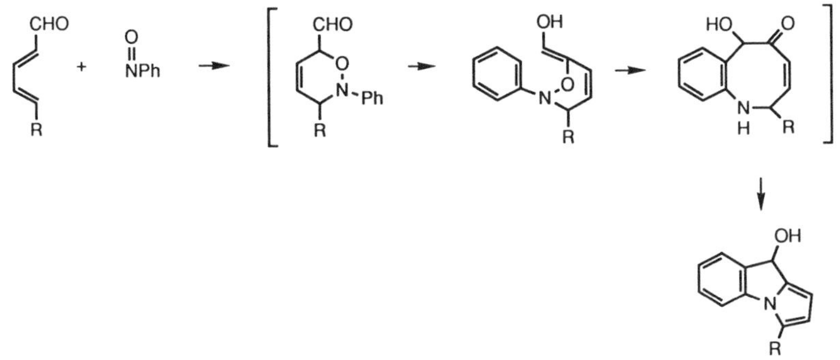

Transacylation is considered as electropilation from the viewpoint of the nucleophile. The very interesting intramolecular tandem transacylation is the "zip reaction" [Kramer, 1978] in which a lactam containing a polyamine sidechain attaching to the ring nitrogen atom incorporates successively into the ring when the lactam is treated with potassium 3-(aminopropyl)amide. A 53-membered ring is formed in reasonably yield.

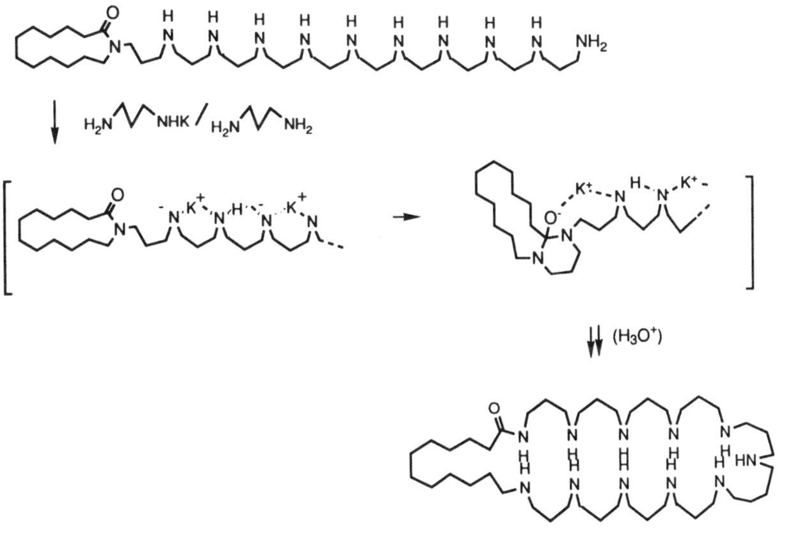

15.3. ELIMINATIONS AND RING OPENINGS

A selection of tandem reactions terminated by an elimination process are discussed in this section.

, Translithiation between 3-iodopropyl vinyl ethers with *t*-butyllithium results in 4-pentenols [W.F. Bailey, 1991]. This reaction is mediated by a 5-*exo-trig* ring closure after the I/Li exchange, but the tetrahydrofuranylmethyllithiums tend to isomerize by a formal elimination reaction.

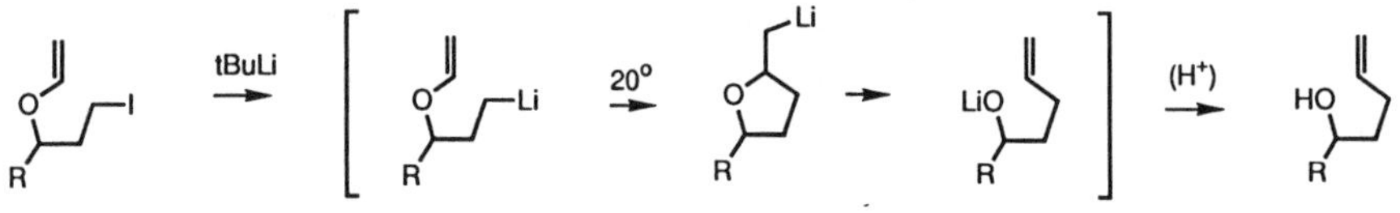

3-Nitropropanoic esters are synthetic equivalents of acrylic ester β-carbanions for Michael reactions owing to the facile in situ elimination of nitrous acid from the adducts [Bakuzis, 1978].

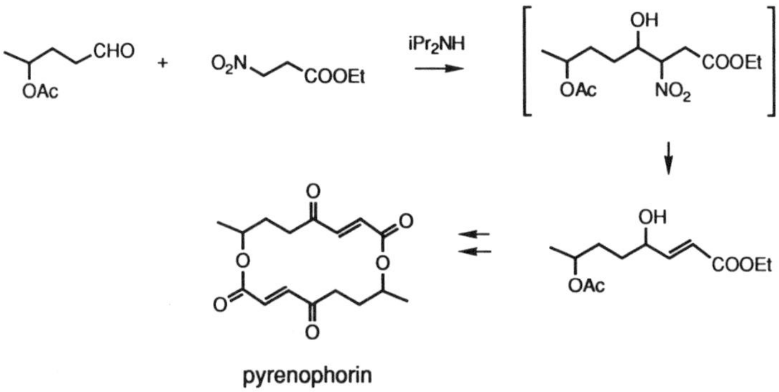

pyrenophorin

The Heck reaction consists of metathesis between C–Pd and C=C bonds and the following [H–Pd] elimination. A combination of an intramolecular Heck reaction with a Pd-ene reaction constitutes a powerful technique for multi-ring formation [Oppolzer, 1991b].

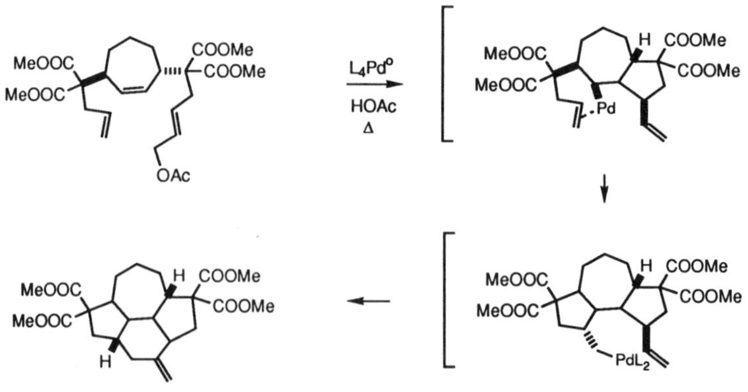

On treatment of monohydroborated products of 1-chloro-2-alkynes with alkali, terminal ellenes are produced [Zweifel, 1970]. 1,2-Elimination is initiated by attack of the hydroxide ion on the borane.

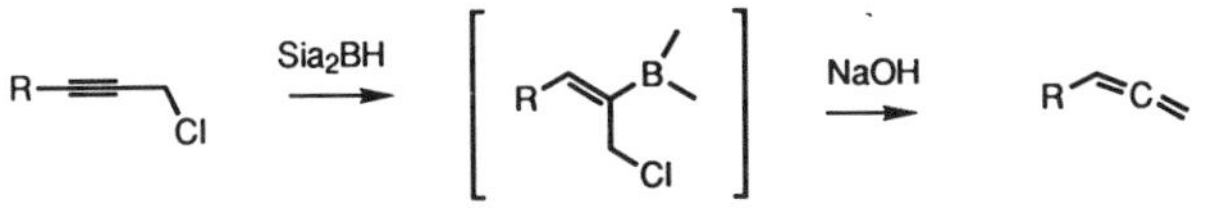

The reaction of allylboranes with enol ethers to give 1,4-dienes [Mikhailov, 1971] manifests the occurrence of an ene reaction and elimination of borate from the adducts.

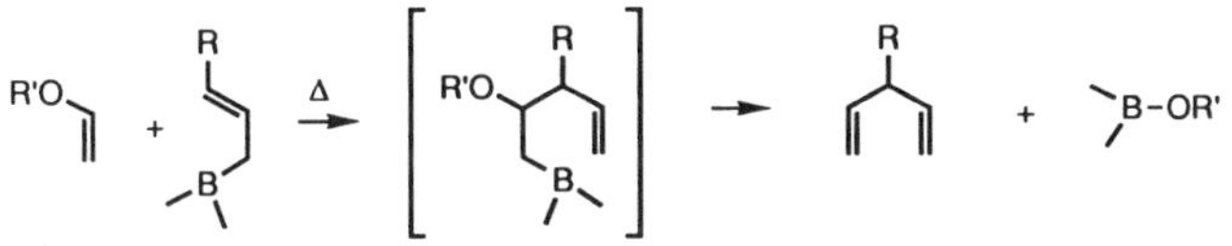

Isoquinolines are formed when o-chloropalladabenzaldehyde t-butylimines are treated with acrylonitrile and heated at 160°C [Girling, 1982]. The Heck reaction products are apparently prone to electrocyclization and subsequent elimination of hydrogen cyanide and isobutene.

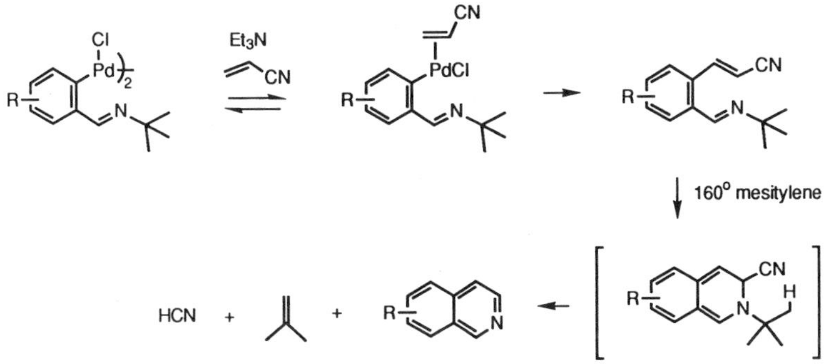

A rather unexpected reaction occurs when the spiroepoxide of 7-methylenebicyclo[3.3.1]nonan-3-one is treated with dimethyloxosulfonium methylide [Yurchenko, 1989]. The reaction pathway apparently consists of intramolecular alkylation, addition of the ylide to the ketone, dehydration, [2.3]-sigmatropic rearrangement, and elimination of methanesulfenic acid.

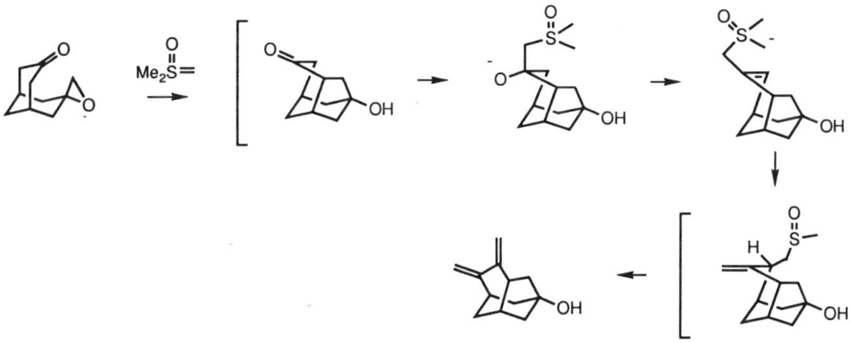

1-Alkoxy-3-siloxy-1,3-butadienes are versatile and reactive dienes for the Diels–Alder reaction. Cycloadducts from these dienes are very readily transformed into cyclohexenones via elimination and hydrolysis. A convenient synthesis of griseofulvin [Danishefsky, 1979] exploited this property as well as the facile elimination of phenylsufenic acid from a spirocyclic intermediate to afford a cross-conjugated cyclohexadienone.

griseofulvin

A direct synthesis of pyridoxine [Firestone, 1967] by a thermal condensation of a 5-alkoxy-4-methyloxazile with 2-butene-1,4-diol indicates in situ elimination of an alcohol molecule from the Diels–Alder adduct. This reaction is similar to the intramolecular reaction of N-alkyl-N-allyl-2-furfurylamines which give dihydroisoindoles [Hernandez, 1988].

pyridoxine

The vinylogous Wolff rearrangement of β',γ'-unsaturated α-diazoketones may proceed via the formation of bicyclo[2.1.0]pentanones [Branca, 1977]. Necessarily such highly strained ketones are prone to undergo ring cleavage to furnish γ,δ-unsaturated ketenes. An application of this method to a synthesis of mayurone is on record.

mayurone

Some controversies still surround the mechanism of olefin metathesis. However, the classification of this process under tandem reactions is appropriate. It has been shown that portions of macrocyclic dienes may be converted into the catenane topoisomers by intramolecular metathesis [Wolovsky, 1970].

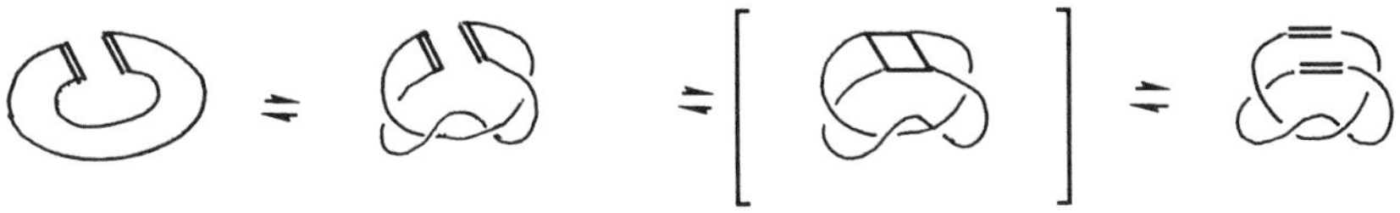

15.4. STITCHING OF ALKENES AND ALKYNES

The mention of the stitching of an alkene and an alkyne to form a (Z)-alkene via hydroboration and iododeboronation [Zweifel, 1968] is most appropriate to connect the last section to the present one. The mechanism for the reaction is shown as the following.

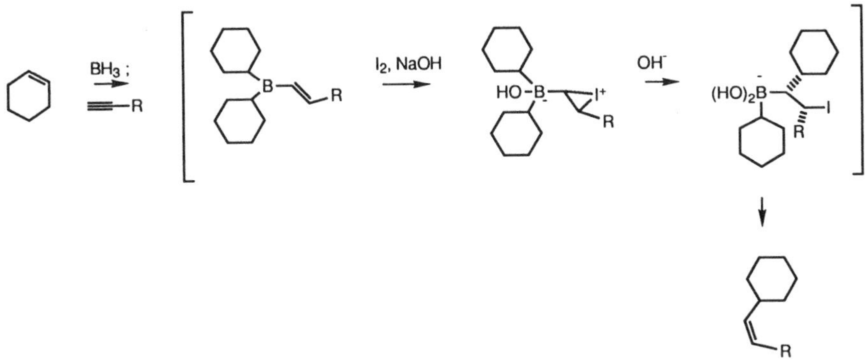

There are several ways to effect ring formation from dienes, but the most convenient method involves hydroboration and carbonylation. To demonstrate its utility we need to point out its incorporation into syntheses of juvabione [Negishi, 1976], δ-coniceine [Garst, 1982], confertin and helenalin [Welch, 1988], and estrone methyl ether [Bryson, 1980].

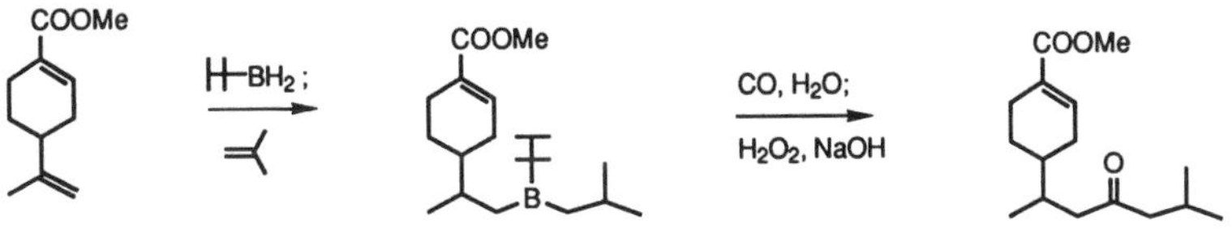

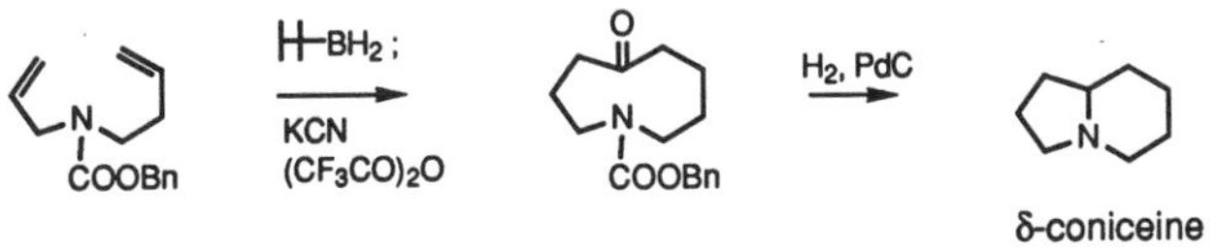

δ-coniceine

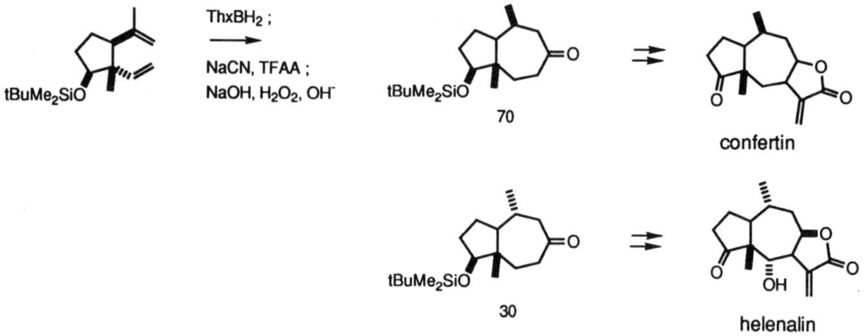

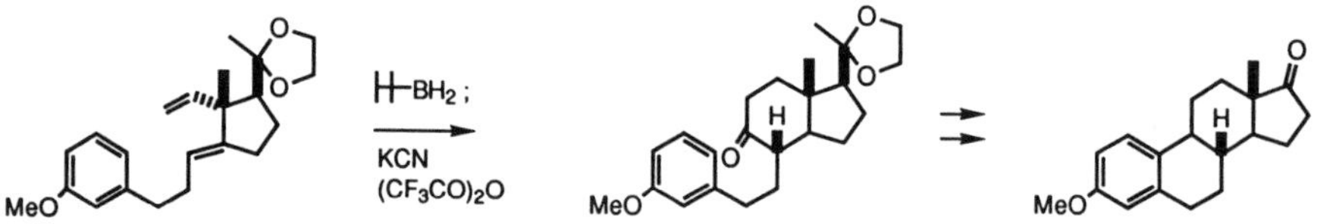

confertin

helenalin

The synthesis of a perhydrophenalenol [H.C. Brown, 1969] from (*E,E,E*)-1,5,9-cyclododecatriene via the hydroboration–carbonylation sequence should be noted.

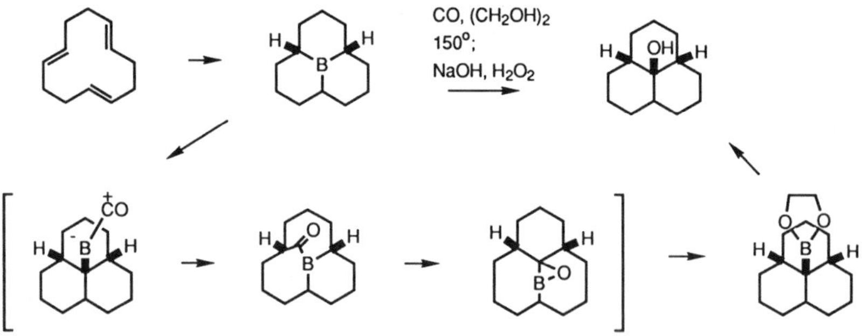

It should also be mentioned the possibility of remote asymmetric induction during the second stage of diene hydroboration [Still, 1980].

Carbonylation followed by transition-metal catalyzed intramolecular coupling with an organometallic moiety also permit a similar type of ring formation. This version is particularly suitable for the synthesis of cyclic enones and dienones, as exemplified by the ring closure leading to jatrophone [Gyorkos, 1990]. (For related carbonylative annulations, see Section 4.3, Pauson–Khand cyclization.)

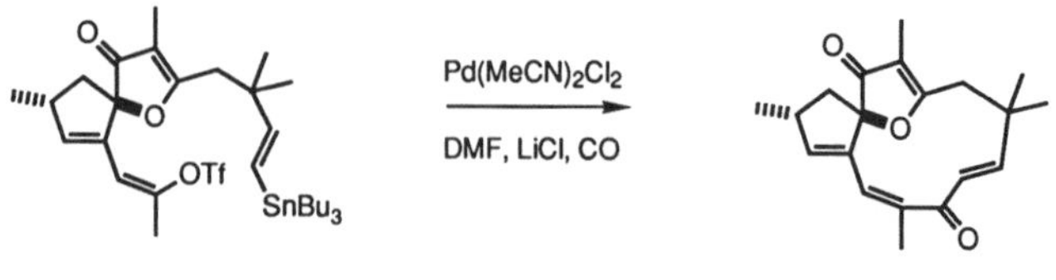

jatrophone

15.5. TANDEM REACTIONS INVOLVING OXIDOREDUCTIONS

Reactions involving redox processes which may or may not be the terminating step ar briefly mentioned here. These few cases serve to illustrate the potential pairing of a redox reaction to form a tandem only.

A very efficient amide synthesis [Mukaiyama, 1977] employing a mixture of triphenylphosphine and bis(α-pyridyl) disulfide as the condensing agent is a redox reaction in that the phosphine is oxidized and the disulfide undergoes reduction. The important amide bond formation involves attack of the carboxylate ion on the phosphonium or pentacovalent phosphorus species derived from the phosphine and the disulfide, the activated "ester" then reacts with the amine.

The Mitsunobu reaction [Mitsunobu, 1981] is based on a similar principle and naturally it is a tandem reaction.

Electrolysis of γ-hydroxycarboxylic acids gives rise to products with a ketone group and an alkene linkage [Corey, 1957; 1959]. An anodic oxidation triggers the fragmentation.

The formation of 3,4-epoxy-3-nitro-1-alkenes by reaction of 3-nitro-4-phenylselenyl-1-alkenes with basic hydrogen peroxide involves a tandem oxidation, elimination, and epoxidation [Najera, 1990].

Reduction that leads to fragmentation has been observed in a synthetic study of vinoxine [Bosch, 1986]. Thus, when a 4-alkylidene-1,4-dihydropyridine was hydrogenated the molecule fell apart, due to the availability of a aromatization pathway leading to a pyridine ring.

The tandem reductive alkylation of aryl carbonyl compounds has been extensively investigated. This process is based on the organometallic reaction of the carbonyl compounds to give benzylic alcohols which are subject to hydrogenolysis by lithium in liquid ammonia [Schumacher, 1981]. Since the two reactions are compatible, that is the metal alkoxides formed in the former reaction can be used directly, a tandem protocol is readily developed.

It is of interest to note that an allene ether in the same molecule is reduced in preference the benzylic alcohol [Weiberth, 1985].

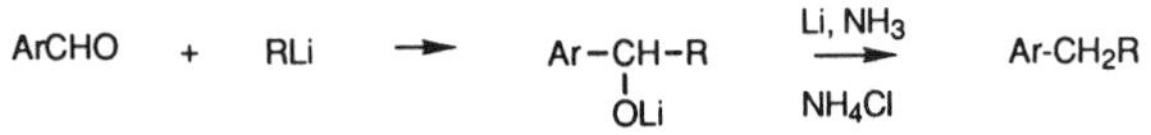

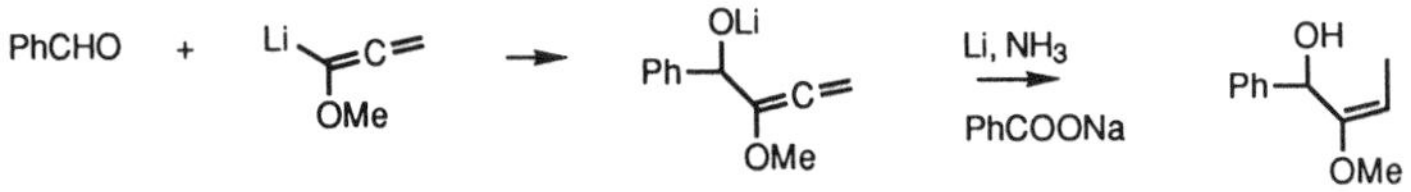

A surprising discovery is the shuffling of a *t*-butyldimethylsilyl group between two oxygen atoms during a cyclic borane-mediated condensation of silyl ketewne acetals with aldehydes [Kiyooka, 1991]. The shuffling enables reduction of the ester group of the intermediates.

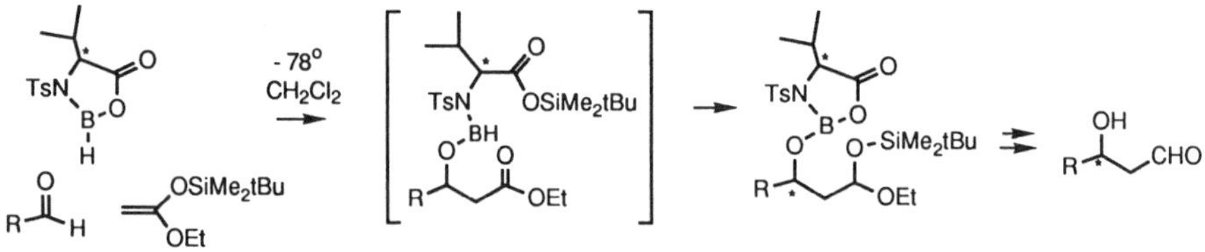

In the conversion of α-cyclopropylbenzyl alcohols into γ-trimethylsilyl-butyrophenone [Hwu, 1985] one of the pathway may involve hydride transfer from the silyl ether [Olah, 1976] and ring opening attack of the cyclopropyl ketone with the silyl anion.

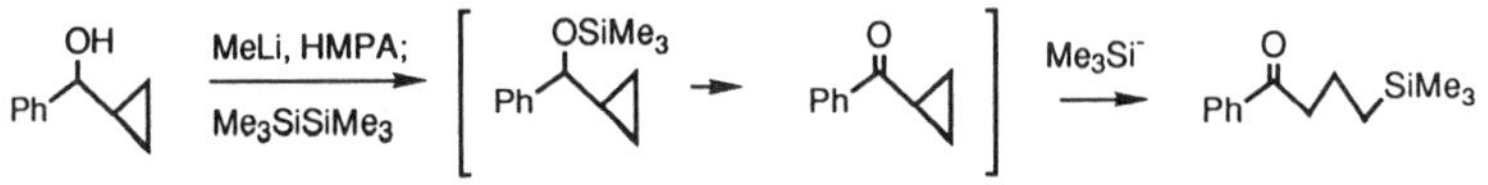

REFERENCES

Abelman, M.M., Overman, L.E. (1988) *J. Am. Chem. Soc.* **110**: 2328.

Abley, P., Byrd, J.E., Halpern, J. (1973) *J. Am. Chem. Soc.* **95**: 2591.

Abrahams, T.S., Baker, R., Exon, C.M., Rao, V.B. (1982) *J. Chem. Soc. Perkin Trans. I* 285.

Achiwa, K., Sekiya, M. (1981) *Chem. Lett.* 1213.

Adam, W., De Lucchi, O. (1980) *J. Org. Chem.* **45**: 4167.

Ager, D.J., Fleming, I., Patel, S.K. (1981) *J. Chem. Soc. Perkin Trans. I* 2520.

Ager, D.J. (1990) *Org. React.* **38**: 1.

Ahmed, Z., Cava, M.P. (1983) *J. Am. Chem. Soc.* **105**: 682.

Alder, A., Bellus, D. (1983) *J. Am. Chem. Soc.* **105**: 6712.

Alder, K., Münz, F. (1949) *Liebigs Ann. Chem.* **565**: 126.

Alexakis, A., Chapdelaine, M.J., Posner, G.H. (1978) *Tetrahedron Lett.* 4209.

Allred, E.L., Beck, B.R. (1973) *J. Am. Chem. Soc.* **95**: 2393.

Allred, E.L., Beck, B.R., Mumford, N.A. (1977) *J. Am. Chem. Soc.* **99**: 2694.

Amri, H., Rambaud, M., Villieras, J. (1989) *Tetrahedron Lett.* **30**: 7381.

Ando, K., Akadegawa, N., Takayama, H. (1991) *Chem. Commun.* 1765.

Ando, M., Büchi, G., Ohnuma, T. (1975) *J. Am. Chem. Soc.* **97**: 6880.

Ando, M., Tajima, K., Takase, K. (1978) *Chem. Lett.* 617.

Ando, R., Sugawara, T., Kuwajima, I. (1983) *Chem. Commun.* 1514; *Tetrahedron Lett.* **24**: 4429.

Andriamialisoa, R.Z., Diatta, L., Rasoanaivo, P., Langlois, N., Potier, P. (1975) *Tetrahedron* **31**: 2347.

Angle, S.R., Turnbull, K.D. (1990) *J. Am. Chem. Soc.* **112**: 3698.

Antczak, K., Kingston, J.F., Fallis, A.G. (1984) *Tetrahedron Lett.* **25**: 2077.

Antczak, K., Kingston, J.F., Fallis, A.G., Hanson, A.W. (1987) *Can. J. Chem.* **65**: 114.

Aoyagi, S., Shishido, Y., Kibayashi, C. (1991) *Tetrahedron Lett.* **32**: 4325.

Arigoni, D. (1962) *Gazz. Chim. Ital.* **92**: 884.

Armstrong, P., Grigg, R., Surendrakumar, S., Warnock, W.J. (1987a) *Chem. Commun.* 1327.

Armstrong, P., Grigg, R., Warnock, W.J. (1987b) *Chem. Commun.* 1325.

Arnold, B.J., Mellows, S.M., Sammes, P.G. (1973) *J. Chem. Soc. Perkin Trans. I* 1266.

Asaoka, M., Mukuta, T., Takei, H. (1981) *Tetrahedron Lett.* **22**: 735.

Atta-ur-Rahman, Beisler, J.A., Harley-Mason, J. (1980) *Tetrahedron* **36**: 1063.

Atta-ur-Rahman, Malik, S., Albert, K. (1986) *Z. Naturforsch.* **B41**: 386.

Attwood, M.R., Churcher, I., Dunsdon, R.M., Hurst, D.N., Jones, P.S. (1991) *Tetrahedron Lett.* **32**: 811.

Bachi, M.D., Denemark, D. (1989) *J. Am. Chem. Soc.* **111**: 1886.

Backenstrass, F., Streith, J., Tschamber, T. (1990) *Tetrahedron Lett.* **31**: 2139.

Bailey, T.R., Garigipati, R.S., Morton, J.A., Weinreb, S.M. (1984) *J. Am. Chem. Soc.* **106**: 3240.

Bailey, W.F., Rossi, K. (1989) *J. Am. Chem. Soc.* **111**: 765.

Bailey, W.F., Khanolkar, A.D. (1990) *Tetrahedron Lett.* **31**: 5993.

Bailey, W.F., Zarcone, L.M.J. (1991) *Tetrahedron Lett.* **32**: 4425.

Bakuzis, P., Bakuzis, M.L.F., Weingartner, T.F. (1978) *Tetrahedron Lett.* 2371.

Balaban, A.T., Nenitzescu, C.D. (1959) *Liebigs Ann. Chem.* **625**: 66.

Balasubramanian, K., John, J.P., Swaminathan, S. (1974) *Synthesis* 51.

Baldwin, J.E., Tzodikov, N.R. (1977) *J. Org. Chem.* **42**: 1878.

Baldwin, J.E., Adlington, R.M., Robertson, J. (1989) *Tetrahedron* **45**: 909.

Baldwin, S.W., Blomquist, H.R. (1982) *J. Am. Chem. Soc.* **104**: 4990.

Ban, T., Wakita, Y., Kanematsu, K. (1980) *J. Am. Chem. Soc.* **102**: 5415.

Ban, Y., Yoshida, K., Goto, J., Oishi, T., Takeda, E. (1983) *Tetrahedron* **39**: 3657.

Banville, J., Grandmaison, J., Lang, G., Brassard, P. (1974) *Can. J. Chem.* **50**: 80.

Banwell, M.G., Onrust, R. (1985) *Tetrahedron Lett.* **26**: 4543.

Banwell, M.G., Lambert, J.N., Gulbis, J.M., Mackay, M.F. (1990) *Chem. Commun.* 1450.

Bao, J., Dragisich, V., Wenglowsky, S., Wulff, W.D. (1991) *J. Am. Chem. Soc.* **113**: 9873.

Baraldi, P.G., Barco, A., Benetti, S., Moroder, F., Pollini, G.P., Simoni, D., Zanirato, V. (1982) *Chem. Commun.* 1265.

Barco, A., Benetti, S., Casolari, A., Pollini, G.P., Spalluto, G. (1990) *Tetrahedron Lett.* **31**: 3039.

Barco, A., Benetti, S., Pollini, G.P., Spalluto, G., Zanirato, V. (1991) *Chem. Commun.* 390.

Barinelli, L.S., Tao, K., Nicholas, K.M. (1986) *Organometallics* **5**: 588.

Barkley, L.B., Knowles, W.S., Raffelson, H., Thompson, Q.E. (1956) *J. Am. Chem. Soc.* **78**: 4111.

Barsi, M.C., Das, B.C., Fourrey, J.-L., Sundaramoorthi, R. (1985) *Chem. Commun.* 88.

Bartlett, P.D., Ando, T. (1970) *J. Am. Chem. Soc.* **92**: 7518.

Barton, D.H.R., Jackman, L.M., Rodriguez-Hahn, L., Sutherland, J.K. (1965) *J. Chem. Soc.* 1772.

Barton, D.H.R. (1970) *Proc. Roy. Soc.* **A319**: 145.

Barton, D.H.R., da Silva, E., Zard, S.Z. (1988) *Chem. Commun.* 285.

Barton, T.J., Burns, G.T. (1978) *J. Am. Chem. Soc.* **100**: 5246.

Bateson, J.H., Smith, C.F., Wilkinson, J.B. (1991) *J. Chem. Soc. Perkin Trans. I* 651.

Bauman, J.G., Hawley, R.C., Rapoport, H. (1984) *J. Org. Chem.* **49**: 3791.

Baxter, A.D., Roberts, S.M., Scheinmann, F., Wakefield, B.J., Newton, R.F. (1983) *Chem. Commun.* 932.

Baxter, E.W., Labaree, D., Chao, S., Mariano, P.S. (1989) *J. Org. Chem.* **54**: 2893.

Baxter, G.J., Brown, R.F.C., McMullen, G.L. (1974) *Aust. J. Chem.* **27**: 2605.

Bayne, D.W., Nicol, A.J., Tennant, G. (1975) *Chem. Commun.* 782.

Bazan, A.C., Edwards, J.M., Weiss, U. (1978) *Tetrahedron* **34**: 3005.

Beak, P., Burg, D.A. (1986) *Tetrahedron Lett.* **27**: 5911.

Beck, A., Knothe, L., Hunkler, D., Prinzbach, H. (1982) *Tetrahedron Lett.* **23**: 2431; (1984) *Tetrahedron Lett.* **25**: 1785.

Becker, D.A., Danheiser, R.L. (1989) *J. Am. Chem. Soc.* **111**: 389.

Becker, H.-D., Becker, H.-C., Sandros, K., Andersson, K. (1985) *Tetrahedron Lett.* **26**: 1589.

Beeley, N.R.A., Cremer, G., Dorlhene, A., Mompon, B., Pascard, C., Dau, E.T.H. (1983) *Chem. Commun.* 1046.

Beereboom, J.J. (1966) *J. Org. Chem.* **31**: 2026.

Begley, M.J., Crombie, L., Slack, D.A., Whiting, D.A. (1977) *J. Chem. Soc. Perkin Trans. I* 2402.

Benson, S.C., Gross, J.L., Snyder, J.K. (1990) *J. Org. Chem.* **55**: 3257.

Benson, W., Winterfeldt, E. (1979) *Chem. Ber.* **112**: 1913.

Bergman, J., Carlsson, R. (1978) *Tetrahedron Lett.* 4055.

Bergman, R., Hansson, T., Sterner, O., Wickberg, B. (1990) *Chem. Commun.* 865.

Bergmann, E.D., Ginsburg, D. (1959a) *Org. React.* **10**: 179.

Bergmann, E.D., Yaroslavsky, S., Weiler-Feilchenfeld, H. (1959b) *J. Am. Chem. Soc.* **81**: 2775.

Beringer, F.M., Galton, S.A. (1963) *J. Org. Chem.* **28**: 3250.

Berson, J.A., Dervan, P.B. (1972a) *J. Am. Chem. Soc.* **94**: 7597.

Berson, J.A., Dervan, P.B., Jenkins, J.A. (1972b) *J. Am. Chem. Soc.* **94**: 7598.

Berubé, G., Fallis, A.G. (1989) *Tetrahedron Lett.* **30**: 4045.

Besselièvre, R., Thal, C., Husson, H.-P., Potier, P. (1975) *Chem. Commun.* 90.

Besselièvre, R., Husson, H.-P. (1976) *Tetrahedron Lett.* 1873.

Bessiere-Chrétien, Y., Grison, C. (1973) *Chem. Commun.* 549.

Best, W.M., Wege, D. (1981) *Tetrahedron Lett.* **22**: 4877.

Bigorra, J., Font, J., Ortuno, R.M., Sanchez-Derrando, F., Florencio, F., Martinez-Carrera, S., Garcia-Blanco, S. (1985) *Tetrahedron* **41**: 5577, 5589.

Billups, W.E., Blakeney, A.J. (1976) *J. Am. Chem. Soc.* **98**: 7817.

Binger, P., Schäfer, B. (1988) *J. Am. Chem. Soc.* **29**: 4539.

Birch, A.J., Hill, J.S. (1966) *J. Chem. Soc. (C)* 419.

Birch, A.J., Mani, N.S., Subba-Rao, G.S.R. (1990) *J. Chem. Soc. Perkin Trans. I* 1423.

Biswas, K.M., Jackson, A.H. (1983) *Chem. Commun.* 85.

Bit, R.A., Davis, P.D., Hill, C.H., Keech, E., Vesey, D.R. (1991) *Tetrahedron* **47**: 4645.

Black, R.M. (1981) *Synthesis* 829.

Bleasdale, D.A., Jones, D.W., Maier, G., Reisenauer, H.P. (1983) *Chem. Commun.* 1095.

Bloch, R., Boivin, F., Bortolussi, M. (1976a) *Chem. Commun.* 371.

Bloch, R., Bortolussi, M. (1976b) *Tetrahedron Lett.* 309.

Bloch, R., Abecassis, J. (1982) *Tetrahedron Lett.* **23**: 3277.

Bloch, R., Hassan, D., Mandard, X. (1983) *Tetrahedron Lett.* **24**: 4691.

Block, E., Revelle, L.K. (1978) *J. Am. Chem. Soc.* **100**: 1630.

Block, E., Putman, D. (1990) *J. Am. Chem. Soc.* **112**: 4072.

Bly, R.S., Bly, R.K., Shibata, T. (1983) *J. Org. Chem.* **48**: 101.

Bobowski, G., Morrison, G.C. (1981) *J. Org. Chem.* **46**: 4927.

Boche, G., Weber, H., Benz, J. (1974) *Angew. Chem. Int. Ed. Engl.* **13**: 207.

Boeckman, R.K. (1973) *J. Am. Chem. Soc.* **95**: 6867.

Boeckman, R.K., Delton, M.H., Nagasaka, T., Watnabe, T. (1977) *J. Org. Chem.* **42**: 2946.

Boeckman, R.S., Ko, S.S. (1980) *J. Am. Chem. Soc.* **102**: 7146.

Boeckman, R.K., Cheon, S.H. (1983) *J. Am. Chem. Soc.* **105**: 4112.

Boekelheide, V., Nottke, J.E. (1969) *J. Am. Chem. Soc.* **34**: 4134.

Boekelheide, V., Ewing, G. (1978) *Tetrahedron Lett.* 4245.

Boger, D.L., Brotherton, C.E. (1984a) *J. Am. Chem. Soc.* **106**: 805; (1984b) *J. Org. Chem.* **49**: 4050.

Boger, D.L., Coleman, R.S. (1984c) *J. Org. Chem.* **49**: 2240.

Boger, D.L., Panek, J.S. (1985) *J. Am. Chem. Soc.* **107**: 5745.

Boger, D.L., Brotherton, C.E. (1986a) *J. Am. Chem. Soc.* **108**: 6695.

Boger, D.L., Coleman, R.S. (1986b) *J. Org. Chem.* **51**: 3250.

Boger, D.L., Patel, M. (1987) *Tetrahedron Lett.* **28**: 2499.

Boger, D.L., Mathvink, R.J. (1990a) *J. Am. Chem. Soc.* **112**: 4003; (1990b) *J. Org. Chem.* **55**: 5442.

Boger, D.L., Zhang, M. (1991) *J. Am. Chem. Soc.* **113**: 4230.

Bond, R.F., Boeyens, J.C.A., Holzapfel, C.W., Steyns, P.S. (1979) *J. Chem. Soc. Perkin Trans. I* 1751.

Bosch, J., Bennasar, M.-L., Zulaica, E. (1986) *J. Org. Chem.* **51**: 2289.

Bourhis, M., Goursolle, M., Leger, J.-M., Duboudin, J.-G., Duboudin, F., Picard, P. (1989) *Tetrahedron* **30**: 4665.

Bowd, A., Turnbull, J.H., Coyle, J.D. (1980) *J. Chem. Res. (S)* 202.

Bowden, B., Cookson, R.C., Davis, H.A. (1973) *J. Chem. Soc. Perkin Trans. I* 2634.

Bowman, E.S., Hugher, G.B., Grutzner, J.B. (1976) *J. Am. Chem. Soc.* **98**: 8273.

Bradbury, R.H., Gilchrist, T.L., Rees, C.W. (1981) *J. Chem. Soc. Perkin Trans. I* 3225.

Brain, E.G., Cassidy, F., Lake, A.W., Cox, P.J., Sim, G.A. (1972) *Chem. Commun.* 497.

Branca, S.J., Lock, R.L., Smith, III, A.B. (1977) *J. Org. Chem.* **42**: 3165.

Brandänge, S., Lundin, C. (1971) *Acta Chem. Scand.* **25**: 2447.

Bratby, D.M., Chadwick, J.C., Fray, G.I., Saxton, R.G. (1977) *Tetrahedron* **33**: 1527.

Braun, M. (1987) *Angew. Chem. Int. Ed. Engl.* **26**: 24.

Bravo, P., Gaudiano, G., Ticozzi, C., Umani-Ronchi, A. (1968) *Tetrahedron Lett.* 4481.

Brennan, T.M., Hill, R.K. (1968) *J. Am. Chem. Soc.* **90**: 5614.

Breslow, R., Olin, S.S., Groves, J.T. (1968) *Tetrahedron Lett.* 1837.

Brieger, G. (1963) *J. Am. Chem. Soc.* **85**: 3783.

Brimacombe, J.S., Zahur-ul-Haque, Murray, A.W. (1974) *Tetrahedron Lett.* 4087.

Brinker, U.H., Wüstler, H., Maas, G. (1985) *Chem. Commun.* 1812.

Brinkley, J.M., Friedman, L. (1972) *Tetrahedron Lett.* 4141.

Britten-Kelly, M.R., Willis, B.J., Barton, D.H.R. (1981) *J. Org. Chem.* **46**: 5027.

Broadhurst, M.J., Hassall, C.H., Thomas, G.J. (1984) *Tetrahedron* **40**: 4649.

Brokatzky, J., Eberbach, W. (1981) *Chem. Ber.* **114**: 384.

Broom, N.J., Sammes, P.G. (1978) *Chem. Commun.* 162.

Brown, H.C., Dickason, W.C. (1969) *J. Am. Chem. Soc.* **91**: 1226.

Brown, R.F.C., Harrington, K.J. (1972) *Chem. Commun.* 1175.

Brown, R.F.C., McMullen, G.L. (1974) *Aust. J. Chem.* **27**: 2385.

Bruce, J.M., Creed, D., Dawes, K. (1971) *J. Chem. Soc. (C)* 2244.

Brückmann, R., Maas, G. (1986) *Chem. Commun.* 1782.

Bryant, R.J., McDonald, E. (1975) *Tetrahedron Lett.* 3841.

Bryson, T.A., Reichel, C.J. (1980) *Tetrahedron Lett.* **21**: 2381.

Bucheister, A., Klemarczyk, P., Rosenblum, M. (1982) *Organometallics* **1**: 1679.

Büchi, G., Coffen, D.L., Kocsis, K., Sonnet, P.E., Ziegler, F.E. (1966) *J. Am. Chem. Soc.* **88**: 3099.

Büchi, G., Kulsa, P., Ogasawara, K., Rosati, R.L. (1970) *J. Am. Chem. Soc.* **92**: 999.

Büchi, G., Wüest, H. (1971) *Helv. Chim. Acta* **54**: 1767.

Büchi, G., Carlson, J.A., Powell, J.E., Tietze, L.-F. (1973) *J. Am. Chem. Soc.* **95**: 540.

Büchi, G., Pawlak, M. (1975) *J. Org. Chem.* **40**: 100.

Büchi, G., Berthet, D., Decorzant, R., Grieder, A., Hauser, A. (1976) *J. Org. Chem.* **41**: 3208.

Büchi, G., Hauser, A., Limacher, J. (1977a) *J. Org. Chem.* **42**: 3323.

Büchi, G., Mak, C.-P. (1977b) *J. Am. Chem. Soc.* **99**: 8073.

Büchi, G., Vogel, D.E. (1983) *J. Org. Chem.* **48**: 5406.

Buchwald, S.L., Lum, R.T., Fisher, R.A., Davis, W.M. (1989) *J. Am. Chem. Soc.* **111**: 9113.

Bucourt, R., Vignau, M., Weill-Raynal, J. (1967) *C.R. Acad. Sci. (Paris)* **265**: 834.

Bunce, R.A., Wamsley, E.J., Pierce, J.D., Schellhammer, A.J., Drumright, R.E. (1987) *J. Org. Chem.* **52**: 464.

Burger, J.J., Chan, T.B.R.A., de Waard, E.R., Huisman, H.O. (1981) *Tetrahedron* **37**: 417.

Burger, U., Etienne, R. (1984) *Helv. Chim. Acta* **67**: 2057.

Burke, S.D., Saunders, J.O., Oplinger, J.A., Murtiashaw, C.W. (1985) *Tetrahedron Lett.* **26**: 1131.

Burke, S.D., Armistead, D.M., Shankaran, K. (1986) *Tetrahedron Lett.* **27**: 6295.

Burke, S.D., Strickland, S.M.S., Organ, H.M., Silks, III, L.A. (1989) *Tetrahedron Lett.* **30**: 6303.

Burkholder, T.P., Fuchs, P.L. (1988) *J. Am. Chem. Soc.* **110**: 2341.

Buschmann, E., Steglich, W. (1974) *Angew. Chem. Int. Ed. Engl.* **13**: 484.

Byrd, J.E., Cassar, L., Eaton, P.E., Halpern, J. (1971) *Chem. Commun.* 40.

Caine, D., Stanhope, B. (1987) *Tetrahedron* **43**: 5545.

Cairns, N., Harwood, L.M., Astles, D.P. (1987) *Chem. Commun.* 400.

Cairns, P.M., Crombie, L., Pattenden, G. (1982) *Tetrahedron Lett.* **23**: 1405.

Callot, H.J., Tschamber, T. (1974) *Tetrahedron Lett.* 3155, 3159.

Cambie, R.C., Crump, D.R., Duvie, R.N. (1969) *Aust. J. Chem.* **22**: 1975.

Cameron, A.G., Hewson, A.T., Osammor, M.I. (1984) *Tetrahedron Lett.* **25**: 2267.

Cano, P., Echavarren, A., Prados, P., Farina, F. (1983) *J. Org. Chem.* **48**: 5373.

Canonne, P., Belanger, D. (1980) *Chem. Commun.* 125.

Canonne, P., Boulanger, R. (1988) *3rd Chem. Congr. N. Am. ORGN* 134.

Cantrell, W.R., Davies, H.M.L. (1991) *J. Org. Chem.* **56**: 723.

Carceller, E., Centellas, V., Moyano, A., Pericas, M.A., Serratos, F. (1985) *Tetrahedron Lett.* **26**: 2475.

Cardwell, K., Hewitt, B., Ladlow, M., Magnus, P. (1988) *J. Am. Chem. Soc.* **110**: 2242.

Cargill, R.L., Pond, D.M., Le Grand, S.O. (1970) *J. Org. Chem.* **35**: 359.

Cargill, R.L., King, T.Y., Sears, A.B., Willcott, M.R. (1971) *J. Org. Chem.* **36**: 1423.

Carpenter, N.E., Kucera, D.J., Overman, L.E. (1989) *J. Org. Chem.* **54**: 5846.

Carruthers, W. (1990) *Cycloaddition Reactions in Organic Synthesis*, Pergamon, Oxford.

Cartier, D., Ouhrani, M., Levy, J. (1989) *Tetrahedron Lett.* **30**: 1951.

Castro, J., Sorensen, H., Riera, A., Morin, C., Moyano, A., Pericas, M.A., Greene, A.E. (1990) *J. Am. Chem. Soc.* **112**: 9388.

Cattanach, C.J., Cohen, A., Heath-Brown, B. (1968) *J. Chem. Soc. (C)* 1235.

Cava, M.P., Ahmad, Z., Benfaremo, N., Murphy, R.A., O'Malley, G.J. (1984) *Tetrahedron* **40**: 4767.

Chan, T.H., Finkenbine, J.R. (1972) *J. Am. Chem. Soc.* **94**: 2880.

Chan, T.H., Li, M.P., Mychajlowskij, W., Harpp, D.N. (1974) *Tetrahedron Lett.* 3511.

Chan, T.H., Brownbridge, P. (1980) *J. Am. Chem. Soc.* **102**: 3534.

Chan, T.H., Kang, G.J. (1982) *Tetrahedron Lett.* **23**: 3011.

Chapdelaine, M.J., Hulce, M. (1990) *Org. React.* **38**: 225.

Chapman, O.L., Lassila, J.D. (1968) *J. Am. Chem. Soc.* **90**: 2449.

Chapman, O.L., Engel, M.R., Springer, J.P., Clardy, J.C. (1971) *J. Am. Chem. Soc.* **93**: 6696.

Chass, D.A., Buddhasukh, D., Magnus, P. (1978) *J. Org. Chem.* **43**: 1750.

Cheng, A.K., Stothers, J.B. (1977) *Can. J. Chem.* **55**: 4184.

Childs, R.F., Grigg, R., Johnson, A.W. (1967) *J. Chem. Soc. (C)* 201.

Childs, R.F., Winstein, S. (1968) *J. Am. Chem. Soc.* **90**: 7146.

Chou, C.-T., Swenton, J.S. (1987) *J. Am. Chem. Soc.* **109**: 6898.

Chow, K., Moore, H.W. (1990a) *J. Org. Chem.* **55**: 370.

Chow, K., Nguyen, N.V., Moore, H.W. (1990b) *J. Org. Chem.* **55**: 3876.

Choy, W. (1990) *Tetrahedron* **46**: 2281.

Cladingboel, D.E., Parson, P.J. (1990) *Chem. Commun.* 1543.

Claisen, L., Tietze, E. (1926) *Liebigs Ann. Chem.* **449**: 89.

Clark, D.A., Bunnell, C.A., Fuchs, P.L. (1978) *J. Am. Chem. Soc.* **100**: 7777.

Clark, D.E., Meredith, R.P.K., Ritchie, A.C., Walker, T. (1962) *J. Chem. Soc.* 2490.

Clarke, T., Stewart, J.D., Ganem, B. (1990) *Tetrahedron* **46**: 731.

Coates, R.M., Hutchins, C.W. (1979) *J. Org. Chem.* **44**: 4742.

Cobb, R.L., Mahan, F.E., Fahey, D.R. (1977) *J. Org. Chem.* **42**: 2601.

Coleman, R.S., Grant, E.B. (1990) *Tetrahedron Lett.* **31**: 3677.

Collie, J.N., Reilly, A.A.B. (1922) *J. Chem. Soc.* **121**: 1984.

Collins, J.L., Grieco, P.A., Gross, R.S. (1990) *J. Org. Chem.* **55**: 5816.

Confalone, P.N., Pizzolato, G. (1981) *J. Am. Chem. Soc.* **103**: 4251.

Confalone, P.N., Pizzolato, G., Confalone, D.L., Uskokovic, M.R. (1980) *J. Am. Chem. Soc.* **102**: 1954.

Confalone, P.N., Huie, E.M. (1984) *J. Am. Chem. Soc.* **106**   7175.

Conia, J.M., LePerchec. P. (1966) *Bull. Soc. Chim. Fr.* 281.

Conover, L.H., Butler, K., Johnston, J.D., Korst, J.J., Woodward, R.B. (1962) *J. Am. Chem. Soc.* **84**: 3222.

Cooke, M.P., Parlman, R.M. (1977) *J. Am. Chem. Soc.* **99**: 5222..

Cooke, M.P. (1979) *Tetrahedron Lett.* 2199.

Cookson, R.C., Rogers, N.R. (1973) *J. Chem. Soc. Perkin Trans. I* 2741.

Corbier, J., Cresson, P. (1970) *C.R. Acad. Sci. (Paris), Ser. C* **270**: 2077.

Cordero, F.M., Brandi, A., Querci, C., Goti, A., DeSarlo, F., Guarna, A. (1990) *J. Org. Chem.* **55**: 1762.

Corey, E.J., Sauers, R.R., Swann, S. (1957) *J. Am. Chem. Soc.* **79**: 5826.

Corey, E.J., Sauers, R.R. (1959) *J. Am. Chem. Soc.* **81**: 1743.

Corey, E.J., Jautelat, M. (1967) *J. Am. Chem. Soc.* **89**: 3912.

Corey, E.J., Broger, E.A. (1969a) *Tetrahedron* 1779.

Corey, E.J., Hegedus, L.S. (1969b) *J. Am. Chem. Soc.* **91**: 4926.

Corey, E.J., Watt, D.S. (1973) *J. Am. Chem. Soc.* **95**: 2304.

Corey, E.J., Balanson, R.D. (1974) *J. Am. Chem. Soc.* **96**: 6516.

Corey, E.J., Tius, M.A., Das, J. (1980) *J. Am. Chem. Soc.* **102**: 1742.

Corey, E.J., Tramantano, A. (1981) *J. Am. Chem. Soc.* **103**: 5599.

Corey, E.J., Desai, M.C. (1985a) *Tetrahedron Lett.* **26**: 3535.

Corey, E.J., Desai, M.C., Engler, T.A. (1985b) *J. Am. Chem. Soc.* **107**: 4339.

Corey, E.J., Su, W. (1987a) *Tetrahedron Lett.* **28**: 5241; (1987b) *J. Am. Chem. Soc.* **109**: 7534.

Corey, E.J., Wess, G., Xiang, Y.B., Singh, A.K. (1987c) *J. Am. Chem. Soc.* **109**: 4717.

Corey, E.J., Carpino, P. (1989) *J. Am. Chem. Soc.* **111**: 5472.

Cornforth, J.W., Robinson, R. (1949) *J. Chem. Soc.* 1861.

Cory, R.M., Chan, D.M.T., Naguib, Y.M.A., Rastall, M.A., Renneboog, R.M. (1980) *J. Org. Chem.* **45**: 1852.

Cory, R.M., Ritchie, B.M. (1983) *Chem. Commun.* 1244.

Cory, R.M., Renneboog, R.M. (1984) *J. Org. Chem.* **49**: 3898.

Cory, R.M., Anderson, P.C., Bailey, M.D., McLaren, F.R., Renneboog, R.M., Yamamoto, B.R. (1985) *Can. J. Chem.* **63**: 2618.

Coxon, J.M., Fong, S.T., Steel, P.J. (1990) *Tetrahedron Lett.* **31**: 7479.

Crandall, J.K., Machleder, W.H. (1968) *J. Am. Chem. Soc.* **90**: 7292, 7346.

Criegee, R., Schweickhardt, C., Knoche, H. (1970) *Chem. Ber.* **103**: 960.

Crimmins, M.T., Nantermet, P.G. (1990) *J. Org. Chem.* **55**: 4235.

Crombie, L., Ponsford, R. (1971) *J. Chem. Soc. (C)* 788.

Cross, B.E., Grove, J.F., Morrison, A. (1961) *J. Chem. Soc.* 2498.

Cross, B.E., Hanson, J.R., Briggs, L.H., Cambie, R.C., Rutledge, P.S. (1963) *Proc. Chem. Soc.* 17.

Cupas, C.A., Kong, M.S., Mullins, M., Heyd, W.E. (1971) *Tetrahedron Lett.* 3157.

Curphey, T.J., Kim, H.L. (1968) *Tetrahedron Lett.* 1441.

Curran, D.P., Chen, M.-H. (1985a) *Tetrahedron Lett.* **26**: 4991.

Curran, D.P., Rakiewics, D.M. (1985b) *J. Am. Chem. Soc.* **107**: 1448.

Curran, D.P., Kuo, S.-C. (1987) *Tetrahedron* **43**: 5653.

Curran, D.P., van Elburg, P.A. (1989) *Tetrahedron Lett.* **30**: 2501.

Curran, D.P. (1990) *Adv. Free Radical Chem.* **1**: 121.

Curran, D.P., Liu, H. (1991a). *J. Am. Chem. Soc.* **113**: 2127.

Curran, D.P., Wolin, R.L. (1991b) *Synlett.* 317.

Cushman, M., Patrick, D.A., Toma, P.H., Byrn, S.R. (1989) *Tetrahedron Lett.* **30**: 7161.

Daeniker, H.U., Hochstetler, A.R., Kaiser, K., Kitchens, G.C. (1972) *J. Org. Chem.* **37**: 6.

Dalacker, V., Hopf, H. (1974) *Tetrahedron Lett.* 15.

Danheiser, R.L., Carini, D.J., Basak, A. (1981a) *J. Am. Chem. Soc.* **103**: 1604.

Danheiser, R.L., Davila, C., Auchus, R.J., Kadonaga, J.T. (1981b) *J. Am. Chem. Soc.* **103**: 1443.

Danheiser, R.L., Gee, S.K., Sard, H. (1982) *J. Am. Chem. Soc.* **104**: 7670.

Danheiser, R.L., Gee, S.K. (1984) *J. Org. Chem.* **49**: 1672.

Danheiser, R.L., Cha, D.D. (1990) *Tetrahedron Lett.* **31**: 1527.

Danieli, B., Lesma, G., Palmisano, G. (1980) *Chem. Commun.* 109.

Danieli, B., Lesma, G., Palmisano, G., Riva, R., Tollari, S. (1984) *J. Org. Chem.* **49**: 547.

Daniewski, A.R., Kowalczyk-Przewloka, T. (1985) *J. Org. Chem.* **50**: 2976.

Danishefsky, S., Migdalof, B.H. (1969) *Tetrahedron Lett.* 4331.

Danishefsky, S., Crawley, L.S., Solomon, D.M., Heggs, P. (1971) *J. Am. Chem. Soc.* **93**: 2356.

Danishefsky, S., Hatch, W.E., Sax, M., Abola, E., Pletcher, J. (1973) *J. Am. Chem. Soc.* **95**: 2410.

Danishefsky, S., Dynak, J., Hatch, E., Yamamoto, M. (1974) *J. Am. Chem. Soc.* **96**: 1256.

Danishefsky, S., Tsai, M.Y., Dynak, J. (1975) *Chem. Commun.* 239.

Danishefsky, S., Schuda, P.F., Caruthers, W. (1977) *J. Org. Chem.* **42**: 2179.

Danishefsky, S., Walker, F.J. (1979) *J. Am. Chem. Soc.* **101**: 7018.

Danishefsky, S., Taniyama, E. (1983) *Tetrahedron Lett.* **24**: 15.

Danishefsky, S., Chackalamannil, S., Harrison, P., Silvestri, M., Cole, P. (1985a) *J. Am. Chem. Soc.* **107**: 2474.

Danishefsky, S.J., Harrison, P.J., Webb, R.R., O'Neil, B.T. (1985b) *J. Am. Chem. Soc.* **107**: 1421.

Das, K.G., Afzal, J., Hazra, B.G., Bhawal, B.M. (1983) *Synth. Commun.* **13**: 787.

Dastur, K.P. (1974) *J. Am. Chem. Soc.* **96**: 2605.

Dauben, W.G., Whalen, D.L. (1966a) *Tetrahedron Lett.* 3743; (1966b) *J. Am. Chem. Soc.* **88**: 4739.

Dauben, W.G., Hart, D.J., Ipaktschi, J., Kozikowski, A. (1973a) *Tetrahedron Lett.* 4425.

Dauben, W.G., Ipaktschi, J. (1973) *J. Am. Chem. Soc.* **95**: 5088.

Dauben, W.G., Hart, D.J. (1977a) *J. Org. Chem.* **42**: 3787.

Dauben, W.G., Hart, D.J. (1977b) *J. Am. Chem. Soc.* **99**: 7307.

Dauben, W.G., Michno, D.M. (1981) *J. Am. Chem. Soc.* **103**: 2284.

Dauben, W.G., Bunce, R.A. (1983) *J. Org. Chem.* **48**: 4642.

Davies, H.M.L., Oldenburg, C.E.M., McAfee, M.J., Nordahl, J.G., Henretta, J.P., Romines, K.R. (1988) *Tetrahedron Lett.* **29**: 975.

Davies, H.W., Saikali, E., Young, W.B. (1991) *J. Org. Chem.* **56**: 5696.

Davies, L.B., Greenberg, S.G., Sammes, P.G. (1981) *J. Chem. Soc. Perkin Trans. I* 1909.

Dawson, B.A., Ghosh, A.K., Jurlina, J.L., Stothers, J.B. (1983) *Chem. Commun.* 204.

Deb, B., Ila, H., Junjappa, H. (1990) *J. Chem. Res. Synop.* 356.

De Camp, M.R., Levin, R.H., Jones, M. (1974) *Tetrahedron Lett.* 3575.

De Capite, P.M., Puar, M.S., Burke, J.M., Scannell, R.T., Stevenson, R. (1990) *199th ACS Nat. Meet.* ORGN 68.

Defoin, A., Fritz, H., Geffroy, G., Streith, J. (1986) *Tetrahedron Lett.* **27**: 3135.

de Groot, A., Jansen, B.J.M. (1984) *J. Org. Chem.* **49**: 2034.

Demole, E., Enggist, P., Borer, C. (1969) *Chem. Commun.* 264.

Denmark, S.E., Dappen, M.S., Sternberg, J.A. (1984) *J. Org. Chem.* **49**: 4741.

Denmark, S.E., Hite, G.A. (1988) *Helv. Chim. Acta* **71**: 195.

Denmark, S.E., Moon, Y.-C., Senanayake, C.B.W. (1990) *J. Am. Chem. Soc.* **112**: 311.

DePuy, C.H., Isaks, M., Eilers, K.L., Morris, G.F. (1964) *J. Org. Chem.* **29**: 3503.

Deslongchamps P. *et al.* (1990) *Can. J. Chem.* **68**: 115, 127, 153, 186.

Differding, E., Ghosez, L. (1985) *Tetrahedron Lett.* **26**: 1647.

Dilthey, W., Quint, F. (1930) *J. Prakt. Chem.* **128**: 139.

Disanayaka, B.W., Weedon, A.C. (1985) *Chem. Commun.* 1282.

Dittami, J.P., Nie, X.Y., Nie, H., Ramanathan, H., Breining, S., Bordner, J., Decosta, D.L., Kiplinger, J., Reiche, P., Ware, R. (1991) *J. Org. Chem.* **56**: 5572.

Djuric, S., Sarkar, T., Magnus, P. (1980) *J. Am. Chem. Soc.* **102**: 6885.

Dodd, J.H., Weinreb, S.M. (1979) *Tetrahedron Lett.* 2263.

Doedens, R.J., Meier, G.P., Overman, L.E. (1988) *J. Org. Chem.* **53**: 685.

Doering, W.v.E., Roth, W.R. (1963) *Tetrahedron* 715.

Doering, W.v.E., Rosenthal, J.W. (1967) *Tetrahedron Lett.* 349.

Dolby, L.J., Nelson, S.J., Senkovich, D. (1972) *J. Org. Chem.* **37**: 3691.

Dolson, M.G., Chenard, B.L., Swenton, J.S. (1981) *J. Am. Chem. Soc.* **103**: 5263.

Dombroski, M.A., Kates, S.A., Snider, B.B. (1990) *J. Am. Chem. Soc.* **112**: 2759

Döpp, D., Weiler, H. (1979) *Chem. Ber.* **112**: 3950.

Dötz, K.-H. (1975) *Angew. Chem. Int. Ed. Engl.* **14**: 644.

Dötz, K.-H. (1983) *Pure Appl. Chem.* **55**: 1689.

Dötz, K.-H. (1984) *Angew. Chem. Int. Ed. Engl.* **23**: 587.

Dötz, K.-H., Popall, M. (1988) *Chem. Ber.* **121**: 665.

Doutheau, A., Sartoretti, J., Gore, J. (1983) *Tetrahedron* **39**: 3059.

Dowd, P., Schappert, R., Garner, P. (1928) *Tetrahedron Lett.* **23**: 7.

Dowd, P., Choi, S.-C. (1989) *Tetrahedron* **45**: 77.

Dowd, P., Zhang, W. (1991) *J. Am. Chem. Soc.* **113**: 9875.

Dower, W.V., Vollhardt, K.P.C. *Tetrahedron* **42**: 1873.

Drouin, J., Leyendecker, F., Conia, J.M. (1975) *Tetrahedron Lett.* 4053.

Duc, D.K.M., Fetizon, M., Lazare, S. (1975) *Chem. Commun.* 282.

Duc, D.K.M., Fetizon, M., Hanna, I., Olesker, A. (1981) *Tetrahedron Lett.* **22**: 3847.

Duddeck, H., Feuerhelm, H.-T., Snatzke, G. (1979) *Tetrahedron Lett.* 829.

Dufour, M., Gramain, J.C., Husson, H.-P., Sinibaldi, M.E., Troin, Y. (1990) *J. Org. Chem.* **55**: 5483.

Dunham, D.J., Lawton, R.G. (1971) *J. Am. Chem. Soc.* **93**: 2075.

Earl, R.A., Vollhardt, K.P.C. (1982) *Heterocycles* **19**: 265.

Earl, R.A., Vollhardt, K.P.C. (1983) *J. Am. Chem. Soc.* **105**: 6991.

East, D.S.R., McMurry, T.B.H., Talekar, R.R. (1974) *Chem. Commun.* 450.

Eaton, P.E., Mueller, R.H., Carlson, G.R., Cullison, D.A., Cooper, G.F., Chou, T.-C., Krebs, E.-P. (1977) *J. Am. Chem. Soc.* **99**: 2751.

Eberbach, W., Laber, N. (1992) *Tetrahedron Lett.* **33**: 57.

Eberlein, T.H., West, F.G., Tester, R.W. (1991) *201st ACS Nat. Meet. ORGN* 194.

Edstrom, E.D., Taylor, T. (1991) *J. Am. Chem. Soc.* **113**: 6690.

Edwards, P.N., Smith, G.F. (1961) *J. Chem. Soc.* 1458.

Ehring, V., Seebach, D. (1975) *Chem. Ber.* **108**: 1961.

Eisenhuth, W., Renfroe, H.B., Schmid, H. (1965) *Helv. Chim. Acta* **48**: 375.

Elix, J.A., Wilson, W.S., Warrener, R.N. (1970) *Tetrahedron Lett.* 1837.

Engler, E.M., Farcasiu, M., Sevin, A., Cense, J.M., Schleyer, P.v.R. (1973) *J. Am. Chem. Soc.* **95**: 5769.

Engler, T.A., Combrink, K.D., Ray, J.E. (1988) *J. Am. Chem. Soc.* **110**: 7931.

Engler, T.A., Combrink, K.D., Reddy, J.P. (1989) *Chem. Commun.* 454.

Enhsen, A., Karabelas, K., Heerding, J.M., Moore, H.W. (1990) *J. Org. Chem.* **55**: 1177.

Eschenmoser, A., Schreiber, J., Julia, S.A. (1953) *Helv. Chim. Acta* **36**: 482.

Eschenmoser, A. (1976) *Chem. Soc. Rev.* **5**: 377.

Estevez, J.C., Estevez, R.J., Guitian, E., Villaverde, M.C., Castedo, L. (1989) *Tetrahedron Lett.* **30**: 5785.

Evans, D.A., Hoffman, J.M. (1976) *J. Am. Chem. Soc.* **98**: 1983.

Evans, D.A., Sims, C.L., Andrews, G.C. (1977) *J. Am. Chem. Soc.* **99**: 5453.

Evans, D.A., Hart, D.J., Koelsch, P.M. (1978) *J. Am. Chem. Soc.* **100**: 4593.

Evans, D.A., Thomas, E.W., Cherpeck, R.E. (1982) *J. Am. Chem. Soc.* **104**: 3695.

Evans, D.A., Weber, A.E. (1987) *J. Am. Chem. Soc.* **109**: 7151.

Exon, C., Magnus, P. (1983) *J. Am. Chem. Soc.* **105**: 2477.

Ezquerra, J., He, W., Paquette, L.A. (1990) *Tetrahedron Lett.* **31**: 6979.

Fagan, P.J., Burns, E.G., Calabrese, J.C. (1988) *J. Am. Chem. Soc.* **110**: 2979.

Falshaw, C.P., Lane, S.A., Ollis, W.D. (1973) *Chem. Commun.* **491.**

Faragher, R., Gilchrist, T.L. (1977) *Chem. Commun.* 252.

Farina, F., Paredes, M.C., Stefani, V. (1986) *Tetrahedron* **42**: 4309.

Farnum, D.G., Carlson, G.R. (1970) *J. Am. Chem. Soc.* **92**: 6700.

Farnum, D.G., Ghandi, M., Raghu, S., Reitz, T. (1982) *J. Org. Chem.* **47**: 2598.

Fayos, J., Clardy, J., Dolby, L.J., Farnham, T. (1977) *J. Org. Chem.* **42**: 1349.

Feldman, K.S., Wu, M.-J., Rotella, D.P. (1989) *J. Am. Chem. Soc.* **111**: 6457.

Felix, D., Muller, R.K., Horn, U., Joos, R., Schreiber, J., Eschenmoser, A. (1972) *Helv. Chim. Acta* **55**: 1276.

Felluga, F., Nitti, P., Pitacco, G., Valentin, E. (1991) *J. Chem. Soc. Perkin Trans. I* 1645.

Fessner, W.-D., Prinzbach, H., Rihs, G. (1983) *Tetrahedron Lett.* **24**: 5857.

Fevig, J.M., Marquix, R.W., Overman, L.E. (1991) *J. Am. Chem. Soc.* **113**: 5085.

Fevig, T.L., Elliott, R.L., Curran, D.P. (1988) *J. Am. Chem. Soc.* **110**: 5064.

Ficini, J., d'Angelo, J., Noire, J. (1974) *J. Am. Chem. Soc.* **96**: 1213.

Ficini, J., Falou, S., d'Angelo, J. (1977) *Tetrahedron Lett.* 1931.

Ficini, J., Revial, G., Genet, J.P. (1981) *Tetrahedron Lett.* **22**: 629, 633.

Finch, A.M.T., Vaughan, W.R. (1969) *J. Am. Chem. Soc.* **91**: 1416.

Findlay, J.A., Desai, D.N., Lonergan, G.C., White, P.S. (1980) *Can. J. Chem.* **58**: 2827.

Firestone, R.A., Harris, E.E., Reuter, W. (1967) *Tetrahedron* **23**: 943.

Fitjer, L., Quaback, U. (1987) *Angew. Chem. Int. Ed. Engl.* **26**: 1023.

Fitjer, L., Kanschik, A., Majewski, M. (1988a) *Tetrahedron Lett.* **29**: 5525.

Fitjer, L., Majewski, M., Kanschik, A. (1988b) *Tetrahedron Lett.* **29**: 1263.

Fitjer, L., Quaback, U. (1989) *Angew. Chem. Int. Ed. Engl.* **28**: 94.

Fleming, I., Pearce, A. (1980) *J. Chem. Soc. Perkin Trans. I* 2485.

Foland, L.D., Decker, O.H.W., Moore, H.W. (1989) *J. Am. Chem. Soc.* **111**: 989.

Forbes, C.F., Michau, J.D., van Ree, T., Wiechers, A., Woudenberg, M. (1976) *Tetrahedron Lett.* 935.

Franck-Neumann, M., Martina, D. (1977) *Tetrahedron Lett.* 2293.

Frater, G. (1975) *Helv. Chim. Acta* **58**: 442.

Frater, G. (1982) *Chem. Commun.* 521.

Freeman, J.P., Plonka, J.H. (1966) *J. Am. Chem. Soc.* **88**: 3662.

Freeman, P.K., Kuper, D.G. (1965) *Chem. Ind.* 424.

Freeman, P.K., Balls, D.M., Blazevich, J.N. (1970) *J. Am. Chem. Soc.* **92**: 2051.

Froborg, J., Magnusson, G. (1978) *J. Am. Chem. Soc.* **100**: 6728.

Fuchs, P.L., Braish, T.F. (1986) *Chem. Rev.* **86**: 903.

Fujioka, H., Yamamoto, H., Kondo, H., Annoura, H., Kita, Y. (1989) *Chem. Commun.* 1509.

Fujioka, H., Miyazaki, M., Yamanaka, T., Yamamoto, H., Kita, Y. (1990) *Tetrahedron Lett.* **31**: 5951.

Fujita, Y., Onishi, T., Nishida, T. (1978) *Synthesis* 532.

Fukamiya, N., Kato, M., Yoshikoshi, A. (1973) *J. Chem. Soc. Perkin Trans. I* 1843.

Fuks, R., King, G.S.D., Viehe, H.G. (1969) *Angew. Chem. Int. Ed. Engl.* **8**: 675.

Fukuyama, T., Yang, L. (1987) *J. Am. Chem. Chem.* **109**: 7881.

Fukuyama, T., Nunes, J.J. (1988) *J. Am. Chem. Soc.* **110**: 5196.

Funita, K., Misumi, A., Mori, A., Ikeda, N., Yamamoto, H. (1984) *Tetrahedron Lett.* **25**: 669.

Funk, R.L., Voohardt, K.P.C. (1977) *J. Am. Chem. Soc.* **99**: 5483; (1979) *J. Am. Chem. Soc.* **101**: 215.

Fusco, R., Sannicolo, F. (1975) *Tetrahedron Lett.* 3351.

Gadamasetti, G., Kuehne, M.E. (1988) in *Enamines* (Cook, A.G., Ed.), 2nd Ed., Dekker, New York: pp. 531–699.

Gajewski, J.J., Shih, C.N. (1972) *J. Org. Chem.* **37**: 64.

Gajewski, J.J., Paul, G.C. (1990) *J. Org. Chem.* **55**: 4575.

Gallagher, T., Magnus, P., Huffman, J.C. (1982) *J. Am. Chem. Soc.* **104**: 1140.

Garigipati, R.S., Freyer, A.J., Whittle, R.R., Weinreb, S.M. (191984) *J. Am. Chem. Soc.* **106**: 7861.

Garratt, P.J., Porter, J.R. (1986) *J. Org. Chem.* **51**: 5450.

Garst, M.E., Bonfiglio, J.N., Marks, J. (1982) *J. Org. Chem.* **47**: 1494.

Gassman, P.G., van Bergen, T.J. (1973) *J. Am. Chem. Soc.* **95**: 590.

Gassman, P.G., Hoye, R.C. (1981) *J. Am. Chem. Soc.* **103**: 2496.

Gawley, R.E. (1976) *Synthesis* 777.

Geiger, R.E., Lalonde, M., Stoller, H., Schleich, K. (1984) *Helv. Chim. Acta* **67**: 1274.

Geissman, T.A., Waiss, A.C. (1962) *J. Org. Chem.* **27**: 139.

Germanas, J., Aubert, C., Vollhardt, K.P.C. (1991) *J. Am. Chem. Soc.* **113**: 4006..

Gervay, J.E., McCapra, F., Money, T., Sharma, G.M. (1966) *Chem. Commun.* 142.

Ghera, E., Maurya, R., Ben-David, Y. (1986) *Tetrahedron Lett.* **27**: 3935.

Ghisalberti, E.L., Jefferies, P.R., Payne, T.G. (1974) *Tetrahedron* **30**: 3099.

Gibbs, R.A., Barteis, K., Lee, R.W.K., Okamura, W.H. (1989) *J. Am. Chem. Soc.* **111**: 3717.

Gibbons, E.G. (1982) *J. Am. Chem. Soc.* **104**: 1767.

Gierer, J., Pettersson, I. (1977) *Can. J. Chem.* **55**: 593.

Giguere, R.J., Namen, A.M., Majetich, G., Defauw, J. (1987) *Tetrahedron Lett.* **28**: 6553.

Gilbert, A., Walsh, R. (1976) *J. Am. Chem. Soc.* **98**: 1606.

Gilbreath, S.G., Harris, C.M., Harris, T.M. (1988) *J. Am. Chem. Soc.* **110**: 6172.

Gilchrist, T.L., Richards, P. (1983) *Synthesis* 153.

Gillon, A., Ovadia, D., Kapon, M., Bien, S. (1982) *Tetrahedron* **38**: 1477.

Gilman, H., Tomasi, R.A. (1962) *J. Org. Chem.* **27**: 3647.

Ginsburg, D. (1974) *Acc. Chem. Res.* **7**: 286.

Ginsburg, D. (1983) *Tetrahedron* **39**: 2095.

Girling, I.R., Widdowson, D.A. (1982) *Tetrahedron Lett.* **23**: 4281.

Gladysz, J.A., Lee, S.J., Tomasello, J.A.V., Yu, Y.S. (1977) *J. Org. Chem.* **42**: 4170.

Gleiter, R., Ginsburg, D. (1979) *Pure Appl. Chem.* **51**: 1301.

Gleiter, R., Karcher, M. (1988) *Angew. Chem. Int. Ed. Engl.* **27**: 840.

Gobao, R.A., Bremmer, M.L., Weinreb, S.M. (1982) *J. Am. Chem. Soc.* **104**: 7065.

Goldschmidt, Z., Gutman, U., Bakal, Y., Worchel, A. (1973) *Tetrahedron Lett.* 3759.

Goldstein, M.J., Gebirtz, A.H. (1965) *Tetrahedron Lett.* 4413.

Goldstein, S., Vannes, P., Honge, C., Frisque-Hesbain, A.M., Wiaux-Zamar, C., Ghosez, L., Germain, G., Declercq, J.P., Van Meersche, M., Arrieta, J.M. (1981) *J. Am. Chem. Soc.* **103**: 4616.

Gommans, L.H.P., Main, L., Nicholson, B.K. (1987) *Chem. Commun.* 761.

Gopalan, A., Magnus, P. (1980) *J. Am. Chem. Soc.* **102**: 1756.

Gordon, H.J., Martin, J.C., McNab, H. (1983) *Chem. Commun.* 957.

Götze, S., Kubuel, B., Steglich, W. (1976) *Chem. Ber.* **109**: 2331.

Govindachari, T.R., Nagarajan, K., Parthasarathy, P.C. (1969) *Chem. Commun.* 823.

Gravett, E.C., Howard, J.A.K., Mackenzie, K., Liu, S.-X., Karadakov, P.B. (1991) *Chem. Commun.* 1763.

Gray, R.W., Dreiding, A.S. (1977) *Helv. Chim. Acta* **60**: 1969.

Grellmann, K.-H., Schmidt, U., Weller, H. (1982) *Chem. Phys. Lett.* **88**: 40.

Greengrass, C.W., Hughman, J.A., Parsons, P.J. (1985) *Chem. Commun.* 889.

Greenlee, W.J., Woodward, R.B. (1976) *J. Am. Chem. Soc.* **98**: 6075.

Gribble, G.W., Switzer, F.L., Soll, R.M. (1988) *J. Org. Chem.* **53**: 3164.

Grieco, P.A., Finkelhor, R.S. (1972) *Tetrahedron Lett.* 3781.

Grieco, P.A., Takigawa, T., Schillinger, W.A. (1980) *J. Org. Chem.* **45**: 2247.

Grieco, P.A., Garner, P., Yoshida, K., Huffman, J.C. (1983) *Tetrahedron Lett.* **24**: 3807.

Grieco, P.A., Nargund, R.P. (1986) *Tetrahedron Lett.* **27**: 4813.

Grieco, P.A., Fobare, W.F. (1987) *Chem. Commun.* 185.

Grieco, P.A., Parker, D.T. (1988) *J. Org. Chem.* **53**: 3658.

Griffin, D.A., Staunton, J. (1975) *Chem. Commun.* 675.

Grigg, R., Heaney, F., Surendrakumar, S., Warnock, W.J. (1989a) *Tetrahedron Lett.* **30**: 609.

Grigg, R., Markandu, J. (1989b) *Tetrahedron Lett.* **30**: 5489.

Grigg, R., Markandu, J., Perrior, T., Surendrakumar, S., Warnick, W.J. (1990) *Tetrahedron Lett.* **31**: 559.

Grimme, W., Mauer, W., Sarter, C. (1985) *Angew. Chem. Int. Ed. Engl.* **24**: 331.

Grob, C.A., Schiess, P. (1958) *Angew. Chem.* **70**: 502.

Grob, C.A., Kiefer, H.R., Lutz, H.J., Wilkens, H.J. (1967) *Helv. Chim. Acta* **50**: 416.

Gronbeck, D.A., Matchett, S.A., Rosenblum, M. (1989) *Tetrahedron Lett.* **30**: 2881.

Gronowitz, S., Nikitidis, G., Halberg, A., Servin, R. (1988) *J. Org. Chem.* **53**: 3351.

Gruber, G.W., Pomerantz, M. (1970) *Tetrahedron Lett.* 3755.

Grundon, M.F., Ramachandran, V.N. (1985) *Tetrahedron Lett.* **26**: 4253.

Gschwend, H.W., Meier, H.P. (1972) *Angew. Chem. Int. Ed. Engl.* **11**: 294.

Gupta, A.K., Fu, X., Snyder, J.P., Cook, J.M. (1991) *Tetrahedron* **47**: 3665.

Gupta, Y.N., Doa, M.J., Houk, K.N. (1982) *J. Am. Chem. Soc.* **104**: 7336.

Guthrie, R.W., Valenta, Z., Wiesner, K. (1966) *Tetrahedron Lett.* 4645.

Gutzwiller, J., Uskokovic, M.R. (1978) *J. Am. Chem. Soc.* **100**: 576.

Guyon, C., Boule, P., Lemaire, J. (1982) *Tetrahedron Lett.* **23**: 1581.

Gyorkos, A.C., Stille, J.K., Hegedus, L.S. (1990) *J. Am. Chem. Soc.* **112**: 8465.

Hafner, K., Haring, J., Jakel, W. (1970) *Angew. Chem. Int. Ed. Engl.* **9**: 159.

Hafner, K., Goltz, M. (1982) *Angew. Chem. Int. Ed. Engl.* **21**: 695.

Hagiwara, H., Uda, H., Kodama, T. (1980) *J. Chem. Soc. Perkin Trans. I* 963.

Hagiwara, H., Okano, A., Uda, H. (1985) *Chem. Commun.* **1047.**

Hagiwara, H., Akama, T., Okano, A., Uda, H. (1988) *Chem. Lett.* 1793.

Hagiwara, K., Akama, T., Uda, H. (1989) *Chem. Lett.* 2067.

Hagiwara, H., Abe, F., Uda, H. (1991) *Chem. Commun.* 1070.

Hajicek, J., Trojanek, J. (1981) *Tetrahedron Lett.* **22**: 2927.

Hajos, Z.G., Parrish, D.R. (1973) *J. Org. Chem.* **38**: 3239, 3244.

Halton, B., Lu, Q., Stang, P.J. (1990) *Aust. J. Chem.* **43**: 1277.

Halverson, A., Keehn, P.M. (1982) *J. Am. Chem. Soc.* **104**: 6125.

Hammond, M.L., Mourino, A., Okamura, W.H. (1978) *J. Am. Chem. Soc.* **100**: 4907.

Hamon, D.P.G., Spurr, P.R. (1982) *Chem. Commun.* 372.

Han, W.C., Takahashi, K., Cook, J.M., Weiss, U., Silverton, J.V. (1982) *J. Am. Chem. Soc.* **104**: 318.

Hantawong, K., Murphy, W.S., Boyd, D.R., Ferguson, G., Parvez, M. (1985) *J. Chem. Soc. Perkin Trans. II* 1577.

Hara, H., Hosaka, M., Hoshino, O., Umezawa, B. (1978) *Tetrahedron Lett.* 3809.

Harada, N., Sugioka, T., Ando, Y., Uda, H., Kuriki, T. (1988) *J. Am. Chem. Soc.* **110**: 8483.

Harano, K., Uchida, K., Izuma, M., Aoki, T., Eto, M., Hisano, T. (1988) *Chem. Pharm. Bull* **36**: 2312.

Harano, K., Ono, K., Nishimoto, M., Eto, M., Hisano, T. (1991) *Tetrahedron Lett.* **32**: 2387.

Harley-Mason, J., Kaplan J. (1967) *Chem. Commun.* 915.

Harling, J.D., Motherwell, W.B. (1988) *Chem. Commun.* 1380.

Harre, M., Winterfeldt, E. (1982) *Chem. Ber.* **115**: 1437.

Harris, T.M., Hauser, C.R. (1964) *J. Org. Chem.* **29**: 1391.

Harris, T.M., Murray, T.P., Harris, C.M., Gumulka, M. (1974) *Chem. Commun.* 362.

Harris, T.M., Webb, A.D., Harris, C.M., Wittek, P.J., Murray, T.P. (1976) *J. Am. Chem. Soc.* **98**: 6065.

Harris, T.M., Harris, C.M., Kuzma, P.C., Lee, J.Y.-C., Mahalingam, S., Gilbreath, S.G. (1988a) *J. Am. Chem. Soc.* **110**: 6186.

Harris, T.M., Harris, C.M., Oster, T.A., Brown, L.E., Lee, J.Y.-C. (1988b) *J. Am. Chem. Soc.* **110**: 6180.

Harrison, R.M., Hobson, J.D., Midgley, A.W. (1973) *J. Chem. Soc. Perkin Trans. I* 1960.

Hart, D.J., Yang, T.-K. (1985) *J. Org. Chem.* **50**: 235.

Hartmann, W., Heine, H.G., Schrader, L. (1974) *Tetrahedron Lett.* 883.

Haruna, M., Ito, K. (1976) *Chem. Commun.* 345.

Harusawa, S., Osaki, H., Kurokawa, T., Fujii, H., Yoneda, R., Kurihara, T. (1991) *Chem. Pharm. Bull* **39**: 1659.

Harvey, D.F., Brown, M.G. (1990) *J. Am. Chem. Soc.* **112**: 7806.

Harvey, D.F., Brown, M.F. (1991a) *Tetrahedron Lett.* **32**: 2871.

Harvey, D.F., Lund, K.P. (1991b) *J. Am. Chem. Soc.* **113**: 5066.

Harwood, L.M., Oxford, A.J., Thomson, C. (1991) *Chem. Commun.* 1303.

Hashimoto, K., Horikawa, M., Shirahama, H. (1990) *Tetrahedron Lett.* **31**: 7047.

Hatanaka, M., Himeda, Y., Ueda, I. (1990) *Chem. Commun.* 526.

Hatanaka, M., Himeda, Y., Ueda, I. (1991) *Tetrahedron Lett.* **32**: 4521.

Hauser, F.M., Rhee, R.P. (1978) *J. Org. Chem.* **43**: 178.

Hayakawa, K., Motohiro, S., Fujii, I., Kanematsu, K. (1981) *J. Am. Chem. Soc.* **103**: 4605.

Hayakawa, K., Kamikawaji, Y., Kanematsu, K. (1982) *Tetrahedron Lett.* **23**: 2171.

Hayakawa, K., Ohsuki, S., Kanematsu, K. (1986a) *Tetrahedron Lett.* **27**: 947.

Hayakawa, K., Ohsuki, S., Kanematsu, K. (1986b) *Tetrahedron Lett.* **27**: 4205.

Hayakawa, K., Yasukouchi, T., Kanematsu, K. (1987) *Tetrahedron Lett.* **28**: 5895.

Hayakawa, K., Nagatsugi, F., Kanematsu, K. (1988) *J. Org. Chem.* **53**: 860.

Hayakawa, Y., Shimizu, F., Noyori, R. (1978) *Tetrahedron Lett.* 993.

Heathcock, C.H. (1981) *Science* **214**: 395.

Heathcock, C.H., Kleinman, E.F., Binkley, E.S. (1982) *J. Am. Chem. Soc.* **104**: 1054.

Heathcock, C.H., Davidsen, S.K., Mills, S., Sanner, M.A. (1986) *J. Am. Chem. Soc.* **108**: 5650.

Heather, J.B., Mittal, R.S.D., Sih, C.J. (1974) *J. Am. Chem. Soc.* **96**: 1976.

Hegedus, L.S., Allen, G.F., Olsen, D.J. (1980) *J. Am. Chem. Soc.* **102**: 3583.

Hegman, J., Christi, M., Peters, K., Peters, E.M., von Schnerling, H.G. (1987) *Tetrahedron Lett.* **28**: 6429; (1988) *Angew. Chem. Int. Ed. Engl.* **27**: 931.

Heimgartner, H., Hansen, H.-J., Schmid, G. (1970) *Helv. Chim. Acta* **53**: 173.

Hendrickson, J.B., Palumbo, P.S. (1985) *J. Org. Chem.* **50**: 2110.

Henrick, C.A., Böhme, E., Edwards, J.A., Fried, J.H. (1968) *J. Am. Chem. Soc.* **90**: 5926.

Hernandez, J.E., Fernandez, S., Arias, G. (1988) *Synth. Commun.* **18**: 2055.

Herndon, J.W., Wu, C., Harp, J.J., Kreutzer, K.A. (1991) *Synlett* 1.

Herrinton, P.M., Hopkins, M.H., Mishra, P., Brown, M.J., Overman, L.E. (1987) *J. Org. Chem.* **52**: 3711.

Herz, W. (1962) *J. Am. Chem. Soc.* **84**: 3857.

Hess, B.A., Bailey, A.S., Bartusek, B., Boekelheide, V. (1969) *J. Am. Chem. Soc.* **91**: 1665.

Hess, J. (1972) *Chem. Ber.* **105**: 441.

Hewson, A.T., MacPherson, D.T. (1985) *J. Chem. Soc. Perkin Trans. I* 2625.

Hickman, D.N., Wallace, T.W., Wardleworth, J.M. (1991) *Tetrahedron Lett.* **32**: 819.

Hikino, H., Agatsume, K., Konno, C., Takemoto, T. (1968) *Tetrahedron Lett.* 4417.

Hill, R.K., Ledford, N.D. (1975) *J. Am. Chem. Soc.* **97**: 666.

Hill, R.K., Bock, M.G. (1978) *J. Am. Chem. Soc.* **100**: 637.

Hillard, III, R.L., Parnell, C.A., Vollhardt, K.P.C. (1983) *Tetrahedron* **39**: 905.

Hirai, Y., Hagiwara, A., Yamazaki, T. (1986) *Heterocycles* **24**: 571.

Hirao, K., Taniguchi, M., Yonemitsu, O., Flippen, J.L., Witkop, B. (1979) *J. Am. Chem. Soc.* **101**: 408.

Hirst, G.C., Howard, P.N., Overman, L.E. (1989) *J. Am. Chem. Soc.* **111**: 1514.

Hiyama, T., Shinoda, M., Nozaki, H. (1979) *Tetrahedron Lett.* 3529.

Hizuka, M., Fang, C., Suemune, H., Sakai, K. (1989) *Chem. Pharm. Bull.* **37**: 1185..

Hlubucek, J.R., Ritchie, E., Taylor, W.C. (1970) *Aust. J. Chem.* **23**: 1881.

Ho, T.-L., Wong, C.M. (1972) *Org. Prep. Prop. Int.* **4**: 265.

Ho, T.-L., Stark, C.J. (1983) *Liebigs Ann. Chem.* 1446.

Ho, T.-L. (1988) *Carbocycle Construction in Terpene Synthesis*, VCH, New York.

Ho, T.-L. (1991) *Polarity Control for Synthesis*, Wiley, New York.

Hobson, J.D., Malpass, J.R. (1967) *J. Chem. Soc. (C)* 1645.

Hoffmann, H.M.R. (1969) *Angew. Chem. Int. Ed. Engl.* **8**: 556.

Hoffman, H.M.R., Rabe, J. (1985) *J. Am. Chem. Soc.* **50**: 3849.

Hoffmann, R.W., Barth, W. (1985) *Chem. Ber.* **118**: 634.

Holmes, B.N., Leonard, N.J. (1976) *J. Am. Chem. Soc.* **41**: 568.

Holmes, T.L., Stevenson, R. (1970) *Tetrahedron Lett.* 199.

Hopf, H., Kirsch, R. (1985) *Tetrahedron Lett.* **26**: 3327.

Hopf, H., Kreuter, M., Jones, P.G. (1991) *Angew. Chem. Int. Ed. Engl.* **30**: 1127.

Hortmann, A.G., Daniel, D.S., Martinelli, J.E. (1973) *J. Org. Chem.* **38**: 728.

Houlihan, W.J., Uike, Y., Parrino, V.A. (1981) *J. Org. Chem.* **46**: 4515.

Hoye, T.R., Dinsmore, C.J., Johnson, D.S., Korkowski, P.F. (1990) *J. Org. Chem.* **55**: 4518.

Hoye, T.R., Dinsmore, C.J. (1992) *Tetrahedron Lett.* **33**: 169.

Huang, C.-G., Shukla, D., Wan, P. (1991) *J. Org. Chem.* **56**: 5437.

Hudlicky, T., Sinai-Zingde, G., Seoane, G. (1987) *Synth. Commun.* **17**: 1155.

Hucklicky, T., Radesca-Kwart, J., Li, L., Bryant, T. (1988) *Tetrahedron Lett.* **29**: 3283.

Hudlicky, T., Heard, N.E., Fleming, A. (1990) *J. Org. Chem.* **55**: 2570.

Hudrlik, P.F., Kulkarni, A.K. (1981) *J. Am. Chem. Soc.* **103**: 6251.

Hug, R., Hansen, H.-J., Schmid, H. (1969) *Chimia* **23**: 108.

Hug, R., Frater, G., Hansen, H.-J., Schmid, H. (1971) *Helv. Chim. Acta* **54**: 306.

Hug, R., Hansen, H.-J., Schmid, H. (1972a) *Helv. Chim. Acta* **55**: 10; (1972b) *Helv. Chim. Acta* **55**: 1675.

Hugel, G., Gourdier, B., Levy, J., LeMen, J. (1974) *Tetrahedron Lett.* 1597.

Hugel, G., Massiot, G., Levy, J., LeMen, J. (1981) *Tetrahedron* **37**: 1369.

Hugel, G., Royer, D., Sigaut, F., Levy, J. (1991) *J. Org. Chem.* **56**: 4631.

Huisgen, R., Moran, J.R. (1985) *Tetrahedron Lett.* **26**: 1057.

Hunter, N.R., Green, N.A., McKinnon, D.M. (1980) *Tetrahedron Lett.* **21**: 4589.

Hussmann, G., Wulff, W.D., Barton, T.J. (1983) *J. Am. Chem. Soc.* **105**: 1263.

Husson, A., Langlois, Y., Riche, C., Husson, H.-P. (1973) *Tetrahedron* **29**: 3095.

Hutchins, R.O., Natale, N.R., Taffer, I.M., Zipkin, R. (1984) *Synth. Commun.* **14**: 445.

Hwang, Y.C., Fowler, F.W. (1985) *J. Org. Chem.* **50**: 2719.

Hwu, J.R. (1985) *Chem. Commun.* 452.

Ichihara, A., Kimura, R., Yamada, S., Sakamura, S. (1980) *J. Am. Chem. Soc.* **102**: 6353.

Ichihara, A. (1987) *Synthesis* 207.

Iguchi, M., Nishiyama, A., Eto, H., Terada, Y., Yamamura, S. (1979) *Chem. Lett.* 1397.

Ihara, M., Toyota, M., Fukumoto, K., Kametani, T. (1986) *J. Chem. Soc. Perkin Trans. I* 2151.

Ihara, M., Kirihara, T., Kawaguchi, A., Tsuruta, M., Fukumoto, K., Kametani, T. (1987) *J. Chem. Soc. Perkin Trans. I* 1719.

Ihara, M., Katogi, M., Fukumoto, K., Kametani, T. (1988a) *J. Chem. Soc. Perkin Trans. I* 2963.

Ihara, M., Suzuki, M., Fukumoto, K., Kametani, T., Kabuto, C. (1988b) *J. Am. Chem. Soc.* **110**: 1963.

Ihara, M., Takahashi, T., Shimizu, N., Ishida, Y., Sudow, I., Fukumoto, K., Kametani, T. (1989) *J. Chem. Soc. Perkin Trans. I* 529.

Ihara, M., Takino, Y., Tomotake, M., Fukumoto, K. (1990) *J. Chem. Soc. Perkin Trans. I* 2287.

Ihara, M., Suzuki, T., Katogi, M., Taniguchi, N., Fukumoto, K. (1991a) *Chem. Commun.* 646.

Ihara, M., Suzuki, S., Taniguchi, N., Fukumoto, K., Kabuto, C. (1991b) *Chem. Commun.* 1168.

Iida, H., Watanabe, Y., Kibayashi, C. (1985) *J. Am. Chem. Soc.* **107**: 5534.

Ireland, R.E., Mueller, R.H. (1972) *J. Am. Chem. Soc.* **94**: 5897.

Ireland, R.E., McKenzie, T.C., Trust, R.I. (1975) *J. Org. Chem.* **40**: 1007.

Ireland, R.E., McGarvey, G.J., Anderson, R.C., Badoud, R., Fitzsimmons, B., Thaisrivongs, S. (1980) *J. Am. Chem. Soc.* **102**: 6178.

Ireland, R.E., Dow, W.C., Godfrey, J.D., Thaisrivongs, S. (1984) *J. Org. Chem.* **49**: 1001.

Irie, H., Katakawa, J., Mizuno, Y., Udaka, S., Taga, T., Osaki, K. (1978) *Chem. Commun.* 717.

Ishibashi, H., Harada, S., Okada, M., Ikeda, M., Ishiyama, K., Yamashita, H., Tamura, Y. (1986) *Synthesis* 847.

Ishibashi, H., So, T.S., Nakatani, H., Minami, K., Ikeda, M. (1988) *Chem. Commun.* 827.

Ishibashi, H., Okano, M., Tamaki, H., Maruyama, K., Yakura, T., Ikeda, M. (1990) *Chem. Commun.* 1436.

Ishikura, M., Terashima, M., Okamura, K., Date, T. (1991) *Chem. Commun.* 1219.

Ito, K., Suzuki, F., Haruna, M. (1978) *Chem. Commun.* 733.

Ito, S. (1991) private communication.

Ito, Y., Nakatsuka, M., Saegusa, T. (1981) *J. Am. Chem. Soc.* **103**: 476.

Ito, Y., Nakajo, E., Saegusa, T. (1983) *Tetrahedron Lett.* **24**: 2881.

Ittah, Y., Shahak, I., Blum, J., Klein, J. (1977) *Synthesis* 678.

Iwasa, S., Yammamoto, M., Kohmoto, S., Yamada, K. (1991) *J. Org. Chem.* **56**: 2849.

Iwata, C., Fusaka, T., Fujiwara, T., Tomita, K., Yamada, M. (1981) *Chem. Commun.* 463.

Jaafar, A., Alilou, E.H., Reglier, M., Waegell, B. (1991) *Tetrahedron Lett.* **32**: 5531.

Jackson, A.H., Shannon, P.V.R., Wilkins, D.J. (1987) *Tetrahedron Lett.* **28**: 4901.

Jackson, A.H., Lertwanawatana, W., Pandey, R.K., Rao, K.R.N. (1989) *J. Chem. Soc. Perkin Trans. I* 374.

Jacobi, P.A., Craig, T.A., Walker, D.G., Arrick, B.A., Frechette, R.F. (1984) *J. Am. Chem. Soc.* **106**: 5585.

Jacobi, P.A., Kaczmarek, C.S.R., Udodong, U.E. (1987) *Tetrahedron* **43**: 5475.

Jacobi, P.A., Kravitz, J.I. (1988) *Tetrahedron Lett.* **29**: 6873.

Jacobi, P.A., Blum, C.A., Desimone, R.W., Udodong, U.E.S. (1989) *Tetrahedron Lett.* **30**: 7173.

Jacobi, P.A., Selnick, H.G. (1990) *J. Org. Chem.* **55**: 202.

Jacobi, P.A., Zheng, W. (1991) *Tetrahedron Lett.* **32**: 1279.

Jacobsen, E.J., Levin, J., Overman, L.E. (1988) *J. Am. Chem. Soc.* **110**: 4329.

Jegham, S., Fourrey, J.-L., Das, B.C. (1989) *Tetrahedron Lett.* **30**: 1959.

Jemison, R.W., Laird, T., Ollis, W.D. (1972) *Chem. Commun.* 556.

Jemison, R.W., Laird, T., Ollis, W.D., Sutherland, I.O. (1980) *J. Chem. Soc. Perkin Trans. I* 1436.

Jeong, N., Yoo, S., Lee, S.J., Lee, S.H., Chung, Y.K. (1991) *Tetrahedron Lett.* **32**: 2137.

Johnson, E.P., Vollhardt, K.P.C. (1991) *J. Am. Chem. Soc.* **113**: 381.

Johnson, W.S., Jensen, N.P., Hooz, J., Leopold, E.J. (1968a) *J. Am. Chem. Soc.* **90**: 5872.

Johnson, W.S., Semmelhack, M.F., Sultanbawa, M.U.S., Dolak, L.A. (1968b) *J. Am. Chem. Soc.* **90**: 2994.

Johnson, W.S., Werthemann, L., Bartlett, W.R., Brocksom, T.J., Li, T.-t., Faulkner, D.J., Petersen, M.R. (1970) *J. Am. Chem. Soc.* **92**: 741.

Johnson, W.S. (1976) *Angew. Chem. Int. Ed. Engl.* **15**: 9.

Johnson, W.S., Brinkmeyer, R.S., Kapoor, V.M., Yarnell, T.M. (1977) *J. Am. Chem. Soc.* **99**: 8341.

Johnson, W.S., Lindell, S.D., Steele, J. (1987) *J. Am. Chem. Soc.* **109**: 5852.

Jones, D.W., Marmon, R.J. (1989a) *Tetrahedron Lett.* **30**: 5467.

Jones, D.W., Thompson, A.M. (1989b) *Chem. Commun.* 1370.

Jones, D.W., Nongrum, F.M. (1990) *J. Chem. Soc. Perkin Trans. I* 3357.

Joseph-Nathan, P., Mendoza, V., Garcia, E. (1977) *Tetrahedron* **33**: 1573.

Joucla, M., Fouchet, B., LeBrun, J., Hamelin, J. (1985) *Tetrahedron Lett.* **26**: 1221.

Julia, M. (1971) *Acc. Chem. Res.* **4**: 386; (1974) *Pure Appl. Chem.* **40**: 553.

Jung, M.E., Lowe, J.A. (1977) *J. Org. Chem.* **42**: 2371.

Jung, M.E., Lowe, J.A. (1978) *Chem. Commun.* 95.

Jung, M.E., Halweg, K.M. (1981) *Tetrahedron Lett.* **22**: 2735.

Jung, M.E., Light, L.A. (1982) *J. Org. Chem.* **47**: 1084.

Jung, M.E., Hatfield, G.L. (1983) *Tetrahedron Lett.* **24**: 2931.

Jung, M.E., Halweg, K.M. (1984) *Tetrahedron Lett.* **25**: 2121.

Jung, M.E., Choi, Y.M. (1991) *J. Org. Chem.* **56**: 6729.

Jutz, C., Wagner, R.M. (1972) *Angew. Chem.* **84**: 299.

Kaczmarek, R., Blechert, S. (1986) *Tetrahedron Lett.* **27**: 2845.

Kakiuchi, K., Ue, M., Tsukahara, H., Shimizu, T., Miyao, T., Tobe, Y., Odaira, Y., Yasuda, M., Shima, K. (1989) *J. Am. Chem. Soc.* **111**: 3707.

Kalvoda, J. (1968) *Helv. Chim. Acta* **51**: 267.

Kalvoda, J., Botta, L. (1972) *Helv. Chim. Acta* **55**: 356.

Kametani, T., Ogasawara, K., Takahashi, T. (1973) *Tetrahedron* **29**: 73.

Kametani, T., Hirai, Y., Takeda, H., Kajiwara, M., Takahashi, T., Satoh, F., Fukumoto, K. (1974) *J. Chem. Soc. Perkin Trans. I* 2141.

Kametani, T., Kajiwara, M., Fukumoto, K. (1974b) *Tetrahedron* **30**: 1053.

Kametani, T., Kato, Y., Fukumoto, K. (1974c) *J. Chem. Soc. Perkin Trans. I* 1712.

Kametani, T., Ichikawa, Y., Suzuki, T., Fukumoto, K. (1975a) *Heterocycles* **3**: 401.

Kametani, T., Kajiwara, M., Takahashi, T., Fukumoto, K. (1975b) *J. Chem. Soc. Perkin Trans. I* 737.

Kametani, T., Fukumoto, K. (1976a) *Acc. Chem. Res.* **9**: 319.

Kametani, T., Higa, T., Fukumoto, K., Koizumi, M. (1976b) *Heterocycles* **4**: 23.

Kametani, T., Kato, Y., Honda, T., Fukumoto, K. (1976c) *J. Am. Chem. Soc.* **98**: 8185.

Kametani, T., Hirai, Y., Satoh, F., Fukumoto, K. (1977a) *Chem. Commun.* 16.

Kametani, T., Tsubuki, M., Shiratori, Y., Kato, Y., Nemoto, H., Ihara, M., Fukumoto, K., Satoh, F., Inoue, H. (1977b) *J. Org. Chem.* **42**: 2672.

Kametani, T., Hirai, Y., Shiratori, Y., Fukumoto, J., Satoh, S. (1978a) *J. Am. Chem. Soc.* **100**: 554.

Kametani, T., Matsumoto, H., Nemoto, H., Fukumoto, K. (1978b) *J. Am. Chem. Soc.* **100**: 6218.

Kametani, T., Enomoto, Y., Takahashi, K., Fukumoto, K. (1979a) *J. Chem. Soc. Perkin Trans. I* 2836.

Kametani, T., Suzuki, K., Nemoto, H., Fukumoto, K. (1979b) *J. Org. Chem.* **44**: 1036.

Kametani, T., Chihiro, M., Honda, T., Fukumoto, K. (1980a) *Chem. Pharm. Bull.* **28**: 2468.

Kametani, T., Honda, T., Shiratori, Y., Fukumoto, K. (1980b) *Tetrahedron Lett.* 1389.

Kametani, T., Matsumoto, H., Honda, T., Nagai, M., Fukumoto, K. (1981a) *Tetrahedron* **37**: 2555.

Kametani, T., Suzuki, K., Nemoto, H. (1981b) *J. Am. Chem. Soc.* **103**: 2890.

Kametani, T., Yukawa, H., Honda, T. (1986) *Chem. Commun.* 651.

Kametani, T., Kawamura, K., Tsubuki, M., Honda, T. (1988) *J. Chem. Soc. Perkin Trans. I* 193.

Kametani, T., Kondoh, H., Tsubuki, M., Honda, T. (1990) *J. Chem. Soc. Perkin Trans. I* 5.

Kanematsu, K., Nagashima, S. (1990) *Chem. Commun.* **1028.**

Kanematsu, K., Soejima, S. (1991) *Heterocycles* **32**: 1483.

Kan-Fan, C., Massiot, G., Ahond, A., Das, B.C., Husson, H.P., Potier, P., Scott, A.I., Wei, C.C. (1974) *Chem. Commun.* 164.

Kano, S., Sugino, E., Hibino, S. (1982) *Heterocycles* **19**: 1673.

Kano, S., Yokomatsu, T., Shibuya, S. (1986) *J. Am. Chem. Soc.* **108**: 6746.

Kanomata, N., Nitta, M. (1989) *Int. Chem. Cong. Pac. Basin Soc. ORGN* 151.

Karabelas, K., Moore, H.W. (1990) *J. Am. Chem. Soc.* **112**: 5372.

Karim, M.R., Sampson, P. (1988) *Tetrahedron Lett.* **29**: 6897.

Kariv-Miller, E., Maeda, H., Lombardo, F. (1989) *J. Org. Chem.* **54**: 4022.

Karpf, M., Dreiding, A.S. (1977) *Helv. Chim. Acta* **60**: 3045.

Kashima, C., Hibi, S., Manuyama, T., Omote, Y. (1986) *Tetrahedron Lett.* **27**: 2131.

Kasturi, T.R., Chandra, R. (1988) *J. Org. Chem.* **53**: 3178.

Kato, M., Sawa, T., Miwa, T. (1971) *Chem. Commun.* 1635.

Kato, M., Funakura, M., Tsuji, M., Miwa, T. (1976) *Chem. Commun.* 63.

Kato, T., Hozumi, T. (1972) *Chem. Pharm. Bull.* **20**: 1574.

Kato, T., Sato, M., Kitagawa, Y. (1978) *J. Chem. Soc. Perkin Trans. I* 352.

Katsube, J., Shimomura, H., Murayama, E., Toki, K., Matsui, M. (1971) *Agri. Biol. Chem.* **35**: 1768.

Kaupp, G., Grüter, H.W. (1978) *Angew. Chem. Int. Ed. Engl.* **17**: 52.

Keay, B.A., Rodrigo, R. (1982) *J. Am. Chem. Soc.* **104**: 4725.

Keck, G.E., Nickell, D.G. (1980) *J. Am. Chem. Soc.* **102**: 3632.

Keck, G.E., Boden, E., Sonnenwald, U. (1981) *Tetrahedron Lett.* **22**: 2615.

Keck, G.E., Webb, R.R., II (1982) *J. Org. Chem.* **47**: 1302.

Keely, S.L., Martinez, A.J., Tahk, F.C. (1970) *Tetrahedron* **26**: 4729.

Kelly, R.B., Zamecnik, J., Beckett, B.A. (1972) *Can. J. Chem.* **50**: 3455.

Kelly, T.R., Gillard, J.W., Goerner, R.N., Lyding, J.M. (1977) *J. Am. Chem. Soc.* **99**: 5513.

Kelly, T.R., Vaya, J., Ananthasubramanian, L. (1980) *J. Am. Chem. Soc.* **102**: 5983.

Kelly, T.R., Ghoshal, M. (1985a) *J. Am. Chem. Soc.* **107**: 3879.

Kelly, T.R., Liu, H. (1985b) *J. Am. Chem. Soc.* **107**: 4998.

Kende, A.S., Curran, D.P., Tsay, Y., Mills, J.E. (1977) *Tetrahedron Lett.* 3537.

Kende, A.S., Demuth, T.P. (1980) *Angew. Chem. Int. Ed. Engl.* **21**: 715.

Kende, A.S., Koch, K. (1986) *Tetrahedron Lett.* **27**: 6051.

Kende, A.S., Heibeisen, P., Newbold, R.C. (1988) *J. Am. Chem. Soc.* **110**: 3315.

Kende, A.S., Kaldor, I. (1989) *Tetrahedron Lett.* **30**: 7329.

Kennedy, M., McKervey, M.A. (1988) *Chem. Commun.* 1028.

Khan, K.M., Knight, D.W. (1991) *Chem. Commun.* 1699.

Khanapure, S.P., Crenshaw, L., Reddy, R.T., Biehl, E.R. (1988) *J. Org. Chem.* **53**: 4915.

Kido, F., Noda, Y., Yoshikoshi, A. (1982) *Chem. Commun.* 1209.

Kido, F., Kawada, Y., Kato, M., Yoshikoshi, A. (1991) *Tetrahedron Lett.* **32**: 6159.

Kienzle, F., Stadlwieser, J. (1991) *Tetrahedron Lett.* **32**: 551.

Kiguchi, T., Kuninobu, N., Naito, T., Ninomiya, I. (1989) *Heterocycles* **28**: 19.

Kilburn, J.D. (1990) *Tetrahedron Lett.* **31**: 2193.

Kilenyi, S.N., Mahaux, J.-M., Van Durme, E. (1991) *J. Org. Chem.* **56**: 2591.

Kim, S., Uh, K.H., Lee, S., Park, J.H. (1991a) *Tetrahedron Lett.* **32**: 3395.

Kim, S., Fuchs, P.L. (1991b) *J. Am. Chem. Soc.* **113**: 9864.

Kitahara, T., Kurata, H., Matsuoka, T., Mori, K. (1985) *Tetrahedron* **41**: 5475.

Kitahara, Y., Funamizu, M. (1958) *Bull. Chem. Soc. Jpn* **31**: 782.

Kitahara, Y., Kato, T. (1964) *Bull. Chem. Soc. Jpn* **37**: 895.

Kitahara, Y., Oda, M., Miyakoshi, S., Nakanishi, S. (1976a) *Tetrahedron Lett.* 2149.

Kitahara, Y., Oda, M., Oda, M. (1976b) *Chem. Commun.* 446.

Kiyooka, S., Kaneko, Y., Komura, M., Matsuo, H., Nakano, M. (1991) *J. Org. Chem.* **56**: 2276.

Klärner, F.G., Dogan, B., Roth, W.R., Hafner, K. (1982) *Angew. Chem. Int. Ed. Engl.* **21**: 708.

Klemm, L.H., Olson, D.R., White, D.V. (1971) *J. Org. Chem.* **36**: 3740.

Klose, W., Nikisch, K., Bohlmann, F. (1980) *Chem. Ber.* **113**: 2694.

Klunder, A.J.H., Bos, W., Verlaak, J.M.M., Zwanenburg, B. (1981) *Tetrahedron Lett.* **22**: 4553.

Knight, J., Parsons, P.J. (1987) *Chem. Commun.* 189.

Koch, K., Lin, J.-M., Fowler, F. (1983) *Tetrahedron Lett.* **24**: 1581.

Kodama, M., Kurihara, T., Sasaki, J., Ito, S. (1979) *Can. J. Chem.* **57**: 3343.

Kodpinid, M.Y., Siwapinyoyos, T., Thebtaranonth, Y. (1984) *J. Am. Chem. Soc.* **106**: 4862.

Koerner, M., Rickborn, B. (1991) *J. Org. Chem.* **56**: 1373.

Koft, E.R., Kotnis, A.S., Broadbent, T.A. (1987) *Tetrahedron Lett.* **28**: 2799.

Kohnke, F.H., Stoddart, J.F. (1989) *Pure Appl. Chem.* **61**: 1581.

Komppa, G. (1903) *Ber* **36**: 4332.

Kondo, K., Ojima, I. (1972) *Chem. Commun.* 860.

Koreeda, M., Mislankar, S.G. (1983) *J. Am. Chem. Soc.* **105**: 7203.

Koreeda, M., Luengo, J.I. (1985) *J. Am. Chem. Soc.* **107**: 5572.

Kostermans, G.B.M., van Dansik, P., de Wolf, W.H., Bickelhaupt, F. (1987) *J. Am. Chem. Soc.* **109**: 7887.

Kowalski, C.J., Lal, G.S. (1988) *J. Am. Chem. Soc.* **110**: 3693.

Kozikowski, A.P., Hasan, N.M. (1977) *J. Org. Chem.* **42**: 2039.

Kozikowski, A.P., Ishida, H. (1980) *J. Am. Chem. Soc.* **102**: 4265.

Kozikowski, A.P., Jung, S.H. (1986) *Tetrahedron Lett.* **27**: 3227.

Kozikowski, A.P., Li, C.-S. (1987) *J. Org. Chem.* **52**: 3541.

Kramer, U., Guggisberg, A., Hesse, M., Schmid, H. (1978) *Angew. Chem. Int. Ed. Engl.* **17**: 200.

Krantz, A., Lin, C.Y. (1973) *J. Am. Chem. Soc.* **95**: 5662.

Kraus, G.A., Sugimoto, H. (1977) *Tetrahedron Lett.* 3929.

Kraus, G.A., Sugimoto, H. (1978) *Tetrahedron Lett.* 2263.

Kraus, G.A., Pezzanite, J.O., Sugimoto, H. (1979) *Tetrahedron Lett.* 853.

Kraus, G.A., Taschner, M.J. (1980) *J. Am. Chem. Soc.* **102**: 1974.

Kraus, G.A., Cho, H., Crowley, S., Roth, B., Sugimoto, H., Prugh, S. (1983) *J. Org. Chem.* **48**: 3439.

Kraus, G.A., Fulton, B.S. (1984) *Tetrahedron* **40**: 4777.

Kraus, G.A., Fulton, B.S. (1985) *J. Org. Chem.* **50**: 1782.

Kraus, G.A., Hon, Y.-S. (1987) *Heterocycles* **25**: 377.

Kretchmer, R.A., Thompson, W.J. (1976) *J. Am. Chem. Soc.* **98**: 3379.

Krief, A., DeVos, M.J. (1985) *Tetrahedron Lett.* **26**: 6115.

Krief, A., Dumont, W., Pasau, P. (1988) *Tetrahedron Lett.* **29**: 1079.

Krohn, K., Tolkiehn, K. (1979) *Chem. Ber.* **112**: 3453.

Krohn, K. (1980) *Tetrahedron Lett.* **21**: 3557.

Krohn, K., Miehe, F. (1985) *Liebigs Ann. Chem.* **1329.**

Kuehne, M.E., Roland, D.M., Hafter, R. (1978) *J. Org. Chem.* **43**: 3705.

Kuehne, M.E., Huebner, J.A., Matsko, T.H. (1979a) *J. Org. Chem.* **44**: 2477.

Kuehne, M.E., Matsko, T.H., Bohnert, J.C., Kirkemo, C.L. (1979b) *J. Org. Chem.* **44**: 1063.

Kuehne, M.E., Kirkemo, C.L., Matsko, T.H., Bohnert, J.C. (1980) *J. Org. Chem.* **45**: 3259.

Kuehne, M.E., Bohnert, J.C. (1981) *J. Org. Chem.* **46**: 3443.

Kuehne, M.E., Okuniewicz, F.J., Kirkemo, C.L., Bohnert, J.C. (1982) *J. Org. Chem.* **47**: 1335.

Kuehne, M.E., Earley, W.G. (1983) *Tetrahedron* **39**: 3707, 3715.

Kuehne, M.E., Reider, P.J. (1985a) *J. Org. Chem.* **50**: 1464.

Kuehne, M.E., Seaton, P.J. (1985b) *J. Org. Chem.* **50**: 4790.

Kuehne, M.E., Bornmann, W.G., Earley, W.G., Marko, I. (1986) *J. Org. Chem.* **51**: 2913.

Kuehne, M.E., Frasier, D.A., Spitzer, T.D. (1991a) *J. Org. Chem.* **56**: 2696.

Kuehne, M.E., Muth, R.S. (1991b) *J. Org. Chem.* **56**: 2701.

Kuhnke, J., Bohlmann, F. (1988) *Liebigs Ann. Chem.* 743.

Kulkarni, Y.S., Snider, B.b. (1985) *J. Org. Chem.* **50**: 2809.

Kumaraswami, K., Shanmugam, P. (1988) *Tetrahedron Lett.* **29**: 2235.

Kündig, E.P., Simmons, D.P. (1983) *Chem. Commun.* 1320.

Kunz, H., Pfrengle, W. (1989) *Angew. Chem. Int. Ed. Engl.* **28**: 1067.

Kupchan, S.M., Liepa, A.J., Kameswaran, V., Bryan, R.F. (1973) *J. Am. Chem. Soc.* **95**: 6861.

Kuroda, S., Asao, T., Funamizu, M., Kurihara, H., Kitahara, Y. (1976) *Tetrahedron Lett.* 251.

Kurney, J.P., Eigendorff, G.K., Matsue, H., Murai, A., Tanaka, K., Sung, W.L., Wada, K., Worth, B.R. (1978) *J. Am. Chem. Soc.* **100**: 938.

Kuzuya, M., Miyake, F., Okuda, T. (1980) *Tetrahedron Lett.* **21**: 1043; (1984) *J. Chem. Soc. Perkin Trans. II* 1471.

Laarhoven, W.H., Lijten, F.A.T., Smits, J.M.M. (1985) *J. Org. Chem.* **50**: 3208.

Labows, J.N., Meinwald, J., Rottele, H., Schroder, G. (1967) *J. Am. Chem. Soc.* 612.

Labuschagne, A.J.H., Meyer, C.J., Spies, H.S.C., Schneider, D.F. (1975) *J. Chem. Soc. Perkin Trans. I* 2129.

LaFontaine, J., Mongrain, M., Sergent-Guay, M., Ruest, L., Deslongchamps, P. (1980) *Can. J. Chem.* **58**: 2460.

Lamberts, J.J.M., Laarhoven, W.H. (1984) *J. Am. Chem. Soc.* **106**: 1736.

Landry, D.W. (1983) *Tetrahedron* **39**: 2761.

Lantos, I., Razgaitis, C., Van Hoeven, H., Loev, B. (1977) *J. Org. Chem.* **42**: 228.

Lasne, M.C., Ripoll, J.L., Thuillier, A. (1982) *J. Chem. Res. Synop.* 214.

Lauer, W.M., Wujciak, D.W. (1956) *J. Am. Chem. Soc.* **78**: 5601.

Lavallée, J.-F., Deslongchamps, P. (1987) *Tetrahedron Lett.* **28**: 3457.

LeBel, N.A., Caprathe, B.W. (1985) *J. Org. Chem.* **50**: 3938.

Lecker, S.H., Nguyen, N.H., Vollhardt, K.P.C. (1986) *J. Am. Chem. Soc.* **108**: 856.

LeDrian, C., Greene, A.C. (1982) *J. Am. Chem. Soc.* **104**: 5473.

Lee, R.A. (1973) *Tetrahedron Lett.* 3333.

Lee, S.Y., Niwa, M., Snider, B.B. (1988) *J. Org. Chem.* **53**: 2356.

Lee, T.V., Channon, J.A., Cregg, C., Porter, J.R., Roden, F.S., Yeoh, H.T.-L., Boucher, R.J., Rockell, C.J.M. (1989) *Tetrahedron* **45**: 5877.

LeGall, T., Lellouche, J.P., Beaucourt, J.P. (1989) *Tetrahedron Lett.* **30**: 6521.

Legseir, B., Henin, J., Massiot, G., Vercauteren, J. (1987) *Tetrahedron Lett.* **28**: 3573.

Lei, X., Doubleday, C.E., Zimmt, M.B., Turro, N.J. (1986) *J. Am. Chem. Soc.* **108**: 2444.

Leimgruber, W. (1974) *Jap. Kokai* 1,254,314.

LeMenez, P., Kunesch, N., Liu, S., Wenkert, E. (1991) *J. Org. Chem.* **56**: 2915.

Lentz, C.M., Posner, G.H. (1978) *Tetrahedron Lett.* 3769.

Leonard, N.J., Frihart, C.R. (1974) *J. Am. Chem. Soc.* **96**: 5894.

Leriverend, P., Conia, J.M. (1970) *Bull. Soc. Chim. Fr.* 1060.

Lewis, C.P., Brookhart, M. (1975) *J. Am. Chem. Soc.* **97**: 651.

Lewis, R.T., Motherwell, W.B., Shipman, M. (1988) *Chem. Commun.* 948.

Leyendecker, F., Drouin, J., Conia, J.M. (1974) *Tetrahedron Lett.* 2931.

Leyendecker, F. (1976) *Tetrahedron* **32**: 349.

Liebeskind, L.S., Iyer, S., Jewell, S., C.F. (1986) *J. Org. Chem.* **51**: 3067.

Lin, C.-T., Chou, T.-C. (1990) *J. Org. Chem.* **55**: 2252.

Lipkowitz, K.B., Scarpone, S., Mundy, B.P., Bornmann, W.G. (1979) *J. Org. Chem.* **44**: 486.

Little, R.D., Muller, G.W., Venegas, M.G., Carroll, G.L., Bukhari, A., Patton, L., Stone, K.J. (1981) *Tetrahedron* **37**: 4371.

Little, R.D., Verhe, R., Monte, W.T., Nugent, S., Dawson, J.R. (1982) *J. Org. Chem.* **47**: 362.

Little, R.D., Carroll, G.L., Petersen, J.L. (1983) *J. Org. Chem.* **105**: 923.

Little, R.D. (1986) *Chem. Rev.* **86**: 875.

Liu, C.-Y., Mareda, J., Houk, K.N., Fronczek, F.R. (1983) *J. Am. Chem. Soc.* **105**: 6714.

Liu, H.-J., Ralitsch, M. (1990) *Chem. Commun.* 997.

Long, N.R., Rathke, M.W. (1981) *Synth. Commun.* **11**: 687.

Longone, D.T., Gladysz, J.A. (1976) *Tetrahedron Lett.* 4559.

Lorenzi-Riatsch, A., Nakashita, Y., Hesse, M. (1981) *Helv. Chim. Acta* **64**: 1854; (1984) *Helv. Chim. Acta* **67**: 249.

Lupi, A., Patamia, M., Arcamone, F. (1990) *Gazz. Chim. Ital.* **120**: 277.

Lutz, R.P. (1984) *Chem. Rev.* **84**: 205.

Lythgoe, B., Roberts, D.A. (1980) *J. Chem. Soc. Perkin Trans. I* 892.

Macdonald, D.I., Durst, T. (1986) *J. Org. Chem.* **51**: 4749.

MacMillan, J., Pryce, R.J. (1967) *J. Chem. Soc. (C)* 740.

Magnus, P.D., Sear, N.L. (1984) *Tetrahedron* **40**: 2795.

Magnus, P., Principe, L.M. (1985) *Tetrahedron Lett.* **26**: 4851.

Magnus, P., Pappalardo, P., Southwell, I. (1986) *Tetrahedron* **42**: 3215.

Magnus, P., Principe, L.M., Slater, M.J. (1987) *J. Org. Chem.* **52**: 1483.

Magnus, P., Coldham, I. (1991) *J. Am. Chem. Soc.* **113**: 672.

Magnus, P., Rodriguez-Lopez, J., Mulholland, K., Matthews, I. (1992) *J. Am. Chem. Soc.* **114**: 382.

Mahalanabis, K.K., Mumtaz, M., Snieckus, V. (1982) *Tetrahedron Lett.* **23**: 3971.

Majetich, G., Hull, K. (1988) *Tetrahedron Lett.* **24**: 2773.

Majumdar, K.C., Thygarajan, B.S. (1972) *Chem. Commun.* 83.

Majumdar, K.C., Chattopadhyay, S.K. (1987) *Chem. Commun.* 524.

Malherbe, R., Bellus, D. (1978) *Helv. Chim. Acta* **61**: 3096.

Mandai, T., Matsumoto, S., Kohama, M., Kawada, M., Tsuji, J. (1990) *J. Org. Chem.* **55**: 5671.

Mandai, T., Murakami, T., Kawada, M., Tsuji, J. (1991a) *Tetrahedron Lett.* **32**: 3399.

Mandai, T., Suzuki, S., Ikawa, A., Murakami, T., Kawada, M., Kawada, M., Tsuji, J. (1991b) *Tetrahedron Lett.* **32**: 7687.

Mander, L.N., Turner, J.V. (1981) *Tetrahedron Lett.* **22**: 3683.

Mander, L.N. (1991) *Synlett.* 134.

Mangeney, P. (1978) *Tetrahedron* **34**: 1359.

Manisse, N., Chuche, J. (1977) *J. Am. Chem. Soc.* **99**: 1272.

Mann, J., Piper, S.E. (1982) *Chem. Commun.* 430.

Marbet, R., Saucy, G. (1967) *Helv. Chim. Acta* **50**: 2091.

Marchand, A.P., Reddy, G.M., Watson, W.H., Nagl, A. (1988) *J. Org. Chem.* **53**: 5969.

Margaretha, P., Tissot, P. (1982) *Helv. Chim. Acta* **65**: 1949.

Marino, J.P., Browne, L.J. (1976) *Tetrahedron Lett.* 3245.

Marino, J.P., Laborde, E. (1985) *J. Am. Chem. Soc.* **107**: 734.

Marino, J.P., Silveira, C., Comasseto, J., Petragnani, N. (1987) *J. Org. Chem.* **52**: 4139.

Marino, J.P., Long, J.K. (1988) *J. Am. Chem. Soc.* **110**: 7916.

Mark, V. (1969) in Thygarajan, B.S. (ed.) *Mechanism of Molecular Migrations* **2**: 319, Wiley, New York.

Märkl, G. (1962) *Angew. Chem. Int. Ed. Engl.* **1**: 511.

Marshall, J.A., Babler, J.H. (1969) *J. Org. Chem.* **34**: 4186.

Marshall, J.A., Ruth, J.A. (1974) *J. Org. Chem.* **39**: 1971.

Marshall, J.A., Conrow, R.E. (1983) *J. Am. Chem. Soc.* **105**: 5679.

Marshall, J.A., DuBay, W.J. (1991a) *J. Org. Chem.* **56**: 1685.

Marshall, J.A., Welmaker, G.S., Gung, B.W. (1991b) *J. Am. Chem. Soc.* **113**: 647.

Marson, C.M., Grabowski, U., Walsgrove, T., Eggleston, D.S., Baures, P.W. (1991) *J. Org. Chem.* **56**: 2603.

Martin, J., Parker, W., Stevart, T., Stevenson, J.R. (1972) *J. Chem. Soc. Perkin Trans. I* 1760.

Martin, J.C., Muchowski, J.M. (1984) *J. Org. Chem.* **49**: 1040.

Martin, R.H., Jespers, J., Defay, N. (1975a) *Helv. Chim. Acta* **58**: 776; (1975b) *Tetrahedron Lett.* 1093.

Martin, S.F., Desai, S.R. (1977) *J. Org. Chem.* **42**: 1664; (1978) *J. Org. Chem.* **43**: 4673.

Martin, S.F., Desai, S.R., Phillips, G.W., Miller, A.C. (1980) *J. Am. Chem. Soc.* **102**: 3294.

Martin, S.F., Tu, C., Kimura, M., Simonsen, S.H. (1992) *J. Org. Chem.* **47**: 3634.

Martin, S.F., Rüeger, H. (1985) *Tetrahedron Lett.* **26**: 5227.

Martin, S.F., Grzejszczak, S., Rüeger, H., Williamson, S.A. (1987) *J. Am. Chem. Soc.* **109**: 6124.

Martin, S.F., Li, W. (1989) *J. Org. Chem.* **54**: 265.

Manuoka, K., Miyazaki, T., Ando, M., Matsumura, Y., Sakane, S., Hattori, K., Yamamoto, H. (1983) *J. Am. Chem. Soc.* **105**: 2831.

Maruoka, S., Sato, J., Banno, H., Yamamoto, H. (1990) *Tetrahedron Lett.* **31**: 377.

Marvel, C.S., Moore, A.C. (1949) *J. Am. Chem. Soc.* **71**: 28.

Marvell, E.N., Anderson, D.R., Ong, J. (1962) *J. Org. Chem.* **27**: 1109.

Marvell, E.N., Titterington, D. (1986) *Tetrahedron Lett.* **21**: 2123.

Maryanoff, B.E., Reitz, A.B. (1989) *Chem. Rev.* **89**: 863.

Mas, J.-M., Gore, J., Malacria, M. (1986) *Tetrahedron Lett.* **27**: 3133.

Masumune, S., Fukumoto, K. (1965) *Tetrahedron Lett.* 4647.

Masamune, S., Cuts, H., Hogben, M.G. (1966) *Tetrahedron Lett.* 1017.

Masamune, S., Chin, C.G., Hojo, K., Seidner, R.T. (1967) *J. Am. Chem. Soc.* **89**: 4804.

Masamune, S., Brooks, D.W., Morio, K., Sobczak, R.L. (1976) *J. Am. Chem. Soc.* **98**: 8277.

Massanet, G.M., Pando, E., Rodriguez-Luis, F., Salva, J. (1987) *Heterocycles* **26**: 1541.

Matlin, A.R., Jin, K. (1989) *Tetrahedron Lett.* **30**: 637.

Matsumura, Y., Fujiwara, J., Maruoka, K., Yamamoto, H. (1983) *J. Am. Chem. Soc.* **105**: 6312.

Maruya, R., Ghera, E. (1990) *Tetrahedron Lett.* **31**: 5179.

May, C., Moody, C.J. (1984) *Chem. Commun.* 926.

Mazza, S., Danishefsky, S., McCurry, P. (1974) *J. Org. Chem.* **39**: 3610.

McClure, K.F., Benbow, J.W., Danishefsky, S.J., Schulte, G.K. (1991) *J. Am. Chem. Soc.* **113**: 8185.

McCullough, D.W., Cohen, T. (1988) *Tetrahedron Lett.* **29**: 27.

McCurry, P.M., Singh, R.K. (1973) *Tetrahedron Lett.* 3325.

McEvoy, F.J., Allen, G.R. (1969) *J. Org. Chem.* **34**: 4199.

McIntosh, J.M., Khalil, H. (1977) *J. Org. Chem.* **42**: 2123.

McIntosh, J.M., Sieler, R.A. (1978) *J. Org. Chem.* **43**: 4431.

McKennis, J.S., Brener, L., Ward, J.S., Pettit, R. (1971) *J. Am. Chem. Soc.* **93**: 4957.

McMurry, J.E., Glass, T.E. (1971) *Tetrahedron Lett.* 2575.

McMurry, J.E., Isser, S.J. (1972) *J. Am. Chem. Soc.* **94**: 7132.

McMurry, J.E., Andrews, A., Ksander, G.M., Musser, J.H., Johnson, M.A. (1979) *J. Am. Chem. Soc.* **101**: 1330.

McMurry, J.E., Swenson, R. (1987) *Tetrahedron Lett.* **28**: 3209.

McNab, H., Monahan, L.C., Gray, T. (1987) *Chem. Commun.* 140.

McNeil, D.W., Kent, M.E., Hedaya, E., D'Angelo, P.F., Schissel, P.O. (1971) *J. Am. Chem. Soc.* **93**: 3817.

Medina, J.C., Cadilla, R., Kyler, K.S. (1987) *Tetrahedron Lett.* **28**: 1059.

Mehta, G., Rao, K.S., Bhadbhade, M.M., Venkatesan, K. (1981) *Chem. Commun.* 755.

Mehta, G., Pramod, K., Subrahmanyam, D. (1986) *Chem. Commun.* 247.

Mehta, G., Reddy, S.H.K. (1991) *Tetrahedron Lett.* **32**: 6403.

Meier, H., Echter, T., Zimmer, O. (1981) *Angew. Chem. Int. Ed. Engl.* **20**: 865.

Meinwald, J., Labana, S.S., Chadha, K.S. (1963) *J. Am. Chem. Soc.* **85**: 582.

Meinwald, J., Schmidt, D. (1969a) *J. Am. Chem. Soc.* **91**: 5877.

Meinwald, J., Tsuruta, H. (1969b) *J. Am. Chem. Soc.* **91**: 5877.

Meister, H. (1963) *Chem. Ber.* **96**: 1688.

Menachery, M.D., Carroll, P., Cava, M.P. (1983) *Tetrahedron Lett.* **24**: 167.

Mestdagh, H., Vollhardt, K.P.C. (1986) *Chem. Commun.* 281.

Meyer, W.L., Brannon, M.J., Merritt, A., Seebach, D. (1986) *Tetrahedron Lett.* **27**: 1449.

Meyers, A.I., Licini, G. (1989) *Tetrahedron Lett.* **30**: 4049.

Michael, A., Ross, J. (1931) *J. Am. Chem. Soc.* **53**: 2394.

Mikami, K., Taya, S., Nakai, T., Fujita, Y. (1981) *J. Org. Chem.* **46**: 5447.

Mikami, K., Nakai, T. (1982) *Chem. Lett.* 1349.

Mikami, K., Kishi, N., Nakai, T. (1983) *Tetrahedron Lett.* **24**: 795.

Mikami, K., Takahashi, K., Nakai, T. (1990) *J. Am. Chem. Soc.* **112**: 4035.

Mikhailov, B.M., Bubnov, Y.N. (1971) *Tetrahedron Lett.* 2127, 2153.

Miller, J.G., Kurz, W., Untch, K.G., Stork, G. (1974) *J. Am. Chem. Soc.* **96**: 6774.

Miller, R.D., Dolce, D. (1973) *Tetrahedron Lett.* 1151.

Miller, R.D., Kaufmann, D., Mayerle, J.J. (1977) *J. Am. Chem. Soc.* **99**: 8511.

Miller, R.D., Theis, W., Heilig, G., Kirchmeyer, S. (1991) *J. Org. Chem.* **56**: 1453.

Minami, T., Nishimura, K., Hirao, I., Suganuma, H., Agawa, T. (1982) *J. Org. Chem.* **47**: 2360.

Minami, T., Yamanouchi, T., Takenaka, S., Mirao, I. (1983) *Tetrahedron Lett.* **24**: 767.

Minami, T., Chikugo, T., Hanamoto, T. (1986a) *J. Org. Chem.* **51**: 2210.

Minimi, T., Kitajima, Y., Chikugo, T. (1986b) *Chem. Lett.* 1229.

Minter, D.E., Fonken, G.J., Cook, F.T. (1979) *Tetrahedron Lett.* 711.

Mirand, C., Massiot, G., Levy, J. (1982) *J. Org. Chem.* **47**: 4169.

Misumi, A., Funita, K., Yamamoto, H. (1984) *Tetrahedron Lett.* **25**: 671.

Mitch, E.L., Dreiding, A.S. (1960) *Chimia* **14**: 424.

Mitra, S., Lawton, R.G. (1979) *J. Am. Chem. Soc.* **101**: 3097.

Mitsunobu, O. (1981) *Synthesis* 1.

Miyagi, Y., Maruyama, K., Tanaka, N., Sato, M., Tomizu, T., Isogawa, Y., Kashiwano, H. (1984) *Bull. Chem. Soc. Jpn* **57**: 791.

Miyakoshi, T., Saito, S., Kumanotani, J. (1982) *Chem. Lett.* 83.

Miyano, M., Stealey, M.A. (1982) *J. Org. Chem.* **47**: 3184.

Miyao, Y., Tanaka, M., Suemune, H., Sakai, K. (1989) *Chem. Commun.* 1535.

Miyashi, T., Kawamoto, H., Mukai, T. (1977) *Tetrahedron Lett.* 4623.

Miyashi, T., Hazato, A., Mukai, T. (1978) *J. Am. Chem. Soc.* **100**: 1008.

Miyashi, T., Kawamoto, H., Nakajo, T., Mukai, T. (1979) *Tetrahedron Lett.* **155.**

Miyashi, T., Hazato, A., Mukai, T. (1982) *J. Am. Chem. Soc.* **104**: 891.

Mizuno, Y., Tomita, M., Irie, H. (1980) *Chem. Lett.* 107.

Moberg, C., Nilsson, M. (1974) *Tetrahedron Lett.* 4521.

Mock, W.L., Sprecher, C.M., Stewart, R.F., Northolt, M.G. (1972) *J. Am. Chem. Soc.* **94**: 2015.

Molander, G.A., Andrews, S.W. (1989) *Tetrahedron Lett.* **30**: 2351.

Molander, G.A., Cameron, K.A. (1991) *J. Org. Chem.* **56.** 2617.

Molina, P., Alajarin, M., Vidal, A., Feneau-Dupont, J., Declercq, J.P. (1990) *Chem. Commun.* 829.

Molina, P., Alajarin, M., Lopez-Leonardo, C. (1991) *Tetrahedron Lett.* **32**: 4041.

Monde, K., Takasugi, M., Katsui, N., Masamune, T. (1990) *Chem. Lett.* 1283.

Mondon, A., Ehrhardt, M. (1966) *Tetrahedron Lett.* 2577.

Monti, S.A., Dean, T.R. (1982) *J. Org. Chem.* **47**: 2679.

Moody, C.J., Shah, P., Knowles, P. (1988) *Tetrahedron Lett.* **29**: 2693.

Moody, C.J., Rahimtoola, K.F. (1990) *Chem. Commun.* 1667.

Moore, H.W. (1964) *J. Am. Chem. Soc.* **86**: 3398.

Moore, H.W. (1979) *Acc. Chem. Res.* **12**: 125.

Mori, N., Takemura, T., Tsuchiya, K. (1988) *Chem. Commun.* 575.

Morita, N., Asao, T., Kitahara, Y. (1974) *Tetrahedron Lett.* 2083.

Morton, D.R., Gravestock, M.B., Parry, R.J., Johnson, W.S. (1973a) *J. Am. Chem. Soc.* **95**: 4417.

Morton, D.R., Johnson, W.S. (1973b) *J. Am. Chem. Soc.* **95**: 4419.

Moursounidis, J., Wege, D. (1986) *Tetrahedron Lett.* **27**: 3045.

Mpango, G.B., Snieckus, V. (1980) *Tetrahedron Lett.* **21**: 4827.

Mucha, B., Hoffmann, H.M.R. (1989) *Tetrahedron Lett.* **30**: 4489.

Muchowski, J.M., Nelson, P.H. (1980) *Tetrahedron Lett.* **21**: 4585.

Mühlemann, H. (1951) *Pharm. Acta Helv.* **26**: 195.

Mukaiyama, T. (1977) *Angew. Chem. Int. Ed. Engl.* **16**: 817.

Mukaiyama, T. (1982) *Org. React.* **28**: 203.

Mukherjee, D., Dunn, L.C., Houk, K.N. (1979) *J. Am. Chem. Soc.* **101**: 251.

Mulzer, J., Bock, H., Eck, W., Buschmann, J., Luger, P. (1991) *Angew. Chem. Int. Ed. Engl.* **30**: 414.

Muquet, M., Kunesch, N., Poisson, J. (1972) *Tetrahedron* **28**: 1363.

Murahashi, S., Yuki, H., Hatada, K., Obokata, T. (1967) *Kobunshi Kagaku* **24**: 309. {(1968) *Chem. Abstr.* **69**: 19610v.}

Murai, A., Sato, S., Masamune, T. (1982) *Chem. Commun.* 511, 513.

Murakami, M., Nishida, S. (1981) *Chem. Lett.* 997.

Murray, W.V., Hadden, S.K. (1991) *J. Chem. Res.* (*S*) 279.

Muxfeldt, H., Bell, J.P., Baker, J.A., Cuntze, U. (1973) *J. Am. Chem. Soc.* 4587.

Muxfeldt, H., Haas, G., Hardtmann, G., Kathawala, F., Moobery, J.B., Vedejs, E. (1979) *J. Am. Chem. Soc.* **101**: 689.

Näf, F., Decorzant, R., Thommen, W. (1975) *Helv. Chim. Acta* **58**: 1808.

Näf, F., Decorzant, R., Thommen, W. (1977) *Helv. Chim. Acta* **60**: 1196.

Nagai, M., Lazor, J., Wilcox, C.S. (1990) *J. Org. Chem.* **55**: 3440.

Nagaoka, H., Kobayashi, K., Matshui, T., Yamada, Y. (1987) *Tetrahedron Lett.* **28**: 2021.

Nagata, R., Yamanaka, H., Okazaki, E., Saito, I. (1989) *Int. Chem. Cong. Pac. Basin Soc. ORGN* 688.

Nagumo, S., Suemune, H., Sakai, K. (1990) *Chem. Commun.* 1778.

Naito, T., Miyata, O., Nimomiya, I. (1979) *Chem. Commun.* 517.

Naito, T., Iida, N., Ninomiya, I. (1986) *J. Chem. Soc. Perkin Trans. I* 99.

Najera, C., Yus, M., Karisson, U., Gogoll, A., Bäckvall, J.-E. (1990) *Tetrahedron Lett.* **31**: 4199.

Nakai, T., Mikami, J. (1986) *Chem. Rev.* **86**: 885.

Nakamura, E., Hashimoto, K., Kuwajima, I. (1977) *J. Org. Chem.* **42**: 4166.

Nakamura, E., Isaka, M., Matsuzawa, S. (1988) *J. Am. Chem. Soc.* **110**: 1297.

Nakayama, J., Yamaoka, S., Nakanishi, T., Hoshino, M. (1988) *J. Am. Chem. Soc.* **110**: 6598.

Nakayama, J., Hasemi, R. (1990) *J. Am. Chem. Soc.* **112**: 5654.

Nalliah, B., Manske, R.H.F., Rodrigo, R., MacLean, D.B. (1973) *Tetrahedron Lett.* 2795.

Narasimhan, N.S., Gokhale, S.M. (1985) *Chem. Commun.* 85.

Naruta, S., Nishimura, H., Kaneko, H. (1975) *Chem. Pharm. Bull.* **23**: 1276.

Naruta, Y., Nishigaichi, Y., Maruyama, K. (1988) *J. Org. Chem.* **53**: 1192.

Naruto, S., Nishimura, H., Kaneko, H. (1975) *Chem. Pharm. Bull.* **23**: 1276.

Natsume, M., Utsunomiya, I. (1984) *Chem. Pharm. Bull.* **32**: 2477.

Neeson, S.J., Stevenson, P.J. (1988) *J. Am. Chem. Soc.* **29**: 813.

Negishi, E., Sabanski, M., Katz, J.J., Brown, H.C. (1976) *Tetrahedron* **32**: 925.

Negishi, E., Holmes, S.J., Tour, J.M., Miller, J.A. (1985) *J. Am. Chem. Soc.* **107**: 2568.

Negishi, E., Wu, G., Tour, J.M. (1988) *Tetrahedron Lett.* **29**: 6745.

Neidlein, R., Kohl, M., Kramer, W. (1989) *Helv. Chim. Acta* **72**: 1311.

Nelson, S.F., Kapp, D.L. (1985) *J. Org. Chem.* **50**: 1339.

Newton, R.F., Pauson, P.L., Taylor, R.G. (1983) *J. Chem. Res. Miniprint* 2201.

Ng, D.K.P., Luh, T.-Y. (1989) *J. Am. Chem. Soc.* **111**: 9119.

Ni, Z.-J., Yang, P.F., Ng, D.K.P., Tzeng, Y.-L., Luh, T.-Y. (1990) *J. Am. Chem. Soc.* **112**: 9356.

Nickon, A., Lambert, J.L. (1962) *J. Am. Chem. Soc.* **84**: 4604.

Nickon, A., Aaronoff, B.R. (1964) *J. Org. Chem.* **29**: 3014.

Nickon, A., Kwasnik, H., Swartz, T., Williams, R.O., DiGiorgio, J.B. (1965) *J. Am. Chem. Soc.* **87**: 1615.

Nicolaou, K.C., Barnette, W.E. (1979) *Chem. Commun.* 1119.

Nicolaou, K.C., Petasis, N.A., Uenishi, J., Zipkin, R.E. (1982a) *J. Am. Chem. Soc.* **104**: 5557.

Nicolaou, K.C., Petasis, N.A., Zipkin, R.E. (1982b) *J. Am. Chem. Soc.* **104**: 5560.

Nicolaou, K.C., Petasis, N.A., Zipkin, R.E., Uenishi, J. (1982c) *J. Am. Chem. Soc.* **104**: 5555.

Nicolaou, K.C., Zuzzarello, G., Petasis, N.A. (1982d) *J. Am. Chem. Soc.* **104**: 5558.

Nicolaou, K.C., Zuccarello, G., Ogawa, Y., Schweiger, E.J., Kumazawa, T. (1988) *J. Am. Chem. Soc.* **110**: 4866.

Nicolaou, K.C., Hwang, C.-K., Smith, A.L., Wendeborn, S.V. (1990) *J. Am. Chem. Soc.* **112**: 7416.

Nickon, A., Lambert, J.L. (1962) *J. Am. Chem. Soc.* **84**: 4604.

Nickon, A., Aaronoff, B.R. (1964) *J. Org. Chem.* **29**: 3014.

Nickon, A., Kwasnik, H., Swartz, T., Williams, R.O., DiGiorgio, J.B. (1965) *J. Am. Chem. Soc.* **87**: 1615.

Nielsen, A.T., Houlihan, W.J. (1968) *Org. React.* **16**: 1.

Ninomiya, I., Naito, T., Mori, T. (1973) *J. Chem. Soc. Perkin Trans. I* 505.

Nishida, A., Takahashi, H., Takeda, H., Takada, N., Yonemitsu, O. (1990) *J. Am. Chem. Soc.* **112**: 902.

Nishiyama, H., Arai, H., Ohki, T., Itoh, K. (1985) *J. Am. Chem. Soc.* **107**: 5310.

Nitta, M., Sekiguchi, A., Koba, H. (1981) *Chem. Lett.* 933.

Niwa, H., Wakamatsu, K., Hida, T., Niiyama, K., Kigoshi, H., Yamada, M., Nagase, H., Suzuki, M., Yamada, K. (1984) *J. Am. Chem. Soc.* **106**: 4547.

Noguchi, S., Morita, K. (1963) *Chem. Pharm. Bull.* **10**: 1235.

Noire, P.D., Franck, R.W. (1982) *Tetrahedron Lett.* **23**: 1031.

Nokami, J., Mandai, T., Watanabe, H., Ohoyama, H., Tsuji, J. (1989a) *J. Am. Chem. Soc.* **111**: 4126.

Nokami, J., Watanabe, H., Mandai, T., Kawada, M., Tsuji, J. (1989b) *Tetrahedron Lett.* **30**: 4829.

Noland, W.E., Wann, S.R. (1979) *J. Org. Chem.* **44**: 4402.

Nomoto, T., Takayama, H. (1985) *Heterocycles* **23**: 2913.

Norman, B.H., Gareua, Y., Padwa, A. (1991) *J. Org. Chem.* **56**: 2154.

Nossin, P.M.M., Speckamp, W.N. (1981) *Tetrahedron Lett.* **22**: 3289.

Novak, M., Novak, J., Salemink, C.A. (1991) *Tetrahedron Lett.* **32**: 4405.

Noyori, R., Baba, Y., Makino, S., Takaya, H. (1973) *Tetrahedron Lett.* 1741.

Noyori, R., Baba, Y., Makino, S., Takaya, H. (1978a) *J. Am. Chem. Soc.* **100**: 1786.

Noyori, R., Sato, T., Hayakawa, Y. (1978b) *J. Am. Chem. Soc.* **100**: 2561.

Noyori, R. (1979a) *Acc. Chem. Res.* **12**: 61.

Noyori, R., Nishizawa, M., Shimizu, F., Hayakawa, Y., Manuoka, K., Hashimoto, S., Yamamoto, H., Nozaki, H. (1979b) *J. Am. Chem. Soc.* **101**: 220.

Nugent, W.A., Hobbs, F.W. (1986) *J. Org. Chem.* **51**: 3376.

Nwaji, M.N., Onyiriuka, O.S. (1974) *Tetrahedron Lett.* 2255.

Oda, M., Breslow, R., Pecoraro, J. (1972) *Tetrahedron Lett.* 4419.

Oda, M., Miyazaki, H., Kitahara, Y. (1976) *Chem. Lett.* 1011.

Oda, M., Okada, K., Motegi, T., Liu, L.-K., Wen, Y.-S. (1990) *Tetrahedron Lett.* **31**: 7341.

O'Dell, D.E., Loper, J.T., Macdonald, T.L. (1988) *J. Org. Chem.* **53**: 5225.

Ogawa, M., Kitagawa, Y., Natsume, M. (1987) *Tetrahedron Lett.* **28**: 3985.

Ogino, T., Awano, K., Fukazawa, Y. (1990) *J. Chem. Soc. Perkin Trans. II* 1735.

Ogura, K., Yahata, N., Minoguchi, M., Ohtsuki, K., Takahashi, K., Iida, H. (1986) *J. Org. Chem.* **51**: 508.

Ohkita, M., Tsuji, T., Nishida, S. (1991) *Chem. Commun.* 37.

Ohloff, G., Uhde, G., Schulte-Elte, K.H. (1967) *Helv. Chim. Acta* **50**: 561.

Ohno, M., Mori, K., Hattori, T., Eguchi, S. (1990) *J. Org. Chem.* **55**: 6086.

Ohnuma, T., Tabe, M., Shiiya, K., Ban, Y., Date, T. (1983) *Tetrahedron Lett.* **24**: 4249.

Oikawa, Y., Yonemitsu, O. (1976) *J. Chem. Soc. Perkin Trans. I* 1479.

Okada, Y., Minimi, T., Yahiro, S., Akinaga, K. (1989) *J. Org. Chem.* **54**: 974.

Okamura, W.H. (1969) *Tetrahedron Lett.* 4717.

Okamura, W.H., Jeganathan, S. (1982) *Tetrahedron Lett.* 4673.

Okamura, W.H., Peter, R., Reischl, W. (1985) *J. Am. Chem. Soc.* **107**: 1034.

Okuda, S., Kamata, H., Tsuda, K. (1963) *Chem. Pharm. Bull.* **11**: 1349.

Olah, G.A., Ho, T.-L. (1976) *Synthesis* 609.

O'Leary, M.A., Wege, D. (1978) *Tetrahedron Lett.* 2811.

Olivier, L., Levy, J., LeMen, J., Janot, M.M., Budzikiewicz, H., Djerassi, C. (1965) *Bull. Soc. Chim. Fr.* 868.

Olsen, H. (1982) *Helv. Chim. Acta* **65**: 1921.

Olsson, T., Rahman, M.T., Ullenius, C. (1977) *Tetrahedron Lett.* 75.

Ong, B.S., Chan, T.H. (1976) *Tetrahedron Lett.* 3257.

Onishi, T., Fujita, Y., Nishida, T. (1980) *Synthesis* 651.
Openshaw, H.T., Whitaker, N. (1963) *J. Chem. Soc.* 1449, 1461.
Oppolzer, W., Keller, K. (1971) *J. Am. Chem. Soc.* **93**: 3836.
Oppolzer, W. (1974) *Tetrahedron Lett.* 1001.
Oppolzer, W., Hauth, H., Pfaffli, P., Wenger, R. (1977a) *Helv. Chim. Acta* **60**: 1801.
Oppolzer, W., Petrzilka, M., Bättig, K. (1977b) *Helv. Chim. Acta* **60**: 2964.
Oppolzer, W., Bättig, K., Petrzilka, M. (1978a) *Helv. Chim. Acta* **61**: 1945.
Oppolzer, W., Godel, T. (1978b) *J. Am. Chem. Soc.* **100**: 2583.
Oppolzer, W., Petrzilka, M. (1978c) *Helv. Chim. Acta* **61**: 2755.
Oppolzer, W., Petrzilka, M. (1978) *Helv. Chim. Acta* **61**: 2757.
Oppolzer, W., Roberts, D.A. (1980) *Helv. Chim. Acta* **63**: 1703.
Oppolzer, W., Francotte, E., Bättig, K. (1981a) *Helv. Chim. Acta* **64**: 478.
Oppolzer, W., Snowden, R.L., Simmons, D.P. (1981b) *Helv. Chim. Acta* **64**: 2002.
Oppolzer, W., Bättig, K. (1982a) *Tetrahedron Lett.* **23**: 4669.
Oppolzer, W., Pitteloud, R. (1982b) *J. Am. Chem. Soc.* **104**: 6578.
Oppolzer, W., Strauss, H.F., Simmons, D.P. (1982c) *Tetrahedron Lett.* **23**: 4673.
Oppolzer, W., Grayson, J.I., Wegmann, H., Urrea, M. (1983a) *Tetrahedron* **39**: 3695.
Oppolzer, W., Robbiani, C. (1983b) *Helv. Chim. Acta* **66**: 1119.
Oppolzer, W., Begley, T., Ashcroft, A. (1984) *Tetrahedron Lett.* **25**: 825.
Oppolzer, W., Cunningham, A.F. (1986a) *Tetrahedron Lett.* **27**: 5467.
Oppolzer, W., Jacobsen, E.J. (1986b) *Tetrahedron Lett.* **27**: 1141.
Oppolzer, W., Nakao, A. (1986c) *Tetrahedron Lett.* **27**: 5471.
Oppolzer, W., Schneider, P. (1986d) *Helv. Chim. Acta* **69**: 1817.
Oppolzer, W., Gaudin, J.-M. (1987) *Helv. Chim. Acta* **70**: 1477.
Oppolzer, W. (1989) *Angew. Chem. Int. Ed. Engl.* **28**: 38.
Oppolzer, W., Bienayme, H., Genevois-Borella, A. (1991a) *J. Am. Chem. Soc.* **113**: 9660.
Oppolzer, W., De Vita, R.J. (1991b) *J. Org. Chem.* **56**: 6256.
Oren, J., Schleifer, L., Shmueli, U., Fuchs, B. (1984) *Tetrahedron Lett.* **25**: 981.
Osterhun, V., Winterfeldt, E. (1977) *Chem. Ber.* **110**: 146.
Overman, L.E., Fukuya, C. (1980a) *J. Am. Chem. Soc.* **102**: 1454.
Overman, L.E., Tsuboi, S., Roos, J.P., Taylor, G.F. (1980b) *J. Am. Chem. Soc.* **102**: 747.
Overman, L.E., Malone, T.C. (1982) *J. Org. Chem.* **47**: 5297.
Overman, L.E., Jacobsen, E.J., Doedens, R.J. (1983a) *J. Org. Chem.* **48**: 3393.
Overman, L.E., Kakimoto, M., Okazaki, M.E., Meier, G.P. (1983b) *J. Am. Chem. Soc.* **105**: 6622.
Overman, L.E., Mendelson, L.T., Jacobsen, E.J. (1983c) *J. Am. Chem. Soc.* **105**: 6629.
Overman, L.E., Sworin, M., Burk, R.M. (1983d) *J. Org. Chem.* **48**: 2685.
Overman, L.E., Angle, S.R. (1985a) *J. Org. Chem.* **50**: 4021.
Overman, L.E., Okazaki, M.E., Jacobsen, E.J. (1985b) *J. Org. Chem.* **50**: 2403.
Overman, L.E., Sugai, S. (1985c) *Helv. Chim. Acta* **68**: 745.
Overman, L.E., Wild, H. (1989a) *Tetrahedron Lett.* **30**: 647.
Overman, L.E., Robertson, G.M., Robichaud, A.J. (1989b) *J. Org. Chem.* **54**: 1236.
Overman, L.E., Sharp, M.J. (1992) *J. Org. Chem.* **57**: 1035.

Paddon-Row, M.N., Warrener, R.N. (1974) *Tetrahedron Lett.* 3797.

Padwa, A., Carlsen, P.J.H. (1978) *J. Org. Chem.* **43**: 3757.

Padwa, A., Fryxell, G.E., Zhi, L. (1990a) *J. Am. Chem. Soc.* **112**: 3100.

Padwa, A., Norman, B.H. (1990b) *J. Org. Chem.* **55**: 4801.

Padwa, A., Krumpe, K.E., Gareau, Y., Chiachhio, U. (1991) *J. Org. Chem.* **56**: 2523.

Pak, H., Canalda, I.I., Fraser-Reid, B. (1990) *J. Org. Chem.* **55**: 3009.

Palensky, F.J., Morrison, H.A. (1977) *J. Am. Chem. Soc.* **99**: 3507.

Palmer, M., Morrison, H. (1980) *J. Org. Chem.* **45**: 798.

Pandit, U.K., Bieraugel, H., Stoit, A.R. (1984) *Tetrahedron Lett.* **25**: 1513.

Pansegrau, P.D., Rieker, W.F., Meyers, A.I. (1988) *J. Am. Chem. Soc.* **110**: 7178.

Papies, O., Grimme, W. (1980) *Tetrahedron Lett.* **21**: 2799.

Paquette, L.A. (1970) *J. Am. Chem. Soc.* **92**: 5765.

Paquette, L.A., Epstein, M.J. (1971a) *J. Am. Chem. Soc.* **93**: 5936.

Paquette, L.A., Philips, J.C., Wingard, R.E. (1971b) *J. Am. Chem. Soc.* **93**: 4516.

Paquette, L.A., Stowell, J.C. (1971c) *J. Am. Chem. Soc.* **93**: 2459.

Paquette, L.A., Meisinger, R.H., Wingard, R.E. (1972a) *J. Am. Chem. Soc.* **94**: 2155.

Paquette, L.A., Wingard, R.E. (1972b) *J. Am. Chem. Soc.* **94**: 4398.

Paquette, L.A., Russell, R.K., Wingard, R.E. (1973) *Tetrahedron Lett.* 1713.

Paquette, L.A., Oku, M. (1974a) *J. Am. Chem. Soc.* **96**: 1219.

Paquette, L.A., Wingard, R.E., Photis, J.M. (1974b) *J. Am. Chem. Soc.* **96**: 5801.

Paquette, L.A. (1977) *Org. React.* **25**: 1.

Paquette, L.A., Wyvratt, M.J., Berk, H.C., Moerck, R.E. (1978) *J. Am. Chem. Soc.* **100**: 5845.

Paquette, L.A., Andrews, D.R., Springer, J.P. (1983a) *J. Org. Chem.* **48**: 1147.

Paquette, L.A., Charumilind, P., Gallucci, J.C. (1983b) *J. Am. Chem. Soc.* **105**: 7364.

Paquette, L.A., Hathaway, S.J., Gallucci, J.C. (1984) *Tetrahedron Lett.* **25**: 2659.

Paquette, L.A., Waykole, L., Jendralla, H., Cottrell, C.E. (1986) *J. Am. Chem. Soc.* **108**: 3739.

Paquette, L.A., DeRussy, D.T., Pegg, N.A., Taylor, R.T., Zydowsky, T.M. (1989a) *J. Org. Chem.* **54**: 4576.

Paquette, L.A., Reagan, J., Schreiber, S.L., Telehar, C.A. (1989b) *J. Am. Chem. Soc.* **111**: 2331.

Paquette, L.A. (1990a) *Angew. Chem. Int. Ed. Engl.* **29**: 609.

Paquette, L.A., Shi, Y.-J. (1990b) *J. Am. Chem. Soc.* **112**: 8478.

Paquette, L.A., Wang, T.Z., Luo, J., Cittrell, C.E., Clough, A.E., Anderson, L.B. (1990c) *J. Am. Chem. Soc.* **112**: 239.

Pariza, R.J., Fuchs, P.L. (1983) *J. Org. Chem.* **48**: 2304.

Parker, K.A., Farmer, J.G. (1986) *J. Org. Chem.* **51**: 4023.

Parker, K.A., Spero, D.M., Van Epp, J. (1988) *J. Org. Chem.* **53**: 4628.

Parsons, P.J., Willis, P.A., Eyley, S.C. (1988) *Chem. Commun.* 283.

Partridge, J.J., Chadha, N.K., Uskokovic, M.R. (1973) *J. Am. Chem. Soc.* **95**: 532.

Pattenden, G., Teague, S.J. (1987) *Tetrahedron* **43**: 5637.

Pauson, P.L. (1985) *Tetrahedron* **41**: 5855.

Pechman, H.v. (1888) *Ber.* **21** : 1419.

Perri, S.T., Foland, L.D., Decker, O.H.W., Moore, H.W. (1986) *J. Org. Chem.* **51** : 3068.

Perri, S.T., Moore, H.W. (1990) *J. Am. Chem. Soc.* **112** : 1897.

Petrzilka, M. (1978) *Helv. Chim. Acta* **61** : 2286.

Piancatelli, G., Scettri, A., D'Auria, M. (1980) *Tetrahedron* **36** : 661.

Piers, E., Nagakura, I., Morton, H.E. (1978) *J. Org. Chem.* **43** : 3630.

Piers, E., Abeysekera, B.F., Herbert, D.J., Suckling, I.D. (1982) *Chem. Commun.* 404.

Pietrusiewicz, K.M., Monkiewicz, J., Bodalski, R. (1983) *J. Org. Chem.* **48** : 788.

Piettre, S., Heathcock, C.H. (1990) *Science* **248** : 1532.

Pindur, U., Abdoust-Housang, E. (1989) *Liebigs Ann. Chem.* 227.

Pirkle, W.H., Turner, W.V. (1975) *J. Org. Chem.* **40** : 1617.

Pirrung, M.C., Werner, J.A. (1986) *J. Am. Chem. Soc.* **108** : 6060.

Porco, J.A., Schoenen, F.J., Stout, T.J., Clardy, J., Schreiber, S.L. (1990) *J. Am. Chem. Soc.* **112** : 7410.

Posner, G.H., Mallamo, J.P., Black, A.Y. (1981) *Tetrahedron* **37** : 3921.

Posner, G.H., Lu, S.-B. (1985) *J. Am. Chem. Soc.* **107** : 1424.

Posner, G.H., Asirvatham, E. (1986a) *Tetrahedron Lett.* **27** : 663.

Posner, G.H., Lu, S.-B., Asirvatham, E., Silversmith, E.F., Shulman, E.M. (1986b) *J. Am. Chem. Soc.* **106** : 511.

Posner, G.H., Webb, K.S., Asirvatham, E., Jew, S., D'Innocenti, A. (1988) *J. Am. Chem. Soc.* **110** : 4754.

Posner, G.H., Shubman-Roskes, E.M. (1989) *J. Org. Chem.* **54** : 3514.

Posner, G.H., Asirvathan, E., Hamill, T.G., Webb, K.S. (1990) *J. Org. Chem.* **55** : 2132.

Powell, K.G., McQuillin, F.J. (1971) *Chem. Commun.* 931.

Prabhakar, S., Lobo, A.M., Oliveira, I.M.C. (1977) *Chem. Commun.* 419.

Praill, P.F.G. (1959) *Chem. Ind.* 1123.

Prelog, V., Metzler, O., Jeger, O. (1947) *Helv. Chim. Acta* **30** : 675.

Prestwich, G.D., Labovitz, J.N. (1974) *J. Am. Chem. Soc.* **96** : 7103.

Priestley, G.M., Warrender, R.N. (1972) *Tetrahedron Lett.* 4295.

Prinzbach, H., Auge, W., Basubudak, M. (1971) *Helv. Chim. Acta* **54** : 759.

Pryde, A., Zsindely, J., Schmid, H. (1974) *Helv. Chim. Acta* **57** : 1598.

Quan, P.M., Karns, T.K.B., Quin, L.D. (1965) *J. Org. Chem.* **30** : 2769.

Quillinan, A.J., Scheinmann, F. (1971) *Chem. Commun.* 966.

Quillinan, A.J., Scheinmann, F. (1972) *J. Chem. Soc. Perkin Trans. I* 1382.

Quinkert, G., Adam, F., Durner, G. (1982a) *Angew. Chem. Int. Ed. Engl.* **21** : 856.

Quinkert, G., Schwartz, U., Stark, H., Weber, W.D., Baier, H., Adam, F., Dürner, G. (1982b) *Liebigs Ann. Chem.* 1999.

Quinkert, G., Schmalz, H.-G., Walzer, E., Kowalczyk-Przewloka, T., Dürner, G., Bats, J.W. (1987) *Angew. Chem. Int. Ed. Engl.* **26** : 61.

Quinkert, G., Heim, N., Glenneberg, J., Döller, U., Eichhorn, M., Billhardt, U.-M., Schwarz, C., Zimmermann, G., Bats, J.W., Dürner, G. (1988) *Helv. Chim. Acta* **71** : 1719.

Radlick, P., Rosen, W. (1967) *J. Am. Chem. Soc.* **89** : 5308.

Rajagopalan, P. (1989) *J. Chem. Soc. Perkin Trans. I* 1691.

Raju, N., Rajagopalan, K., Swaminathan, S., Schoolery, J.N. (1980) *Tetrahedron Lett.* **21**: 1577.

Ramage, R., Sattar, A. (1970) *Chem. Commun.* 173.

Ranasinghe, M.G., Fuchs, P.L. (1989) *J. Am. Chem. Soc.* **111**: 779.

Ranganathan, S., Ranganathan D., Iyengar, R. (1976) *Tetrahedron* **32**: 961.

Ranu, B.C., Sarkar, D.C. (1988) *Chem. Commun.* 245.

Rao, C.S.S., Kumar, G., Rajagopalan, K., Swaminathan, S. (1982) *Tetrahedron* **38**: 2195.

Raucher, S., Chi, K.-W., Hwang, K.-J., Burks, J.E. (1986) *J. Org. Chem.* **51**: 5503.

Rawal, V.H., Newton, R.C., Krishnamurthy, V. (1990) *J. Org. Chem.* **55**: 5181.

Rebek, J., Shue, Y.K. (1980) *J. Am. Chem. Soc.* **102**: 5427.

Rebek, J., Shaber, S.H., Shue, Y.-K., Gehret, J.-C., Zimmerman, S. (1984) *J. Org. Chem.* **49**: 5164.

Reddy, G.S., Bhatt, M.V. (1980) *Tetrahedron Lett.* 3627.

Reetz, M.T., Schmitz, A., Holdgrün, X. (1989) *Tetrahedron Lett.* **30**: 5421.

Reinaud, O., Capdevielle, P., Maumy, M. (1988) *Synthesis* 293.

Reuter, J.M., Salomon, R.G. (1977) *J. Org. Chem.* **42**: 3360.

Richter, F., Otto, H.H. (1987) *Tetrahedron Lett.* **28**: 2945.

Ried, W., Anthofer, F. (1953) *Angew. Chem.* **65**: 601.

Ried, W., Schmidt, A.H., Kuhn, W., Bierendempfel, A. (1972) *Tetrahedron Lett.* 3885.

Riemann, J.M., Trahanovsky, W.S. (1977) *Tetrahedron Lett.* 1863.

Ripoll, J.L., Rouessac, A., Rouessac, F. (1978) *Tetrahedron* **34**: 19.

Risch, N. (1983) *Chem. Commun.* 532.

Risch, N., Langhais, M., Hohberg, T. (1991) *Tetrahedron Lett.* **32**: 4465.

Roberts, B.W., Ross, M., Wong, J. (1980) *Chem. Commun.* 428.

Roberts, M.R., Schlessinger, R.H. (1981) *J. Am. Chem. Soc.* **103**: 724.

Roberts, S.M. (1974) *Chem. Commun.* 948.

Robichaud, A.J., Meyers, A.I. (1991) *J. Org. Chem.* **56**: 2607.

Robinson, M.J.T. (1957) *Tetrahedron* **1**: 60.

Robinson, R. (1917) *J. Chem. Soc.* 762.

Robinson, R., Walker, J. (1937) *J. Chem. Soc.* 60.

Rodrigo, R. (1980) *J. Org. Chem.* **45**: 4538.

Rosenblum, M., Watkins, J.C. (1990) *J. Am. Chem. Soc.* **112**: 6316.

Rosenmund, P., Hosseini-Merescht, M. (1990) *Tetrahedron Lett.* **31**: 647.

Roush, W.R., Gillis, H.R., Essenfeld, A.P. (1984) *J. Org. Chem.* **49**: 4674.

Rowley, M., Tsukamoto, M., Kishi, Y. (1989) *J. Am. Chem. Soc.* **111**: 2735.

Rubottom, G.M., Krueger, D.S. (1977) *Tetrahedron Lett.* 611.

Ruggeri, R.B., Hansen, M.M., Heathcock, C.H. (1988) *J. Am. Chem. Soc.* **110**: 8734.

Russell, R.A., Evans, D.A.C., Warrener, R.N. (1984) *Aust. J. Chem.* **37**: 1699.

Ryckman, D.M., Stevens, R.V. (1987) *J. Am. Chem. Soc.* **109**: 4940.

Saá, C., Guitian, E., Castedo, L., Suau, R., Saá, J.M. (1986) *J. Org. Chem.* **51**: 2781.

Saito, K., Omura, Y., Maekawa, E., Gassman, P.G. (1984) *Tetrahedron Lett.* **25**: 2573.

Saito, S., Hirohara, Y., Narahara, O., Moriwake, T. (1989) *J. Am. Chem. Soc.* **111**: 4533.

Sakai, T., Miyata, K., Ishikawa, M., Takeda, A. (1985) *Tetrahedron Lett.* **26**: 4727.

Sakane, S., Matsumura, Y., Yamamura, Y., Ishida, Y., Maruoka, K., Yamamoto, H. (1983) *J. Am. Chem. Soc.* **105**: 672.

Saksena, A.K., Green, M.J., Shue, H.-J., Wong, J.K., McPhail, A.T. (1985) *Tetrahedron Lett.* **26**: 551.

Sakurai, H., Shirahata, A., Hosomi, A. (1979) *Angew. Chem. Int. Ed. Engl.* **18**: 163.

Salomon, R.G., Burns, J.R., Dominic, W.J. (1976) *J. Org. Chem.* **41**: 2918.

Sammes, P.G., Street, L.J. (1983) *Chem. Commun.* 666.

Sampath, V., Lund, E.C., Knudsen, M.J., Olmstead, M.M., Schore, N.E. (1987) *J. Org. Chem.* **52**: 3595.

Sardina, F.J., Howard, M.H., Koskinen, A.M.P., Rapoport, H. (1989) *J. Org. Chem.* **54**: 4654.

Sathyamoorthi, G., Thangaraj, K., Srinivasan, P.C., Swaminathan, S. (1989) *Tetrahedron Lett.* **30**: 4427; (1990) *Tetrahedron* **46**: 3559.

Saward, C.J., Vollhardt, K.P.C. (1975) *Tetrahedron Lett.* 4539.

Scanio, C.J.V., Starrett, R.M. (1971) *J. Am. Chem. Soc.* **93**: 1539.

Schaefer, J.P., Bloomfield, J.J. (1967) *Org. React.* **15**: 1.

Schaffer, O., Dimroth, K. (1975) *Angew. Chem. Int. Ed. Engl.* **14**: 112.

Schank, K., Moell, N. (1969) *Chem. Ber.* **102**: 71.

Schell, F.M., Smith, A.N. (1983) *Tetrahedron Lett.* **24**: 1883.

Schiess, P., Seeger, R., Suter, C. (1970) *Helv. Chim. Acta* **53**: 1713.

Schiess, P., Wisson, M. (1974) *Helv. Chim. Acta* **57**: 1692.

Schiess, P., Dinkel, R. (1975) *Tetrahedron Lett.* 2503.

Schiess, P., Fünfschilling, P. (1976) *Helv. Chim. Acta* **59**: 1756.

Schiess, P., Huys-Franchotte, M., Vogel, C. (1985) *Tetrahedron Lett.* **26**: 3959.

Schinzer, D., Kalesse, M. (1989) *Synlett* 34.

Schinzer, D., Bo, Y. (1991) *Angew. Chem. Int. Ed. Engl.* **30**: 687.

Schleigh, W.R. (1969) *Tetrahedron Lett.* 1405.

Schmid, K., Fahrni, P., Schmid, H. (1956) *Helv. Chim. Acta* **39**: 708.

Schmid, R., Huesmann, P.L., Johnson, W.S. (1980) *J. Am. Chem. Soc.* **102**: 5122.

Schmitthenner, H.F., Weinreb, S.M. (1980) *J. Org. Chem.* **45**: 3372.

Schöpf, C., Lehmann, G., Arnold, W. (1937) *Angew. Chem.* **50**: 783.

Schore, N.E., Najdi, S.D. (1987) *J. Org. Chem.* **52**: 5296.

Schore, N.E. (1988a) *Chem. Rev.* **88**: 1081.

Schore, N.E., Rowley, E.G. (1988b) *J. Am. Chem. Soc.* **110**: 5224.

Schreiber, J., Leimgruber, W., Pesaro, M., Schudel, P., Threlfall, T., Eschenmoser, A. (1961) *Helv. Chim. Acta* **44**: 540.

Schreiber, S.L., Kelly, S.E. (1984) *Tetrahedron Lett.* **25**: 1757.

Schröder, G. (1963) *Angew. Chem. Int. Ed. Engl.* **2**: 481.

Schultz, A.G., Shen, M. (1979) *Tetrahedron Lett.* 2969.

Schultz, A.G., Sha, C.-K. (1980) *J. Org. Chem.* **45**: 2040.

Schultz, A.G., Shen, M. (1981) *Tetrahedron Lett.* **22**: 1775.

Schultz, A.G., Dittami, J.P., Eng., K.K. (1984) *Tetrahedron Lett.* 1255.

Schultz, A.G., Puig, S., Wang, Y. (1985) *Chem. Commun.* 785.

Schumacher, D.P., Hal., S.S. (1981) *J. Org. Chem.* **46**: 5060.

Schumann, D., Müller, H.-J., Naumann, A. (1982) *Liebigs Ann. Chem.* 1700.

Schumann, D., Naumann, A. (1983) *Liebigs Ann. Chem.* 220.

Schumann, D., Schmidd, H. (1963) *Helv. Chim. Acta* **46**: 1996.

Schuster, H., Sichert, H., Sauer, J. (1983) *Tetrahedron Lett.* **24**: 1485.

Schwartz, C.E., Curran, D.P. (1990) *J. Am. Chem. Soc.* **112**: 9272.

Schweizer, E.E. (1964) *J. Am. Chem. Soc.* **86**: 2744.

Scott, L.T., Hashemi, M.M., Meyer, D.T., Warren, H.B. (1991) *J. Am. Chem. Soc.* **113**: 7082.

Sciacovelli, O., v. Philipsborn, W., Amith, C., Ginsburg, D. (1970) *Tetrahedron* **26**: 4589.

Semmelhack, M.F., Stauffer, R.D., Rogerson, T.D. (1973) *Tetrahedron Lett.* 4519.

Semmelhack, M.F., Seufert, W., Keller, L. (1980) *J. Am. Chem. Soc.* **102**: 6584.

Semmelhack, M.F., Brickner, S.J. (1981) *J. Am. Chem. Soc.* **103**: 3945, 6460.

Semmelhack, M.F., Herndon, J.W., Springer, J.P. (1983) *J. Am. Chem. Soc.* **105**: 2497.

Semmelhack, M.F., Bozell, J.J., Keller, L., Sato, T., Spiess, E.J., Wulff, W., Zask, A. (1985a) *Tetrahedron* **41**: 5803.

Semmelhack, M.F., Keller, L., Sato, E.J., Wulff, W. (1985b) *J. Org. Chem.* **50**: 5566.

Seto, H., Fujimoto, Y., Tatsuno, T., Yoshioka, H. (1985) *Synth. Commun.* **15**: 1217.

Seyferth, D., Blank, D.R., Evnin, A.B. (1967) *J. Am. Chem. Soc.* **89**: 4793.

Sha, C.-K., Tsou, C.-P. (1991) *J. Chin. Chem. Soc.* **38**: 183.

Shambayati, S., Crowe, W.E., Schreiber, S.L. (1990) *Tetrahedron Lett.* **31**: 5289.

Shamma, M., Jones, C.D. (1969) *J. Am. Chem. Soc.* **91**: 4009.

Sharpless, K.B., Lauer, R.F. (1972) *J. Am. Chem. Soc.* **94**: 7154.

Shea, K.J., Wada, E. (1982) *Tetrahedron Lett.* **23**: 1523.

Shea, K.J., Burke, L.D., England, W.P. (1988) *Tetrahedron Lett.* **29**: 407.

Shea, K.J., Lease, T.G., Ziller, J.W. (1990) *J. Am. Chem. Soc.* **112**: 8627.

Sheehan, J.C., Wilson, R.M., Oxford, A.W. (1971) *J. Am. Chem. Soc.* **93**: 7222.

Shen, K.-W. (1971) *Chem. Commun.* 391.

Shih, C., Swenton, J.S. (1982) *J. Org. Chem.* **47**: 2668.

Shimizu, I., Ohashi, Y., Tsuji, J. (1984) *Tetrahedron Lett.* **25**: 5183.

Shimizu, I., Ohashi, Y., Tsuji, J. (1985) *Tetrahedron Lett.* **26**: 3825.

Shishido, K., Shitara, E., Komatsu, H., Hiroya, K., Fukumoto, K., Kametani, T. (1986) *J. Org. Chem.* **51**: 3007.

Shishido, K., Hiroya, K., Komatsu, H., Fukumoto, K., Kametani, T. (1987) *J. Chem. Soc. Perkin Trans. I* 2491.

Shishido, K., Hiroya, K., Fukumoto, K., Kametani, T. (1989a) *Heterocycles* **28**: 39.

Shishido, K., Hiroya, K., Fukumoto, K., Kametani, K., Kabuto, C. (1989b) *J. Chem. Soc. Perkin Trans. I* 1443.

Shishido, K., Komatsu, H., Fukumoto, K., Kametani, T. (1989c) *Heterocycles* **28**: 43.

Shishido, K., Yamashita, A., Hiroya, K., Fukumoto, K., Kametani, T. (1989d) *Tetrahedron Lett.* **30**: 111.

Shishido, K., Yamashita, A., Hiroya, K., Fukumoto, K., Kametani, T. (1990) *J. Chem. Soc. Perkin Trans. I* 469.

Shizuri, Y., Shigemori, H., Okuno, Y., Yamamura, S. (1986a) *Chem. Lett.* **2097.**

Shizuri, Y., Suyama, K., Yamamura, S. (1986b) *Chem. Commun.* 63.

Shizuri, Y., Okuno, Y., Shigemori, H., Yamamura, S. (1987) *Tetrahedron Lett.* **28**: 6661.

Shizuri, Y., Ohkubo, M., Yamamura, S. (1989) *Tetrahedron Lett.* **30**: 3797.

Shizuri, Y., Maki, S., Ohkubo, M., Yamamura, S. (1990) *Tetrahedron Lett.* 7167.

Sigrist, R., Rey, M., Dreiding, A.S. (1986) *Chem. Commun.* 944.

Sih, C.J., Heather, H.B., Sood, R., Price, P., Perruzzotti, G., Lee, L.F.H., Lee, S.S. (1975) *J. Am. Chem. Soc.* **97**: 865.

Singh, A.K., Bakshi, R.K., Corey, E.J. (1987) *J. Am. Chem. Soc.* **109**: 6187.

Singleton, D.A., Huval, C.C., Church, K.M. (1991) *J. Am. Chem. Soc.* **32**: 5765.

Sliwa, H., Le Bot, Y. (1977) *Tetrahedron Lett.* 4129.

Smit, A., Kok, J.G.J., Geluk, H.W. (1975) *Chem. Commun.* 513.

Smith, III, A.B., Dieter, R.K. (1977) *J. Org. Chem.* **42**: 396.

Smith, R., Livinghouse, T. (1983) *J. Org. Chem.* **48**: 1554.

Smith, S., Elango, V., Shamma, M. (1984) *J. Org. Chem.* **49**: 581.

Snider, B.B., Hrib, N.J. (1977) *Tetrahedron Lett.* **1725.**

Snider, B.B. (1980) *Acc. Chem. Res.* **13**: 426.

Snider, B.B., Deutsch, E.A. (1983a) *J. Org. Chem.* **48**: 1822.

Snider, B.B., Kirk, T.C. (1983b) *J. Am. Chem. Soc.* **105**: 2364.

Snider, B.B., Phillips, G.B., Cordova, R. (1983c) *J. Org. Chem.* **48**: 3003.

Snider, B.B., Beal, R.B. (1988) *J. Org. Chem.* **53**: 4508.

Snider, B.B., Kwon, T. (1990) *J. Org. Chem.* **55**: 4786.

Snider, B.B., Wan, B.Y.-F., Buckman, B.O., Foxman, B.M. (1991) *J. Org. Chem.* **56**: 328.

Snowden, R.L. (1986) *Tetrahedron* **42**: 3277.

South, M.S., Liebeskind, L.S. (1984) *J. Am. Chem. Soc.* **106**: 4181.

Spangler, C.W. (1976) *Chem. Rev.* **76**: 187.

Speckamp, W.N., Hiemstra, H. (1985) *Tetrahedron* **41**: 4367.

Spengler, R.J., Beckman, B.G., Kim, J.H. (1977) *J. Org. Chem.* **42**: 2989.

Spitzner, D., Engler, A., Liese, T., Splettstosser, G., de Meijere, A. (1982) *Angew. Chem. Int. Ed. Engl.* **21**: 791.

Spitzner, D., Wagner, P., Simon, A., Peters, K. (1989) *Tetrahedron Lett.* **30**: 547.

Srinivasan, R. (1971) *Tetrahedron Lett.* 4551.

Srinivasan, R. (1972) *Tetrahedron Lett.* 4537.

Stach, H., Hesse, M. (1986) *Helv. Chim. Acta* **69**: 85, 1614.

Stamos, I.K., Howie, G.A., Manni, P.E., Haws, W.J., Byrn, S.R., Cassady, J.M. (1977) *J. Org. Chem.* **42**: 1703.

Staretz, M.E., Hastie, S.B. (1991) *J. Org. Chem.* **56**: 428.

Steglich, W. (1984) *IUPAC Chemistry for the Future* (Grunewald, H., Ed.) Pergamon, Oxford: pp. 211–218.

Sternberg, E.D., Vollhardt, K.P.C. (1982) *J. Org. Chem.* **47**: 3447.

Stetter, H., Thomas, H.G. (1968) *Chem. Ber.* **101**: 1115.

Stetter, H., Elfert, K. (1974) *Synthesis* 36.

Stevens, R.V., Wentland, M.P. (1968) *J. Am. Chem. Soc.* **90**: 5580.

Stevens, R.V., DuPree, L.E., Loewenstein, P.L. (1972) *J. Org. Chem.* **37**: 977.

Stevens, R.V., Christensen, C.G., Cory, R.M., Thorsett, E. (1975a) *J. Am. Chem. Soc.* **97**: 5940.

Stevens, R.V., Lesko, P.M., Lapalme, R. (1975b) *J. Org. Chem.* **40**: 3495.

Stevens, R.V., Lee, A.W.M. (1979) *J. Am. Chem. Soc.* **101**: 7032.

Stevens, R.V., Kenney, P.M. (1983a) *Chem. Commun.* 384.

Stevens, R.V., Pruitt, J. (1983b) *Chem. Commun.* 1425.

Still, W.C., Darst, K.P. (1980) *J. Am. Chem. Soc.* **102**: 7385.

Still, W.C., Gennari, C., Noguez, J.A., Pearson, D.A. (1984) *J. Am. Chem. Soc.* **106**: 260.

Stille, J.R., Grubbs, R.H. (1986) *J. Am. Chem. Soc.* **108**: 855.

Stockinger, H., Schmidt, U. (1976) *Liebigs Ann. Chem.* 1617.

Stork, G., Loewenthal, H.J.E., Mukharji, P.C. (1956) *J. Am. Chem. Soc.* **78**: 501.

Stork, G., Worrall, W.S., Pappas, J.J. (1960) *J. Am. Chem. Soc.* **82**: 4315.

Stork, G., Brizzolara, A., Landesman, H., Szmuszkovicz, J., Terrell, R. (1963a) *J. Am. Chem. Soc.* **85**: 207.

Stork, G., Dolfini, J.E. (1963b) *J. Am. Chem. Soc.* **85**: 2872.

Stork, G., Tomasz, M. (1964) *J. Am. Chem. Soc.* **86**: 471.

Stork, G., Grieco, P.A. (1969) *J. Am. Chem. Soc.* **91**: 2407.

Stork, G., Guthikonda, R.N. (1972) *J. Am. Chem. Soc.* **94**: 5109.

Stork, G., Clark, G., Shiner, C.S. (1981) *J. Am. Chem. Soc.* **103**: 4948.

Stork, G., Atwal, K.S. (1982a) *Tetrahedron Lett.* **23**: 2023.

Stork, G., Shiner, C.S., Winkler, J.D. (1982b) *J. Am. Chem. Soc.* **104**: 310.

Stork, G., Winkler, J.D., Shiner, C.S. (1982c) *J. Am. Chem. Soc.* **104**: 3767.

Stork, G., Mook, R. (1983a) *J. Am. Chem. Soc.* **105**: 3720.

Stork, G., Sher, P.M. (1983b) *J. Am. Chem. Soc.* **105**: 6765.

Stork, G., Sher, P.M., Chen, H.L. (1986) *J. Am. Chem. Soc.* **108**: 6384.

Stork, G. (1989) *Pure Appl. Chem.* **61**: 439.

Stork, G. (1990a) *199th ACS Nat. Meet. ORGN* 269.

Stork, G., Zhao, K. (1990b) *J. Am. Chem. Soc.* **112**: 5875.

Strauss, M.J., Schran, H.F., Bard, R.R. (1973) *J. Org. Chem.* **38**: 3394.

Strauss, M.J. (1974) *Acc. Chem. Res.* **7**: 181.

Strickler, H., Ohloff, G., Kovats, E.s. (1967) *Helv. Chim. Acta* **50**: 759.

Sugawara, F., Takahashi, N., Strobel, G.A., Strobel, S.A., Lu, H.S.M., Clardy, J. (1988) *J. Am. Chem. Soc.* **110**: 4086.

Sugita, T., Koyama, J., Tagahara, K., Suzuta, Y. (1986) *Heterocycles* **24**: 29.

Sunitha, K., Balasubramanian, K.K., Rajagopalan, K. (1985) *Tetrahedron Lett.* **26**: 4393.

Suri, S.C., Rodgers, S.L., Lauderdale, W.J. (1988) *Tetrahedron Lett.* **29**: 4031.

Sustmann, R. (1974) *Pure Appl. Chem.* **40**: 569.

Suzuki, S., Fujita, Y., Nishida, T. (1983) *Tetrahedron Lett.* **24**: 5737.

Swarbrick, T.M., Marko, I.E., Kennard, L. (1991) *Tetrahedron Lett.* **32**: 2549.

Swenton, J.S., Anderson, D.K., Coburn, C.E., Haag, A.P. (1984) *Tetrahedron* **40**: 4633.

Swindell, C.S., Patel, B.P. (1990) *J. Org. Chem.* **55**: 3.

Sworin, M., Lin, K.C. (1987) *J. Org. Chem.* **52**: 5640.

Szantay, C., Take, L., Honty, K. (1965) *Tetrahedron Lett.* 1665.

Taber, D.F., Gunn, B.P. (1979) *J. Am. Chem. Soc.* **101**: 3992.

Tada, M. (1982) *Chem. Lett.* 441.

Takahashi, K., Mikami, K., Nakai, T. (1988a) *Tetrahedron Lett.* **29**: 5277.

Takahashi, K., Namekata, N., Fukazawa, Y., Takase, K., Mikami, E. (1988b) *Tetrahedron Lett.* **29**: 4123.

Takahashi, T., Hori, K., Tsuji, J. (1981a) *Tetrahedron Lett.* **22**: 119.

Takahashi, T., Naito, Y., Tsuji, J. (1981b) *J. Am. Chem. Soc.* **103**: 5261.

Takahata, H., Yamabe, K., Suzuki, T., Yamazaki, T. (1986) *Chem. Pharm. Bull.* **34**: 4523.

Takano, S., Sasaki, M., Kanno, H., Shishido, K., Ogasawara, K. (1978a) *J. Org. Chem.* **43**: 4169.

Takano, S., Shishido, K., Sato, M., Yuta, K., Ogasawara, K. (1978b) *Chem. Commun.* 943.

Takano, S., Hatakeyama, S., Ogasawara, K. (1979a) *J. Am. Chem. Soc.* **101**: 6414.

Takano, S., K., Hatakeyama, S., Ogasawara, K. (1979b) *Tetrahedron Lett.* 369.

Takano, S., Ogawa, N., Ogasawara, K. (1981) *Heterocycles* **16**: 915.

Takano, S., Ohkawa, T., Tamori, S., Satoh, S., Ogasawara, K. (1988a) *Chem. Commun.* 189.

Takano, S., Sugihara, T., Satoh, S., Ogasawara, K. (1988b) *J. Am. Chem. Soc.* **110**: 6467.

Takano, S., Kijima, A., Sugihara, T., Satoh, S., Ogasawara, K. (1989) *Chem. Lett.* 87.

Takeuchi, N., Handa, S., Koyama, K., Kamata, K., Goto, K., Tobinaga, S. (1991) *Chem. Pharm. Bull.* **39**: 1655.

Talley, J.J. (1985) *J. Org. Chem.* **50**: 1695.

Tamao, K., Kobayashi, K., Ito, Y. (1988) *J. Am. Chem. Soc.* **110**: 1286.

Tamaru, Y., Harada, T., Yoshida, Z. (1980) *J. Am. Chem. Soc.* **102**: 2392.

Tamura, Y., Maeda, H., Akai, S., Ishibashi, H. (1982) *Tetrahedron Lett.* **23**: 2209.

Tamura, Y., Miyamoto, T., Nishimura, T., Kita, Y. (1973) *Tetrahedron Lett.* 2351.

Tamura, Y., Mohri, S., Maeda, H., Tsugoshi, T., Sasho, M., Kita, Y. (1984a) *Tetrahedron Lett.* **25**: 309.

Tamura, Y., Sasho, M., Nakagawa, K., Tsugoshi, T., Kita, Y. (1984b) *J. Org. Chem.* **49**: 473.

Tanaka, K., Uchiyama, F., Sakamoto, K., Inubushi, Y. (1982) *J. Am. Chem. Soc.* **104**: 4965.

Tanaka, M., Suemune, H., Sakai, K. (1988) *Tetrahedron Lett.* **29**: 1733.

Tanaka, M., Sakai, K. (1991) *Tetrahedron Lett.* **32**: 5581.

Tanida, H., Hata, Y. (1965) *J. Org. Chem.* **30**: 1985.

Tanner, D., Wennerström, O., Olsson, T. (1983) *Tetrahedron Lett.* **24**: 5407.

Tarnchompoo, B., Thebtaranonth, C., Thebtaranonth, Y. (191987) *Tetrahedron Lett.* **28**: 6671.

Taylor, E.C., Robey, R.L., Liu, K.-T., Favre, H.T., Bozimo, R.A., Conley, R.A., Chiang, C.-S., McKillop, A., Ford, M.E. (1976) *J. Am. Chem. Soc.* **98**: 3037.

Tegmo-Larsson, I.-M., Houk, K.N. (1978) *Tetrahedron Lett.* 941.

Ternansky, R.J., Balogh, D.W., Paquette, L.A. (1982) *J. Am. Chem. Soc.* **104**: 4503.

Thanupran, C., Thebtaranonth, C., Thebtaranonth, Y. (1986) *Tetrahedron Lett.* **27**: 2295.

Theuns, H.G., LaVos, G.F., tenNoever de Brauw, M.C., Salemink, C.A. (1984) *Tetrahedron Lett.* **25**: 4161.

Thomas, A.F. (1969) *J. Am. Chem. Soc.* **91**: 3281.

Thomas, A.F., Ozainne, M. (1970) *J. Chem. Soc.* (*C*) 220.

Thyagarajan, B.S., Balasubramanian, K.K., Rao, R.B. (1967) *Chem. Ind.* 401.

Thyagarajan, B.S., Hillard, J.B., Reddy, K.V., Majumdar, K.C. (1974) *Tetrahedron Lett.* 1999.

Tice, C.M., Heathcock, C.H. (1981) *J. Org. Chem.* **46**: 9.

Tietze, L.-F., von Kiedrowski, G., Berger, B. (1982) *Angew. Chem. Int. Ed. Engl.* **21**: 221; (1982) *Synthesis* 683.

Tobe, Y., Kishimura, T., Kakiuchi, K., Odaira, Y. (1983) *J. Org. Chem.* **48**: 551.

Tobe, Y., Yamashita, D., Takahashi, T., Inata, M., Sato, J., Kakiuchi, J., Kobiro, K., Odaira, Y. (1990) *J. Am. Chem. Soc.* **112**: 775.

Toda, M., Takaoka, S., Konno, M., Okuyama, S., Hayashi, M., Hamanaka, N., Iwashita, T. (1982) *Tetrahedron Lett.* **23**: 1477.

Torii, S., Tanaka, H., Nagai, Y. (1977) *Bull. Chem. Soc. Jpn* **50**: 2825.

Toth, J.E., Hamann, P.R., Fuchs, P.L. (1988) *J. Org. Chem.* **53**: 4694.

Townsend, C.A., Davis, S.G., Christensen, S.B., Link, J.C., Lewis, C.P., III (1981) *J. Am. Chem. Soc.* **103**: 6885.

Trah, S., Weidmann, K., Fritz, H., Prinzbach, H. (1987a) *Tetrahedron Lett.* **28**: 4403. (1987b) *Tetrahedron Lett.* **28**: 4399.

Trahanovsky, W.S., Mullen, P.W. (1972) *J. Am. Chem. Soc.* **94**: 5911.

Trahanovsky, W.S., Park, M.G. (1973) *J. Am. Chem. Soc.* **95**: 5412.

Trepanier, D.L., Wang, S., Moppett, C.E. (1973) *Chem. Commun.* 642.

Trost, B.M., Bogdanowicz, M.J. (1972) *J. Am. Chem. Soc.* **94**: 4777.

Trost, B.M., Bridges, A.J. (1976) *J. Am. Chem. Soc.* **98**: 5017.

Trost, B.M., Godleski, S.A., Ippen, J. (1978) *J. Org. Chem.* **43**: 4559.

Trost, B.M., Runge, T.A. (1981) *J. Am. Chem. Soc.* **103**: 7550.

Trost, B.M., McDougal, P.G. (1982a) *J. Am. Chem. Soc.* **104**: 6110.

Trost, B.M., Renaut, P. (1982b) *J. Am. Chem. Soc.* **104**: 6668.

Trost, B.M., Warner, R.W. (1982c) *J. Am. Chem. Soc.* **104**: 6112.

Trost, B.M., Adams, B.R. (1983) *J. Am. Chem. Soc.* **105**: 4849.

Trost, B.M., Nanninga, T.N. (1985) *J. Am. Chem. Soc.* **107**: 1293.

Trost, B.M. (1986a) *Angew. Chem. Int. Ed. Engl.* **25**: 1.

Trost, B.M., Mignani, S.M. (1986b) *Tetrahedron Lett.* **27**: 4137.

Trost, B.M., Tometzki, G.B., Hung, M.-H. (1987) *J. Am. Chem. Soc.* **109**: 2176.

Trost, B.M., Lee, D.C. (1988) *J. Am. Chem. Soc.* **110**: 6556.

Trost, B.M., Ohmori, M., Boyd, S.A., Okawara, H., Brickner, S.J. (1989) *J. Am. Chem. Soc.* **111**: 8281.

Trost, B.M., Urabe, H. (1990) *J. Am. Chem. Soc.* **112**: 4982.

Trost, B.M., Lautens, M., Chan, C., Jebaratnam, D.J., Mueller, T. (1991a) *J. Am. Chem. Soc.* **113**: 636.

Trost, B.M., Shi, Y. (1991b) *J. Am. Chem. Soc.* **113**: 701.

Trost, B.M., Trost, M.K. (1991c) *J. Am. Chem. Soc.* **113**: 1850.

Tsai, T.Y.R., Tsai, C.S.J., Sy, W.W., Shanbhag, M.N., Liu, W.C., Lee, S.F., Wiesner, K. (1977) *Heterocycles* **7**: 217.

Tsai, Y.-M., Cheng, C.-D. (1991) *Tetrahedron Lett.* **32**: 3515.

Tsuda, T., Chujo, Y., Nishi, S., Tawara, K., Saegusa, T. (1980) *J. Am. Chem. Soc.* **102**: 6381.

Tsuge, O., Wada, E., Kanemasa, S. (1983) *Chem. Lett.* 239.

Tsuji, J., Kobayashi, Y., Shimizu, I. (1979) *Tetrahedron Lett.* 39.

Tsuji, J., Okumoto, H., Kobayashi, Y., Takahashi, T. (1981) *Tetrahedron Lett.* **22**: 1357.

Tsuji, J., Minami, I., Shimizu, I. (1984) *Chem. Lett.* 1721.

Tsuruta, H., Kuraybayashi, K., Mukai, T. (1968) *J. Am. Chem. Soc.* **90**: 7167.

Tsuzuki, K., Hashimoto, H., Shirahama, H., Matsumoto, T. (1977) *Chem. Lett.* 1469.

Tufariello, J.J., Mullen, G.B., Tegler, J.J., Trybulski, E.J., Wong, S.C., Ali, S.A. (1979) *J. Am. Chem. Soc.* **101**: 2435.

Ulrich, H., Rao, D.V. (1977) *J. Org. Chem.* **42**: 3444.

Uma, R., Rajagopalan, K., Swaminathan, S. (1986) *Tetrahedron* **42**: 2757.

Uno, H., Naruta, Y., Maruyama, K. (1984) *Tetrahedron* **40**: 4725.

Uno, H., Goto, K., Watanabe, N., Suzuki, H. (1989) *J. Chem. Soc. Perkin Trans. I* 289.

Uyehara, T., Suzuki, I., Yamamoto, Y. (1990) *Tetrahedron Lett.* **31**: 3753.

Uyehara, T., Chiba, N., Suzuki, I., Yamamoto, Y. (1991) *Tetrahedron Lett.* **32**: 4371.

Vanderzande, D.J., Ceustermans, R.A., Martens, H.J., Toppet, S.M., Hoornaert, G.J. (1983) *J. Org. Chem.* **48**: 2188.

Van Higfte, L., Little, R.D. (1985) *J. Org. Chem.* **50**: 3940.

van Tamelen, E.E., Foltz, R.L. (1960) *J. Am. Chem. Soc.* **82**: 2400.

van Tamelen, E.E., Oliver, L.K. (1970) *J. Am. Chem. Soc.* **92**: 2136.

van Tamelen, E.E., Hwu, J.R. (1983) *J. Am. Chem. Soc.* **105**: 2490.

Vedejs, E. (1971) *Chem. Commun.* 536.

Vedejs, E., Steiner, R.P., Wu, E.S.C. (1974) *J. Am. Chem. Soc.* **96**: 4040.

Vedejs, E., Singer, S.P. (1978) *J. Org. Chem.* **43**: 4884.

Vedejs, E., Martinez, G.R. (1980) *J. Am. Chem. Soc.* **102**: 7993.

Vedejs, E., Dolphin, J.M., Mastalerz, H. (1983) *J. Am. Chem. Soc.* **105**: 127.

Vedejs, E., Eberlein, T.H., Wilde, R.G. (1988a) *J. Org. Chem.* **53**: 2220.

Vedejs, E., Rodgers, J.D., Wittenberger, S.J. (1988b) *J. Am. Chem. Soc.* **110**: 4822.

Vedejs, E., Dax, S.L. (1989) *Tetrahedron Lett.* **30**: 2627.

Veenstra, S.J., Speckamp, W.N. (1981) *J. Am. Chem. Soc.* **103**: 4645.

Vincent, T., Chuang, C., Scott, R.B. (1969) *Chem. Commun.* 758.

Vinick, F.J., Desai, M.C., Jung, S., Thadeio, P. (1989) *Tetrahedron Lett.* **30**: 787.

Visnick, M., Battiste, M.A. (1985) *Chem. Commun.* 1621.

Visser, G.W., Verboom, W., Reinhoudt, D.N., Harkema, S., van Hummel, G.J. (1982) *J. Am. Chem. Soc.* **104**: 6842.

Vogel, E., Grimme, W., Korte, S. (1965) *Tetrahedron Lett.* 3625.

Vogel, E., Weyres, F., Lepper, H., Rautenstrauch, V. (1966) *Angew. Chem.* **78**: 754.

Vogel, E., Engels, H.-W., Huber, W., Lex, J., Mullen, K. (1982) *J. Am. Chem. Soc.* **104**: 3729.

Vogel, E. (1983) in *Current Trends in Organic Synthesis* (Nozaki, H., Ed.) Pergamon Press, Oxford: 379.

Vögtle, E., Erb, R., Lenz, G., Bothner-By, A. (1965) *Liebigs Ann. Chem.* **682**: 1.

Volkmann, R., Danishefsky, S., Eggler, J., Solomon, D.M. (1971) *J. Am. Chem. Soc.* **93**: 5576.

Volkmann, R.A., Andrews, G.C., Johnson, W.S. (1975) *J. Am. Chem. Soc.* **97**: 4777.

Vorländer, E., Erig, J. (1897) *Liebigs Ann. Chem.* **294**: 314.

Wakselman, C., Mondon, M. (1973) *Tetrahedron Lett.* 4285.

Wallquist, O., Rey, M., Dreiding, A.S. (1983) *Helv. Chim. Acta* **66**: 1891.

Wang, D., Chan, T.-H. (1984) *Chem. Commun.* 1273.

Wang, S., Gates, B.D., Swenton, J.S. (1991) *J. Org. Chem.* **56**: 1979.

Wang, S.L.B., Wulff, W.D. (1990) *J. Am. Chem. Soc.* **112**: 4550.

Wang, Y., Mukherjee, D., Birney, D., Houk, K.N. (1990) *J. Org. Chem.* **55**: 4504.

Warner, P., Chu, I.-S., Boulanger, W. (1983) *Tetrahedron Lett.* **24**: 4165.

Warrener, R.N., Russell, R.A. (1978) *Tetrahedron Lett.* 4447.

Wasserman, H.H., Doumaux, A.R. (1962) *J. Am. Chem. Soc.* **84**: 4611.

Wasserman, H.H., Kitzing, R. (1966) *Tetrahedron Lett.* 3343.

Wasserman, H.H., McCarthy, K.E., Prowse, K.S. (1986) *Chem. Rev.* **86**: 845.

Wasserman, H.H., Lombardo, L.J. (1989) *Tetrahedron Lett.* **30**: 1725.

Wasserman, H.H., Cook, J.D., Vu, C.B. (1990a) *J. Org. Chem.* **55**: 1701.

Wasserman, H.H., van Duzer, J.H., Vu, C.B. (1990b) *Tetrahedron Lett.* **31**: 1609.

Watabe, T., Hosoda, Y., Okada, K., Oda, M. (1987) *Bull. Chem. Soc. Jpn* **60**: 3801.

Watabe, T., Okada, K., Oda, M. (1988) *J. Org. Chem.* **53**: 216.

Watanabe, M., Snieckus, V. (1980) *J. Am. Chem. Soc.* **102**: 1457.

Watkins, J.C., Rosenblum, M. (1984) *Tetrahedron Lett.* **25**: 2097.

Watt, D., Corey, E.J. (1972) *Tetrahedron Lett.* 4651.

Wehle, D., Fitjer, L. (1987) *Angew. Chem. Int. Ed. Engl.* **26**: 130.

Weiberth, F.J., Hall, S.S. (1985) *J. Org. Chem.* **46**: 5060.

Weinreb, S.M., Khatri, N.A., Shringarpure, J. (1979) *J. Am. Chem. Soc.* **101**: 5073.

Welch, M.C., Bryson, T.A. (1988) *Tetrahedron Lett.* **29**: 521.

Weller, T., Seebach, D. (1982) *Tetrahedron Lett.* **23**: 935.

Wender, P.A., Hillemann, C.L., Szymonifka, M.J. (1980a) *Tetrahedron Lett.* **21**: 2205.

Wender, P.A., Hubbs, J.C. (1980b) *J. Org. Chem.* **45**: 365.

Wender, P.A., Letendre, L.J. (1980c) *J. Org. Chem.* **45**: 367.

Wender, P.A., Dreyer, G.B. (1981a) *Tetrahedron* **37**: 4445.

Wender, P.A., Howbert, J.J. (1981b) *J. Am. Chem. Soc.* **103**: 688.

Wender, P.A., Dreyer, G.B. (1982a) *J. Am. Chem. Soc.* **104**: 5805.

Wender, P.A., Eck, S.L. (1982b) *Tetrahedron Lett.* **23**: 1871.

Wender, P.A., Howbert, J.J. (1982c) *Tetrahedron Lett.* **23**: 3985.

Wender, P.A., Dreyer, G.B. (1983a) *Tetrahedron Lett.* **24**: 4543.

Wender, P.A., Holt, D.A., Sieburth, S.M. (1983b) *J. Am. Chem. Soc.* **105**: 3348.

Wender, P.A. (1984) in *Selectivity—A Goal for Synthetic Efficiency* (Bartmann, W., Trost, B.M., Eds) VCH, New York.

Wender, P.A., Holt, D.A. (1985a) *J. Am. Chem. Soc.* **107**: 7771.

Wender, P.A., Ternansky, R.J. (1985b) *Tetrahedron Lett.* **26**: 2625.

Wender, P.A., Ternansky, R.J., Sieburth, S.M. (1985c) *Tetrahedron Lett.* **26**: 4319.

Wender, P.A., Fisher, K. (1986) *Tetrahedron Lett.* **27**: 1857.

Wender, P.A., Brighty, K. (1988a) *Tetrahedron Lett.* **29**: 6741.

Wender, P.A., White, A.W. (1988b) *J. Am. Chem. Soc.* **110**: 2218.

Wender, P.A., von Geldern, T.W., Levine, B.H. (1988c) *J. Am. Chem. Soc.* **110**: 4858.

Wender, P.A., Lee, H.Y., Wilhelm, R.S., Williams, P.D. (1989a) *J. Am. Chem. Soc.* **111**: 8954.

Wender, P.A., Siggel, L., Nuss, J.M. (1989b) *Org. Photochem.* **10**: 357.

Wender, P.A., McDonald, F.E. (1990a) *Tetrahedron Lett.* **31**: 3691.

Wender, P.A., Singh, S.K. (1990b) *Tetrahedron Lett.* **31**: 2517.

Wendler, N.L., Taub, D., Kuo, H. (1960) *J. Am. Chem. Soc.* **82**: 5701.

Wenkert, E., Dave, K.G. (1962) *J. Am. Chem. Soc.* **84**: 94.

Wenkert, E., Mueller, R.A., Reardon, E.J., Sathe, S.S., Scharf, D.J., Tosi, G. (1970) *J. Am. Chem. Soc.* **92**: 7428.

Wenkert, E., Bindra, J.S., Chauncy, B. (1972) *Synth. Commun.* **2**: 285.

Wenkert, E., Buckwalter, B.L., Craveiro, A.A., Sanchez, E.L., Sathe, S.S. (1978) *J. Am. Chem. Soc.* **100**: 1267.

Wenkert, E., Piettre, S.R. (1988) *J. Org. Chem.* **53**: 5850.

Wenkert, E., Decorzant, R., Näf, F. (1989) *Helv. Chim. Acta* **72**: 756.

Wenkert, E., Guo, M., Lavilla, R., Porter, R., Porter, B., Ramachandran, K., Sheu, J.-H. (1990) *J. Org. Chem.* **55**: 6203.

Westberg, H.H., Cain, E.N., Masamune, S. (1969) *J. Am. Chem. Soc.* **91**: 7512.

Wharton, P.S., Sundin, C.E., Johnson, D.W., Kluender, H.C. (1972) *J. Org. Chem.* **37**: 34.

White, D.L., Seyferth, D. (1972) *J. Org. Chem.* **37**: 3545.

White, D.R. (1975) *Chem. Commun.* 95.

White, J.D., Skeean, R.W. (1978) *J. Am. Chem. Soc.* **100**: 6292.

White, J.D., Ruppert, J.F., Avery, M.A., Torii, S., Nokami, J. (1981) *J. Am. Chem. Soc.* **103**: 1813.

Whitlock, H.W., Siefkin, M.W. (1968) *J. Am. Chem. Soc.* **90**: 4929.

Wiberg, K.B., Burgmaier, G.J., Warner, P.J. (1971) *J. Am. Chem. Soc.* **93**: 246.

Wiberg, K.B., Olli, L.K., Golembeski, N., Adams, R.D. (1980) *J. Am. Chem. Soc.* **102**: 7467.

Wick, A.E., Felix, D., Steen, K., Eschenmoser, A. (1964) *Helv. Chim. Acta* **47**: 2425.

Wiechert, R. (1970) *Angew. Chem.* **82**: 219.

Wiesner, K., Götz, M., Simmons, D.L., Fowler, L.R. (1963) *Coll. Czech. Chem. Commun.* **28**: 2462.

Wilkens, J., Kühling, A., Blechert, S. (1987) *Tetrahedron* **43**: 3237.

Williams, J.R., Callahan, J.F. (1980) *J. Org. Chem.* **45**: 4479.

Williams, J.R., Lin, C. (1981) *Chem. Commun.* 752.

Wilson, R.M., Musser, A.K. (1980) *J. Am. Chem. Soc.* **102**: 1720.

Wilson, R.M. (1984) *Proc. SPIE, Int. Soc. Opt. Eng.* **458**: 58.

Wilson, R.M. (1985) *Org. Photochem.* **7**: 339.

Wilson, S.R., Phillips, L.R., Natalie, K.J. (1979) *J. Am. Chem. Soc.* **101**: 3340.

Winkler, J.D., Muller, C.L., Scott, R.D. (1988) **110**: 4831.

Winkler, J.D., Scott, R.D., Williard, P.G. (1990) *J. Am. Chem. Soc.* **112**: 8971.

Winterfeldt, E., Gaskell, A.J., Korth, T., Radunz, H.E., Walkowiak, M. (1969) *Chem. Ber.* **102**: 3558.

Wiseman, J.R., Pendery, J.J., Otto, C.A., Chiong, K.G. (1980) *J. Org. Chem.* **45**: 516.

Wittek, P.J., Harris, T.M. (1973) *J. Am. Chem. Soc.* **95**: 6865.

Wolf, A.D., Jones, M. (1973) *J. Am. Chem. Soc.* **95**: 8209.

Wolovsky, R., Ben-Efraim, D.A., Batich, C., Wasserman, E. (1970) *J. Am. Chem. Soc.* **92**: 2133.

Wong, C.-M., Singh, R., Singh, K., Lam, H.Y.P. (1979) *Can. J. Chem.* **57**: 3304.

Woodward, R.B., Doering, W.v.E. (1945) *J. Am. Chem. Soc.* **67**: 860.

Woodward, R.B., Eastman, R.H. (1946) *J. Am. Chem. Soc.* **68**: 2229.

Woodward, R.B., Singh, T. (1950a) *J. Am. Chem. Soc.* **72**: 494; (1950b) *J. Am. Chem. Soc.* **72**: 499.

Woodward, R.B. (1961) *Pure Appl. Chem.* **2**: 383.

Woodward, R.B., Yates, P. (1963) *J. Am. Chem. Soc.* **85**: 551,553.

Woodward, R.B., Ayer, W.A., Beaton, J.M., Bickelhaupt, F., Bonnett, R., Buchschacher, P., Closs, G.L., Dutler, H., Hannah, J., Hauck, F.P., Ito, S., Langemann, A., LeGoff, E., Leimgruber, W., Lwowski, W., Sauer, J., Valenta, Z., Volz, H. (1990) *Tetrahedron* **46**: 7599.

Wu, G., Shimoyama, I., Negishi, E. (1991) *J. Org. Chem.* **56**: 6506.

Wu, H.-J., Pan, K. (1987) *Chem. Commun.* 898.

Wu, K.-M., Midland, M.M., Okamura, W.H. (1990a) *J. Org. Chem.* **55**: 4381.

Wu, K.-M., Okamura, W.H. (1990b) *J. Org. Chem.* **55**: 4025.

Wu, T.-C., Mareda, J., Gupta, Y.N., Houk, K.N. (1983) *J. Am. Chem. Soc.* **105**: 6996.

Wulff, W.D., Tang, P.-C. (1984) *J. Am. Chem. Soc.* **106**: 434.

Wulff, W.D., Kaesler, R.W. (1985) *Organometallics* **4**: 1461.

Wulff, W.D., Xu, Y.-C. (1988) *J. Am. Chem. Soc.* **110**: 2312.

Wuonola, M.A., Smallheer, J.M., Read, J.M., Calabrese, J.C. (1991) *Tetrahedron Lett.* **32**: 5481.

Wynberg, H., Helder, R. (1971) *Tetrahedron Lett.* 4317.

Xie, Z.-F., Suemune, H., Sakai, K. (1988) *Chem. Commun.* 612; (1989) *Synth. Commun.* **19**: 987.

Xie, Z.-F., Sakai, K. (1990) *J. Org. Chem.* **55**: 820.

Xiong, H., Rieke, R.D. (1991) *Tetrahedron Lett.* **32**: 5269.

Xu, S.L., Moore, H.W. (1989) *J. Org. Chem.* **54**: 6018.

Xu, Y.-C., Cantin, M., Deslongchamps, P. (1990) *Can. J. Chem.* **68**: 2137.

Yamada, K., Kyotani, Y., Manabe, S., Suzuki, M. (1979) *Tetrahedron* **35**: 293.

Yamada, Y., Miljkovic, D., Wehrli, P., Golding, B., Löliger, P., Keese, R., Müller, K., Eschenmoser, A. (1969) *Angew. Chem. Int. Ed. Engl.* **8**: 343.

Yamaguchi, M., Tsukamoto, M., Hirao, I. (1985) *Tetrahedron Lett.* **26**: 1723.

Yamaguchi, M., Hasabe, K., Tanaka, S., Minami, T. (1986) *Tetrahedron Lett.* **27**: 959.

Yamaguchi, M., Hasabe, K., Hiyashi, H., Uchida, M., Irie, A., Minami, T. (1990) *J. Org. Chem.* **55**: 1611.

Yamaguchi, R., Hamasaki, T., Utimoto, K., Kozima, S., Takaya, H. (1990) *Chem. Lett.* 2161.

Yamaguchi, Y., Hayakawa, K., Kanematsu, K. (1987) *Chem. Commun.* 515.

Yamamoto, I., Fujimoto, T., Ohta, K., Matsuzaki, K. (1987) *J. Chem. Soc. Perkin Trans. I* 1537.

Yamamoto, T., Eki, T., Nagumo, S., Suemune, H., Sakai, K. (1991) *Tetrahedron Lett.* **32**: 515.

Yamamoto, Y., Furuta, T. (1990) *J. Org. Chem.* **55**: 3971.

Yang, N.C., Lin, L.C., Shani, A., Yang, S.S. (1969) *J. Org. Chem.* **34**: 1845.

Yasukouchi, T., Kanematsu, K. (1989) *Chem. Commun.* 953.

Yates, P., Abrams, G.D., Betts, M.J., Goldstein, S. (1971) *Can. J. Chem.* **49**: 2850.

Yokomori, Y., Tamura, R. (1991) *J. Chem. Soc. Perkin Trans. II* 159.

Young, S.D., Borden, W.T. (1980) *J. Org. Chem.* **45**: 724.

Yurchenko, A.G., Melnik, N.N., Likhotvorik, I.R. (1989) *Tetrahedron Lett.* **30**: 3653.

Zhang, Y., Negishi, E. (1989) *J. Am. Chem. Soc.* **111**: 3454.

Zhang, Y., Wu, G., Agnel, G., Negishi, E. (1990) *J. Am. Chem. Soc.* **112**: 8590.

Zheng, G.-C., Kakisawa, H. (1989) *Bull. Chem. Soc. Jpn* **62**: 602.

Ziegler, F.E., Schwartz, J.A. (1975) *Tetrahedron Lett.* 4643.

Ziegler, F.E., Wang, T.-F. (1981) *Tetrahedron Lett.* **22**: 1179.

Ziegler, F.E., Lim, H. (1982a) *J. Org. Chem.* **47**: 5229.

Ziegler, F.E., Piwinski, J.J. (1982b) *J. Am. Chem. Soc.* **104**: 7181.

Ziegler, F.E., Mencel, J.J. (1984) *Tetrahedron Lett.* **25**: 123.

Ziegler, F.E. (1988) *Chem. Rev.* **88**: 1423.

Zimmerman, H.E., Grunewald, G.L. (1964) *J. Am. Chem. Soc.* **86**: 1434.

Zimmerman, H.E., Crumrine, D.S. (1968) *J. Am. Chem. Soc.* **90**: 5612.

Zimmerman, H.E., Grunewald, G.L., Paufler, R.M., Sherwin, M.A. (1969a) *J. Am. Chem. Soc.* **91**: 2330.

Zimmerman, H.E., Robbins, J.D., Schantl, J. (1969b) *J. Am. Chem. Soc.* **91**: 5878.

Zschiesche, R., Reissig, H.-U. (1987) *Liebigs Ann. Chem.* 387.

Zschiesche, R., Reissig, H.-U. (1988) *Liebigs Ann. Chem.* 1165.

Zweifel, G., Polston, N.L., Whitney, C.C. (1968) *J. Am. Chem. Soc.* **90**: 6243.

Zweifel, G., Horng, A., Snow, J.T. (1970) *J. Am. Chem. Soc.* **92**: 1427.

INDEX